ENDORSED BY

International Mathematics for Cambridge IGCSE® (0607)

Extended

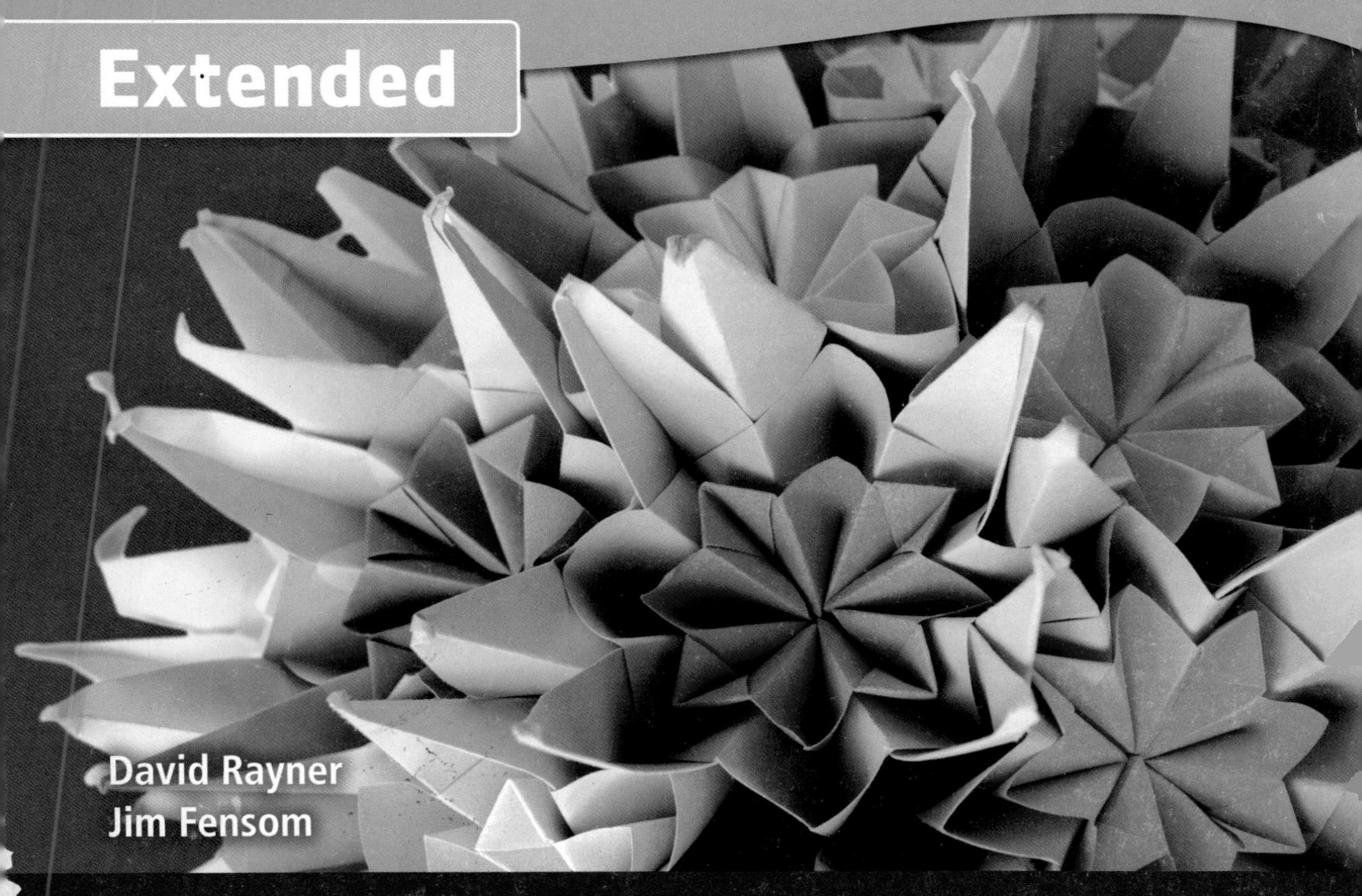

David Rayner
Jim Fensom

Oxford and Cambridge
leading education together

OXFORD
UNIVERSITY PRESS

OXFORD
UNIVERSITY PRESS

Great Clarendon Street, Oxford OX2 6DP

Oxford University Press is a department of the University of Oxford. It furthers the University's objective of excellence in research, scholarship, and education by publishing worldwide in

Oxford New York

Auckland Cape Town Dar es Salaam Hong Kong Karachi Kuala Lumpur Madrid Melbourne Mexico City Nairobi New Delhi Shanghai Taipei Toronto

With offices in

Argentina Austria Brazil Chile Czech Republic France Greece Guatemala Hungary Italy Japan Poland Portugal Singapore South Korea Switzerland Thailand Turkey Ukraine Vietnam

First published 2013

British Library Cataloguing in Publication Data

Data available

ISBN: 978-0-19-838918-7
10 9 8 7 6 5 4

Printed in Malaysia by Vivar Printing Sdn. Bhd.

Acknowledgements

The publisher would like to thank the following for their permission to reproduce photographs:

P1: Texas Instruments; **P1:** Texas Instruments; **P1:** Casio Electronics Co. Ltd; **P38:** Lapoo/Shutterstock; **P46:** Grzym/Shutterstock; **P55:** Floridastock/Shutterstock; **P55:** Rodho/Shutterstock; **P56:** Leonello Calvetti/Shutterstock; **P59:** Fstop/Alamy; **P61:** Dean Bertoncelj/Shutterstock; **P61:** Milovad/Shutterstock; **P63:** M. Unal Ozmen/Shutterstock; **P63:** Ikongraphics/Alamy; **P64:** Serg64/Shutterstock; **P67:** Mike Flippo/Shutterstock; **P68:** Miroslav Hlavko/Shutterstock; **P69:** Oleksiy Mark/Shutterstock; **P72:** Alexussk/Shutterstock; **P73:** Gorilla/Shutterstock; **P74:** Maxim Petrichuk/Shutterstock; **P75:** Musicman/Shutterstock; **P76:** Joerg Beuge/Shutterstock; **P81:** Fedor Selivanov/Shutterstock; **P84:** Nicku/Shutterstock.Com; **P85:** Rich Carey/Shutterstock; **P112:** Aleksei Ruzhin/Shutterstock; **P113:** Michael Weseman/Shutterstock; **P127:** Classic Image/Alamy; **P129:** Sandys/Shutterstock; **P135:** Brocreative/Shutterstock; **P144:** Temiropix/Shutterstock; **P149:** Pichugin Dmitry/Shutterstock; **P154:** Evgeny Atamanenko/Shutterstock; **P154:** Clearviewstock/Shutterstock; **P156:** Georgios Kollidas/Shutterstock; **P177:** Emilio Segre Visual Archives/American Institute Of Physics/Science Photo Library; **P186:** Offscreen/Shutterstock; **P204:** Turtix/Shutterstock; **P204:** Images.Etc/Shutterstock; **P205:** Lee Yiu Tung/Shutterstock; **P205:** Lee Yiu Tung/Shutterstock; **P206:** Dan Breckwoldt/Shutterstock; **P214:** Volodymyr Goinyk/Shutterstock; **P227:** Science Source/Science Photo Library; **P228:** Robert Kneschke/Shutterstock; **P230:** Alexander Tolstykh/Shutterstock; **P230:** TV/Shutterstock; **P239:** Mopic/Shutterstock; **P239:** Matthew Benoit/Shutterstock; **P260:** Artsilense/Shutterstock; **P267:** Anatolym/Shutterstock; **P269:** Khoroshunova Olga/Shutterstock; **P269:** Andrey Pavlov/Shutterstock; **P276:** Antonio Abrignani/Shutterstock; **P308:** New York Public Library/Science Photo Library; **P320:** Danielbotha/Shutterstock; **P321:** Manuel Fernandes/Shutterstock; **P322:** Coprid/Shutterstock; **P342:** David Fowler/Shutterstock.Com; **P374:** Bettmann/Corbis; **P385:** Georgios Kollidas/Shutterstock; **P388:** Pedro Nogueira/Shutterstock; **P399:** Leventegyori/Shutterstock; **P402:** Pictorial Press Ltd/Alamy; **P413:** Ljupco Smokovski/Shutterstock; **P435:** Zoran Vukmanov Simokov/Shutterstock; **P444:** Tatiana Popova/Shutterstock; **P449:** Professor Peter Goddard/Science Photo Library; **P455:** Katatonia82/Shutterstock.Com
Cover image: Oksix/Dreamstime.com

Contents

Introduction

About this book

This book has been written to cover the **Cambridge IGCSE® International Mathematics (0607)** course, and is fully aligned to the syllabus.

In addition to the main curriculum content, you will find:

- **Two chapters devoted entirely to investigations and mathematical modelling**, which will help you develop fundamental techniques for solving problems and open-ended questions.
- **Comprehensive support for the use of a graphic display calculator**, an integral part of the course. Chapter 1 will show you the basic skills you need to learn, and there are examples throughout the rest of the book of how to put these into practice.

Throughout the book, you will encounter worked examples and a host of rigorous exercises. The examples show you the important techniques required to tackle questions. The exercises are carefully graded, starting from a basic level and going up to exam standard, allowing you plenty of opportunities to practise your skills. Together, the examples and exercises put maths in a real-world context, with a truly international focus.

At the start of each chapter, you will see a list of objectives that are covered in that chapter. These objectives are drawn from the Cambridge IGCSE syllabus.

About the authors

David Rayner is a highly experienced author. He has taught and examined mathematics for over thirty years and has published a number of leading textbooks on the subject.

Jim Fensom has many years of experience teaching and examining mathematics at secondary level. He is currently Mathematics Coordinator at Nexus International School in Singapore.

Special thanks to James Nicholson for his contribution to the resources on the CD.

What's on the CD?

The CD that accompanies the book contains a wealth of material to help solidify your understanding of the Cambridge IGCSE International Mathematics course, and to aid revision for your examinations:

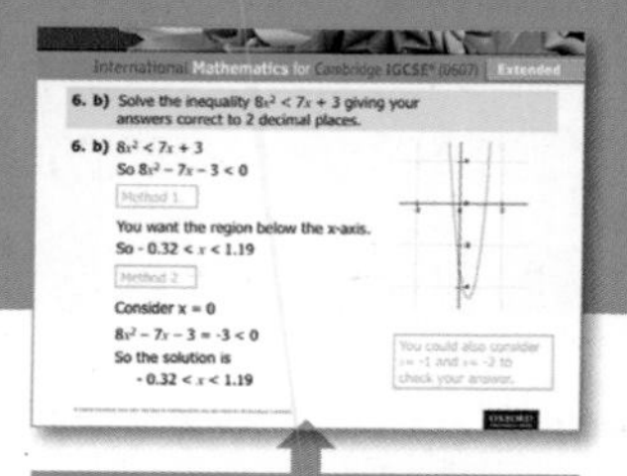

Worked solutions are available for selected questions from the examination exercises in the book. These are in the form of PowerPoint presentations, giving clear, step-by-step guidance on how to approach exam-style questions.

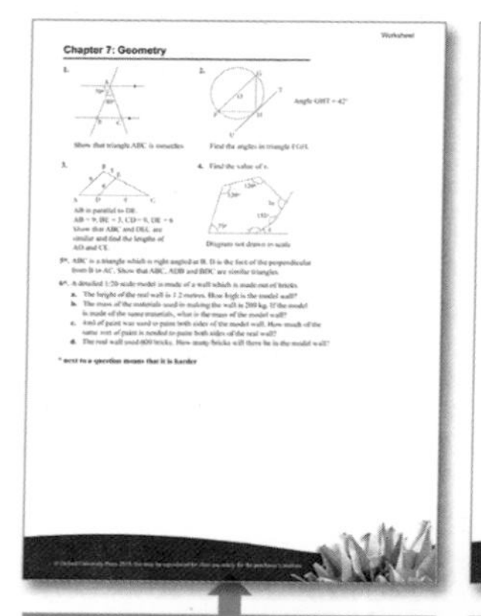
Chapter 7: Geometry

Worksheets are provided for further practice. They accompany each of the main chapters in the book, and some questions are intended to be more difficult, to challenge you.

Chapter 10: Trigonometry

Revision checklists will help you track your progress as you consolidate your knowledge of the topics on the Cambridge IGCSE International Mathematics course.

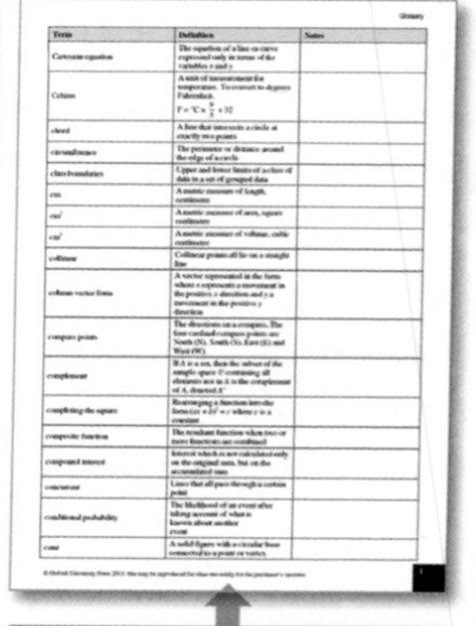

A glossary offers detailed coverage of mathematical terminology, and can be edited to include your own notes and definitions.

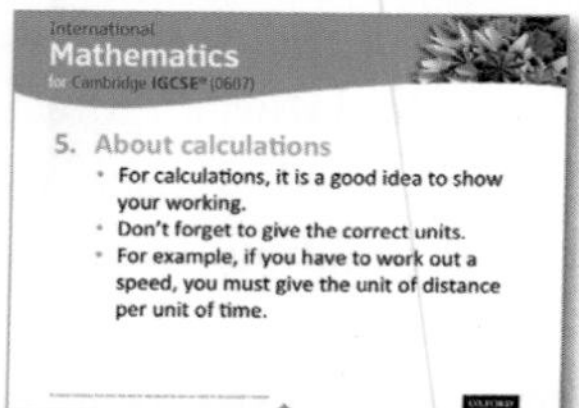

Exam preparation is given in a range of materials, covering advice for revision, the language used in question papers, and tips for avoiding commonly-made errors.

What's on the website?

Full worked solutions to each of the examination exercises in the book can be found online, at: **www.oxfordsecondary.co.uk/0607**

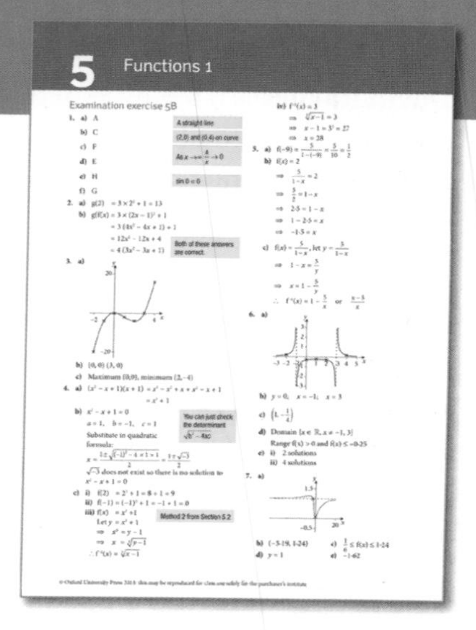
5 Functions 1

Examination exercise 5B

1 Using your graphic display calculator

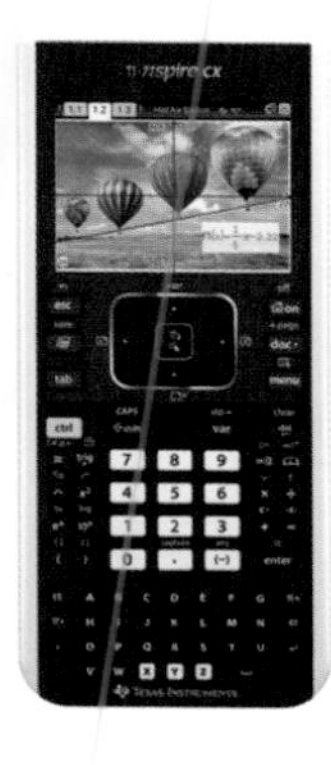
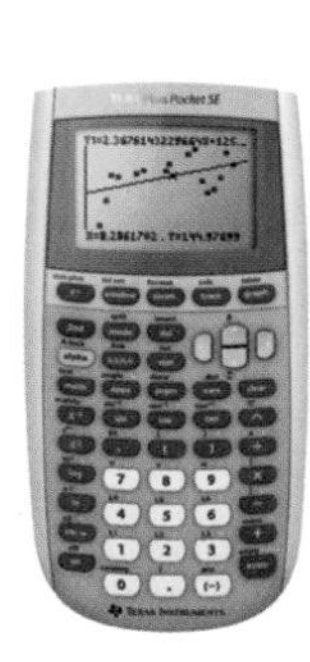

To get the most out of the International Mathematics course and the examination, you will need to use a graphic display calculator (GDC).

Instructions in this book have been written for the TI-Nspire™ CX Handheld (but will work with any of the TI-Nspire™ family), for the TI-84 Plus Pocket SE (but will work with any other TI-84 Plus) and for the Casio PRIZM fx-CG10/20 (but will also work with the fx-9860 GII and other similar models).

However, there are a number of popular manufactures and models available, which may be used for your Cambridge IGCSE course. All leading brands of calculator are capable of performing the same mathematical functions, although their menu structures and the exact name of a function may vary from one to another.

The International Mathematics course requires that you should be able to do the following using a graphics calculator:

- Sketch a graph.
- Produce a table of values for a function.
- Find zeros and local maxima or minima of a function.
- Find the intersection point of two graphs.
- Find mean, median, quartiles.
- Find the linear regression equation.

Other existing in-built applications should *not* be used and will gain no credit.

Any other applications and programs from external sources are *not* permitted.

This chapter is split into three sections and covers the use of a GDC for each of the above. Instructions are given side-by-side for each of the models shown. Being an efficient user of your GDC is very important in the examination. The GDC will not do the mathematics for you; it will help you to do your mathematics. If you use it resourcefully, it will give you more time in the examination to do other things.

As well as the instructions in this chapter, you will also find examples throughout the textbook of applications where you can use the GDC.

1 Using templates in basic calculations

GDCs use a method known as MathPrint™ or Natural Input and Display™ to show mathematics on their screens. The most efficient way to use the calculator is to enter a calculation in the same way that you see it written on paper. By using expression templates, you can do this easily.
You can move around templates using the arrow keys on your GDC.

1.1 Fractions

Consider the calculation $\frac{4.2+1.75}{3.63-2.14}$

Here are the instructions for doing this with a scientific calculator from a textbook written in 2000:

Find the bottom line first:

So, after you calculated the bottom line, you had to store it in the calculator's memory (M), clear the screen (C), calculate the top line and finally divide the result by the value that you stored in the memory (MR). There was another way you could do the same thing, but using brackets. This was to type:

Neither of these methods looks much like the original. The use of templates for things like fractions, exponents and roots has made a huge difference to the use of the calculator. In mathematics examinations in the past, the questions on the quadratic formula and the cosine rule (both fairly complicated formulas) were very often answered incorrectly. This was due to the difficulty of entering the calculation. Now with a template and your GDC, you have no excuse for getting the answer wrong.

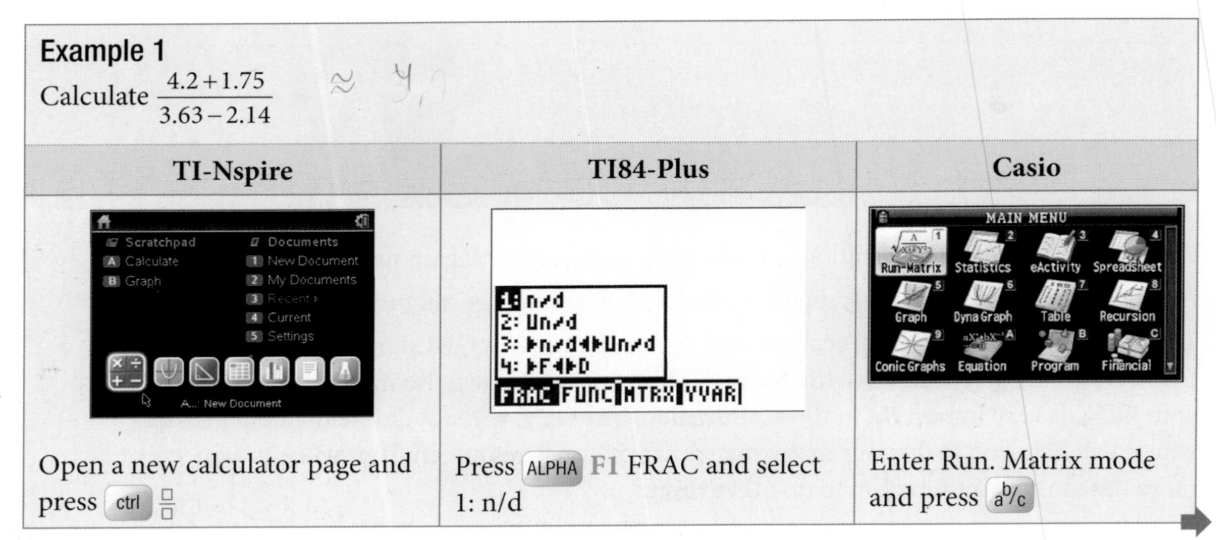

Example 1

Calculate $\frac{4.2+1.75}{3.63-2.14}$

TI-Nspire	TI84-Plus	Casio
Open a new calculator page and press ctrl $\frac{\square}{\square}$	Press ALPHA F1 FRAC and select 1: n/d	Enter Run. Matrix mode and press a^b/c

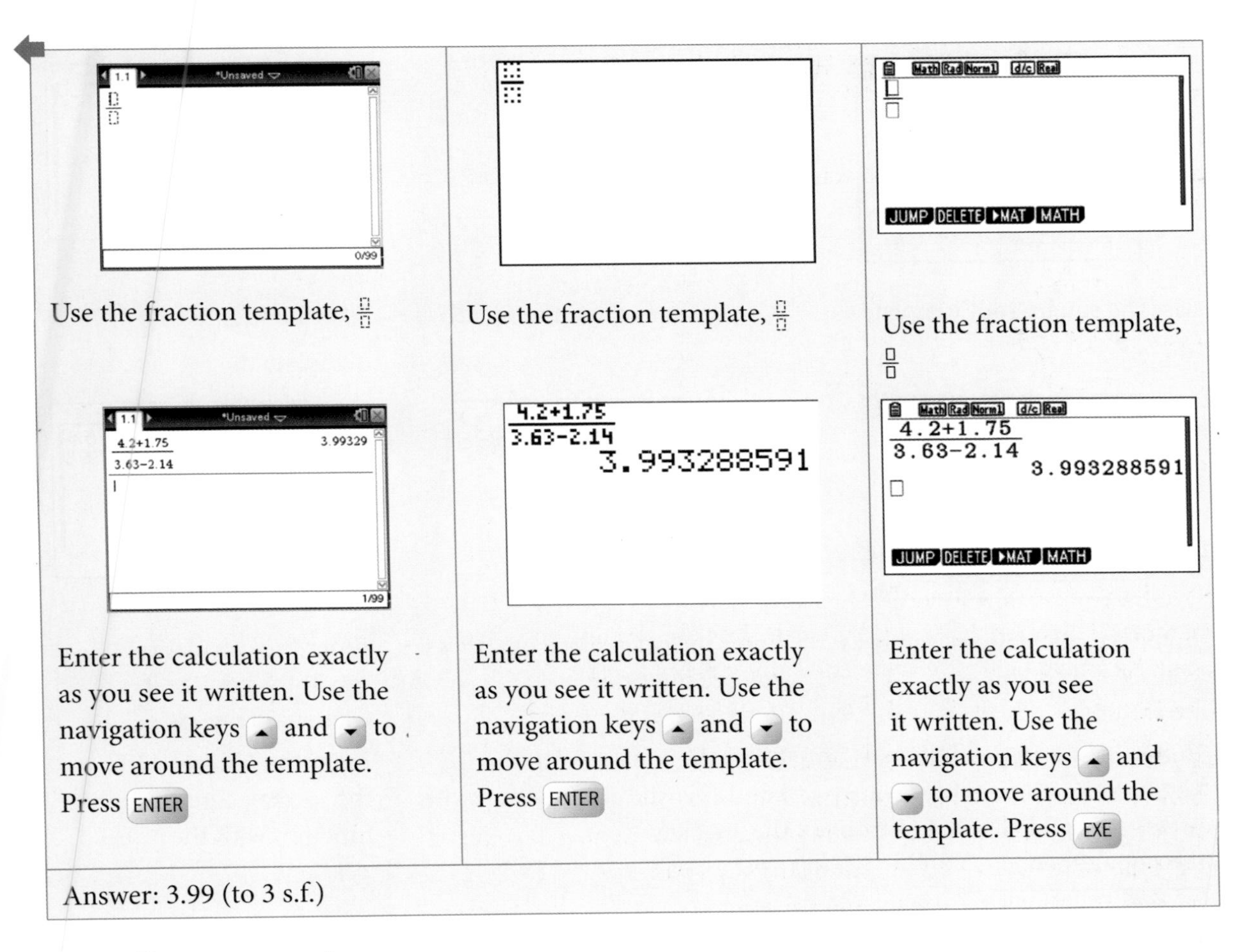

Use the fraction template, $\frac{\square}{\square}$	Use the fraction template, $\frac{\square}{\square}$	Use the fraction template, $\frac{\square}{\square}$
Enter the calculation exactly as you see it written. Use the navigation keys ▲ and ▼ to move around the template. Press ENTER	Enter the calculation exactly as you see it written. Use the navigation keys ▲ and ▼ to move around the template. Press ENTER	Enter the calculation exactly as you see it written. Use the navigation keys ▲ and ▼ to move around the template. Press EXE

Answer: 3.99 (to 3 s.f.)

1.2 Square roots

Another template that you will need to use is the square root template – for example, when you are doing a calculation for the cosine rule.

Example 2

Calculate $\sqrt{8.1^2 + 7.5^2 - 2\times 8.1\times 7.5\times \cos(62.3)}$

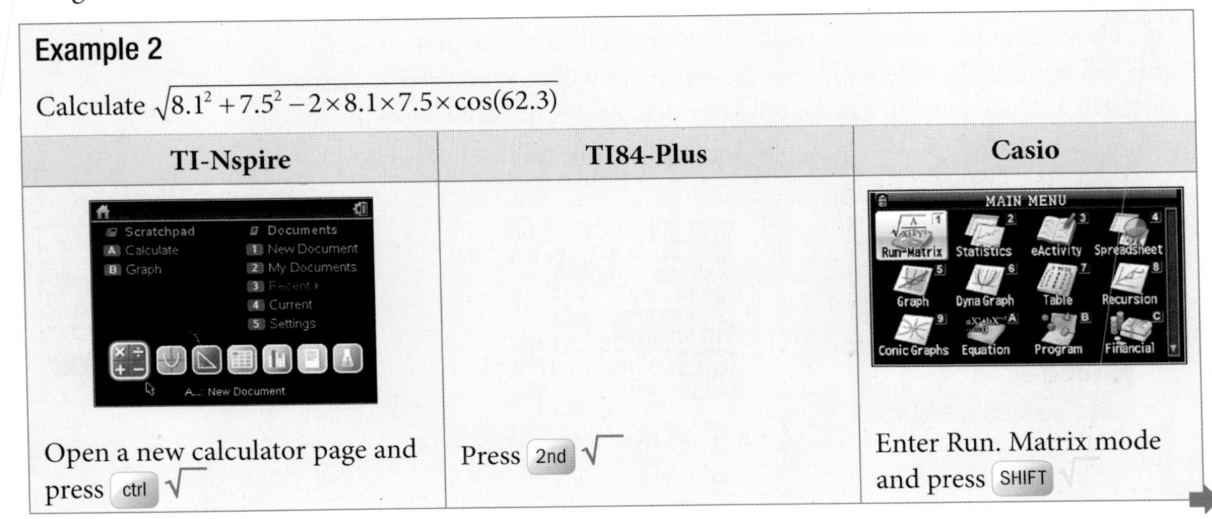

TI-Nspire	TI84-Plus	Casio
Open a new calculator page and press ctrl $\sqrt{\ }$	Press 2nd $\sqrt{\ }$	Enter Run. Matrix mode and press SHIFT

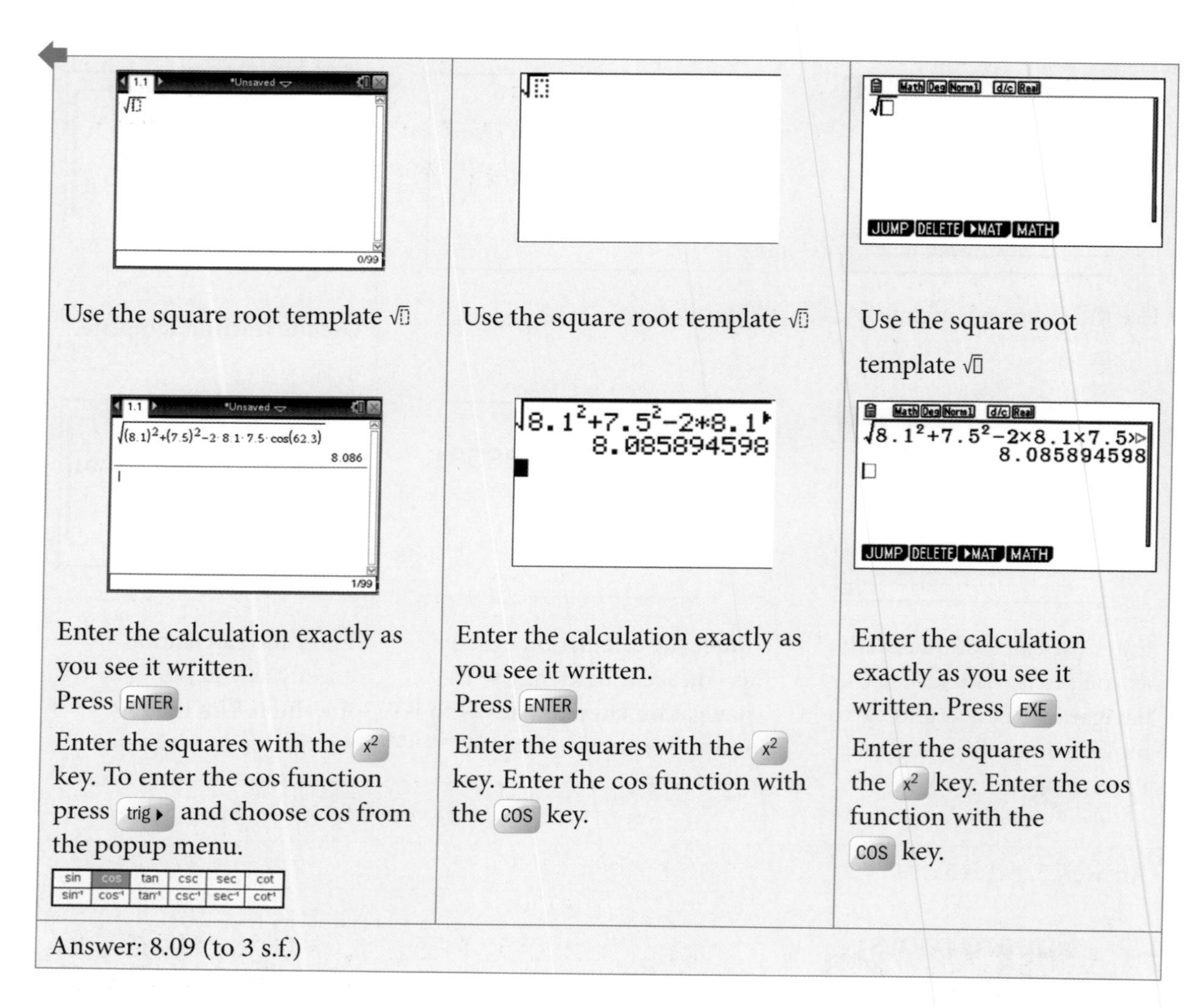

Use the square root template √▯	Use the square root template √▯	Use the square root template √▯
Enter the calculation exactly as you see it written. Press ENTER. Enter the squares with the x^2 key. To enter the cos function press trig ▸ and choose cos from the popup menu.	Enter the calculation exactly as you see it written. Press ENTER. Enter the squares with the x^2 key. Enter the cos function with the COS key.	Enter the calculation exactly as you see it written. Press EXE. Enter the squares with the x^2 key. Enter the cos function with the COS key.

Answer: 8.09 (to 3 s.f.)

Using degrees for angles

In the above example, you will need to set your calculator in **degree** mode.
In the International Mathematics course, all angles will be measured in degrees.
Your GDC is likely to be in **radian** mode by default. Here is how to set the mode.

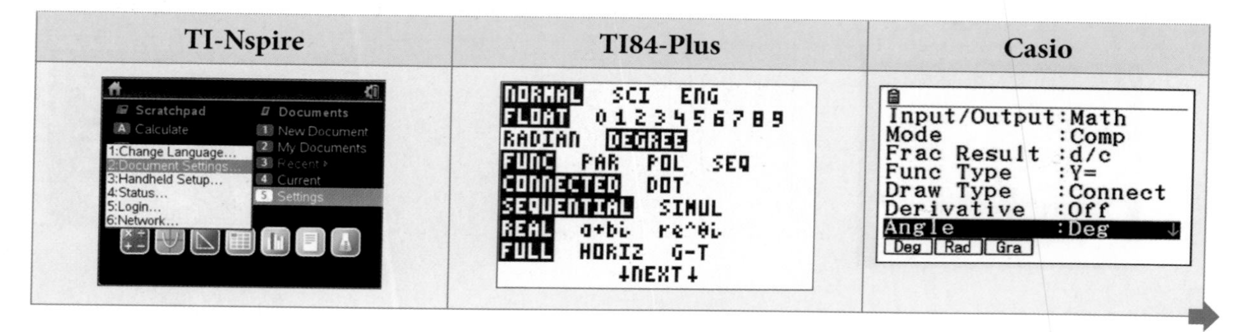

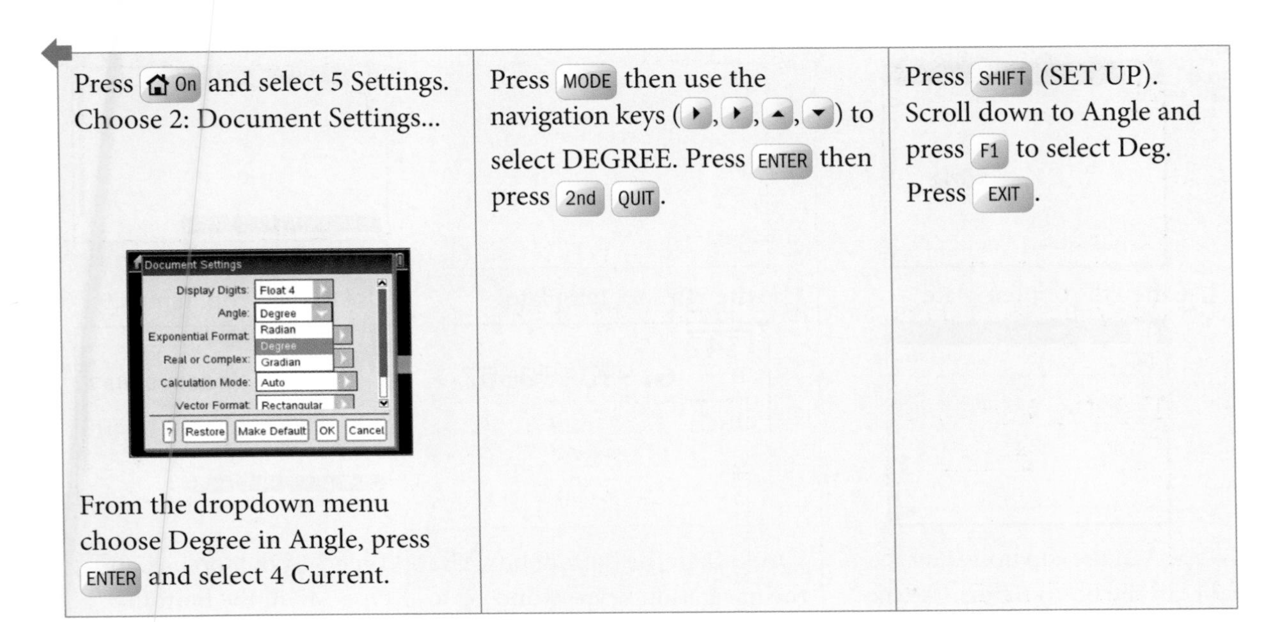

Press [On] and select 5 Settings. Choose 2: Document Settings... From the dropdown menu choose Degree in Angle, press [ENTER] and select 4 Current.	Press [MODE] then use the navigation keys (▸, ▸, ▴, ▾) to select DEGREE. Press [ENTER] then press [2nd] [QUIT].	Press [SHIFT] (SET UP). Scroll down to Angle and press [F1] to select Deg. Press [EXIT].

1.3 *n*th roots

For finding other roots apart from the square root there is another template for finding the *n*th roots of a number. The TI-84 Plus and the Casio also have a cube root function.

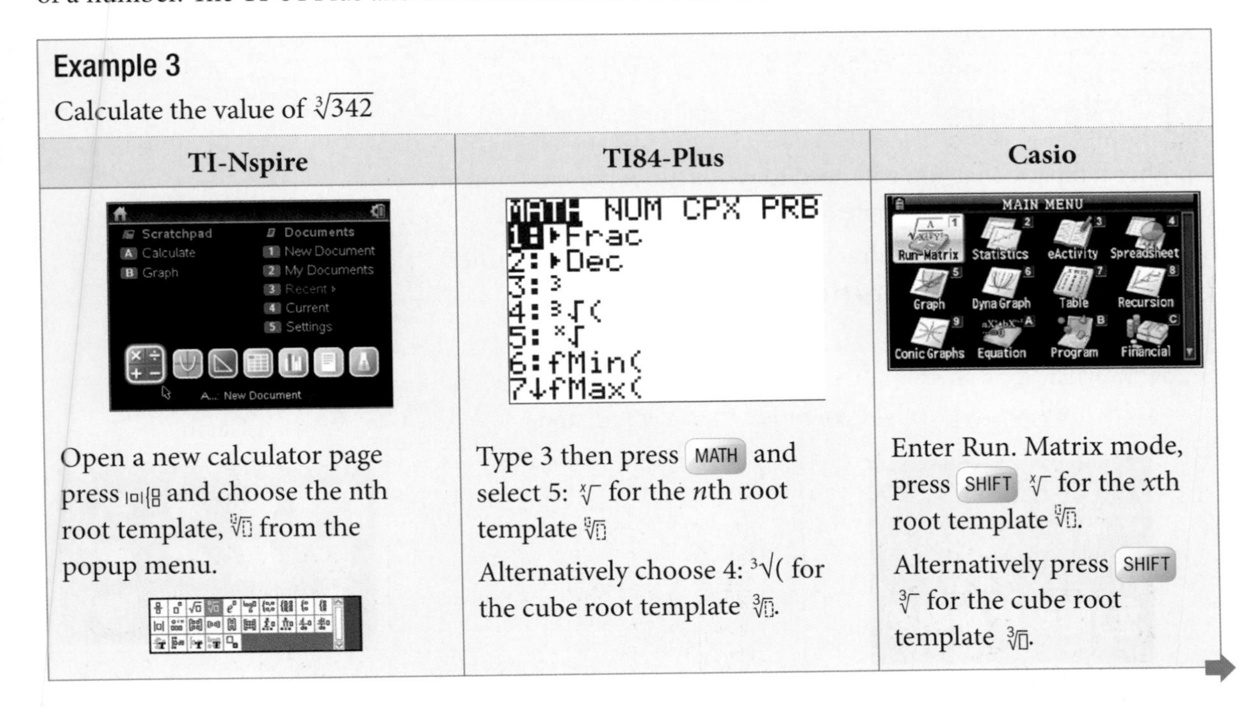

Example 3

Calculate the value of $\sqrt[3]{342}$

TI-Nspire	TI84-Plus	Casio
Open a new calculator page press [templates] and choose the nth root template, $\sqrt[\square]{\square}$ from the popup menu.	Type 3 then press [MATH] and select 5: $\sqrt[x]{}$ for the *n*th root template $\sqrt[\square]{\square}$ Alternatively choose 4: $^3\sqrt{}($ for the cube root template $\sqrt[3]{\square}$.	Enter Run. Matrix mode, press [SHIFT] $\sqrt[x]{}$ for the *x*th root template $\sqrt[\square]{\square}$. Alternatively press [SHIFT] $\sqrt[3]{}$ for the cube root template $\sqrt[3]{\square}$.

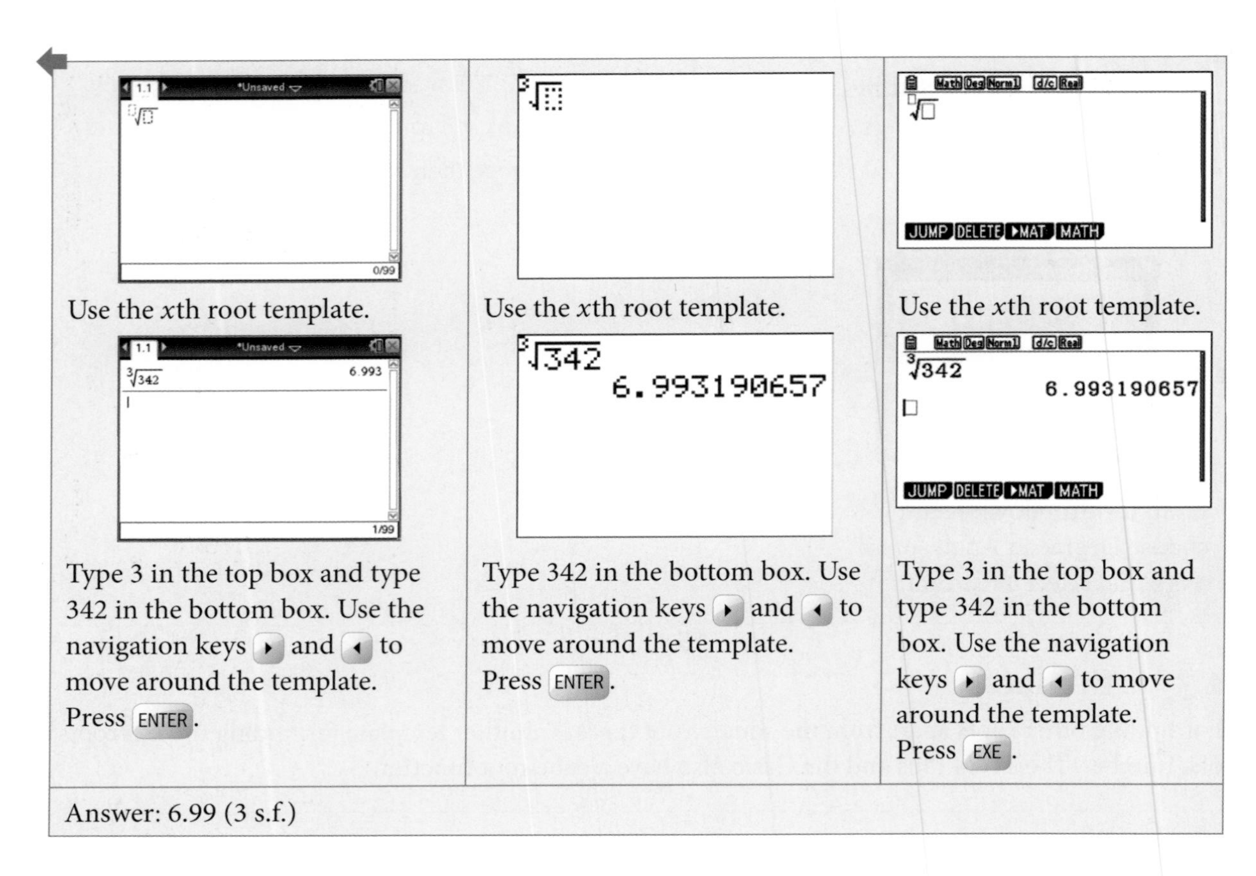

Use the xth root template.	Use the xth root template.	Use the xth root template.
Type 3 in the top box and type 342 in the bottom box. Use the navigation keys ▸ and ◂ to move around the template. Press ENTER.	Type 342 in the bottom box. Use the navigation keys ▸ and ◂ to move around the template. Press ENTER.	Type 3 in the top box and type 342 in the bottom box. Use the navigation keys ▸ and ◂ to move around the template. Press EXE.
Answer: 6.99 (3 s.f.)		

1.4 Exponents

Another template that you will need to use is the exponent template. Consider the following example.

You will also need to use the fraction template as described earlier.

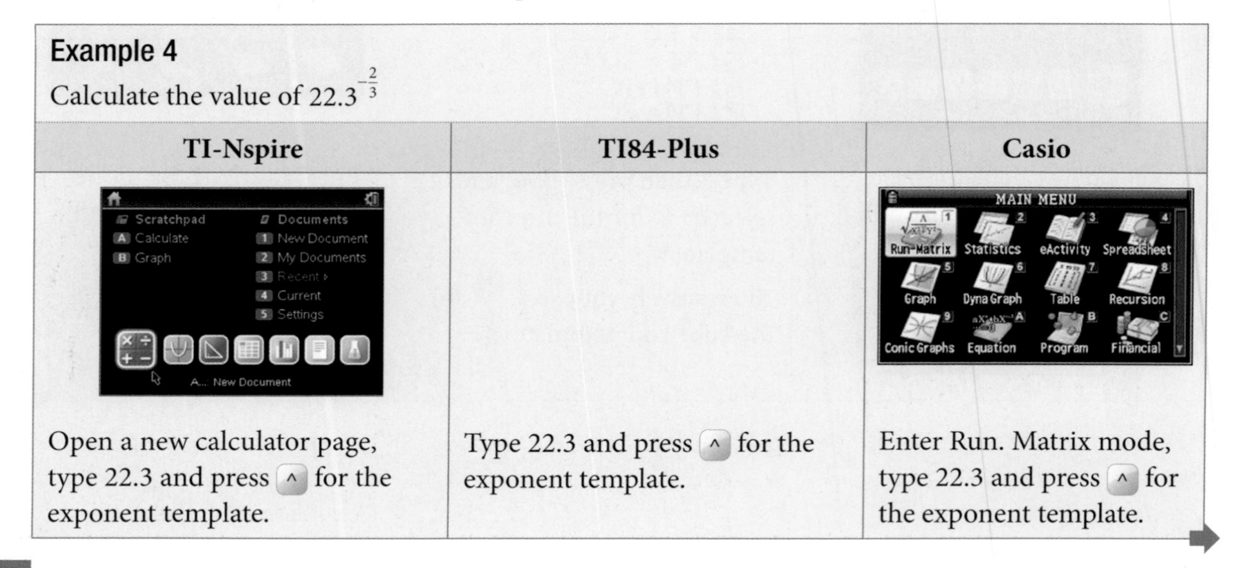

Example 4

Calculate the value of $22.3^{-\frac{2}{3}}$

TI-Nspire	TI84-Plus	Casio
Open a new calculator page, type 22.3 and press ^ for the exponent template.	Type 22.3 and press ^ for the exponent template.	Enter Run. Matrix mode, type 22.3 and press ^ for the exponent template.

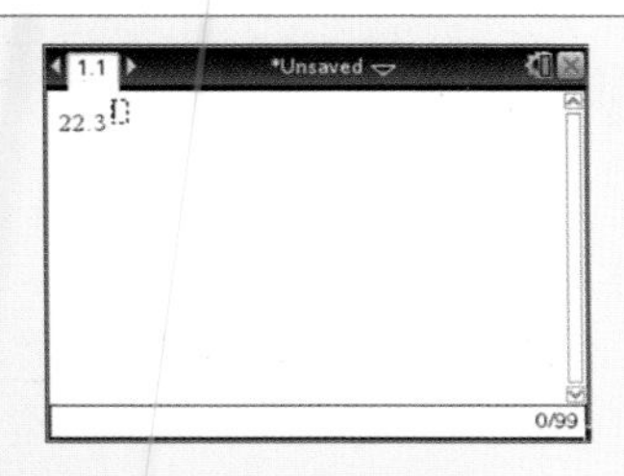	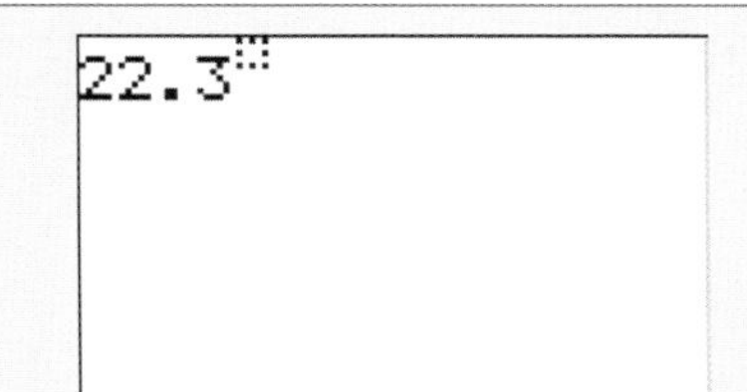	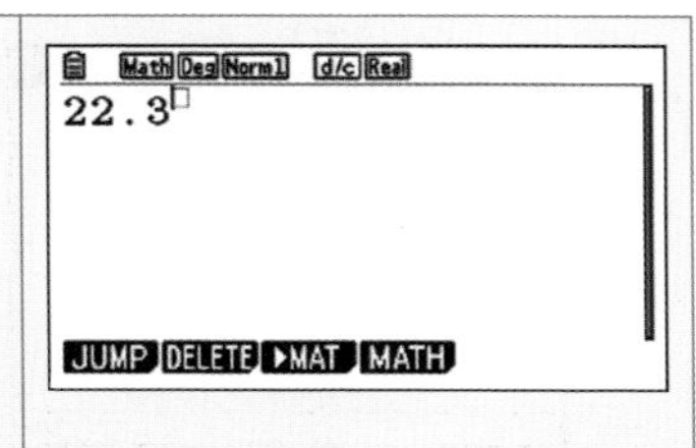
Use the exponent template.	Use the exponent template.	Use the exponent template.
	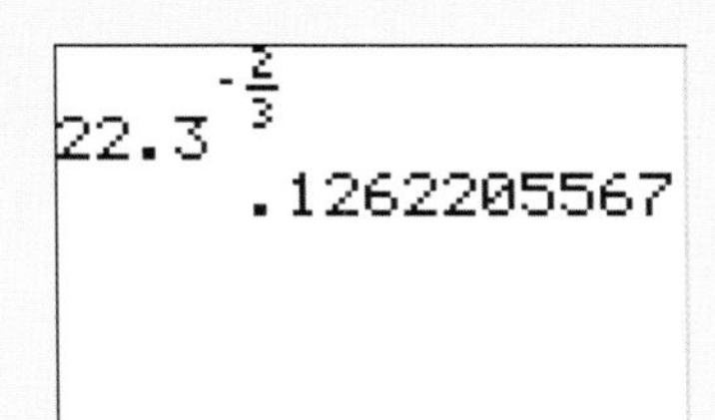	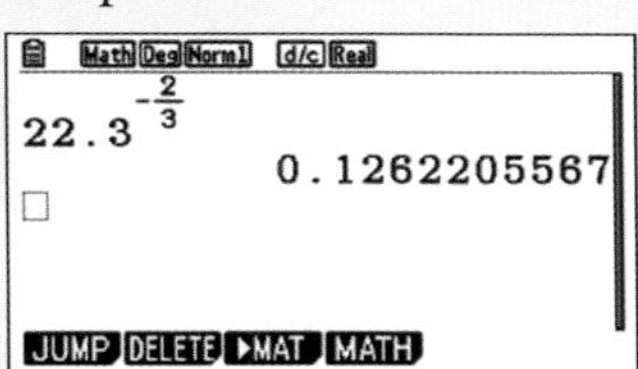
Press (-) to enter the negative symbol then use the fraction template, $\frac{\square}{\square}$ to enter $\frac{2}{3}$. (The brackets are inserted automatically). Use the navigation keys ▲, ▼, ► and ◄ to move around the templates. Press ENTER.	Press (-) to enter the negative symbol then use the fraction template, $\frac{\square}{\square}$ to enter $\frac{2}{3}$. Use the navigation keys ▲, ▼, ► and ◄ to move around the templates. Press ENTER.	Press (-) to enter the negative symbol then use the fraction template, $\frac{\square}{\square}$ to enter $\frac{2}{3}$. Use the navigation keys ▲, ▼, ► and ◄ to move around the templates. Press EXE.

Answer: 0.126 (3 s.f.)

When you are using the exponent template as a part of a longer calculation, make sure that you exit the template before you enter the rest of the calculation. Consider the following example.

Example 5

Calculate the value of $4^3 + 5^3$

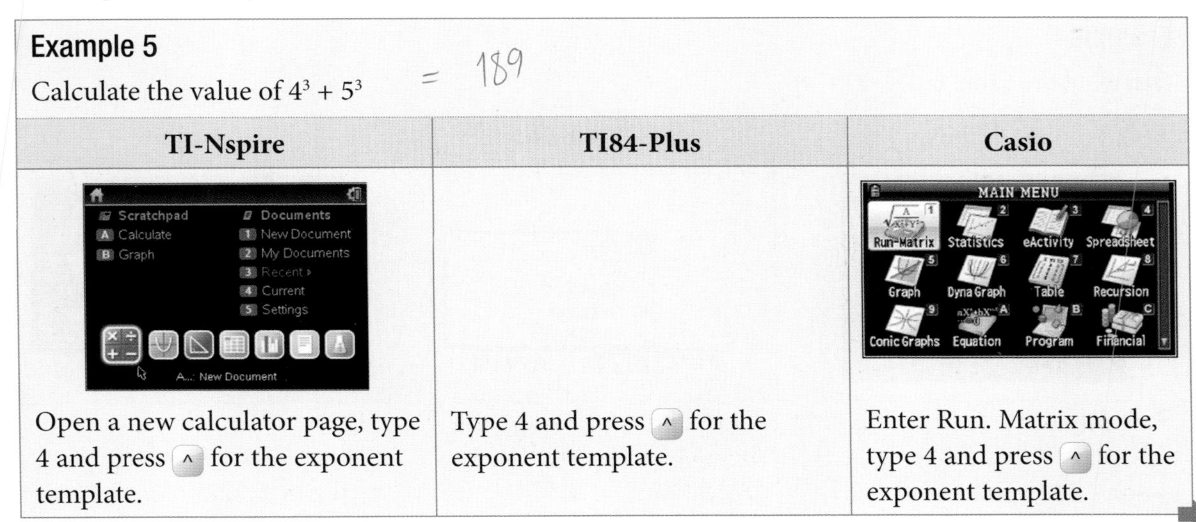

TI-Nspire	TI84-Plus	Casio
Open a new calculator page, type 4 and press ^ for the exponent template.	Type 4 and press ^ for the exponent template.	Enter Run. Matrix mode, type 4 and press ^ for the exponent template.

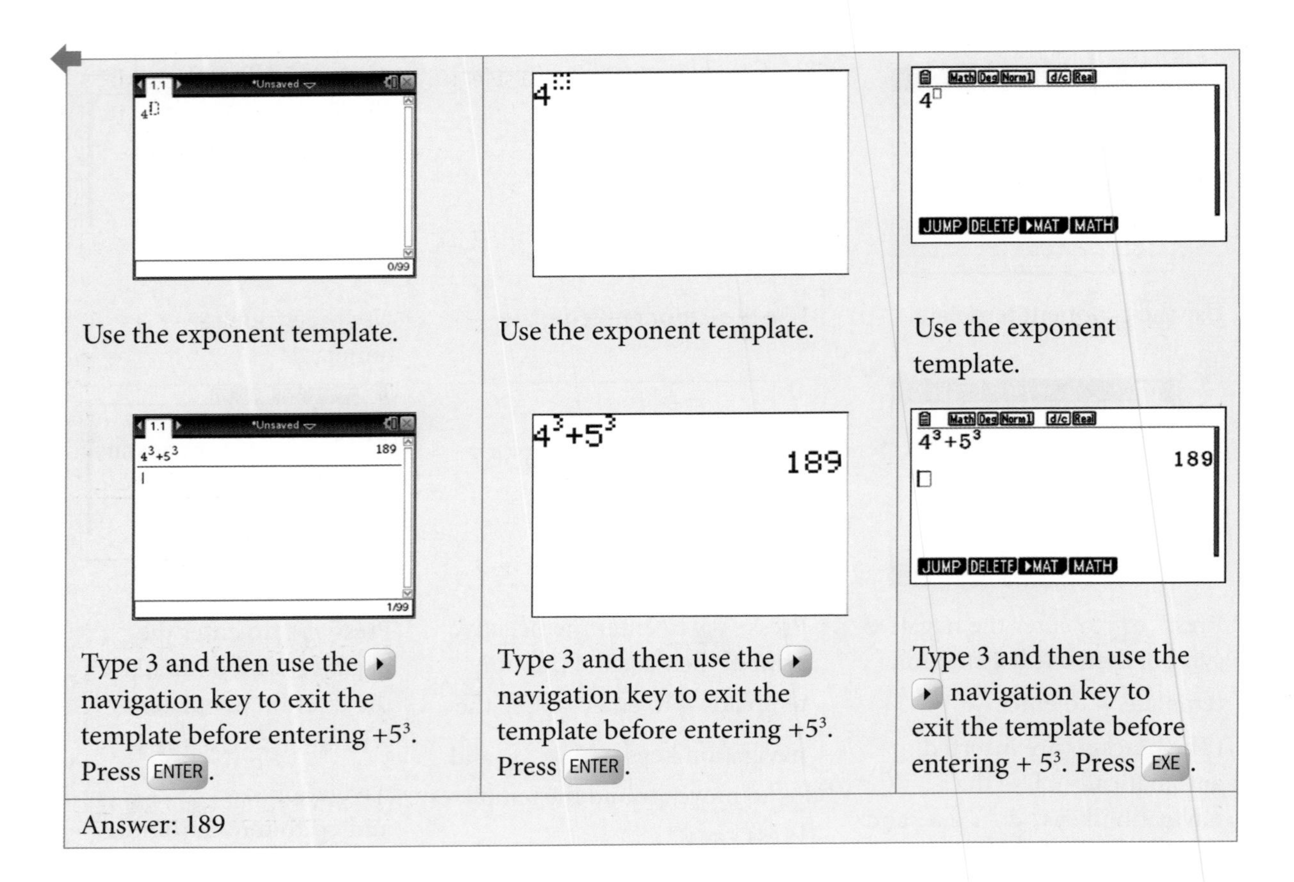

Use the exponent template.	Use the exponent template.	Use the exponent template.
Type 3 and then use the ▸ navigation key to exit the template before entering $+5^3$. Press ENTER.	Type 3 and then use the ▸ navigation key to exit the template before entering $+5^3$. Press ENTER.	Type 3 and then use the ▸ navigation key to exit the template before entering $+ 5^3$. Press EXE.

Answer: 189

1.5 Absolute value

Another function from the International Mathematics syllabus that uses a template is the absolute value function.

Example 6

Calculate the value of $|3(-2) + 4|$

TI-Nspire	TI84-Plus	Casio

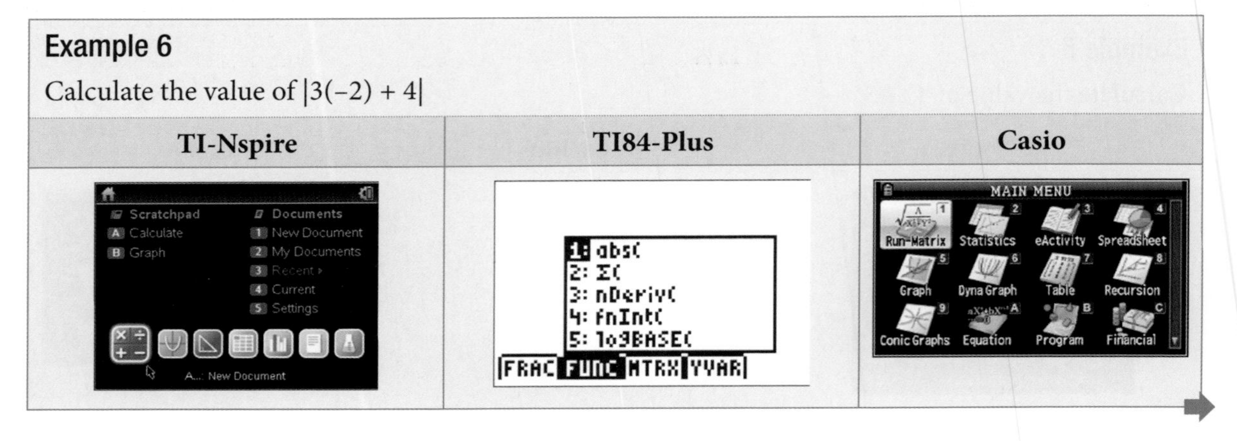

Open a new calculator page press [key] and choose the absolute value template, $|\square|$ from the popup menu.  	Press ALPHA π FUNC and select 1: abs(. Alternatively press MATH | NUM | 1:abs(. 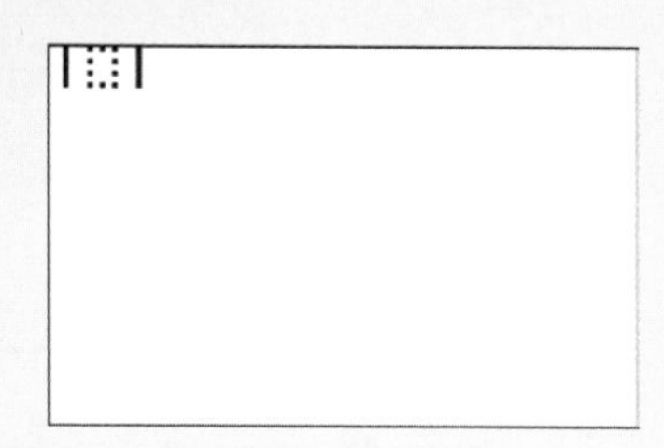	Enter Run. Matrix mode, Press F4 MATH F3 Abs for the absolute value template, $|\square|$. 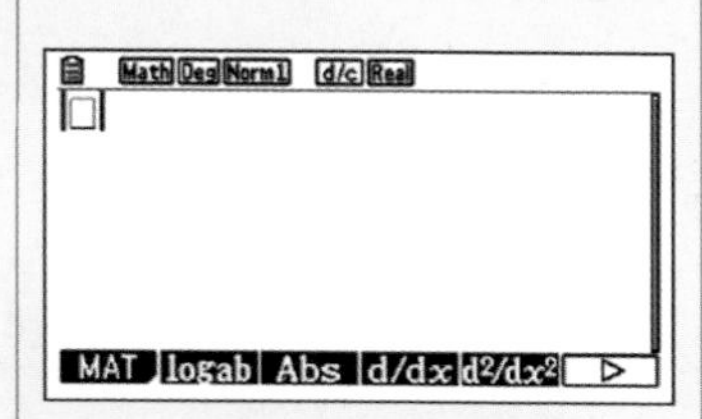
Use the absolute value template. 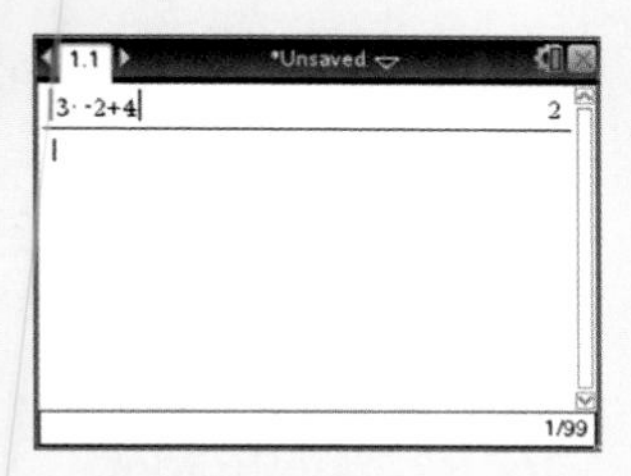	Use the absolute value template. 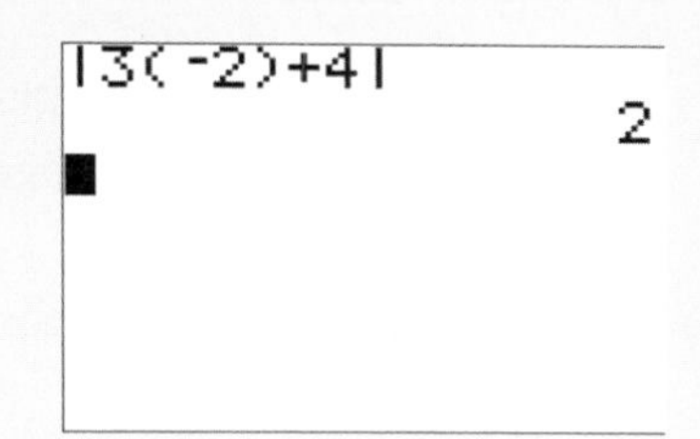	Use the absolute value template. 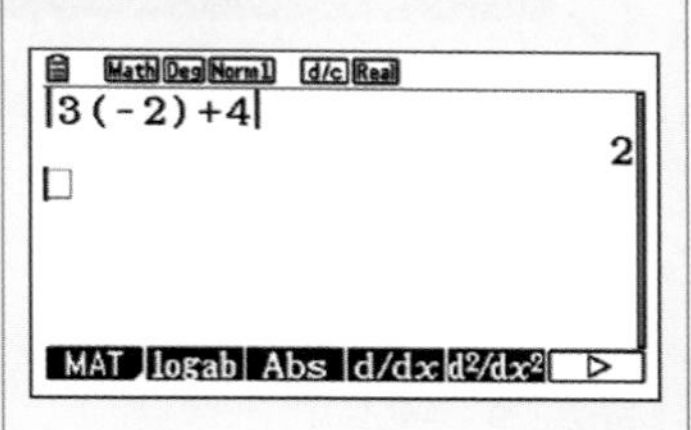
Type 3 ((-) 2) + 4. You can use the navigation key ▸ to move outside the template. Press ENTER. Note that the GDC automatically removes the brackets.	Type 3 ((-) 2) + 4. You can use the navigation key ▸ to move outside the template. Press ENTER.	Type 3 ((-) 2) + 4. You can use the navigation key ▸ to move outside the template. Press EXE.
Answer: 2		

1.6 Logarithms

There are two or three ways that a GDC will calculate logarithms. Logarithms can be found to base 10, base e or to any base you choose. Natural logarithms to base e are not a part of the International Mathematics course. You should not use the key marked ln for calculating with logarithms.

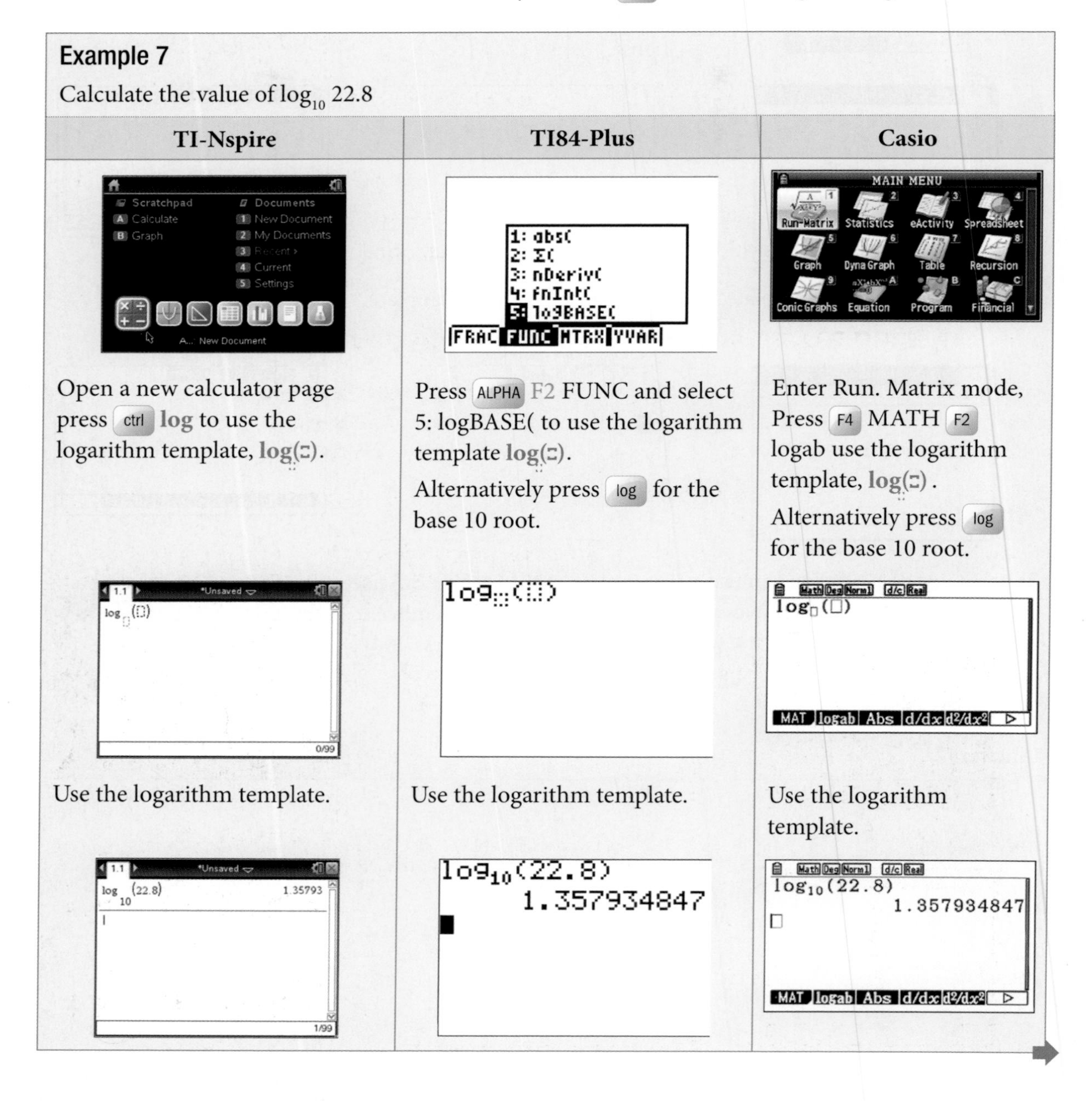

Example 7

Calculate the value of $\log_{10} 22.8$

TI-Nspire	TI84-Plus	Casio
Open a new calculator page press ctrl **log** to use the logarithm template, **log(□)**.	Press ALPHA F2 FUNC and select 5: logBASE(to use the logarithm template **log(□)**. Alternatively press log for the base 10 root.	Enter Run. Matrix mode, Press F4 MATH F2 logab use the logarithm template, **log(□)**. Alternatively press log for the base 10 root.
Use the logarithm template.	Use the logarithm template.	Use the logarithm template.

Type 10 in the bottom box and type 22.8 in the top box. Use the navigation keys ▸ and ◂ to move around the templates. Press ENTER.	Type 10 in the bottom box and type 22.8 in the top box. Use the navigation keys ▸ and ◂ to move around the templates. Press ENTER.	Type 10 in the bottom box and type 22.8 in the top box. Use the navigation keys ▸ and ◂ to move around the templates. Press EXE.
Answer: 1.36 (3 s.f.)		

2 Working with graphs

Of all the things that a GDC can do, drawing graphs is one of the main features. You must learn to use your GDC to sketch graphs. The calculator does a lot for you, but you will need to know how to get the most out of the graphing application.

2.1 Entering a function and choosing a window

To display the graph of a function successfully you will need to choose the right domain and range to define the window that you will use to view it. Sometimes both of these will be given to you, but sometimes you will have just the domain.

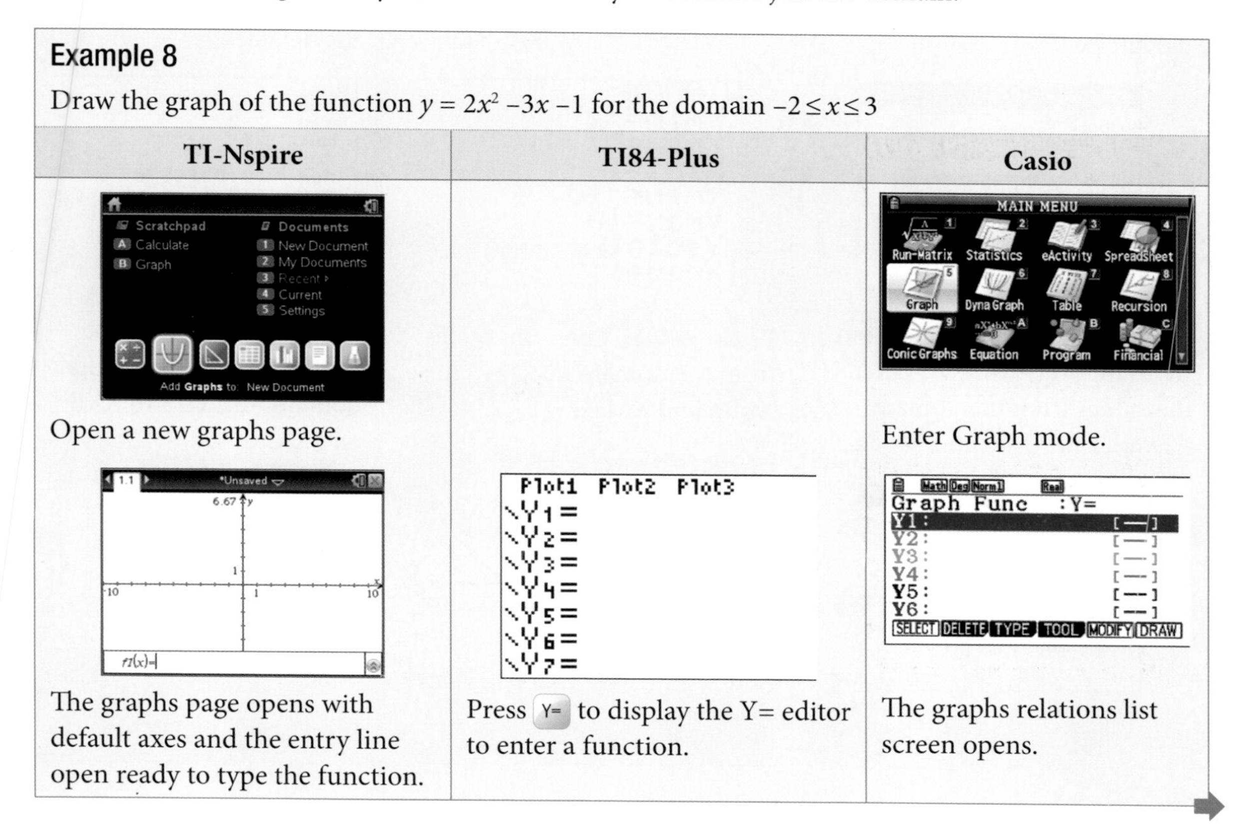

Example 8

Draw the graph of the function $y = 2x^2 - 3x - 1$ for the domain $-2 \leq x \leq 3$

TI-Nspire	TI84-Plus	Casio
Open a new graphs page.		Enter Graph mode.
The graphs page opens with default axes and the entry line open ready to type the function.	Press Y= to display the Y= editor to enter a function.	The graphs relations list screen opens.

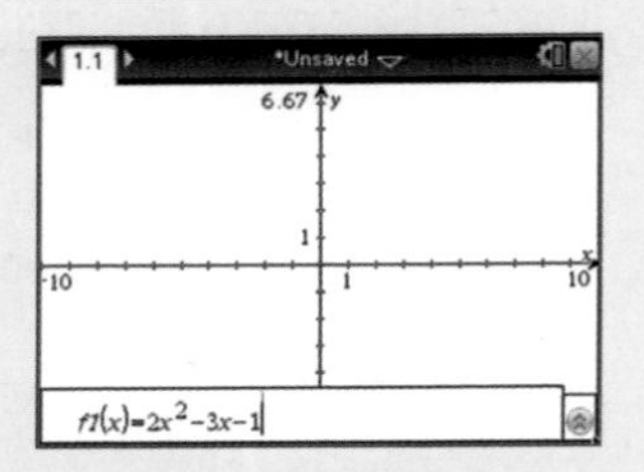

Enter the function exactly as you see it written, press ENTER.

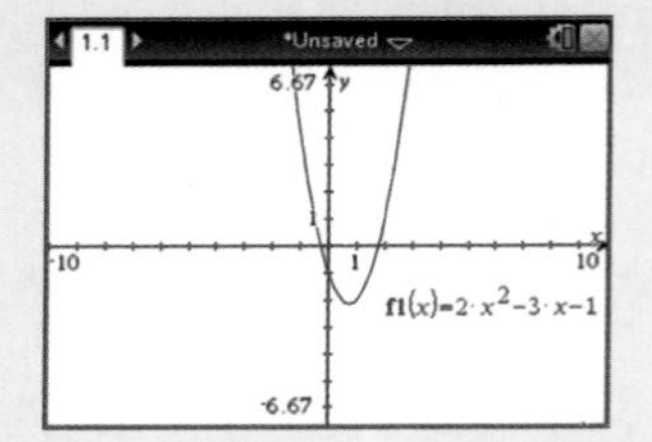

Press ENTER and you will see the function displayed with the default axes.

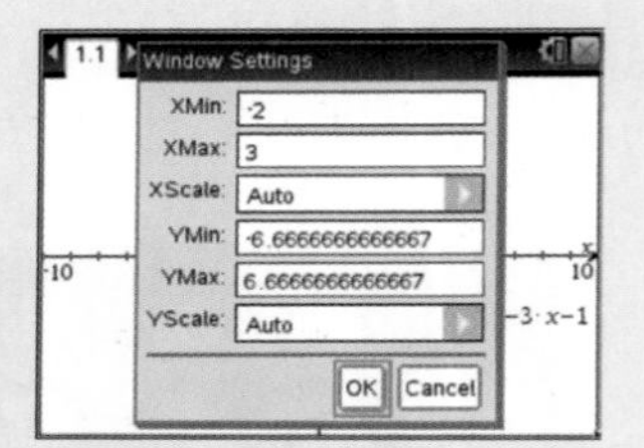

Press MENU 4: Window / Zoom | 1: Window Settings… Enter the values from the domain $-2<x<3$ in XMin and XMax.

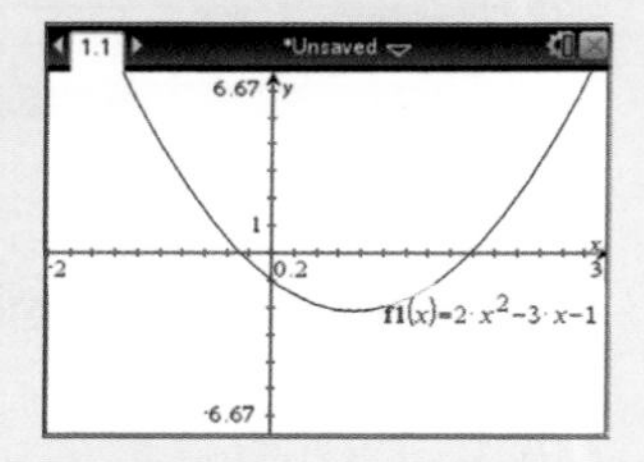

Press ENTER.

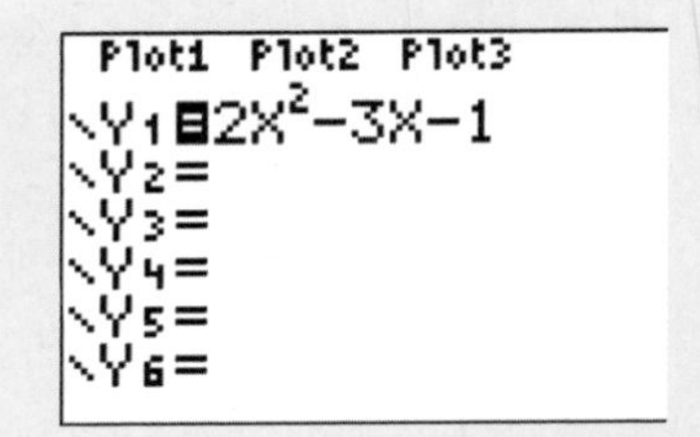

Enter the function exactly as you see it written, press ENTER. (Use the X,T,Θ,n key to enter x).

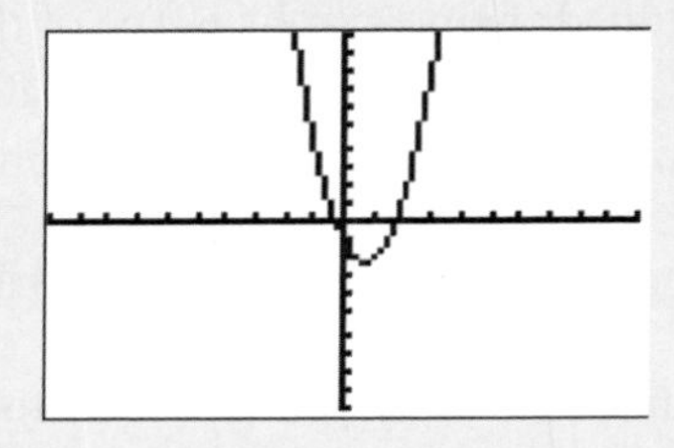

Press GRAPH to display the graph with the default axes.

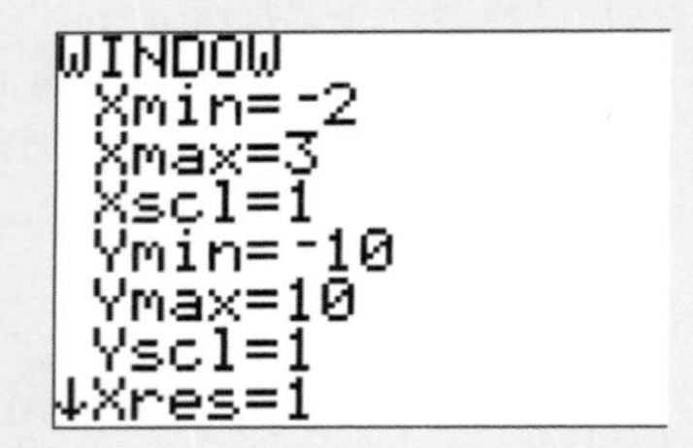

Press WINDOW. Enter the values from the domain $-2\leq x\leq 3$ in XMin and XMax.

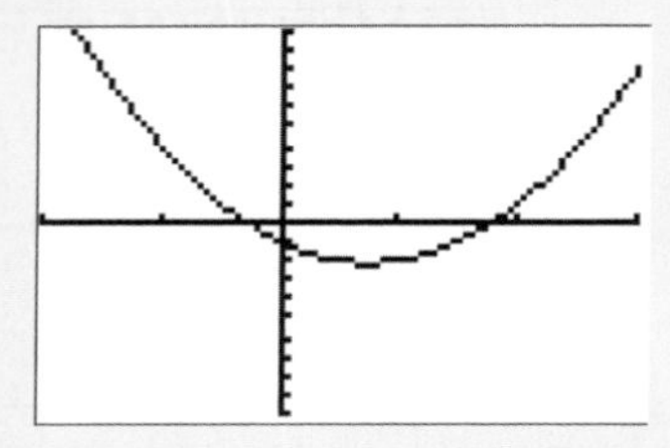

Press GRAPH.

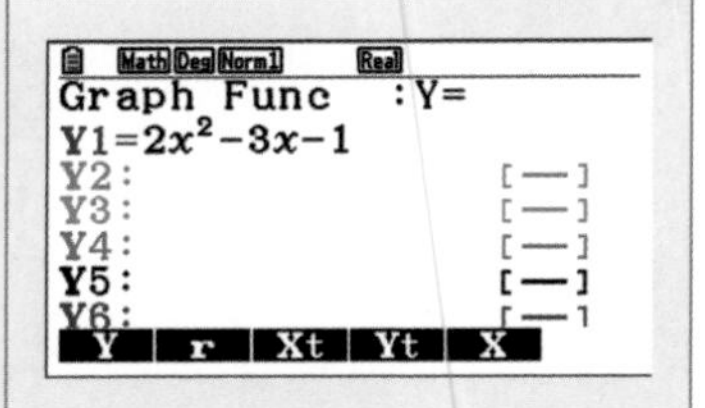

Enter the calculation exactly as you see it written, press EXE. (Use the X,θ,T key to enter x).

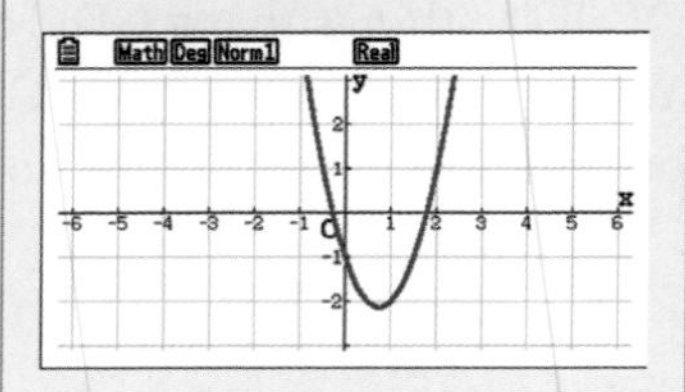

Press F6 DRAW to display the graph with the default axes.

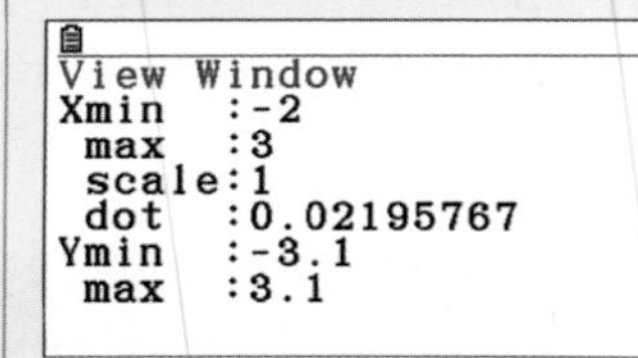

Press SHIFT F3 V-WIN. Enter the values from the domain $-2\leq x\leq 3$ in XMin and XMax.

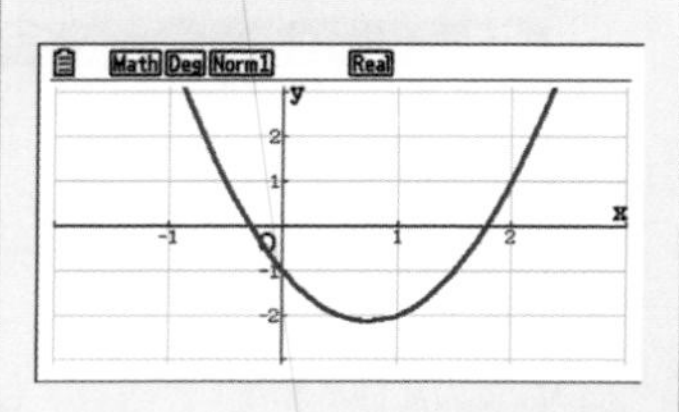

Press EXIT F6 DRAW.

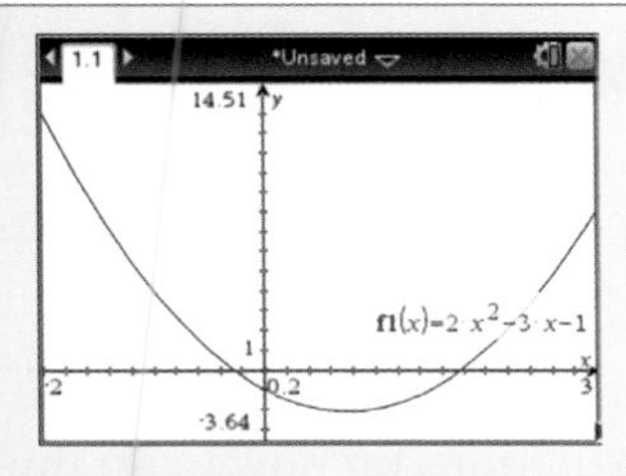

A quick way to choose suitable values for the range is to press MENU 4: Window / Zoom | A: Zoom – Fit.

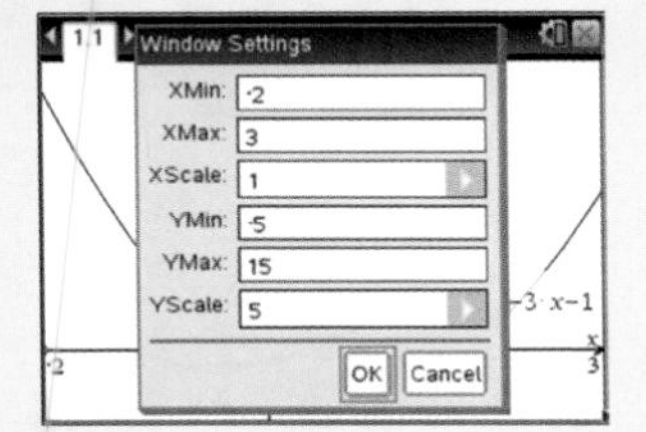

To improve on the automated axes, choose a scale and some round figures for the range.

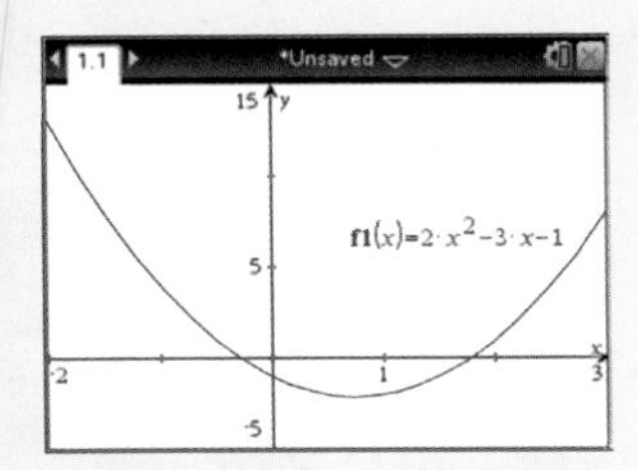

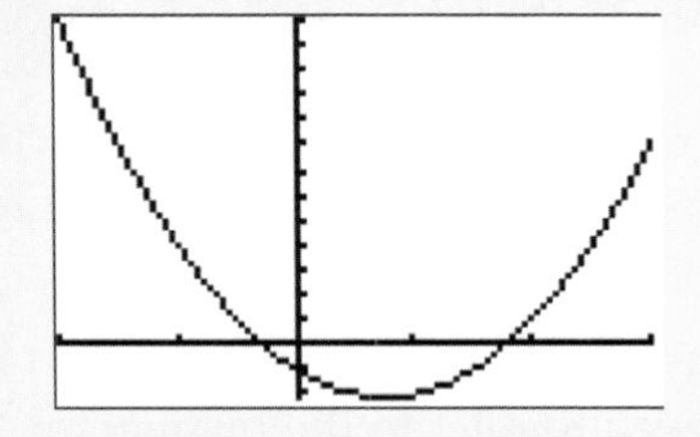

A quick way to choose suitable values for the range is to press ZOOM 0:ZoomFit.

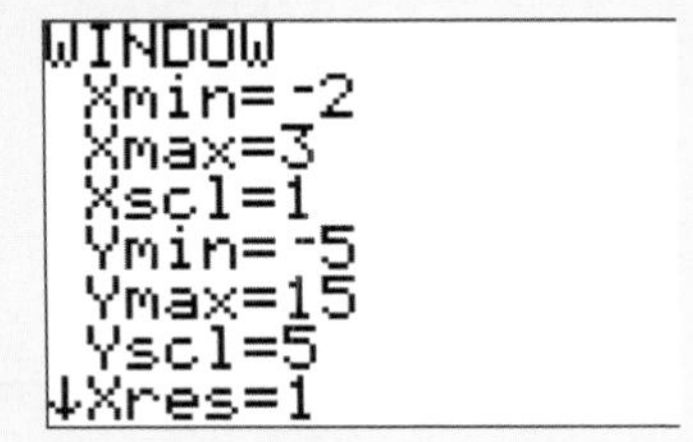

To improve on the automated axes, choose a scale and some round figures for the range.

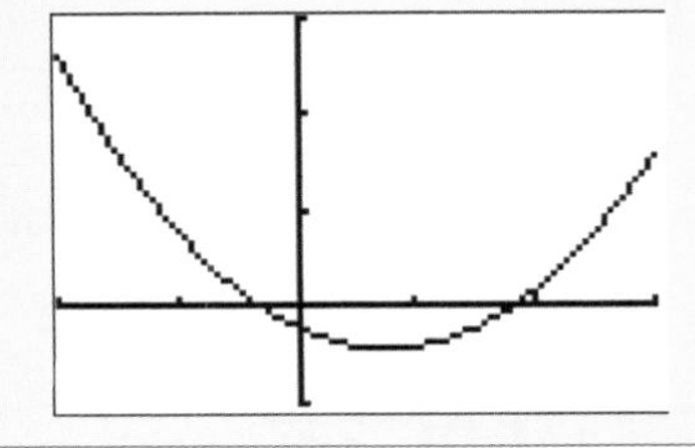

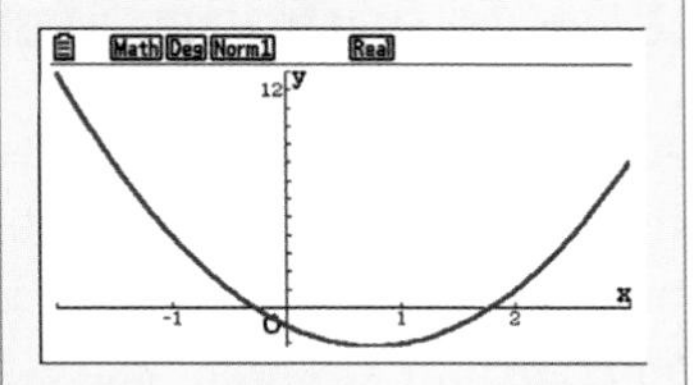

A quick way to choose suitable values for the range is to press SHIFT F2 ZOOM F5, AUTO.

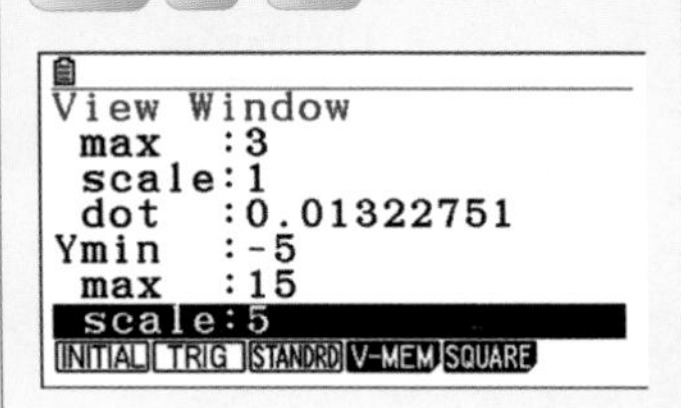

To improve on the automated axes, choose a scale and some round figures for the range.

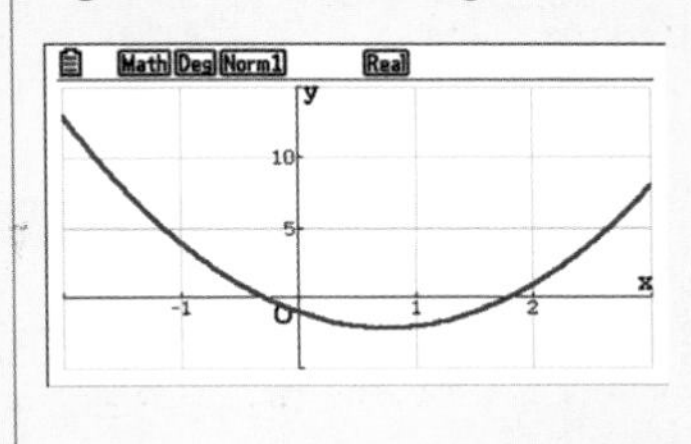

2.2 Producing a table of values for a function

A table of values is useful if you want to make a more accurate sketch of the graph of a function than a freehand sketch. A table is also a useful way to find the range of a function, to find turning points on graphs or to find asymptotes.

TI GDCs allow you to view the graph of the function and the table side by side. Casio GDCs will do something similar, but you will need to refer to the manual for full instructions on how to do this.

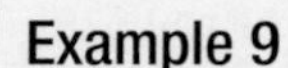

Example 9

Make a table of values for the function $y = 2x^2 - 3x - 1$ for the domain $-2 \leq x \leq 3$

TI-Nspire	TI84-Plus	Casio
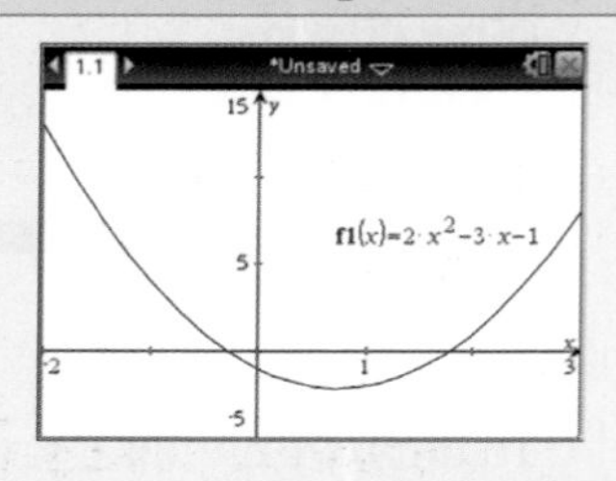	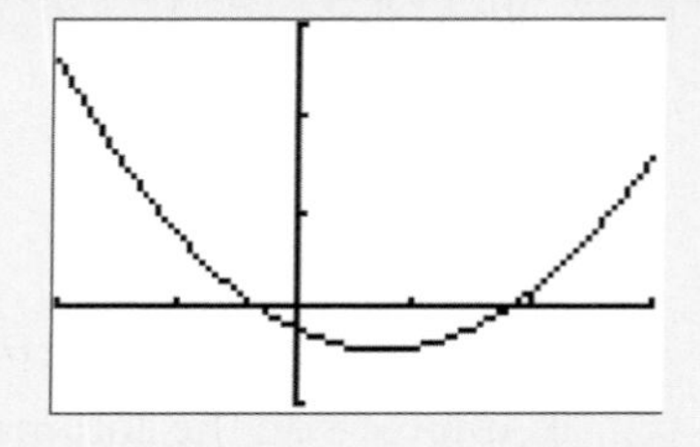	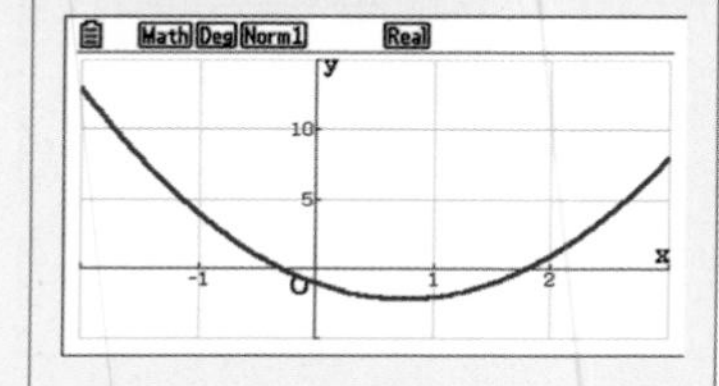
Open a new graphs page and draw the graph of the function $y = 2x^2 - 3x - 1$ for the domain $-2 \leq x \leq 3$ as in Example 8.	Draw the graph of the function $y = 2x^2 - 3x - 1$ for the domain $-2 \leq x \leq 3$ as in Example 8.	Enter Graph mode and draw the graph of the function $y = 2x^2 - 3x - 1$ for the domain $-2 \leq x \leq 3$ as in Example 8.
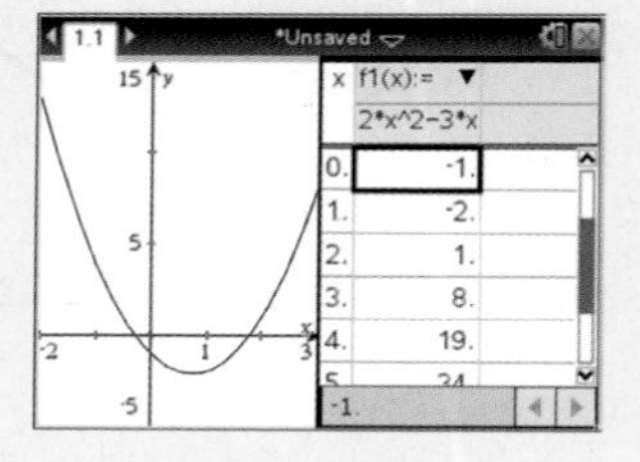	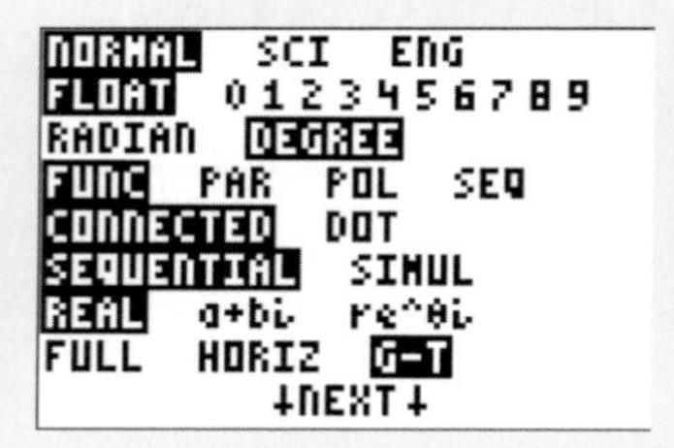	
Press MENU 7: Table | 1: Split-screen Table or press ctrl T to show the graph and a table side by side.	Press MODE, scroll down to G-T (Graph-Table) and press ENTER.	Press MENU 7 to enter Table mode.
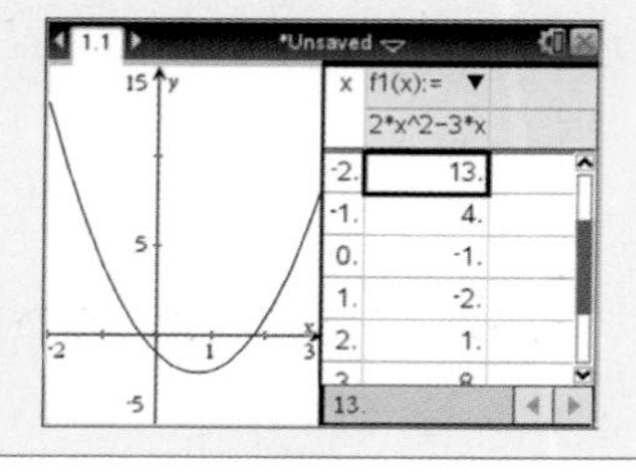	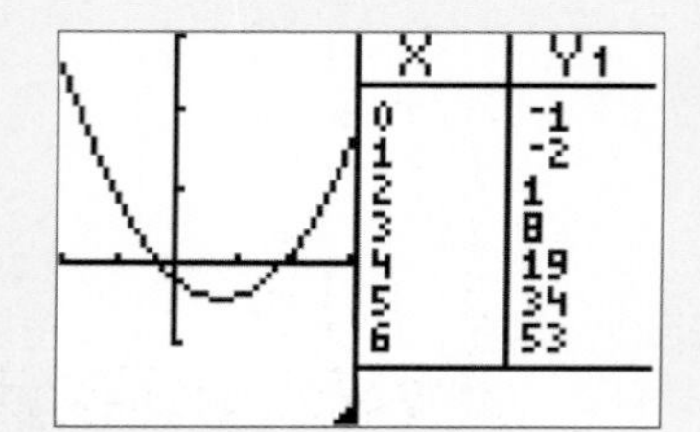	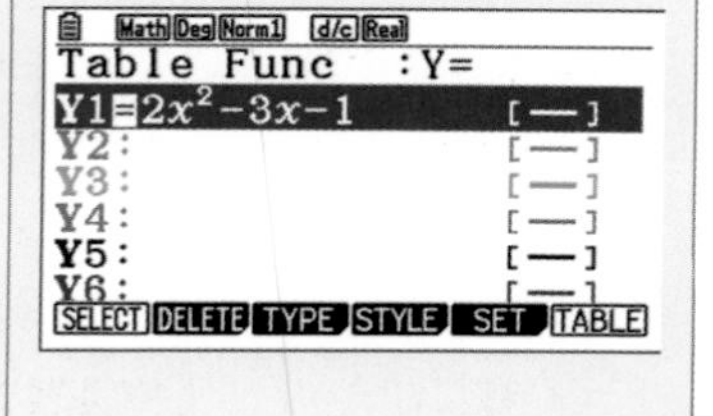

Use the navigation keys ▲ ▼ to scroll through the table.

To change the step size in the table, press MENU 2: Table | 5: Edit Table Settings… Navigate to Table Step and enter a new value.

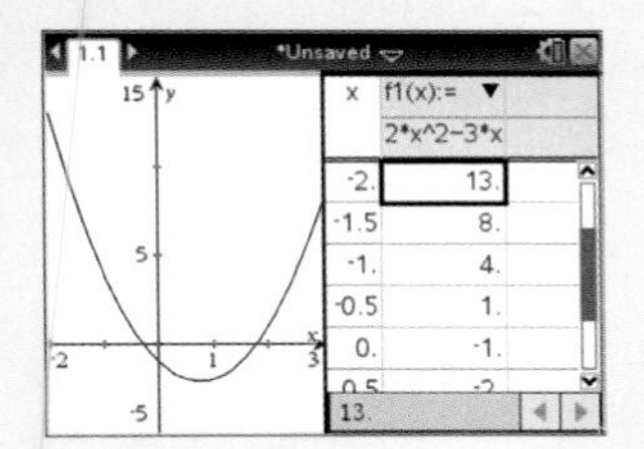

Press ENTER.

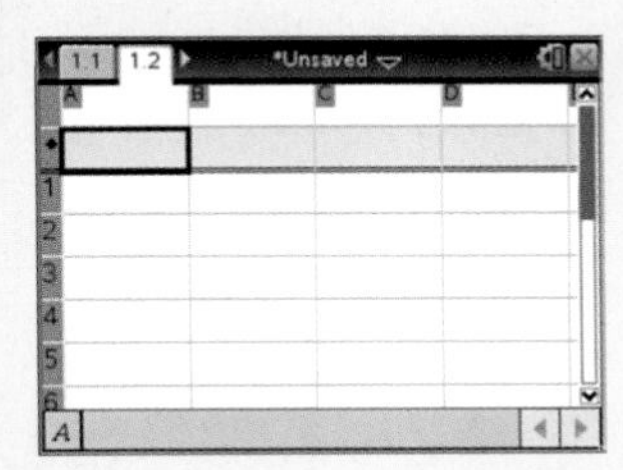

To show the table on its own press ctrl +page and press 4: Add Lists & Spreadsheet

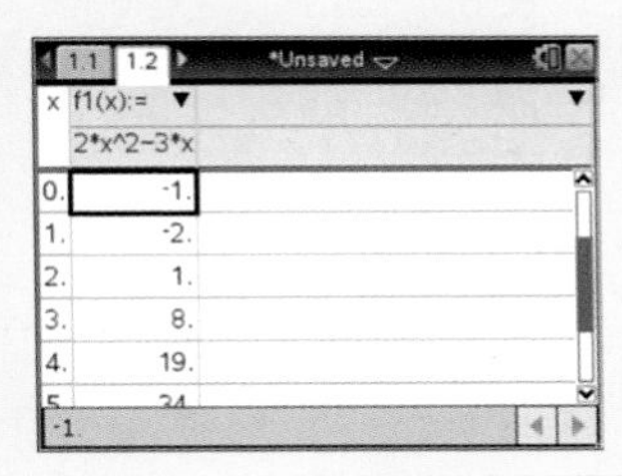

Press GRAPH to show the graph and a table side by side.

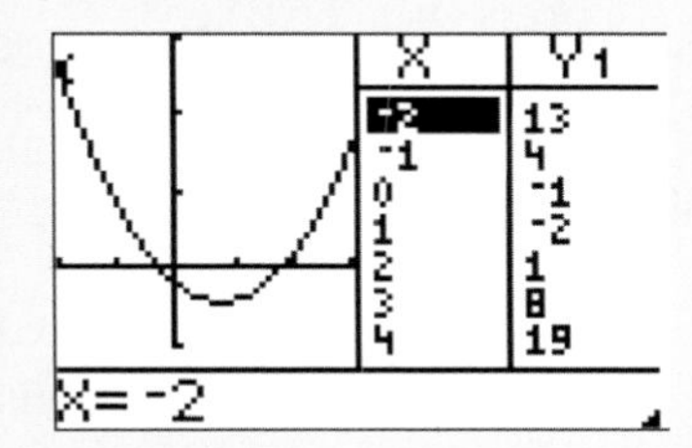

Press 2nd TABLE to switch to the table and use the navigation keys ▲ ▼ to scroll through the table.

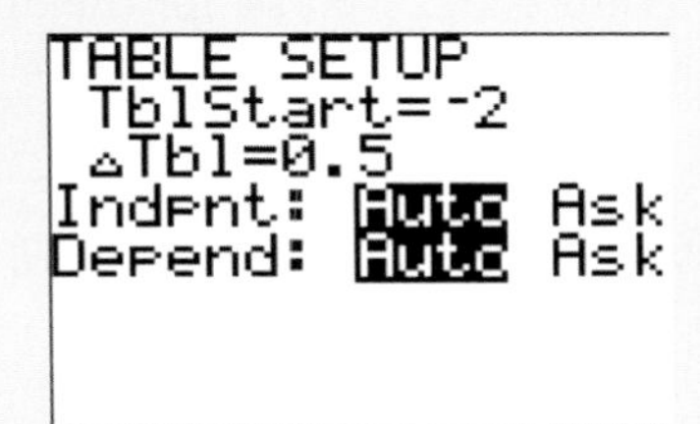

To change the step size in the table, press 2nd TBLSET. Navigate to ΔTbl and enter a new value.

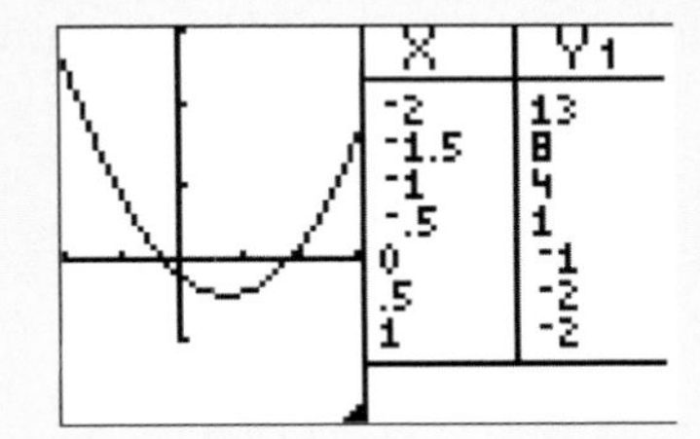

Press GRAPH.

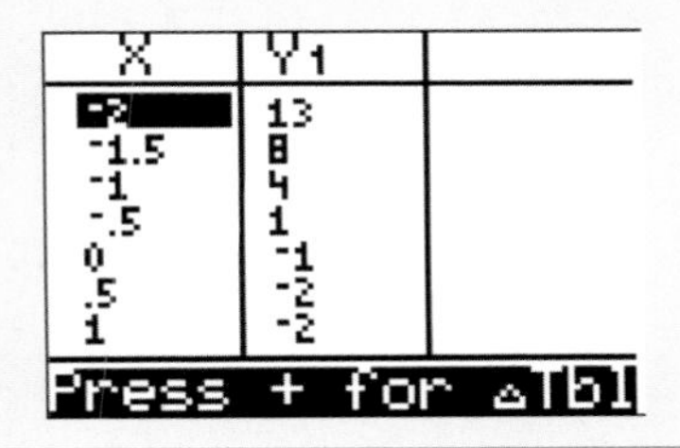

Press EXIT F5 SET to enter the domain for the table.

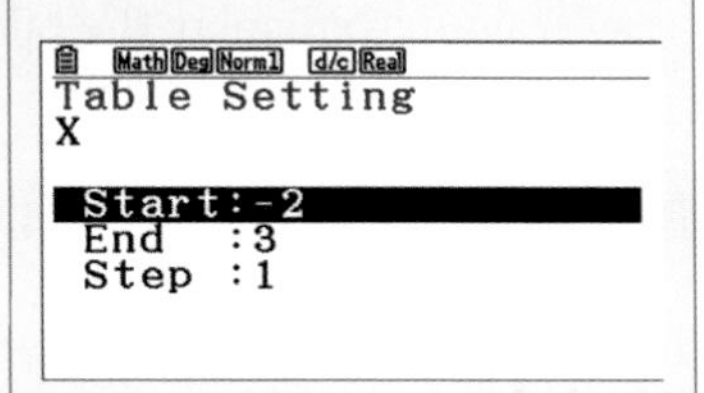

Enter the values for the domain $-2 \leq x \leq 3$ as the Start and End values.

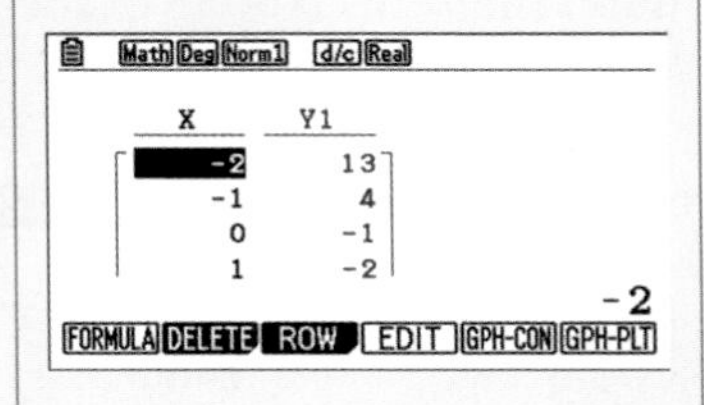

Press EXIT F6 TABLE.

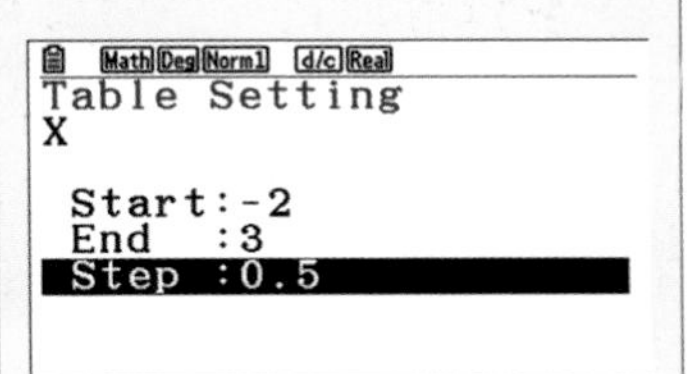

To change the step size in the table, Press EXIT F5 SET. Navigate to Step and enter a new value.

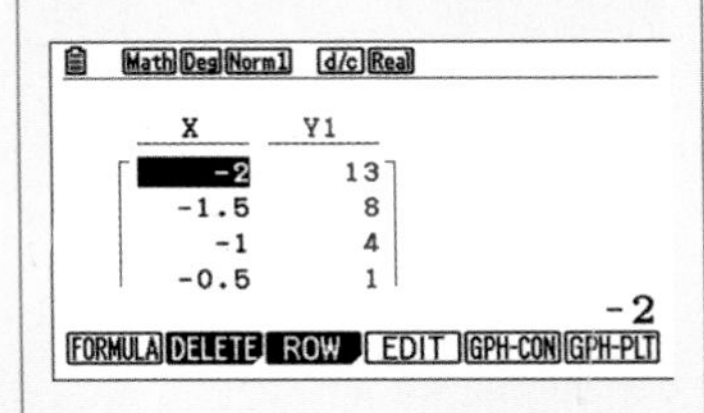

Press MENU 5: Table | 1: Switch to Table (or press ctrl T). Press ENTER again to display the table.	To show the table on its own, change MODE to FULL and press 2nd **TABLE**	Press EXIT F6 TABLE.

2.3 Finding zeros (roots)

The points where the graph of a function crosses the x-axis are known as the **zeros** or **roots** of the function. The zeros (roots) of the function f(x) are the solutions of the equation f(x) = 0.
The GDC provides a quick, efficient way to find these zeros (roots). For some functions there is also an algebraic method to do this. For others, the GDC is the only way for you to do this.

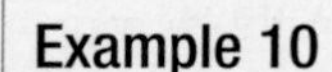

Example 10

Find the zeros (roots) of the function $y = 2x^2 - 3x - 1$ for the domain $-2 \le x \le 3$

TI-Nspire	TI84-Plus	Casio
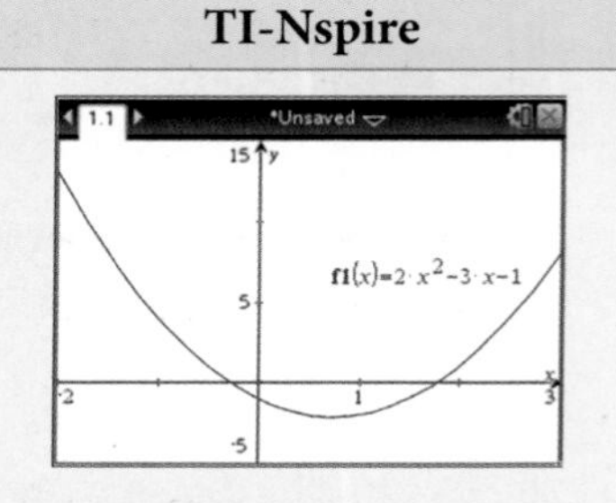	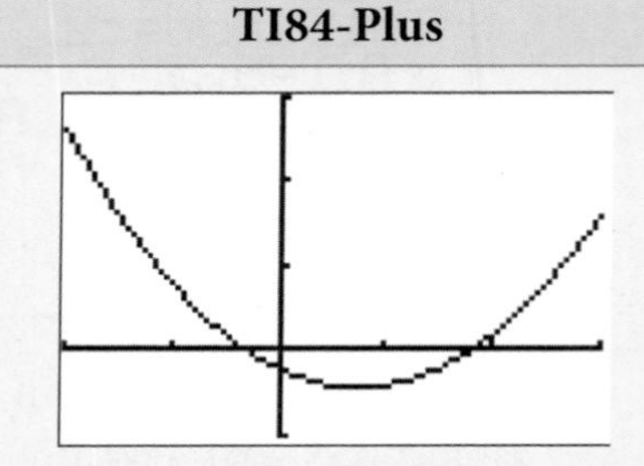	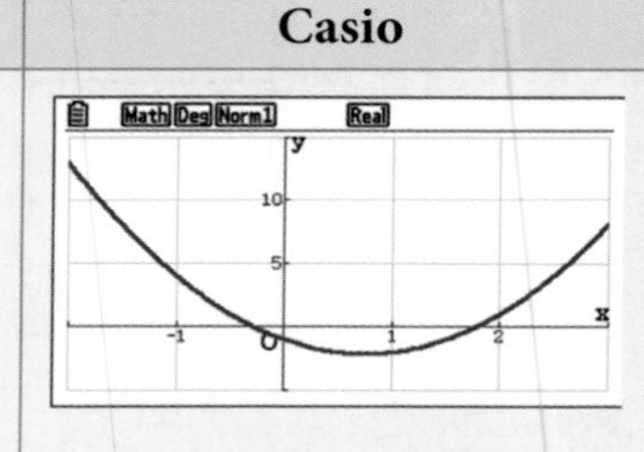
Open a new graphs page and draw the graph of the function $y = 2x^2 - 3x - 1$ for the domain $2 \le x \le 3$ as in Example 8.	Draw the graph of the function $y = 2x^2 - 3x - 1$ for the domain $2 \le x \le 3$ as in Example 8.	Enter Graph mode and draw the graph of the function $y = 2x^2 - 3x - 1$ for the domain $2 \le x \le 3$ as in Example 8.
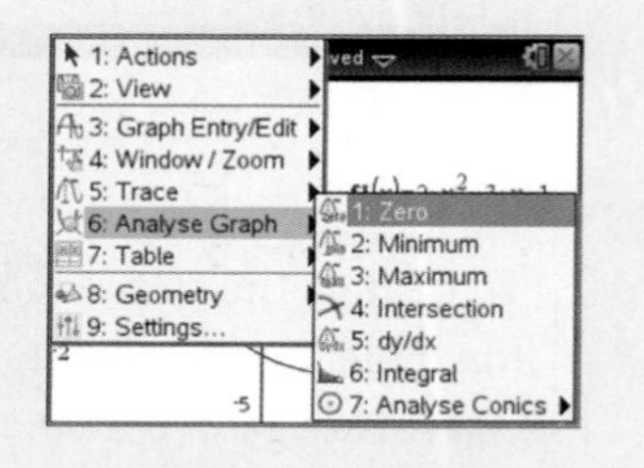	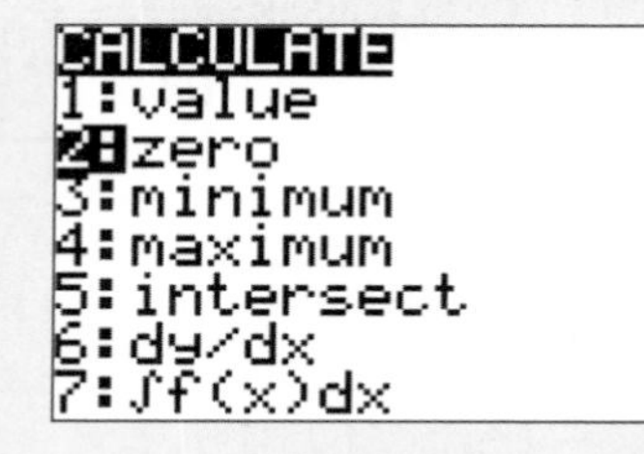	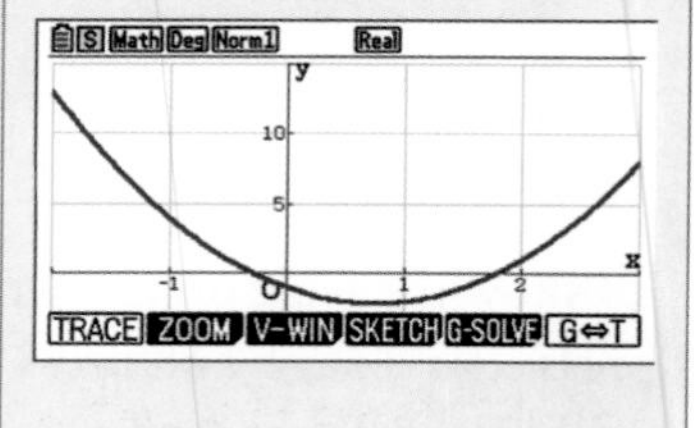
Press MENU 6: Analyse Graph | 1: Zero…	Press 2nd **CALC** 2: zero	Press SHIFT F5 G-SOLVE.

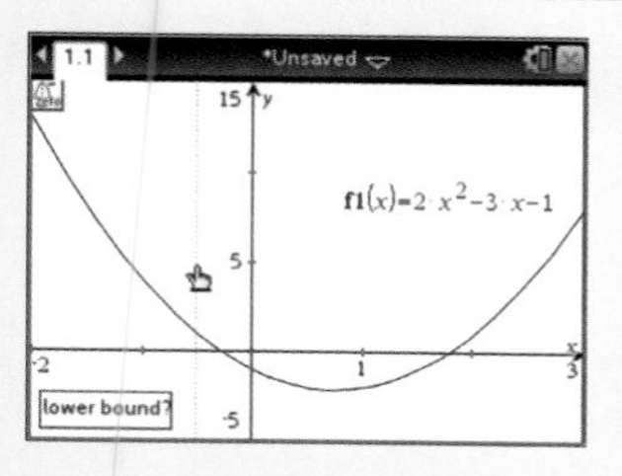

Select a point to the left of the zero as the lower bound and press ENTER. (Use the ▸ ◂ keys or type a value).

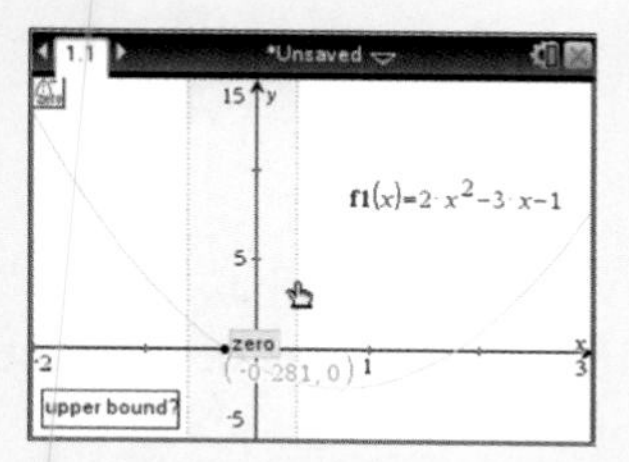

Select another point to the right of the zero as the upper bound and press ENTER. (Use the ▸ ◂ keys, the touchpad or type a value).

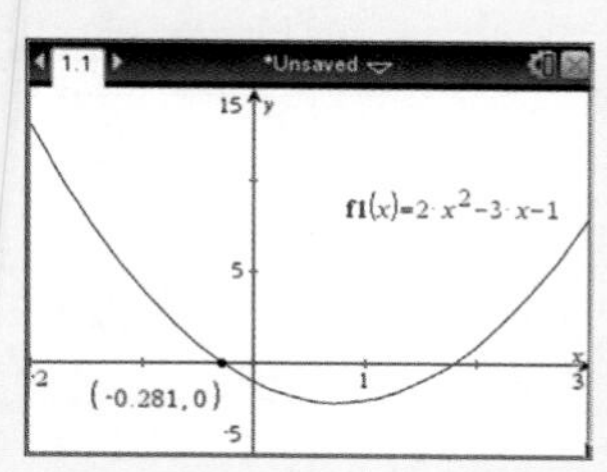

The GDC displays the coordinates of the zero.

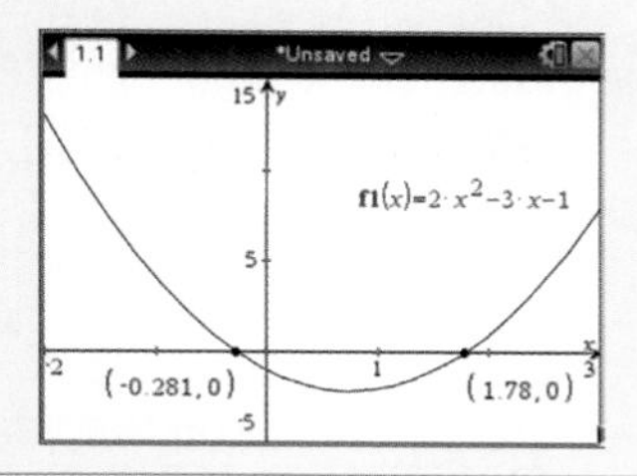

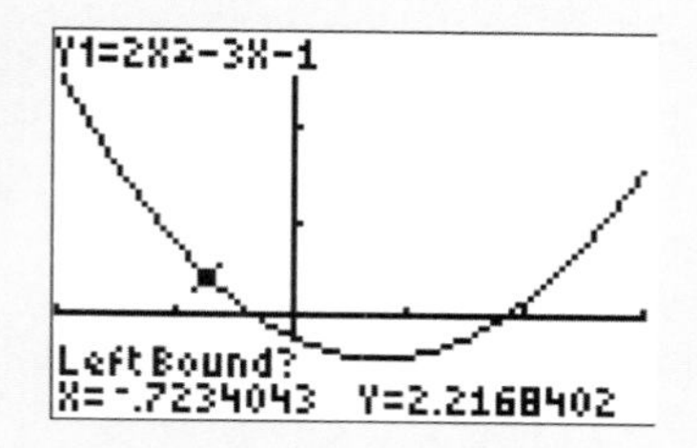

Select a point to the left of the zero as the left bound and press ENTER. (Use the ▸ ◂ keys or type a value).

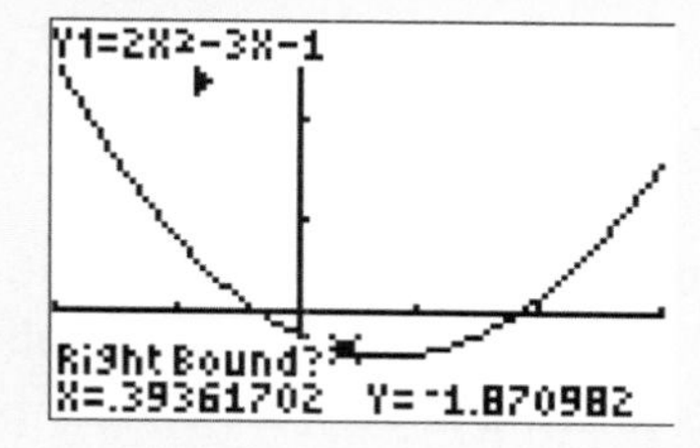

Select another point to the right of the zero as the upper bound and press ENTER. (Use the ▸ ◂ keys or type a value).

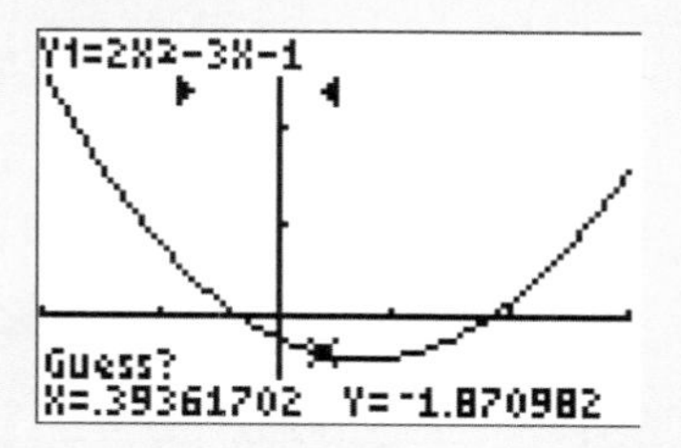

Press ENTER again.

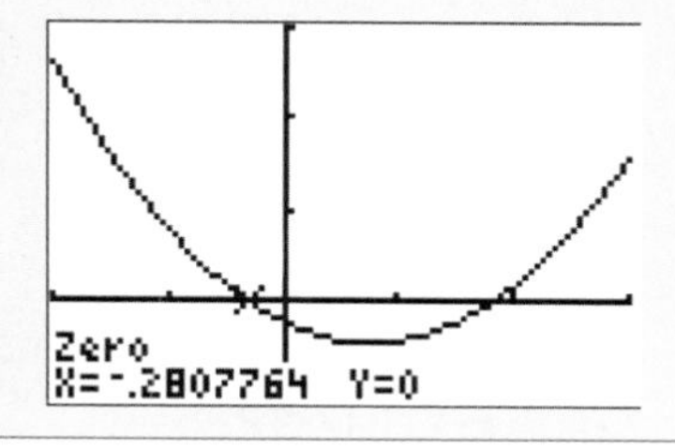

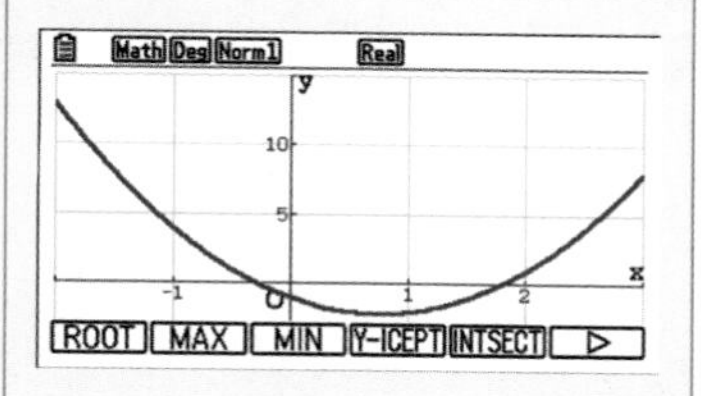

Press F1 ROOT.

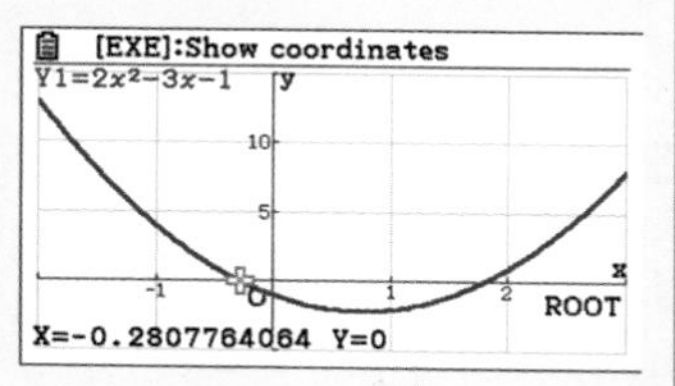

The GDC displays the coordinates of the root.

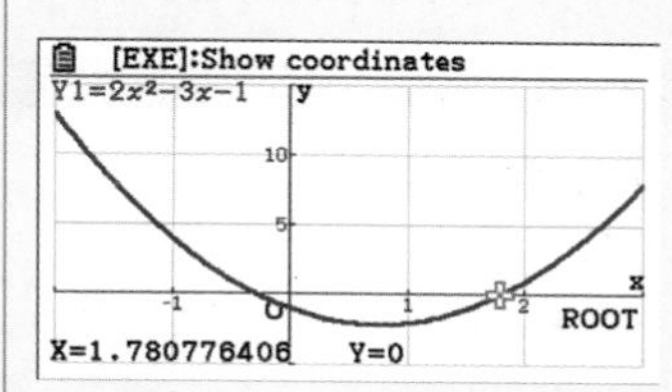

Press ▸ to find the second root. The GDC displays the coordinates of the root.

Repeat the process for the second zero.	The GDC displays the coordinates of the zero. 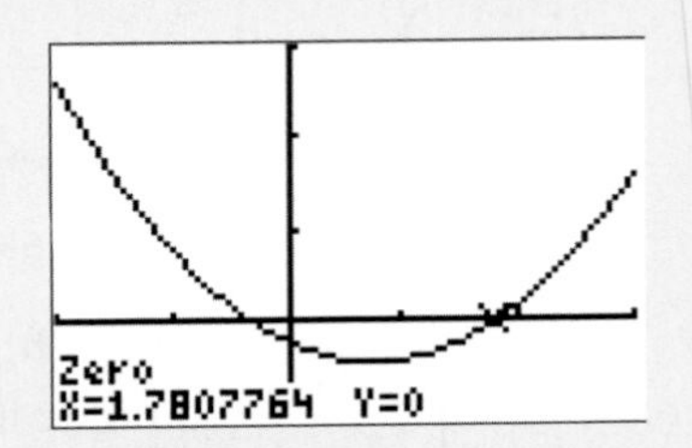 Repeat the process for the second zero.	

2.4 Finding a local minimum

The point $(x, f(x_0))$ is a **local minimum** of the function $f(x)$ if, for all values of x in the neighbourhood of x_0, $f(x_0) < f(x)$. This point may not be the absolute minimum value of the function. Points like this are also known as **turning points, stationary points** or **extrema**.

Example 11

Find the local minimum point of the function $y = 2x^2 - 3x - 1$ for the domain $-2 \le x \le 3$

TI-Nspire	TI84-Plus	Casio
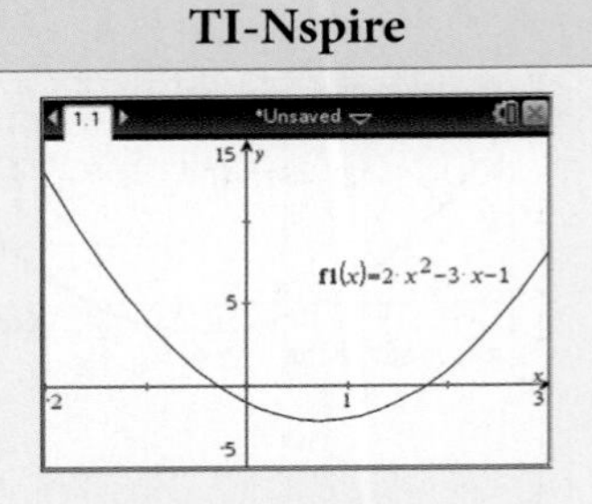 Open a new graphs page and draw the graph of the function $y = 2x^2 - 3x - 1$ for the domain $-2 \le x \le 3$ as in Example 8.	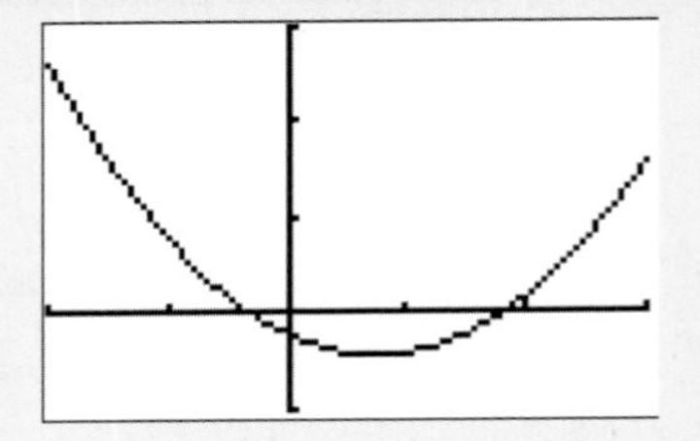 Draw the graph of the function $y = 2x^2 - 3x - 1$ for the domain $-2 \le x \le 3$ as in Example 8.	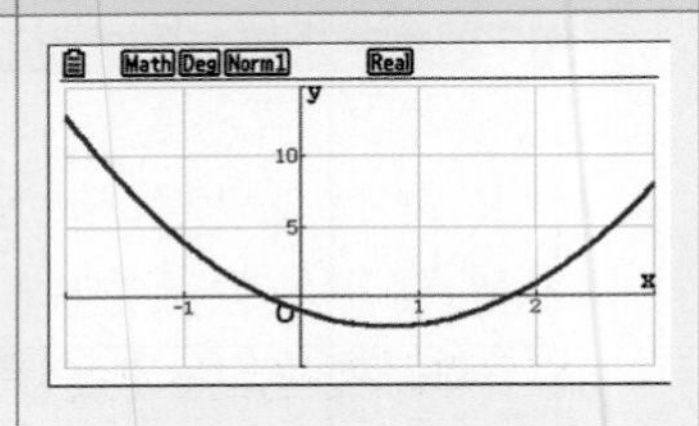 Enter Graph mode and draw the graph of the function $y = 2x^2 - 3x - 1$ for the domain $-2 \le x \le 3$ as in Example 8.
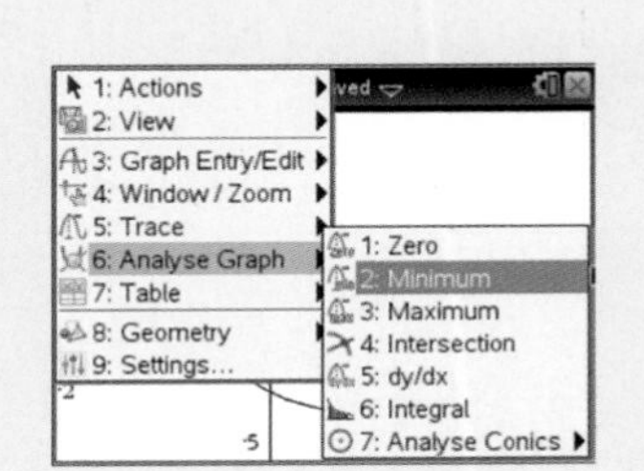 	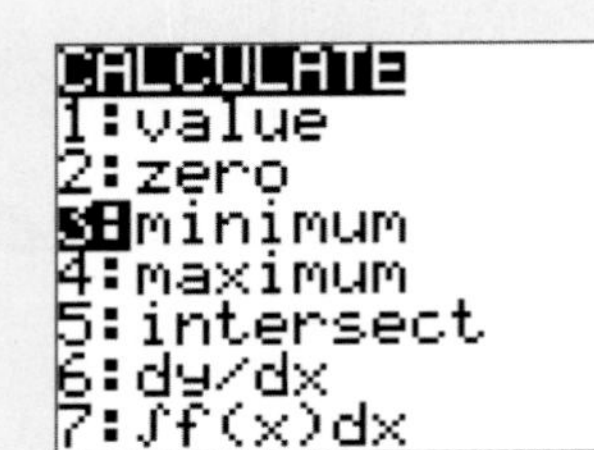 	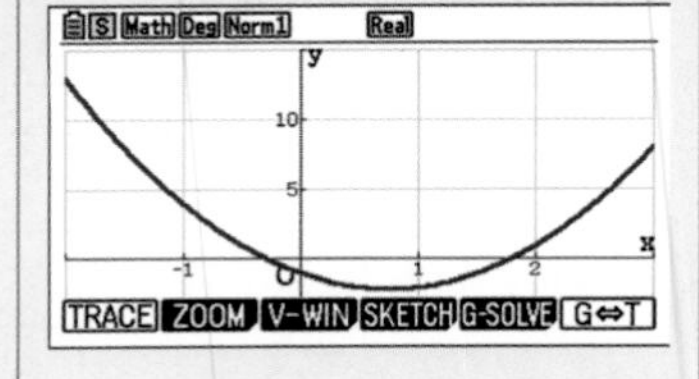

Press MENU 6: Analyse Graph | 2: Minimum…

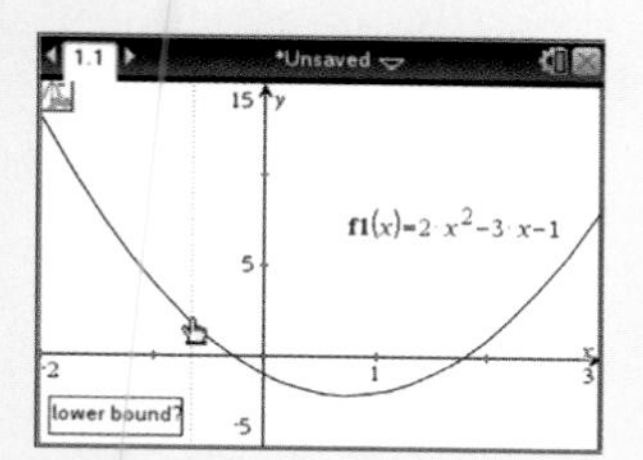

Select a point to the left of the local minimum as the lower bound and press ENTER. (Use the ▶ ◀ keys, the touchpad or type a value).

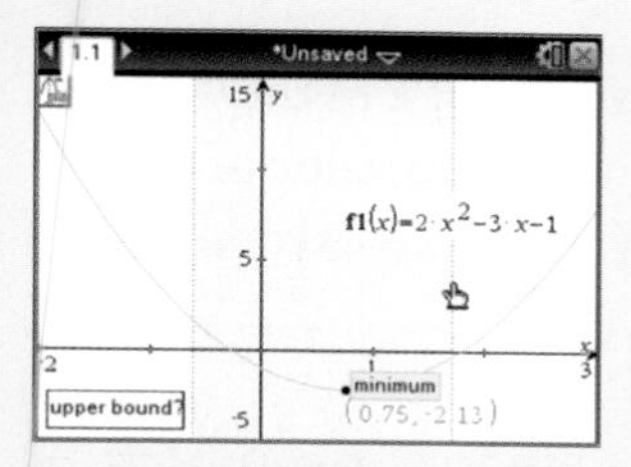

Select another point to the right of the local minimum as the upper bound and press ENTER. (Use the ▶ ◀ keys, the touchpad or type a value).

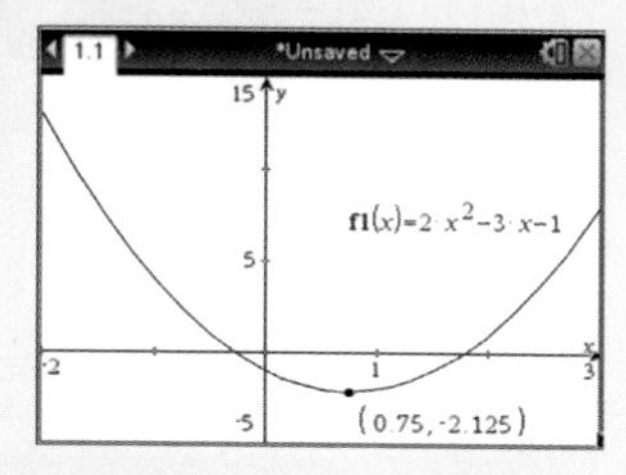

The GDC displays the coordinates of the local minimum point.

Press 2nd CALC 3: minimum.

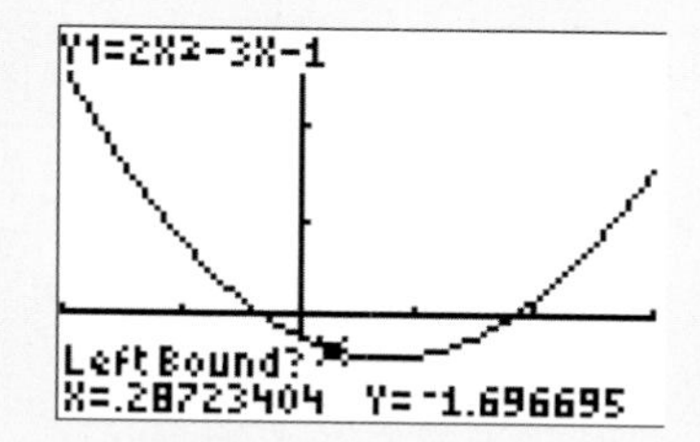

Select a point to the left of the local minimum as the left bound and press ENTER. (Use the ▶ ◀ keys or type a value).

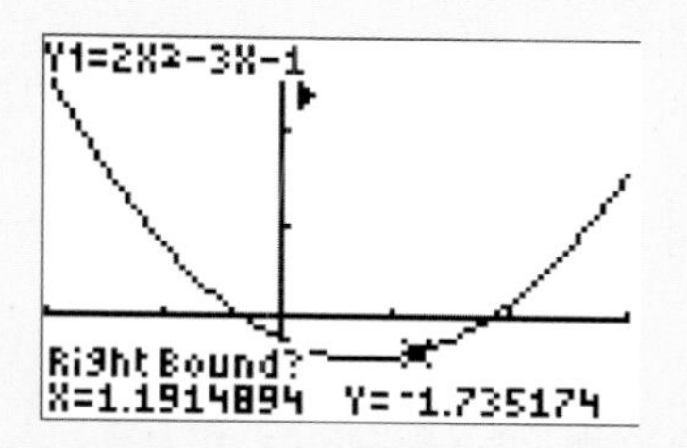

Select a point to the right of the local minimum as the right bound and press ENTER. (Use the ▶ ◀ keys or type a value).

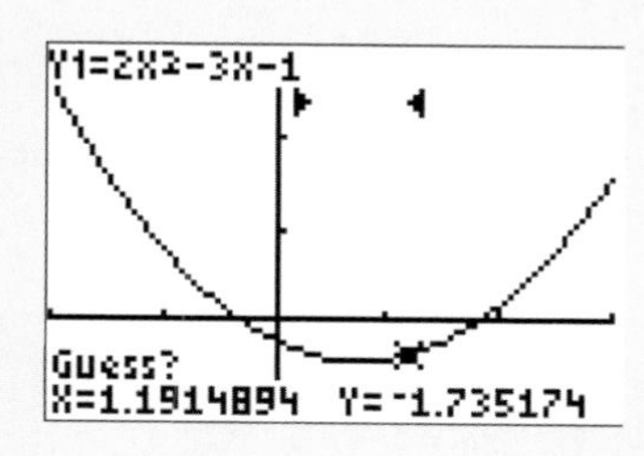

Press ENTER again.

Press SHIFT F5 G-SOLVE.

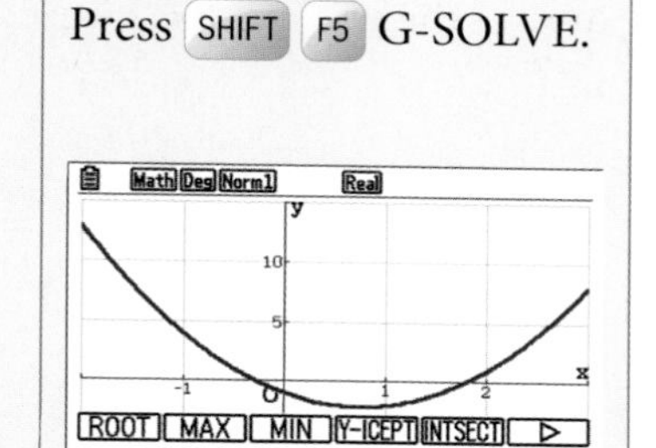

Press F3 MIN.

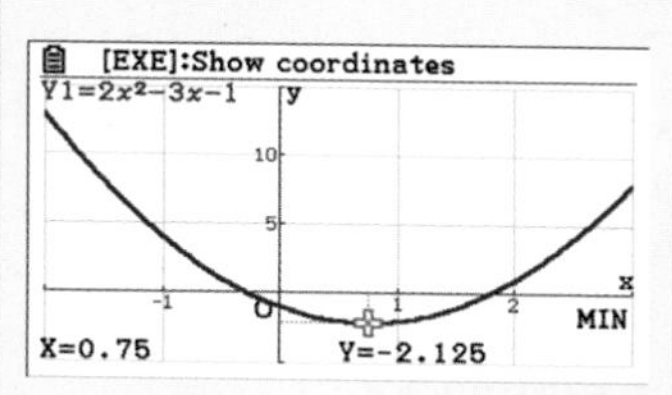

The GDC displays the coordinates of the local minimum point.

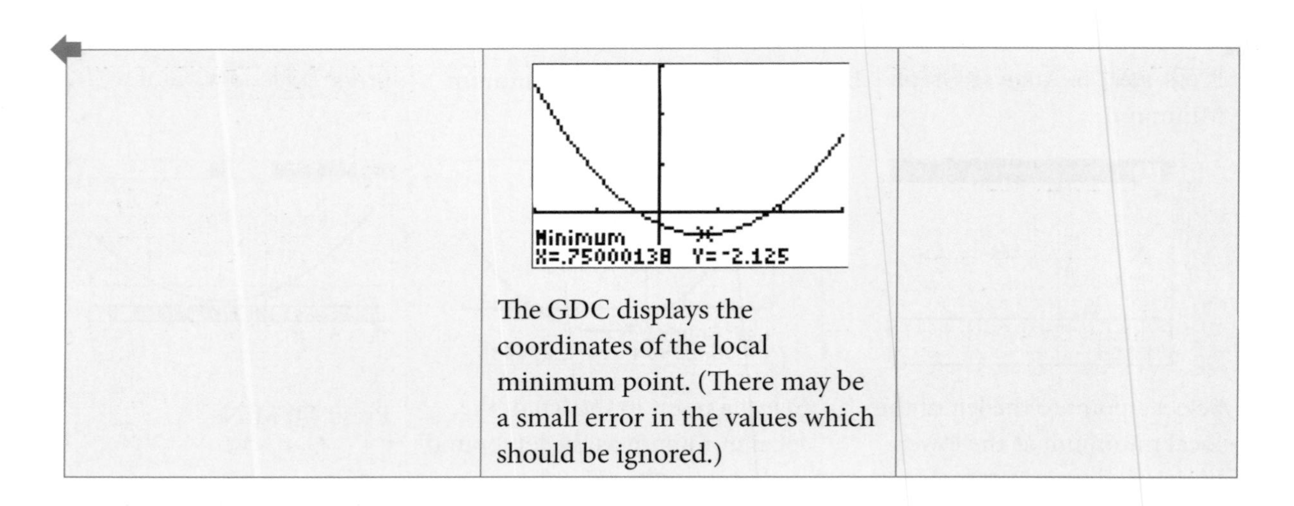

The GDC displays the coordinates of the local minimum point. (There may be a small error in the values which should be ignored.)

2.5 Finding a local maximum

The point $(x, f(x_0))$ is a **local maximum** of the function $f(x)$ if, for all values of x in the neighbourhood of x_0, $f(x_0) > f(x)$. This point may not be the absolute maximum value of the function. Points like this are also known as **turning points**, **stationary points** or **extrema**.

Example 12

Find the local maximum point of the function $y = 3 - 2x - 3x^2$ for the domain $-2 \le x \le 2$

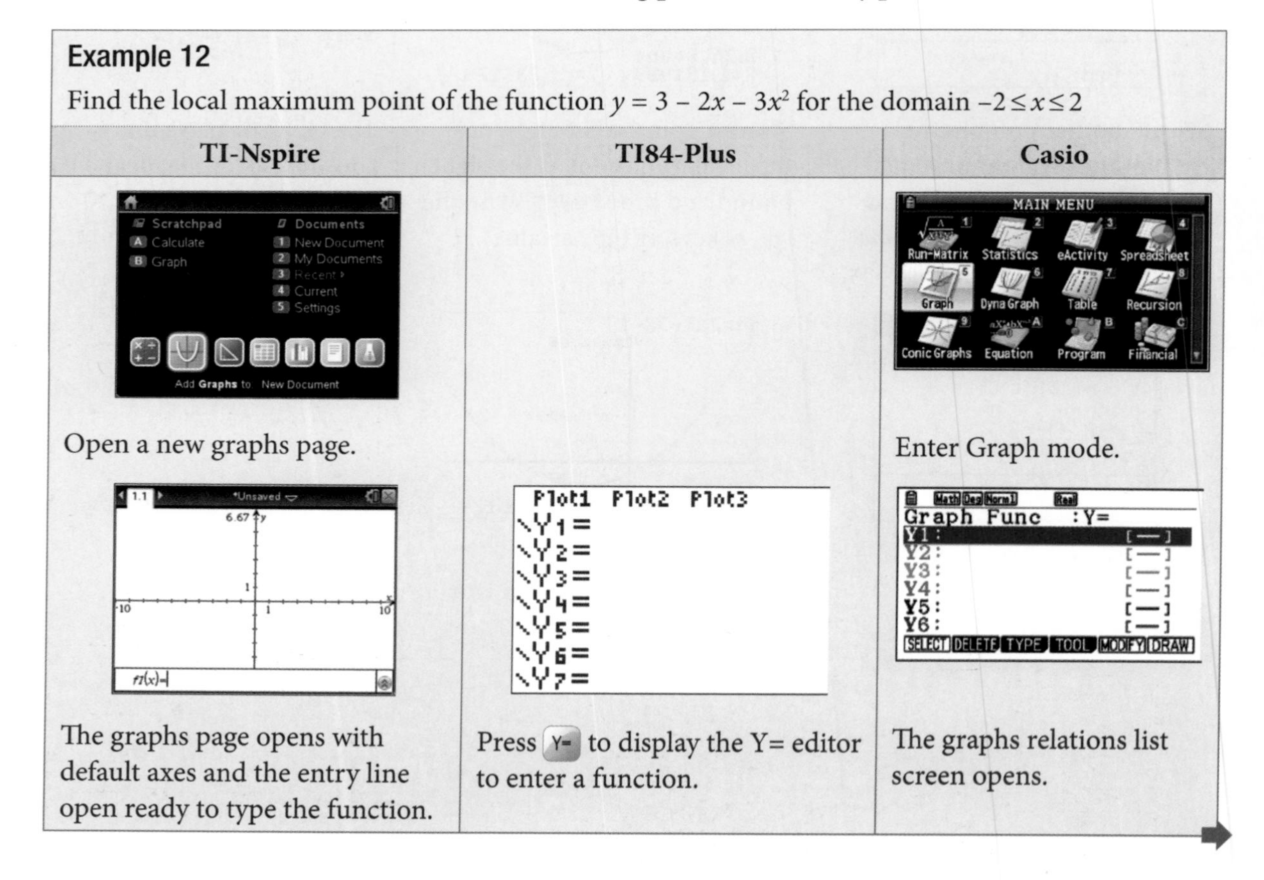

TI-Nspire	TI84-Plus	Casio
Open a new graphs page.		Enter Graph mode.
The graphs page opens with default axes and the entry line open ready to type the function.	Press Y= to display the Y= editor to enter a function.	The graphs relations list screen opens.

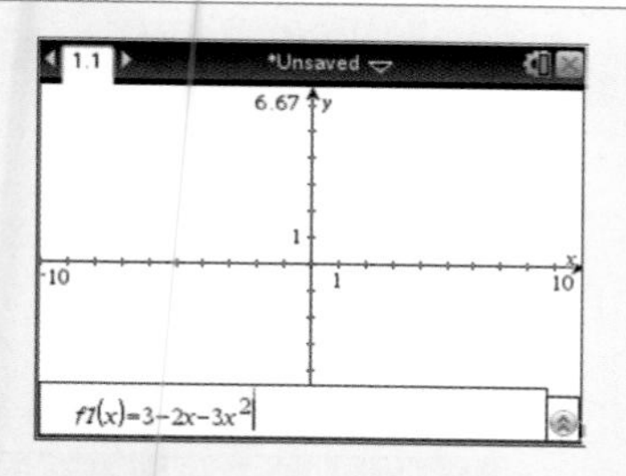

Enter the function exactly as you see it written, press ENTER.

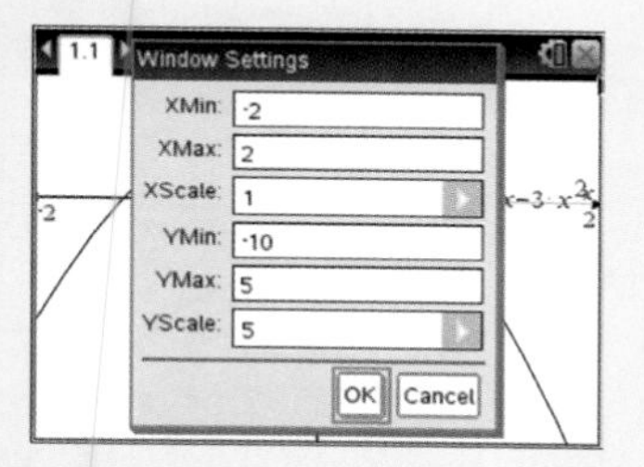

Press MENU 4: Window / Zoom | 1: Window Settings…

Enter the values from the domain $-2 \le x \le 2$ in XMin and XMax.

Enter the values from the range $-10 \le y \le 5$ in YMin and YMax, with a scale of 5.

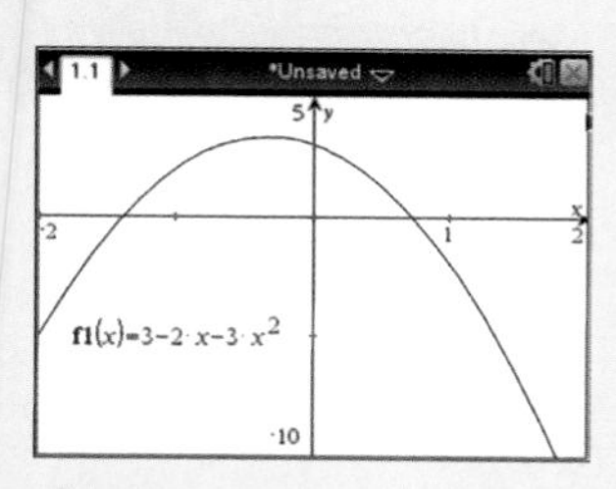

Press ENTER.

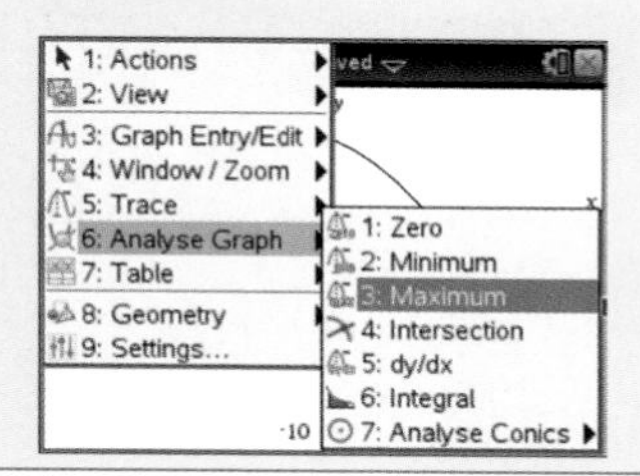

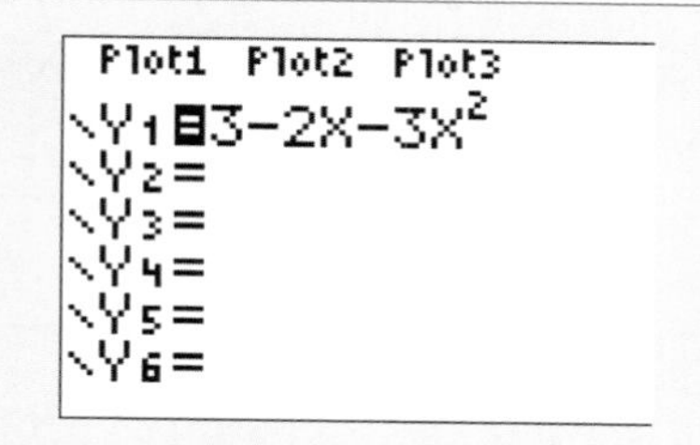

Enter the function exactly as you see it written, press ENTER. (Use the X,T,Θ,n key to enter x).

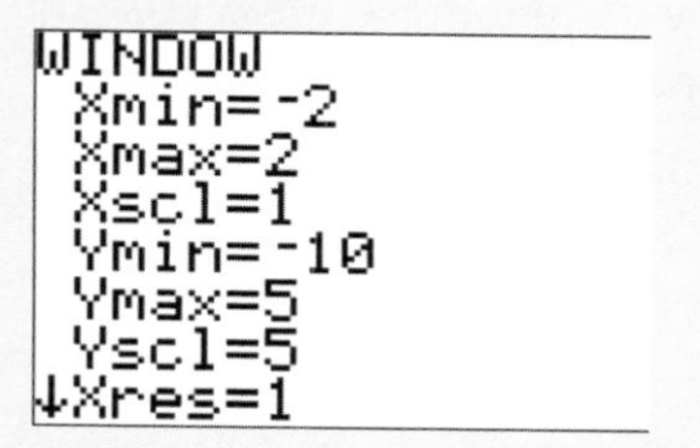

Press WINDOW.

Enter the values from the domain $-2 \le x \le 2$ in Xmin and Xmax.

Enter the values from the range $-10 \le y \le 5$ in Ymin and Ymax, with a scale of 5.

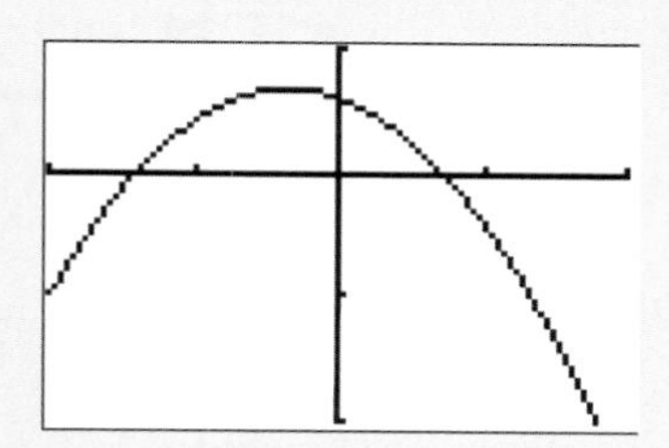

Press GRAPH.

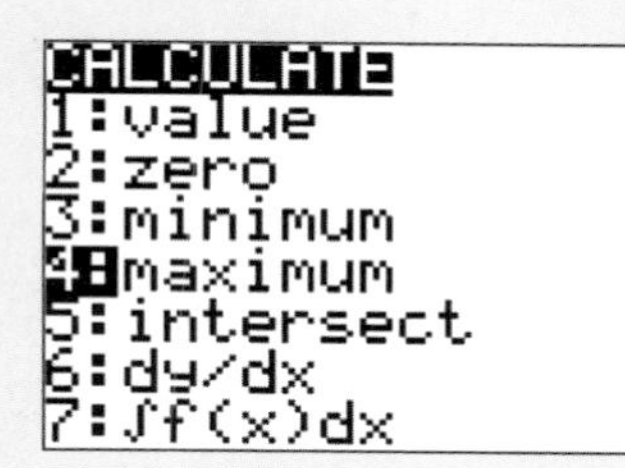

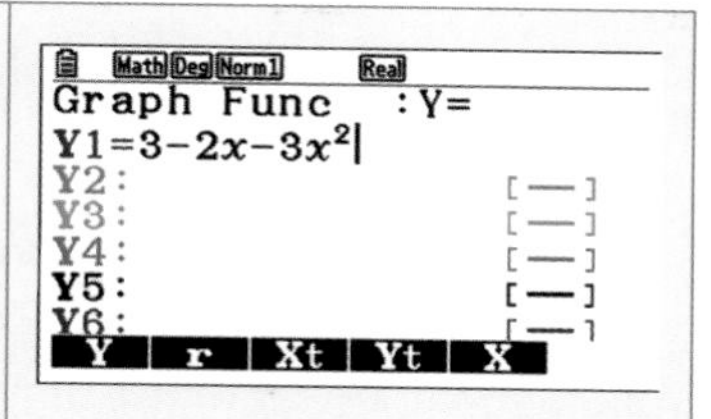

Enter the calculation exactly as you see it written, press EXE. (Use the X,θ,T key to enter x).

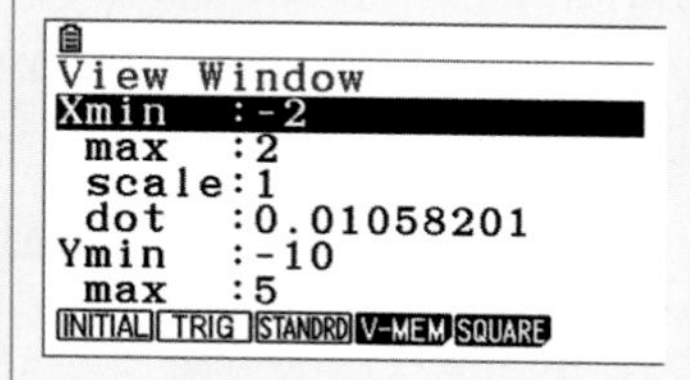

Press SHIFT F3 V-WIN.

Enter the values from the domain $-2 \le x \le 2$ in Xmin and max.

Enter the values from the range $-10 \le y \le 5$ in Ymin and max, with a scale of 5.

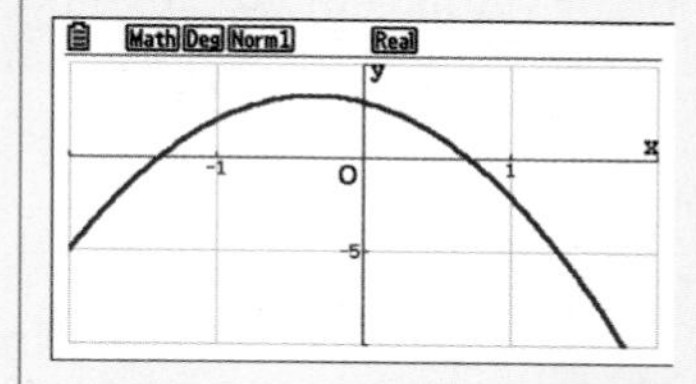

Press EXIT F6 DRAW.

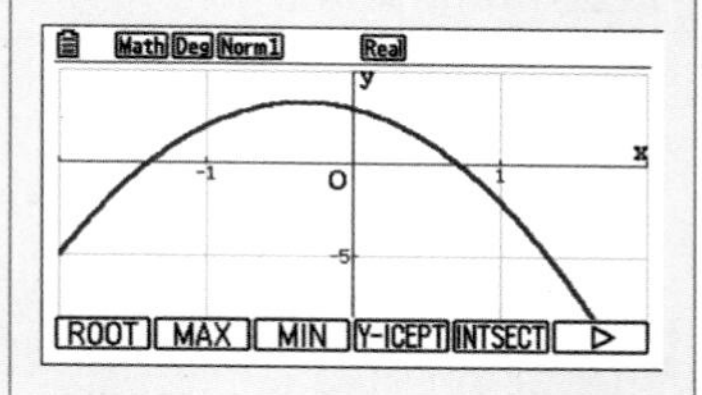

Press MENU 6: Analyse Graph | 3: Maximum.

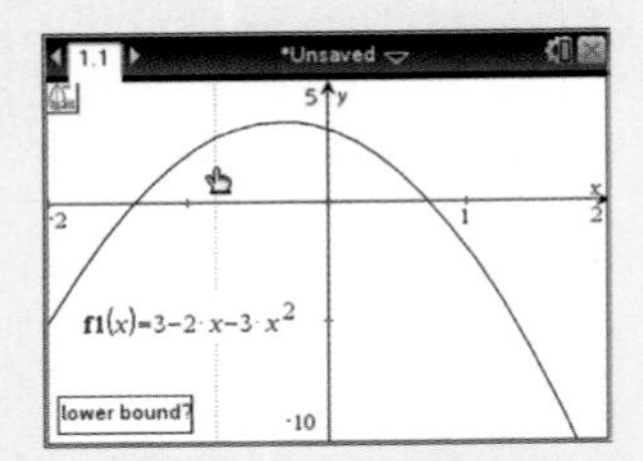

Select a point to the left of the local maximum as the lower bound and press ENTER. (Use the ▶ ◀ keys, the touchpad or type a value).

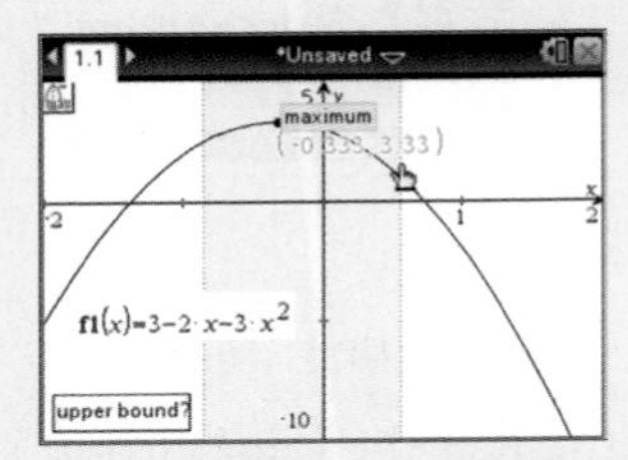

Select another point to the right of the local maximum as the upper bound and press ENTER. (Use the ▶ ◀ keys or type a value).

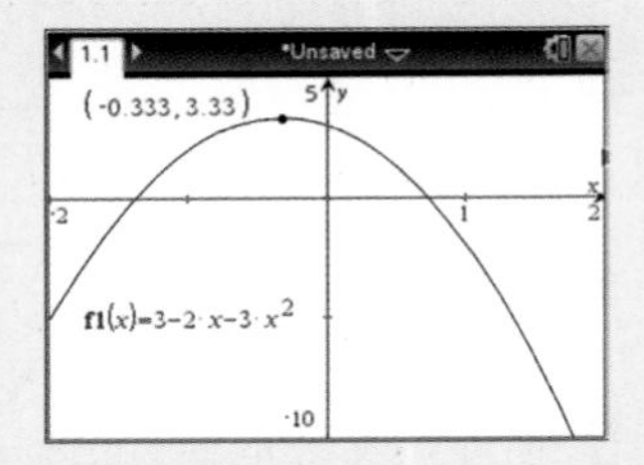

The GDC displays the coordinates of the local maximum point.

Press 2nd CALC 4: maximum.

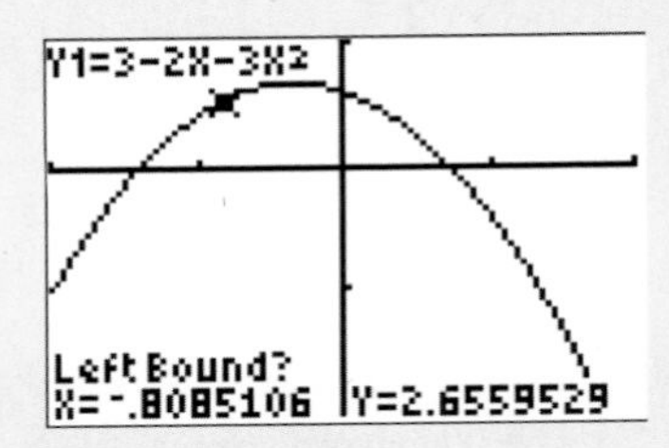

Select a point to the left of the local maximum as the left bound and press ENTER. (Use the ▶ ◀ keys or type a value).

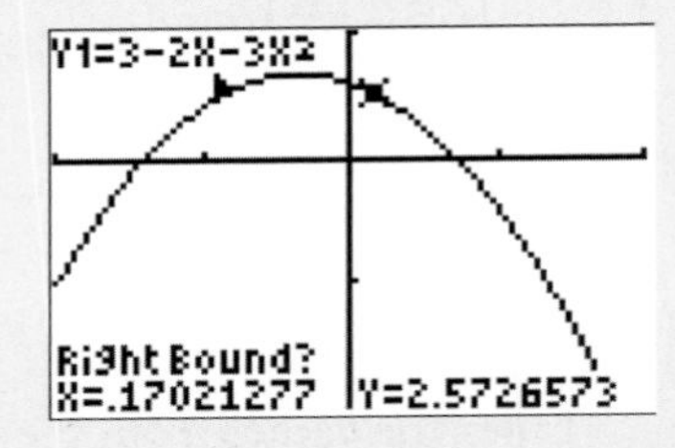

Select a point to the right of the local maximum as the right bound and press ENTER. (Use the ▶ ◀ keys or type a value).

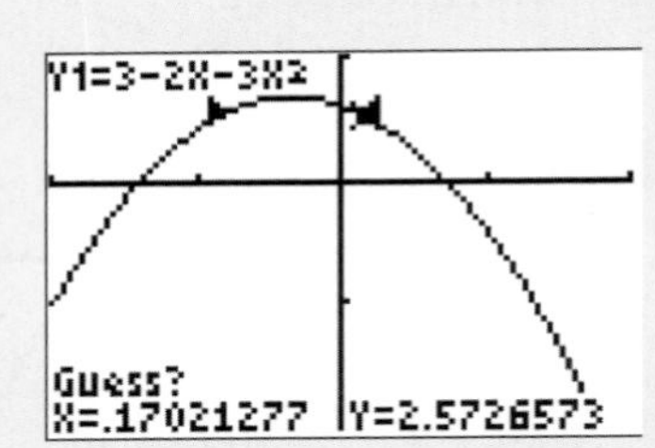

Press ENTER again.

Press SHIFT F5 G-SOLVE.

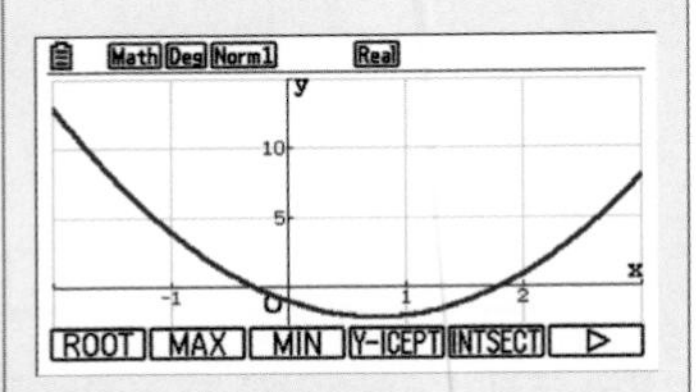

Press F2 MAX.

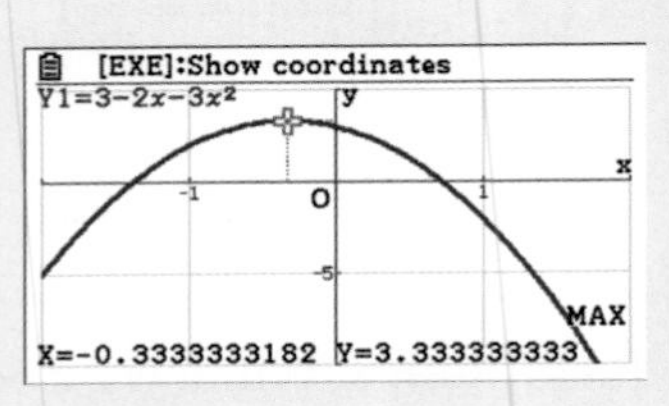

The GDC displays the coordinates of the local maximum point.

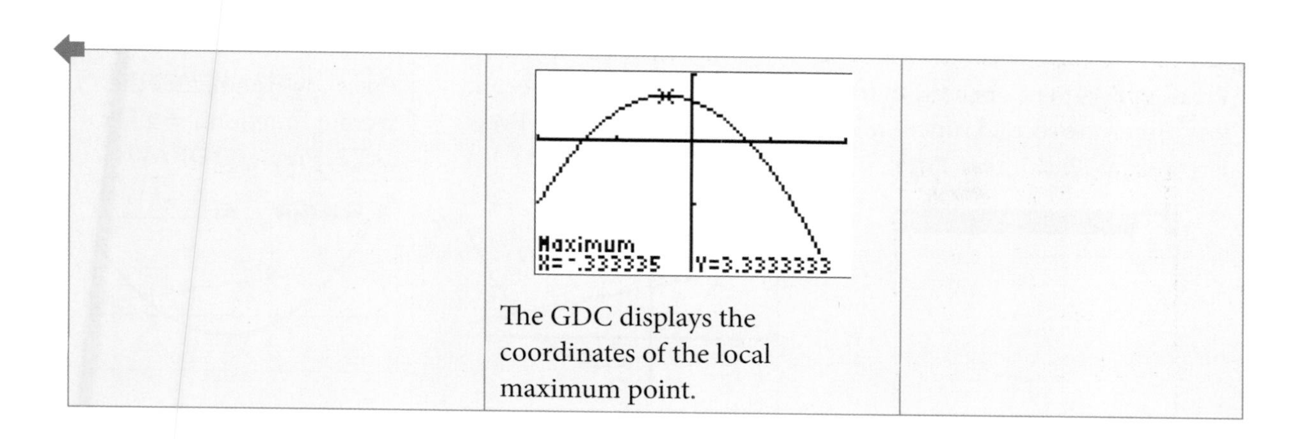

The GDC displays the coordinates of the local maximum point.

2.6 Finding the intersection point of two graphs

The point where two graphs intersect can be found quickly and efficiently using the GDC. For some functions there is an algebraic method to do this as well. For others, the GDC is almost the only way to do this.

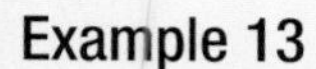

Example 13

Find the intersection points of the function $y = 2x^2 - 3x - 1$ and $y = 4 - x$ for the domain $-2 \le x \le 3$

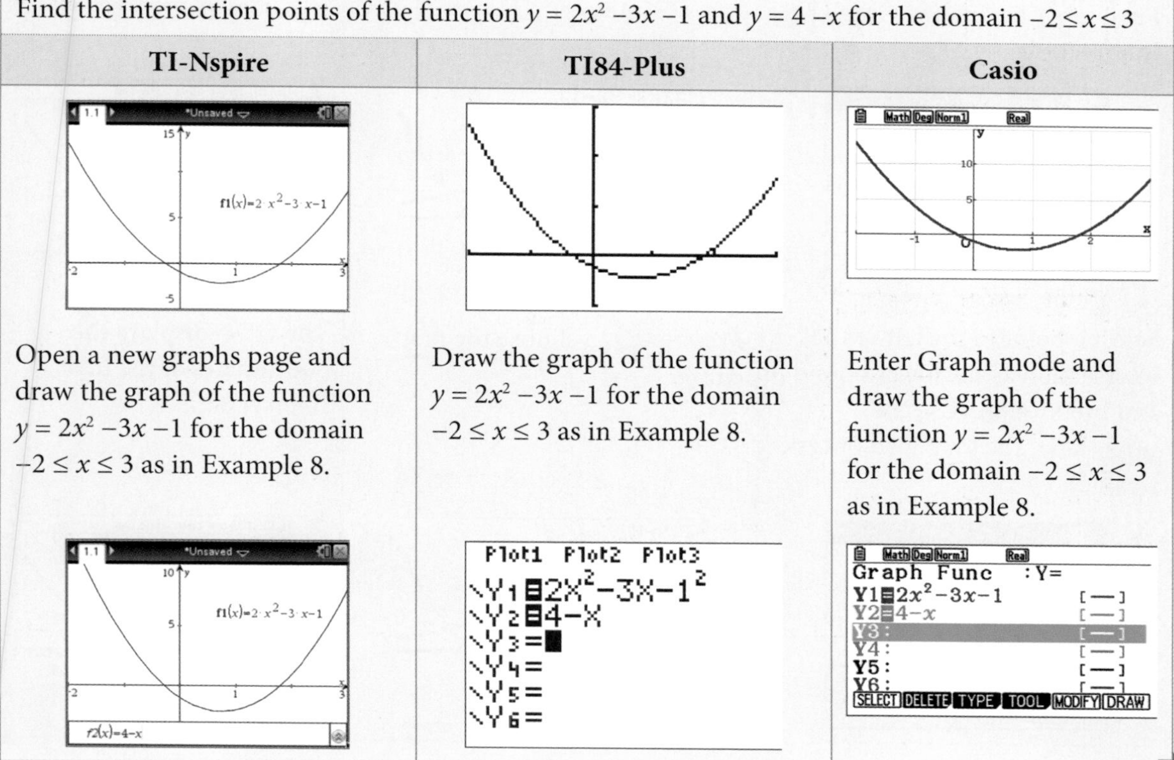

TI-Nspire	TI84-Plus	Casio
Open a new graphs page and draw the graph of the function $y = 2x^2 - 3x - 1$ for the domain $-2 \le x \le 3$ as in Example 8.	Draw the graph of the function $y = 2x^2 - 3x - 1$ for the domain $-2 \le x \le 3$ as in Example 8.	Enter Graph mode and draw the graph of the function $y = 2x^2 - 3x - 1$ for the domain $-2 \le x \le 3$ as in Example 8.

Press ctrl G to re-open the entry line. Enter the second function $y = 4 - x$ as f2(x). Press ENTER.

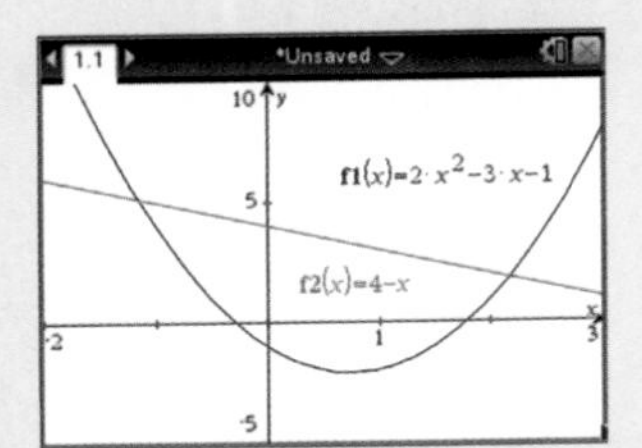

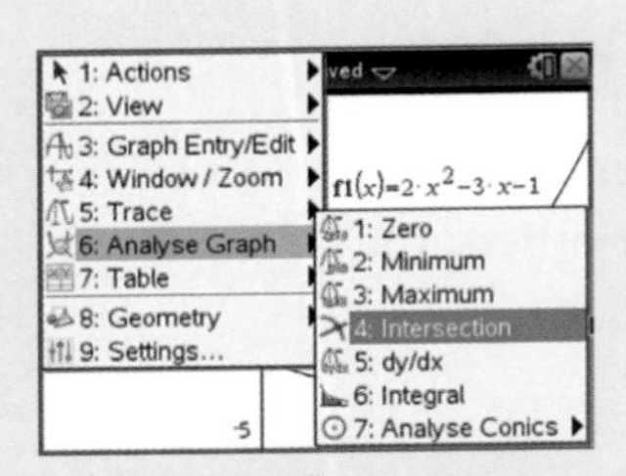

Press MENU 6: Analyse Graph | 4: Intersection.

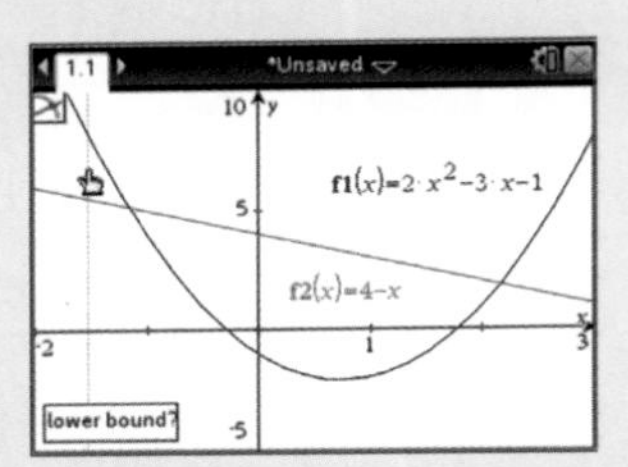

Select a point to the left of the intersection as the lower bound and press ENTER. (Use the ▶ ◀ keys, the touchpad or type a value)

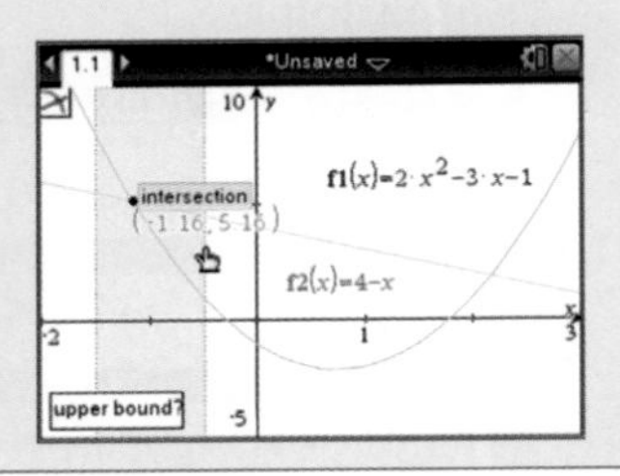

Press Y= and enter the second function $y = 4 - x$ as Y_2. Press GRAPH.

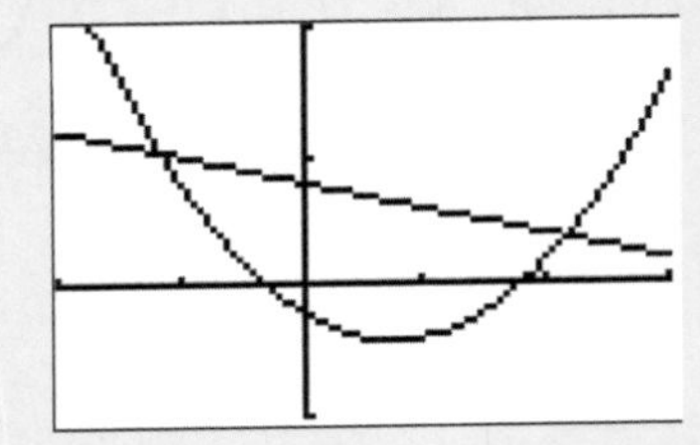

Press 2nd CALC.

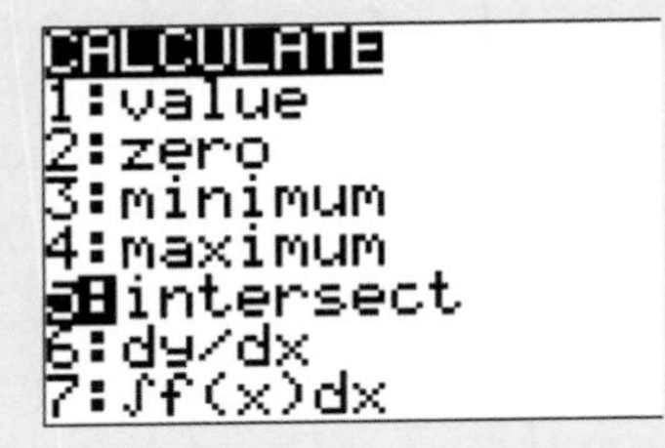

Press 5: intersect

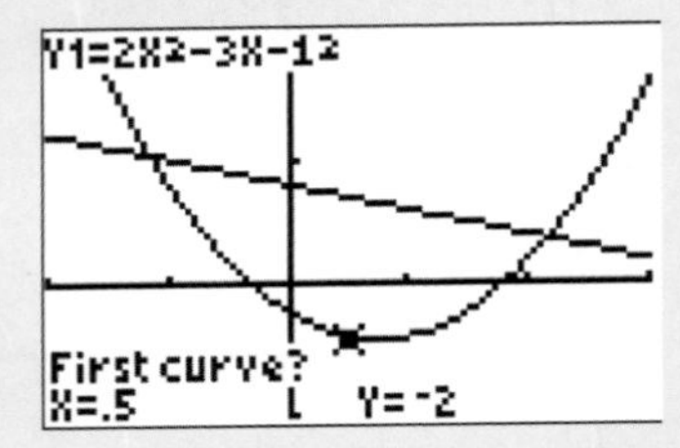

Press ENTER to choose the first curve.

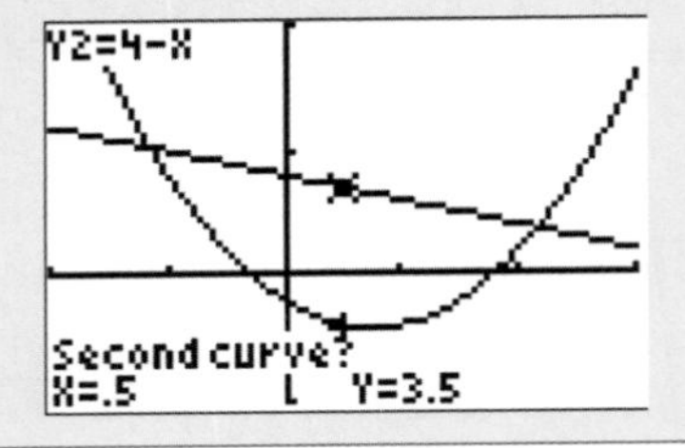

Press EXIT and enter the second function $y = 4 - x$ as Y2. Press F6 DRAW.

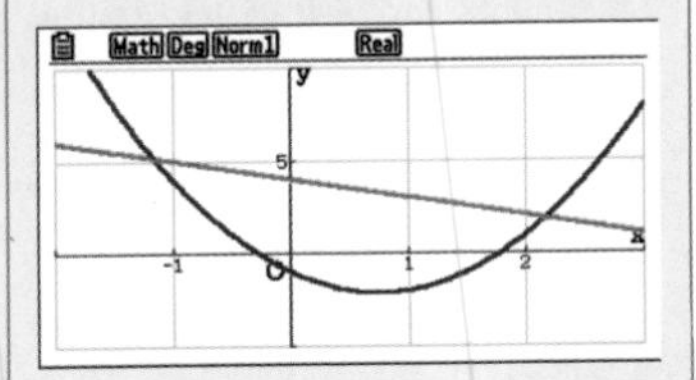

Press SHIFT F5 G-SOLVE.

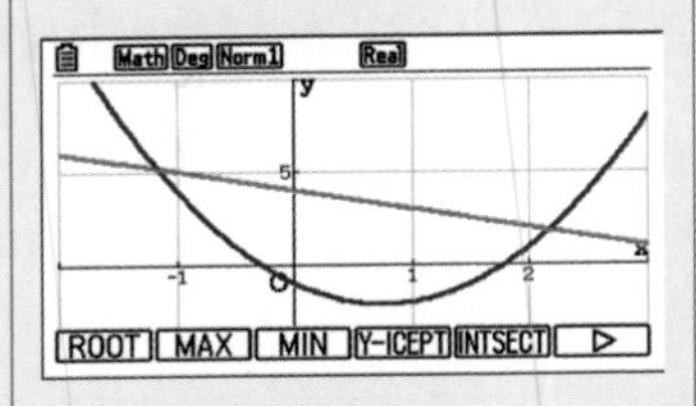

Press F5 INTSECT.

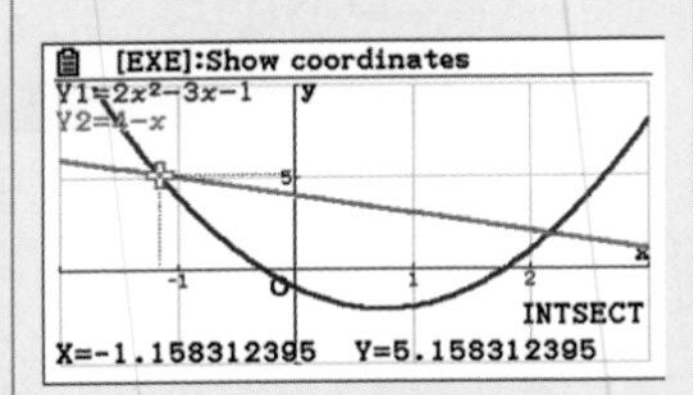

The GDC displays the coordinates of the first intersection.

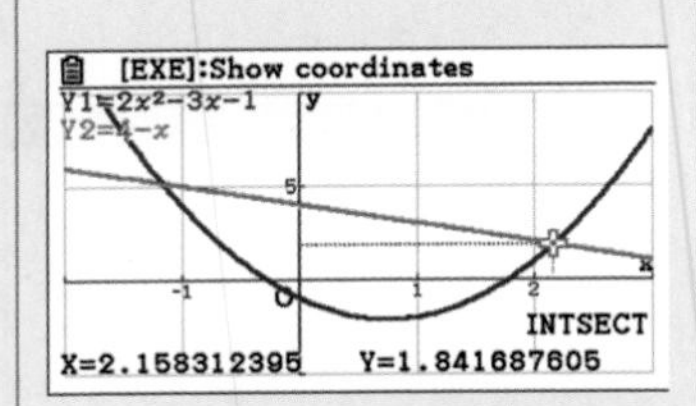

Select another point to the right of the intersection as the upper bound and press ENTER. (Use the ▶ ◀ keys or type a value)

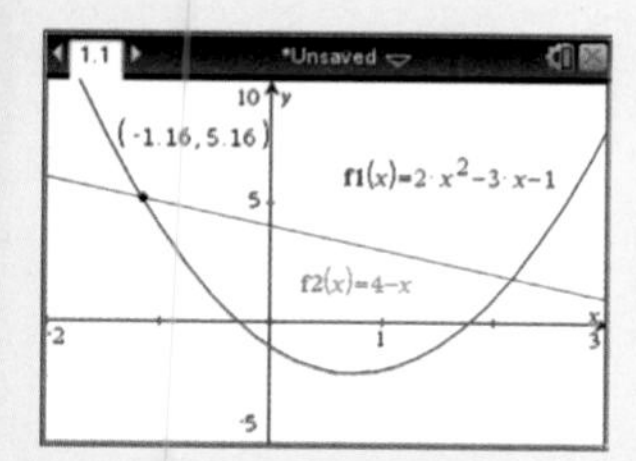

The GDC displays the coordinates of the intersection point.

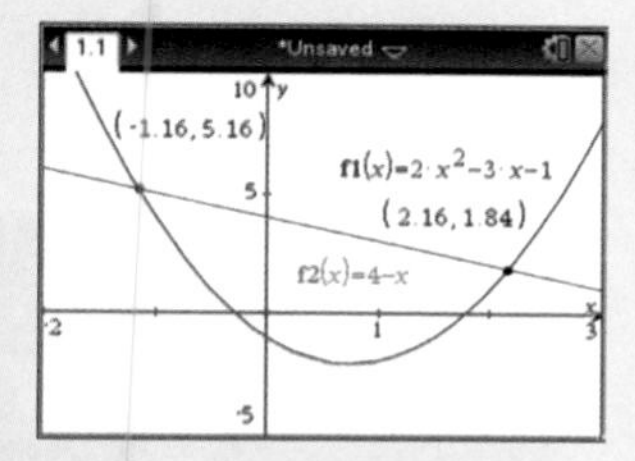

Repeat the process for the second point of intersection.

Press ENTER again to choose the second curve. (It does not matter that the curve is a straight line.)

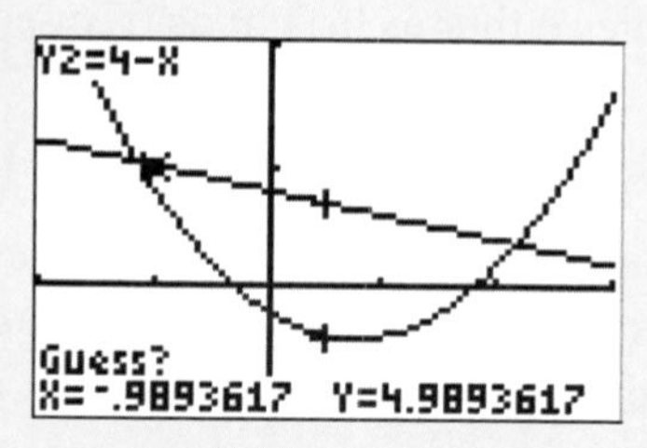

Use the ▶ ◀ keys or type a value to select a point close to the first intersection point and press ENTER.

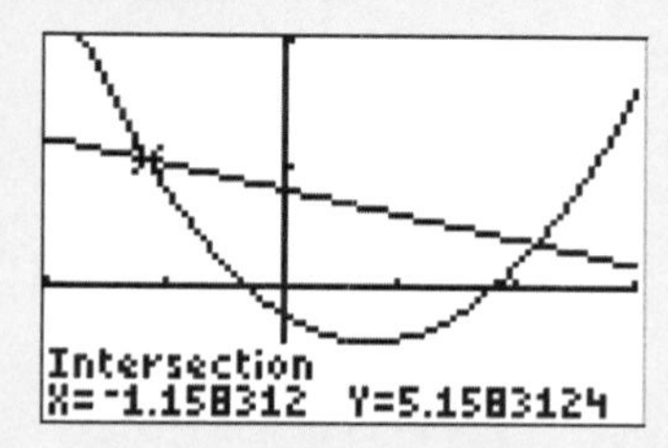

The GDC displays the coordinates of the intersection.

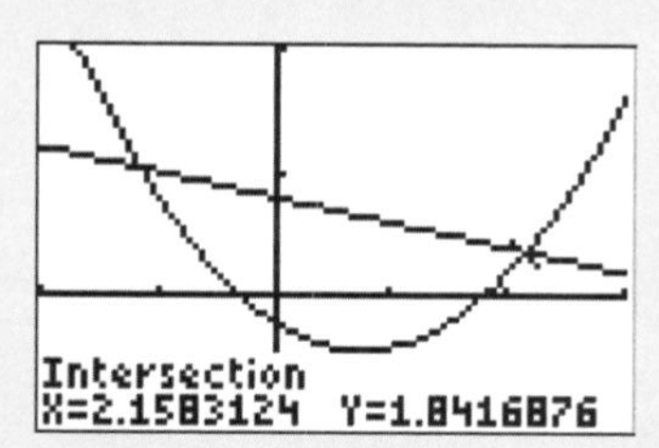

Repeat the process for the second point of intersection.

Press ▶ to find the second intersection. The GDC displays the coordinates of the intersection.

3 Working with data

The GDC can perform many statistical calculations and draw a number of statistical graphs. You will need to know how to find the mean, median, quartiles of data and how to find linear regression equations. The GDC can work with data entered in tables either as lists or as frequency tables.

3.1 Calculation of basic statistics from a list

Data is first entered into a table as a list and then basic statistics can be calculated. The values found are stared as variables that can then be used in further calculations if necessary.

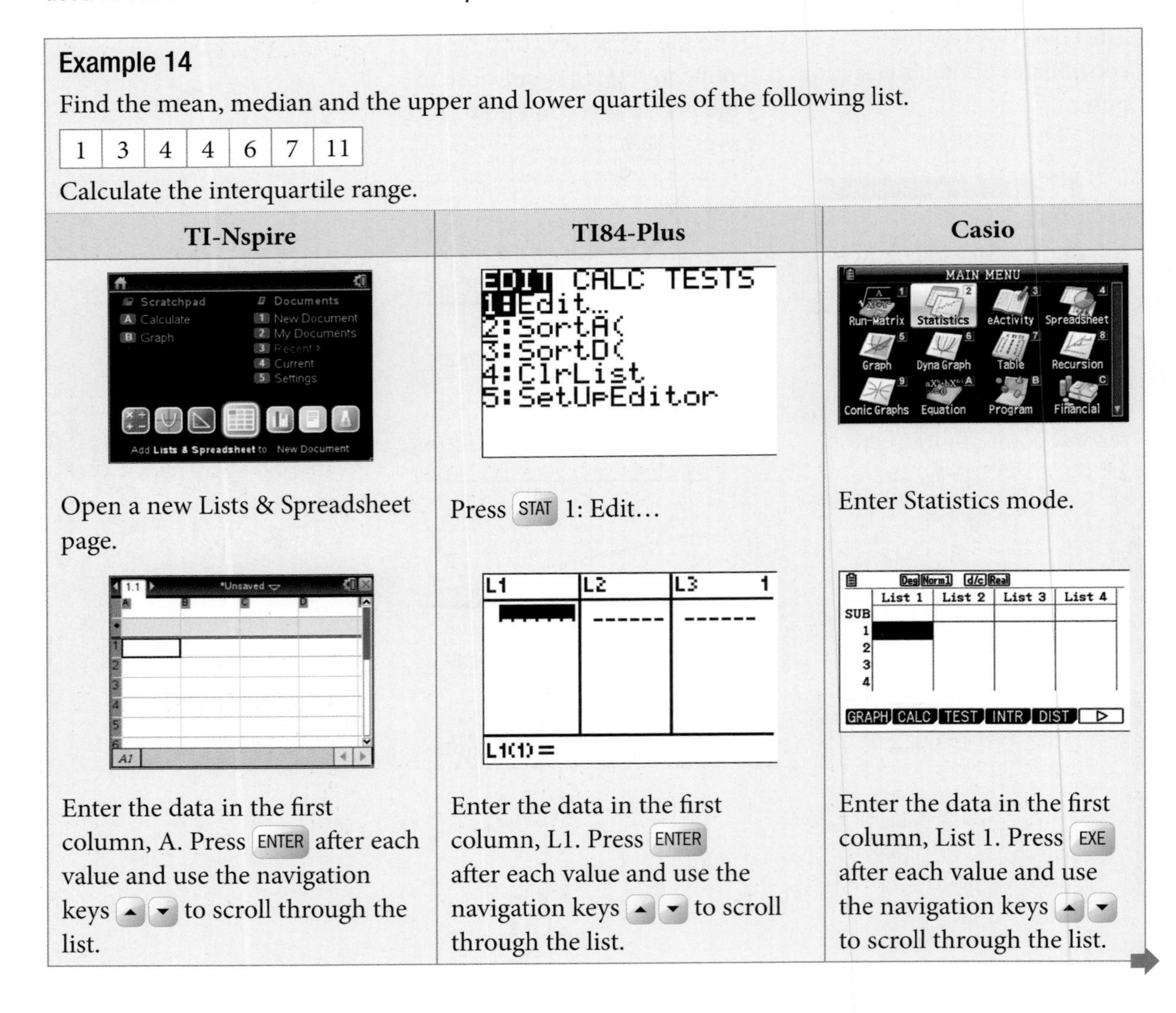

Example 14

Find the mean, median and the upper and lower quartiles of the following list.

1	3	4	4	6	7	11

Calculate the interquartile range.

TI-Nspire	TI84-Plus	Casio
Open a new Lists & Spreadsheet page.	Press STAT 1: Edit…	Enter Statistics mode.
Enter the data in the first column, A. Press ENTER after each value and use the navigation keys ▲ ▼ to scroll through the list.	Enter the data in the first column, L1. Press ENTER after each value and use the navigation keys ▲ ▼ to scroll through the list.	Enter the data in the first column, List 1. Press EXE after each value and use the navigation keys ▲ ▼ to scroll through the list.

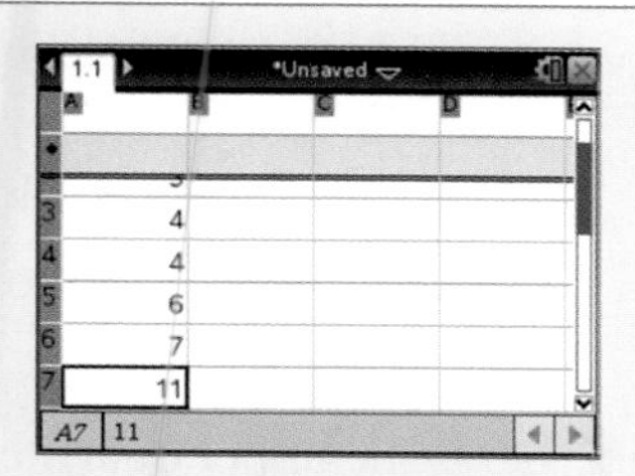

Press MENU

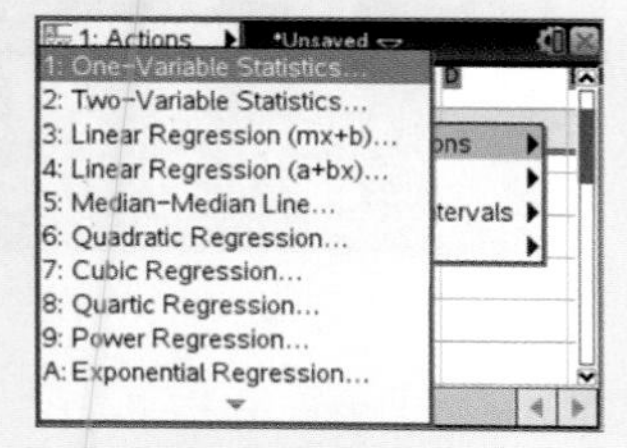

Select 4: Statistics | 1: Stat Calculations | 1:One-Variable Statistics.

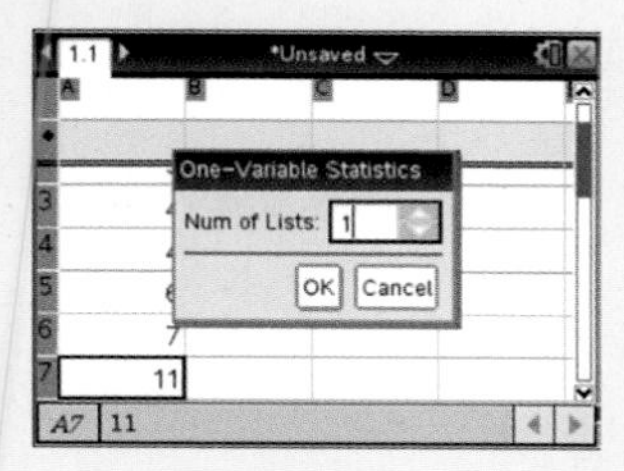

Choose 1 list and press ENTER.

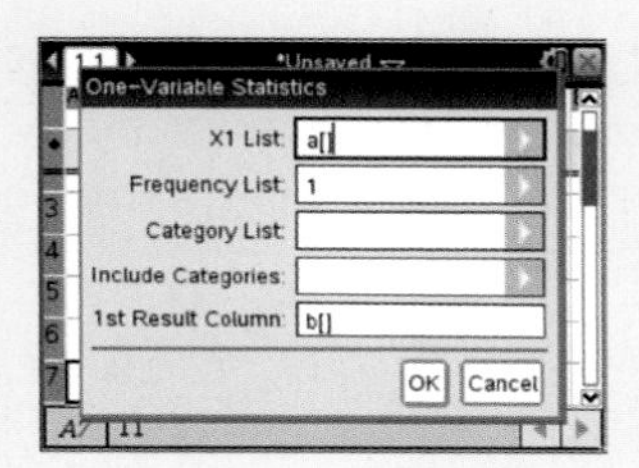

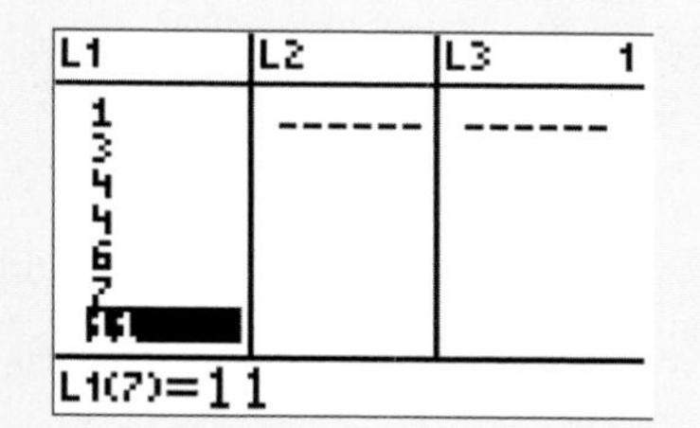

Press STAT CALC.

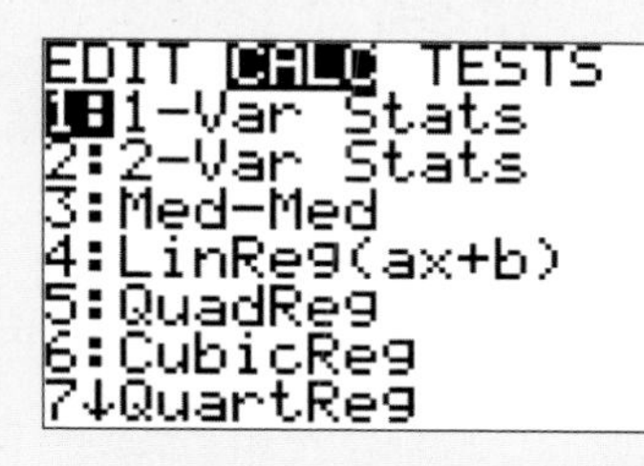

Select 1: 1-Var Stats. Press ENTER.

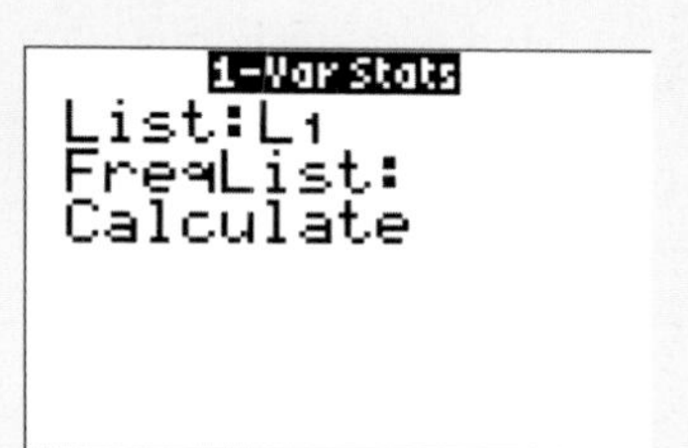

Select the default values for this dialogue. Press ENTER ENTER ENTER.

```
1-Var Stats
x̄=5.142857143
Σx=36
Σx²=248
Sx=3.236694375
σx=2.996596709
↓n=7
```

```
      List 1  List 2  List 3  List 4
SUB
4        4
5        6
6        7
7       11
                                  11
GRAPH CALC TEST INTR DIST ▷
```

Press F2 CALC

```
      List 1  List 2  List 3  List 4
SUB
4        4
5        6
6        7
7       11
                                  11
1-VAR 2-VAR REG                  SET
```

Press F1 1-VAR

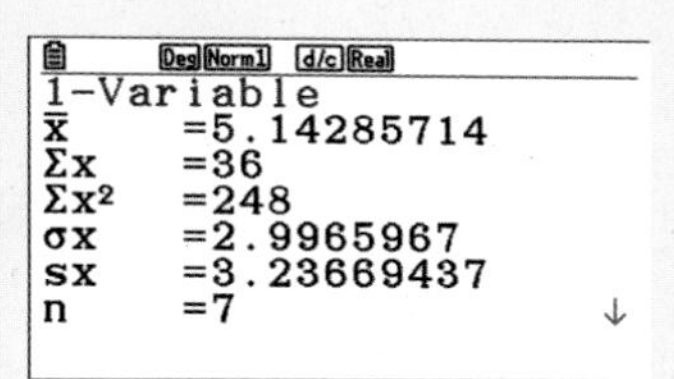

The GDC displays a list of basic statistics.

$\bar{x}$ is the mean

Σx is the sum of all the values

n is the number of items.

```
1-Variable
minX =1
Q1   =3
Med  =4
Q3   =7
maxX =11
Mod  =4
```

Select the default values in this dialogue and press ENTER.

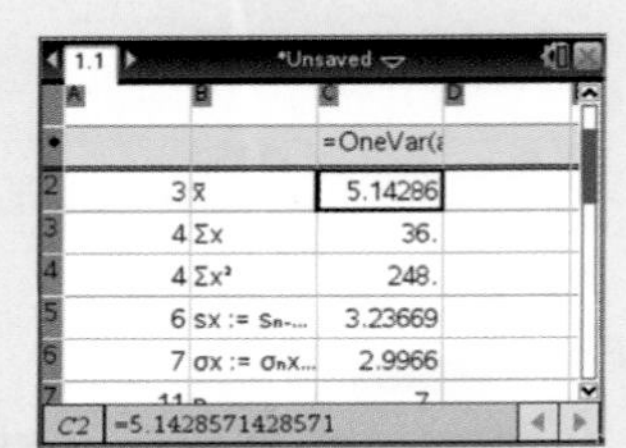

The GDC displays a list of basic statistics.

$\bar{x}$ is the mean.

Σx is the sum of all the values.

Scrolling down the table using the navigation keys ▲ ▼ shows further values.

n is the number of items

Q_1X is the lower quartile

MedianX… is the median

Q_3X is the upper quartile

The GDC displays a list of basic statistics.

$\bar{x}$ is the mean

Σx is the sum of all the values

n is the number of items.

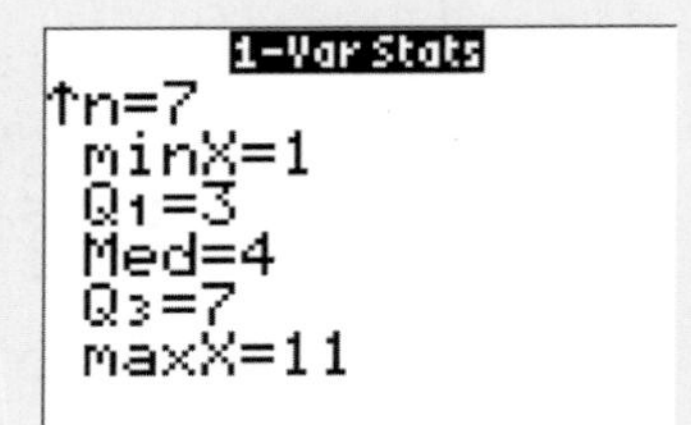

Scrolling down the table using the navigation keys ▲ ▼ shows further values.

Q_1 is the lower quartile.

Med is the median.

Q_3 is the upper quartile.

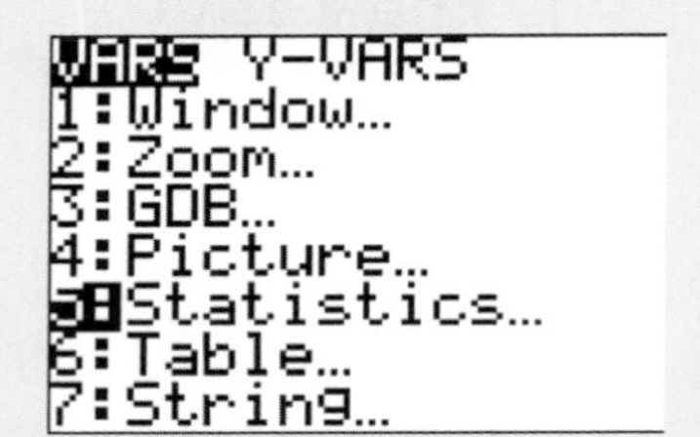

Press VARS and choose 5:Statistics…

Scrolling down the table using the navigation keys ▲ ▼ shows further values.

Q1 is the lower quartile

Med is the median

Q3 is the upper quartile

Press MENU and select Run-Matrix mode.

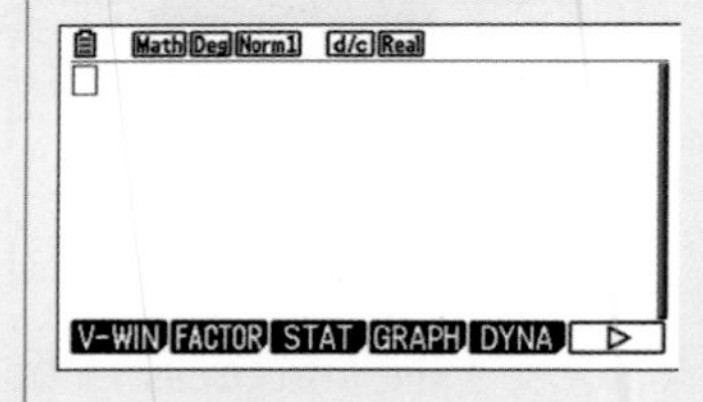

Press VARS then press F3 STAT and F3 GRAPH.

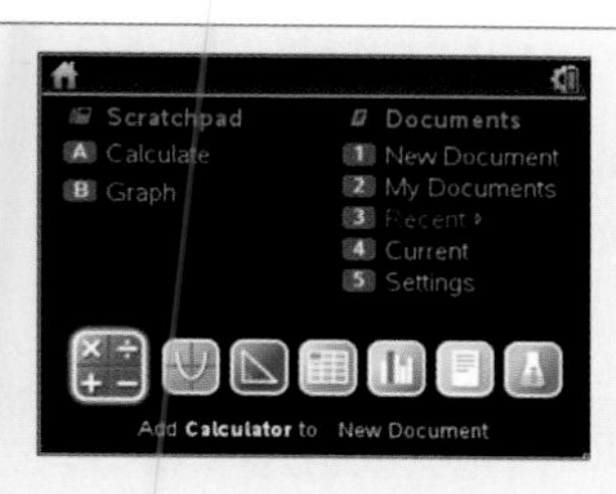

Press On and add a calculator page.

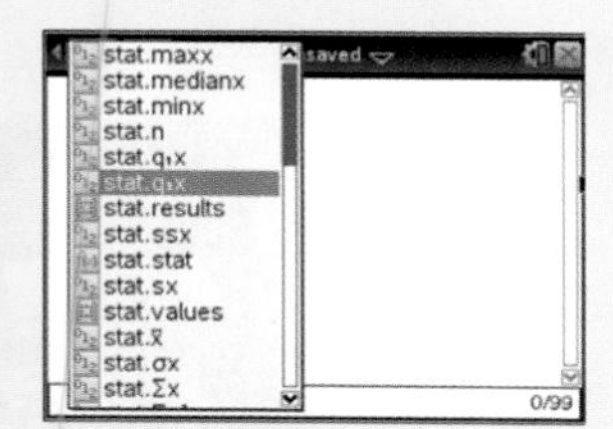

Press h and select stat.q_3x press – and then Press h and select stat.q_1x. Press ENTER.

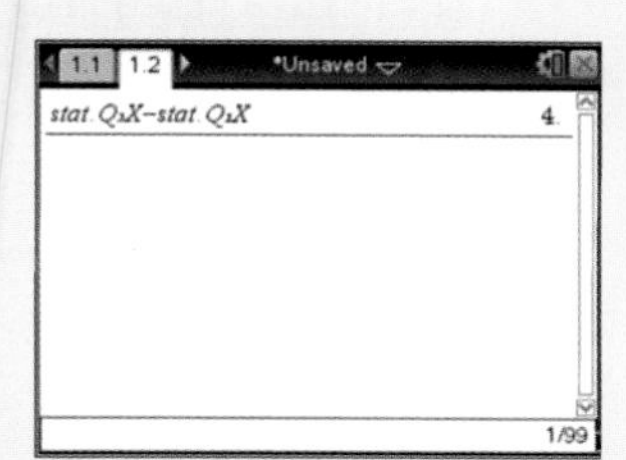

The GDC calculates the interquartile range.

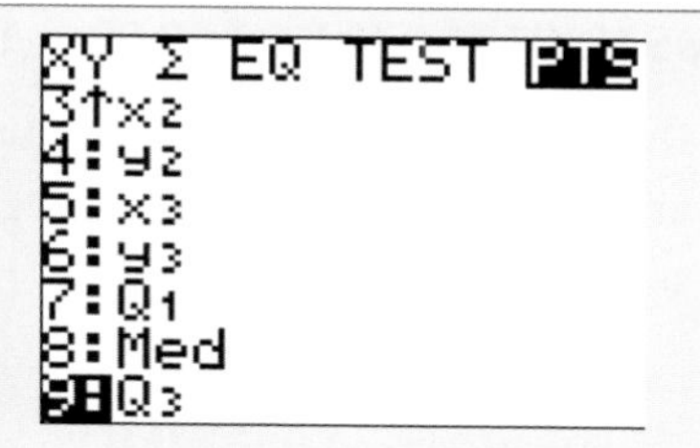

Select PTS using the navigation keys ▶ ◀. Select 9:Q_3. Press – and then VARS | 5:Statistics… | PTS | 7:Q_1. Press ENTER.

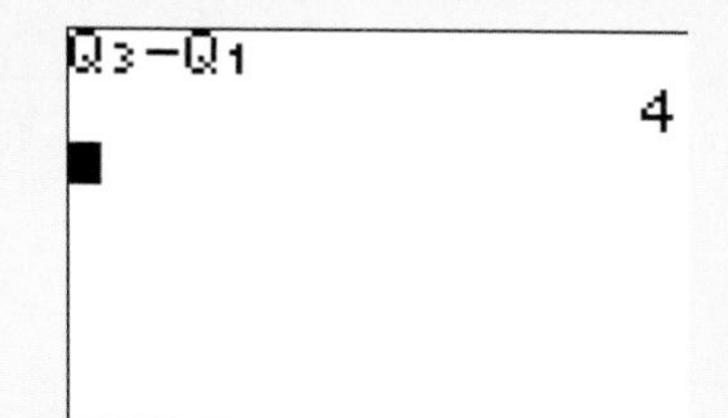

The GDC calculates the interquartile range.

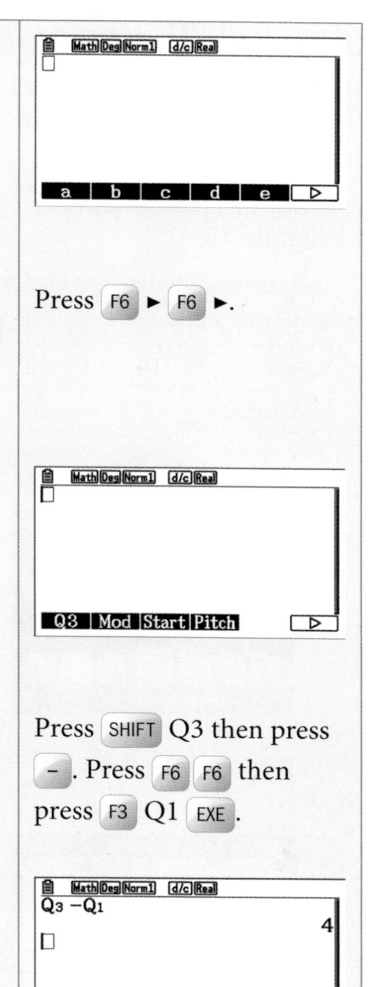

Press F6 ► F6 ►.

Press SHIFT Q3 then press –. Press F6 F6 then press F3 Q1 EXE.

The GDC calculates the interquartile range.

3.2 Calculation of basic statistics from a frequency table

Data is first entered into a table in two lists and then basic statistics can be calculated. The values found are stared as variables that can then be used in further calculations if necessary.

Example 15

Find the mean, median and the upper and lower quartiles from the following frequency table.

Value	10	11	12	13	14
Frequency	3	9	18	12	8

TI-Nspire

Open a new Lists & Spreadsheet page.

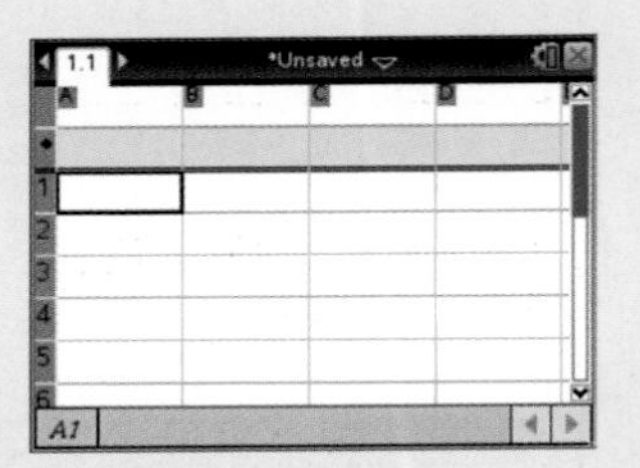

Enter the values in the first column, A. Press ENTER after each value and use the navigation keys ▲ ▼ to scroll through the list. Enter the frequencies in the second column B.

TI84-Plus

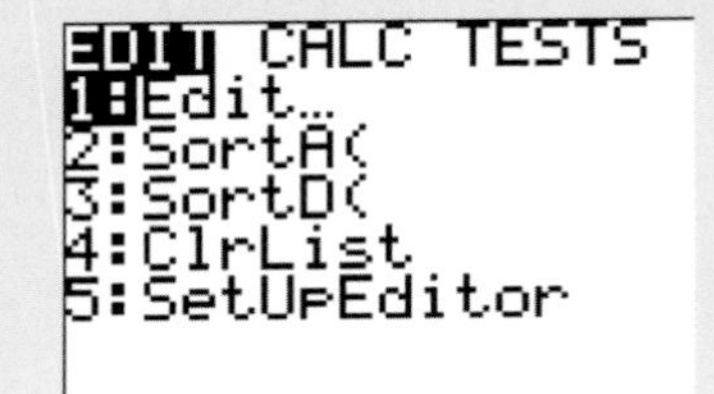

Press STAT 1: Edit…

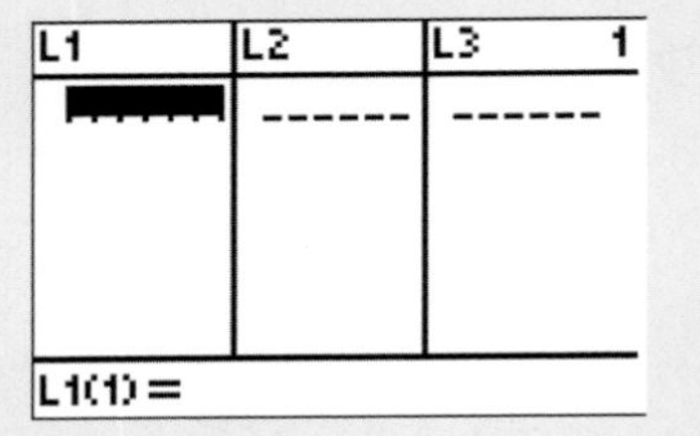

Enter the values in the first column, L1. Press ENTER after each value and use the navigation keys ▲ ▼ to scroll through the list. Enter the frequencies in the second column L2.

Casio

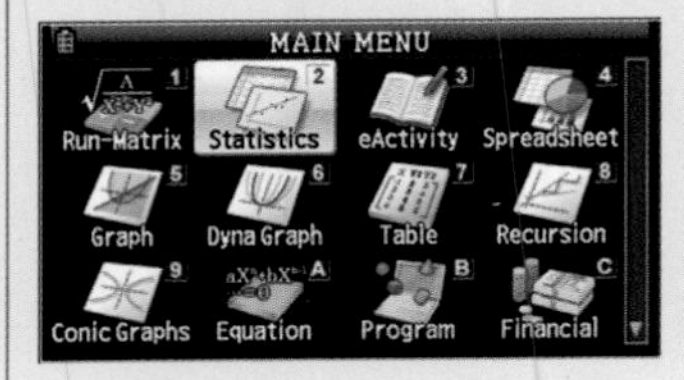

Enter Statistics mode.

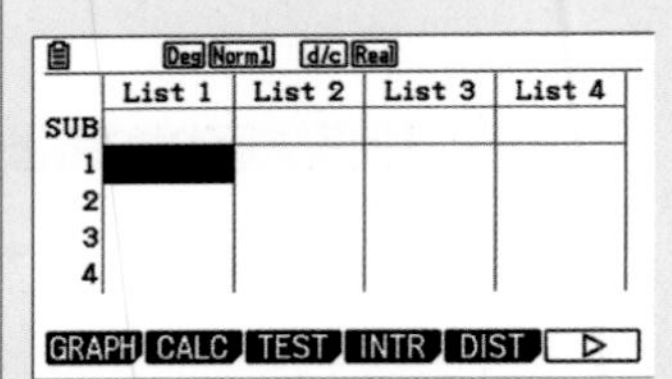

Enter the values in the first column, List 1. Press EXE after each value and use the navigation keys ▲ ▼ to scroll through the list. Enter the frequencies in the second column List 2.

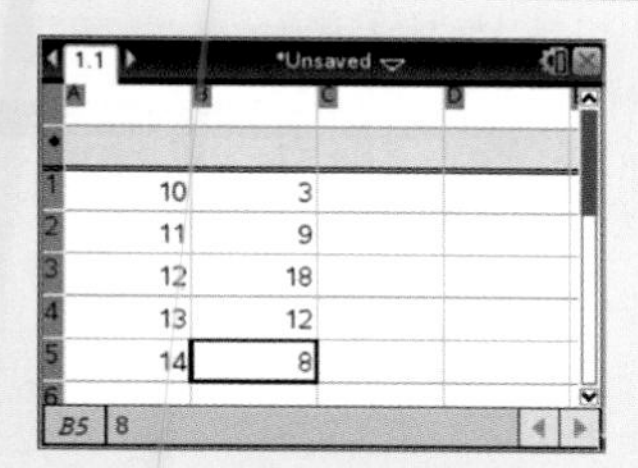

Press MENU

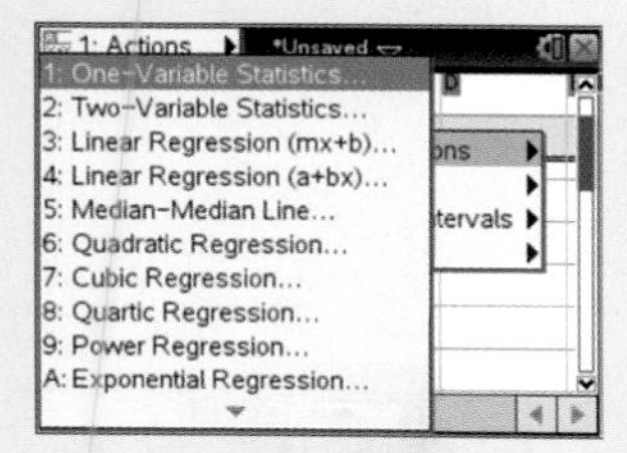

Select 4: Statistics | 1: Stat Calculations | 1: One-Variable Statistics.

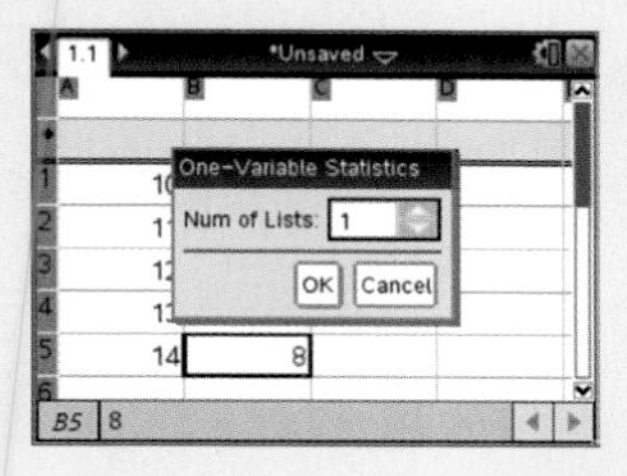

Choose 1 list and press ENTER.

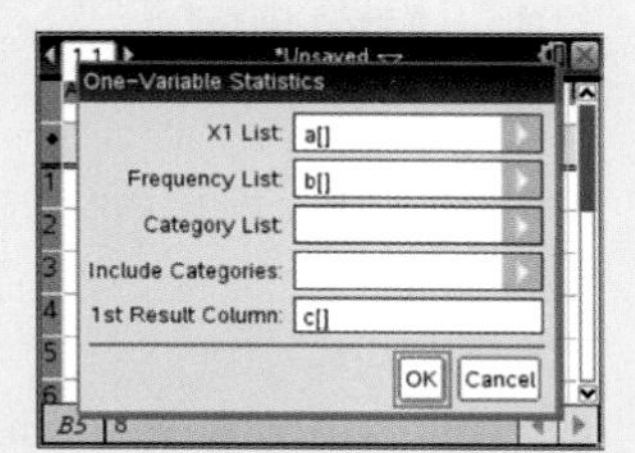

Delete the default values in this dialogue, enter a[] in X1 List, b[] in Frequency List, leave c[]as the 1st Result Column and press ENTER.

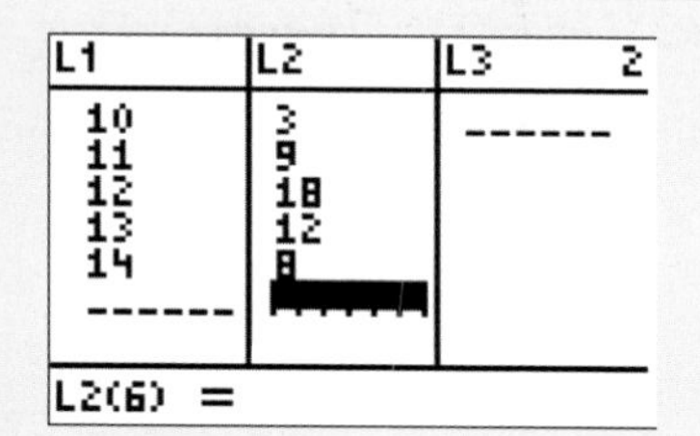

Press STAT CALC.

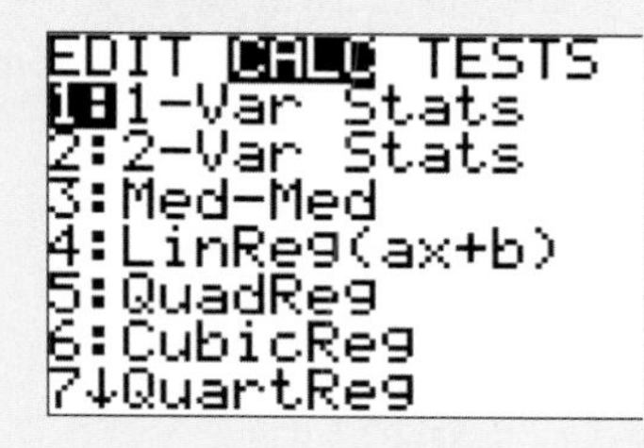

Select 1: 1-Var Stats. Press ENTER.

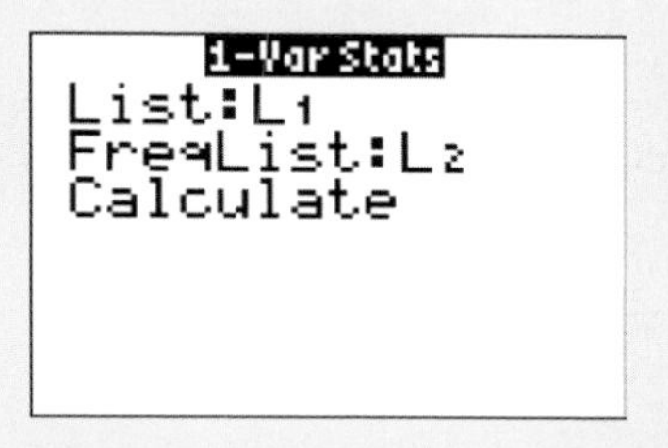

Leave the default value for List and enter 2nd L2 in FreqList.

Press ENTER ENTER.

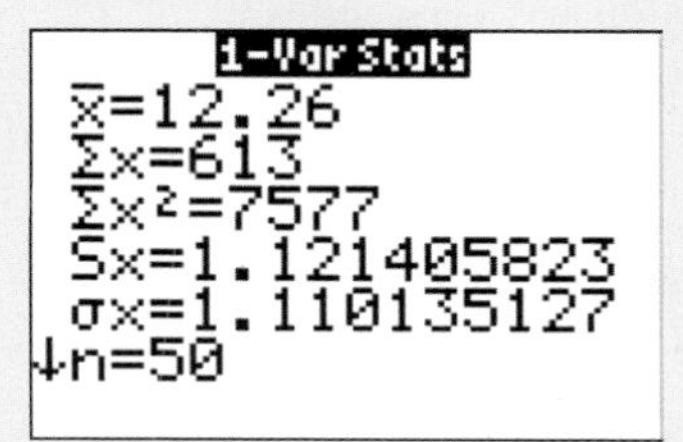

The GDC displays a list of basic statistics.

$\bar{x}$ is the mean.

Σx is the sum of all the values.

n is the number of items.

Press F2 CALC

Press F6 SET

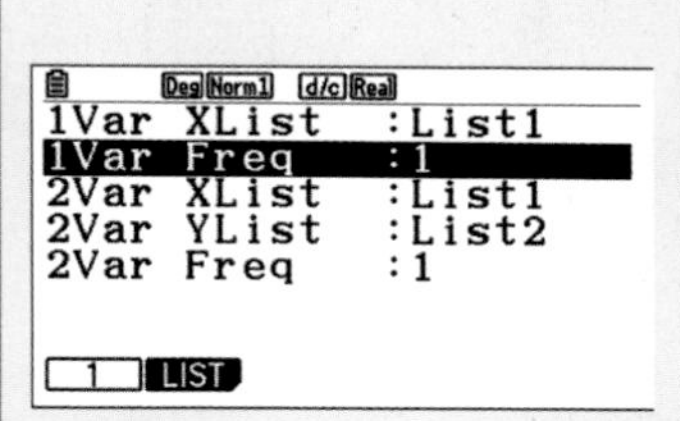

Navigate to 1Var Freq. Press F2 LIST.

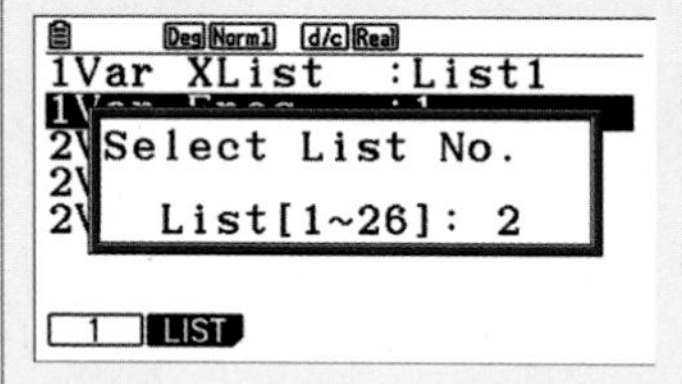

Type 2 and press EXE.

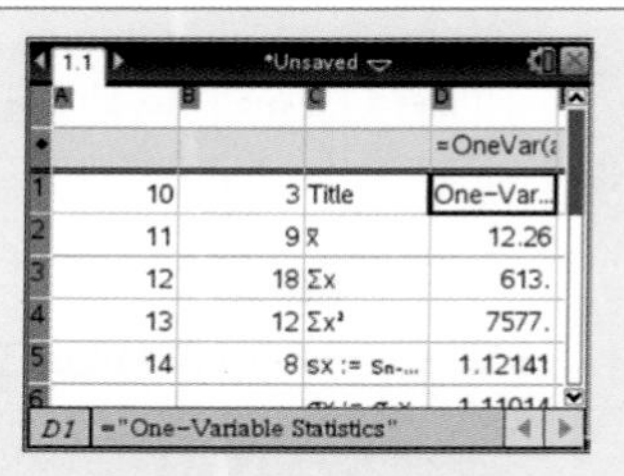

The GDC displays a list of basic statistics.

$\bar{x}$ is the mean.

Σx is the sum of all the values.

```
1.1  *Unsaved
=OneVar(
7   n          50.
8   MinX       10.
9   Q₁X        12.
10  MedianX... 12.
11  Q₃X        13.
C11 ="Q₃X"
```

Scrolling down the table using the navigation keys ▲ ▼ shows further values.

n is the number of items.

Q_1X is the lower quartile.

MedianX… is the median.

Q_3X is the upper quartile.

```
1-Var Stats
↑n=50
 minX=10
 Q₁=12
 Med=12
 Q₃=13
 maxX=14
```

Scrolling down the table using the navigation keys ▲ ▼ shows further values.

Q_1 is the lower quartile.

Med is the median.

Q_3 is the upper quartile.

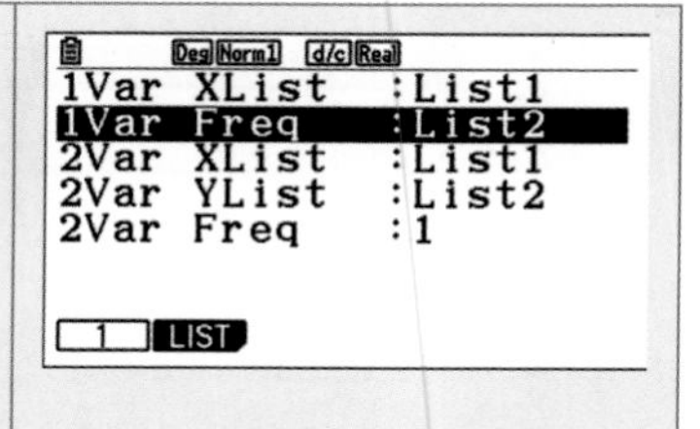

Press EXIT then F1 1-VAR.

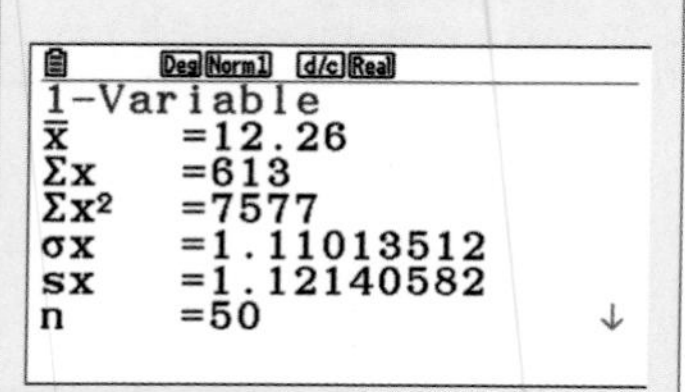

The GDC displays a list of basic statistics.

$\bar{x}$ is the mean.

Σx is the sum of all the values.

n is the number of items.

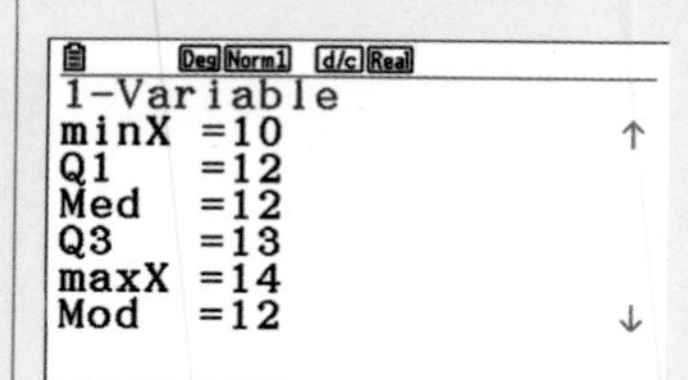

Scrolling down the table using the navigation keys ▲ ▼ shows further values.

Q1 is the lower quartile.

Med is the median.

Q3 is the upper quartile.

3.3 Finding a linear regression equation

The GDC can be used to plot a scatter graph, then find and plot the equation of linear regression between two variables. Although you will only need to find the equation with your GDC, it is also useful to know how to plot the scatter graph and the show the equation.

Example 16

The lengths in cm (x) and weights in grams (y) of the fish in the school pond are shown in this table.

Length	16.3	21.2	22.3	33.6	36.1	38.0
Weight	32.3	45.5	42.8	72.2	75.9	82.2

Find the equation of linear regression of y on x.

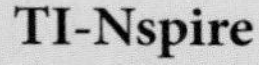

TI-Nspire

Open a new Lists & Spreadsheet page.

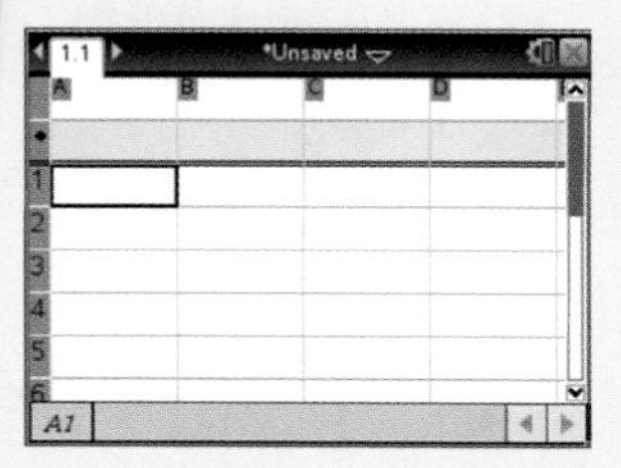

Enter the values for length in the first column, A. Press ENTER after each value and use the navigation keys ▲ ▼ to scroll through the list. Enter the values for weight in the second column B.

TI84-Plus

Press STAT 1: Edit...

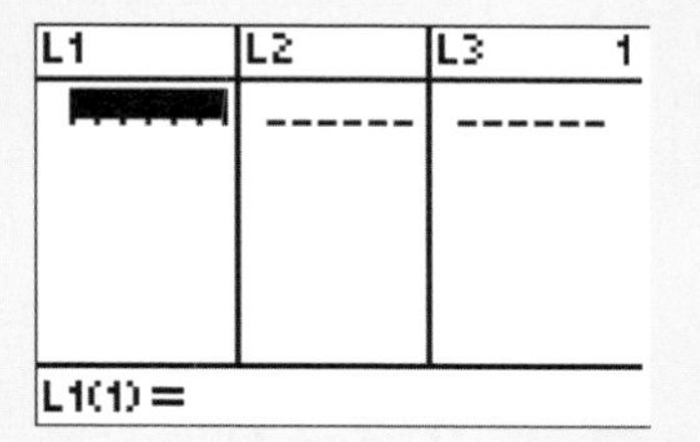

Enter the values for length in the first column, L1. Press ENTER after each value and use the navigation keys ▲ ▼ to scroll through the list. Enter the values for weight in the second column L2.

Casio

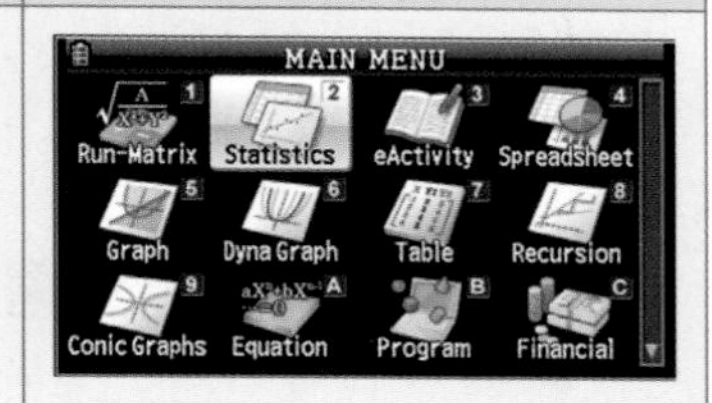

Enter Statistics mode.

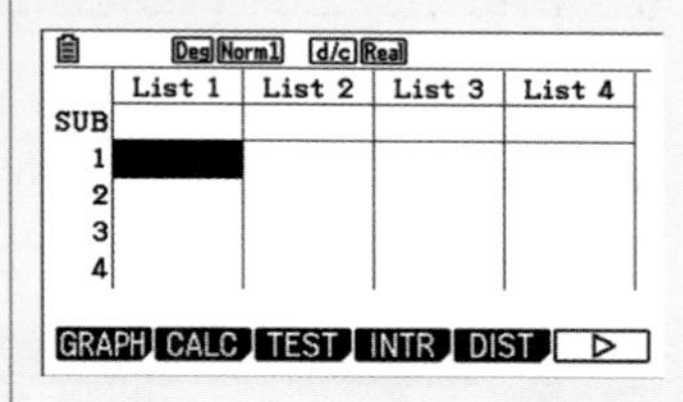

Enter the data for length in the first column, List 1. Press EXE after each value and use the navigation keys ▲ ▼ to scroll through the list. Enter the values for weight in the second column List 2.

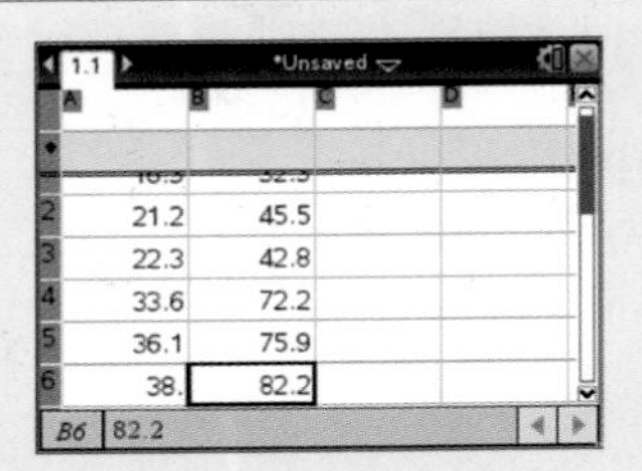

Press MENU

Press 4: Statistics | 1: Stat Calculations | 3: Linear Regression(mx+b)…

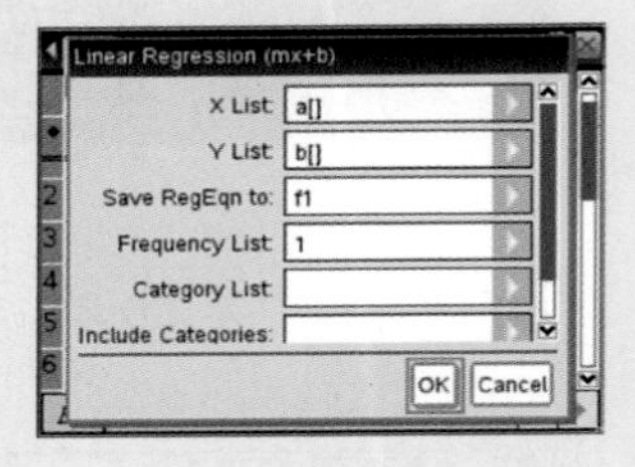

Delete the default values in this dialogue, enter a[] in X List, b[] in Y List and press ENTER.

The equation of linear regression is $y = 2.27x - 5.06$

Press STAT

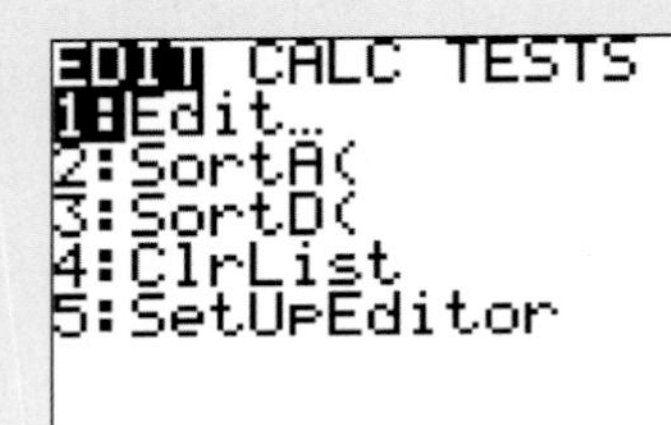

Select CALC using the ▸ ◂ navigation keys.

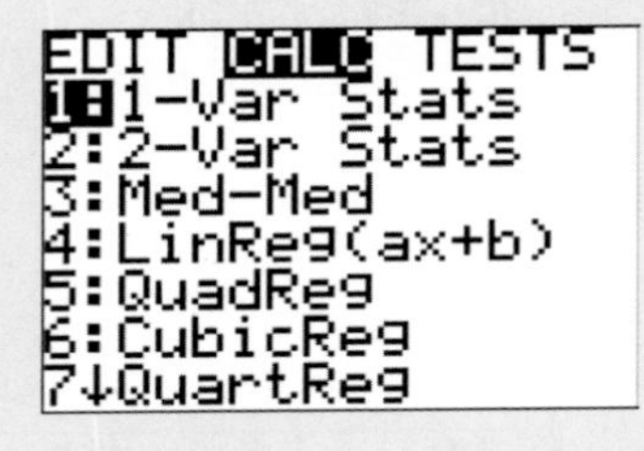

Select 4: LinReg(ax+b).

Select L_1 as Xlist and L_2 as Ylist. The reqList should be clear. Press ENTER.

Press F2 CALC.

Press F3 REG. (You can use SET to change the two lists).

Press F1 X.

Press F1 ax + b.

	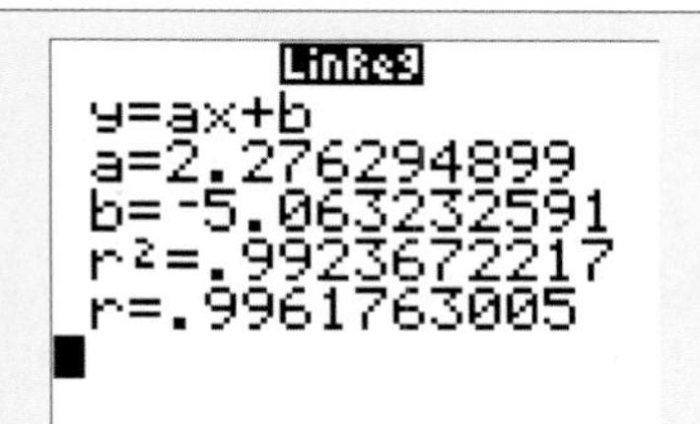 The equation of linear regression is $y = 2.27x - 5.06$ (Depending on settings the values of r and r^2 may not be shown).	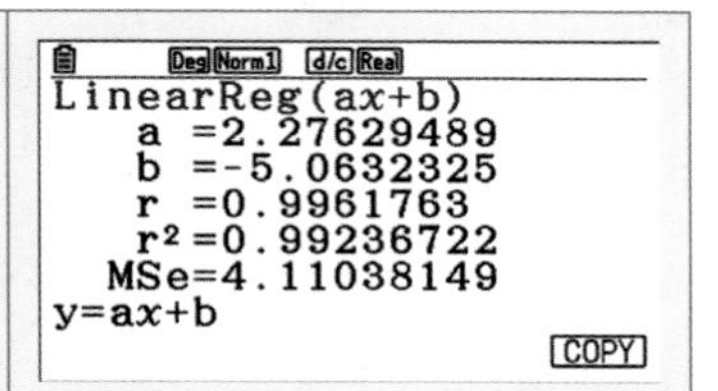 The equation of linear regression is $y = 2.27x - 5.06$

3.4 Drawing a scatter graph and the graph of the equation of linear regression

Example 17

The lengths in cm (x) and weights in grams (y) of the carp in the school pond are shown in this table.

Length	16.3	21.2	22.3	33.6	36.1	38.0
Weight	32.3	45.5	42.8	72.2	75.9	82.2

Plot a scatter graph of y against x and show the graph equation of linear regression of y on x.

TI-Nspire	TI84-Plus	Casio
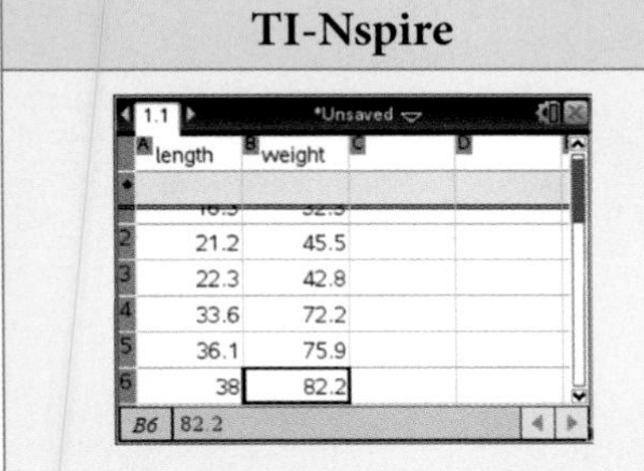 Enter the data as in Example 16 but name the columns length and weight as shown.	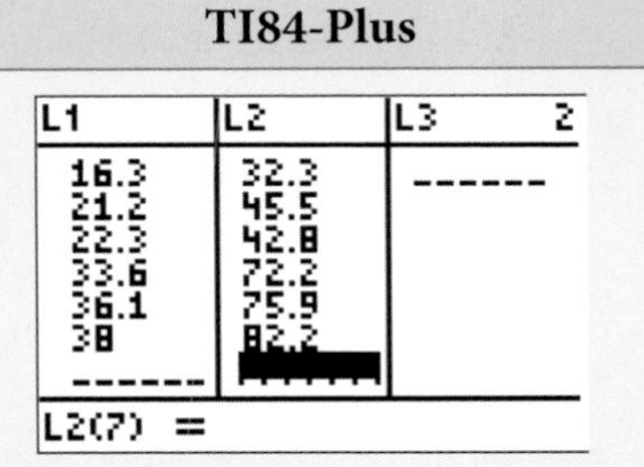 Enter the data as in Example 16	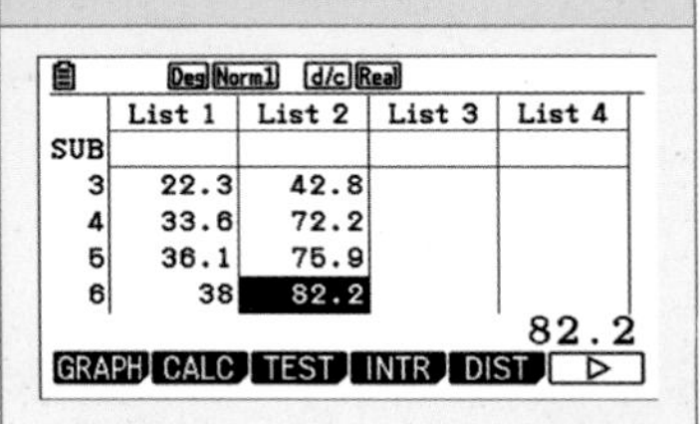 Enter the data as in Example 16
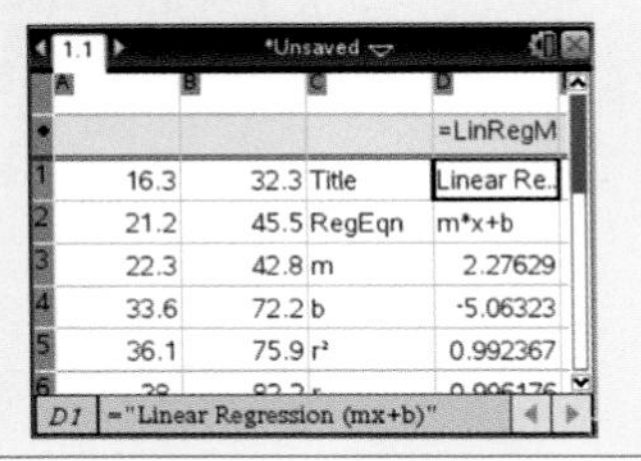	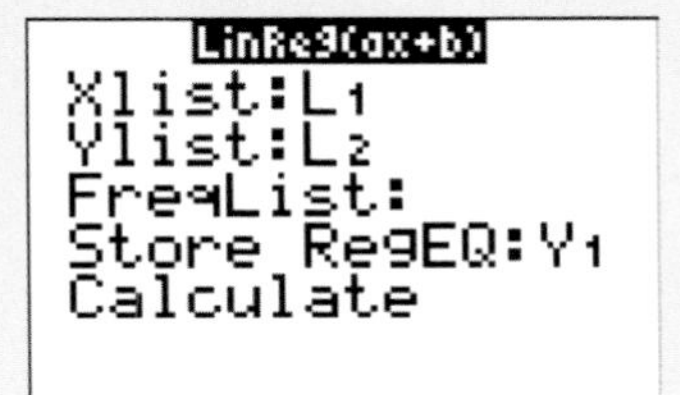	

Calculate the equation of linear regression as in Example 16 choosing the variables length and weight for X List and Y List.

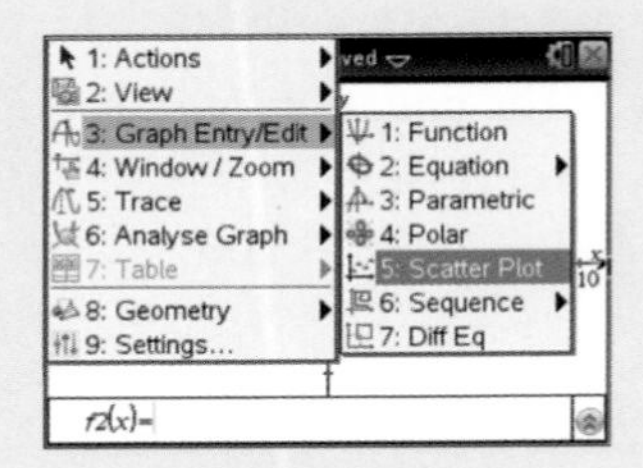

Press On ans add a Graphs page. Press MENU 3: Graph Entry/ Edit | 5: Scatter Plot

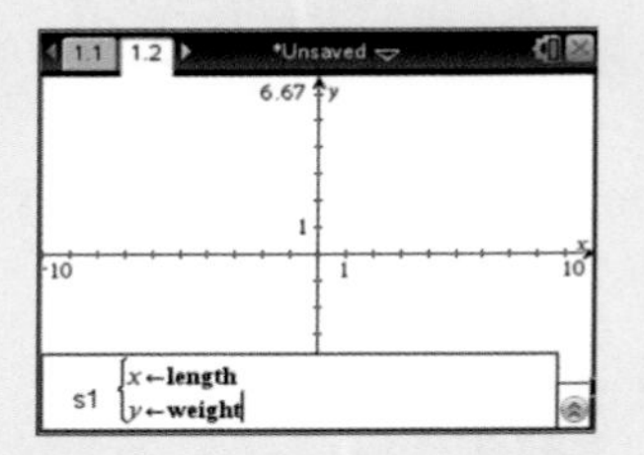

Use VAR to select length and weight from the variables to enter as x and y. Press ENTER.

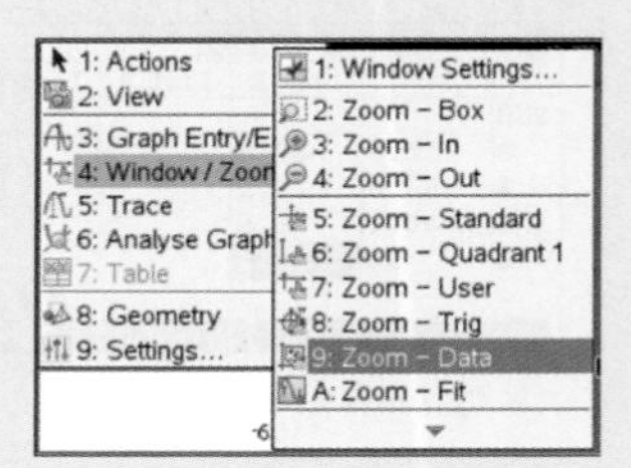

Press MENU 4: Window/Zoom | 9: Zoom-Data

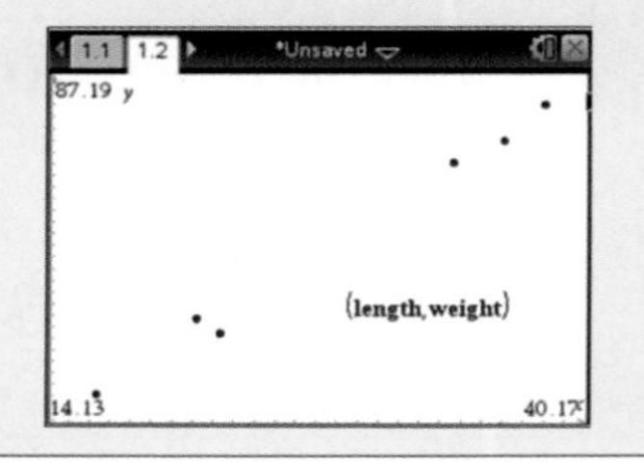

Calculate the equation of linear regression as in Example 16.

When entering the lists, enter Y_1 in Store RegEQ. (Press ALPHA F4 and select Y_1.)

LinReg
y=ax+b
a=2.276294899
b=-5.063232591
r²=.9923672217
r=.9961763005

Press 2nd STAT PLOT

Press ENTER.

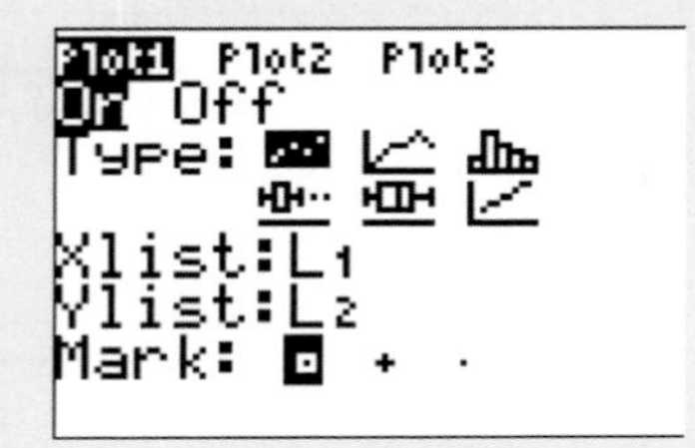

Select On, the scatter graph icon in Type, L_1 as Xlist and L_2 as Ylist.

ZOOM MEMORY
3↑Zoom Out
4:ZDecimal
5:ZSquare
6:ZStandard
7:ZTrig
8:ZInteger
9↓ZoomStat

Press EXIT until you see this screen. Press F1 GRAPH and F1 GRAPH1.

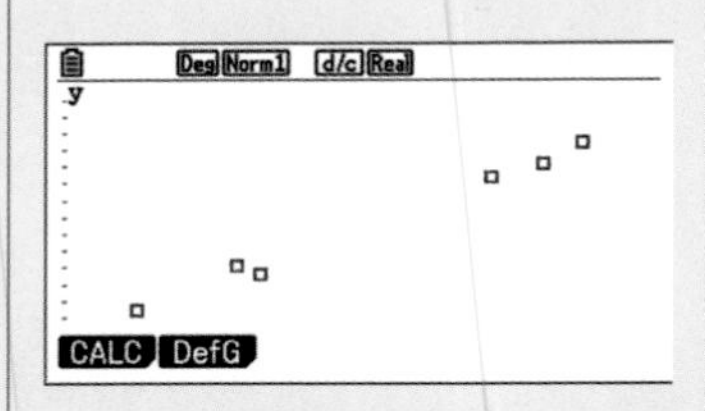

Press F1 CALC | F2 X | F1 ax+b.

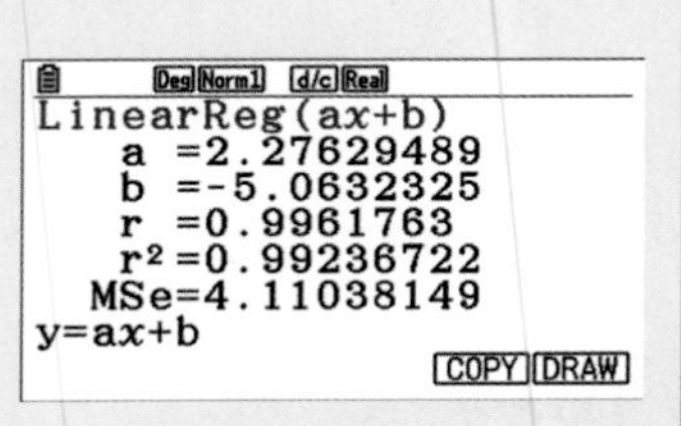

Press F6 DRAW.

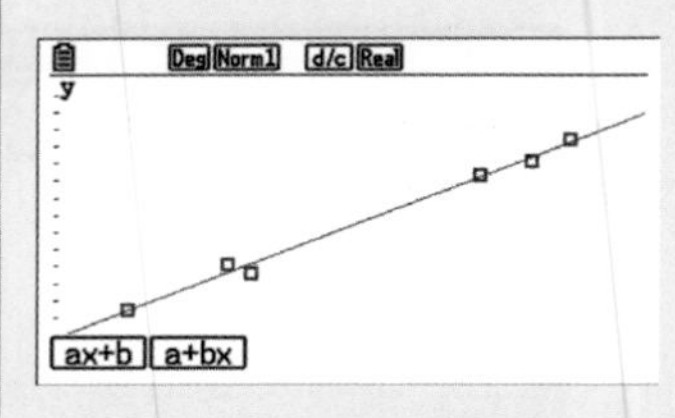

The GDC displays the scatter graph and the line of regression.

Press MENU

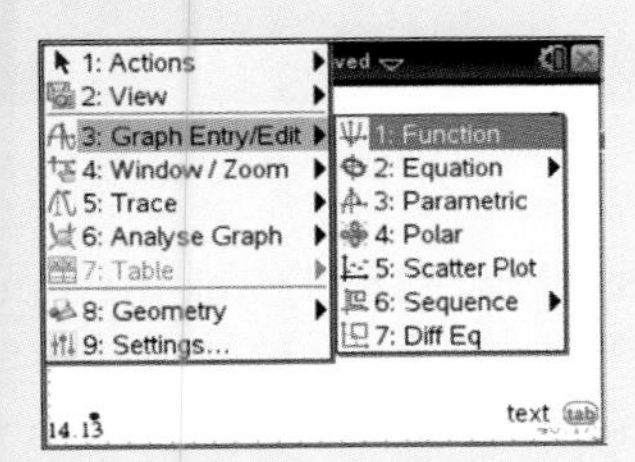

Press 3: Graph Entry/Edit | 1: Function

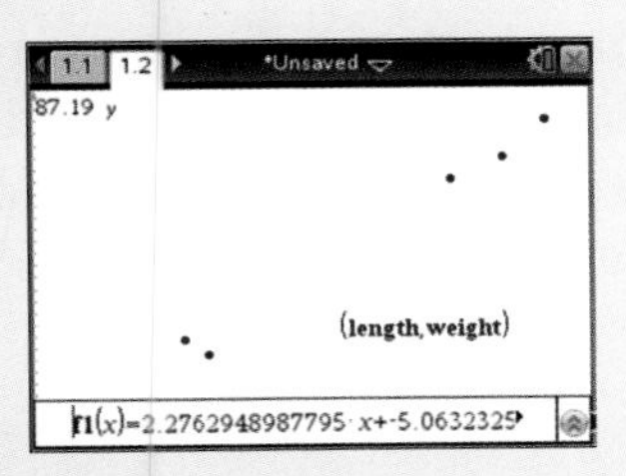

Use the ▲ ▼ keys to navigate to f1(x) and press ENTER.

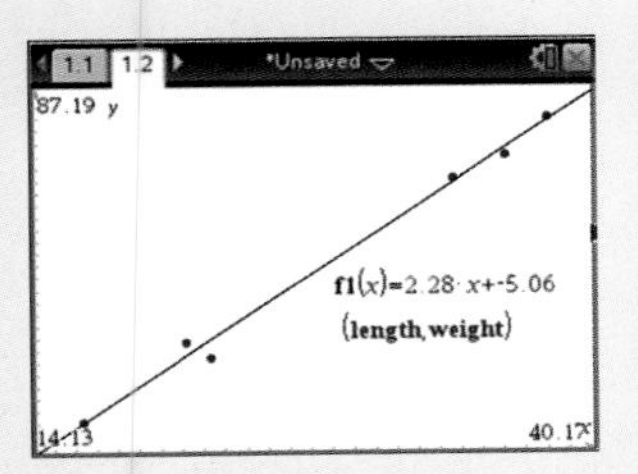

The GDC displays the scatter graph and the line of regression.

Press ZOOM 9: ZoomStat.

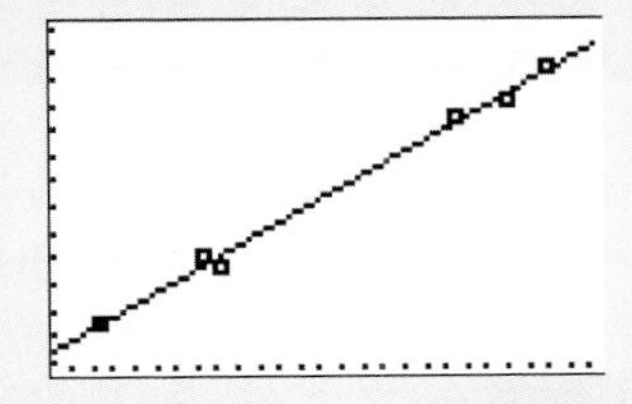

The GDC displays the scatter graph and the line of regression.

2 Number

Karl Friedrich Gauss (1777–1855) was the son of a German labourer and is thought by many to have been the greatest all-round mathematician ever. He considered that his finest discovery was the method for constructing a regular seventeen-sided polygon. This was not of the slightest use outside the world of mathematics, but was a great achievement of the human mind. Gauss would not have understood the modern view held by many that mathematics must somehow be 'useful' to be worthy of study.

1.1 Vocabulary and notation for different sets of numbers: natural numbers $\mathbb{N}$, primes, squares, cubes, integers $\mathbb{Z}$, rational numbers $\mathbb{Q}$, irrational numbers, real numbers $\mathbb{R}$, triangle numbers
$\mathbb{N} = \{0, 1, 2, \ldots\}$

1.2 Use of the four operations and brackets

1.3 Highest common factor, lowest common multiple

1.4 Calculation of powers and roots

1.5 Ratio and proportion including use of e.g. map scales

1.6 Absolute value $|x|$

1.7 Equivalences between decimals, fractions, ratios and percentages

1.8 Percentages including applications such as interest and profit, includes both simple and compound interest and also percentiles

1.9 Meaning of exponents (powers, indices) in $\mathbb{Q}$
Standard Form $a \times 10^n$ where $1 \leq a < 10$ and $n \in \mathbb{Z}$
Rules for exponents

1.10 Surds (radicals), simplification of square root expressions
Rationalisation of the denominator
e.g. $\dfrac{1}{\sqrt{3} - 1}$

1.11 Estimating, rounding, decimal places and significant figures

1.12 Calculations involving time: second (s), minutes (min), hours (h), days, months, years including the relation between consecutive units

1.13 Problems involving speed, distance and time problems

2.1 Vocabulary and notation for sets of numbers

- **Natural** numbers are the whole numbers 0, 1, 2, . . . which are numbers used for counting.

  The set of natural numbers, $\mathbb{N} = \{0, 1, 2, 3, \ldots\}$
- **Integers** are positive and negative whole numbers.

  We write $\mathbb{Z} = \{\ldots -4, -3, -2, -1, 0, 1, 2, 3, 4, \ldots\}$
- A **prime** number is divisible only by itself and by 1, e.g. 2, 3, 5, 7, 11 . . .
- The **square** numbers are 1, 4, 9, 16, . . . n^2. . .
- The **cube** numbers are 1, 8, 27, 64, . . . n^3. . .
- A **rational** number can be written in the form $\frac{m}{n}$, where m and n are whole numbers.

  The set of rational numbers is denoted by $\mathbb{Q}$.
- **Irrational** numbers cannot be written in the form $\frac{m}{n}$ as above.

  It can be shown that $\sqrt{2}$, $\sqrt{3}$, $\sqrt{5}$ are irrational numbers.
- The set of rational and irrational numbers form the set of **real** numbers $\mathbb{R}$.
- **Triangle** numbers 1, 3, 6, 10, 15, 21, . . . can be represented by objects or dots that can form an equilateral triangle as below.

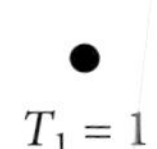

$T_1 = 1$

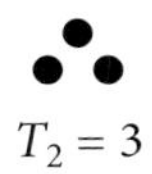

$T_2 = 3$

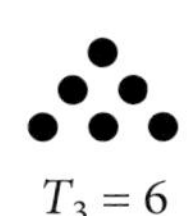

$T_3 = 6$

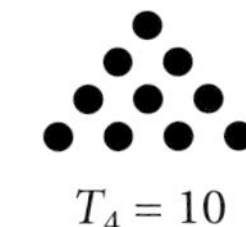

$T_4 = 10$

$T_5 = 15$

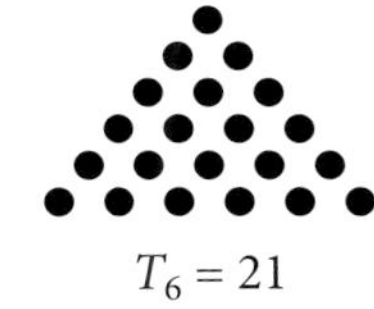

$T_6 = 21$

The **absolute** value (or modulus) $|n|$ of a real number is the numerical value of n not regarding its *sign* (+ or −). For example, $|-5| = 5$.
The absolute value of a number can be thought of as its *distance* from zero.

2.2 Arithmetic

Decimals

Example 1

Evaluate: **a)** $7{\cdot}6 + 19$ **b)** $3{\cdot}4 - 0{\cdot}24$ **c)** $7{\cdot}2 \times 0{\cdot}21$ **d)** $0{\cdot}84 \div 0{\cdot}2$ **e)** $3{\cdot}6 \div 0{\cdot}004$

a)
$$\begin{array}{r} 7{\cdot}6 \\ +\,19{\cdot}0 \\ \hline 26{\cdot}6 \\ \hline \end{array}$$

b)
$$\begin{array}{r} 3{\cdot}40 \\ -\,0{\cdot}24 \\ \hline 3{\cdot}16 \\ \hline \end{array}$$

c)
$$\begin{array}{r} 7{\cdot}2 \\ \times\,0{\cdot}21 \\ \hline 72 \\ 1440 \\ \hline 1{\cdot}512 \\ \hline \end{array}$$

d) $0{\cdot}84 \div 0{\cdot}2 = 8{\cdot}4 \div 2$

$$2\overline{)8{\cdot}4} = 4{\cdot}2$$

e) $3{\cdot}6 \div 0{\cdot}004 = 3600 \div 4$
$= 900$

No decimal points in the working. 3 figures after the points in the question and in the answer to part **c)**.

In part **d)**, multiply both numbers by 10 so that we can divide by a whole number.

.xercise 2.1

Evaluate the following without a calculator:

1. $7.6 + 0.31$	**2.** $15 + 7.22$	**3.** $7.004 + 0.368$	**4.** $0.06 + 0.006$
5. $4.2 + 42 + 420$	**6.** $3.84 - 2.62$	**7.** $11.4 - 9.73$	**8.** $4.61 - 3$
9. $17 - 0.37$	**10.** $8.7 + 19.2 - 3.8$	**11.** $25 - 7.8 + 9.5$	**12.** $3.6 - 8.74 + 9$
13. $20.4 - 20.399$	**14.** 2.6×0.6	**15.** 0.72×0.04	**16.** 27.2×0.08
17. 0.1×0.2	**18.** $(0.01)^2$	**19.** 2.1×3.6	**20.** 2.31×0.34
21. 0.36×1000	**22.** $0.34 \times 100\,000$	**23.** $3.6 \div 0.2$	**24.** $0.592 \div 0.8$
25. $0.1404 \div 0.06$	**26.** $3.24 \div 0.002$	**27.** $0.968 \div 0.11$	**28.** $600 \div 0.5$
29. $0.007 \div 4$	**30.** $2640 \div 200$	**31.** $1100 \div 5.5$	**32.** $(11 + 2.4) \times 0.06$
33. $(0.4)^2 \div 0.2$	**34.** $77 \div 1000$	**35.** $(0.3)^2 \div 100$	**36.** $(0.1)^4 \div 0.01$
37. $\frac{92 \times 4.6}{2.3}$	**38.** $\frac{180 \times 4}{36}$	**39.** $\frac{0.55 \times 0.81}{4.5}$	**40.** $\frac{63 \times 600 \times 0.2}{360 \times 7}$

Exercise 2.2

1. A maths teacher bought 40 calculators at \$8·20 each and a number of other calculators costing \$2·95 each. In all she spent \$387. How many of the cheaper calculators did she buy?

2. At a temperature of 20°C the common amoeba reproduces by splitting in half every 24 hours. If we start with a single amoeba how many will there be after

a) 8 days

b) 16 days?

3.

1.5 litres
£2·90

The prices of various amounts of freshly squeezed orange juice are shown above.
Lian needs to buy 5 litres of this orange juice.
Which cartons should she buy so that she spends the least amount of money?

4. A bank is selling euros at the rate of 1·147 euros for one pound sterling. Find the cost in euros for £20 000.

5. A lake is 2·5 cm long on a map of scale 1:10 000. What is the actual length of the lake?

6. Which is larger: 0.1×0.2 or $0.5 \div 10$?

7. Copy and complete.

$3^2 + 4^2 + 12^2 = 13^2$

$5^2 + 6^2 + 30^2 = 31^2$

$6^2 + 7^2 + \square = \square$

$x^2 + \square + \square = \square$

8. Find all the missing digits in these multiplications.

a) $\begin{array}{r} 5\,* \\ \times\, 9 \\ \hline *\,*\,6 \\ \hline \end{array}$ **b)** $\begin{array}{r} *\,7 \\ \times\, * \\ \hline 4\,*\,6 \\ \hline \end{array}$ **c)** $\begin{array}{r} 5\,* \\ \times\, * \\ \hline 1\,*\,4 \\ \hline \end{array}$

9. Pages 6 and 27 are on the same (double) sheet of a newspaper. What are the page numbers on the opposite side of the sheet? How many pages are there in the newspaper altogether?

10. Use the numbers 1, 2, 3, 4, 5, 6, 7, 8, 9 once each and in their natural order to obtain an answer of 100. You may use only the operations $+, -, \times, \div$.

11. The ruler below has eleven marks and can be used to measure lengths from one unit to twelve units.

Design a ruler which can be used to measure all the lengths from one unit to twelve units but this time put the minimum possible number of marks on the ruler.

12. Each packet of washing powder carries a token and four tokens can be exchanged for a free packet. How many free packets will I receive if I buy 64 packets?

13. Put three different numbers in the circles so that when you add the numbers at the end of each line you always get a square number.

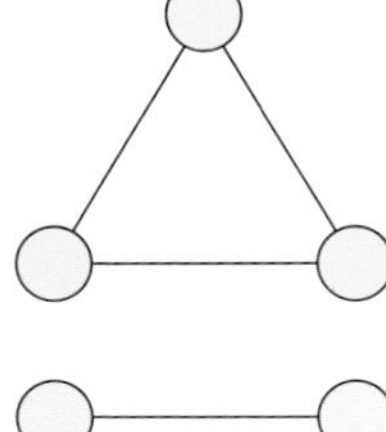

14. Put four different numbers in the circles so that when you add the numbers at the end of each line you always get a square number.

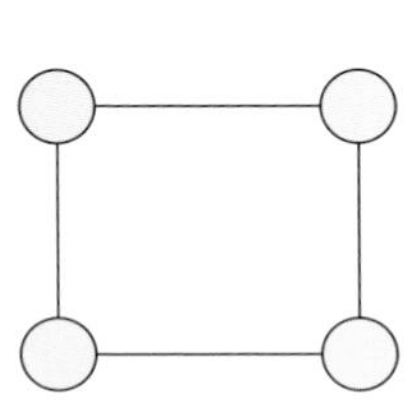

15. A group of friends share a bill for \$13·69 equally between them. How many were in the group?

Fractions

Common fractions are added or subtracted from one another directly only when they have a common denominator.

Example 2

Evaluate: **a)** $\frac{3}{4}+\frac{2}{5}$ **b)** $2\frac{3}{8}-1\frac{5}{12}$ **c)** $\frac{2}{5}\times\frac{6}{7}$ **d)** $2\frac{2}{5}\div 6$

a) $\frac{3}{4}+\frac{2}{5}=\frac{15}{20}+\frac{8}{20}$
$=\frac{23}{20}$
$=1\frac{3}{20}$

b) $2\frac{3}{8}-1\frac{5}{12}=\frac{19}{8}-\frac{17}{12}$
$=\frac{57}{24}-\frac{34}{24}$
$=\frac{23}{24}$

c) $\frac{2}{5}\times\frac{6}{7}=\frac{12}{35}$

d) $2\frac{2}{5}\div 6=\frac{12}{5}\div\frac{6}{1}$
$=\frac{{}^{2}\cancel{12}}{5}\times\frac{1}{\cancel{6}_{1}}=\frac{2}{5}$

Remember
The order of operations follows the BODMAS rule:
Brackets then
Powers Of then
Divide then
Multiply then
Add then
Subtract

Exercise 2.3

Evaluate, and simplify your answer.

1. $\frac{3}{4}+\frac{4}{5}$ **2.** $\frac{1}{3}+\frac{1}{8}$ **3.** $\frac{5}{6}+\frac{6}{9}$ **4.** $\frac{3}{4}-\frac{1}{3}$ **5.** $\frac{3}{5}-\frac{1}{3}$

6. $\frac{1}{2}-\frac{2}{5}$ **7.** $\frac{2}{3}\times\frac{4}{5}$ **8.** $\frac{1}{7}\times\frac{5}{6}$ **9.** $\frac{5}{8}\times\frac{12}{13}$ **10.** $\frac{1}{3}\div\frac{4}{5}$

11. $\frac{3}{4}\div\frac{1}{6}$ **12.** $\frac{5}{6}\div\frac{1}{2}$ **13.** $\frac{3}{8}+\frac{1}{5}$ **14.** $\frac{3}{8}\times\frac{1}{5}$ **15.** $\frac{3}{8}\div\frac{1}{5}$

16. $1\frac{3}{4}-\frac{2}{3}$ **17.** $1\frac{3}{4}\times\frac{2}{3}$ **18.** $1\frac{3}{4}\div\frac{2}{3}$ **19.** $3\frac{1}{2}+2\frac{3}{5}$ **20.** $3\frac{1}{2}\times 2\frac{3}{5}$

21. $3\frac{1}{2}\div 2\frac{3}{5}$ **22.** $\left(\frac{3}{4}-\frac{2}{3}\right)\div\frac{3}{4}$ **23.** $\left(\frac{3}{5}+\frac{1}{3}\right)\times\frac{5}{7}$ **24.** $\dfrac{\frac{3}{8}-\frac{1}{5}}{\frac{7}{10}-\frac{2}{3}}$ **25.** $\dfrac{\frac{2}{3}+\frac{1}{5}}{\frac{3}{4}-\frac{1}{3}}$

26. Arrange the fractions in order of size.

a) $\frac{7}{12},\frac{1}{2},\frac{2}{3}$ **b)** $\frac{3}{4},\frac{2}{3},\frac{5}{6}$ **c)** $\frac{1}{3},\frac{17}{24},\frac{5}{8},\frac{3}{4}$ **d)** $\frac{5}{6},\frac{8}{9},\frac{11}{12}$

27. Find the fraction which is mid-way between the two fractions given.

a) $\frac{2}{5},\frac{3}{5}$ **b)** $\frac{5}{8},\frac{7}{8}$ **c)** $\frac{2}{3},\frac{3}{4}$

d) $\frac{1}{3},\frac{4}{9}$ **e)** $\frac{4}{15},\frac{1}{3}$ **f)** $\frac{3}{8},\frac{11}{24}$

28. In the equation below all the numbers in the boxes are the same number. What is the number?

$$\frac{\square}{\square} - \frac{\square}{6} = \frac{\square}{30}$$

29. Each card has a fraction and a letter. Find the cards which contain equivalent fractions and arrange the letters to make the name of a capital city.

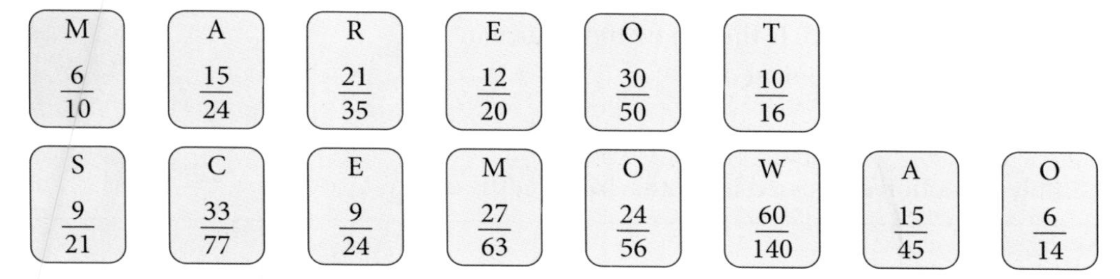

30. Draw a 4×6 rectangle. Divide it into three parts using three different fractions, each with numerator 1.

31. Draw a 4×5 rectangle and divide it into four parts using four different fractions, each with numerator 1.

32. A girl read $\frac{3}{8}$ of her book one day and $\frac{2}{5}$ the next day. How much was there left to read?

33. A blackbird shares 3 g of worms between her three chicks. Emily receives $\frac{3}{4}$ g and Brian receives $\frac{5}{8}$ g. How much is left for Lucky, the third chick?

34. A wooden pole is painted in four colours: red, yellow, blue and green. If $\frac{1}{8}$ is red, $\frac{5}{12}$ is yellow and $\frac{1}{16}$ is blue, what fraction of the pole is painted green?

35. Work out:

a) $\left[\frac{1}{2} \text{ of } \left(\frac{1}{3} \times \frac{1}{4}\right)\right] \div \frac{1}{5}$ **b)** $\dfrac{\frac{1}{2} + \frac{1}{3} + \frac{1}{4}}{\frac{1}{2} \times \frac{1}{3} \times \frac{1}{4}}$ **c)** $\sqrt{\left(\frac{2}{3} \times \frac{3}{4} \times \frac{4}{5} \times \ldots \times \frac{99}{100}\right) \times \left(\frac{9}{50}\right)}$

36. Copy and complete:

$$\left(\frac{4}{7} - \frac{1}{2}\right) \times \frac{\square}{\square} + \frac{1}{9} = \frac{1}{6}$$

37. Two whole numbers a and b are chosen such that $a < b$ and $b < 7$. How many *different* fractions are there of the form $\frac{a}{b}$?

38. When it hatches from its egg, the shell of a certain crab is 1 cm across.
When fully grown, the shell is approximately 10 cm across.
Each new shell is one-third bigger than the previous one.
How many shells does a fully grown crab have during its life?

39. Glass A contains 100 ml of water and glass B contains 100 ml of juice.

A 10 ml spoonful of juice is taken from glass B and mixed thoroughly with the water in glass A. A 10 ml spoonful of the mixture from A is returned to B. Is there now more juice in the water or more water in the juice?

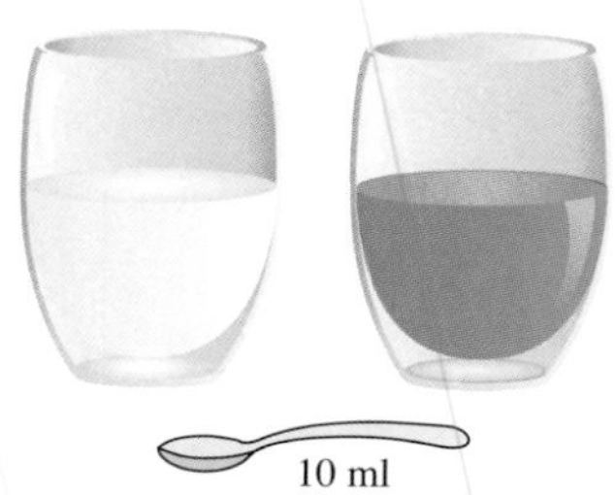

Fractions and decimals

A decimal is simply a fraction expressed in tenths, hundredths etc.

Example 3

Change **a)** $\frac{7}{8}$ to a decimal **b)** 0·35 to a fraction **c)** $\frac{1}{3}$ to a decimal

a) For $\frac{7}{8}$, divide 8 into 7:

$$8\overline{)7{\cdot}000}\quad\text{gives } 0{\cdot}875$$

$\frac{7}{8} = 0{\cdot}875$

b) $0{\cdot}35 = \frac{35}{100} = \frac{7}{20}$

c) For $\frac{1}{3}$, divide 3 into 1:

$$3\overline{)1{\cdot}0^{1}0^{1}0^{1}000}\quad\text{gives } 0{\cdot}3333$$

$\frac{1}{3} = 0{\cdot}\dot{3}$ (0·3 recurring)

Exercise 2.4

In questions **1** to **24**, change the fractions to decimals.

1. $\frac{1}{4}$ **2.** $\frac{2}{5}$ **3.** $\frac{4}{5}$ **4.** $\frac{3}{4}$ **5.** $\frac{1}{2}$ **6.** $\frac{3}{8}$ **7.** $\frac{9}{10}$ **8.** $\frac{5}{8}$

9. $\frac{5}{12}$ **10.** $\frac{1}{6}$ **11.** $\frac{2}{3}$ **12.** $\frac{5}{6}$ **13.** $\frac{2}{7}$ **14.** $\frac{3}{7}$ **15.** $\frac{4}{9}$ **16.** $\frac{5}{11}$

17. $1\frac{1}{5}$ **18.** $2\frac{5}{8}$ **19.** $2\frac{1}{3}$ **20.** $1\frac{7}{10}$ **21.** $2\frac{3}{16}$ **22.** $2\frac{2}{7}$ **23.** $2\frac{6}{7}$ **24.** $3\frac{19}{100}$

In questions **25** to **40**, change the decimals to fractions and simplify.

25. 0·2 **26.** 0·7 **27.** 0·25 **28.** 0·45 **29.** 0·36 **30.** 0·52

31. 0·125 **32.** 0·625 **33.** 0·84 **34.** 2·35 **35.** 3·95 **36.** 1·05

37. 3·2 **38.** 0·27 **39.** 0·007 **40.** 0·000 11

Evaluate, giving the answer to 2 decimal places:

41. $\frac{1}{4} + \frac{1}{3}$ **42.** $\frac{2}{3} + 0{\cdot}75$ **43.** $\frac{8}{9} - 0{\cdot}24$ **44.** $\frac{7}{8} + \frac{5}{9} + \frac{2}{11}$

45. $\frac{1}{3} \times 0{\cdot}2$ **46.** $\frac{5}{8} \times \frac{1}{4}$ **47.** $\frac{8}{11} \div 0{\cdot}2$ **48.** $\left(\frac{4}{7} - \frac{1}{3}\right) \div 0{\cdot}4$

Arrange the numbers in order of size (smallest first).

49. $\frac{1}{3}$, 0·33, $\frac{4}{15}$ **50.** $\frac{2}{7}$, 0·3, $\frac{4}{9}$ **51.** 0·71, $\frac{7}{11}$, 0·705 **52.** $\frac{4}{13}$, 0·3, $\frac{5}{18}$

Recurring decimals

We have seen that fractions can be converted into decimals by dividing the numerator by the denominator. [E.g. $\frac{3}{4} = 3 \div 4 = 0{\cdot}75$]
Sometimes the division gives a decimal which recurs, for example:

$\frac{1}{3} = 1 \div 3 = 0{\cdot}333 \ldots.$

It is important to realise that all recurring decimals can be written as exact fractions. Here is a method for converting recurring decimals to fractions:

a) Change 0·7777... to a fraction.

Let $r = 0{\cdot}7777\ldots$

$10r = 7{\cdot}7777\ldots$ (multiply by 10)

$9r = 7$ (subtract: $10r - r$)

$r = \frac{7}{9}$

So $0{\cdot}7777\ldots = \frac{7}{9}$

b) Change 0·373737 ... to a fraction.

Let $r = 0{\cdot}373737\ldots$

$100r = 37{\cdot}373737\ldots$

$100r - r = 99r$

$99r = 37$

$r = \frac{37}{99}$

so $0{\cdot}\dot{3}\dot{7} = \frac{37}{99}$

Exercise 2.5

1. Copy and complete to change 0·4444 ... to a fraction.

Let $r = 0{\cdot}4444\ldots$

$10r = \square$ (multiply both sides by 10)

$9r = \square$ (subtract)

$r = \square$

2. Copy and complete to change 0·28282828 ... to a fraction.

Let $r = 0{\cdot}28282828\ldots$

$100r = \square$

$99r = \square$

$r = \square$

In questions **3** to **10** change the recurring decimals to fractions.

3. $0{\cdot}2222\ldots$ **4.** $0{\cdot}737373\ldots$ **5.** $0{\cdot}\dot{5}\dot{1}(= 0{\cdot}515151\ldots)$ **6.** $0{\cdot}\dot{2}\dot{9}$

7. $0{\cdot}245245245\ldots$ **8.** $0{\cdot}\dot{3}2\dot{6}$ **9.** $0{\cdot}\dot{4}1\dot{7}$ **10.** $0{\cdot}\dot{8}\dot{2}$

11. Change these fractions to recurring decimals.

a) $\frac{1}{6}$ **b)** $\frac{8}{9}$ **c)** $\frac{2}{7}$ **d)** $\frac{5}{13}$

2.3 Number facts and sequences

Number facts

- An **integer** is a whole number. e.g. 2, −3,...
- A **prime** number is divisible only by itself and by 1.
  e.g. 2, 3, 5, 7, 11, 13,...
- The **multiples** of 12 are 12, 24, 36, 48,...
- The **factors** of 12 are 1, 2, 3, 4, 6, 12.
- A **square number** is the result of multiplying a number by itself.
  e.g. $5 \times 5 = 25$ so 25 is a square number.
- A **cube number** is the result of multiplying a number by itself three times. e.g. $5 \times 5 \times 5 = 125$, so 125 is a cube number.

Exercise 2.6

1. Which of the following are prime numbers?

3, 11, 15, 19, 21, 23, 27, 29, 31, 37, 39, 47, 51, 59, 61, 67, 72, 73, 87, 99

2. Write down the first five multiples of the following numbers:

a) 4 **b)** 6 **c)** 10 **d)** 11 **e)** 20

3. Write down the first six multiples of 4 and of 6. What are the first two *common* multiples of 4 and 6? [i.e. multiples of both 4 and 6]

4. Write down the first six multiples of 3 and of 5. What is the lowest common multiple of 3 and 5?

5. Write down all the factors of the following:

a) 6 **b)** 9 **c)** 10 **d)** 15 **e)** 24 **f)** 32

6. **a)** Is 263 a prime number?

By how many numbers do you need to divide 263 so that you can find out?

b) Is 527 a prime number?

c) Suppose you used a computer to find out if 1147 was a prime number. Which numbers would you tell the computer to divide by?

7. Make six prime numbers using the digits 1, 2, 3, 4, 5, 6, 7, 8, 9 once each.

8. Answer 'true' or 'false':

   a) All prime numbers are odd.

   b) If the product of two numbers is zero, then one of the numbers must be zero.

   c) Every positive integer greater than 20 has an even number of factors.

9. The reciprocal of 3 is $\frac{1}{3}$. The reciprocal of x is $\frac{1}{x}$.

   a) Find the reciprocal of $1\frac{1}{2}$.

   b) Find the reciprocal of [the reciprocal of 3 + the reciprocal of 4].

10. Find the square root of the reciprocal of $\left(\frac{71}{121} - \frac{2}{11}\right)$.

11. If three fifths of a number is 45, what is two thirds of it?

12. The number in the square is the product of the two numbers on either side of it.

    Copy and complete:

    a)
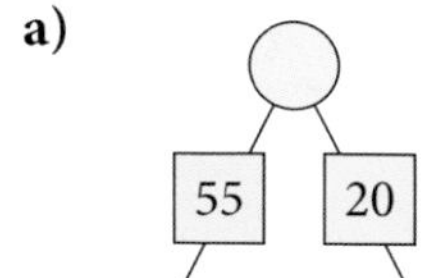

    b)
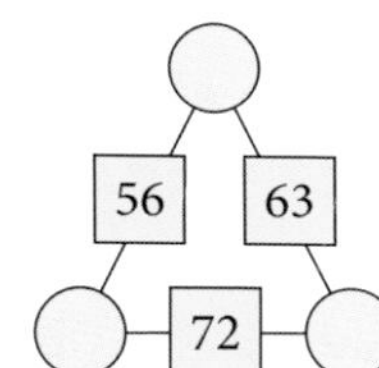

    c)
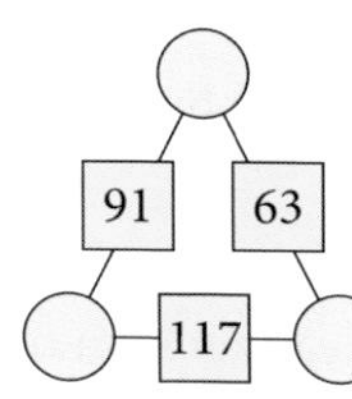

13. Write the following sets of numbers in order of magnitude (size).

    a) $\frac{7}{5}$, $\sqrt{25}$, -2, 3^2, $|-11|$, 1^3, $-\sqrt{9}$

    b) $|x|$ when $x = -10$, $\sqrt{400}$, $(1 \div 0{\cdot}01)$, $\left(\frac{1}{10}\right)^2$, $-(0{\cdot}2)^2$

Rational and irrational numbers

- A rational number can always be written exactly in the form $\frac{a}{b}$ where a and b are whole numbers.

> $\frac{3}{7}$ $\quad 1\frac{1}{2} = \frac{3}{2}$ $\quad 5{\cdot}14 = \frac{257}{50}$ $\quad 0{\cdot}\dot{6} = \frac{2}{3}$
>
> All of these are rational numbers.

- An irrational number cannot be written in the form $\frac{a}{b}$.
  $\sqrt{2}$, $\sqrt{5}$, π, $\sqrt[3]{2}$ are all irrational numbers.
- In general $\sqrt{n}$ is irrational unless n is a square number.

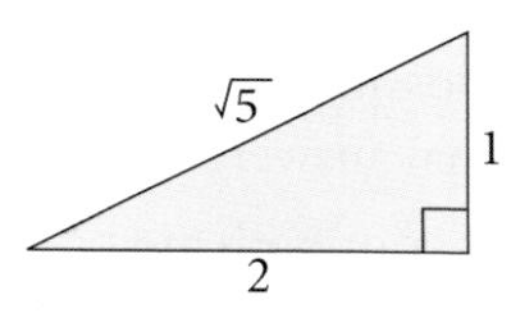

In this triangle the length of the hypotenuse is *exactly* $\sqrt{5}$.
On a calculator, $\sqrt{5} = 2{\cdot}236068$. This value of $\sqrt{5}$ is *not* exact and is correct only to 6 decimal places.

Exercise 2.7

1. Which of the following numbers are rational?

$\frac{\pi}{2}$ $\sqrt{5}$ $(\sqrt{17})^2$ $\sqrt{3}$

$3{\cdot}14$ $\frac{\sqrt{12}}{\sqrt{3}}$ π^2 $3^{-1} + 3^{-2}$

$7^{-\frac{1}{2}}$ $\frac{22}{7}$ $\sqrt{2} + 1$ $\sqrt{2{\cdot}25}$

2. **a)** Write down any rational number between 4 and 6.

b) Write down any irrational number between 4 and 6.

c) Find a rational number between $\sqrt{2}$ and $\sqrt{3}$.

d) Write down any rational number between π and $\sqrt{10}$.

3. **a)** For each triangle use Pythagoras' theorem to calculate the length x.

b) For each triangle state whether the *perimeter* is rational or irrational.

c) For each triangle state whether the *area* is rational or irrational.

d) In which triangle is $\sin\theta$ an irrational number?

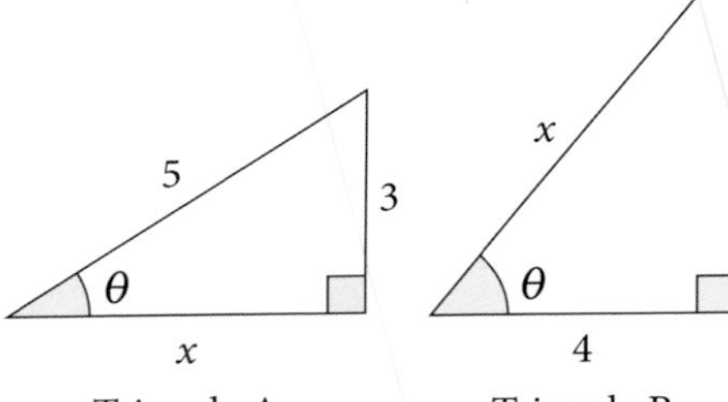

4. The diagram shows a circle of radius 3 cm drawn inside a square. Write down the exact value of the following and state whether the answer is rational or not:

a) the circumference of the circle

b) the diameter of the circle

c) the area of the square

d) the area of the circle

e) the shaded area.

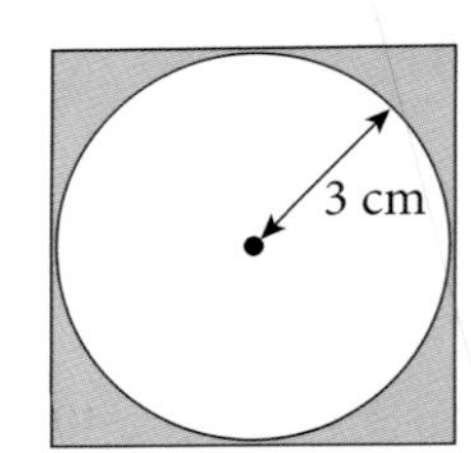

5. Answer 'true' or 'false':

a) $\sqrt{5} + \sqrt{5} = 2\sqrt{5}$ **b)** $\sqrt{12} = 2\sqrt{3}$ **c)** $\frac{3}{\sqrt{3}} = \sqrt{3}$

6. Think of two *irrational* numbers x and y such that $\frac{x}{y}$ is a *rational* number.

7. Explain the difference between a rational number and an irrational number.

8. **a)** Is it possible to multiply a rational number by an irrational number to give an answer which is rational?

b) Is it possible to multiply two irrational numbers together to give a rational answer?

c) If either or both are possible, give an example.

9. a) Write down two rational numbers, both between 0 and 1, one of which is equal to a non-recurring decimal and the other which is equal to a recurring decimal.

   b) Add together the two numbers you found above and show that the answer is rational.

10. n is a positive integer such that $\sqrt{n} = 8{\cdot}6$, correct to 1 decimal place.

   a) Find the value of n.

   b) State whether $\sqrt{n}$ is rational or irrational, giving your reason.

11. The number N is rational and is not zero. Decide whether $\frac{1}{N}$ is rational or irrational.

12. Write down a number between 5 and 6 that has a rational square root.

13. Adding or subtracting two irrational numbers usually gives an irrational result. The result can, however, be rational, for example:

$$(\sqrt{3}+2)+(5-\sqrt{3})=7 \quad \text{or} \quad (3-\sqrt{2})+\sqrt{2}=3$$

   Find two examples of adding two irrational numbers so that the result is a rational number.

14. Find an irrational number which, when multiplied by the number below, gives a rational number.

   a) $\sqrt{7}$ b) $\frac{1}{\sqrt{3}}$ c) $5\sqrt{2}$ d) $\frac{4}{\pi}$

Surds

- Numbers like $\sqrt{4}, \sqrt{3}, \sqrt{7}$ are called **surds**. The following rules apply.
- $\sqrt{a} \times \sqrt{b} = \sqrt{ab}$
- $\frac{\sqrt{a}}{\sqrt{b}} = \sqrt{\frac{a}{b}}$

$$\sqrt{2} \times \sqrt{3} = \sqrt{6}$$
$$\sqrt{80} = \sqrt{16 \times 5} = \sqrt{16} \times \sqrt{5} = 4\sqrt{5}$$

$$\frac{\sqrt{27}}{\sqrt{3}} = \sqrt{\frac{27}{3}} = \sqrt{9} = 3$$

- The fraction $\frac{6}{\sqrt{3}}$ can be written with a rational denominator by multiplying numerator and denominator by $\sqrt{3}$:

$$\frac{6}{\sqrt{3}} = \frac{6 \times \sqrt{3}}{\sqrt{3} \times \sqrt{3}} = \frac{6\sqrt{3}}{3} = 2\sqrt{3}$$

- A common mistake occurs with surds.

  $\sqrt{4}$ means 'the positive square root of 4' So $\sqrt{4} = 2$ only and *not* ± 2.

  Notice that the solutions of the equation $x^2 = 4$ are $x = 2, -2$

  You *can* write $x = +\sqrt{4}$ or $-\sqrt{4}$.

Look back at sequences **A**, **B** and **C** on the previous page.
For sequence **A**, the nth term $= 2n + b$ [the terms go up by 2]
For sequence **B**, the nth term $= 20n + b$ [the terms go up by 20]
For sequence **C**, the nth term $= -3n + b$ [the terms go up by -3]
Look at each sequence and find the value of b in each case.

For example in sequence **A**, when $n = 1$: $2 \times 1 + b = 5$
so $b = 3$

The nth term in sequence **A** is $2n + 3$.

Exercise 2.10

In questions **1** to **18** find a formula for the nth term of the sequence.

1. 5, 9, 13, 17, …
2. 7, 10, 13, 16, …
3. 4, 9, 14, 19, …
4. 6, 10, 14, 18, …
5. 5, 8, 11, 14, …
6. 25, 22, 19, 16, …
7. 5, 10, 15, 20, …
8. 2, 4, 8, 16, 32, …
9. $(1 \times 3), (2 \times 4), (3 \times 5), \ldots$
10. $\frac{1}{2}, \frac{2}{3}, \frac{3}{4}, \frac{4}{5}, \ldots$
11. 7, 14, 21, 28, …
12. 1, 4, 9, 16, 25, …
13. $\frac{5}{1^2}, \frac{5}{2^2}, \frac{5}{3^2}, \frac{5}{4^2}, \ldots$
14. $\frac{3}{1}, \frac{4}{2}, \frac{5}{3}, \frac{6}{4}, \ldots$
15. 3, 7, 11, 15, …
16. 5, 7, 9, 11, …
17. 7, 5, 3, 1, …
18. $-5, -1, 3, 7, \ldots$

19. Write down each sequence and then find the nth term.
a) 8, 10, 12, 14, 16, … **b)** 3, 7, 11, 15, … **c)** 8, 13, 18, 23, …

20. Write down each sequence and write the nth term.
a) 11, 19, 27, 35, … **b)** $2\frac{1}{2}, 4\frac{1}{2}, 6\frac{1}{2}, 8\frac{1}{2}, \ldots$ **c)** $-7, -4, -1, 2, 5, \ldots$

21. Here is a sequence of shapes made from sticks:

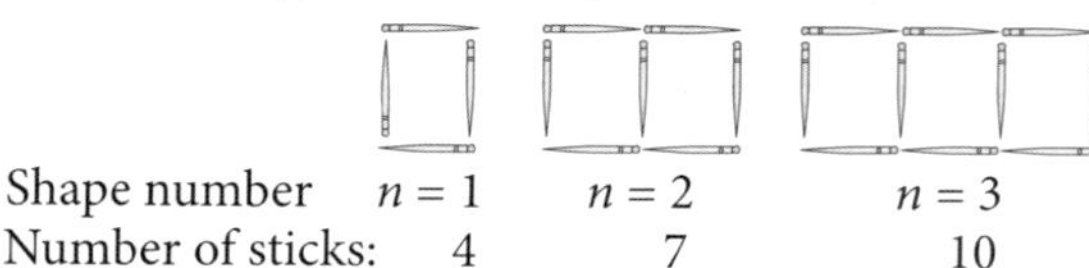

Shape number	$n = 1$	$n = 2$	$n = 3$
Number of sticks:	4	7	10

The number of sticks makes the sequence 4, 7, 10, 13, …
a) Find an expression for the nth term in the sequence.
b) How many sticks are there in shape number 1000?

22. Write down each sequence and select the correct expression for the nth term from the list given.
a) 4, 8, 12, 16, … **b)** 3, 5, 7, 9, …
c) $\frac{1}{2}, \frac{2}{3}, \frac{3}{4}, \frac{4}{5}, \ldots$ **d)** 4, 7, 10, 13, …

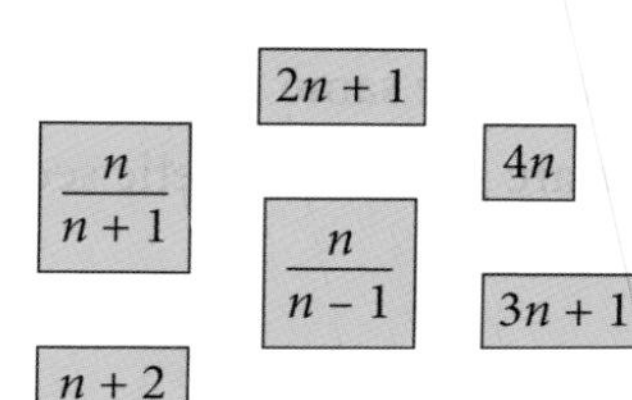

23. The nth term of a sequence is $n^2 + 1$
Write down **a)** the 5th term **b)** the 100th term.

2.4 Approximations and estimation

You can estimate answers by rounding to the nearest whole number or to a number of significant figures or decimal places.

a) 7·8126 = 8 to the nearest whole number
↑ This figure is '5 or more'.

b) 7·8126 = 7·81 to three significant figures
↑ This figure is not '5 or more'.

c) 7·8126 = 7·813 to three decimal places
↑ This figure is '5 or more'.

d) 0·078 126 = 0·0781 to three significant figures.
↑ 7 is the first significant figure.

e) 3596 = 3600 to two significant figures.
↑ This figure is '5 or more'.

Exercise 2.11

Write the following numbers correct to

a) the nearest whole number **b)** three significant figures **c)** two decimal places.

1. 8·174 **2.** 19·617 **3.** 20·041 **4.** 0·814 52 **5.** 311·14

6. 0·275 **7.** 0·007 47 **8.** 15·62 **9.** 900·12 **10.** 3·555

11. 5·454 **12.** 20·961 **13.** 0·0851 **14.** 0·5151 **15.** 3·071

Write the following numbers correct to one decimal place.

16. 5·71 **17.** 0·7614 **18.** 11·241 **19.** 0·0614 **20.** 0·0081 **21.** 11·12

Estimation

You should check that the answer to a calculation is 'about the right size'.

Example 4

Estimate the value of $\frac{57{\cdot}2 \times 110}{2{\cdot}146 \times 46{\cdot}9}$, correct to one significant figure.

We have approximately, $\frac{\cancel{50} \times 100}{2 \times \cancel{50}} \approx 50$ ←

On a calculator the value is 62·52 (to 4 significant figures).

Exercise 2.12

In this exercise there are 25 questions, each followed by three possible answers. Decide (by estimating) which answer is correct.

1.	$7{\cdot}2 \times 9{\cdot}8$	52·16	98·36	70·56
2.	$2{\cdot}03 \times 58{\cdot}6$	118·958	87·848	141·116
3.	$23{\cdot}4 \times 19{\cdot}3$	213·32	301·52	451·62
4.	$313 \times 107{\cdot}6$	3642·8	4281·8	33 678·8
5.	$6{\cdot}3 \times 0{\cdot}098$	0·6174	0·0622	5·98
6.	$1200 \times 0{\cdot}89$	722	1068	131
7.	$0{\cdot}21 \times 93$	41·23	9·03	19·53
8.	$88{\cdot}8 \times 213$	18 914·4	1693·4	1965·4
9.	$0{\cdot}04 \times 968$	38·72	18·52	95·12
10.	$0{\cdot}11 \times 0{\cdot}089$	0·1069	0·0959	0·009 79
11.	$13{\cdot}92 \div 5{\cdot}8$	0·52	4·2	2·4
12.	$105{\cdot}6 \div 9{\cdot}6$	8·9	11	15
13.	$8405 \div 205$	4·6	402	41
14.	$881{\cdot}1 \div 99$	4·5	8·9	88
15.	$4{\cdot}183 \div 0{\cdot}89$	4·7	48	51
16.	$6{\cdot}72 \div 0{\cdot}12$	6·32	21·2	56
17.	$20{\cdot}301 \div 1010$	0·0201	0·211	0·0021
18.	$0{\cdot}288\,96 \div 0{\cdot}0096$	312	102·1	30·1
19.	$0{\cdot}143 \div 0{\cdot}11$	2·3	1·3	11·4
20.	$159{\cdot}65 \div 515$	0·11	3·61	0·31
21.	$(5{\cdot}6 + 0{\cdot}21) \times 39$	389·21	210·21	20·51
22.	$\dfrac{17{\cdot}5 \times 42}{2{\cdot}5}$	294	504	86
23.	$(906 + 4{\cdot}1) \times 0{\cdot}31$	473·21	282·131	29·561
24.	$\dfrac{543 + 472}{18{\cdot}1 + 10{\cdot}9}$	65	35	85
25.	$\dfrac{112{\cdot}2 + 75{\cdot}9}{6{\cdot}9 + 5{\cdot}1}$	242	15.675	25·2

Exercise 2.13

In questions **1** to **5** give your answer correct to one significant figure. Do not use a calculator.

1. A fireman is paid a salary of $98 750 per year. Work out a rough estimate for his weekly pay.

2. Estimate the mean weight of items with the following weights.

4·9 kg, 0·21 kg, 0·72 kg, 25·1 kg, 0·11 kg

3. In 2012 Suki's pay was $38 970 per year. In 2013 she receives a pay increase of 19·3%. Estimate the *monthly increase* in her pay.

4. Estimate the length of the diagonals of a square of side 7·2 cm.

5. An eagle flies at an average speed of 18·7 km/h from 0710 until 1117. Roughly how far does it fly?

6. Give an estimate for each of the following calculations. Show your working.

a) $\frac{82 \cdot 4 \times \sqrt{907 \cdot 4}}{2 \cdot 824}$ **b)** $\frac{2848 \cdot 7 - 1 \cdot 94}{0 \cdot 32 + 39 \cdot 83}$ **c)** 52% of 0·394 kg

d) $\frac{3 \cdot 15^2 + 30 \cdot 63^2}{0 \cdot 104^2}$ **e)** $\frac{7}{15}$ of £3918·25 **f)** $\frac{207 \cdot 5 + 4 \cdot 21 + 0 \cdot 63}{109 \cdot 4 + 293 \cdot 2}$

g) $\frac{5 \cdot 13 \times 18 \cdot 777}{0.952}$ **h)** $\pi \times 9{\cdot}73^2$ **i)** $\frac{17}{31}$ of 12% of £2057

7. The population of the Earth is about 7 billion. Estimate how many people share the same birthday as you.

8. Find an estimate for the radius of a circular pool of area 150 m^2.

2.5 Standard form

When dealing with either very large or very small numbers, it is not convenient to write them out in full in the normal way. It is better to use standard form. Most calculators represent large and small numbers in this way.

The number $a \times 10^n$ is in standard form when $1 \leq a < 10$ and n is a positive or negative integer.

Example 5

Write the following numbers in standard form.

a) 2000 b) 150 c) 0·0004

a) $2000 = 2 \times 1000 = 2 \times 10^3$

b) $150 = 1{\cdot}5 \times 100 = 1{\cdot}5 \times 10^2$

c) $0{\cdot}0004 = \times \frac{1}{10000} = 4 \times 10^{-4}$

Exercise 2.14

Write the following numbers in standard form.

1. 4000	**2.** 500	**3.** 70 000
4. 60	**5.** 2400	**6.** 380
7. 46 000	**8.** 46	**9.** 900 000
10. 2560	**11.** 0·007	**12.** 0·0004
13. 0·0035	**14.** 0·421	**15.** 0·000 055
16. 0·01	**17.** 564 000	**18.** 19 million

19. The population of China is estimated at 1100 000 000. Write this in standard form.

20. The mass of a hydrogen atom is 0·000 000 000 000 000 000 000 001 67 grams. Write this mass in standard form.

21. The area of the surface of the Earth is about 510 000 000 km². Express this in standard form.

22. An atom is 0·000 000 000 25 cm in diameter. Write this in standard form.

23. Avogadro's number is 602 300 000 000 000 000 000 000. Express this in standard form.

24. The speed of light is 300 000 km/s. Express this speed in cm/s in standard form.

1 km = 1000 m
1 m = 100 cm

25. A very rich man leaves his fortune of $\$3{\cdot}6 \times 10^8$ to be divided between his 100 grandchildren. How much does each child receive? Give the answer in standard form.

GDC help

There are three different ways of entering a number in standard form.
For example, to enter 2.4×10^{12} press the keys shown below.
These screenshots show also how numbers in standard form are displayed

TI-Nspire	TI84-Plus	Casio
2 . 4 × 1 0 ^ 1 2	2 . 4 × 1 0 ^ 1 2	2 . 4 × 1 0 ^ 1 2
2 . 4 × 10ˣ 1 2	2 . 4 × 2nd 10ˣ 1 2	2 . 4 × SHIFT 10ˣ 1 2
2 . 4 EE 1 2	2 . 4 2nd EE 1 2	2 . 4 EXP 1 2 EXE

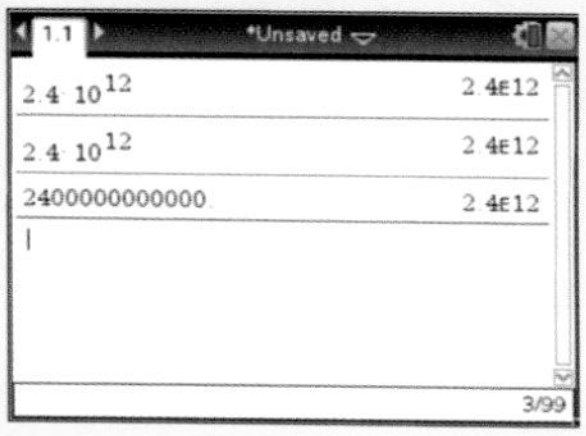

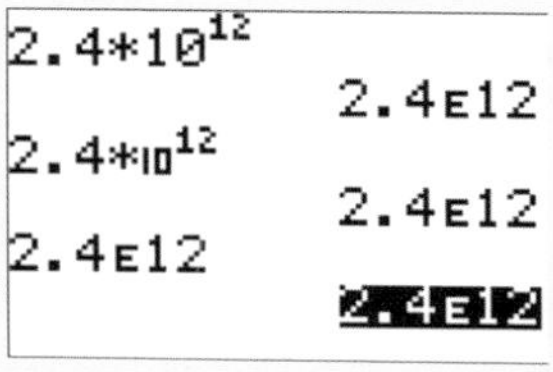

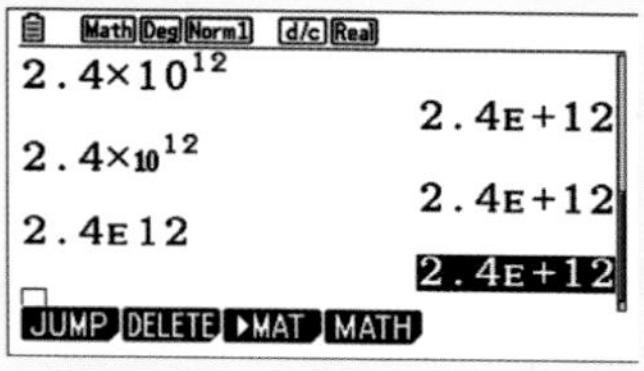

Note: never copy the display from the calculator. You should write your answer in correct standard form.

Example 6

Work out $1500 \times 8\,000\,000$

$$1500 \times 8\,000\,000 = (1{\cdot}5 \times 10^3) \times (8 \times 10^6)$$
$$= 12 \times 10^9$$
$$= 1{\cdot}2 \times 10^{10}$$

Notice that we multiply the numbers and the powers of 10 separately.

Exercise 2.15

In questions **1** to **12** give the answer in standard form.

1. 5000×3000 **2.** $60\,000 \times 5000$ **3.** $0{\cdot}000\,07 \times 400$

4. $0{\cdot}0007 \times 0{\cdot}000\,01$ **5.** $8000 \div 0.004$ **6.** $(0{\cdot}002)^2$

7. $150 \times 0{\cdot}0006$ **8.** $0{\cdot}000\,033 \div 500$ **9.** $0{\cdot}007 \div 20\,000$

10. $(0{\cdot}0001)^4$ **11.** $(2000)^3$ **12.** $0{\cdot}005\,92 \div 8000$

13. If $a = 512 \times 10^2$, $b = 0{\cdot}478 \times 10^6$ and $c = 0{\cdot}0049 \div 10^7$, arrange a, b and c in order of size (smallest first).

14. If the number $2{\cdot}74 \times 10^{15}$ is written out in full, how many zeros follow the 4?

15. If the number $7{\cdot}31 \times 10^{-17}$ is written out in full, how many zeros would there be between the decimal point and the first significant figure?

16. If $x = 2 \times 10^5$ and $y = 3 \times 10^{-3}$ correct to one significant figure, find the greatest and least possible values of

a) xy **b)** $\frac{x}{y}$

Remember

The limits of accuracy of 2 to one significant figure are 1·5 to 2·5.

17. Oil flows through a pipe at a rate of $40\text{m}^3/\text{s}$. How long will it take to fill a tank of volume $1{\cdot}2 \times 10^5\text{m}^3$?

18. Given that $L = 2\sqrt{\frac{a}{k}}$, find the value of L in standard form when $a = 4{\cdot}5 \times 10^{12}$ and $k = 5 \times 10^7$.

19. a) The number 10 to the power 100 (10 000 sexdecillion) is called a 'Googol'! If it takes $\frac{1}{5}$ second to write a 'zero' and $\frac{1}{10}$ second to write a 'one', how long would it take to write the number 100 'Googols' in full?

b) The number 10 to the power of a 'Googol' is called a 'Googolplex'. Using the same speed of writing, how long in years would it take to write 1 'Googolplex' in full? You may assume that your pen has enough ink.

2.6 Ratio and proportion

The word **ratio** is used to describe a fraction. If the ratio of a boy's height to his father's height is 4 : 5, then he is $\frac{4}{5}$ as tall as his father.

Example 7

Change the ratio 2:5 into the form

a) $1:n$ **b)** $m:1$

a) $2:5 = 1:\frac{5}{2}$
$= 1:2{\cdot}5$

b) $2:5 = \frac{2}{5}:1$
$= 0{\cdot}4:1$

Example 8

Divide \$60 between two people A and B in the ratio 5 : 7.

Consider \$60 as 12 equal parts (i.e. 5 + 7).
Then A receives 5 parts and B receives 7 parts.

$\therefore$ A receives $\frac{5}{12}$ of \$60 = \$25

B receives $\frac{7}{12}$ of \$60 = \$35

Example 9

Divide 200 kg in the ratio 1:3:4.

The parts are $\frac{1}{8}$, $\frac{3}{8}$ and $\frac{4}{8}$ (of 200 kg), i.e. 25 kg, 75 kg and 100 kg.

Exercise 2.16

In questions **1** to **8** express the ratios in the form $1:n$.

1. 2:6 **2.** 5:30 **3.** 2:100 **4.** 5:8

5. 4:3 **6.** 8:3 **7.** 22:550 **8.** 45:360

In questions **9** to **12** express the ratios in the form $n:1$.

9. 12:5 **10.** 5:2 **11.** 4:5 **12.** 2:100

In questions **13** to **18** divide the quantity in the ratio given.

13. $40; (3:5) **14.** $120; (3:7)

15. 250 m; (14:11) **16.** $117; (2:3:8)

17. 180 kg; (1:5:6) **18.** 184 minutes; (2:3:3)

19. When $143 is divided in the ratio 2:4:5, what is the difference between the largest share and the smallest share?

20. Divide 180 kg in the ratio 1:2:3:4.

21. Divide $4000 in the ratio 2:5:5:8.

22. If $\frac{5}{8}$ of the children in a school are boys, what is the ratio of boys to girls?

23. A man and a woman share a prize of $1000 between them in the ratio 1:4. The woman shares her part between herself, her mother and her daughter in the ratio 2:1:1. How much does her daughter receive?

24. A man and a woman share a sum of money in the ratio 3:2. If the sum of money is doubled, in what ratio should they divide it so that the man still receives the same amount?

25. In the photo what is the ratio of blue umbrellas: not blue umbrellas?

26. In a herd of x cattle, the ratio of the number of bulls to cows is 1:6. Find the number of bulls in the herd in terms of x.

27. If $x:3 = 12:x$, calculate the positive value of x.

28. If $y:18 = 8:y$, calculate the positive value of y.

29. $400 is divided between Kas, Jaspar and Jae so that Kas has twice as much as Jaspar and Jaspar has three times as much as Jae. How much does Jaspar receive?

30. A cake of mass 550 g has three ingredients: flour, sugar and raisins. There is twice as much flour as sugar and one and a half times as much sugar as raisins. How much flour is there?

31. A brother and sister share out their collection of 5000 stamps in the ratio 5 : 3. The brother then shares his stamps with two friends in the ratio 3 : 1 : 1, keeping most for himself. How many stamps do each of his friends receive?

Proportion

The majority of problems where proportion is involved are usually solved by finding the value of a unit quantity.

Example 10

If a wire of length 2 metres costs $10, find the cost of a wire of length 35 cm.

200 cm costs 1000 cents

$\therefore$ 1 cm costs $\frac{1000}{200}$ cents = 5 cents

$\therefore$ 35 cm costs 5×35 cents = 175 cents

= $1·75

Example 11

Eight men can dig a hole in 4 hours. How long will it take five men to dig the same size hole?

8 men take 4 hours

1 man would take 32 hours

5 men would take $\frac{32}{5}$ hours = 6 hours 24 minutes

Exercise 2.17

1. Five cans of cola cost $1·20. Find the cost of seven cans.
2. A man earns $420 in a 5-day week. What is his pay for 3 days?
3. Three people build a wall in 10 days. How long would it take five people?
4. Nine fruit juice bottles contain $4\frac{1}{2}$ litres of fruit juice between them. How much juice do five bottles hold?

5. A car uses 10 litres of petrol in 75 km.
   How far will it go on 8 litres?
6. A wire 11 cm long has a mass of 187 g.
   What is the mass of 7 cm of this wire?
7. A shopkeeper can buy 36 toys for \$20·52.
   What will he pay for 120 toys?
8. A ship has sufficient food to supply 600 passengers for 3 weeks. How long would the food last for 800 people?
9. The cost of a phone call lasting 3 minutes 30 seconds was 52·5 cents. At this rate, what was the cost of a call lasting 5 minutes 20 seconds?
10. 80 machines can produce 4800 identical pens in 5 hours. At this rate
    a) how many pens would one machine produce in one hour?
    b) how many pens would 25 machines produce in 7 hours?
11. Three men can build a wall in 10 hours. How many men would be needed to build the wall in $7\frac{1}{2}$ hours?
12. If it takes 6 men 4 days to dig a hole 3 metres deep, how long will it take 10 men to dig a hole 7 metres deep?
13. Find the cost of 1 km of pipe at 7 cents for every 40 cm.
14. A motorcycle wheel turns through 90 revolutions per minute.
    How many degrees does it turn through in 1 second?

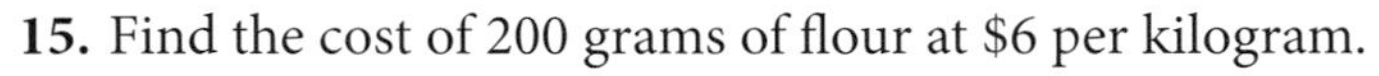

15. Find the cost of 200 grams of flour at \$6 per kilogram.
16. The height of One Kansas City Place is 623 feet. Express this height to the nearest metre using the conversion 1m = 3·281 feet.
17. A floor is covered by 800 tiles of side 10 cm. How many square tiles of side 8 cm would be needed to cover the same floor?
18. A battery has enough energy to operate eight toy bears for 21 hours. For how long could the battery operate 15 toy bears?
19. An engine has enough fuel to operate at full power for 20 minutes. For how long could the engine operate at 35% of full power?
20. A large drum, when full, contains 260 kg of oil of density 0·9 g/cm^3. What mass of petrol, of density 0·84 g/cm^3, can be contained in the drum?
21. A wall can be built by 6 men working 8 hours per day in 5 days. How many days will it take 4 men to build the wall if they work only 5 hours per day?

Foreign exchange

Money is changed from one currency into another using the method of proportion.

Exchange rate for US dollars ($):

Country	*Rate of exchange*
Argentina (pesos)	$1 = 3·79 ARS
Australia (dollars)	$1 = 1·13 AUD
Euro (euros)	$1 = €0·70 EUR
India (rupees)	$1 = 46·50 INR
Japan (yen)	$1 = 91·20 JPY
Kuwait (dinar)	$1 = 0·29 KWD
UK (pounds)	$1 = £0·63 GBP

Example 12

Convert: **a)** $22·50 to dinars **b)** €300 to dollars.

a) $1 = 0·29 dinars (KWD)

so $22·50 = 0·29 × 22·50 KWD

= 6·53 KWD

b) €0·70 = $1

so €1 = $\frac{1}{0 \cdot 70}$

so €300 = $\$\frac{1}{0 \cdot 70} \times 300$

= $428·57

Exercise 2.18

Give your answers correct to two decimal places. Use the exchange rates given in the table.

1. Change the number of dollars into the foreign currency stated.

a) $20 [euros] **b)** $70 [pounds] **c)** $200 [pesos]

d) $1·50 [rupees] **e)** $2.30 [yen] **f)** 90c [dinars]

2. Change the amount of foreign currency into dollars.

a) €500 **b)** £2500 **c)** 7·50 rupees

d) 900 dinars **e)** 125·24 pesos **f)** 750 AUD

3. A CD costs £9·50 in London and $9·70 in Chicago. How much cheaper, in British money, is the CD when bought in the US?

4. A DVD costs €20·46 in Spain and £12·60 in the UK. Which is the cheaper in dollars, and by how much?

5. The monthly rent of a flat in New Delhi is 32 860 rupees. How much is this in euros?
6. A handbag is sold in several countries at the prices given below.

   Kuwait   150 dinars

   France   550 euros

   Japan   92 000 yen

   Write out in order a list of the prices converted into GBP.

7. An Australian man on holiday in Germany finds that his wallet contains 700 AUD. If he changes the money at a bank how many euros will he receive?

Map scales

You can use proportion to work out map scales. First you need to know these metric equivalents:

$1\text{ km} = 1000\text{ m}$

$1\text{ m} = 100\text{ cm}$

$1\text{ cm} = 10\text{ mm}$

km means kilometre
m means metre
cm means centimetre
mm means millimetre

Example 13

A map is drawn to a scale of 1 to 50 000.
Calculate

a) the length of a road which appears as 3 cm long on the map
b) the length on the map of a lake which is 10 km long.

a) 1 cm on the map is equivalent to 50 000 cm on the Earth.

$\therefore\ 1\text{ cm} \equiv 50\,000\text{ cm}$

$\therefore\ 1\text{ cm} \equiv 500\text{ m}$

$\therefore\ 1\text{ cm} \equiv 0{\cdot}5\text{ km}$

so $3\text{ cm} \equiv 3 \times 0{\cdot}5\text{ km} = 1{\cdot}5\text{ km}$.

The road is 1·5 km long.

b) $0{\cdot}5\text{ km} \equiv 1\text{ cm}$

$\therefore\ 1\text{ km} \equiv 2\text{ cm}$

$\therefore\ 10\text{ km} \equiv 2 \times 10\text{ cm}$

$= 20\text{ cm}$

The lake appears 20 cm long on the map.

Exercise 2.19

1. Find the actual length represented on a drawing by

   a) 14 cm **b)** 3·2 cm **c)** 0·71 cm **d)** 21·7 cm

   when the scale is 1 cm to 5 m.

2. Find the length on a drawing that represents

   a) 50 m **b)** 35 m **c)** 7·2 m **d)** 28·6 m

   when the scale is 1 cm to 10 m.

3. If the scale is 1 : 10 000, what length will 45 cm on the map represent

   a) in cm **b)** in m **c)** in km?

4. On a map of scale 1 : 100 000, the distance between De'aro and Debeka is 12·3 cm. What is the actual distance in km?

5. On a map of scale 1 : 15 000, the distance between Noordwijk aan Zee and Katwijk aan Zee is 31·4 cm. What is the actual distance in km?

6. The diameter of a globe is 60 cm. The circumference of the earth is about 40 000 km. Estimate the scale of this globe. Give your answer in the form 1 : *n*.

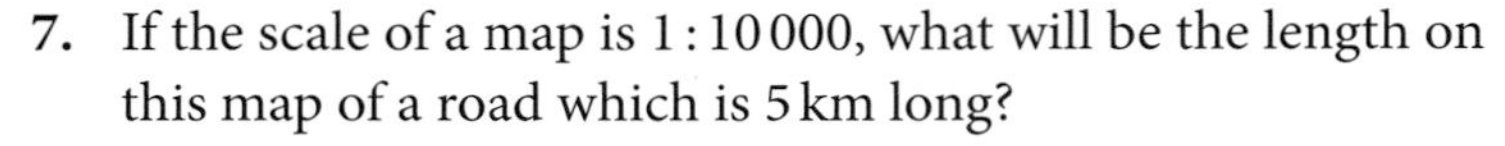

7. If the scale of a map is 1 : 10 000, what will be the length on this map of a road which is 5 km long?

8. The distance from Hong Kong to Shenzhen is 32 km. How far apart will they be on a map of scale 1 : 50 000?

9. The 17th hole at the famous St Andrews golf course is 420 m in length. How long will it appear on a plan of the course of scale 1 : 8000?

An area involves two dimensions multiplied together and hence the scale is multiplied *twice*.

For example, if the linear scale is $\frac{1}{100}$, then the area scale is $\frac{1}{100} \times \frac{1}{100} = \frac{1}{10000}$.

You can use a diagram to help:

If a scale is 1 : 50 000 then 2 cm ≡ 1 km

An area of 6 cm² can be thought of as:

3 cm

6 cm² | 2 cm

so the equivalent area using the scale is:

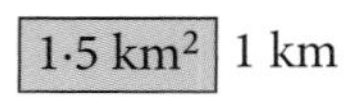

Exercise 2.20

1. The scale of a map is 1 : 1000. What are the actual dimensions of a rectangle which appears as 4 cm by 3 cm on the map? What is the area on the map in cm^2? What is the actual area in m^2?
2. The scale of a map is 1 : 100. What area does 1 cm^2 on the map represent? What area does 6 cm^2 represent?
3. The scale of a map is 1 : 20 000. What area does 8 cm^2 represent?
4. The scale of a map is 1 : 1000. What is the area, in cm^2, on the map of a lake of area 5000 m^2?
5. The scale of a map is 1 cm to 5 km. A farm is represented by a rectangle measuring 1·5 cm by 4 cm. What is the actual area of the farm?
6. On a map of scale 1 cm to 250 m the area of a car park is 3 cm^2. What is the actual area of the car park in hectares? [1 hectare = 10 000 m^2]
7. The area of the playing surface at the Olympic Stadium in Beijing is $\frac{3}{5}$ of a hectare. What area will it occupy on a plan drawn to a scale of 1 : 500?
8. On a map of scale 1 : 20 000 the area of a forest is 50 cm^2. On another map the area of the forest is 8 cm^2. Find the scale of the second map.

2.7 Percentages

Percentages are simply a convenient way of expressing fractions or decimals. '50% of \$60' means $\frac{50}{100}$ of \$60, or more simply $\frac{1}{2}$ of \$60. Percentages are used very frequently in everyday life and are misunderstood by a large number of people. For example, what are the implications if 'inflation falls from 10% to 8%'? Does this mean prices will fall?

Example 14

a) Change 80% to a fraction. **b)** Change $\frac{3}{8}$ to a percentage. **c)** Change 8% to a decimal.

a) $80\% = \frac{80}{100} = \frac{4}{5}$

b) $\frac{3}{8} = \left(\frac{3}{8} \times \frac{100}{1}\right)\% = 37\frac{1}{2}\%$

c) $8\% = \frac{8}{100} = 0{\cdot}08$

Exercise 2.21

1. Change to fractions:

   a) 60% b) 24% c) 35% d) 2%

2. Change to percentages:

   a) $\frac{1}{4}$ b) $\frac{1}{10}$ c) $\frac{7}{8}$

   d) $\frac{1}{3}$ e) 0·72 f) 0·31

3. Change to decimals:

   a) 36% b) 28% c) 7%

   d) 13·4% e) $\frac{3}{5}$ f) $\frac{7}{8}$

4. Arrange in order of size (smallest first):

   a) $\frac{1}{2}$; 45%; 0·6 b) 0·38; $\frac{6}{16}$; 4%

   c) 0·111; 11%; $\frac{1}{9}$ d) 32%; 0·3; $\frac{1}{3}$

5. The following are marks obtained in various tests. Convert them to percentages.

   a) 17 out of 20 b) 31 out of 40 c) 19 out of 80

   d) 112 out of 200 e) $2\frac{1}{2}$ out of 25 f) $7\frac{1}{2}$ out of 20

6. Work out the following correct to 2 decimal places.

   a) $\frac{7}{12} - 0{\cdot}22$ b) $0{\cdot}\dot{4}5\dot{2} + \frac{2}{9}$ c) $\frac{3}{5}$ of 0·713

   d) $\left(\frac{3}{4} \text{ of } 0.8\right) - \left(15\% \text{ of } \frac{4}{5}\right)$ e) $0{\cdot}\dot{4} - (22\% \text{ of } 0{\cdot}5\dot{1})$

   f) $\left(\frac{1}{12} + \frac{1}{7} - \frac{1}{13}\right) \times 0{\cdot}73$ g) $85\% \text{ of } \left(0.8\dot{1}\dot{2} - \frac{4}{5}\right)$

7. Satheesh claims that 35% of £40 is the same as 40% of £35. Is he correct?

Example 15

A second hard car costing \$2400 is reduced in price by 10%. Find the new price.

$$10\% \text{ of } \$2400 = \frac{10}{100} \times \frac{2400}{1}$$
$$= \$240$$

New price of car = \$(2400 – 240)

= \$2160

Example 16

After a price increase of 10% a television costs \$286.

What was the price before the increase?

The price before the increase is 100%.

$\therefore$ 110% of old price = \$286

$\therefore$ 1% of old price = $\$\frac{286}{110}$

$\therefore$ 100% of old price = $\$\frac{286}{110} \times \frac{100}{1}$

Old price of TV = \$260

Exercise 2.22

1. Calculate

   a) 30% of \$50 b) 45% of 2000 kg

   c) 4% of \$70 d) 2·5% of 5000 people

2. In a sale, a jacket costing \$40 is reduced by 20%. What is the sale price?

3. The charge for a telephone call costing 12 cents is increased by 10%. What is the new charge?

4. In peeling potatoes 4% of the mass of the potatoes is lost as 'peel'. How much is left for use from a bag containing 55 kg?

5. Work out, to the nearest cent:

   a) 6·4% of \$15·95 b) 11·2% of \$192·66

   c) 8·6% of \$25·84 d) 2·9% of \$18·18

6. Find the total bill:

   - 5 golf clubs at \$18·65 each
   - 60 golf balls at \$16·50 per dozen
   - 1 pair of golf shoes at \$35·80

   Sales tax at 15% is added to the total cost.

7. In 2013 a club has 250 members who each pay an annual subscription of \$95. In 2014 the membership increases by 4% and the annual subscription is increased by 6%. What is the total income from subscriptions in 2014?

8. In Thailand the population of a town is 48 700 men and 1600 women. What percentage of the total population are men?

9. In South Korea there are 21 280 000 licensed vehicles on the road. Of these, 16 486 000 are private cars. What percentage of the licensed vehicles are private cars?

10. A quarterly telephone bill consists of \$19·15 rental plus 4·7 cents for each dialled unit. Tax is added at 15%. What is the total bill for Bryndis, who used 915 dialled units?

11. 70% of Hassan's collection of goldfish died. If he has 60 survivors, how many did he have originally?

12. The average attendance at Parma football club fell by 7% in 2013. If 2030 fewer people went to matches in 2013, how many went in 2012?

13. An iron bar expands by 0·2% when heated. If the increase in length is 1 cm, what is the original length of the bar?

14. In the last two weeks of a sale, prices are reduced first by 30% and then by a *further* 40% of the new price. What is the final sale price of a T-shirt which originally cost \$15?

15. During a Grand Prix car race, the tyres on a car are reduced in mass by 3%. If their mass is 388 kg at the end of the race, what was their mass at the start?

16. Over a period of 6 months, a colony of rabbits increases in number by 25% and then by a further 30%. If there were originally 200 rabbits in the colony, how many were there at the end?

17. A dress costs \$270.25 including 15% tax. How much of the cost is tax?

18. The cash price for a car was \$7640. Gurtaj bought the car on the following terms: 'A deposit of 20% of the cash price and 36 monthly payments of \$191·60'. Calculate the total amount Gurtaj paid.

Percentage increase or decrease

In the next exercise use the formulae:

$$\text{Percentage profit} = \frac{\text{Actual profit}}{\text{Original price}} \times \frac{100}{1}$$

$$\text{Percentage loss} = \frac{\text{Actual loss}}{\text{Original price}} \times \frac{100}{1}$$

Example 17

A radio is bought for \$16 and sold for \$20. What is the percentage profit?

Actual profit = \$4

$\therefore$ Percentage profit $= \frac{4}{16} \times \frac{100}{1} = 25\%$

The radio is sold at a 25% profit.

Example 18

A car is sold for \$2280, at a loss of 5% on the cost price. Find the cost price.

Do *not* calculate 5% of \$2280!

The loss is 5% of the cost price.

$\therefore$ 95% of cost price = \$2280

1% of cost price $= \$\frac{2280}{95}$

$\therefore$ 100% of cost price $= \$\frac{2280}{95} \times \frac{100}{1}$

Cost price = \$2400

Exercise 2.23

c is the symbol for cents.
100c = \$1

1. The first figure is the cost price and the second figure is the selling price. Calculate the percentage profit or loss in each case.
   a) \$20, \$25 **b)** \$400, \$500 **c)** \$60, \$54
   d) \$9000, \$10 800 **e)** \$460, \$598 **f)** \$512, \$550·40
   g) \$45, \$39·60 **h)** 50c, 23c
2. A car dealer buys a car for \$500, and then sells it for \$640. What is the percentage profit?
3. A damaged carpet which cost \$180 when new, is sold for \$100. What is the percentage loss?
4. During the first four weeks of her life, a baby girl increases her mass from 3·2kg to 4·7 kg. What percentage increase does this represent? (Give your answer to 3 s.f.)
5. When tax is added to the cost of a camera, its price increases from \$165 to \$184·80. What is the rate at which tax is charged?

6. The price of a sports car is reduced from \$30 000 to \$28 400. What percentage reduction is this?

7. Find the cost price of the following:

   a) selling price \$55, profit 10% b) selling price \$558, profit 24%

   c) selling price \$680, loss 15% d) selling price \$11·78, loss 5%

8. A sari is sold for \$60, thereby making a profit of 20% on the cost price. What was the cost price?

9. A pair of jeans is sold for \$15, thereby making a profit of 25% on the cost price. What was the cost price?

10. A book is sold for \$5·40, at a profit of 8% on the cost price. What was the cost price?

11. A can of worms is sold for 48c, incurring a loss of 20%. What was the cost price?

12. A car was sold for \$1430, thereby making a loss of 35% on the cost price. What was the cost price?

13. If an employer reduces the working week from 40 hours to 35 hours, with no loss of weekly pay, calculate the percentage increase in the hourly rate of pay.

14. The cost of insuring a mobile phone changed from \$80 per year to \$8 per month. What is the percentage increase in the yearly cost?

15. A greengrocer sells a melon at a profit of $37\frac{1}{2}\%$ on the price he pays for it. What is the ratio of the cost price to the selling price?

16. Given that $G = ab$, find the percentage increase in G when both a and b increase by 10%.

17. Given that $T = \frac{kx}{y}$, find the percentage increase in T when k, x and y all increase by 20%.

Simple interest

When a sum of money $\$P$ is invested for T years at $R\%$ interest per annum (each year), then the interest gained I is given by:

$$I = \frac{P \times R \times T}{100}$$

This is known as **simple interest**.

Example 19

Joel invests \$400 for 6 months at 5%.

Work out the simple interest gained.

$P = \$400 \quad R = 5 \quad T = 0{\cdot}5$ (6 months is half a year)

so $I = \frac{400 \times 5 \times 0{\cdot}5}{100}$

$I = \$10$

Exercise 2.24

1. Calculate:
   - **a)** the simple interest on $1200 for 3 years at 6% per annum
   - **b)** the simple interest on $700 at 8·25% per annum for 2 years
   - **c)** the length of time for $5000 to earn $1000 if invested at 10% per annum
   - **d)** the length of time for $400 to earn $160 if invested at 8% per annum.
2. Khalid invests $6750 at 8·5% per annum. How much interest has he earned and what is the total amount in his account after 4 years?
3. Petra invests $10 800. After 4 years she has earned $3240 in interest. At what annual rate of interest did she invest her money?

Compound interest

Suppose a bank pays a fixed interest of 10% on money in deposit accounts. A man puts $500 in the bank.

After one year he has

$500 + 10\% \text{ of } 500 = \$550$

After two years he has

$550 + 10\% \text{ of } 550 = \$605$

[Check that this is $1{\cdot}10^2 \times 500$]

After three years he has

$605 + 10\% \text{ of } 605 = \$665{\cdot}50$

[Check that this is $1{\cdot}10^3 \times 500$]

In general after n years the money in the bank will be $\$(1{\cdot}10^n \times 500)$

Exercise 2.25

1. A bank pays interest of 9% on money in deposit accounts. Carme puts $2000 in the bank. How much has she after

   a) one year **b)** two years **c)** three years?
2. A bank pays interest of 11%. Mamuru puts $5000 in the bank. How much has he after

   a) one year **b)** three years **c)** five years?

3. A student gets a grant of \$10 000 a year. Assuming her grant is increased by 7% each year, what will her grant be in four years' time?

4. Isoke's salary in 2014 is \$30 000 per year. Every year her salary is increased by 5%.

   In 2015 her salary will be $30000 \times 1{\cdot}05$ = \$31 500

   In 2016 her salary will be $30000 \times 1{\cdot}05 \times 1{\cdot}05$ = \$33 075

   In 2017 her salary will be $30000 \times 1{\cdot}05 \times 1{\cdot}05 \times 1{\cdot}05$ = \$34 728·75

   And so on.

   a) What will her salary be in 2018?

   b) What will her salary be in 2020?

5. The rental price of a dacha was \$9000. At the end of each month the price is increased by 6%.

   a) Find the price of the house after 1 month.

   b) Find the price of the house after 3 months.

   c) Find the price of the house after 10 months.

6. Assuming an average inflation rate of 8%, work out the probable cost of the following items in 10 years:

   a) motorbike \$6500

   b) iPod \$340

   c) car \$50 000

7. A new scooter is valued at \$15 000. At the end of each year its value is reduced by 15% of its value at the start of the year. What will it be worth after 3 years?

8. The population of an island increases by 10% each year. After how many years will the original population be doubled?

9. A bank pays interest of 11% on \$6000 in a deposit account. After how many years will the money have trebled?

10. A tree grows in height by 21% per year. It is 2 m tall after one year. After how many more years will the tree be over 20 m tall?

11. Which is the better investment over ten years:

    \$20 000 at 12% compound interest

    or \$30 000 at 8% compound interest?

2.8 Speed, distance and time

Calculations involving these three quantities are simpler when the speed is *constant*. The formulae connecting the quantities are as follows:

- distance = speed × time
- speed = $\dfrac{\text{distance}}{\text{time}}$
- time = $\dfrac{\text{distance}}{\text{speed}}$

A helpful way of remembering these formulae is to write the letters D, S and T in a triangle,

thus:

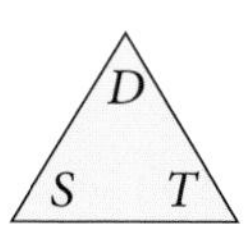

to find D, cover D and we have ST

to find S, cover S and we have $\dfrac{D}{T}$

to find T, cover T and we have $\dfrac{D}{S}$

Great care must be taken with the units in these questions.

Example 20

A girl is skiing at a speed of 8 km/h for a distance of 5200 metres.

Find the time taken in minutes.

5200 metres = 5·2 km

time taken in hours = $\left(\dfrac{D}{S}\right) = \dfrac{5{\cdot}2}{8}$

= 0·65 hours

time taken in minutes = 0·65 × 60

= 39 minutes

Example 21

Change the units of a speed of 54 km/h into metres per second.

54 km/hour = 54 000 metres/hour

= $\dfrac{54000}{60}$ metres/minute

= $\dfrac{54000}{60 \times 60}$ metres/second

= 15 m/s

Exercise 2.26

1. Find the time taken for the following journeys:

a) 100 km at a speed of 40 km/h

b) 250 miles at a speed of 80 miles per hour

c) 15 metres at a speed of 20 cm/s (answer in seconds)

d) 10^4 metres at a speed of 2·5 km/h

2. Change the units of the following speeds as indicated:

a) 72 km/h into m/s

b) 108 km/h into m/s

c) 300 km/h into m/s

d) 30 m/s into km/h

e) 22 m/s into km/h

f) 0·012 m/s into cm/s

g) 9000 cm/s into m/s

h) 600 miles/day into miles per hour

i) 2592 miles/day into miles per second

3. Find the speeds of the bodies which move as follows:

a) a distance of 600 km in 8 hours

b) a distance of 31·64 km in 7 hours

c) a distance of 136·8m in 18 seconds

d) a distance of 4×10^4 m in 10^{-2} seconds

e) a distance of 5×10^5 cm in 2×10^{-3} seconds

f) a distance of 10^8 mm in 30 minutes (in km/h)

g) a distance of 500 m in 10 minutes (in km/h)

4. Find the distance travelled (in metres) in the following:

a) at a speed of 55 km/h for 2 hours

b) at a speed of 40 km/h for $\frac{1}{4}$ hour

c) at a speed of 338·4 km/h for 10 minutes

d) at a speed of 15 m/s for 5 minutes

e) at a speed of 14 m/s for 1 hour

f) at a speed of 4×10^3 m/s for 2×10^{-2} seconds

g) at a speed of 8×10^5 cm/s for 2 minutes

5. A car travels 60 km at 30 km/h and then a further 180 km at 160 km/h. Find:

a) the total time taken

b) the average speed for the whole journey.

6. A cyclist travels 25 kilometres at 20 km/h and then a further 80 kilometres at 25 km/h. Find:

a) the total time taken

b) the average speed for the whole journey.

7. A swallow flies at a speed of 50 km/h for 3 hours and then at a speed of 40 km/h for a further 2 hours. Find the average speed for the whole journey.
8. A runner ran two laps around a 400 m track. She completed the first lap in 50 seconds and then decreased her speed by 5% for the second lap. Find:
   a) her speed on the first lap
   b) her speed on the second lap
   c) her total time for the two laps
   d) her average speed for the two laps.
9. An aeroplane flies 2000 km at a speed of 1600 km/h and then returns due to bad weather at a speed of 1000 km/h. Find the average speed for the whole trip.
10. A train travels from A to B, a distance of 100 km, at a speed of 20 km/h. If it had gone two and a half times as fast, how much earlier would it have arrived at B?

11. Two men running towards each other at 4 m/s and 6 m/s respectively are one kilometre apart. How long will it take before they meet?
12. A car travelling at 90 km/h is 500 m behind another car travelling at 70 km/h in the same direction. How long will it take the first car to catch the second?
13. How long is a train which passes a signal in twenty seconds at a speed of 108 km/h?
14. A train of length 180 m approaches a tunnel of length 620 m. How long will it take the train to pass completely through the tunnel at a speed of 54 km/h?
15. An earthworm of length 15 cm is crawling along at 2 cm/s. An ant overtakes the worm in 5 seconds. How fast is the ant walking?
16. A train of length 100 m is moving at a speed of 50 km/h. A horse is running alongside the train at a speed of 56 km/h. How long will it take the horse to overtake the train?
17. A car completes a journey at an average speed of 40 km/h. At what speed must it travel on the return journey if the average speed for the complete journey (out and back) is 60 km/h?

Exercise 2.27

1. Fill in the blank spaces in the table so that each row contains equivalent values.

Fraction	Decimal	Percentage
	0·28	
		64%
$\frac{5}{8}$		

2. An engine pulls four identical carriages. The engine is $\frac{2}{3}$ the length of a carriage and the total length of the train is 86·8 m. Find the length of the engine.

3. A cake is made from the ingredients listed below.

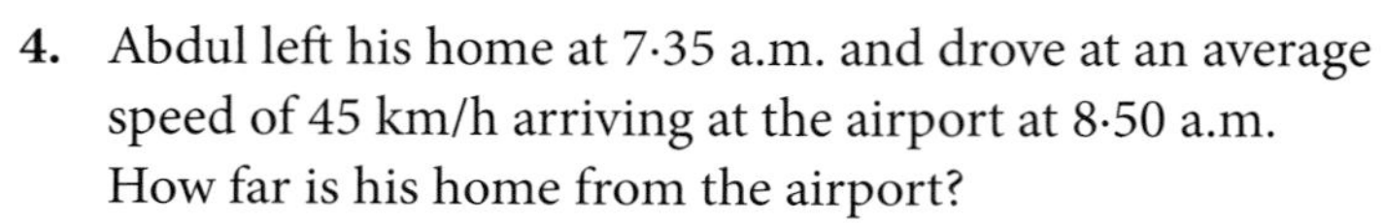
500 g flour, 450 g butter, 470 g sugar,
1·8 kg mixed fruit, 4 eggs (70 g each)

The cake loses 12% of its mass during cooking.
What is its final mass?

4. Abdul left his home at 7·35 a.m. and drove at an average speed of 45 km/h arriving at the airport at 8·50 a.m. How far is his home from the airport?

5. Tuwile's parents have agreed to lend him 60% of the cost of buying a car. If Tuwile still has to find $328 himself, how much does the car cost?

6. Which bag of potatoes is the better value:
   Bag A, 6 kg for $4·14 or
   Bag B, 2·5 kg for $1·80?

7. An aeroplane was due to take off from Madrid airport at 18:42 but it was 35 minutes late. During the flight, thanks to a tail wind, the plane made up the time and in fact landed 16 min before its scheduled arrival time of 00:05.

   a) What time did the aeroplane take off?

   b) What time did it land?

   (Assume that the plane did not cross any time zones on its journey.)

8. A 20 cent coin is 1·2 mm thick. What is the value of a pile of 20 cent coins which is 21·6 cm high?

9. Work out $\frac{3}{5} + 0{\cdot}12 + 6\%$ of 10.

Exercise 2.28

1. Find the distance travelled by light in one hour, given that the speed of light is 300 000 kilometres per second. Give the answer in kilometres in standard form.
2. When the lid is left off an ink bottle, the ink evaporates at a rate of 2.5×10^{-6} cm³/s. A full bottle contains 36 cm³ of ink. How long, to the nearest day, will it take for all the ink to evaporate?
3. Convert 3·35 hours into hours and minutes.

Remember
There are 60 minutes in 1 hour.

4. When I think of a number, multiply it by 6 and subtract 120, my answer is −18. What was my original number?
5. The cost of advertising in a local paper for one week is shown below.

   28 cents per word plus 75 cents

   a) What is the cost of an advertisement of 15 words for one week?
   b) What is the greatest number of words in an advertisement costing up to $8 for one week?
   c) If an advertisement is run for two weeks, the cost for the second week is reduced by 30%. Calculate the total cost for an advertisement of 22 words for two weeks.
6. Bronze is made up of zinc, tin and copper in the ratio 1 : 4 : 95. A bronze statue contains 120 g of tin. Find the quantities of the other two metals required and the total mass of the statue.

Exercise 2.29

1. In the diagram $\frac{5}{6}$ of the circle is red and $\frac{2}{3}$ of the triangle is red. What is the ratio of the area of the circle to the area of the triangle?

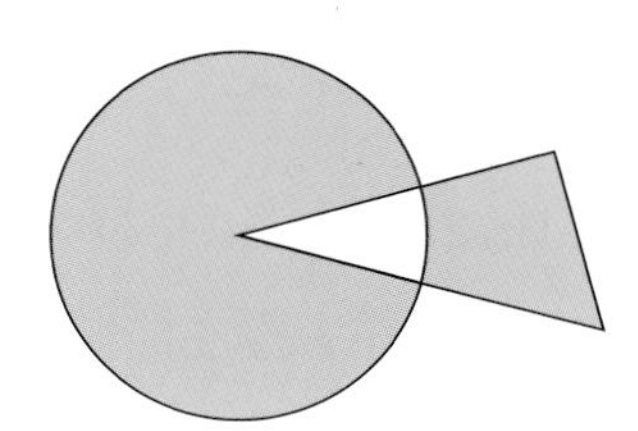

2. Find the exact answer to the following by first working out a rough answer and then using the information given.
   Do *not* use a calculator.
   a) If $142{\cdot}3 \times 98{\cdot}5 = 14\,016{\cdot}55$ find $140{\cdot}1655 \div 14{\cdot}23$
   b) If $76{\cdot}2 \times 8{\cdot}6 = 655{\cdot}32$ find $6553{\cdot}2 \div 86$
   c) If $22{\cdot}3512 \div 0{\cdot}268 = 83{\cdot}4$ find $8340 \times 26{\cdot}8$
   d) If $1{\cdot}6781 \div 17{\cdot}3 = 0{\cdot}097$ find $9700 \times 0{\cdot}173$

3. A sales manager reports an increase of 28% in sales this year compared to last year.

   The increase was \$70 560.

   What were the sales last year?

4. Small cubes of side 1 cm are stuck together to form a large cube of side 4 cm. Opposite faces of the large cube are painted the same colour, but adjacent faces are different colours. The three colours used are red, blue and green.

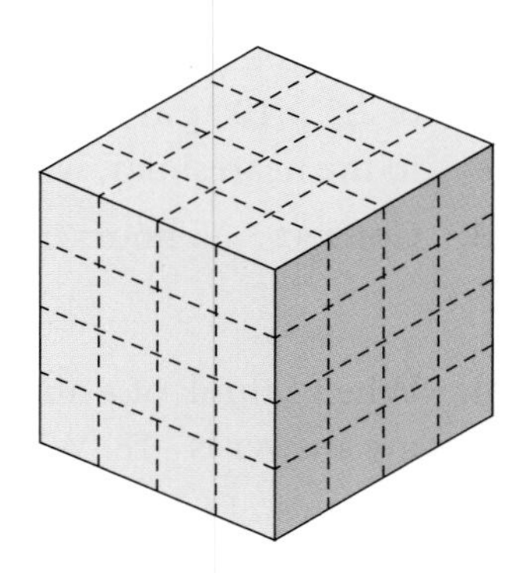

   a) How many small cubes have just one red and one green face?

   b) How many small cubes are painted on one face only?

   c) How many small cubes have one red, one green and one blue face?

   d) How many small cubes have no faces painted?

5. A bullet travels at a speed of 3×10^4 cm/s. Work out the length of time in seconds taken for the bullet to hit a target 54m away.

6. A sewing machine cost \$162·40 after a price increase of 16%. Find the price before the increase.

7. To get the next number in a sequence you double the previous number and subtract two.

   The fifth number in the sequence is 50.

   Find the first number.

8. A code uses 1 for A, 2 for B, 3 for C and so on, up to 26 for Z. Coded words are written without spaces to confuse the enemy, so 18 could be AH or R. Decode the following message.

   208919 919 1 2251825 199121225 31545

9. A coach can take 47 passengers. How many coaches are needed to transport 1330 passengers?

Checking answers

Here are five calculations, followed by sensible checks.

a) $22{\cdot}2 \div 6 = 3{\cdot}7$ check $3{\cdot}7 \times 6 = 22{\cdot}2$

b) $31{\cdot}7 - 4{\cdot}83 = 26{\cdot}87$ check $26{\cdot}87 + 4{\cdot}83 = 31{\cdot}7$

c) $42{\cdot}8 \times 30 = 1284$ check $1284 \div 30 = 42{\cdot}8$

d) $\sqrt{17} = 4{\cdot}1231$ check $4{\cdot}1231^2$

e) $3{\cdot}7 + 17{\cdot}6 + 13{\cdot}9$ check $13{\cdot}9 + 17{\cdot}6 + 3{\cdot}7$ (add in reverse order)

Calculations can also be checked by rounding numbers to a given number of significant figures.

f) $\frac{6 \cdot 1 \times 32 \cdot 6}{19 \cdot 3} = 10{\cdot}3$ (to 3 s.f.)

Check this answer by rounding each number to 1 significant figure and estimating.

$$\frac{6.1 \times 32.6}{19.3} \approx \frac{6 \times 30}{20} = \frac{180}{20} = 9$$

This is close to 10·3
so the actual answer probably is 10·3

'≈' means 'approximately equal to'

Exercise 2.30

1. Use a calculator to work out the following then check the answers as indicated.

a) $92{\cdot}5 \times 20 = \square$ Check $\square \div 20 = \square$
b) $14 \times 328 = \square$ Check $\square \div 328 = \square$
c) $63 - 12{\cdot}6 = \square$ Check $\square + 12{\cdot}6 = \square$
d) $221{\cdot}2 \div 7 = \square$ Check $\square \times 7 = \square$
e) $384{\cdot}93 \div 9{\cdot}1 = \square$ Check $\square \times 9{\cdot}1 = \square$
f) $13{\cdot}71 + 25{\cdot}8 = \square$ Check $\square - 25{\cdot}8 = \square$
g) $95{\cdot}4 \div 4{\cdot}5 = \square$ Check $\square \times 4{\cdot}5 = \square$
h) $8{\cdot}2 + 3{\cdot}1 + 19{\cdot}6 + 11{\cdot}5 = \square$ Check $11{\cdot}5 + 19{\cdot}6 + 3{\cdot}1 + 8{\cdot}2 = \square$
i) $\sqrt{39} = \square$ Check $\square^2 = 39$
j) $3{\cdot}17 + 2{\cdot}06 + 8{\cdot}4 + 16 = \square$ Check $16 + 8{\cdot}4 + 2{\cdot}06 + 3{\cdot}17 = \square$

2. Ten elastic bands weigh 11 grams. Estimate the weight of 985 elastic bands.

3. The numbers below are rounded to 1 significant figure to *estimate* the answer to each calculation. Match each question below to the correct estimated answer.

A	$21{\cdot}9 \times 1{\cdot}01$	P	10
B	$\frac{19{\cdot}82^2}{(18{\cdot}61 + 22{\cdot}3)}$	Q	5
C	$7{\cdot}8 \times 1{\cdot}01$	R	0·5
D	$\frac{\sqrt{98{\cdot}7}}{8{\cdot}78 + 11{\cdot}43}$	S	8
E	$\frac{21{\cdot}42 + 28{\cdot}6}{18{\cdot}84 - 8{\cdot}99}$	T	20

4. $281 \times 36 = 10\,116$

Work out:

a) $10\,116 \div 36$ **b)** $10\,116 \div 281$ **c)** $28{\cdot}1 \times 3{\cdot}6$

(Do *not* use a calculator.)

5. Mavis is paid a salary of \$49 620 per year. Work out a rough estimate for her weekly pay.

(Give your answer correct to one significant figure.)

6. In 2012, the population of France was 61 278 514 and the population of Greece was 9 815 972.
Roughly how many times bigger is the population of France compared to the population of Greece?

Round the numbers to one significant figure.

In questions **7** and **8** there are six calculations and six answers. Write down each calculation and insert the correct answer from the list given. Use estimation.

7. **a)** $5{\cdot}9 \times 6{\cdot}1$ **b)** $19{\cdot}8 \div 5$ **c)** $32 \times 9{\cdot}9$
d) $0{\cdot}89 + 14{\cdot}7$ **e)** $4{\cdot}5 \times 44$ **f)** $4141 \div 40$

Answers:	198	35·99	103·5	15·59	316·8	3·96

8. **a)** $102{\cdot}8 \div 5$ **b)** $11{\cdot}2 \div 98{\cdot}6$ **c)** $3 \times 0{\cdot}41$
d) $34 \times 2{\cdot}9$ **e)** $51 \times 3{\cdot}9$ **f)** $238{\cdot}6 \div 4{\cdot}7$

Answers:	50·76	20·56	1·23	198·9	98·6	0·114

9. Estimate, correct to 1 significant figure:

a) $41{\cdot}56 \div 7{\cdot}88$ **b)** $\dfrac{5{\cdot}13 \times 18{\cdot}777}{0{\cdot}952}$ **c)** $\dfrac{1}{5}$ of \$14 892

d) $\dfrac{0{\cdot}0974 \times \sqrt{104}}{1{\cdot}03}$ **e)** 52% of 0·394 kg **f)** $\dfrac{6{\cdot}84^2 + 0{\cdot}983}{5{\cdot}07^2}$

g) $\dfrac{2848{\cdot}7 + 1024{\cdot}8}{51{\cdot}2 - 9{\cdot}98}$ **h)** $\dfrac{2}{3}$ of £3124 **i)** $18{\cdot}13 \times (3{\cdot}96^2 + 2{\cdot}07^2)$

Revision exercise 2A

1. Evaluate, without a calculator:

a) $148 \div 0{\cdot}8$ **b)** $0{\cdot}024 \div 0{\cdot}000\,16$

c) $(0{\cdot}2)^2 \div (0{\cdot}1)^3$ **d)** $2 - \dfrac{1}{2} - \dfrac{1}{3} - \dfrac{1}{4}$

e) $1\dfrac{3}{4} \times 1\dfrac{3}{5}$ **f)** $\dfrac{1\frac{1}{6}}{1\frac{2}{3} + 1\frac{1}{4}}$

2. On each bounce, a ball rises to $\dfrac{4}{5}$ of its previous height. To what height will it rise after the third bounce, if dropped from a height of 250 cm?

3. A man spends $\dfrac{1}{3}$ of his salary on accommodation and $\dfrac{2}{5}$ of the remainder on food. What fraction is left for other purposes?

4. If $a = \frac{1}{2}$, $b = \frac{1}{4}$, which one of the following has the greatest value?

i) ab **ii)** $a + b$ **iii)** $\frac{a}{b}$

iv) $\frac{b}{a}$ **v)** $(ab)^2$

5. Express 0·05473

a) correct to three significant figures

b) correct to three decimal places

c) in standard form.

6. Evaluate $\frac{2}{3} + \frac{4}{7}$, correct to three decimal places.

7. Labour costs, totalling \$472·50, account for 63% of a car repair bill. Calculate the total bill.

8. **a)** \$143 is divided in the ratio 2 : 3 : 6. Calculate the smallest share.

b) A prize is divided between three people X, Y and Z. If the ratio of X's share to Y's share is 3 : 1 and Y's share to Z's share is 2 : 5, calculate the ratio of X's share to Z's share.

c) If $a : 3 = 12 : a$, calculate the positive value of a.

9. Evaluate the following and give the answer in standard form.

a) $3600 \div 0{\cdot}00012$ **b)** $\frac{3{\cdot}33 \times 10^4}{9 \times 10^{-1}}$

c) $(30\,000)^3$

10. **a)** Convert to percentages:

i) 0·572 **ii)** $\frac{7}{8}$

b) Express 2·6 kg as a percentage of 6·5 kg.

c) In selling a red herring for 92c, a fishmonger makes a profit of 15%. Find the cost price of the fish.

Revision exercise 2B

1. The length of a rectangle is decreased by 25% and the width is increased by 40%. Calculate the percentage change in the area of the rectangle.

2. **a)** What sum of money, invested at 9% interest per year, is needed to provide an income of \$45 per year?

b) A particle increases its speed from 8×10^5 m/s to $1{\cdot}1 \times 10^6$ m/s. What is the percentage increase?

3. A family on holiday in France exchanged \$450 for euros when the exchange rate was 1·41 euros to the dollar. They spent 500 euros and then changed the rest back into dollars, by which time the exchange rate had become 1·46 euros to the dollar. How much did the holiday cost? (Answer in dollars.)

4. Given that

$$t = 2\pi\sqrt{\left(\frac{l}{g}\right)},$$

find the value of t, to three significant figures, when $l = 2{\cdot}31$ and $g = 9{\cdot}81$.

5. A map is drawn to a scale of 1 : 10 000. Find:
   a) the distance between two railway stations which appear on the map 24 cm apart.
   b) the area, in square kilometres, of a lake which has an area of 100 cm^2 on the map.
6. A map is drawn to a scale of 1 : 2000. Find:
   a) the actual distance between two points, which appear 15 cm apart on the map.
   b) the length on the map of a road, which is 1·2 km in length.
   c) the area on the map of a field, with an actual area of 60 000m^2.
7. a) On a map, the distance between two points is 16 cm. Calculate the scale of the map if the actual distance between the points is 8 km.
   b) On another map, two points appear 1·5 cm apart and are in fact 60 km apart. Calculate the scale of the map.
8. a) A house is bought for \$20 000 and sold for \$24 400. What is the percentage profit?
   b) A piece of fish, initially of mass 2·4 kg, is cooked and subsequently has mass 1·9 kg. What is the percentage loss in mass?
   c) An article is sold at a 6% loss for \$225·60. What was the cost price?
9. 2015 people share the cost of hiring a cruise boat. Roughly how much does each person pay, if the total cost was half a million dollars?
10. a) Calculate the speed (in metres per second) of a slug which moves a distance of 30 cm in 1 minute.
    b) Calculate the time taken for a bullet to travel 8 km at a speed of 5000 m/s.
    c) Calculate the distance flown, in a time of four hours, by a pigeon which flies at a speed of 12 m/s.

Revision exercise 2C

1. A motorist travelled 200 km in five hours. Her average speed for the first 100 km was 50 km/h. What was her average speed for the second 100 kilometres?
2. 1 3 8 9 10
   a) From these numbers, write down
      i) the prime number
      ii) a multiple of 5
      iii) two square numbers
      iv) two factors of 32.

   1 is not a prime number

   b) Find two numbers m and n from the list such that $m = \sqrt{n}$ and $n = \sqrt{81}$.
   c) If each of the numbers in the list can be used once, find p, q, r, s, t such that $(p + q)\,r = 2(s + t) = 36$.
3. The value of t is given by

   $$t = 2\pi\sqrt{\left(\frac{2\cdot31^2 + 0\cdot9^2}{2\cdot31 \times 9\cdot81}\right)}$$

   Without using a calculator, and using suitable approximate values for the numbers in the formula, find an estimate for the value of t.
   (To earn the marks in this question you must show the various stages of your working.)
4. Throughout his life Baichu's heart has beaten at an average rate of 72 beats per minute. Baichu is sixty years old. How many times has his heart beaten during his life? Give the answer in standard form correct to two significant figures.
5. Estimate the answer correct to one significant figure. Do not use a calculator.
   a) $(612 \times 52) \div 49\cdot2$
   b) $(11\cdot7 + 997\cdot1) \times 9\cdot2$
   c) $\sqrt{\left(\frac{91\cdot3}{10\cdot1}\right)}$
   d) $\pi\sqrt{(5\cdot2^2 + 18\cdot2)}$

6. Evaluate the following using a calculator. Give your answers to four significant figures.

a) $\frac{0\cdot74}{0\cdot81\times1\cdot631}$ b) $\sqrt{\left(\frac{9\cdot61}{8\cdot34-7\cdot41}\right)}$

c) $\left(\frac{0\cdot741}{0\cdot8364}\right)^4$ d) $\frac{8\cdot4-7\cdot642}{3\cdot333-1\cdot735}$

7. Evaluate the following and give the answers to three significant figures:

a) $\sqrt[3]{(9\cdot61\times0\cdot0041)}$ b) $\left(\frac{1}{9\cdot5}-\frac{1}{11\cdot2}\right)^3$

c) $\frac{15\cdot6\times0\cdot714}{0\cdot0143\times12}$ d) $\sqrt[4]{\left(\frac{1}{5\times10^3}\right)}$

8. The edges of a cube are all increased by 10%. What is the percentage increase in the volume?

Examination exercise 2D

1. a) A car uses fuel at a rate of 5.6 litres per 100 km.

   Calculate the distance travelled when the car has used 14 litres of fuel. [2]

   b) The car passes a post at a speed of 72 km/h.

   i) Change 72 km/h into m/s. [2]

   ii) The car has a length of 4.5 metres. Calculate, in seconds, the time the car takes to pass the post completely. [2]

Cambridge IGCSE International Mathematics 0607 Paper 4 Q6 May/June 2010

2. a) In 2009 the height of a tree was 25.2 m. A year later, the height was 28 m. Calculate the percentage increase in height. [3]

   b) The height of 25.2 m was a 20% increase of the height in 2008. Calculate the height in 2008. [3]

   c) The height of the tree is expected to increase by 5% of its value **each** year. The height is now 30 m.

   i) Calculate the expected height in 3 years time. [3]

   ii) Calculate the number of years it will take for the tree to reach a height of 40m. [2]

Cambridge IGCSE International Mathematics 0607 Paper 41 Q1 May/June 2011

3. a) Simplify $\sqrt{75}$. [1]

   b) Simplify $\frac{2}{5-\sqrt{3}}$ by rationalising the denominator. [2]

Cambridge IGCSE International Mathematics 0607 Paper 21 Q1 May/June 2011

4. A train from Picton to Christchurch leaves Picton at 1300.

   The length of the journey is 340 km.

   a) The train arrives at Christchurch at 1821. Show that the average speed is 63.55 km/h, correct to 2 decimal places. [4]

   b) One day the weather is bad and the average speed of 63.55 km/h is reduced by 15%.

   i) Calculate the new average speed. [2]

   ii) Calculate the new time of arrival at Christchurch. Give your answer to the nearest minute. [3]

Cambridge IGCSE International Mathematics 0607 Paper 4 Q1 October/November 2010

3 Algebra 1

Isaac Newton (1642–1727) is thought by many to have been one of the greatest intellects of all time. He went to Trinity College Cambridge in 1661, and by the age of 23 he had made three major discoveries: the nature of colours, calculus and the law of gravitation. He used his version of calculus to give the first satisfactory explanation of the motion of the sun, the moon and the stars. Because he was extremely sensitive to criticism, Newton was always very secretive, but he was eventually persuaded to publish his discoveries in 1687.

2.3 Solution of linear equations including those with fractional expressions
2.6 Solution of simultaneous linear equations in two variables
2.7 Expansion of brackets, including the square of a binomial
2.8 Factorisation:
common factor e.g. $6x^2 + 9x = 3x\,(2x + 3)$
difference of squares e.g. $9x^2 + 16y^2 = (3x - 4y)\,(3x + 4y)$
trinomial e.g. $6x^2 + 11x - 10 = (3x - 2)\,(2x + 5)$
four term e.g. $xy - 3x + 2y - 6 = (x + 2)\,(y - 3)$
2.10 Solution of quadratic equations:
by factorisation
using a graphics calculator
using the quadratic formula
2.11 Use of a graphics calculator to solve equations, including those which may be unfamiliar e.g. $2^x - 1 = \frac{1}{x^3}$
Syllabus objectives 2.1, 2.2, 2.4, 2.5, 2.9, 2.12 and 2.13 are covered in Chapter 8.

3.1 Negative numbers

- If the weather is very cold and the temperature is 3 degrees below zero, it is written $-3°$.
- If a golfer is 5 under par for his round, the scoreboard will show -5.
- On a bank statement if someone is $55 overdrawn (or 'in the red') it would appear as $-$55.

The above are examples of the use of negative numbers.

An easy way to begin calculations with negative numbers is to think about changes in temperature:

a) Suppose the temperature is $-2°$ and it rises by $7°$.
The new temperature is $5°$.
We can write $-2 + 7 = 5$.

b) Suppose the temperature is $-3°$ and it falls by $6°$.
The new temperature is $-9°$.
We can write $-3 - 6 = -9$.

GDC help

When you enter a negative number, use the key marked (–) for the negative sign. See below for an example.

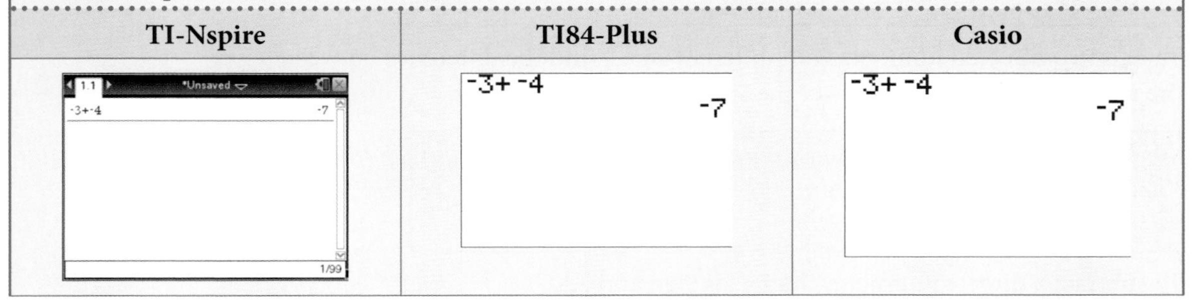

Exercise 3.1

In Questions **1** to **10** move up or down a thermometer as described, and state the new temperature.

1. The temperature is $+8°$ and it falls by $3°$.

2. The temperature is $-8°$ and it rises by $4°$.

3. The temperature is $+4°$ and it falls by $5°$.

4. The temperature is $-3°$ and it rises by $7°$.

5. The temperature is $+2°$ and it falls by $6°$.

6. The temperature is $+4°$ and it rises by $8°$.

7. The temperature is $-1°$ and it falls by $6°$.

8. The temperature is $+9°$ and it falls by $14°$.

9. The temperature is $-5°$ and it rises by $1°$.

10. The temperature is $-13°$ and it rises by $13°$.

11. Some land in Bangladesh is below sea level.
Here are the heights, above sea level, of five villages.

A 1 m **B** −4 m **C** 21 m **D** −2 m **E** −1·5 m

a) Which village is safest from flooding?
b) Which village is most at risk from serious flooding?

12. A diver is below the surface of the water at −15 m.
He dives down by 6 m, then rises 4 m.
Where is he now?

3.2 Directed numbers

To add two directed numbers with the same sign, find the sum of the numbers and give the answer the same sign.

- $+3 + (+5) = +3 + 5 = +8$
- $-7 + (-3) = -7 - 3 = -10$
- $-9{\cdot}1 + (-3{\cdot}1) = 9{\cdot}1 - 3{\cdot}1 = -12{\cdot}2$
- $-2 + (-1) + (-5) = (-2 - 1) - 5$
  $= -3 - 5$
  $= -8$

To add two directed numbers with different signs, find the difference between the numbers and give the answer the sign of the larger number.

- $+7 + (-3) = +7 - 3 = +4$
- $+9 + (-12) = +9 - 12 = -3$
- $-8 + (+4) = -8 + 4 = -4$

To subtract a directed number, change its sign and add.

- $+7 - (+5) = +7 - 5 = +2$
- $+7 - (-5) = +7 + 5 = +12$
- $-8 - (+4) = -8 - 4 = -12$
- $-9 - (-11) = -9 + 11 = +2$

Exercise 3.2

1. $+7 + (+6)$
2. $+11 + (+200)$
3. $-3 + (-9)$
4. $-7 + (-24)$
5. $-5 + (-61)$
6. $+0.2 + (+5{\cdot}9)$
7. $+5 + (+4.1)$
8. $-8 + (-27)$
9. $+17 + (+1{\cdot}7)$
10. $-2 + (-3) + (-4)$
11. $-7 + (+ 4)$
12. $+7 + (-4)$
13. $-9 + (+7)$
14. $+16 + (-30)$
15. $+14 + (-21)$
16. $-7 + (+10)$
17. $-19 + (+200)$
18. $+7{\cdot}6 + (-9{\cdot}8)$
19. $-1{\cdot}8 + (+10)$
20. $-7 + (+24)$
21. $+7 - (+5)$
22. $+9 - (+15)$
23. $-6 - (+9)$
24. $-9 - (+5)$
25. $+8 - (+10)$
26. $-19 - (-7)$
27. $-10 - (+70)$
28. $-5{\cdot}1 - (+8)$
29. $-0{\cdot}2 - (+4)$
30. $+5{\cdot}2 - (-7{\cdot}2)$
31. $-4 + (-3)$
32. $+6 - (-2)$
33. $+8 + (-4)$
34. $-4 - (+6)$
35. $+7 - (-4)$
36. $+6 + (-2)$
37. $+10 - (+30)$
38. $+19 - (+11)$
39. $+4 + (-7) + (-2)$
40. $-3 - (+2) + (-5)$
41. $-17 - (-1) + (-10)$
42. $-5 + (-7) - (+9)$

43. $+9+(-7)-(-6)$

44. $-7-(-8)$

45. $-10{\cdot}1+(-10{\cdot}1)$

46. $-75-(-25)$

47. $-204-(+304)$

48. $-7+(-11)-(+11)$

49. $+17-(+17)$

50. $-6+(-7)-(+8)$

51. $+7+(-7{\cdot}1)$

52. $-11-(-4)+(+3)$

53. $-2-(-8{\cdot}7)$

54. $+7+(-11)+(+5)$

55. $-610+(-240)$

56. $-7-(-3)-(-8)$

57. $+9-(-6)+(-9)$

58. $-1-(-5)+(-8)$

59. $-2{\cdot}1+(-9{\cdot}9)$

60. $-47-(-16)$

When two directed numbers with the same sign are multiplied together, the answer is positive.

- $+7\times(+3)=+21$
- $-6\times(-4)=+24$

When two directed numbers with different signs are multiplied together, the answer is negative.

- $-8\times(+4)=-32$
- $+7\times(-5)=-35$
- $-3\times(+2)\times(+5)=-6\times(+5)=-30$

When dividing directed numbers, the rules are the same as in multiplication.

- $-70\div(-2)=+35$
- $+12\div(-3)=-4$
- $-20\div(+4)=-5$

Exercise 3.3

1. $+2\times(-4)$

2. $+7\times(+4)$

3. $-4\times(-3)$

4. $-6\times(-4)$

5. $-6\times(-3)$

6. $+5\times(-7)$

7. $-7\times(-7)$

8. $-4\times(+3)$

9. $+0{\cdot}5\times(-4)$

10. $-1\frac{1}{2}\times(-6)$

11. $-8\div(+2)$

12. $+12\div(+3)$

13. $+36\div(-9)$

14. $-40\div(-5)$

15. $-70\div(-1)$

16. $-56\div(+8)$

17. $-12\div(-2)$

18. $-3\div(+5)$

19. $+0{\cdot}1\div(-10)$

20. $-0{\cdot}02\div(-100)$

21. $-11\times(-11)$

22. $-6\times(-1)$

23. $+12\times(-50)$

24. $-\frac{1}{2}\div\left(+\frac{1}{2}\right)$

25. $-600\div(+30)$

26. $-5{\cdot}2\div(+2)$

27. $+7\times(-100)$

28. $-6\div\left(-\frac{1}{3}\right)$

29. $100\div(-0{\cdot}1)$

30. -8×-80

31. $-3\times(-2)\times(-1)$

32. $+3\times(-7)\times(+2)$

33. $+0{\cdot}4\div(-1)$

34. $-16\div(+40)$

35. $+0{\cdot}2\times(-1000)$

36. $-7\times(-5)\times(-1)$

37. $-14\div(+7)$

38. $-7\div(-14)$

39. $+1\frac{1}{4}\div(-5)$

40. $-6\times\left(-\frac{1}{2}\right)\times(-30)$

Exercise 3.4

1. $-7 + (-3)$ **2.** $-6 - (-7)$ **3.** $-4 \times (-3)$ **4.** $-4 \times (+7)$

5. $4 - (+6)$ **6.** $-4 \times (-4)$ **7.** $+6 \div (-2)$ **8.** $+8 - (-6)$

9. $-7 \times (+4)$ **10.** $-8 \div (-2)$ **11.** $+10 \div (-60)$ **12.** $(-3)^2$

13. $40 - (+70)$ **14.** $-6 \times (-4)$ **15.** $(-1)^5$ **16.** $-8 \div (+4)$

17. $+10 \times (-3)$ **18.** $-7 \times (-1)$ **19.** $+10 + (-7)$ **20.** $+12 - (-4)$

21. $+100 + (-7)$ **22.** $-60 \times (-40)$ **23.** $-20 \div (-2)$ **24.** $(-1)^{20}$

25. $6 - (+10)$ **26.** $-6 \times (+4) \times (-2)$ **27.** $+8 \div (-8)$ **28.** $0 \times (-6)$

29. $(-2)^3$ **30.** $+100 - (-70)$ **31.** $+18 \div (-6)$ **32.** $(-1)^{12}$

33. $-6 - (-7)$ **34.** $(-2)^2 + (-4)$ **35.** $+8 - (-7)$ **36.** $+7 + (-2)$

37. $-6 \times (+0{\cdot}4)$ **38.** $-3 \times (-6) \times (-10)$ **39.** $(-2)^2 + (+1)$ **40.** $+6 - (+1000)$

41. $(-3)^2 - 7$ **42.** $-12 \div \frac{1}{4}$ **43.** $-30 \div -12$ **44.** $5 - (+7) + (-0{\cdot}5)$

45. $(-2)^5$ **46.** $0 \div \left(-\frac{1}{5}\right)$ **47.** $(-0{\cdot}1)^2 \times (-10)$ **48.** $3 - (+19)$

49. $2{\cdot}1 + (-6{\cdot}4)$ **50.** $\left(-\frac{1}{2}\right)^2 \div (-4)$

51. In a 'magic' square the sum of the numbers in each row, column and the main diagonals is the same.

Copy and complete these magic squares.

		3
	0	
−3	2	

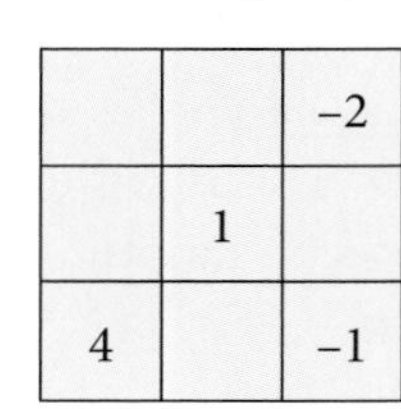

		−2
	1	
4		−1

	1	
2	−3	−2

Find the letters

Perform each calculation and write down the corresponding letter from the list on the next page, to make a sentence.

A. $5 - 8$; $-3 - 2$; $(-2)^2$; $3^2 - 20$; $6 \div (-6)$; $(-2)^2 + 3$; $-5\,(-3)$; $(-49) \div (-7)$; $-3 - (-5)$; $(-5) \times (-1)$; $-7 + 11$; $4^2 - 4$; $-1 + 13$; $(-1)^2 \times 4$; $30 \div (-10)$; $-2 + 9$; $\left(-3\frac{1}{2}\right) \times 2$; $(-8) \div (-2)$; $(-1) \div \frac{1}{10}$; $8 - 11$; $(-7)^2 + (-1)^2$; $-6 - 5$; $(-10) \times \frac{1}{2}$; $-3 + 10$; $(-2)^3 - 2$; $3 - (-5)$; $(-3) \times (-4)$; $(-16) \div (-2)$; $2 \div (-4)$? $(-6 - 2) \times (-1)$; $2 \times (-5)^2$; $(-6)^2 \div 3$; $(-2) \times (-2) \times 2$; $-5 + 13$; $-12 - (-2)$.

B. $-11+8$; $-3+(-2)$; $(-2)^2$; $1-12$; $3\times(-1)$; $1-(-6)$; $-2+4$; $(-3)\times(-4)$; $6\div(-6)$;

$(-16)\div 2$; $\left(-3\frac{1}{2}\right)\times(-2)$; 2^2-2; $1\div\left(\frac{1}{2}\right)$; $-7-(-7)$; $(-3)^2-1$; $-3-8$; $(-14)\div(-2)$;

$-3+8$; $-8+15$; $(-1)^2+(-1)^2$; $(-5)\times 2$; $-1-10$; $(-2)\div\left(-\frac{1}{2}\right)$; $-2-2-1$; 2^3;

$4+(-6)$; $(-1)^5$; $-3+10$; $(-6)\div(-1)$; $2-(-3)$; $(-3-4)\times(-1)$; $19-22$; $(-5)\times 0$?

$(-2)^3\div(-2)$; $-2+7$; $1-(-6)$; $3\times(-1)$; $(-50)\div(-10)$; $0{\cdot}1\times 20$; $3^2+2^2-1^2$;

$(-1)\div\left(-\frac{1}{4}\right)$; $(-2)^3-3$; $\frac{1}{7}\times 49$; 4^3-66.

W	O	U	D	E	C
−3	7	2	−1	8	5

Y	A	H	N	T	L	M
−8	4	−5	−10	−11	12	−7

S	I	R	H	F	G
0	50	−2	−6	6	$-\frac{1}{2}$

3.3 Formulae and expressions

When a calculation is repeated many times it is often helpful to use a formula. Publishers use a formula to work out the selling price of a book based on the production costs and the expected sales of the book. An **expression** has no equals sign. For example, $2x-y$ and x^2+7x+1 are expressions.

Exercise 3.5

1. The final speed v of a car is given by the formula $v=u+at$.
   [u = initial speed, a = acceleration, t = time taken]
   Find v when $u=15$ m/s, $a=0{\cdot}2$ m/s^2, $t=30$ s.
2. The time period T of a simple pendulum is given by the formula
   $T=2\pi\sqrt{\left(\frac{l}{g}\right)}$, where l is the length of the pendulum and g is the gravitational acceleration. Find T when $l=0{\cdot}65$ m, $g=9{\cdot}81$ m/s^2 and $\pi=3{\cdot}142$.
3. The total surface area A of a cone is related to the radius r and the slant height l by the formula $A=\pi r(r+l)$. Find A when $r=7$ cm and $l=11$ cm.
4. The sum S of the squares of the integers from 1 to n is given by $S=\frac{1}{6}n(n+1)(2n+1)$. Find S when $n=12$.

5. The acceleration a of a train is found using the formula $a = \frac{v^2 - u^2}{2s}$. Find a when $v = 20$ m/s, $u = 9$ m/s and $s = 2{\cdot}5$ m.

6. Einstein's famous equation relating energy, mass and the speed of light is $E = mc^2$. Find E when $m = 0{\cdot}0001$ kg and $c = 3 \times 10^8$ m/s.

7. The distance s travelled by an accelerating rocket is given by $s = ut + \frac{1}{2}at^2$. Find s when $u = 3$ m/s, $t = 100$ s and $a = 0{\cdot}1$ m/s^2.

8. Find a formula for the area of the shape below, in terms of a, b and c.

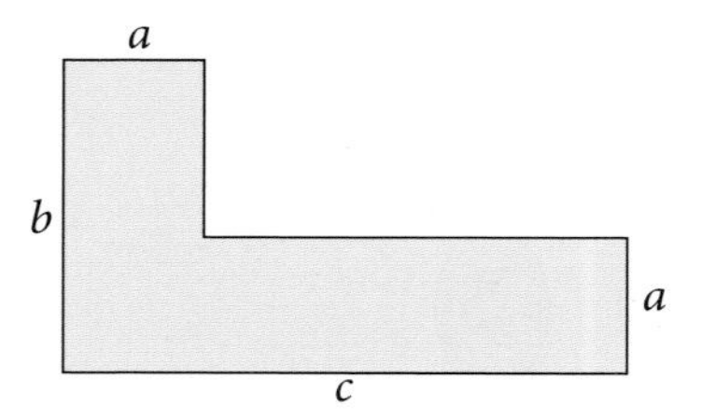

9. Find a formula for the length of the blue part below, in terms of p, q and r.

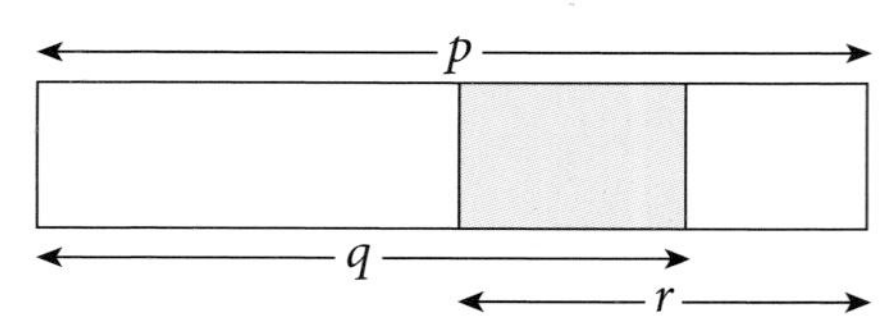

10. A fish lays brown eggs or white eggs and it likes to lay them in a certain pattern. Each brown egg is surrounded by six white eggs. Here there are 3 brown eggs and 14 white eggs.

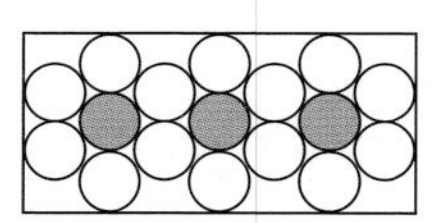

a) How many eggs does it lay altogether if it lays 200 brown eggs?

b) How many eggs does it lay altogether if it lays n brown eggs?

11. In the diagrams below the rows of red tiles are surrounded by white tiles.

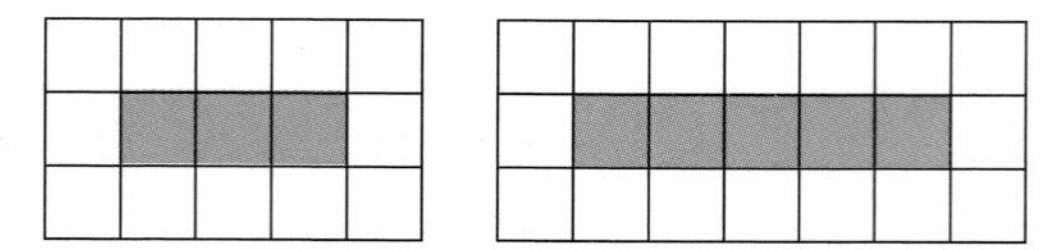

Find a formula for the number of white tiles which would be needed to surround a row of n red tiles.

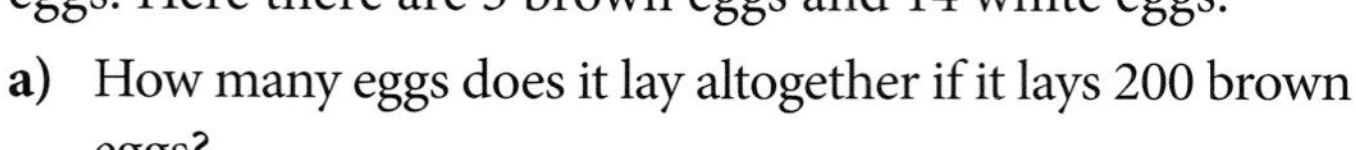

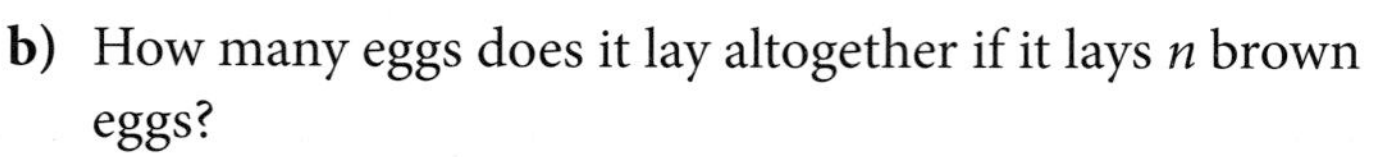

Exercise 3.6

1. Four rods P, Q, R and S have lengths, in cm, as shown.

P	Q	R	S
x	$x + 3$	$2x - 1$	$x - 1$

In each diagram find the length l, in terms of x. Give your answers in their simplest form. [The diagrams are not drawn to scale.]

a) 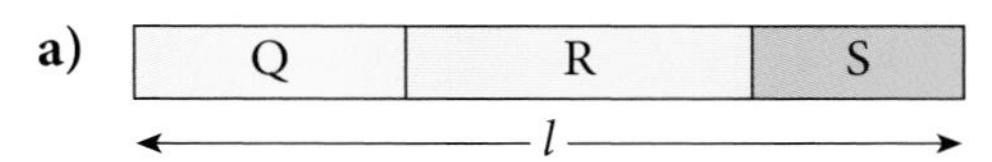

b) R R, l

c) 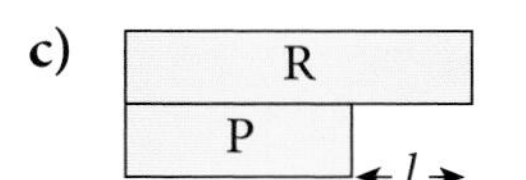

d) 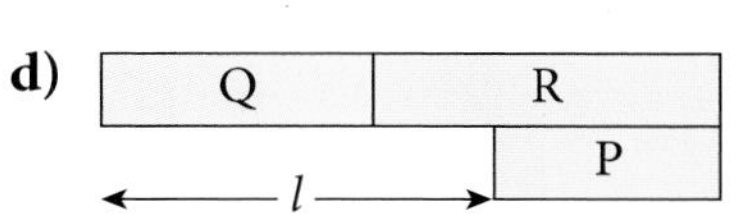

e) 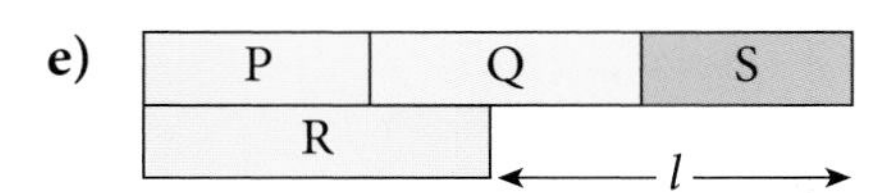

f) 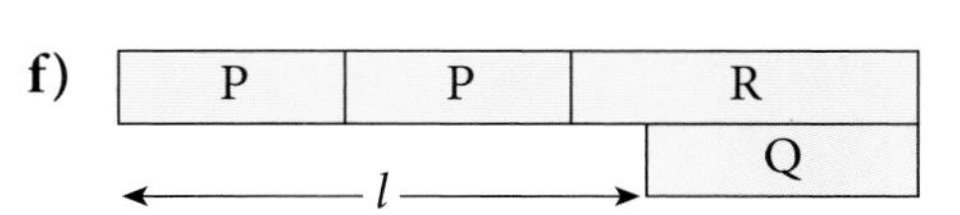

g) 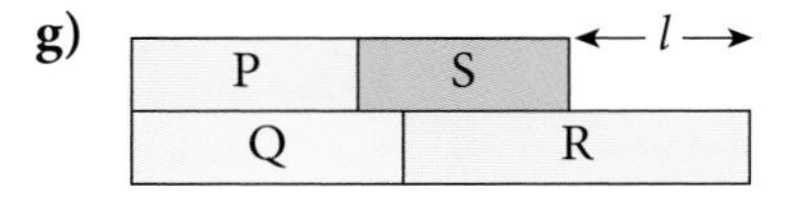

h) 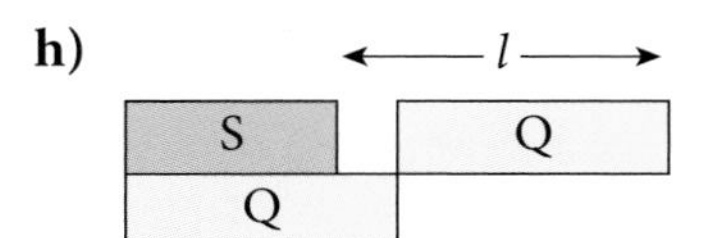

i) 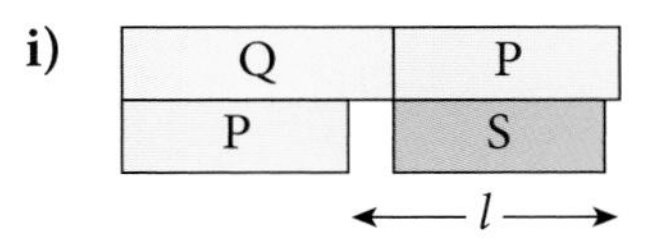

j) 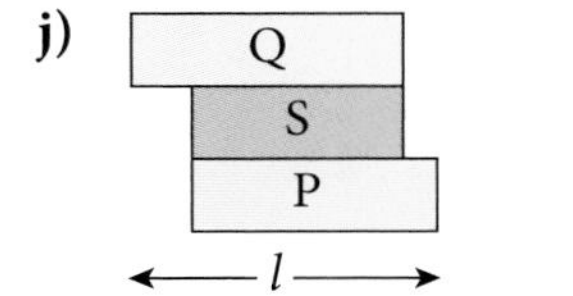

k) 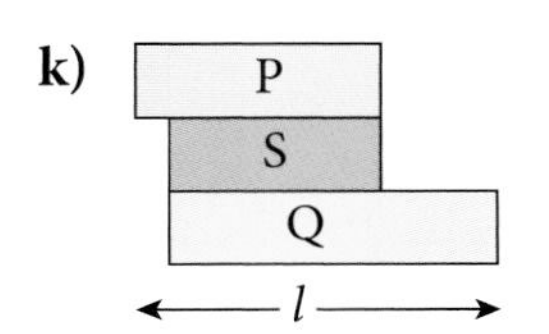

l) 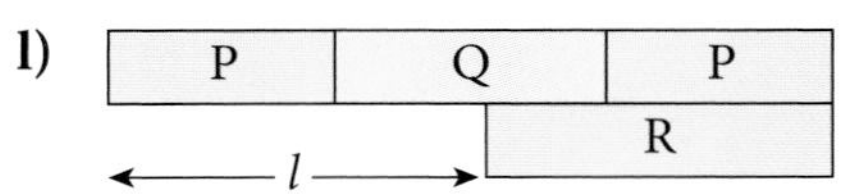

Substituting in expressions

Example 1

When $a = 3$, $b = -2$, $c = 5$, find the value of:

a) $3a + b$ **b)** $ac + b^2$ **c)** $\frac{a+c}{b}$ **d)** $a(c - b)$

a) $3a + b = (3 \times 3) + (-2)$
$= 9 - 2$
$= 7$

b) $ac + b^2 = (3 \times 5) + (-2)^2$
$= 15 + 4$
$= 19$

c) $\frac{a+c}{b} = \frac{3+5}{-2}$
$= \frac{8}{-2}$
$= -4$

d) $a(c - b) = 3[5 - (-2)]$
$= 3[7]$
$= 21$

Notice that working *down* the page is often easier to follow.

Exercise 3.7

Evaluate the following expressions.

For questions **1** to **12**, $a = 3$, $c = 2$, $e = 5$.

1. $3a - 2$ **2.** $4c + e$ **3.** $2c + 3a$ **4.** $5e - a$

5. $e - 2c$ **6.** $e - 2a$ **7.** $4c + 2e$ **8.** $7a - 5e$

9. $c - e$ **10.** $10a + c + e$ **11.** $a + c - e$ **12.** $a - c - e$

For questions **13** to **24**, $h = 3$, $m = -2$, $t = -3$.

13. $2m - 3$ **14.** $4t + 10$ **15.** $3h - 12$ **16.** $6m + 4$

17. $9t - 3$ **18.** $4h + 4$ **19.** $2m - 6$ **20.** $m + 2$

21. $3h + m$ **22.** $t - h$ **23.** $4m + 2h$ **24.** $3t - m$

For questions **25** to **36**, $x = -2$, $y = -1$, $k = 0$.

25. $3x + 1$ **26.** $2y + 5$ **27.** $6k + 4$ **28.** $3x + 2y$

29. $2k + x$ **30.** xy **31.** xk **32.** $2xy$

33. $2(x + k)$ **34.** $3(k + y)$ **35.** $5x - y$ **36.** $3k - 2x$

$2x^2$ means $2(x^2)$
$(2x)^2$ means 'work out $2x$ and *then* square it'
$-7x$ means $-7(x)$
$-x^2$ means $-1(x^2)$

Example 2

When $x = -2$, find the value of:

a) $2x^2 - 5x$ **b)** $(3x)^2 - x^2$

a) $2x^2 - 5x = 2(-2)^2 - 5(-2)$
$= 2(4) + 10$
$= 18$

b) $(3x)^2 - x^2 = (3 \times -2)^2 - 1(-2)^2$
$= (-6)^2 - 1(4)$
$= 36 - 4$
$= 32$

Exercise 3.8

If $x = -3$ and $y = 2$, evaluate the following:

1. x^2 **2.** $3x^2$ **3.** y^2 **4.** $4y^2$

5. $(2x)^2$ **6.** $2x^2$ **7.** $10 - x^2$ **8.** $10 - y^2$

9. $20 - 2x^2$ **10.** $20 - 3y^2$ **11.** $5 + 4x$ **12.** $x^2 - 2x$

13. $y^2 - 3x^2$ **14.** $x^2 - 3y$ **15.** $(2x)^2 - y^2$ **16.** $4x^2$

17. $(4x)^2$ **18.** $1 - x^2$ **19.** $y - x^2$ **20.** $x^2 + y^2$

21. $x^2 - y^2$ **22.** $2 - 2x^2$ **23.** $(3x)^2 + 3$ **24.** $11 - xy$

25. $12 + xy$ **26.** $(2x)^2 - (3y)^2$ **27.** $2 - 3x^2$ **28.** $y^2 - x^2$

29. $x^2 + y^3$ **30.** $\frac{x}{y}$ **31.** $10 - 3x$ **32.** $2y^2$

33. $25 - 3y$ **34.** $(2y)^2$ **35.** $-7 + 3x$ **36.** $-8 + 10y$

37. $(xy)^2$ **38.** xy^2 **39.** $-7 + x^2$ **40.** $17 + xy$

41. $-5 - 2x^2$ **42.** $10 - (2x)^2$ **43.** $x^2 + 3x + 5$ **44.** $2x^2 - 4x + 1$ **45.** $\frac{x^2}{y}$

Example 3

When $a = -2$, $b = 3$, $c = -3$, evaluate:

a) $\frac{2a\left(b^2 - a\right)}{c}$ **b)** $\sqrt{\left(a^2 + b^2\right)}$

a)

$$(b^2 - a) = 9 - (-2)$$
$$= 11$$
$$\therefore \frac{2a\left(b^2 - a\right)}{c} = \frac{2\times(-2)\times(11)}{-3}$$
$$= 14\frac{2}{3}$$

b)

$$a^2 + b^2 = (-2)^2 + (3)^2$$
$$= 4 + 9$$
$$= 13$$
$$\sqrt{\left(a^2 + b^2\right)} = \sqrt{13}$$

GDC help

Using a GDC

	TI-Nspire	TI84-Plus	Casio
a)	2·-2·((-3)²--2) / -3 → 44/3	2*(-2)*((-3)²-(-2)) / -3 → 44/3; Ans▸n/d◂▸Un/d → 14 2/3	2×(-2)×((-3)²-(-2)) / -3 → 14 2/3
b)	√((-2)²+3²) → 3.60555	√((-2)²+3² → 3.605551275	√((-2)²+3² → √13

Exercise 3.9

Evaluate the following:

In questions **1** to **16**, $a = 4$, $b = -2$, $c = -3$.

1. $a(b + c)$

2. $a^2(b - c)$

3. $2c(a - c)$

4. $b^2(2a + 3c)$

5. $c^2(b - 2a)$

6. $2a^2(b + c)$

7. $2(a + b + c)$

8. $3c(a - b - c)$

9. $b^2 + 2b + a$

10. $c^2 - 3c + a$

11. $2b^2 - 3b$

12. $\sqrt{(a^2 + b^2)}$

13. $\sqrt{(ab + c^2)}$

14. $\sqrt{(c^2 - b^2)}$

15. $\frac{b^2}{a} + \frac{2c}{b}$

16. $\frac{c^2}{b} + \frac{4b}{a}$

In questions **17** to **32**, $k = -3$, $m = 1$, $n = -4$.

17. $k^2(2m - n)$

18. $5m\sqrt{(k^2 + n^2)}$

19. $\sqrt{(kn + 4m)}$

20. $kmn(k^2 + m^2 + n^2)$

21. $k^2m^2(m - n)$

22. $k^2 - 3k + 4$

23. $m^3 + m^2 + n^2 + n$

24. $k^3 + 3k$

25. $m(k^2 - n^2)$

26. $m\sqrt{(k - n)}$

27. $100k^2 + m$

28. $m^2(2k^2 - 3n^2)$

29. $\frac{2k + m}{k - n}$

30. $\frac{kn - k}{2m}$

31. $\frac{3k + 2m}{2n - 3k}$

32. $\frac{k + m + n}{k^2 + m^2 + n^2}$

In questions **33** to **48**, $w = -2$, $x = 3$, $y = 0$, $z = -\frac{1}{2}$.

33. $\frac{w}{z} + x$

34. $\frac{w + x}{z}$

35. $y\left(\frac{x + z}{w}\right)$

36. $x^2(z + wy)$

37. $x\sqrt{(x + wz)}$

38. $w^2\sqrt{(x^2 + y^2)}$

39. $2(w^2 + x^2 + y^2)$

40. $2x(w - z)$

41. $\frac{z}{w} + x$

42. $\frac{z + w}{x}$

43. $\frac{x + w}{z^2}$

44. $\frac{y^2 - w^2}{x^2}$

45. $z^2 + 4z + 5$

46. $\frac{1}{w} + \frac{1}{z} + \frac{1}{x}$

47. $\frac{4}{z} - \frac{10}{w}$

48. $\frac{yz - xw}{xz - w}$

49. Find $K = \sqrt{\left(\frac{a^2 + b^2 + c^2 - 2c}{a^2 + b^2 + 4c}\right)}$ if $a = 3$, $b = -2$, $c = -1$.

50. Find $W = \frac{kmn(k + m + n)}{(k + m)(k + n)}$ if $k = \frac{1}{2}$, $m = -\frac{1}{3}$, $n = \frac{1}{4}$.

3.4 Brackets and simplifying

A term outside a bracket multiplies each of the terms inside the bracket. This is the **distributive law** and is demonstrated in the following examples.

$3(x - 2y) = 3x - 6y$

$2x(x - 2y + z) = 2x^2 - 4xy + 2xz$

$7y - 4(2x - 3) = 7y - 8x + 12$

In general, numbers can be added to numbers

- x's can be added to x's
- y's can be added to y's
- x^2's can be added to x^2's

But they must not be mixed.

$2x + 3y + 3x^2 + 2y - x = x + 5y + 3x^2$

$7x + 3x(2x - 3) = 7x + 6x^2 - 9x$
$= 6x^2 - 2x$

$4(x + 3) - (x - 2)$
Write this as $4(x + 3) - 1(x - 2)$
Multiply out: $4x + 12 - x + 2 = 3x + 14$

Introduce '–1' to replace '–'

Exercise 3.10

Simplify as far as possible:

1. $3x + 4y + 7y$
2. $4a + 7b - 2a + b$
3. $3x - 2y + 4y$
4. $2x + 3x + 5$
5. $7 - 3x + 2 + 4x$
6. $5 - 3y - 6y - 2$
7. $5x + 2y - 4y - x^2$
8. $2x^2 + 3x + 5$
9. $2x - 7y - 2x - 3y$
10. $4a + 3a^2 - 2a$
11. $7a - 7a^2 + 7$
12. $x^2 + 3x^2 - 4x^2 + 5x$
13. $\frac{3}{a} + b + \frac{7}{a} - 2b$
14. $\frac{4}{x} - \frac{7}{y} + \frac{1}{x} + \frac{2}{y}$
15. $\frac{m}{x} + \frac{2m}{x}$
16. $\frac{5}{x} - \frac{7}{x} + \frac{1}{2}$
17. $\frac{3}{a} + b + \frac{2}{a} + 2b$
18. $\frac{n}{4} - \frac{m}{3} - \frac{n}{2} + \frac{m}{3}$
19. $x^3 + 7x^2 - 2x^3$
20. $(2x)^2 - 2x^2$
21. $(3y)^2 + x^2 - (2y)^2$
22. $(2x)^2 - (2y)^2 - (4x)^2$
23. $5x - 7x^2 - (2x)^2$
24. $\frac{3}{x^2} + \frac{5}{x^2}$

Remove the brackets and collect like terms:

25. $3x + 2(x + 1)$
26. $5x + 7(x - 1)$
27. $7 + 3(x - 1)$
28. $9 - 2(3x - 1)$
29. $3x - 4(2x + 5)$
30. $5x - 2x(x - 1)$
31. $7x + 3x(x - 4)$
32. $4(x - 1) - 3x$
33. $5x(x + 2) + 4x$
34. $3x(x - 1) - 7x^2$
35. $3a + 2(a + 4)$
36. $4a - 3(a - 3)$
37. $3ab - 2a(b - 2)$
38. $3y - y(2 - y)$
39. $3x - (x + 2)$
40. $7x - (x - 3)$
41. $5x - 2(2x + 2)$
42. $3(x - y) + 4(x + 2y)$
43. $x(x - 2) + 3x(x - 3)$
44. $3x(x + 4) - x(x - 2)$
45. $y(3y - 1) - (3y - 1)$
46. $7(2x + 2) - (2x + 2)$
47. $7b(a + 2) - a(3b + 3)$
48. $3(x - 2) - (x - 2)$

49. Write down and simplify an expression for

a) the area of this shape

b) the perimeter of this shape.

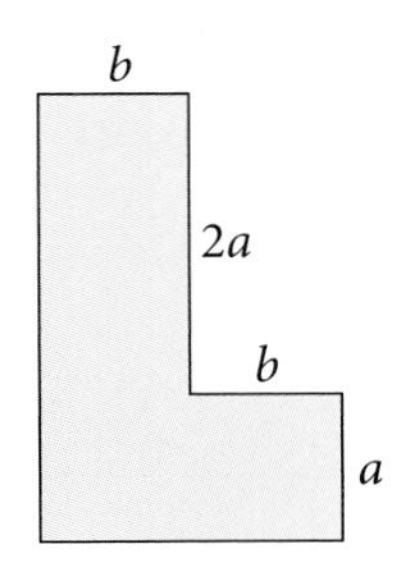

50. In the expression $2(3n - 5)$, three operations are performed in the following order:

$n \rightarrow$ [$\times 3$] $\rightarrow$ [-5] $\rightarrow$ [$\times 2$] $\rightarrow$

Draw similar diagrams to show the correct order of operations for the following expressions:

a) $6n + 1$ **b)** $3(5n - 2)$ **c)** $\frac{4n+5}{3}$ **d)** $2(7n + 3)$

e) $n^2 + 7$ **f)** $(n - 7)^2$ **g)** $(2n - 3)^2 + 10$ **h)** $\frac{(3n+1)^2 + 5}{7}$

51. Find the value of n in each case.

a) $n \rightarrow$ [$\times 4$] $\rightarrow$ [$+92$] $\rightarrow$ [$\div 3$] $\rightarrow 34$

b) $n \rightarrow$ [$\times 15$] $\rightarrow$ [$\sqrt{\ }$] $\rightarrow$ [$\times 4$] $\rightarrow$ [$\div 12$] $\rightarrow 5$

c) $n \rightarrow$ [square] $\rightarrow$ [$+51$] $\rightarrow$ [$\div 1000$] $\rightarrow$ [$+0{\cdot}9$] $\rightarrow 1$

Two brackets

The following examples demonstrate how to multiply out two sets of brackets.

$$\begin{aligned}(x + 5)(x + 3) &= x(x + 3) + 5(x + 3)\\ &= x^2 + 3x + 5x + 15\\ &= x^2 + 8x + 15\end{aligned}$$

$$\begin{aligned}(2x - 3)(4y + 3) &= 2x(4y + 3) - 3(4y + 3)\\ &= 8xy + 6x - 12y - 9\end{aligned}$$

$$\begin{aligned}3(x + 1)(x - 2) &= 3[x(x - 2) + 1(x - 2)]\\ &= 3[x^2 - 2x + x - 2]\\ &= 3x^2 - 3x - 6\end{aligned}$$

Exercise 3.11

Remove the brackets and simplify:

1. $(x + 1)(x + 3)$ **2.** $(x + 3)(x + 2)$ **3.** $(y + 4)(y + 5)$

4. $(x - 3)(x + 4)$ **5.** $(x + 5)(x - 2)$ **6.** $(x - 3)(x - 2)$

7. $(a - 7)(a + 5)$ **8.** $(z + 9)(z - 2)$ **9.** $(x - 3)(x + 3)$

10. $(k - 11)(k + 11)$ **11.** $(2x + 1)(x - 3)$ **12.** $(3x + 4)(x - 2)$

13. $(2y - 3)(y + 1)$ **14.** $(7y - 1)(7y + 1)$ **15.** $(3x - 2)(3x + 2)$

16. $(3a + b)(2a + b)$ **17.** $(3x + y)(x + 2y)$ **18.** $(2b + c)(3b - c)$

19. $(5x - y)(3y - x)$ **20.** $(3b - a)(2a + 5b)$ **21.** $2(x - 1)(x + 2)$

22. $3(x-1)(2x+3)$ **23.** $4(2y-1)(3y+2)$ **24.** $2(3x+1)(x-2)$

25. $4(a+2b)(a-2b)$ **26.** $x(x-1)(x-2)$ **27.** $2x(2x-1)(2x+1)$

28. $3y(y-2)(y+3)$ **29.** $x(x+y)(x+z)$ **30.** $3z(a+2m)(a-m)$

Be careful with an expression like $(x-3)^2$. It is not x^2-9 or even x^2+9.

$$\begin{aligned}(x-3)^2 &= (x-3)(x-3)\\ &= x(x-3)-3(x-3)\\ &= x^2-6x+9\end{aligned}$$

Another common mistake occurs with an expression like $4-(x-1)^2$.

$$\begin{aligned}4-(x-1)^2 &= 4-1(x-1)(x-1)\\ &= 4-1(x^2-2x+1)\\ &= 4-x^2+2x-1\\ &= 3+2x-x^2\end{aligned}$$

Exercise 3.12

Remove the brackets and simplify:

1. $(x+4)^2$ **2.** $(x+2)^2$ **3.** $(x-2)^2$ **4.** $(2x+1)^2$

5. $(y-5)^2$ **6.** $(3y+1)^2$ **7.** $(x+y)^2$ **8.** $(2x+y)^2$

9. $(a-b)^2$ **10.** $(2a-3b)^2$ **11.** $3(x+2)^2$ **12.** $(3-x)^2$

13. $(3x+2)^2$ **14.** $(a-2b)^2$ **15.** $(x+1)^2+(x+2)^2$ **16.** $(x-2)^2+(x+3)^2$

17. $(x+2)^2+(2x+1)^2$ **18.** $(y-3)^2+(y-4)^2$ **19.** $(x+2)^2-(x-3)^2$ **20.** $(x-3)^2-(x+1)^2$

21. $(y-3)^2-(y+2)^2$ **22.** $(2x+1)^2-(x+3)^2$ **23.** $3(x+2)^2-(x+4)^2$ **24.** $2(x-3)^2-3(x+1)^2$

25. a) A solid rectangular block measures x cm by x cm by $(x+3)$ cm. Find a simplified expression for its surface area in cm².

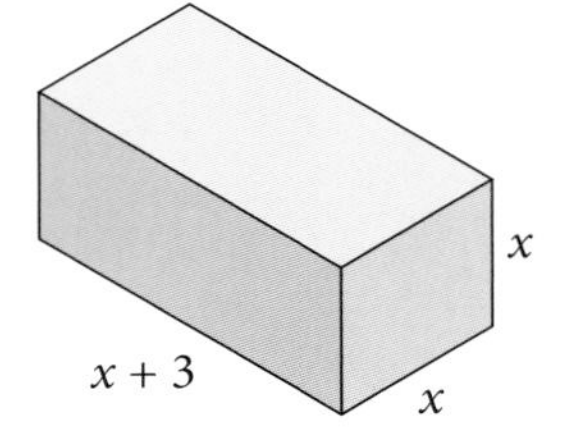

b) The block above is divided into two rectangular blocks by cutting as shown in the diagram. Find a simplified expression for the total surface area of the two blocks formed, in cm².

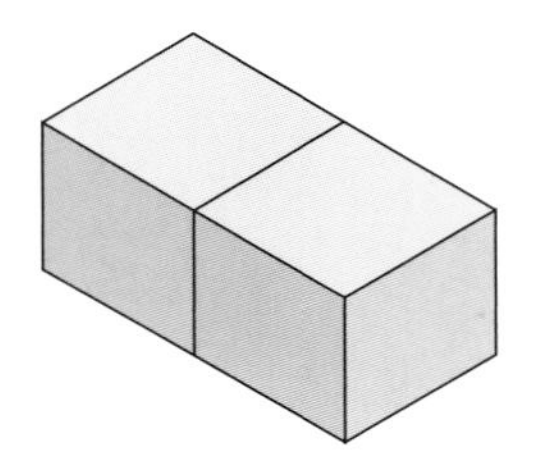

3.5 Linear equations

- If the x term is negative, take it to the other side, where it becomes positive.

$$4 - 3x = 2$$
$$4 = 2 + 3x$$
$$2 = 3x$$
$$\frac{2}{3} = x$$

- If there are x terms on both sides, collect them on one side.

$$2x - 7 = 5 - 3x$$
$$2x + 3x = 5 + 7$$
$$5x = 12$$
$$x = \frac{12}{5} = 2\frac{2}{5}$$

- If there is a fraction in the x term, multiply out to simplify the equation.

$$\frac{2x}{3} = 10$$
$$2x = 30$$
$$x = \frac{30}{2} = 15$$

Exercise 3.13

Solve the following equations:

1. $2x - 5 = 11$
2. $3x - 7 = 20$
3. $2x + 6 = 20$
4. $5x + 10 = 60$
5. $8 = 7 + 3x$
6. $12 = 2x - 8$
7. $-7 = 2x - 10$
8. $3x - 7 = -10$
9. $12 = 15 + 2x$
10. $5 + 6x = 7$
11. $\frac{x}{5} = 7$
12. $\frac{x}{10} = 13$
13. $7 = \frac{x}{2}$
14. $\frac{x}{2} = \frac{1}{3}$
15. $\frac{3x}{2} = 5$
16. $\frac{4x}{5} = -2$
17. $7 = \frac{7x}{3}$
18. $\frac{3}{4} = \frac{2x}{3}$
19. $\frac{5x}{6} = \frac{1}{4}$
20. $-\frac{3}{4} = \frac{3x}{5}$
21. $\frac{x}{2} + 7 = 12$
22. $\frac{x}{3} - 7 = 2$
23. $\frac{x}{5} - 6 = -2$
24. $4 = \frac{x}{2} - 5$
25. $10 = 3 + \frac{x}{4}$
26. $\frac{a}{5} - 1 = -4$
27. $100x - 1 = 98$
28. $7 = 7 + 7x$
29. $\frac{x}{100} + 10 = 20$
30. $1000x - 5 = -6$
31. $-4 = -7 + 3x$
32. $2x + 4 = x - 3$
33. $x - 3 = 3x + 7$
34. $5x - 4 = 3 - x$
35. $4 - 3x = 1$
36. $5 - 4x = -3$
37. $7 = 2 - x$
38. $3 - 2x = x + 12$
39. $6 + 2a = 3$
40. $a - 3 = 3a - 7$
41. $2y - 1 = 4 - 3y$
42. $7 - 2x = 2x - 7$
43. $7 - 3x = 5 - 2x$
44. $8 - 2y = 5 - 5y$

45. $x - 16 = 16 - 2x$ **46.** $x + 2 = 3{\cdot}1$ **47.** $-x - 4 = -3$

48. $-3 - x = -5$ **49.** $-\frac{x}{2} + 1 = -\frac{1}{4}$ **50.** $-\frac{3}{5} + \frac{x}{10} = -\frac{1}{5} - \frac{x}{5}$

Example 4

Solve this equation:

$x - 2(x - 1) = 1 - 4(x + 1)$

$$x - 2(x - 1) = 1 - 4(x + 1)$$
$$x - 2x + 2 = 1 - 4x - 4$$
$$x - 2x + 4x = 1 - 4 - 2$$
$$3x = -5$$
$$x = -\frac{5}{3}$$

Exercise 3.14

Solve the following equations:

1. $x + 3(x + 1) = 2x$

2. $1 + 3(x - 1) = 4$

3. $2x - 2(x + 1) = 5x$

4. $2(3x - 1) = 3(x - 1)$

5. $4(x - 1) = 2(3 - x)$

6. $4(x - 1) - 2 = 3x$

7. $4(1 - 2x) = 3(2 - x)$

8. $3 - 2(2x + 1) = x + 17$

9. $4x = x - (x - 2)$

10. $7x = 3x - (x + 20)$

11. $5x - 3(x - 1) = 39$

12. $3x + 2(x - 5) = 15$

13. $7 - (x + 1) = 9 - (2x - 1)$

14. $10x - (2x + 3) = 21$

15. $3(2x + 1) + 2(x - 1) = 23$

16. $5(1 - 2x) - 3(4 + 4x) = 0$

17. $7x - (2 - x) = 0$

18. $3(x + 1) = 4 - (x - 3)$

19. $3y + 7 + 3(y - 1) = 2(2y + 6)$

20. $4(y - 1) + 3(y + 2) = 5(y - 4)$

21. $4x - 2(x + 1) = 5(x + 3) + 5$

22. $7 - 2(x - 1) = 3(2x - 1) + 2$

23. $10(2x + 3) - 8(3x - 5) + 5(2x - 8) = 0$

24. $2(x + 4) + 3(x - 10) = 8$

25. $7(2x - 4) + 3(5 - 3x) = 2$

26. $10(x + 4) - 9(x - 3) - 1 = 8(x + 3)$

27. $5(2x - 1) - 2(x - 2) = 7 + 4x$

28. $6(3x - 4) - 10(x - 3) = 10(2x - 3)$

29. $3(x - 3) - 7(2x - 8) - (x - 1) = 0$

30. $5 + 2(x + 5) = 10 - (4 - 5x)$

31. $6x + 30(x - 12) = 2\left(x - 1\frac{1}{2}\right)$

32. $3\left(2x - \frac{2}{3}\right) - 7(x - 1) = 0$

33. $5(x - 1) + 17(x - 2) = 2x + 1$

34. $6(2x - 1) + 9(x + 1) = 8(x - 1\frac{1}{4})$

35. $7(x + 4) - 5(x + 3) + (4 - x) = 0$

36. $0 = 9(3x + 7) - 5(x + 2) - (2x - 5)$

37. $10(2{\cdot}3 - x) - 0{\cdot}1(5x - 30) = 0$

38. $8\left(2\frac{1}{2}x - \frac{3}{4}\right) - \frac{1}{4}(1 - x) = \frac{1}{2}$

39. $(6 - x) - (x - 5) - (4 - x) = -\frac{x}{2}$

40. $10\left(1 - \frac{x}{10}\right) - (10 - x) - \frac{1}{100}(10 - x) = 0{\cdot}05$

Example 5

Solve this equation:

$$(x + 3)^2 = (x + 2)^2 + 3^2$$

$$(x + 3)^2 = (x + 2)^2 + 3^2$$
$$(x + 3)(x + 3) = (x + 2)(x + 2) + 9$$
$$x^2 + 6x + 9 = x^2 + 4x + 4 + 9$$
$$6x + 9 = 4x + 13$$
$$2x = 4$$
$$x = 2$$

Exercise 3.15

Solve the following equations:

1. $x^2 + 4 = (x + 1)(x + 3)$
2. $x^2 + 3x = (x + 3)(x + 1)$
3. $(x + 3)(x - 1) = x^2 + 5$
4. $(x + 1)(x + 4) = (x - 7)(x + 6)$
5. $(x - 2)(x + 3) = (x - 7)(x + 7)$
6. $(x - 5)(x + 4) = (x + 7)(x - 6)$
7. $2x^2 + 3x = (2x - 1)(x + 1)$
8. $(2x - 1)(x - 3) = (2x - 3)(x - 1)$
9. $x^2 + (x + 1)^2 = (2x - 1)(x + 4)$
10. $x(2x + 6) = 2(x^2 - 5)$
11. $(x + 1)(x - 3) + (x + 1)^2 = 2x(x - 4)$
12. $(2x + 1)(x - 4) + (x - 2)^2 = 3x(x + 2)$
13. $(x + 2)^2 - (x - 3)^2 = 3x - 11$
14. $x(x - 1) = 2(x - 1)(x + 5) - (x - 4)^2$
15. $(2x + 1)^2 - 4(x - 3)^2 = 5x + 10$
16. $2(x + 1)^2 - (x - 2)^2 = x(x - 3)$
17. The area of the rectangle shown exceeds the area of the square by 2 cm². Find x.

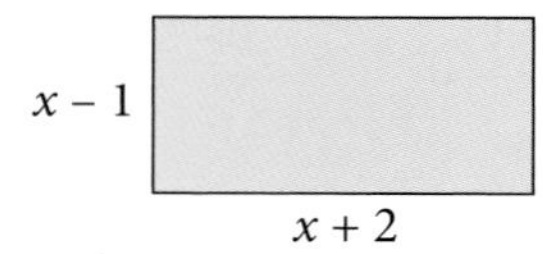

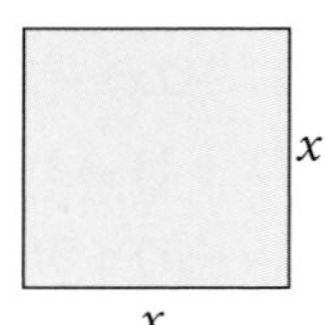

18. The area of the square exceeds the area of the rectangle by 13 m². Find y.

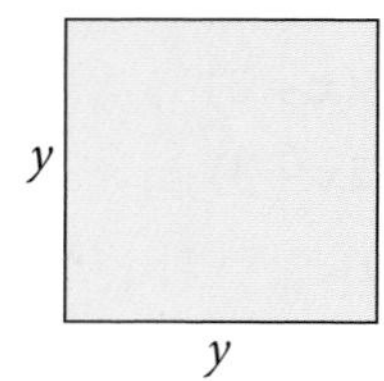

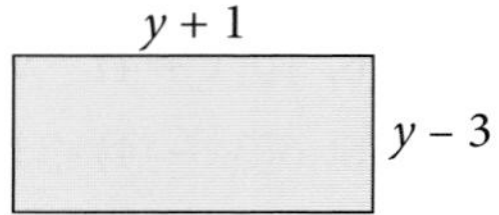

Remember

For a rectangle:
area = length × width

19. The area of the square is half the area of the rectangle. Find x.

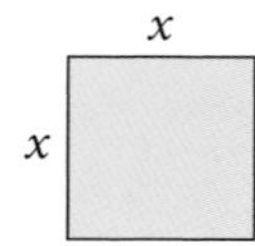

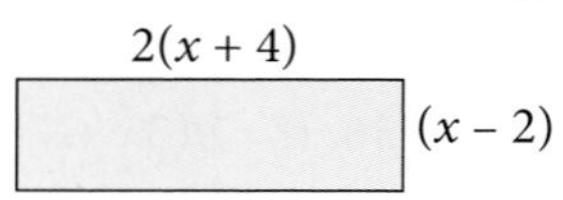

Equations involving fractions

When solving equations involving fractions, multiply both sides of the equation by a suitable number to eliminate the fractions.

$\frac{5}{x} = 2$

$5 = 2x$ (multiply both sides by x)

$\frac{5}{2} = x$

$\frac{x+4}{4} = \frac{2x-1}{3}$ . . . [A]

$12\frac{(x+3)}{4} = 12\frac{(2x-1)}{3}$

(multiply both sides by 12)

$3(x + 3) = 4(2x - 1)$ . . . [B]

$\therefore \quad 3x + 9 = 8x - 4$

$13 = 5x$

$\frac{13}{5} = x$

$x = 2\frac{3}{5}$

Note: It is possible to go straight from line [A] to line [B] by 'cross-multiplying'.

$\frac{5}{(x-1)} + 2 = 12$

$\frac{5}{(x-1)} = 10$

$5 = 10(x - 1)$

$5 = 10x - 10$

$15 = 10x$

$\frac{15}{10} = x$

$x = 1\frac{1}{2}$

Exercise 3.16

Solve the following equations:

1. $\frac{7}{x} = 21$

2. $30 = \frac{6}{x}$

3. $\frac{5}{x} = 3$

4. $\frac{9}{x} = -3$

5. $11 = \frac{5}{x}$

6. $-2 = \frac{4}{x}$

7. $\frac{x}{4} = \frac{3}{2}$

8. $\frac{x}{3} = 1\frac{1}{4}$

9. $\frac{x+1}{3} = \frac{x-1}{4}$

10. $\frac{x+3}{2} = \frac{x-4}{5}$

11. $\frac{2x-1}{3} = \frac{x}{2}$

12. $\frac{3x+1}{5} = \frac{2x}{3}$

13. $\frac{8-x}{2} = \frac{2x+2}{5}$

14. $\frac{x+2}{7} = \frac{3x+6}{5}$

15. $\frac{1-x}{2} = \frac{3-x}{3}$

16. $\frac{2}{x-1} = 1$

17. $\frac{x}{3} + \frac{x}{4} = 1$

18. $\frac{x}{3} + \frac{x}{2} = 4$

19. $\frac{x}{2} - \frac{x}{5} = 3$

20. $\frac{x}{3} = 2 + \frac{x}{4}$

21. $\frac{5}{x-1} = \frac{10}{x}$

Exercise 3.17

Solve the following equations:

1. $\frac{12}{2x-3} = 4$

2. $2 = \frac{18}{x+4}$

3. $\frac{5}{x+5} = \frac{15}{x+7}$

4. $\frac{9}{x} = \frac{5}{x-3}$

5. $\frac{4}{x-1} = \frac{10}{3x-1}$

6. $\frac{-7}{x-1} = \frac{14}{5x+2}$

7. $\frac{4}{x+1} = \frac{7}{3x-2}$

8. $\frac{x-1}{2} + \frac{x-1}{3} = \frac{1}{6}$

9. $\frac{1}{3}(x+2) = \frac{1}{5}(3x+2)$

10. $\frac{1}{2}(x-1) - \frac{1}{6}(x+1) = 0$

11. $\frac{1}{4}(x+5) - \frac{2x}{3} = 0$

12. $\frac{4}{x} + 2 = 3$

13. $\frac{6}{x} - 3 = 7$

14. $\frac{9}{x} - 7 = 1$

15. $-2 = 1 + \frac{3}{x}$

16. $4 - \frac{4}{x} = 0$

17. $5 - \frac{6}{x} = -1$

18. $7 - \frac{3}{2x} = 1$

19. $4 + \frac{5}{3x} = -1$

20. $\frac{9}{2x} - 5 = 0$

21. $\frac{x-1}{5} - \frac{x-1}{3} = 0$

22. $\frac{x-1}{4} - \frac{2x-3}{5} = \frac{1}{20}$

23. $\frac{4}{1-x} = \frac{3}{1+x}$

24. $\frac{x-1}{4} - \frac{x}{3} = \frac{1}{12}$

25. $\frac{2x+1}{8} - \frac{x-1}{3} = \frac{5}{24}$

3.6 Problems solved by linear equations

- Let the unknown quantity be x (or any other letter) and state the units (where appropriate).
- Express the given statement in the form of an equation.
- Solve the equation for x and give the answer in *words*. (Do not finish by writing '$x = 3$'.)
- Check your solution using the problem (not your equation).

Example 6

The sum of three consecutive whole numbers is 78. Find the numbers.

a) Let the smallest number be x; then the other numbers are $(x + 1)$ and $(x + 2)$

b) Form an equation:
$x + (x + 1) + (x + 2) = 78$

c) Solve: $3x = 75$
$x = 25$

In words:
The three numbers are 25, 26 and 27.

d) Check: $25 + 26 + 27 = 78$

Example 7

The length of a rectangle is three times the width. If the perimeter is 36 cm, find the width.

a) Let the width of the rectangle be x cm.
Then the length of the rectangle is $3x$ cm.

x

$3x$

b) Form an equation:
$x + 3x + x + 3x = 36$

c) Solve: $8x = 36$

$$x = \frac{36}{8}$$

$$x = 4{\cdot}5$$

In words:
The width of the rectangle is 4·5 cm.

d) Check: If width = 4·5 cm
length = 13·5 cm
perimeter = 36 cm

Exercise 3.18

Solve each problem by forming an equation. The first questions are easy but should still be solved using an equation, in order to practise the method.

1. The sum of three consecutive numbers is 276. Find the numbers.

2. The sum of four consecutive numbers is 90. Find the numbers.

3. The sum of three consecutive odd numbers is 177. Find the numbers.

4. Find three consecutive even numbers which add up to 1524.

5. When a number is doubled and then added to 13, the result is 38. Find the number.

6. When a number is doubled and then added to 24, the result is 49. Find the number.

7. When 7 is subtracted from three times a certain number, the result is 28. What is the number?

8. The sum of two numbers is 50. The second number is five times the first. Find the numbers.

9. Two numbers are in the ratio 1 : 11 and their sum is 15. Find the numbers.

10. The length of a rectangle is twice the width. If the perimeter is 20 cm, find the width.

11. The width of a rectangle is one third of the length. If the perimeter is 96 cm, find the width.

12. If AB is a straight line, find x. (The angles on a straight line add to 180°.)

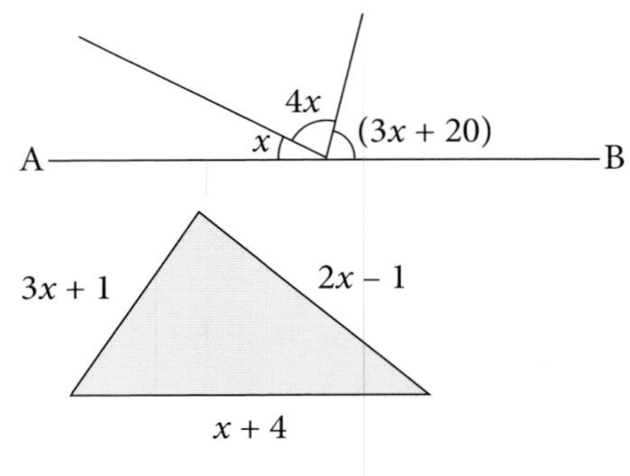

13. If the perimeter of the triangle is 22 cm, find the length of the shortest side.

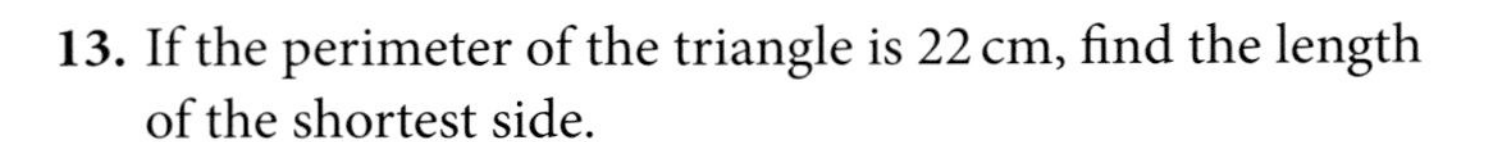

14. If the perimeter of the rectangle is 34 cm, find x.

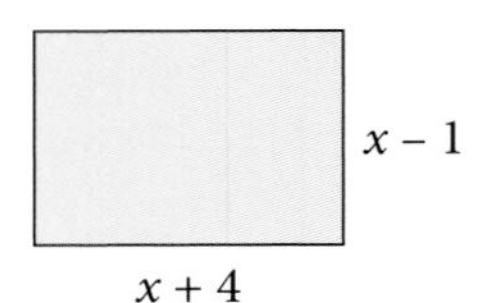

15. The difference between two numbers is 9. Find the numbers if their sum is 46.

Exercise 3.19

1. The three angles in a triangle are in the ratio 1 : 3 : 5. Find them.
2. The three angles in a triangle are in the ratio 3 : 4 : 5. Find them.
3. The product of two consecutive odd numbers is 10 more than the square of the smaller number. Find the smaller number.
4. The product of two consecutive even numbers is 12 more than the square of the smaller number. Find the numbers.
5. The sum of three numbers is 66. The second number is twice the first and six less than the third. Find the numbers.
6. The sum of three numbers is 28. The second number is three times the first and the third is 7 less than the second. What are the numbers?
7. Find the area of the rectangle if the perimeter is 52 cm. [The perimeter is the distance around the edge of the rectangle.]

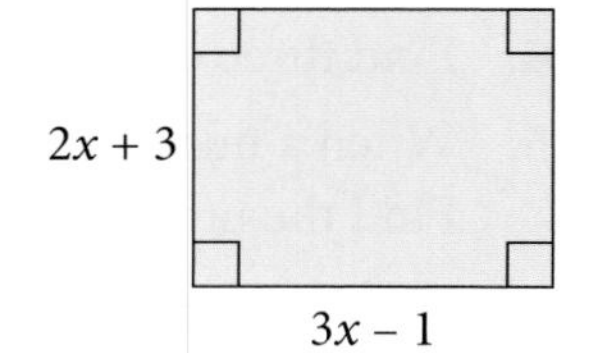

8. David's mass is 5 kg less than John's, who in turn is 8 kg lighter than Paul. If their total mass is 197 kg, how heavy is each person?
9. Nilopal is 2 years older than Devjan who is 7 years older than Sucha. If their combined age is 61 years, find the age of each person.
10. Kimiya has four times as many marbles as Ramneet. If Kimiya gave 18 to Ramneet they would have the same number. How many marbles has each?
11. Mukat has five times as many books as Usha. If Mukat gave 16 books to Usha, they would each have the same number. How many books did each girl have?

12. The result of trebling a number is the same as adding 12 to it. What is the number?

13. The area of the rectangle is 20 square units. Find the length and width of the rectangle.

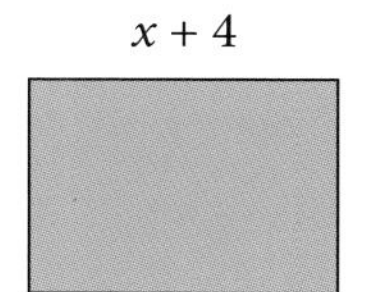

14. The result of multiplying a number by 3 and subtracting 5 is the same as doubling the number and adding 9. What is the number?

15. Two girls have \$76 between them. If the first gave the second \$7 they would each have the same amount of money. How much did each girl have?

16. A tennis racquet costs \$12 more than a hockey stick. If the price of the two is \$31, find the cost of the tennis racket.

17. Find the value of x so that the areas of the pink rectangles are equal.

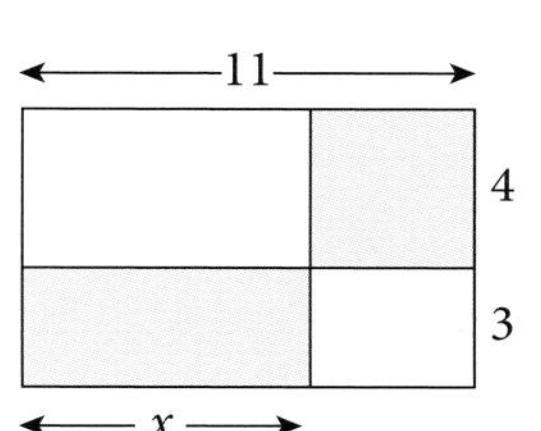

Example 8

A man goes out at 16:42 and arrives at a post box, 6 km away, at 17:30. He walked part of the way at 5 km/h and then, realising the time, he ran the rest of the way at 10 km/h. How far did he have to run?

- Let the distance he ran be x km. Then the distance he walked $= (6 - x)$ km
- Time taken to walk $(6 - x)$ km at 5 km/h $= \frac{(6-x)}{5}$ hours

  Time taken to run x km at 10 km/h $= \frac{x}{10}$ hours

  Total time taken = 48 minutes $= \frac{4}{5}$ hour

  $$\therefore \quad \frac{(6-x)}{5} + \frac{x}{10} = \frac{4}{5}$$

- Multiply by 10:

  $2(6 - x) + x = 8$

  $12 - 2x + x = 8$

  $4 = x$

  He ran a distance of 4 km

- Check:

  Time to run 4 km $= \frac{4}{10} = \frac{2}{5}$ hour

  Time to walk 2 km $= \frac{2}{5}$ hour

  Total time taken $= \left(\frac{2}{5} + \frac{2}{5}\right)$ hour $= \frac{4}{5}$ hour

Exercise 3.20

1. Every year a man is paid \$500 more than the previous year. If he receives \$17 800 over four years, what was he paid in the first year?
2. Samir buys x cans of soda at 30 cents each and $(x + 4)$ cans of soda at 35 cents each. The total cost was \$3·35. Find x.
3. The length of a straight line ABC is 5 m. If AB : BC = 2 : 5, find the length of AB.
4. The opposite angles of a cyclic quadrilateral are $(3x + 10)°$ and $(2x + 20)°$. Find the angles.

Opposite angles of a cyclic quadrilateral add to 180°

5. The interior angles of a hexagon are in the ratio 1 : 2 : 3 : 4 : 5 : 9. Find the angles. This is an example of a concave hexagon. Try to sketch the hexagon.

Interior angles of a hexagon add to 720°

6. A man is 32 years older than his son. Ten years ago he was three times as old as his son was then. Find the present age of each.
7. Mahmoud runs to a marker and back in 15 minutes. His speed on the way to the marker is 5 m/s and his speed on the way back is 4 m/s. Find the distance to the marker.
8. A car completes a journey in 10 minutes. For the first half of the distance the speed was 60 km/h and for the second half the speed was 40 km/h. How far is the journey?
9. A lemming runs from a point A to a cliff at 4 m/s, jumps over the edge at B and falls to C at an average speed of 25 m/s. If the total distance from A to C is 500 m and the time taken for the journey is 41 seconds, find the height BC of the cliff.
10. A bus is travelling with 48 passengers. When it arrives at a stop, x passengers get off and 3 get on. At the next stop half the passengers get off and 7 get on. There are now 22 passengers. Find x.
11. A bus is travelling with 52 passengers. When it arrives at a stop, y passengers get off and 4 get on. At the next stop one-third of the passengers get off and 3 get on. There are now 25 passengers. Find y.
12. Mr Lee left his fortune to his 3 sons, 4 daughters and his wife. Each son received twice as much as each daughter and his wife received \$6000, which was a quarter of the money. How much did each son receive?
13. In a regular polygon with n sides each interior angle is $180 - \frac{360}{n}$ degrees. How many sides does a polygon have if each interior angle is 156°?
14. Two angles of an isosceles triangle are $a°$ and $(a + 10)°$. Find the two possible values of a.
15. A sparrow flies to see a friend at a speed of 4 km/h. His friend is out, so the sparrow immediately returns home at a speed of 5 km/h. The complete journey took 54 minutes. How far away does his friend live?

16. Consider the equation $an^2 = 182$ where a is any number between 2 and 5 and n is a positive integer. What are the possible values of n?

17. Consider the equation $\frac{k}{x} = 12$ where k is any number between 20 and 65, and x is a positive integer. What are the possible values of x?

3.7 Simultaneous equations

To find the value of two unknowns in a problem, two *different* equations must be given that relate the unknowns to each other. These two equations are called **simultaneous** equations.

Substitution method

This method is used when one equation contains a unit quantity of one of the unknowns, as in equation [2] of the example below.

Example 9

Solve these simultaneous equations.

$3x - 2y = 0$ ... [1]

$2x + y = 7$ ... [2]

a) Label the equations so that the working is made clear.

b) In *this case*, write y in terms of x from equation [2].

c) Substitute this expression for y in equation [1] and solve to find x.

d) Find y from equation [2] using this value of x.

$2x + y = 7$ ... [2]

$y = 7 - 2x$

Substituting in [1]:

$3x - 2(7 - 2x) = 0$

$3x - 14 + 4x = 0$

$7x = 14$

$x = 2$

Substituting in [2]:

$2 \times 2 + y = 7$

$y = 3$

The solutions are $x = 2$, $y = 3$

These values of x and y are the only pair which simultaneously satisfy *both* equations.

Exercise 3.21

Use the substitution method to solve the following:

1. $2x + y = 5$
$x + 3y = 5$

2. $x + 2y = 8$
$2x + 3y = 14$

3. $3x + y = 10$
$x - y = 2$

4. $2x + y = -3$
$x - y = -3$

5. $4x + y = 14$
$x + 5y = 13$

6. $x + 2y = 1$
$2x + 3y = 4$

7. $2x + y = 5$
$3x - 2y = 4$

8. $2x + y = 13$
$5x - 4y = 13$

9. $7x + 2y = 19$
$x - y = 4$

10. $b - a = -5$
$a + b = -1$

11. $a + 4b = 6$
$8b - a = -3$

12. $a + b = 4$
$2a + b = 5$

13. $3m = 2n - 6\frac{1}{2}$
$4m + n = 6$

14. $2w + 3x - 13 = 0$
$x + 5w - 13 = 0$

15. $x + 2(y - 6) = 0$
$3x + 4y = 30$

16. $2x = 4 + z$
$6x - 5z = 18$

17. $3m - n = 5$
$2m + 5n = 7$

18. $5c - d - 11 = 0$
$4d + 3c = -5$

It is useful at this point to revise the operations of addition and subtraction with negative numbers.

Example 10

Simplify:

a) $-7 + -4$ **b)** $-3x + (-4x)$
c) $4y - (-3y)$ **d)** $3a + (-3a)$

a) $-7 + -4 = -7 - 4 = -11$

b) $-3x + (-4x) = -3x - 4x = -7x$

c) $4y - (-3y) = 4y + 3y = 7y$

d) $3a + (-3a) = 3a - 3a = 0$

Exercise 3.22

Evaluate:

1. $7 + (-6)$ **2.** $8 + (-11)$ **3.** $5 - (+7)$ **4.** $6 - (-9)$ **5.** $-8 + (-4)$

6. $-7 - (-4)$ **7.** $10 + (-12)$ **8.** $-7 - (+4)$ **9.** $-10 - (+11)$ **10.** $-3 - (-4)$

11. $4 - (+4)$ **12.** $8 - (-7)$ **13.** $-5 - (+5)$ **14.** $-7 - (-10)$ **15.** $16 - (+10)$

16. $-7 - (+4)$ **17.** $-6 - (-8)$ **18.** $10 - (+5)$ **19.** $-12 + (-7)$ **20.** $7 + (-11)$

Simplify:

21. $3x + (-2x)$ **22.** $4x + (-7x)$ **23.** $6x - (+2x)$ **24.** $10y - (+6y)$ **25.** $6y - (-3y)$

26. $7x + (-4x)$ **27.** $-5x + (-3x)$ **28.** $-3x - (-7x)$ **29.** $5x - (+3x)$ **30.** $-7y - (-10y)$

Elimination method

Use this method when the first method is unsuitable (some prefer to use it for every question).

Example 11

Solve these simultaneous equations.

$x + 2y = 8$ ... [1]

$2x + 3y = 14$ ... [2]

a) Label the equations so that the working is made clear.

b) Choose an unknown in one of the equations and multiply the equations by a factor or factors so that this unknown has the same coefficient in both equations.

c) Eliminate this unknown from the two equations by subtracting them, then solve for the remaining unknown.

d) Substitute in the first equation and solve for the eliminated unknown.

$x + 2y = 8$ ... [1]

[1] × 2 $2x + 4y = 16$ ... [3]

$2x + 3y = 14$ ... [2]

Subtract [2] from [3]:

$y = 2$

Substituting in [1]:

$x + 2 \times 2 = 8$

$x = 8 - 4$

$x = 4$

The solutions are $x = 4$, $y = 2$.

Example 12

Solve these simultaneous equations.

$2x + 3y = 5$ ... [1]

$5x - 2y = -16$ ... [2]

[1] × 5 $10x + 15y = 25$ ... [3]

[2] × 2 $10x - 4y = -32$ ... [4]

[3] −4 $15y - (-4y) = 25 - (-32)$

$19y = 57$

$y = 3$

Substitute in [1]:

$2x + 3 - 3 = 5$

$2x = 5 - 9 = -4$

$x = -2$

The solutions are $x = -2$, $y = 3$.

Exercise 3.23

Use the elimination method to solve the following:

1. $2x + 5y = 24$
$4x + 3y = 20$

2. $5x + 2y = 13$
$2x + 6y = 26$

3. $3x + y = 11$
$9x + 2y = 28$

4. $x + 2y = 17$
$8x + 3y = 45$

5. $3x + 2y = 19$
$x + 8y = 21$

6. $2a + 3b = 9$
$4a + b = 13$

7. $2x + 3y = 11$
$3x + 4y = 15$

8. $3x + 8y = 27$
$4x + 3y = 13$

9. $2x + 7y = 17$
$5x + 3y = -1$

10. $5x + 3y = 23$
$2x + 4y = 12$

11. $7x + 5y = 32$
$3x + 4y = 23$

12. $3x + 2y = 4$
$4x + 5y = 10$

13. $3x + 2y = 11$
$2x - y = -3$

14. $3x + 2y = 7$
$2x - 3y = -4$

15. $x + 2y = -4$
$3x - y = 9$

16. $5x - 7y = 27$
$3x - 4y = 16$

17. $3x - 2y = 7$
$4x + y = 13$

18. $x - y = -1$
$2x - y = 0$

19. $y - x = -1$
$3x - y = 5$

20. $x - 3y = -5$
$2y + 3x + 4 = 0$

21. $x + 3y - 7 = 0$
$2y - x - 3 = 0$

22. $3a - b = 9$
$2a + 2b = 14$

23. $3x - y = 9$
$4x - y = -14$

24. $x + 2y = 4$
$3x + y = 9\frac{1}{2}$

25. $2x - y = 5$
$\frac{x}{4} + \frac{y}{3} = 2$

26. $3x - y = 17$
$\frac{x}{5} + \frac{y}{2} = 0$

27. $3x - 2y = 5$
$\frac{2x}{3} + \frac{y}{2} = -\frac{7}{9}$

28. $2x = 11 - y$
$\frac{x}{5} - \frac{y}{4} = 1$

29. $4x - 0{\cdot}5y = 12{\cdot}5$
$3x + 0{\cdot}8y = 8{\cdot}2$

30. $0{\cdot}4x + 3y = 2{\cdot}6$
$x - 2y = 4{\cdot}6$

3.8 Problems solved by simultaneous equations

Example 13

A motorist buys 24 litres of petrol and 5 litres of oil for \$10·70, while another motorist buys 18 litres of petrol and 10 litres of oil for \$12·40.
Find the cost of 1 litre of petrol and 1 litre of oil at this garage.

Let cost of 1 litre of petrol be x cents.
Let cost of 1 litre of oil be y cents.
We have, $24x + 5y = 1070$ ... [1]
$18x + 10y = 1240$ ... [2]

a) Multiply [1] by 2:
$48x + 10y = 2140$ ... [3]

b) Subtract [2] from [3]:
$30x = 900$
$x = 30$

c) Substitute $x = 30$ into equation [2]:
$18(30) + 10y = 1240$
$10y = 1240 - 540$
$10y = 700$
$y = 70$

1 litre of petrol costs 30 cents and
1 litre of oil costs 70 cents.

Exercise 3.24

Solve each problem by forming a pair of simultaneous equations.

1. Find two numbers with a sum of 15 and a difference of 4.
2. Twice one number added to three times another gives 21. Find the numbers, if the difference between them is 3.
3. The average of two numbers is 7, and three times the difference between them is 18. Find the numbers.
4. The line, with equation $y + ax = c$, passes through the points (1, 5) and (3, 1). Find a and c.

For the point (1, 5) put $x = 1$ and $y = 5$ into $y + ax = c$ and solve.

5. The line $y = mx + c$ passes through (2, 5) and (4, 13). Find m and c.
6. The curve $y = ax^2 + bx$ passes through (2, 0) and (4, 8). Find a and b.

7. A gardener buys fifty carrot seeds and twenty lettuce seeds for \$1·10 and her mother buys thirty carrot seeds and forty lettuce seeds for \$1·50. Find the cost of one carrot seed and one lettuce seed.

8. A shop owner can buy either two televisions and three DVD players for \$1750 or four televisions and one DVD player for \$1250. Find the cost of one of each.

9. Half the difference between two numbers is 2. The sum of the greater number and twice the smaller number is 13. Find the numbers.

10. Angle A is 12° greater than angle C. Find the angles of the triangle.

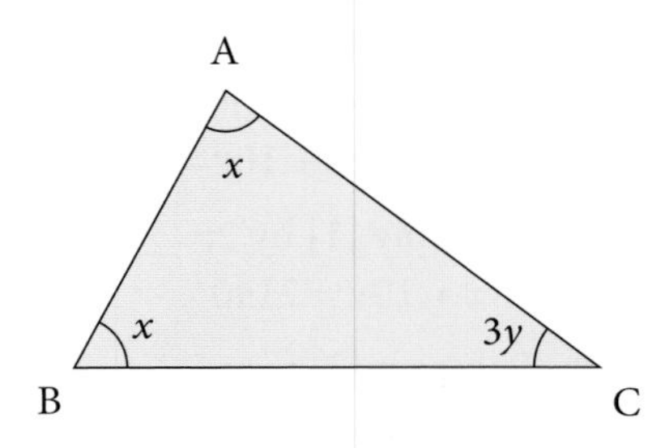

11. A bird can lay either white or brown eggs. Three white eggs and two brown eggs have a mass of 13 grams, while five white eggs and four brown eggs have a mass of 24 grams. Find the mass of a brown egg and of a white egg.

12. A tortoise makes a journey in two parts; it can either walk at 4 cm/s or crawl at 3 cm/s. If the tortoise walks the first part and crawls the second, it takes 110 seconds. If it crawls the first part and walks the second, it takes 100 seconds. Find the lengths of the two parts of the journey.

13. A girl on a snow mobile completes a journey of 500 m in 22 seconds, travelling part of the way at 10 m/s and the remainder at 50 m/s. How far does she travel at each speed?

14. A bag contains forty coins, all of them either 2 cent or 5 cent coins. If the value of the money in the bag is \$1·55, find the number of each type of coin.

15. A slot machine takes only 10 cent and 50 cent coins and contains a total of twenty-one coins altogether. If the value of the coins is \$4·90, find the number of coins of each value.

16. Thirty tickets were sold for a concert, some at €6 and the rest at €10. If the total raised was €220, how many people had the cheaper tickets?

Exercise 3.25

1. The weekly wage bill for five teachers and six doctors is \$6700, while the bill for eight teachers and three doctors is \$6100. Find the weekly wage for a teacher and the wage for a doctor.
2. A fish can swim at 14 m/s in the direction of the current and at 6 m/s against it. Find the speed of the current and the speed of the fish in still water.
3. If the numerator and denominator of a fraction are both decreased by 1 the fraction becomes $\frac{2}{3}$. If the numerator and denominator are both increased by 1 the fraction becomes $\frac{3}{4}$. Find the original fraction.
4. The denominator of a fraction is 2 more than the numerator. If both denominator and numerator are increased by 1 the fraction becomes $\frac{2}{3}$. Find the original fraction.
5. In three years' time a pet snake will be as old as his owner was four years ago. Their present ages total 13 years. Find the age of each now.

6. Find two numbers where three times the smaller number exceeds the larger by 5, and the sum of the numbers is 11.
7. A straight line passes through the points (2, 4) and (−1, −5). Find its equation.
8. A spider can walk at a certain speed and run at another speed. If she walks for 10 seconds and runs for 9 seconds she travels 85 m. If she walks for 30 seconds and runs for 2 seconds she travels 130 m. Find her walking and running speeds.
9. A wallet containing \$40 has three times as many \$1 notes as \$5 notes. Find the number of each type of note.
10. At the present time a man is four times as old as his son. Six years ago he was 10 times as old. Find their present ages.
11. In the square shown the sum of the numbers in each row and column is given. Find the values of A, B and C.

A	B	A	17
B	A	A	17
C	B	B	15
13	19	17	

12. A submarine can travel at 25 knots with the current and at 16 knots against it. Find the speed of the current and the speed of the submarine in still water.

The knot is a unit of speed for ships and aircraft.

13. The curve $y = ax^2 + bx + c$ passes through the points (1, 8), (0, 5) and (3, 20). Find the values of a, b and c and hence the equation of the curve.

14. The curve $y = ax^2 + bx + c$ passes through the points (1, 4), (−2, 19) and (0, 5). Find the equation of the curve.

15. The curve $y = ax^2 + bx + c$ passes through (1, 8), (−1, 2) and (2, 14). Find the equation of the curve.

16. The curve $y = ax^2 + bx + c$ passes through (2, 5), (3, 12) and (−1, −4). Find the equation of the curve.

3.9 Factorising

Earlier in this section we expanded expressions such as $x(3x - 1)$ to give $3x^2 - x$. The reverse of this process is called **factorising**.

Example 14

Factorise: a) $4x + 4y$ b) $x^2 + 7x$
c) $3y^2 - 12y$ d) $6a^2b - 10ab^2$

a) 4 is common to $4x$ and $4y$
$\therefore \quad 4x + 4y = 4(x + y)$

b) x is common to x^2 and $7x$:
$\therefore \quad x^2 + 7x = x(x + 7)$
The factors are x and $(x + 7)$.

c) $3y$ is common
$\therefore \quad 3y^2 - 12y = 3y(y - 4)$

d) $2ab$ is common
$\therefore \quad 6a^2b - 10\,ab^2 = 2ab(3a - 5b)$

Exercise 3.26

Factorise the following expressions completely:

1. $5a + 5b$	**2.** $7x + 7y$	**3.** $7x - x^2$	**4.** $y^2 + 8y$
5. $2y^2 + 3y$	**6.** $6y^2 - 4y$	**7.** $3x^2 - 21x$	**8.** $16a - 2a^2$
9. $6c^2 - 21c$	**10.** $15x - 9x^2$	**11.** $56y - 21y^2$	**12.** $ax + bx + 2cx$
13. $x^2 + xy + 3xz$	**14.** $x^2y + y^3 + z^2y$	**15.** $3a^2b + 2ab^2$	**16.** $x^2y + xy^2$
17. $6a^2 + 4ab + 2ac$	**18.** $ma + 2bm + m^2$	**19.** $2kx + 6ky + 4kz$	**20.** $ax^2 + ay + 2ab$
21. $x^2k + xk^2$	**22.** $ab^3 + 2ab^2$	**23.** $abc - 3b^2c$	**24.** $2a^2e - 5ae^2$
25. $a^3b + ab^3$	**26.** $x^3y + x^2y^2$	**27.** $6xy^2 - 4x^2y$	**28.** $3ab^3 - 3a^3b$
29. $2a^3b + 5a^2b^2$	**30.** $ax^2y - 2ax^2z$	**31.** $2abx + 2ab^2 + 2a^2b$	**32.** $ayx + yx^3 - 2y^2x^2$

Example 15

Factorise $ah + ak + bh + bk$

a) Divide into pairs, $ah + ak + bh + bk$:

a is common to the first pair.
b is common to the second pair.
$a(h + k) + b(h + k)$

b) $(h + k)$ is common to both terms.
Thus we have $(h + k)(a + b)$

Example 16

Factorise $6mx - 3nx + 2my - ny$

$6mx - 3nx + 2my - ny$

$= 3x(2m - n) + y(2m - n)$

$= (2m - n)(3x + y)$

Exercise 3.27

Factorise the following expressions:

1. $ax + ay + bx + by$

2. $ay + az + by + bz$

3. $xb + xc + yb + yc$

4. $xh + xk + yh + yk$

5. $xm + xn + my + ny$

6. $ah - ak + bh - bk$

7. $ax - ay + bx - by$

8. $am - bm + an - bn$

9. $hs + ht + ks + kt$

10. $xs - xt + ys - yt$

11. $ax - ay - bx + by$

12. $xs - xt - ys + yt$

13. $as - ay - xs + xy$

14. $hx - hy - bx + by$

15. $am - bm - an + bn$

16. $xk - xm - kz + mz$

17. $2ax + 6ay + bx + 3by$

18. $2ax + 2ay + bx + by$

19. $2mh - 2mk + nh - nk$

20. $2mh + 3mk - 2nh - 3nk$

21. $6ax + 2bx + 3ay + by$

22. $2ax - 2ay - bx + by$

23. $x^2a + x^2b + ya + yb$

24. $ms + 2mt^2 - ns - 2nt^2$

Quadratic expressions

Example 17

Factorise $x^2 + 6x + 8$

a) Find two numbers which multiply to give 8 and add up to 6. In this case the numbers are 4 and 2.

b) Put these numbers into brackets:
$x^2 + 6x + 8 = (x + 4)(x + 2)$

Example 18

Factorise **a)** $x^2 + 2x - 15$

b) $x^2 - 6x + 8$

a) Two numbers which multiply to give -15 and add up to $+2$ are -3 and 5.

$\therefore \quad x^2 + 2x - 15 = (x - 3)(x + 5)$

b) Two numbers which multiply to give $+8$ and add up to -6 are -2 and -4.

$\therefore \quad x^2 - 6x + 8 = (x - 2)(x - 4)$

Exercise 3.28

Factorise the following:

1. $x^2 + 7x + 10$

2. $x^2 + 7x + 12$

3. $x^2 + 8x + 15$

4. $x^2 + 10x + 21$

5. $x^2 + 8x + 12$

6. $y^2 + 12y + 35$

7. $y^2 + 11y + 24$

8. $y^2 + 10y + 25$

9. $y^2 + 15y + 36$

10. $a^2 - 3a - 10$

11. $a^2 - a - 12$

12. $z^2 + z - 6$

13. $x^2 - 2x - 35$

14. $x^2 - 5x - 24$

15. $x^2 - 6x + 8$

16. $y^2 - 5y + 6$

17. $x^2 - 8x + 15$

18. $a^2 - a - 6$

19. $a^2 + 14a + 45$

20. $b^2 - 4b - 21$

21. $x^2 - 8x + 16$

22. $y^2 + 2y + 1$

23. $y^2 - 3y - 28$

24. $x^2 - x - 20$

25. $x^2 - 8x - 240$

26. $x^2 - 26x + 165$

27. $y^2 + 3y - 108$

28. $x^2 - 49$

29. $x^2 - 9$

30. $x^2 - 16$

Example 19

Factorise $3x^2 + 13x + 4$

a) Find two numbers which multiply to give 12 and add up to 13. In this case the numbers are 1 and 12.

b) Split the '$13x$' term:

$3x^2 + x + 12x + 4$

c) Factorise in pairs:

$x(3x + 1) + 4(3x + 1)$

d) $(3x + 1)$ is common:

$(3x + 1)(x + 4)$

The '12' comes from 3×4 and the '13' is the coefficient of x.

Example 20

Factorise $6x^2 + x - 2$

a) Find the numbers which multiply to give -12 [i.e. $6 \times (-2)$] and add up to 1 (the coeffecient of x).
The numbers are 4 and -3.

b) Split the '$1x$' term:
$6x^2 + 4x - 3x - 2$

c) Factorise in pairs:
$2x(3x + 2) - 1(3x + 2)$

d) $(3x + 2)$ is common: we have $(3x + 2)(2x - 1)$

Exercise 3.29

Factorise the following:

1. $2x^2 + 5x + 3$ **2.** $2x^2 + 7x + 3$ **3.** $3x^2 + 7x + 2$

4. $2x^2 + 11x + 12$ **5.** $3x^2 + 8x + 4$ **6.** $2x^2 + 7x + 5$

7. $3x^2 - 5x - 2$ **8.** $2x^2 - x - 15$ **9.** $2x^2 + x - 21$

10. $3x^2 - 17x - 28$ **11.** $6x^2 + 7x + 2$ **12.** $12x^2 + 23x + 10$

13. $3x^2 - 11x + 6$ **14.** $3y^2 - 11y + 10$ **15.** $4y^2 - 23y + 15$

16. $6y^2 + 7y - 3$ **17.** $6x^2 - 27x + 30$ **18.** $10x^2 + 9x + 2$

19. $6x^2 - 19x + 3$ **20.** $8x^2 - 10x - 3$ **21.** $12x^2 + 4x - 5$

22. $16x^2 + 19x + 3$ **23.** $4a^2 - 4a + 1$ **24.** $12x^2 + 17x - 14$

25. $15x^2 + 44x - 3$ **26.** $48x^2 + 46x + 5$ **27.** $64y^2 + 4y - 3$

28. $120x^2 + 67x - 5$ **29.** $9x^2 - 1$ **30.** $4a^2 - 9$

The difference of two squares

$$x^2 - y^2 = (x - y)(x + y)$$

Remember this result.

Example 21

Factorise **a)** $4a^2 - b^2$ **b)** $3x^2 - 27y^2$

a) $4a^2 - b^2 = (2a)^2 - b^2$
$= (2a - b)(2a + b)$

b) $3x^2 - 27y^2 = 3(x^2 - 9y^2)$
$= 3[x^2 - (3y)^2]$
$= 3(x - 3y)(x + 3y)$

Exercise 3.30

Factorise the following:

1. $y^2 - a^2$
2. $m^2 - n^2$
3. $x^2 - t^2$
4. $y^2 - 1$
5. $x^2 - 9$
6. $a^2 - 25$
7. $x^2 - \frac{1}{4}$
8. $x^2 - \frac{1}{9}$
9. $4x^2 - y^2$
10. $a^2 - 4b^2$
11. $25x^2 - 4y^2$
12. $9x^2 - 16y^2$
13. $x^2 - \frac{y^2}{4}$
14. $9m^2 - \frac{4}{9}n^2$
15. $16t^2 - \frac{4}{25}s^2$
16. $4x^2 - \frac{z^2}{100}$
17. $x^3 - x$
18. $a^3 - ab^2$
19. $4x^3 - x$
20. $8x^3 - 2xy^2$
21. $12x^3 - 3xy^2$
22. $18m^3 - 8mn^2$
23. $5x^2 - 1\frac{1}{4}$
24. $50a^3 - 18ab^2$
25. $12x^2 y - 3yz^2$
26. $36a^3 b - 4ab^3$
27. $50a^5 - 8a^3b^2$
28. $36x^3y - 225xy^3$

Evaluate the following:

29. $81^2 - 80^2$
30. $102^2 - 100^2$
31. $225^2 - 215^2$
32. $1211^2 - 1210^2$
33. $723^2 - 720^2$
34. $3{\cdot}8^2 - 3{\cdot}7^2$
35. $5{\cdot}24^2 - 4{\cdot}76^2$
36. $1234^2 - 1235^2$
37. $3{\cdot}81^2 - 3{\cdot}8^2$
38. $540^2 - 550^2$
39. $7{\cdot}68^2 - 2{\cdot}32^2$
40. $0{\cdot}003^2 - 0{\cdot}002^2$

3.10 Quadratic equations

So far, we have met linear equations which have one solution only. Quadratic equations always have an x^2 term, and often an x term and a number term, and generally have two different solutions.

Solution by factors

Consider the equation $a \times b = 0$, where a and b are numbers. The product $a \times b$ can only be zero if either a or b (or both) is equal to zero. Can you think of other possible pairs of numbers which multiply together to give zero?

Example 22

Solve the equation $x^2 + x - 12 = 0$

Factorising, $(x - 3)(x + 4) = 0$

either $x - 3 = 0$ or $x + 4 = 0$

$x = 3$ $\quad$ $x = -4$

Example 23

Solve the equation $6x^2 + x - 2 = 0$

Factorising, $(2x - 1)(3x + 2) = 0$

either $2x - 1 = 0$ or $3x + 2 = 0$

$2x = 1$ $\quad$ $3x = -2$

$x = \frac{1}{2}$ $\quad$ $x = -\frac{2}{3}$

Exercise 3.31

Solve the following equations:

1. $x^2 + 7x + 12 = 0$
2. $x^2 + 7x + 10 = 0$
3. $x^2 + 2x - 15 = 0$
4. $x^2 + x - 6 = 0$
5. $x^2 - 8x + 12 = 0$
6. $x^2 + 10x + 21 = 0$
7. $x^2 - 5x + 6 = 0$
8. $x^2 - 4x - 5 = 0$
9. $x^2 + 5x - 14 = 0$
10. $2x^2 - 3x - 2 = 0$
11. $3x^2 + 10x - 8 = 0$
12. $2x^2 + 7x - 15 = 0$
13. $6x^2 - 13x + 6 = 0$
14. $4x^2 - 29x + 7 = 0$
15. $10x^2 - x - 3 = 0$
16. $y^2 - 15y + 56 = 0$
17. $12y^2 - 16y + 5 = 0$
18. $y^2 + 2y - 63 = 0$
19. $x^2 + 2x + 1 = 0$
20. $x^2 - 6x + 9 = 0$
21. $x^2 + 10x + 25 = 0$
22. $x^2 - 14x + 49 = 0$
23. $6a^2 - a - 1 = 0$
24. $4a^2 - 3a - 10 = 0$
25. $z^2 - 8z - 65 = 0$
26. $6x^2 + 17x - 3 = 0$
27. $10k^2 + 19k - 2 = 0$
28. $y^2 - 2y + 1 = 0$
29. $36x^2 + x - 2 = 0$
30. $20x^2 - 7x - 3 = 0$

Example 24

Solve the equation $x^2 - 7x = 0$

Factorising, $x(x - 7) = 0$

either $x = 0$ or $x - 7 = 0$

$x = 7$

The solutions are $x = 0$ and $x = 7$.

You can check your factorisation by multiplying out the brackets and checking that the result matches the original equation.

Example 25

Solve the equation $4x^2 - 9 = 0$

a) Factorising, $(2x - 3)(2x + 3) = 0$

either $2x - 3 = 0$ or $2x + 3 = 0$

$2x = 3$ $\quad$ $2x = -3$

$x = \frac{3}{2}$ $\quad$ $x = -\frac{3}{2}$

b) Alternative method:

$4x^2 - 9 = 0$

$4x^2 = 9$

$x^2 = \frac{9}{4}$

$x = +\frac{3}{2}$ or $-\frac{3}{2}$

You must give both the solutions. A common error is to give only the positive square root.

Exercise 3.32

Solve the following equations:

1. $x^2 - 3x = 0$

2. $x^2 + 7x = 0$

3. $2x^2 - 2x = 0$

4. $3x^2 - x = 0$

5. $x^2 - 16 = 0$

6. $x^2 - 49 = 0$

7. $4x^2 - 1 = 0$

8. $9x^2 - 4 = 0$

9. $6y^2 + 9y = 0$

10. $6a^2 - 9a = 0$

11. $10x^2 - 55x = 0$

12. $16x^2 - 1 = 0$

13. $y^2 - \frac{1}{4} = 0$

14. $56x^2 - 35x = 0$

15. $36x^2 - 3x = 0$

16. $x^2 = 6x$

17. $x^2 = 11x$

18. $2x^2 = 3x$

19. $x^2 = x$

20. $4x = x^2$

21. $3x - x^2 = 0$

22. $4x^2 = 1$

23. $9x^2 = 16$

24. $x^2 = 9$

25. $12x = 5x^2$

26. $1 - 9x^2 = 0$

27. $x^2 = \frac{x}{4}$

28. $2x^2 = \frac{x}{3}$

29. $4x^2 = \frac{1}{4}$

30. $\frac{x}{5} - x^2 = 0$

Solution by formula

The solutions of the quadratic equation $ax^2 + bx + c = 0$ are given by the formula:

$$x = \frac{-b \pm \sqrt{b^2 - 4ac}}{2a}$$

Use this formula only after trying (and failing) to factorise.

Example 26

Solve the equation $2x^2 - 3x - 4 = 0$.

In this case $a = 2$, $b = -3$, $c = -4$.

$$x = \frac{-(-3) \pm \sqrt{[(-3)^2 - (4 \times 2 \times -4)]}}{2 \times 2}$$

$$x = \frac{3 \pm \sqrt{[9 - 32]}}{4} = \frac{3 \pm \sqrt{41}}{4} = \frac{3 \pm 6{\cdot}403}{4}$$

Either $x = \frac{3 + 6{\cdot}403}{4}$ or $x = \frac{3 - 6{\cdot}403}{4}$

$= 2{\cdot}35$ (2 d.p.) $= -0{\cdot}85$ (2 d.p.)

Exercise 3.33

Solve the following, giving answers to two decimal places where necessary:

1. $2x^2 + 11x + 5 = 0$
2. $3x^2 + 11x + 6 = 0$
3. $6x^2 + 7x + 2 = 0$
4. $3x^2 - 10x + 3 = 0$
5. $5x^2 - 7x + 2 = 0$
6. $6x^2 - 11x + 3 = 0$
7. $2x^2 + 6x + 3 = 0$
8. $x^2 + 4x + 1 = 0$
9. $5x^2 - 5x + 1 = 0$
10. $x^2 - 7x + 2 = 0$
11. $2x^2 + 5x - 1 = 0$
12. $3x^2 + x - 3 = 0$
13. $3x^2 + 8x - 6 = 0$
14. $3x^2 - 7x - 20 = 0$
15. $2x^2 - 7x - 15 = 0$
16. $x^2 - 3x - 2 = 0$
17. $2x^2 + 6x - 1 = 0$
18. $6x^2 - 11x - 7 = 0$
19. $3x^2 + 25x + 8 = 0$
20. $3y^2 - 2y - 5 = 0$
21. $2y^2 - 5y + 1 = 0$
22. $\frac{1}{2}y^2 + 3y + 1 = 0$
23. $2 - x - 6x^2 = 0$
24. $3 + 4x - 2x^2 = 0$
25. $1 - 5x - 2x^2 = 0$
26. $3x^2 - 1 + 4x = 0$
27. $5x - x^2 + 2 = 0$
28. $24x^2 - 22x - 35 = 0$
29. $36x^2 - 17x - 35 = 0$
30. $20x^2 + 17x - 63 = 0$
31. $x^2 + 2{\cdot}5x - 6 = 0$
32. $0{\cdot}3y^2 + 0{\cdot}4y - 1{\cdot}5 = 0$
33. $10 - x - 3x^2 = 0$
34. $x^2 + 3{\cdot}3x - 0{\cdot}7 = 0$
35. $12 - 5x^2 - 11x = 0$
36. $5x - 2x^2 + 187 = 0$

The solution to a problem can involve an equation which does not at first appear to be quadratic. The terms in the equation may need to be rearranged.

Example 27

Solve $2x(x - 1) = (x + 1)^2 - 5$

$$2x^2 - 2x = x^2 + 2x + 1 - 5$$
$$2x^2 - 2x - x^2 - 2x - 1 + 5 = 0$$
$$x^2 - 4x + 4 = 0$$
$$(x - 2)(x - 2) = 0$$
$$x = 2$$

In this example the quadratic has a repeated solution of $x = 2$

Exercise 3.34

Solve the following, giving answers to two decimal places where necessary:

1. $x^2 = 6 - x$
2. $x(x + 10) = -21$
3. $3x + 2 = 2x^2$
4. $x^2 + 4 = 5x$
5. $6x(x + 1) = 5 - x$
6. $(2x)^2 = x(x - 14) - 5$
7. $(x - 3)^2 = 10$
8. $(x + 1)^2 - 10 = 2x(x - 2)$
9. $(2x - 1)^2 = (x - 1)^2 + 8$
10. $3x(x + 2) - x(x - 2) + 6 = 0$
11. $x = \frac{15}{x} - 22$
12. $x + 5 = \frac{14}{x}$
13. $4x + \frac{7}{x} = 29$
14. $10x = 1 + \frac{3}{x}$

15. $2x^2 = 7x$

16. $16 = \dfrac{1}{x^2}$

17. $2x + 2 = \dfrac{7}{x} - 1$

18. $\dfrac{2}{x} + \dfrac{2}{x+1} = 3$

19. $\dfrac{3}{x-1} + \dfrac{3}{x+1} = 4$

20. $\dfrac{2}{x-2} + \dfrac{4}{x+1} = 3$

21. One of the solutions published by the Italian Renaissance mathematician Cardan in 1545 for the solution of cubic equations is given below.
For an equation in the form $x^3 + px = q$,

$$x = \sqrt[3]{\left[\sqrt{\left(\frac{p}{3}\right)^3 + \left(\frac{q}{2}\right)^2} + \frac{q}{2}\right]} - \sqrt[3]{\left[\sqrt{\left(\frac{p}{3}\right)^3 + \left(\frac{q}{2}\right)^2} - \frac{q}{2}\right]}$$

Use the formula to solve the following equations, giving answers to 4 s.f. where necessary.

a) $x^3 + 7x = -8$

b) $x^3 + 6x = 4$

c) $x^3 + 3x = 2$

d) $x^3 + 9x - 2 = 0$

3.11 Problems solved by quadratic equations

Example 28

The area of rectangle A is $16\,\text{cm}^2$ greater than the area of rectangle B. Find the height of rectangle A.

7 cm

A (x + 3) cm

B (x – 1) cm

(x + 2) cm

Area of rectangle A = $7(x + 3)$
Area of rectangle B = $(x + 2)(x - 1)$
We are given $(x + 3)(x - 1) + 16 = 7(x + 3)$
Solve this equation $x^2 + 2x - x - 2 + 16 = 7x + 21$

$$x^2 + x + 14 = 7x + 21$$
$$x^2 - 6x - 7 = 0$$
$$(x - 7)(x + 1) = 0$$
$$x = 7 \text{ (x cannot be negative)}$$

The height of rectangle A, $x + 3$, is 10 cm.

Example 29

A man bought a certain number of golf balls for \$20. If each ball had cost 20 cents less, he could have bought five more for the same money. How many golf balls did he buy?

Let the number of balls bought be x.

Cost of each ball $= \frac{2000}{x}$ cents

If five more balls had been bought

Cost of each ball now $= \frac{2000}{(x+5)}$ cents

The new price is 20 cents less than the original price.

$$\therefore \quad \frac{2000}{x} - \frac{2000}{(x+5)} = 20 \text{ (multiply by } x)$$

$$x \cdot \frac{2000}{x} - x \cdot \frac{2000}{(x+5)} = 20x \text{ (multiply by } (x+5))$$

$$2000(x+5) - x\frac{2000}{(x+5)}(x+5) = 20x(x = 5)$$

$$2000x + 10\,000 - 2000x = 20x^2 + 100x$$

$$20x^2 + 100x - 10\,000 = 0$$

$$x^2 + 5x - 500 = 0$$

$$(x - 20)(x + 25) = 0$$

$$x = 20 \text{ or } x = -25$$

We discard $x = -25$ as meaningless.

The number of balls bought = 20.

Exercise 3.35

Solve by forming a quadratic equation:

1. Two numbers, which differ by 3, have a product of 88. Find them.
2. The product of two consecutive odd numbers is 143. Find the numbers.

If the first odd number is x, what is the next odd number?

3. The length of a rectangle exceeds the width by 7 cm. If the area is 60 cm^2, find the length of the rectangle.
4. The length of a rectangle exceeds the width by 2 cm. If the diagonal is 10 cm long, find the width of the rectangle.
5. The area of the rectangle exceeds the area of the square by 24m^2. Find x.

$(x - 3)$m

$(x - 6)$m

$(x - 7)$m

$(x - 7)$m

Questions 4, 6 and 7 use Pythagoras' Theorem. For more information on Pythagoras' Theorem see page 193 in Chapter 7.

6. The perimeter of a rectangle is 68 cm. If the diagonal is 26 cm, find the dimensions of the rectangle.

7. Sang Jae walks a certain distance due North and then the same distance plus a further 7 km due East. If the final distance from the starting point is 17 km, find the distances he walks North and East.

8. A farmer makes a profit of x cents on each of the $(x + 5)$ eggs her hen lays. If her total profit was 84 cents, find the number of eggs the hen lays.

9. A rectangle is 7 cm longer than it is wide. The largest possible circle is cut out of the rectangle and the remaining area is $40\,\text{cm}^2$. [Call the width of the rectangle $2x$.] Find the dimensions of the rectangle.

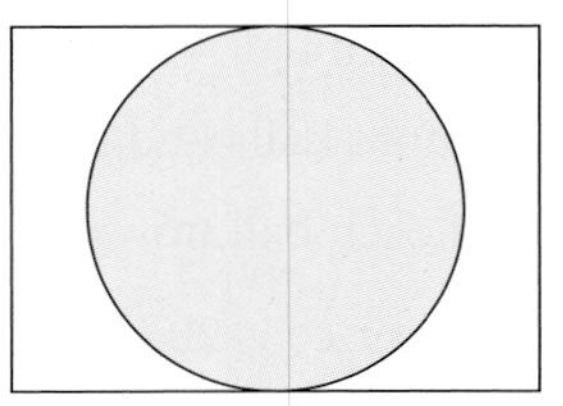

10. Sirak buys x eggs at $(x - 8)$ cents each and $(x - 2)$ bread rolls at $(x - 3)$ cents each. If the total bill is \$1·75, how many eggs does he buy?

11. A number exceeds four times its reciprocal by 3. Find the number.

12. Two positive numbers differ by 3. The sum of their reciprocals is $\frac{7}{10}$. Find the numbers.

13. A cyclist travels 40 km at a speed of x km/h. Find the time taken in terms of x. Find the time taken when his speed is reduced by 2 km/h. If the difference between the times is 1 hour, find the original speed x.

14. An increase of speed of 4 km/h on a journey of 32 km reduces the time taken by 4 hours. Find the original speed.

15. A train normally travels 240 km at a certain speed. One day, due to bad weather, the train's speed is reduced by 20 km/h so that the journey takes two hours longer. Find the normal speed.

16. The speed of a swallow is x km/h in still air. When the wind is blowing at 1 km/h, the swallow takes 5 hours to fly 12 kilometres to her nest and 12 kilometres back again. She goes out directly into the wind and returns with the wind behind her. Find her speed in still air.

17. An aircraft flies a certain distance on a bearing of 135° and then twice the distance on a bearing of 225°. Its distance from the starting point is then 350 km. Find the length of the first part of the journey.

A **bearing** is a clockwise angle measured from North. For more information about bearings see page 317 in Chapter 10.

18. In the diagram, ABCD is a rectangle with AB = 12 cm and BC = 7 cm. AK = BL = CM = DN = x cm. If the area of KLMN is $54\,\text{cm}^2$, find x.

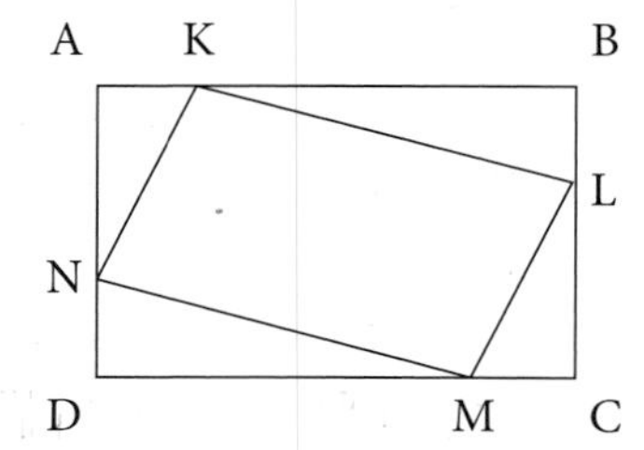

19. In the diagram, AB = 14 cm, BC = 11 cm and AK = BL = CM = DN = x cm. If the area of KLMN is now $97\,\text{cm}^2$, find x.

20. The numerator of a fraction is 1 less than the denominator. When both numerator and denominator are increased by 2, the fraction is increased by $\frac{1}{12}$. Find the original fraction.

21. The perimeters of a square and a rectangle are equal. One side of the rectangle is 11 cm and the area of the square is 4 cm^2 more than the area of the rectangle. Find the side of the square.

Revision exercise 3A

1. Solve the equations:

a) $x + 4 = 3x + 9$ **b)** $9 - 3a = 1$

c) $y^2 + 5y = 0$ **d)** $x^2 - 4 = 0$

e) $3x^2 + 7x - 40 = 0$

2. Given $a = 3$, $b = 4$ and $c = -2$, evaluate:

a) $2a^2 - b$ **b)** $a(b - c)$ **c)** $2b^2 - c^2$

3. Factorise completely:

a) $4x^2 - y^2$ **b)** $2x^2 + 8x + 6$

c) $6m + 4n - 9km - 6kn$ **d)** $2x^2 - 5x - 3$

4. Solve the simultaneous equations:

a) $3x + 2y = 5$, $2x - y = 8$ **b)** $2m - n = 6$, $2m + 3n = -6$

c) $3x - 4y = 19$, $x + 6y = 10$ **d)** $3x - 7y = 11$, $2x - 3y = 4$

5. Given that $x = 4$, $y = 3$, $z = -2$, evaluate:

a) $2x(y + z)$ **b)** $(xy)^2 - z^2$

c) $x^2 + y^2 + z^2$ **d)** $(x + y)(x - z)$

e) $\sqrt{[x(1-4z)]}$ **f)** $\frac{xy}{z}$

6. **a)** Simplify $3(2x - 5) - 2(2x + 3)$.

b) Factorise $2a - 3b - 4xa + 6xb$.

c) Solve the equation $\frac{x-11}{2} - \frac{x-3}{5} = 2$.

7. Write a number or an expression in each of the six cards so that you obtain the expression

$5a + 10b + 3n + 12m$ from

□(□ + □) + □(□ + □)

8. Solve the equations:

a) $5 - 7x = 4 - 6x$ **b)** $\frac{7}{x} = \frac{2}{3}$

c) $2x^2 - 7x = 0$ **d)** $x^2 + 5x + 6 = 0$

e) $\frac{1}{x} + \frac{1}{4} = \frac{1}{3}$

9. Factorise completely:

a) $z^3 - 16z$ **b)** $x^2y^2 + x^2 + y^2 + 1$

c) $2x^2 + 11x + 12$

10. Find the value of $\frac{2x-3y}{5x+2y}$ when $x = 2a$ and $y = -a$.

11. Solve the simultaneous equations:

a) $7c + 3d = 29$, $5c - 4d = 33$ **b)** $2x - 3y = 7$, $2y - 3x = -8$

c) $5x = 3(1 - y)$, $3x + 2y + 1 = 0$ **d)** $5s + 3t = 16$, $11s + 7t = 34$

12. Solve the equations:

a) $4(y + 1) = \frac{3}{1-y}$ **b)** $4(2x - 1) - 3(1 - x) = 0$

c) $\frac{x+3}{x} = 2$ **d)** $x^2 = 5x$

13. Solve the following, giving your answers correct to two decimal places.

a) $2x^2 - 3x - 1 = 0$ **b)** $x^2 - x - 1 = 0$

c) $3x^2 + 2x - 4 = 0$ **d)** $x + 3 = \frac{7}{x}$

14. Find x by forming a suitable equation.

a)

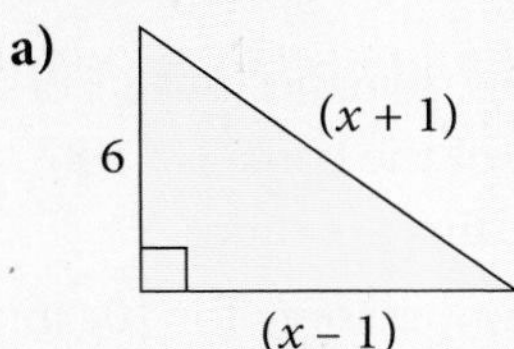

b)

(x + 14)

x

26

15. Given that $m = -2$ and $n = 4$, evaluate:

a) $5m + 3n$ b) $5 + 2m - m^2$

c) $m^2 + 2n^2$ d) $(2m + n)(2m - n)$

e) $(n - m)^2$ f) $n - mn - 2m^2$

16. A car travels for x hours at a speed of $(x + 2)$ km/h. If the distance travelled is 15 km, write down an equation for x and solve it to find the speed of the car.

17. ABCD is a rectangle, where AB = x cm and BC is 1·5 cm less than AB.

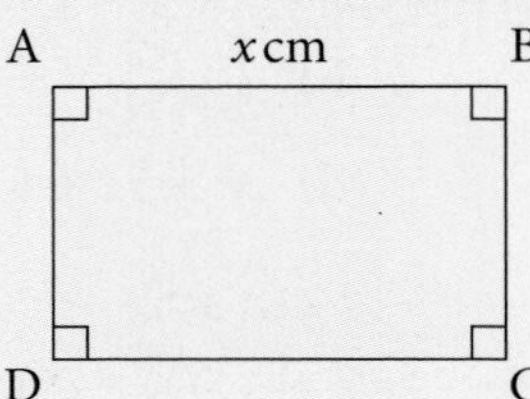

If the area of the rectangle is 52 cm², form an equation in x and solve it to find the dimensions of the rectangle.

18. Solve these equations:

a) $(2x + 1)^2 = (x + 5)^2$ b) $\frac{x+2}{2} - \frac{x-1}{3} = \frac{x}{4}$

c) $x^2 - 7x + 5 = 0$,

giving your answers correct to two decimal places.

19. Solve the equation:

$$\frac{x}{x+1} - \frac{x+1}{3x-1} = \frac{1}{4}$$

20. Given that $a + b = 2$ and that $a^2 + b^2 = 6$, prove that $2ab = -2$. Find also the value of $(a - b)^2$.

21. The sides of a right-angled triangle have lengths $(x - 3)$ cm, $(x + 11)$ cm and $2x$ cm, where $2x$ is the hypotenuse. Find x.

22. The six numbers around each pink hexagon add up to 100.

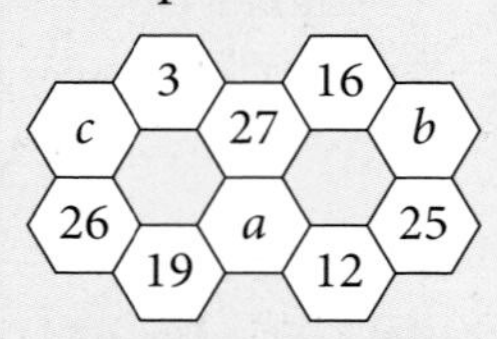

Find a, b and c, given that $b + c = 29$.

23. A jar contains 50 coins, all either 2 cents or 5 cents. The total value of the coins is \$1·87. How many 2 cents coins are there?

24. Stefan bought 45 stamps, some for 10c and some for 18c. If he spent \$6·66 altogether, how many 10c stamps did he buy?

25. Angle P is 20° greater than angle Q. Angle R is half the sum of angles P and Q.

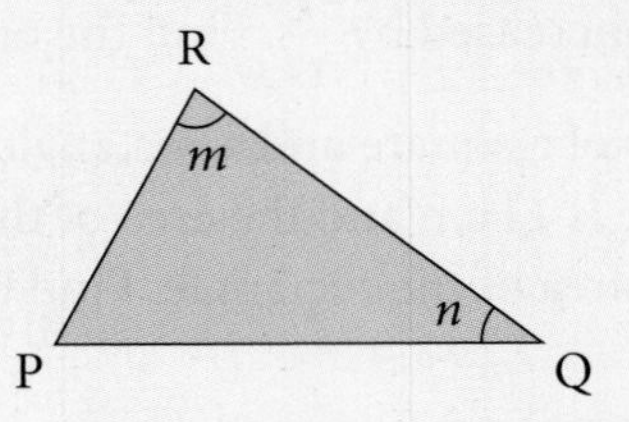

Find the value of m and n.

26. When each edge of a cube is decreased by 1 cm, its volume is decreased by 91 cm³. Find the length of a side of the original cube.

27. One solution of the equation $2x^2 - 7x + k = 0$ is $x = -\frac{1}{2}$. Find the value of k.

Examination exercise 3B

1. Solve the simultaneous equations. [4]

$3x + 2y = 7$

$5x + 3y = 12$

Cambridge IGCSE International Mathematics 0607 Specimen Paper 2 Q4 2010

2. Solve the equation $2x^2 + 11 = x + 21$. [4]

Cambridge IGCSE International Mathematics 0607 5 Specimen Paper 2 Q5 2010

3. Factorise completely. [2]

$2ac - 5bc + 6a - 15b$

Cambridge IGCSE Mathematics 0607 Paper 2 Q2 October/November 2010

4.

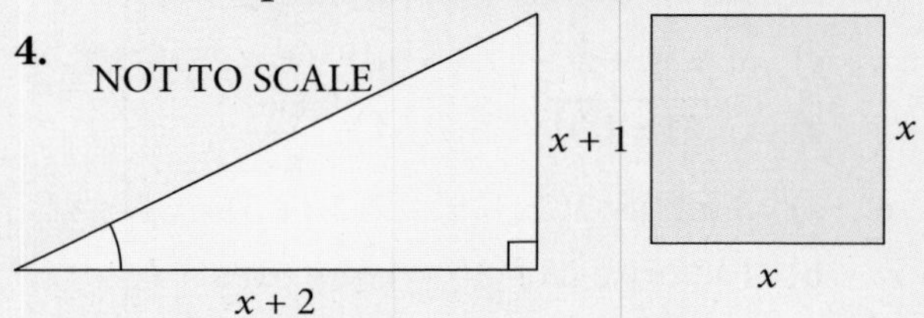

The area of the right-angled triangle is equal to the area of the square.

a) Show that $x^2 - 3x - 2 = 0$. [3]

b) Solve the equation $x^2 - 3x - 2 = 0$. [3]
Give your answer correct to 2 decimal places.

Cambridge IGCSE International Mathematics 0607 Paper 41 Q13 (a) and (b) May/June 2011

4 Mensuration

Archimedes of Samos (287–212 BCE) studied at Alexandria as a young man. One of the first to apply scientific thinking to everyday problems, he was a practical man of common sense. He gave proofs for finding the area, the volume and the centre of gravity of circles, spheres, conics and spirals. By drawing polygons with many sides, he arrived at a value of π between $3\frac{10}{71}$ and $3\frac{10}{70}$. He was killed in the siege of Syracuse at the age of 75.

6.1 Units: mm, cm, m, km

mm^2, cm^2, m^2, ha, km^2

mm^3, cm^3, m^3

ml, cl, l

g, kg, t

convert between units

6.2 Perimeter and area of rectangle, triangle and compound shapes derived from these

6.3 Circumference and area of a circle

Arc length and area of sector

6.4 Surface area and volume of prism and pyramid (in particular, cuboid, cylinder and cone)

Surface area and volume of sphere and hemisphere

(formulae given for curved surface areas of cylinder, cone and sphere; volume of pyramid, cone, cylinder and sphere)

6.5 Areas and volumes of compound shapes

4.1 Area

Rectangle

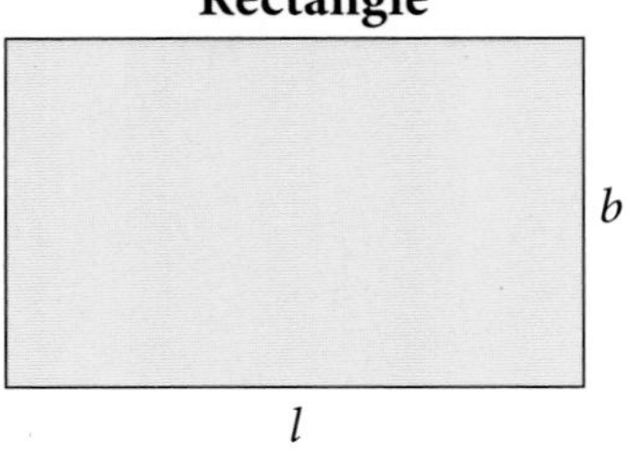

area = $l \times b$

Trapezium

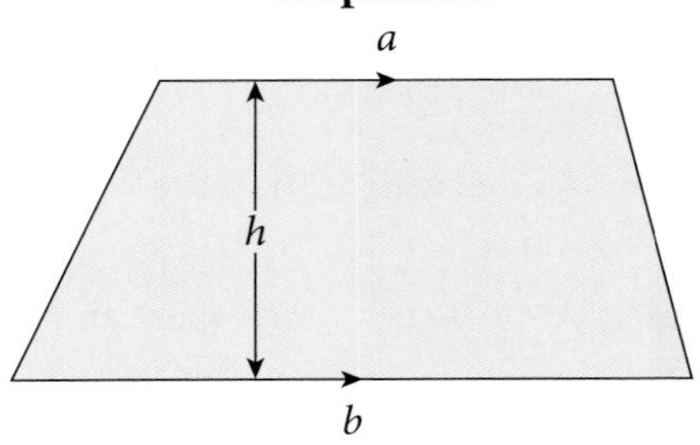

area = $\frac{1}{2}(a + b)\,h$

Kite

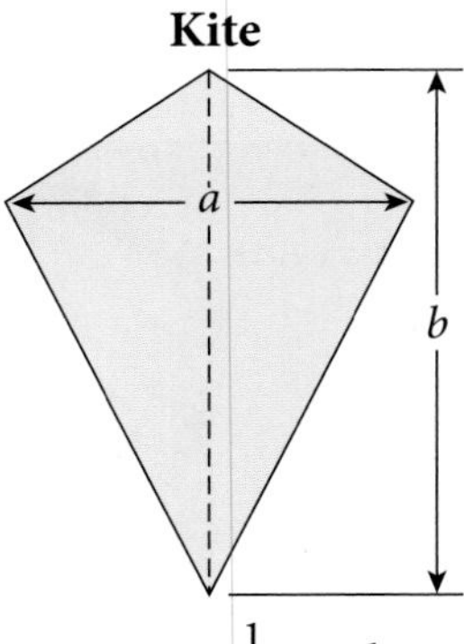

area = $\frac{1}{2}\,a \times b$

$= \frac{1}{2} \times$ (product of diagonals)

Exercise 4.1

For questions **1** to **7**, find the area of each shape.
Decide which information to use – you may not need all of it.

1.

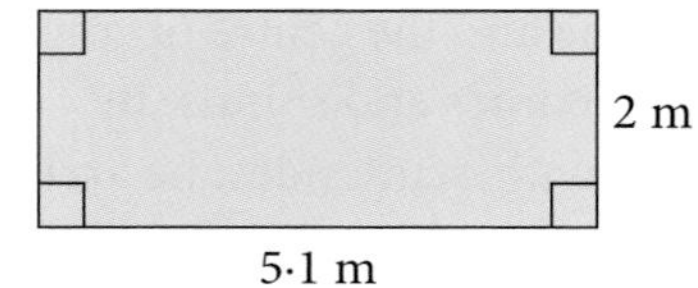

2.

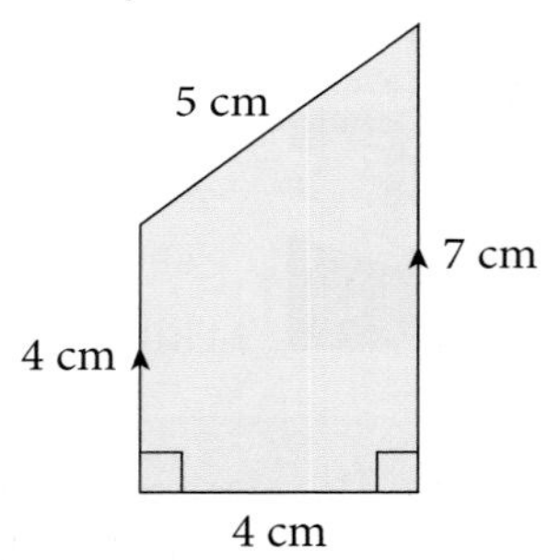

3.

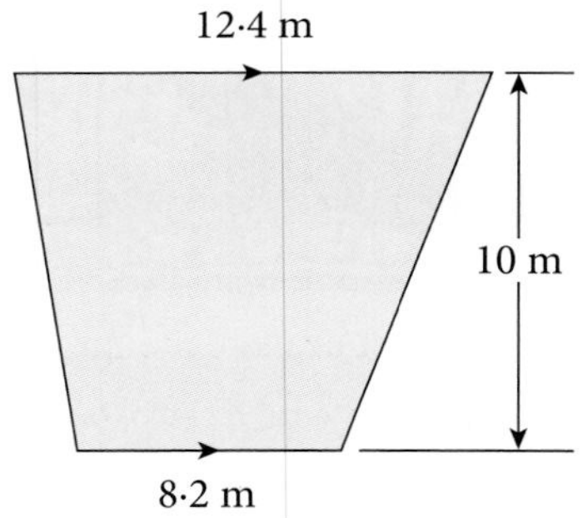

4.

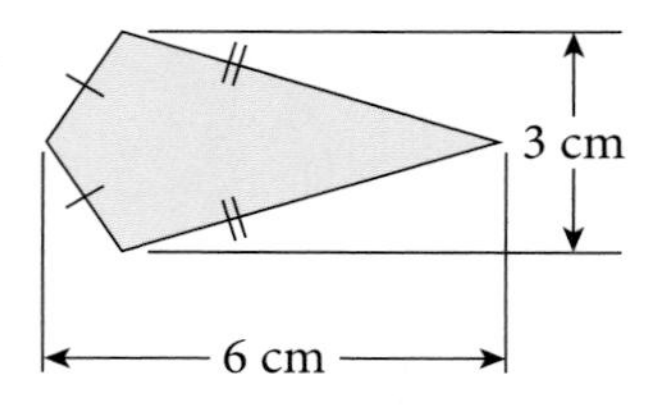

5.

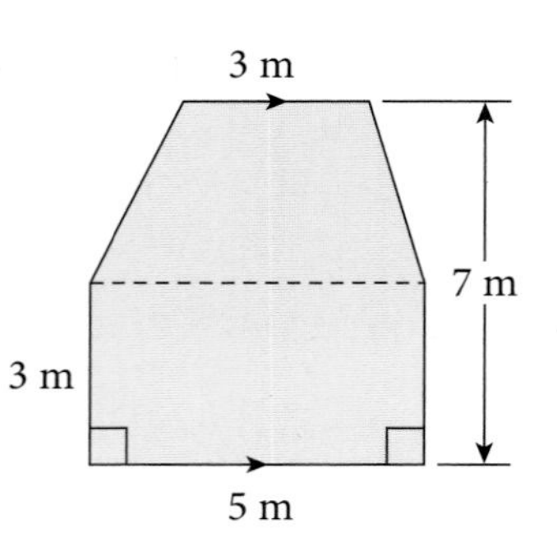

6.

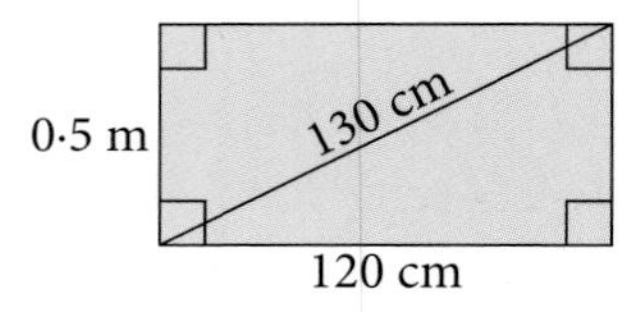

7.

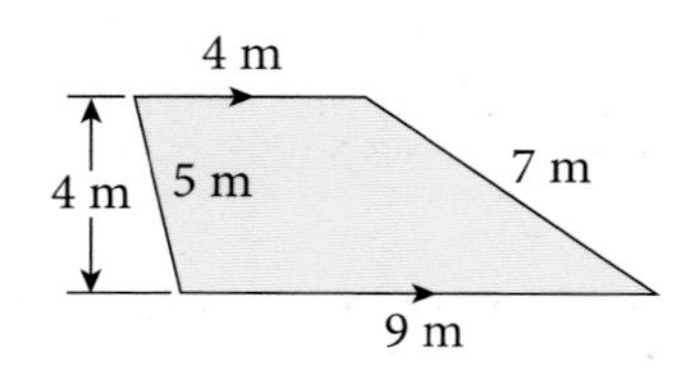

8. Find the area shaded pink.

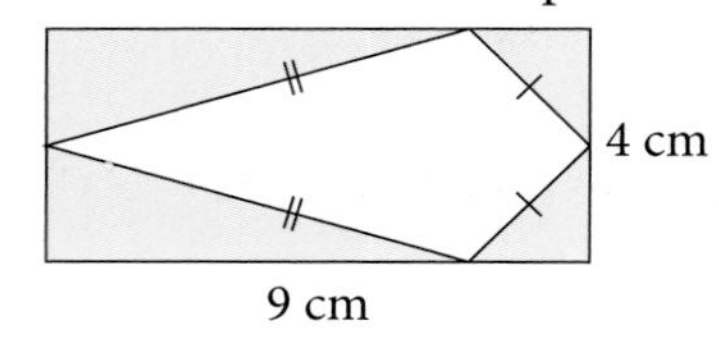

9. Find the area shaded blue.

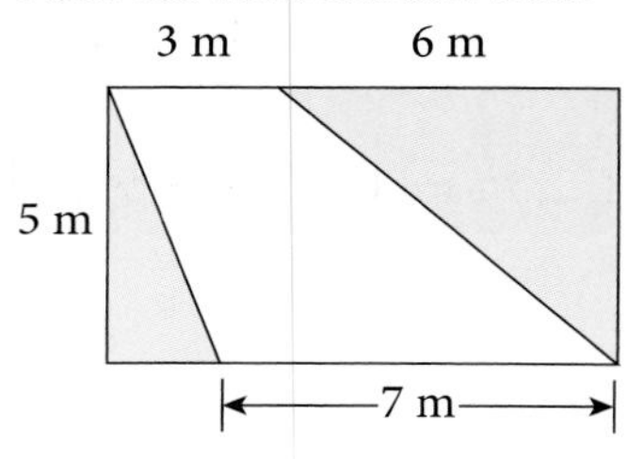

10. A rectangle has an area of $117\,m^2$ and a width of 9 m. Find its length.

11. A trapezium of area $105\,cm^2$ has parallel sides of length 5 cm and 9 cm. How far apart are the parallel sides?

12. A kite of area $252\,m^2$ has one diagonal of length 9 m. Find the length of the other diagonal.

13. A kite of area $40\,m^2$ has one diagonal 2 m longer than the other. Find the lengths of the diagonals.

14. A trapezium of area $140\,cm^2$ has parallel sides 10 cm apart and one of these sides is 16 cm long. Find the length of the other parallel side.

15. A floor 5 m by 20 m is covered by square tiles of side 20 cm. How many tiles are needed?

16. 

The area of a forest is $2{\cdot}5\,km^2$. The land is sold at a price of \$4200 per hectare. How much does it cost?

1 hectare = $10\,000\,m^2$
Hectare can be abbreviated to 'ha'.

17. On squared paper draw the triangle with vertices at (1, 1), (5, 3), (3, 5). Find the area of the triangle.

18. Draw the quadrilateral with vertices at (1, 1), (6, 2), (5, 5), (3, 6). Find the area of the quadrilateral.

19. A square wall is covered with square tiles. There are 85 tiles altogether along the two diagonals. How many tiles are there on the whole wall?

20. On squared paper draw a 7×7 square. Divide it up into nine smaller squares.

21. A rectangular field, 400 m long, has an area of 6 hectares. Calculate the perimeter of the field.

Area of a triangle and a parallelogram

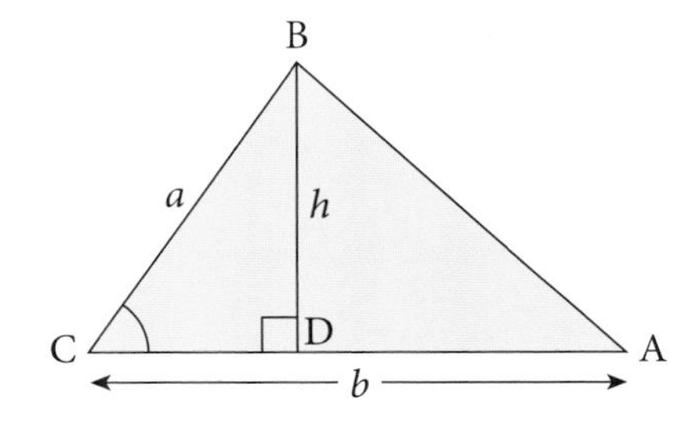

In triangle BCD, $\sin C = \frac{h}{a}$

$\therefore \quad h = a \sin C$

$$\therefore \quad \text{area of triangle ABC} = \frac{1}{2} \times b \times a \sin C$$

This formula is useful when *two sides* and the *included angle* are known.

Area of a triangle $= \frac{1}{2} \times b \times h$

Example 1

Find the area of the triangle shown.

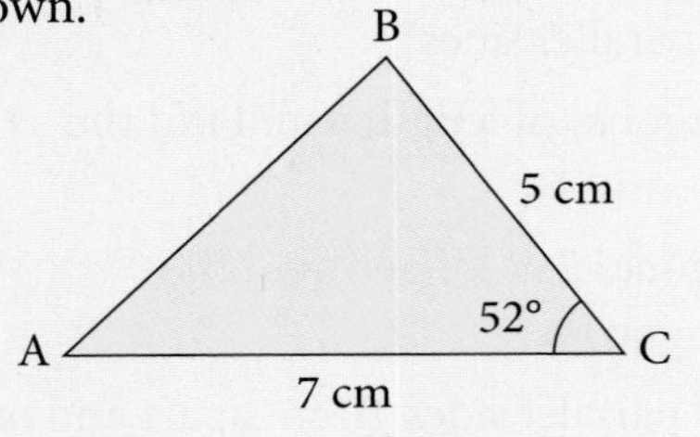

$$\text{Area} = \frac{1}{2}ab\sin C$$

$$= \frac{1}{2} \times 5 \times 7 \times \sin 52° \quad = 13{\cdot}8\,\text{cm}^2 \text{ (1 d.p.)}$$

There are two ways of finding the area of a parallelogram.

Parallelogram

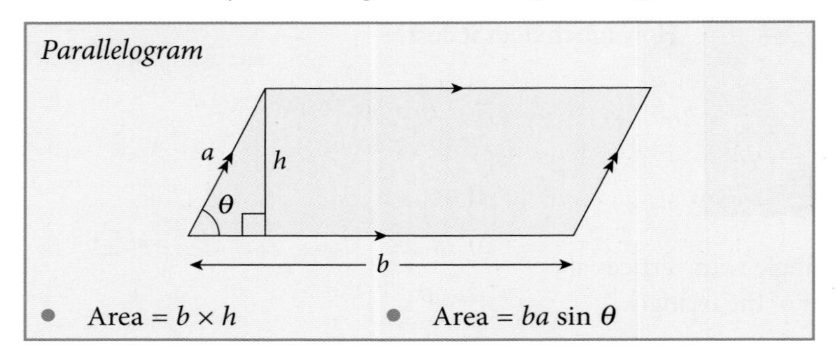

- Area = $b \times h$
- Area = $ba \sin \theta$

You can find out more about $\sin \theta$ on page 309 in Chapter 10.

Exercise 4.2

In questions **1** to **12** find the area of ΔABC where AB = c, AC = b and BC = a. (Sketch the triangle in each case.) You will need some basic trigonometry (see page 309 in Chapter 10).

1. $a = 7\,\text{cm}$, $b = 14\,\text{cm}$, $\hat{C} = 80°$.

2. $b = 11\,\text{cm}$, $a = 9\,\text{cm}$, $\hat{C} = 35°$.

3. $c = 12\,\text{m}$, $b = 12\,\text{m}$, $\hat{A} = 67{\cdot}2°$.

4. $a = 5\,\text{cm}$, $c = 6\,\text{cm}$, $\hat{B} = 11{\cdot}8°$.

5. $b = 4{\cdot}2\,\text{cm}$, $a = 10\,\text{cm}$, $\hat{C} = 120°$.

6. $a = 5\,\text{cm}$, $c = 8\,\text{cm}$, $\hat{B} = 142°$.

7. $b = 3{\cdot}2\,\text{cm}$, $c = 1{\cdot}8\,\text{cm}$, $\hat{B} = 10°$, $\hat{C} = 65°$.

8. $a = 7\,\text{m}$, $b = 14\,\text{m}$, $\hat{A} = 32°$, $\hat{B} = 100°$.

9. $a = b = c = 12\,\text{m}$.

10. $a = c = 8\,\text{m}$, $\hat{B} = 72°$.

11. $b = c = 10\,\text{cm}$, $\hat{B} = 32°$.

12. $a = b = c = 0{\cdot}8\,\text{m}$.

In questions **13** to **20**, find the area of each shape.

13.

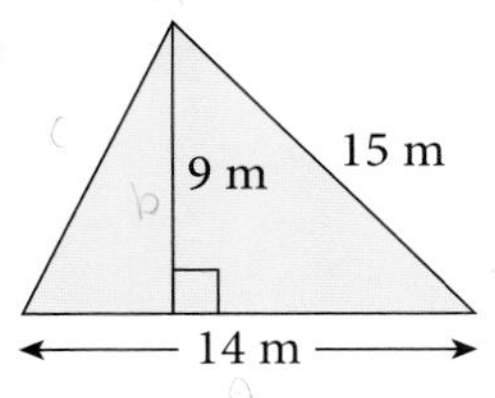

14.

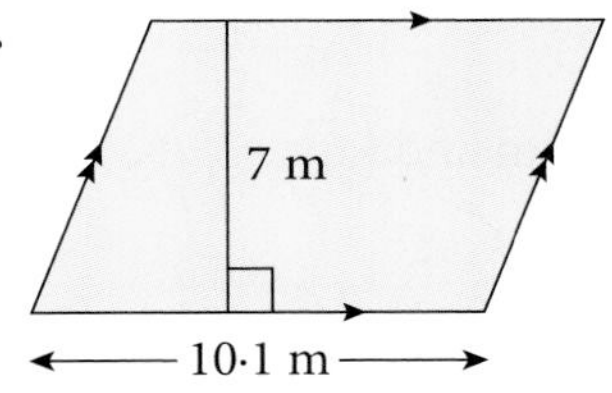

15.

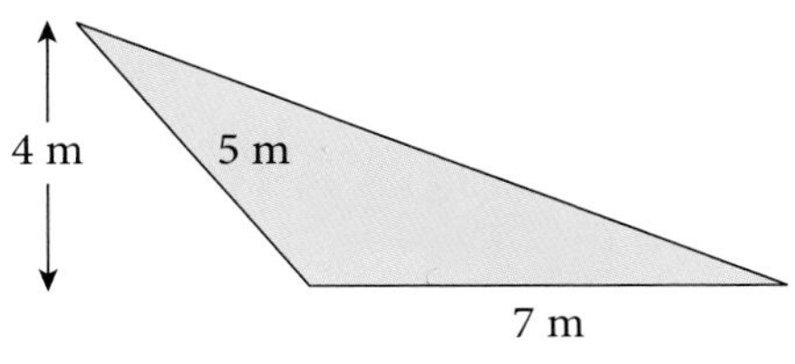

16.

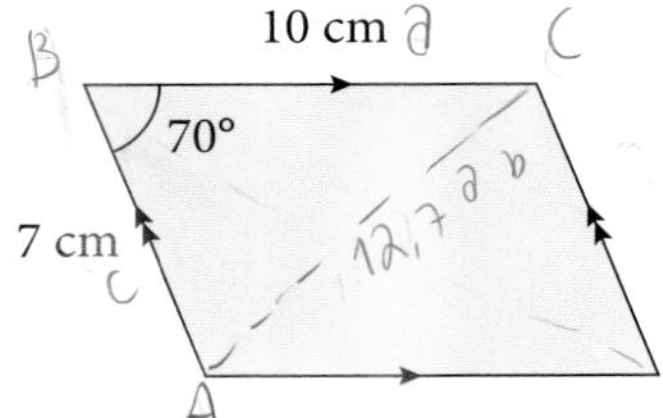

17.

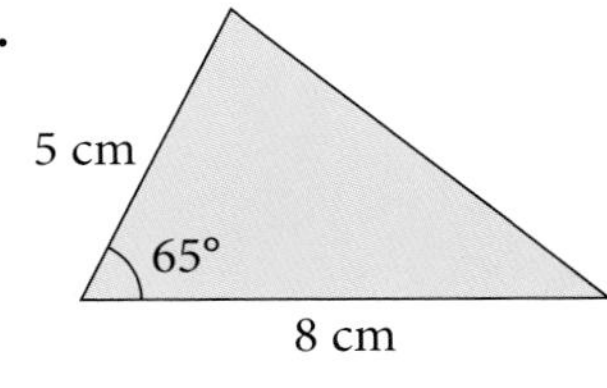

18.

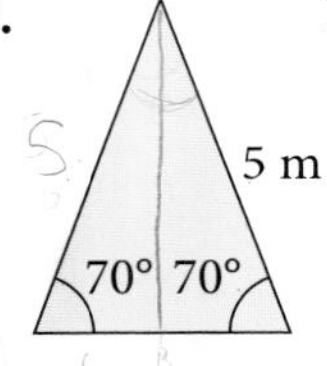

19.

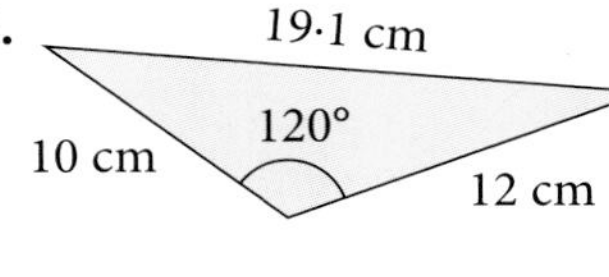

Exercise 4.3

1. Find the area of a parallelogram ABCD with AB = 7 m, AD = 20 m and $B\hat{A}D = 62°$.
2. The area of an equilateral triangle ABC is $50\,\text{cm}^2$. Find AB.
3. The area of a triangle ABC is $64\,\text{cm}^2$. Given AB = 11 cm and BC = 15 cm, Find $A\hat{B}C$.
4. Find the length of a side of an equilateral triangle of area $10{\cdot}2\,\text{m}^2$.
5. A rhombus has an area of $40\,\text{cm}^2$ and adjacent angles of 50° and 130°. Find the length of a side of the rhombus.
6. A regular hexagon is circumscribed by a circle of radius 3 cm with centre O.
   - **a)** What is angle EOD?
   - **b)** Find the area of triangle EOD and hence find the area of the hexagon ABCDEF.

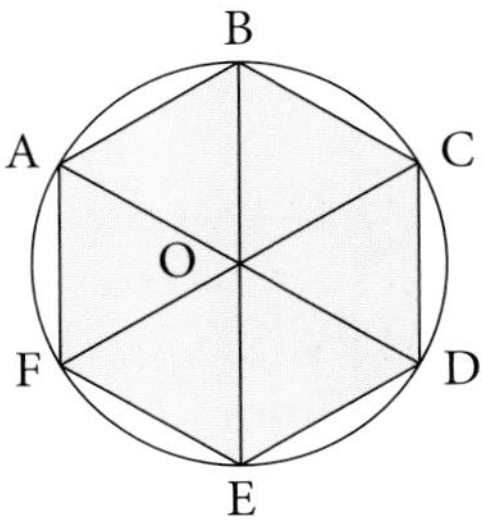

7. A triangle and a square are drawn on dotty paper with dots 1 cm apart. What is the area of the region coloured red?
8. Find the area of a regular pentagon of side 8 cm.
9. The area of a regular pentagon is $600\,\text{cm}^2$. Calculate the length of one side of the pentagon.

Converting units

Length

The standard units of length are mm, cm, m and km.

10 mm = 1 cm, 100 cm = 1 m, 1000 m = 1 km

We can easily convert length from cm to m, km to m, and so on.

Example:
- 580 cm = 5·8 m
- 3·2 km = 3200 m
- 2870 mm = 2·87 m

Areas

Areas are also sometimes given in different units and we need to be able to convert from one to another.

Here is a square with sides of 1 m:

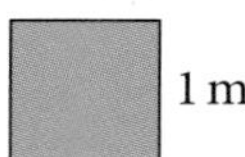

… and the same square with sides of 100 cm:

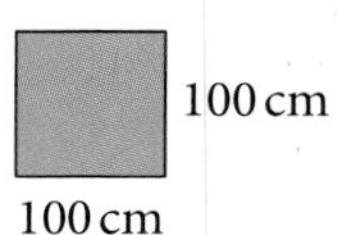

The area of each square is the same, so

$1\ m^2 = 100 \times 100\ cm^2$
$1\ m^2 = 10\,000\ cm^2$

For larger areas, such as fields, we use **hectares**.

A figure of side 100 m has an area of 1 hectare.

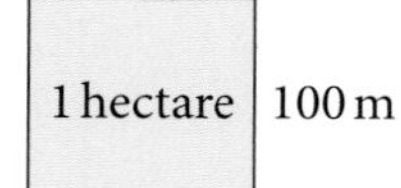

$1 \text{ hectare} = 100 \times 100\ m^2$
$1 \text{ hectare} = 10\,000\ m^2$

Exercise 4.4

1. Copy and complete:

a) 57 cm = □ mm **b)** 3·7 m = □ cm **c)** 21 km = □ m **d)** 99 mm = □ cm
e) 0·84 m = □ cm **f)** 200 mm = □ m **g)** 0·45 km = □ m **h)** 7 km = □ m = □ mm

2. Find the area of this triangle in

a) cm^2 **b)** mm^2

4 cm
6 cm

3. A closed card board box has dimensions 1 m × 2 m × 3 m

Find the total surface area of the box:

a) in m^2 **b)** in cm^2

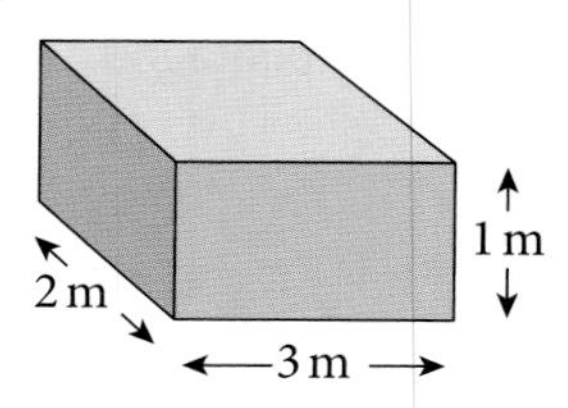

4. Copy and complete:

a) $1\ m^2 =$ □ cm^2 **b)** $80\,000\ cm^2 =$ □ cm^2
c) $0{\cdot}005\ m^2 =$ □ cm **d)** 260 000 mm = □ km **e)** $0{\cdot}02\ m^2 =$ □ cm^2

4.2 The circle

For any circle, the ratio $\left(\dfrac{\text{circumference}}{\text{diameter}}\right)$ is equal to π.

The value of π is usually taken to be 3·14, but this is not an exact value. Through the centuries, mathematicians have been trying to obtain a better value for π.

For example, in the third century CE, the Chinese mathematician Liu Hui obtained the value 3·14159 by considering a regular polygon having 3072 sides! Ludolph van Ceulen (1540–1610) worked even harder to produce a value correct to 35 significant figures. He was so proud of his work that he had this value of π engraved on his tombstone.

Computers are now able to calculate the value of π to many thousands of figures, but its value is still not exact. It was shown in 1761 that π is an **irrational number** which, like $\sqrt{2}$ or $\sqrt{3}$ cannot be expressed exactly as a fraction.

GDC help

To enter π on your calculator, press π▸ + enter, 2nd [π] or SHIFT (π).

The first fifteen significant figures of π can be remembered from the number of letters in each word of the following sentence:

> "How I need a drink, cherryade of course, after the silly lectures involving Italian kangaroos."

There remain a lot of unanswered questions concerning π, and many mathematicians today are still working on them.

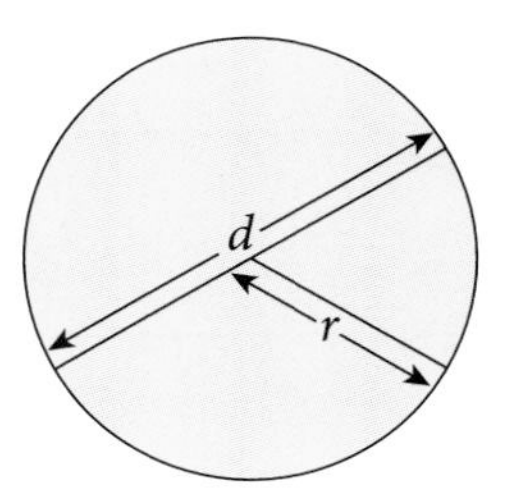

The following formulae should be memorised:

- circumference = πdr or $2\pi r$
- area = πr^2

Example 2

Find the circumference and area of a circle of diameter 8 cm. (Take $\pi = 3{\cdot}142$)

Circumference = πd
$= 3{\cdot}142 \times 8$
$= 25{\cdot}1$ cm (1 d.p.)

Area = πr^2
$= 3{\cdot}142 \times 4^2$
$= 50{\cdot}3\ \text{cm}^2$ (1 d.p.)

Exercise 4.5

For each shape find

a) the perimeter

b) the area.

All lengths are in cm. Use the π button on a calculator or take $\pi = 3{\cdot}142$.

All the arcs are either semi-circles or quarter-circles.

1.

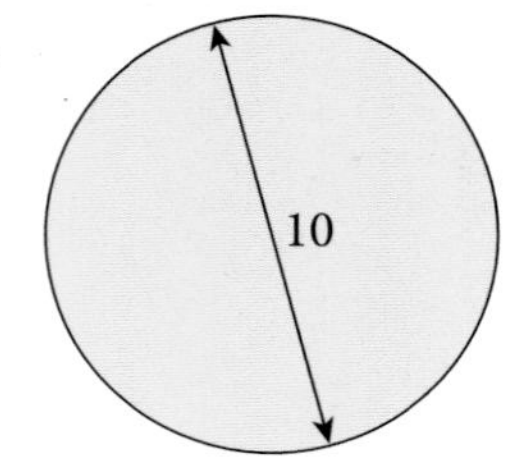

2.

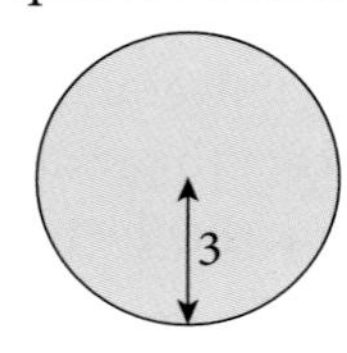

3.

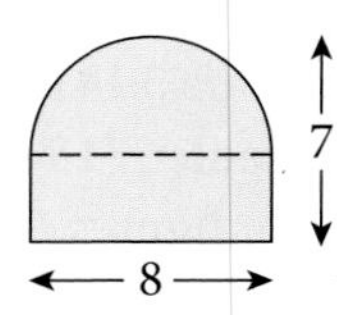

4.

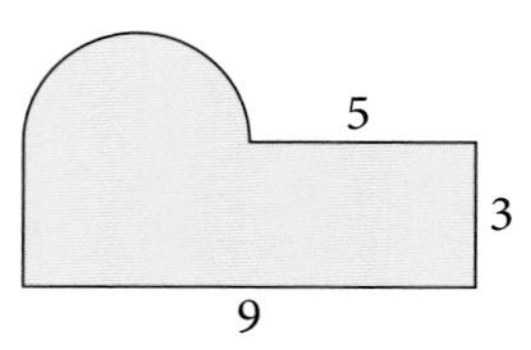

5.

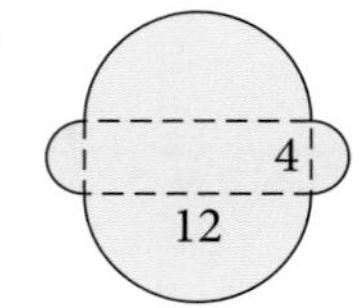

6.

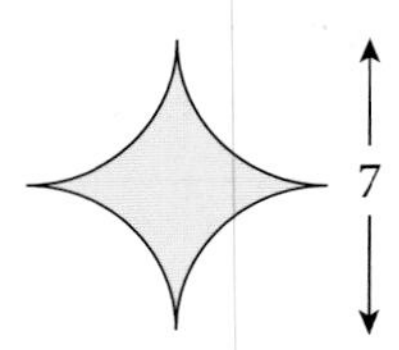

7.

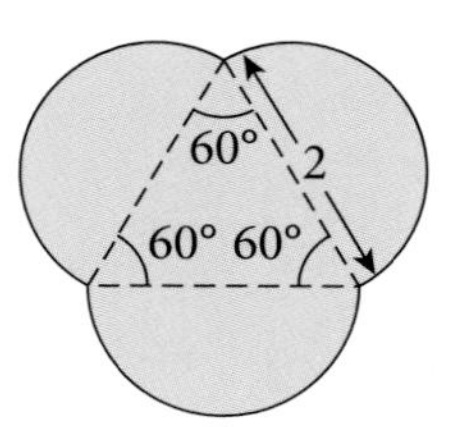

8.

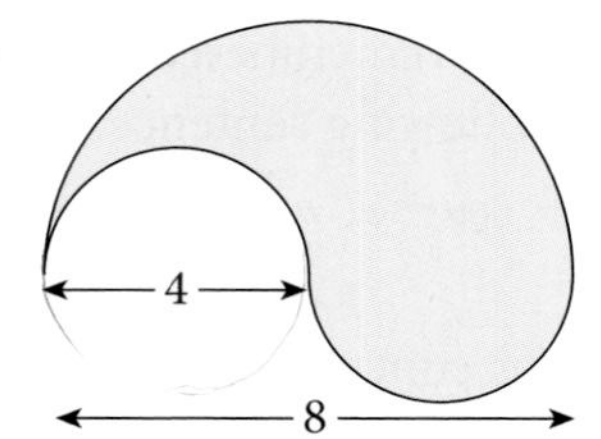

9.

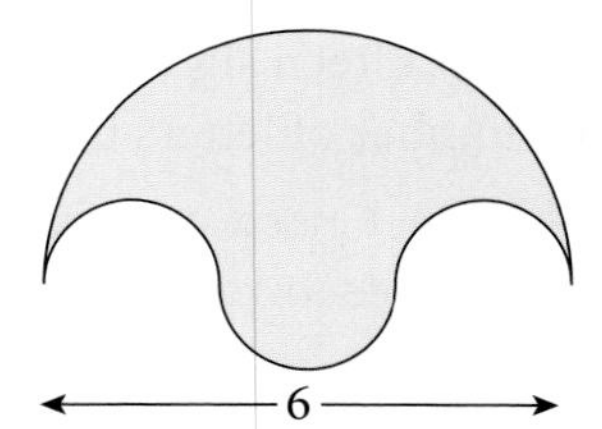

Example 3

A circle has a circumference of 20 m.
Find the radius of the circle.

Let the radius of the circle be r m.

Circumference $= 2\pi r$

$\therefore \quad 2\pi r = 20$

$\therefore \quad r = \dfrac{20}{2\pi} = 3{\cdot}18$

The radius of the circle is 3·18 m (3 s.f.)

Example 4

A circle has an area of 45 cm^2.
Find the radius of the circle.

Let the radius of the circle r cm.

$\pi r^2 = 45$

$r^2 = \dfrac{45}{\pi}$

$r = \sqrt{\left(\dfrac{45}{\pi}\right)} = 3{\cdot}78$ (3 s.f.)

The radius of the circle is 3·78 cm.

▶ Continued on next page

GDC help

Using a GDC

TI-Nspire	TI84-Plus	Casio
1.1 *Unsaved $\sqrt{\frac{45}{\pi}}$ 3.7847 1/99	$\sqrt{\frac{45}{\pi}}$ 3.784698783	Math Deg Norm1 d/c Real $\sqrt{\frac{45}{\pi}}$ 3.784698783 JUMP DELETE ▸MAT MATH

Exercise 4.6

Use the π button on a calculator and give answers to 3 s.f.

1. A circle has an area of $15\,\text{cm}^2$. Find its radius.
2. A circle has a circumference of 190 m. Find its radius.
3. Find the radius of a circle of area $22\,\text{km}^2$.
4. Find the radius of a circle of circumference 58.6 cm.
5. A circle has an area of $16\,\text{mm}^2$. Find its circumference.
6. A circle has a circumference of 2500 km. Find its area.
7. Discs of radius 4 cm are cut from a rectangular plastic sheet of length 84 cm and width 24 cm.
   a) How many complete discs can be cut out?
   Find:
   b) the total area of the discs cut
   c) the area of the sheet wasted.
8. A circular pond of radius 6 m is surrounded by a path of width 1 m.
   a) Find the area of the path.
   b) The path is resurfaced with astroturf which is bought in packs each containing enough to cover an area of $7\,\text{m}^2$. How many packs are required?

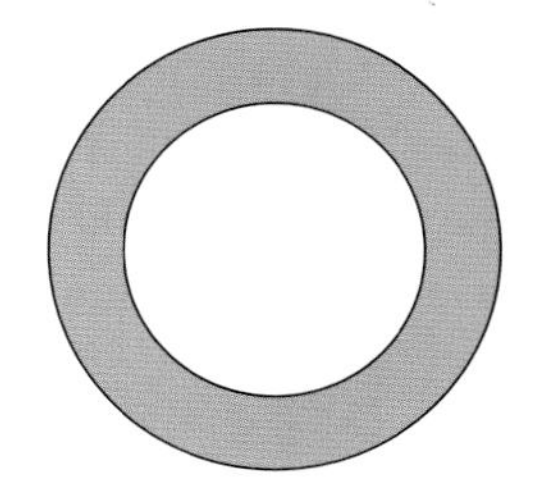

9. The tyre of a car wheel has an outer diameter of 30 cm. How many times will the wheel rotate on a journey of 5 km?
10. When cycling 2.4 km to the beach, the wheels on Sue's bike rotated 1255 times. Find the diameter of the wheels.

11. A golf ball of diameter 1·68 inches rolls a distance of 4 m in a straight line. How many times does the ball rotate completely? [1 inch = 2·54 cm]

Exercise 4.7

1. A rectangular metal plate has a length of 65 cm and a width of 35 cm. It is melted down and recast into circular discs of the same thickness. How many complete discs can be formed if
   a) the radius of each disc is 3 cm
   b) the radius of each disc is 10 cm?

2. A square is inscribed in a circle of radius 7 cm. Find:
   a) the area of the square
   b) the area shaded blue.

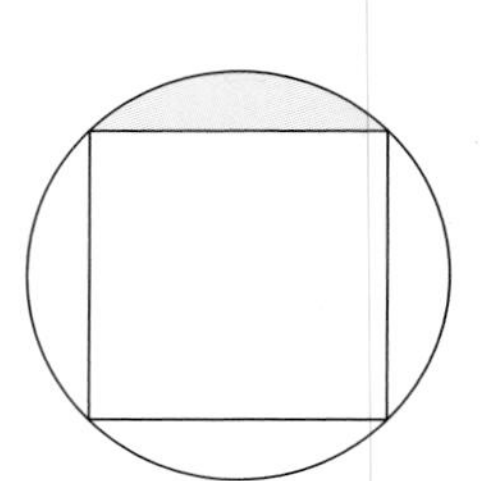

3. The diameter of a circle is given as 10 cm, correct to the nearest cm. Calculate:
   a) the maximum possible circumference
   b) the minimum possible area of the circle consistent with this data.

4. Calculate the radius of a circle whose area is equal to the sum of the areas of three circles of radii 2 cm, 3 cm and 4 cm respectively.

5. A farmer has 100 m of wire fencing. What area can she enclose if she makes a circular enclosure for her sheep?

6. The diagram shows a semicircle inside a rectangle. The pink shaded area is 84 cm². Find x.

7. The semi-circle and the isosceles triangle have the same base AB and the same area. Find the angle x.

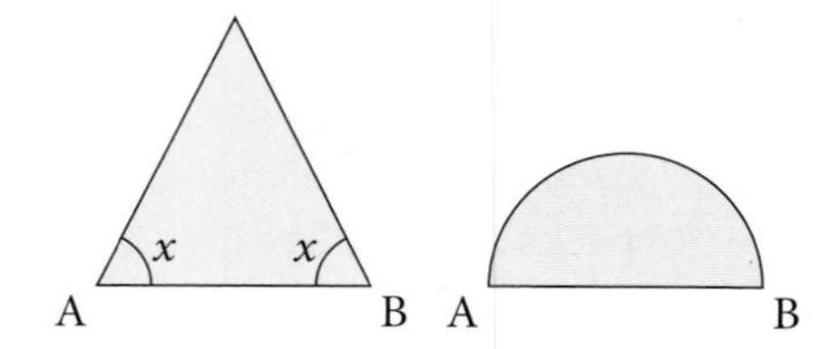

8. Lakmini decided to measure the circumference of the Earth using a very long tape measure. She held the tape measure 1 m from the surface of the (perfectly spherical) Earth all the way round. When she had finished her friend said that her measurement gave too large an answer and suggested taking off 6 m. Was her friend correct?
[Take the radius of the Earth to be 6400 km (if you need it).]

4.3 Arc length and sector area

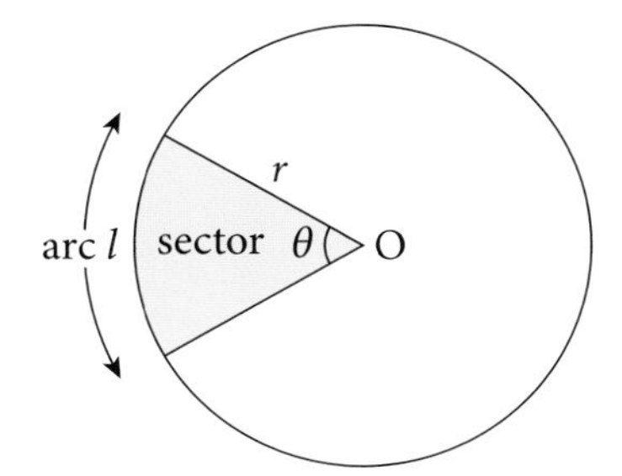

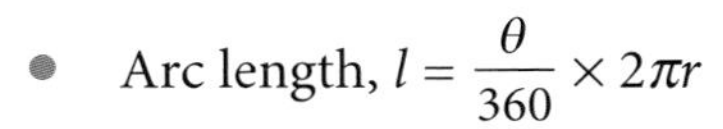

- Arc length, $l = \dfrac{\theta}{360} \times 2\pi r$

We take a fraction of the whole circumference depending on the angle at the centre of the circle.

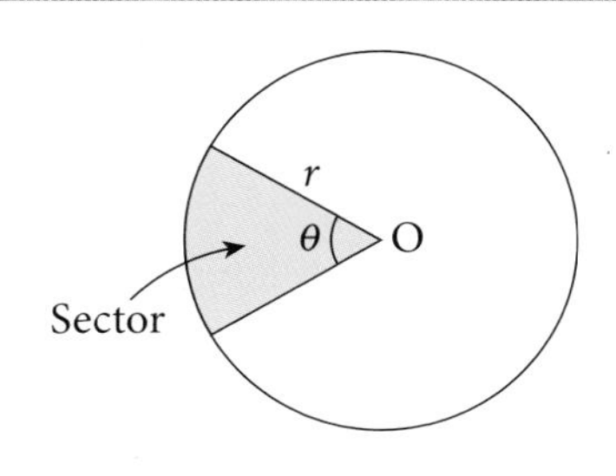

- Sector area, $A = \dfrac{\theta}{360} \times \pi r^2$

We take a fraction of the whole area depending on the angle at the centre of the circle.

Example 5

Find the length of an arc which subtends an angle of 140° at the centre of a circle of radius 12 cm.

Arc length, $l = \dfrac{140}{360} \times 2 \times \pi \times 12$

$= \dfrac{28}{3}\pi$

$= 29\dfrac{1}{3}$ cm

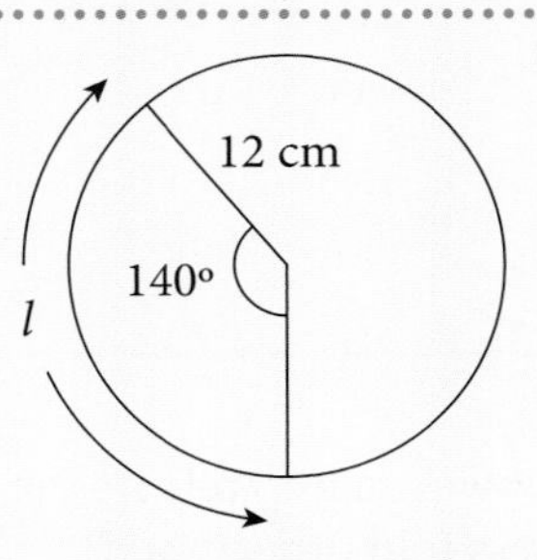

GDC help

Using a GDC

TI-Nspire	TI84-Plus	Casio
$\frac{140}{360} \cdot 2 \cdot \pi \cdot 12$ 29.3215	$\frac{140}{360}$*2*π*12 29.32153143	$\frac{140}{360}\times 2\times\pi\times 12$ $9\frac{1}{3}\pi$

Example 6

A sector of a circle of radius 10 cm has an area of 25 cm².
Find the angle at the centre of the circle.

Let the angle at the centre of the circle be θ.

$$\frac{\theta}{360} \times \pi \times 10^2 = 25$$

$$\therefore \quad \theta = \frac{25 \times 360}{\pi \times 100}$$

$$\theta = 28{\cdot}6° \text{ (3 s.f.)}$$

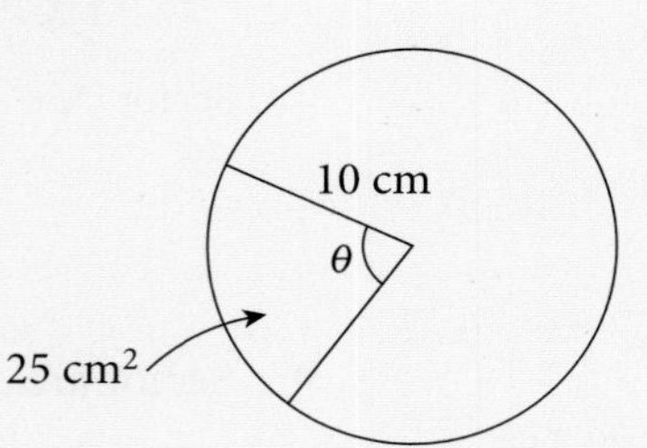

The angle at the centre of the circle is 28·6°.

GDC help

Using a GDC

TI-Nspire	TI84-Plus	Casio
1.1 *Unsaved $\frac{25 \cdot 360}{\pi \cdot 10^2}$ 28.6479 1/99	$\frac{25*360}{\pi*10^2}$ 28.64788976	Math Deg Norm1 d/c Real $\frac{25\times360}{\pi\times10^2}$ 28.64788976 JUMP DELETE ▸MAT MATH

Exercise 4.8

[Use the π button on your calculator unless told otherwise.]

1. Arc AB subtends an angle θ at the centre of circle radius r.
Find the arc length and sector area when:

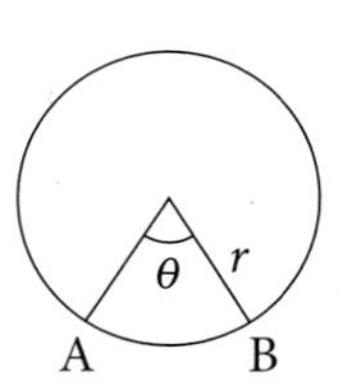

a) $r = 4$ cm, $\theta = 30°$ **b)** $r = 10$ cm, $\theta = 45°$

c) $r = 2$ cm, $\theta = 235°$

In questions **2** and **3** find the total area of the shape.

2.

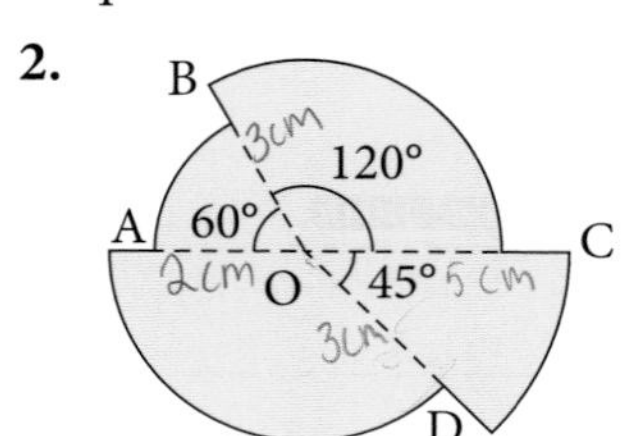

OA = 2 cm, OB = 3 cm, OC = 5 cm, OD = 3 cm.

3.

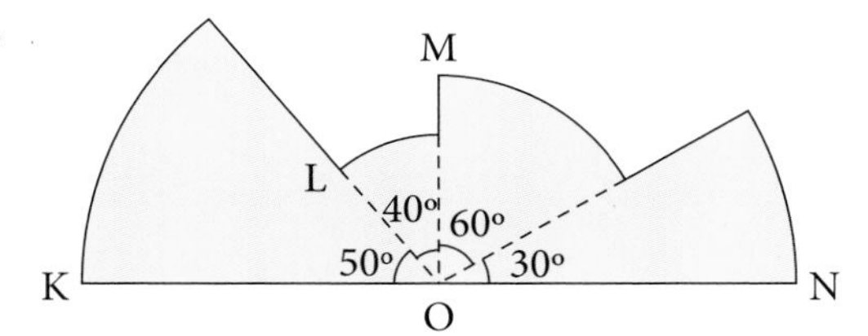

ON = 6 cm, OM = 3 cm, OL = 2 cm, OK = 6 cm.

4. In the diagram the arc length is l and the sector area is A.

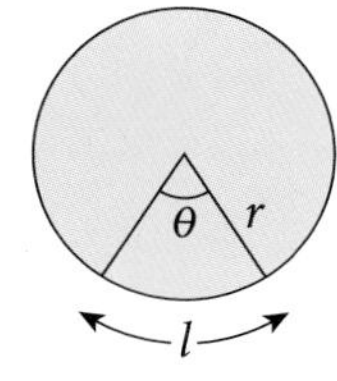

a) Find θ, when $r = 5$ cm and $l = 7.5$ cm.

b) Find θ, when $r = 2$ m and $A = 2\,\text{cm}^2$.

c) Find r, when $\theta = 55°$ and $l = 6$ cm.

5. The length of the minor arc AB of a circle, centre O, is 2π cm and the length of the major arc is 22π cm. Find:

a) the radius of the circle

b) the acute angle AOB.

6. The lengths of the minor and major arcs of a circle are 5·2 cm and 19·8 respectively. Find:

a) the radius of the circle

b) the angle subtended at the centre by the minor arc.

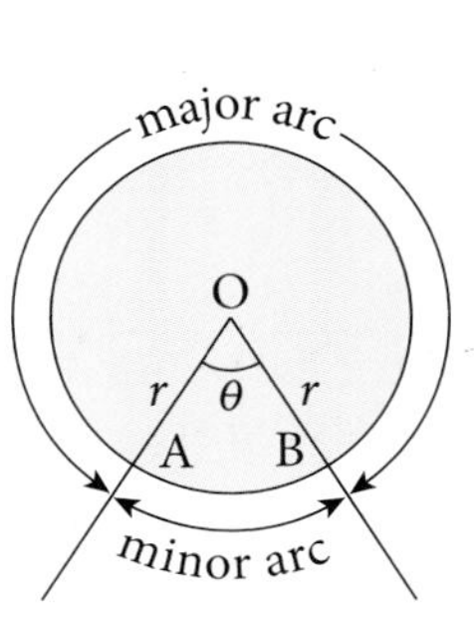

7. A wheel of radius 10 cm is turning at a rate of 5 revolutions per minute. Calculate:

a) the angle through which the wheel turns in 1 second

b) the distance moved by a point on the rim in 2 seconds.

8. The length of an arc of a circle is 12 cm. The corresponding sector area is $108\,\text{cm}^2$. Find:

a) the radius of the circle

b) the angle subtended at the centre of the circle by the arc.

9. The length of an arc of a circle is 7·5 cm. The corresponding sector area is $37{\cdot}5\,\text{cm}^2$. Find:

a) the radius of the circle

b) the angle subtended at the centre of the circle by the arc.

10. A long time ago Dulani found an island shaped like a triangle with three straight shores of length 3 km, 4 km and 5 km. He said nobody could come within 1 km of his shore. What was the area of his exclusion zone?

11. 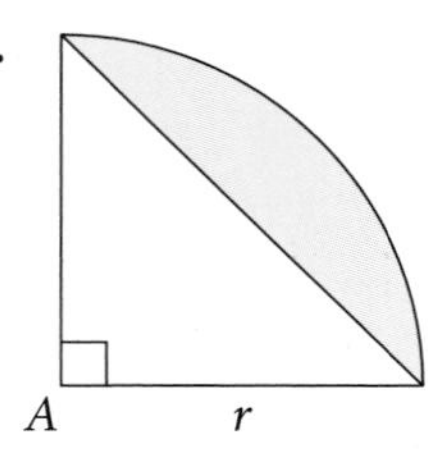

The diagram consists of a quarter circle with centre A and a triangle. The pink shaded area is $200\,\text{cm}^2$. Find the radius r.

4.4 Chord of a circle

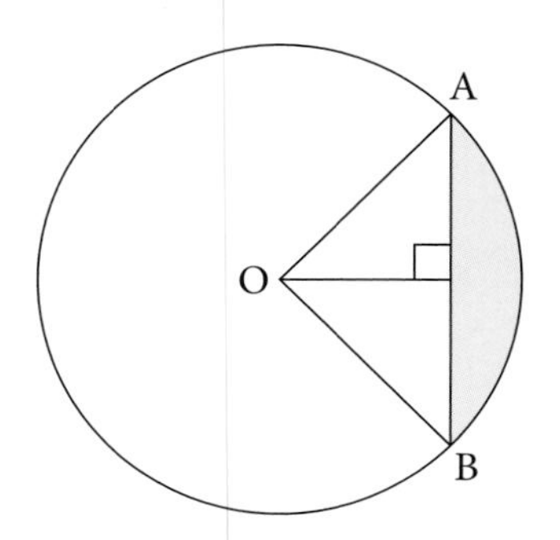

The line AB is a chord. The area of a circle cut off by a chord is called a **segment**. In the diagram the **minor** segment is shaded and the **major** segment is all of the unshaded region.

- The line from the centre of a circle to the mid-point of a chord **bisects** the chord at right angles.
- The line from the centre of a circle to the mid-point of a chord bisects the angle subtended by the chord at the centre of the circle.

Example 7

XY is a chord of length 12 cm of a circle of radius 10 cm, centre O.

Calculate:

a) the angle XOY

b) the area of the minor segment cut off by the chord XY.

You can find more about trigonometry in Chapter 10.

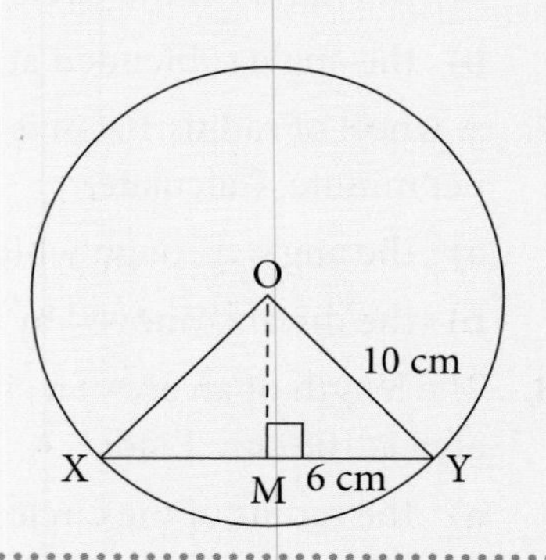

a) Let the mid-point of XY be M.

$\therefore \quad MY = 6\,\text{cm}$

$\sin M\widehat{O}Y = \dfrac{6}{10}$

$\therefore \quad M\widehat{O}Y = 36{\cdot}87°$

$\therefore \quad X\widehat{O}Y = 2 \times 36{\cdot}87$

$= 73{\cdot}74°$

b) Area of minor segment = area of sector XOY – area of ΔXOY

Area of sector XOY $= \dfrac{73{\cdot}74}{360} \times \pi \times 10^2$

$= 64{\cdot}32\,\text{cm}^2$.

Area of ΔXOY $= \dfrac{1}{2} \times 10 \times 10 \times \sin 73{\cdot}74°$

$= 48{\cdot}00\,\text{cm}^2$

$\therefore \quad$ Area of minor segment $= 64{\cdot}32 - 48{\cdot}00$

$= 16{\cdot}3\,\text{cm}^2$ (3 s.f.)

▶ Continued on next page

GDC help

Using a GDC

TI-Nspire	TI84-Plus	Casio
$2\cdot\sin^{-1}\left(\frac{6}{10}\right)$ 73.7398 $\frac{73.739795291688}{360}\cdot\pi\cdot 10^2$ 64.3501 $\frac{1}{2}\cdot 10\cdot 10\cdot\sin(73.739795291688)$ 48. 64.350110879327−48. 16.3501	2*sin⁻¹(6/10) 73.73979529 73.73979529/360 *π*10² 64.35011088 1/2*10*10*sin(73.▸ 48 64.35011088−48 16.35011088	2×sin⁻¹(6/10) 73.73979529 Ans/360×π×10² 64.35011088 JUMP DELETE ▸MAT MATH 2×sin⁻¹(6/10) 73.73979529 1/2×10×10×sin (Ans) 48 DEL-LINE DEL-ALL 64.35011088−48 16.35011088 JUMP DELETE ▸MAT MATH

Exercise 4.9

For this exercise, you will need to use the π button on your calculator. You will also need basic trigonometry (page 309 in Chapter 10).

1. The chord AB subtends an angle of 130° at the centre O. The radius of the circle is 8 cm. Find:
   a) the length of AB
   b) the area of sector OAB
   c) the area of triangle OAB
   d) the area of the minor segment (shown shaded).

2. 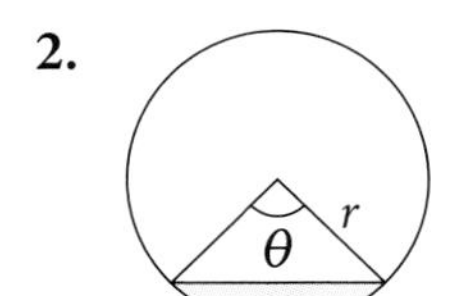

   Find the shaded area of the circle on the left when:
   a) $r = 6\,\text{cm}$, $\theta = 70°$
   b) $r = 14\,\text{cm}$, $\theta = 104°$
   c) $r = 5\,\text{cm}$, $\theta = 80°$

3. Find θ and hence the blue shaded area when:
   a) $\text{AB} = 10\,\text{cm}$, $r = 10\,\text{cm}$
   b) $\text{AB} = 8\,\text{cm}$, $r = 5\,\text{cm}$

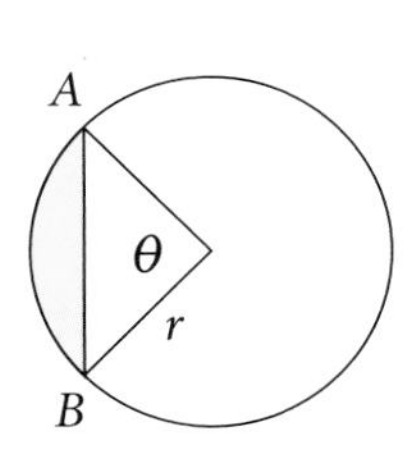

4. How far is a chord of length 8 cm from the centre of a circle of radius 5 cm?

5. How far is a chord of length 9 cm from the centre of a circle of radius 6 cm?

6. The diagram on the right shows the cross-section of a cylindrical pipe with water lying in the bottom.

   a) If the maximum depth of the water is 2 cm and the radius of the pipe is 7 cm, find the area shaded.

   b) What is the *volume* of water in a length of 30 cm?

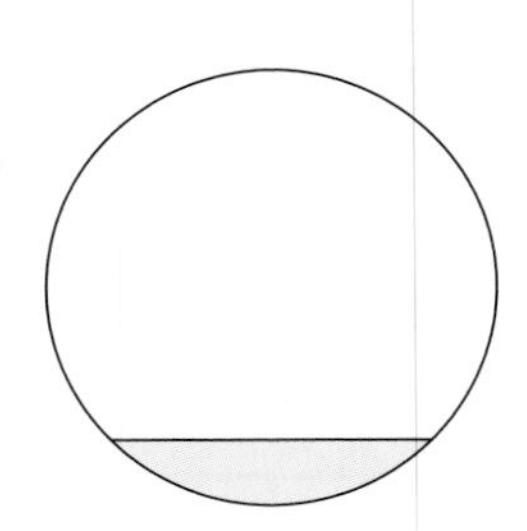

7. 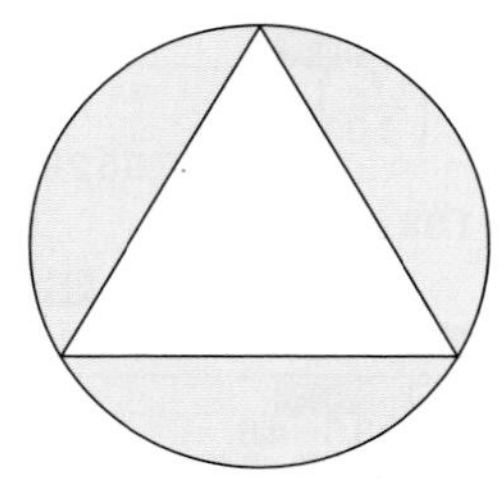

An equilateral triangle is inscribed in a circle of radius 10 cm. Find:

   a) the area of the triangle

   b) the area shaded green.

8. A regular hexagon is circumscribed by a circle of radius 6 cm. Find the area shaded green.

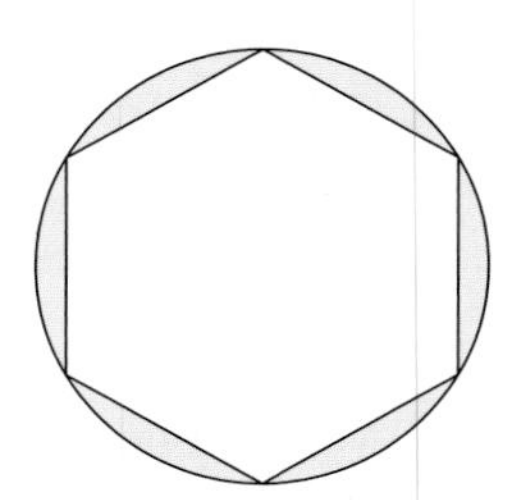

9. A regular octagon is circumscribed by a circle of radius r cm. Find the area enclosed between the circle and the octagon. (Give the answer in terms of r.)

10. Find the radius of the circle:

   a) when $\theta = 90°$, $A = 20\,\text{cm}^2$

   b) when $\theta = 30°$, $A = 35\,\text{cm}^2$

   c) when $\theta = 150°$, $A = 114\,\text{cm}^2$

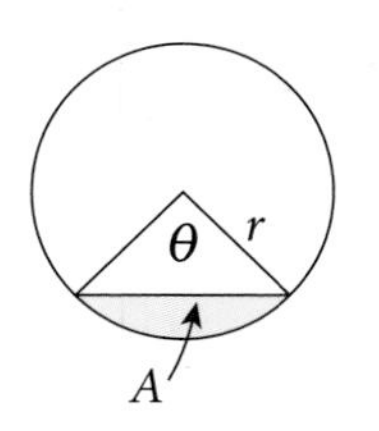

4.5 Volume

Volume of a prism

A **prism** is an object with the same cross-section throughout its length.

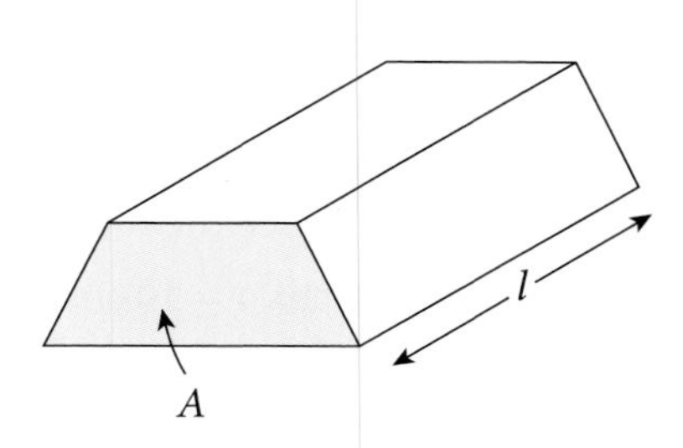

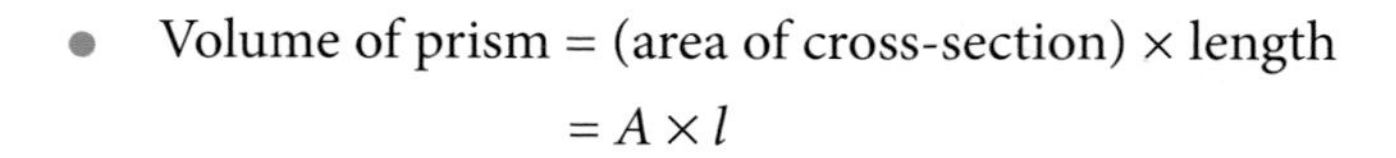

- Volume of prism = (area of cross-section) × length

$$= A \times l$$

A **cuboid** is a prism whose six faces are all rectangles.
A **cube** is a special case of a cuboid in which all six faces are squares.
A **cylinder** is a prism whose cross-section is a circle.

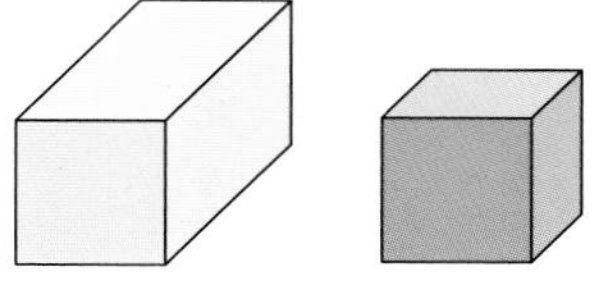

radius = r

height = h

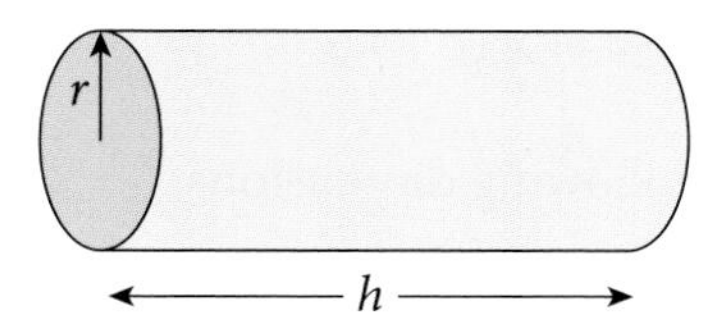

- Volume of cylinder = (area of cross-section) × length

  Volume = $\pi r^2 h$

Example 8

Calculate the height of a cylinder of volume 500 cm³ and base radius 8 cm. Let the height of the cylinder be h cm.

$$\pi r^2 h = 500$$

$$3{\cdot}142 \times 8^2 \times h = 500$$

$$h = \frac{500}{3{\cdot}142 \times 64}$$

$$h = 2{\cdot}49 \text{ (3 s.f.)}$$

The height of the cylinder is 2·49 cm.

GDC help

Using a GDC

TI-Nspire	TI84-Plus	Casio
1.1 *Unsaved; $\frac{500}{\pi \cdot 64}$ 2.4868; 1/99	$\frac{500}{\pi * 64}$ 2.486795986	Math Deg Norm1 d/c Real; $\frac{500}{\pi \times 64}$ 2.486795986; JUMP DELETE ▶MAT MATH

Exercise 4.10

1. Calculate the volume of the prisms. All lengths are in cm.

a)

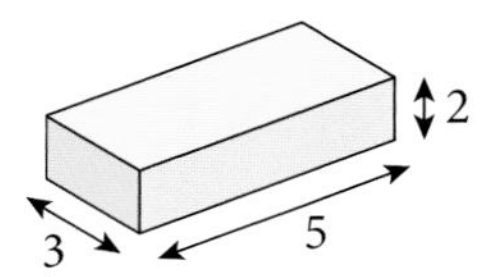

b)

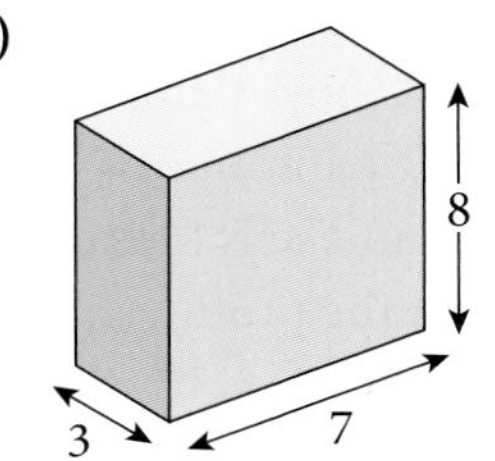

c)

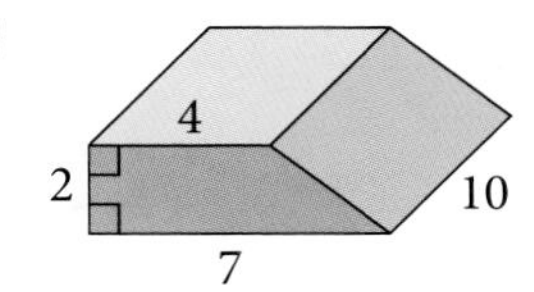

d)

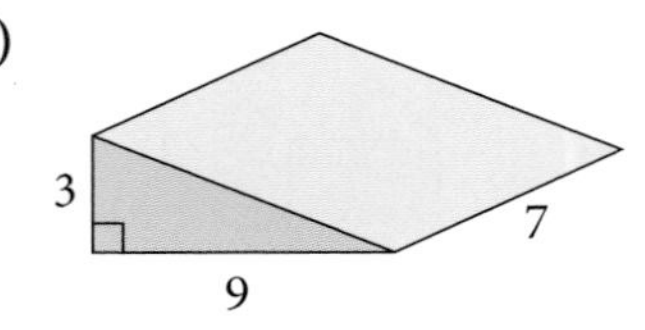

e)

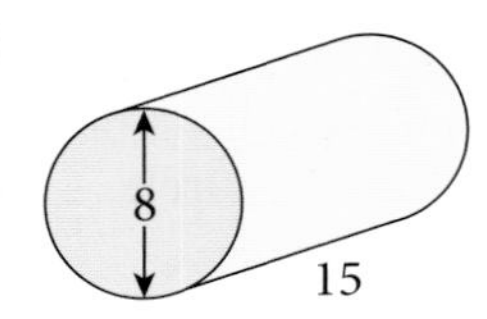

f)

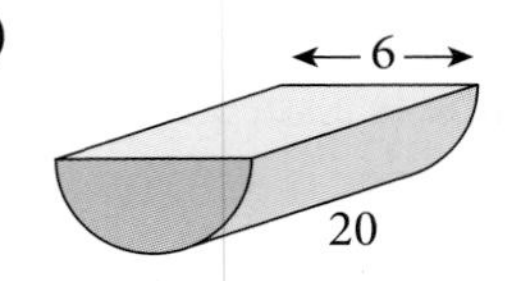

2. Calculate the volume of the cylinders with the following dimensions:
   a) $r = 4\,\text{cm}$, $h = 10\,\text{cm}$ b) $r = 11\,\text{m}$, $h = 2\,\text{m}$
   c) $r = 2{\cdot}1\,\text{cm}$, $h = 0{\cdot}9\,\text{cm}$
3. Find the height of a cylinder of volume $200\,\text{cm}^3$ and radius 4 cm.
4. Find the length of a cylinder of volume 2 litres and radius 10 cm.
5. Find the radius of a cylinder of volume $45\,\text{cm}^3$ and length 4 cm.
6. The total surface area of cube is $121{\cdot}5\,\text{cm}^2$. What is its volume?
7. When 3 litres of oil is removed from an upright cylindrical can, the level falls by 10 cm. Find the radius of the can.

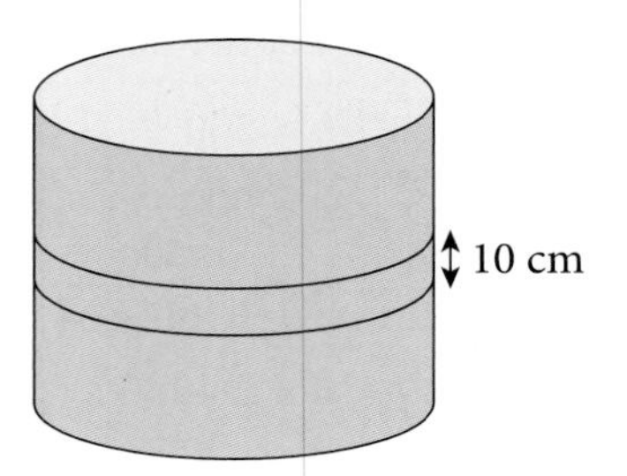

8. A solid cylinder of radius 4 cm and length 8 cm is melted down and recast into a solid cube. Find the side of the cube.
9. A solid rectangular block of copper 5 cm by 4 cm by 2 cm is drawn out to make a cylindrical wire of diameter 2 mm. Calculate the length of the wire.
10. Water flows through a circular pipe of internal diameter 3 cm at a speed of 10 cm/s. If the pipe is full, how much water flows from the pipe in one minute? (Answer in litres.)
11. Water flows from a hose-pipe of internal diameter 1 cm at a rate of 5 litres per minute. At what speed is the water flowing through the pipe?
12. A cylindrical metal pipe has external diameter of 6 cm and internal diameter of 4 cm. Calculate the volume of metal in a pipe of length 1 m. If $1\,\text{cm}^3$ of the metal has a mass of 8 g, find the mass of the pipe.
13. For two cylinders A and B, the ratio of lengths is 3 : 1 and the ratio of diameters is 1 : 2. Calculate the ratio of their volumes.
14. A prism has volume $100\,\text{cm}^3$ and length 8 cm. If the cross-section is an equilateral triangle, find the length of a side of the triangle.
15. 

    A business buys paint at \$30 000 for $20\,\text{m}^3$. It puts the paint in tins of capacity 0·8 litres and sells them at \$5·95 each. Work out the profit.

16. Natalia decided to build a garage and began by calculating the number of bricks required. The garage was to be 6 m by 4 m and 2·5 m in height. Each brick measures 22 cm by 10 cm by 7 cm. Natalia estimated that she would need about 40 000 bricks. Is this a reasonable estimate?

17. A cylindrical can of internal radius 20 cm stands upright on a flat surface. It contains water to a depth of 20 cm. Calculate the rise in the level of the water when a brick of volume 1500 cm³ is immersed in the water.

18. A cylindrical tin of height 15 cm and radius 4 cm is filled with sand from a rectangular box. How many times can the tin be filled if the dimensions of the box are 50 cm by 40 cm by 20 cm?

19. A rectangular piece of card 12 cm by 20 cm is rolled up to make a tube (with no overlap). Find the volume of the tube if

 a) the long sides are joined

 b) the short sides are joined.

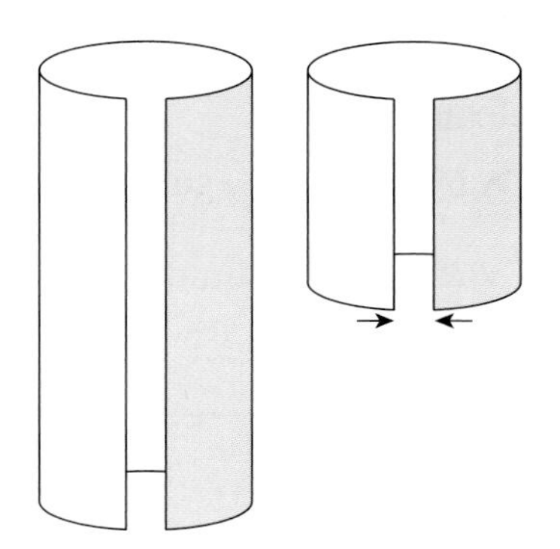

20. Rain which falls onto a flat rectangular surface of length 6 m and width 4 m is collected in a cylinder of internal radius 20 cm. What is the depth of water in the cylinder after a storm in which 1 cm of rain fell?

Volume of other shapes

- Volume of a **pyramid** $= \frac{1}{3}$ (base area) × height

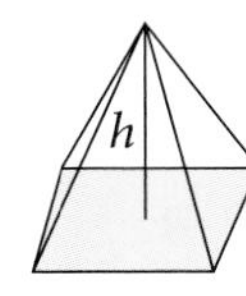

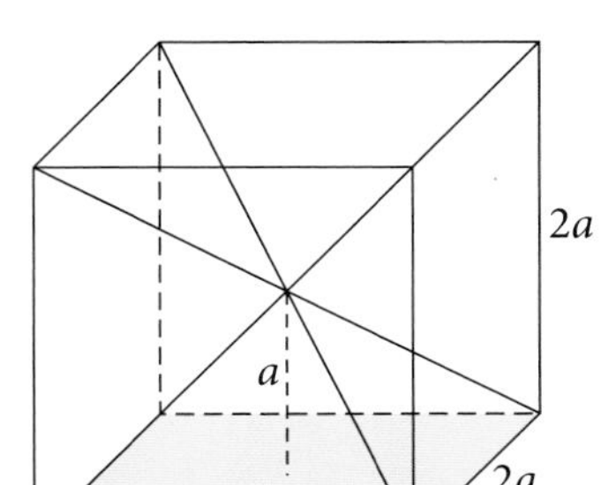

Figure 1

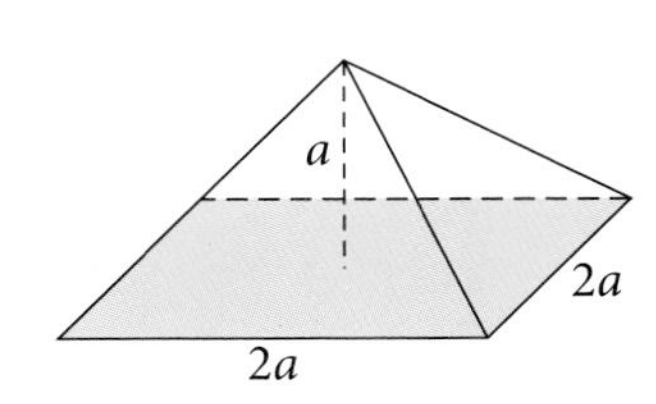

Figure 2

Figure 1 shows a cube of side $2a$ broken down into six pyramids of height a as given in Figure 2.

If the volume of each pyramid is V,

then $6V = 2a \times 2a \times 2a$

$V = \frac{1}{6} \times (2a)^2 \times 2a$

So $\quad V = \frac{1}{3} \times (2a)^2 \times a$

$V = \frac{1}{3}$ (base area) $\times$ height

- Volume of a **cone** $= \frac{1}{3}\pi r^2 h$ (note the similarity with the pyramid)

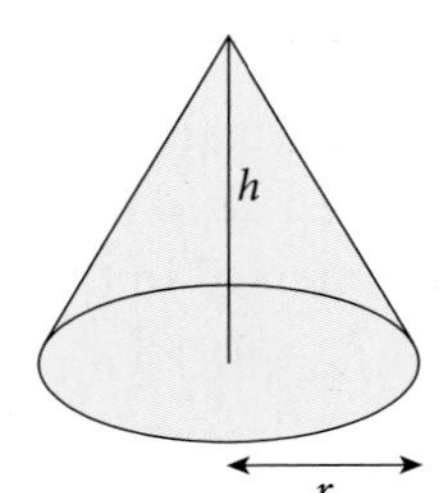

- Volume of a **sphere** $= \frac{4}{3}\pi r^3$

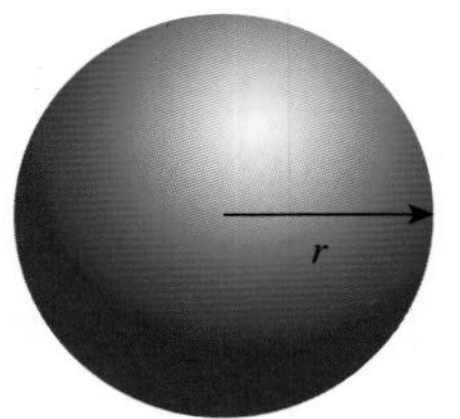

Example 9

A pyramid has a square base of side 5 m and vertical height 4 m. Find its volume.

Volume of pyramid $= \frac{1}{3}(5 \times 5) \times 4$

$= 33\frac{1}{3}\text{ m}^3$

GDC help

Using a GDC

TI-Nspire	TI84-Plus	Casio
1.1 *Unsaved $\frac{1}{3}\cdot 5\cdot 5\cdot 4$ $\quad \frac{100}{3}$ $\frac{1}{3}\cdot 5\cdot 5\cdot 4$ $\quad$ 33.3333 2/99	$\frac{1}{3}$*5*5*4 $\frac{100}{3}$ Ans▸n/d◂▸Un/d $33\frac{1}{3}$	Math Deg Norm1 d/c Real $\frac{1}{3}\times5\times5\times4$ $33\frac{1}{3}$ JUMP DELETE ▸MAT MATH

Example 10

Calculate the radius of a sphere of volume 500 cm³.

Let the radius of the sphere be r cm

$$\frac{4}{3}\pi r^3 = 500$$

$$r^3 = \frac{3 \times 500}{4\pi}$$

$$r = \sqrt[3]{\left(\frac{3 \times 500}{4\pi}\right)} = 4\cdot 92 \text{ (3 s.f.)}$$

The radius of the sphere is 4·92 cm.

▶ Continued on next page

GDC help

Using a GDC

TI-Nspire	TI84-Plus	Casio
$\sqrt[3]{\frac{3\cdot 500}{4\cdot \pi}}$ 4.92373	$\sqrt[3]{\frac{3*500}{4\pi}}$ 4.923725109	$\sqrt[3]{\frac{3\times 500}{4\pi}}$ 4.923725109

Exercise 4.11

Find the volumes of the following objects.

1. Cone: height = 5 cm, radius = 2 cm
2. Sphere: radius = 5 cm
3. Sphere: radius = 10 cm
4. Cone: height = 6 cm, radius = 4 cm
5. Sphere: diameter = 8 cm
6. Cone: height = x cm, radius = $2x$ cm
7. Sphere: radius = 0.1 m
8. Cone: height = $\frac{1}{\pi}$ cm, radius = 3 cm
9. Pyramid: rectangular base 7 cm by 8 cm; height = 5 cm
10. Pyramid: square base of side 4 m, height = 9 m
11. Pyramid: equilateral triangular base of side = 8 cm, height = 10 cm
12. Find the volume of a hemisphere of radius 5 cm.
13. A cone is attached to a hemisphere of radius 4 cm. If the total height of the object is 10 cm, find its volume.

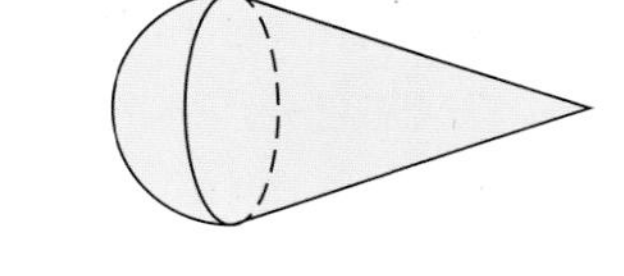
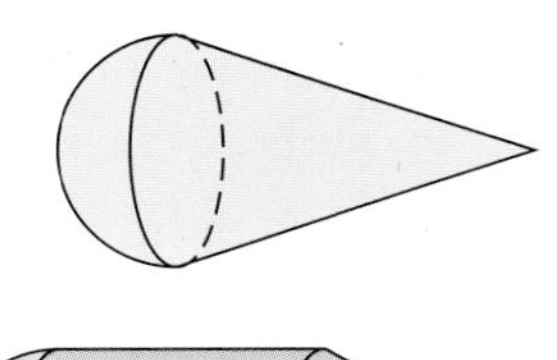

14. A toy consists of a cylinder of diameter 6 cm 'sandwiched' between a hemisphere and a cone of the same diameter. If the cone is of height 8 cm and the cylinder is of height 10 cm, find the total volume of the toy.

15. Find the height of a pyramid of volume 20 m^3 and base area 12 m^2.
16. Find the radius of a sphere of volume 60 cm^3.
17. Find the height of a cone of volume 2·5 litres and radius 10 cm.
18. Six square-based pyramids fit exactly onto the six faces of a cube of side 4 cm. If the volume of the object formed is 256 cm^3, find the height of each of the pyramids.
19. A solid metal cube of side 6 cm is recast into a solid sphere. Find the radius of the sphere.
20. A hollow spherical vessel has internal and external radii of 6 cm and 6·4 cm respectively. Calculate the mass of the vessel if it is made of metal of density 10 g/cm^3.

Exercise 4.12

1. Water is flowing into an inverted cone of diameter and height 30 cm, at a rate of 4 litres per minute. How long, in seconds, will it take to fill the cone?

2. A solid metal sphere is recast into many smaller spheres. Calculate the number of the smaller spheres if the initial and final radii are as follows

   a) Initial radius = 10 cm, final radius = 2 cm

   b) Initial radius = 7 cm, final radius = $\frac{1}{2}$ cm

   c) Initial radius = 1 m, final radius = $\frac{1}{3}$ cm

3. A spherical ball is immersed in water contained in a vertical cylinder. Assuming the water covers the ball, calculate the rise in the water level if:

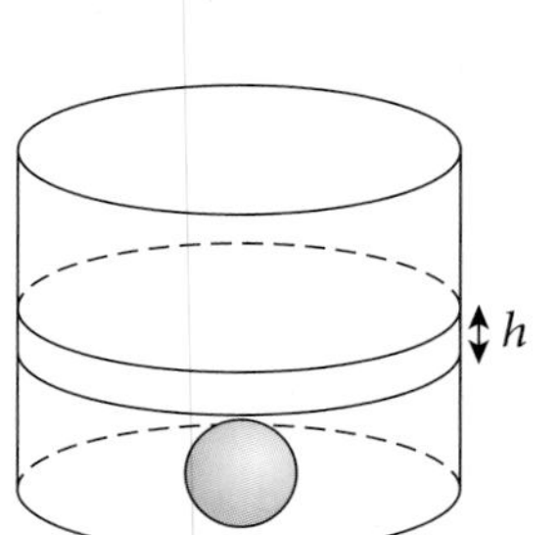

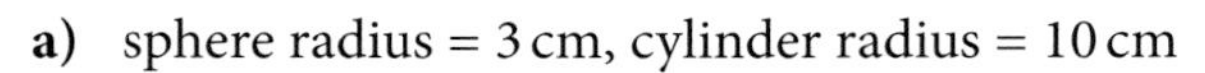

   a) sphere radius = 3 cm, cylinder radius = 10 cm

   b) sphere radius = 2 cm, cylinder radius = 5 cm.

4. A spherical ball is immersed in water contained in a vertical cylinder. The rise in water level is measured in order to calculate the radius of the spherical ball. Calculate the radius of the ball in the following cases.

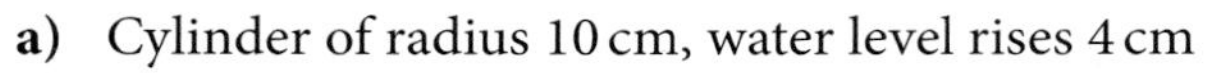

   a) Cylinder of radius 10 cm, water level rises 4 cm

   b) Cylinder of radius 100 cm, water level rises 8 cm

5. One corner of a solid cube of side 8 cm is removed by cutting through the mid-points of three adjacent sides. Calculate the volume of the piece removed.

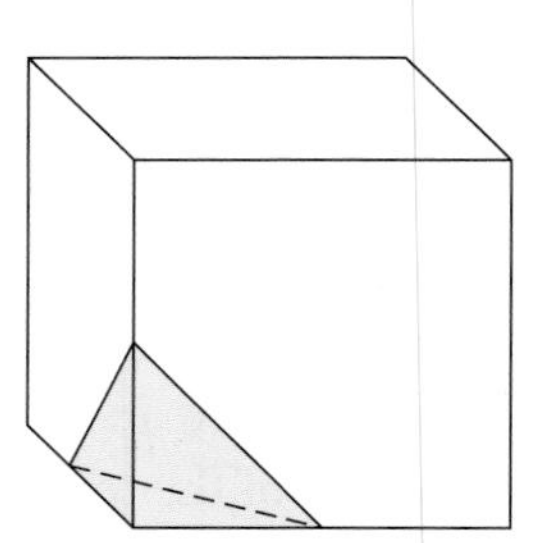

6. The cylindrical end of a pencil is sharpened to produce a perfect cone at the end with no overall loss of length. If the diameter of the pencil is 1 cm, and the cone is of length 2 cm, calculate the volume of the shavings.

7. Metal spheres of radius 2 cm are packed into a rectangular box of internal dimensions 16 cm × 8 cm × 8 cm. When 16 spheres are packed the box is filled with a preservative liquid. Find the volume of this liquid.

8. An inverted cone of height 10 cm and base radius 6.4 cm contains water to a depth of 5 cm, measured from the vertex. Calculate the volume of water in the cone.

9. An inverted cone of height 15 cm and base radius 4 cm contains water to a depth of 10 cm. Calculate the volume of water in the cone.

10. An inverted cone of height 12 cm and base radius 6 cm contains 20 cm^3 of water. Calculate the depth of water in the cone, measured from the vertex.

11. The volume of paint in a tube is 150 cm^3. If this paint is squirted out in a straight line through a circular hole of diameter 5 mm, how long will this line be?

12. A **frustum** is a cone with 'the end chopped off'. A bucket in the shape of a frustum as shown has diameters of 10 cm and 4 cm at its ends and a depth of 3 cm. Calculate the volume of the bucket.

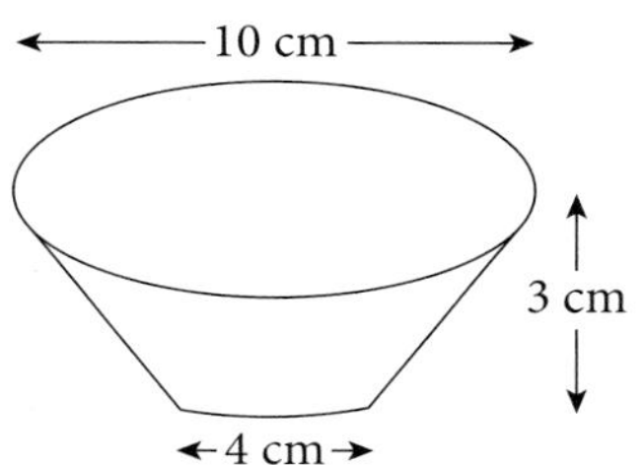

13. Find the volume of a frustum with end diameters of 60 cm and 20 cm and a depth of 40 cm.

14. The diagram shows a sector of a circle of radius 10 cm.

    a) Find, as a multiple of π, the arc length of the sector.

    The straight edges are brought together to make a cone. Calculate:

    b) the radius of the base of the cone,

    c) the vertical height of the cone.

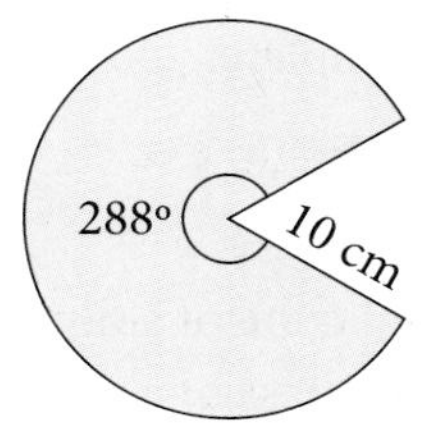

15. Calculate the volume of a regular octahedron whose edges are all 10 cm.

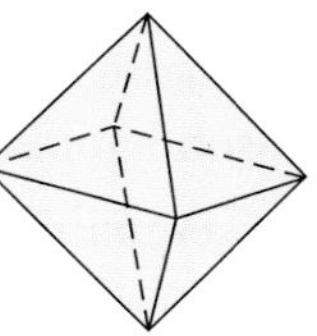

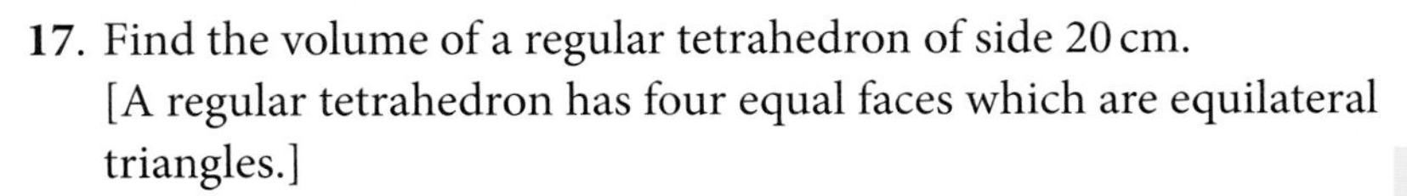

16. A sphere passes through the eight corners of a cube of side 10 cm. Find the volume of the sphere.

17. Find the volume of a regular tetrahedron of side 20 cm. [A regular tetrahedron has four equal faces which are equilateral triangles.]

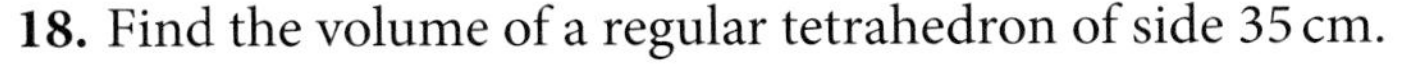

18. Find the volume of a regular tetrahedron of side 35 cm.

Question **17** is harder than the rest of the exercise.

Converting units

Volume

Here is a cube with sides of 1 cm

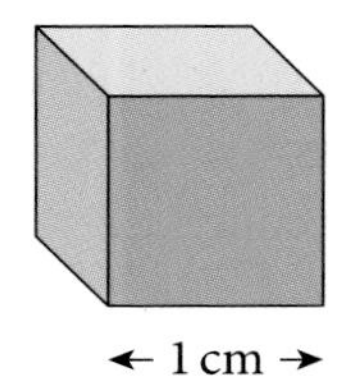

. . . and the same cube with sides of 10 mm

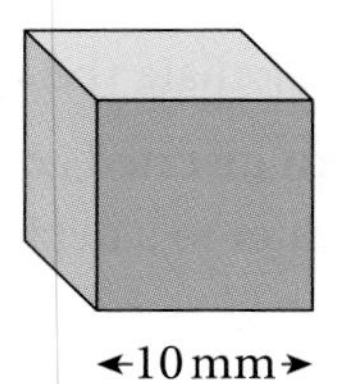

The volume of each cube in the same,

so $\quad 1\text{ cm}^3 = 10 \times 10 \times 10\text{ mm}^3$

$\quad\quad 1\text{ cm}^3 = 1000\text{ mm}^3$

Similarly $\quad 1\text{ cm}^3 = 1\,000\,000\text{ cm}^3$

Mass The standard units of mass are mg, g, kg, and tonnes.

1000 mg = 1 g, 1000 g = 1 kg, 1000 kg = 1 tonne [1t]

Examples:
- 8000 mg = 8 kg
- 2200 g = 2·2 kg
- 600 kg = 0·6 tonne

Exercise 4.13

1. Copy and complete:

a) $7\text{ cm}^3 = \square\text{ mm}^3$ **b)** $0{\cdot}6\text{ cm}^3 = \square\text{ mm}^3$ **c)** $1000\text{ mm}^3 = \square\text{ cm}^3$

d) $1\text{ m}^3 = \square\text{ cm}^3$ **e)** $3\text{ tonne} = \square\text{ kg}$ **f)** $8500\text{ kg} = \square\text{ tonne}$

2. Copy and complete:

a) $55\text{ g} = \square\text{ mg}$ **b)** $13\text{ kg} = \square\text{ g}$ **c)** $200\text{ mg} = \square\text{ g}$

d) $0{\cdot}6\text{ g} = \square\text{ mg}$ **e)** $0{\cdot}02\text{ tonne} = \square\text{ kg}$ **f)** $1\text{ mm} = \square\text{ mg}$

3. A solid metal cube of side 1·2 cm has a mass of 15 grams. Calculate the mass of a cube of side 5 cm made of the same metal.

4. The total surface area of a cube is $121{\cdot}5\text{ cm}^2$. Work out the volume of the cube.

5. Ten gold coins of radius 0·6 cm and thickness 0·3 cm are melted down to form a solid gold cube of side x cm. Calculate the value of x.

6. An ice rink with semicircular ends is to be filled to a depth of 10 cm with ice. If the straight sides are 60 m long and 30 m apart, how many cubic metres of ice will be needed?

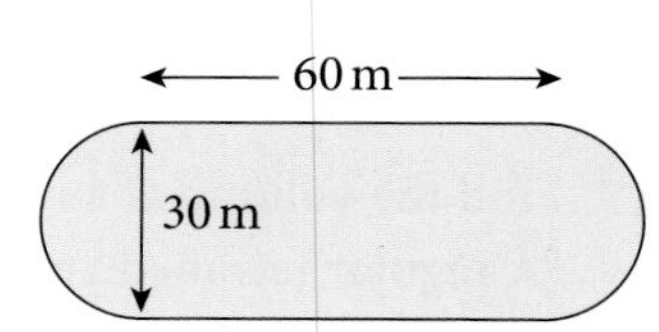

4.6 Surface area

We are concerned here with the surface areas of the *curved* parts of cylinders, spheres and cones. The areas of the plane faces are easier to find.

- **Cylinder**
  Curved surface area = $2\pi rh$

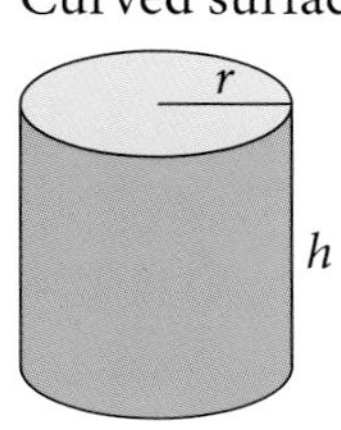

- **Sphere**
  Surface area = $4\pi r^2$

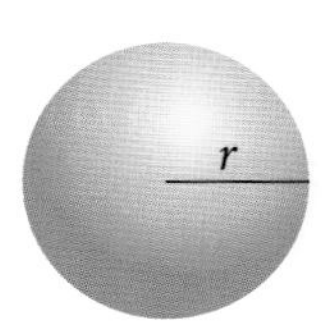

- **Cone**
  Curved surface area = πrl
  where l is the slant height

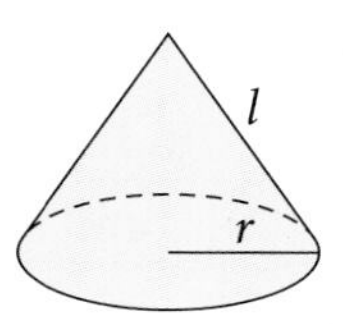

Example 11

Find the *total* surface area of a solid cone of radius 4 cm and vertical height 3 cm.

Let the slant height of the cone be l cm.

$l^2 = 3^2 + 4^2$ (by Pythagoras' theorem)
$l = 5$

Curved surface area $= \pi \times 4 \times 5$
$= 20\pi \text{ cm}^2$

Area of end face $= \pi \times 4^2 = 16\pi \text{ cm}^2$

$\therefore$ Total surface area $= 20\pi + 16\pi$
$= 36\pi \text{ cm}^2$
$= 113 \text{ cm}^2$ to 3 s.f.

Exercise 4.14

For this exercise, use the π button on your calculator unless otherwise instructed.

1. Copy the table and find the quantities marked *.
(Leave π in your answers.)

	Solid object	Radius	Vertical height	Curved surface area	Total surface area
a)	sphere	3 cm		*	
b)	cylinder	4 cm	5 cm		*
c)	cone	6 cm	8 cm	*	
d)	cylinder	0.7 m	1 m		*
e)	sphere	10 m		*	
f)	cone	5 cm	12 cm	*	
g)	cylinder	6 mm	10 mm		*
h)	cone	2·1 cm	4·4 cm	*	
i)	sphere	0·01 m		*	
j)	hemisphere	7 cm		*	*

2. Find the radius of a sphere of surface area 34 cm^2.

3. Find the slant height of a cone of curved surface area 20 cm^2 and radius 3 cm.

4. Find the height of a solid cylinder of radius 1 cm and *total* surface area 28 cm^2.

5. Copy the table and find the quantities marked $*$. (Take $\pi = 3$)

	Object	Radius	Vertical height	Curved surface area	Total surface area
a)	cylinder	4 cm	*	72 cm^2	
b)	sphere	*		192 cm^2	
c)	cone	4 cm	*	60 cm^2	
d)	sphere	*		0·48 m^2	
e)	cylinder	5 cm	*		330 cm^2
f)	cone	6 cm	*		225 cm^2
g)	cylinder	2 m	*		108 m^2

6. A solid wooden cylinder of height 8 cm and radius 3 cm is standing vertically. It is then cut in two along a vertical plane of symmetry. Calculate the total surface area of the two pieces.

7. A tin of paint covers a surface area of 60 m^2 and costs \$4.50. Find the cost of painting the outside surface of a hemispherical dome of radius 50 m. (Just the curved part.)

8. A solid cylinder of height 10 cm and radius 4 cm is to be plated with material costing \$11 per cm^2. Find the cost of the plating.

9. Find the volume of a sphere of surface area 100 cm^2.

10. Find the surface area of a sphere of volume 28 cm^3.

11. An inverted cone of vertical height 12 cm and base radius 9 cm contains water to a depth of 4 cm. Find the area of the interior surface of the cone not in contact with the water.

12. A circular piece of paper of radius 20 cm is cut in half and each half is made into a hollow cone by joining the straight edges. Find the slant height and base radius of each cone.

13. A golf ball has a diameter of 4·1 cm and the surface has 150 dimples of radius 2 mm. Calculate the total surface area which is exposed to the surroundings. (Assume the dimples are hemispherical.)

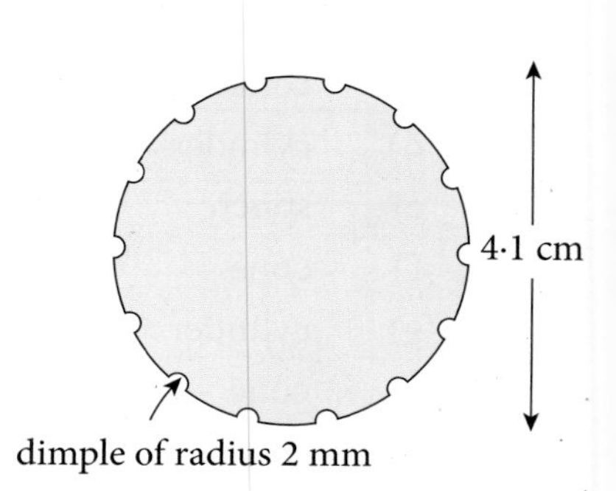

14. A cone of radius 3 cm and slant height 6 cm is cut into four identical pieces. Calculate the total surface area of the four pieces.

Revision exercise 4A

1. Find the area of the following shapes.

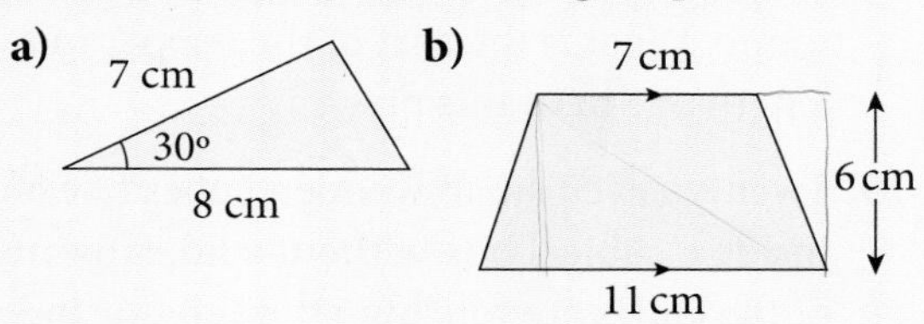

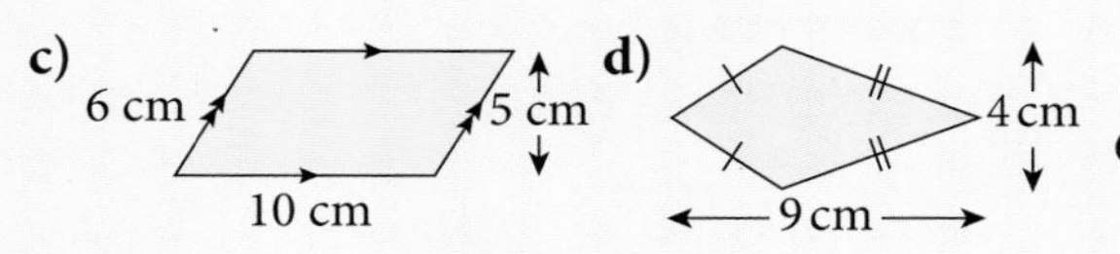

2. a) A circle has radius 9 m. Find its circumference and area.
   b) A circle has circumference 34 cm. Find its diameter.
   c) A circle has area 50 cm^2. Find its radius.

3. A target consists of concentric circles of radius 3 cm and 9 cm.

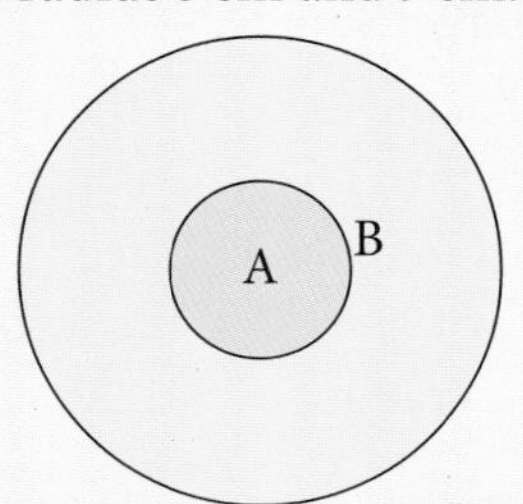

   a) Find the area of A, in terms of π.
   b) Find the ratio $\dfrac{\text{area of B}}{\text{area of A}}$.

4. In Figure 1 a circle of radius 4 cm is inscribed in a square. In Figure 2 a square is inscribed in a circle of radius 4 cm.
   Calculate the shaded area in each diagram.

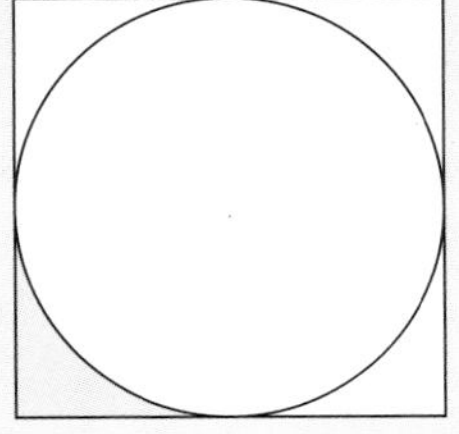

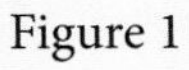

Figure 1

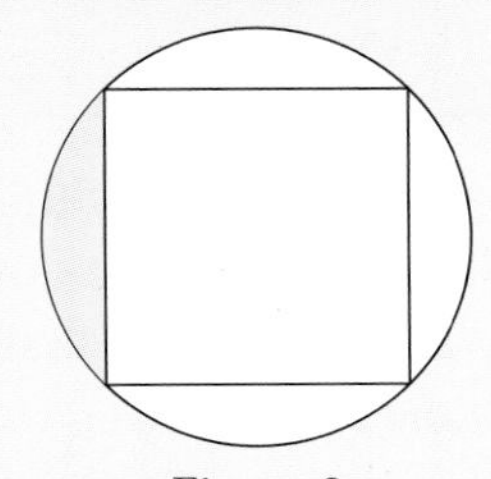

Figure 2

5. Given that OA = 10 cm and $A\widehat{O}B = 70°$ (where O is the centre of the circle), calculate:

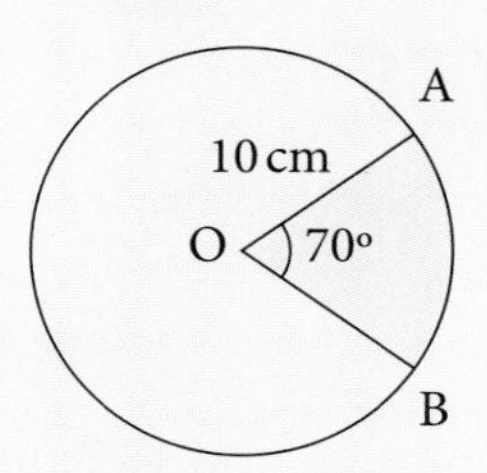

   a) the arc length AB
   b) the area of minor sector AOB.

6. The points X and Y lie on the circumference of a circle. The circle has centre O and radius 8 cm, and $X\widehat{O}Y = 80°$. Calculate:
   a) the length of the minor arc XY
   b) the length of the chord XY
   c) the area of sector XOY
   d) the area of triangle XOY
   e) the area of the minor segment of the circle cut off by XY.

7. Given that ON = 10 cm and minor arc MN = 18 cm, calculate the angle $M\widehat{O}N$ (shown as $x°$).

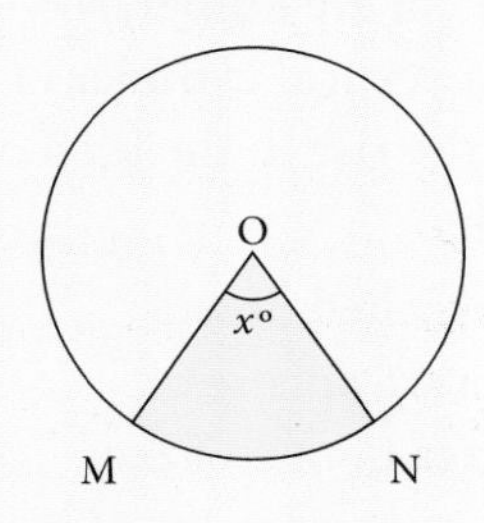

8. A cylinder of radius 8 cm has a volume of 2 litres. Calculate the height of the cylinder.

9. Calculate:
   a) the volume of a sphere of radius 6 cm
   b) the radius of a sphere whose volume is 800 cm^3.

10. A sphere of radius 5 cm is melted down and made into a solid cube.
    Find the length of a side of the cube.

11. The curved surface area of a solid circular cylinder of height 8 cm is 100 cm^2. Calculate the volume of the cylinder.

12.

The perimeter of a lemon measures 18·8 cm. How many of these lemons could be put in a single layer in a box measuring 42 cm by 36 cm?

13. Calculate the radius of a hemispherical solid whose total surface area is $48\pi\,cm^2$.

14. Calculate:

a) the area of an equilateral triangle of side 6 cm

b) the area of a regular hexagon of side 6 cm

c) the volume of a regular hexagonal prism of length 10 cm, where the side of the hexagon is 12 cm.

15. Ten spheres of radius 1 cm are immersed in liquid contained in a vertical cylinder of radius 6 cm. Calculate the rise in the level of the liquid in the cylinder.

16. A cube of side 10 cm is melted down and made into ten identical spheres. Calculate the surface area of one of the spheres.

17.

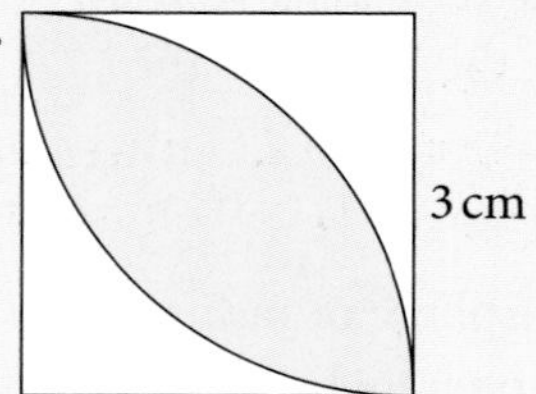

The square has sides of length 3 cm and the arcs have centres at the corners. Find the area coloured blue.

18. A copper pipe has external diameter 18 mm and thickness 2 mm. The density of copper is 9 g/cm^3 and the price of copper is \$150 per tonne. What is the cost of the copper in a length of 5 m of this pipe?

19. Twenty-seven small wooden cubes fit exactly inside a cubical box without a lid. How many of the cubes are touching the sides or the bottom of the box?

20.

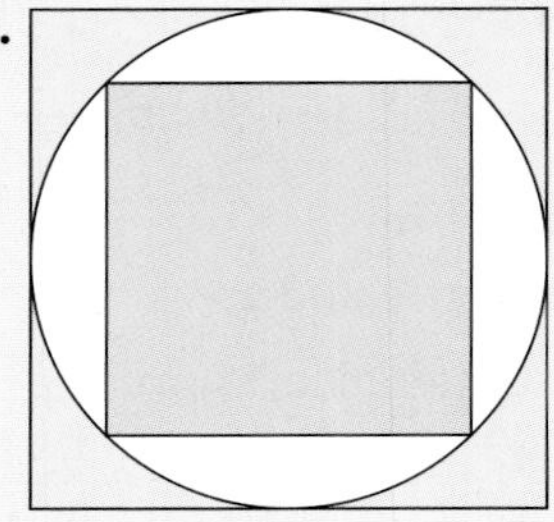

In the diagram the area of the smaller square is 10 cm^2.
Find the area of the larger square.

21. A tiny island in the Pacific Ocean is the only place on Earth where the Gibson Oak grows. Each specimen of this magnificent tree requires a land area of 195 m^2 and the wood from each tree can be sold for \$550 000.

The diameter of this circular island is 1·8 km. Calculate the value of the wood on the island. Assume that the island is covered by the trees.

Examination exercise 4B

1.

NOT TO SCALE

40 cm

70 cm

The diagram shows a water tank which is open at the top.
The tank is in the shape of a cylinder with radius 40 cm and height 70 cm.

a) i) Calculate the total surface area of the cylinder. [2]

ii) The cylinder is made of metal which costs \$2.40 per **square metre**.
Calculate the cost of the metal. [3]

b) i) Calculate the volume of the cylinder. [2]

ii) Water is poured into the empty cylinder at a rate of 8 cm^3 per second.
Calculate the time taken to fill the cylinder.
Give your answer in hours and minutes, correct to the nearest minute. [4]

Cambridge IGCSE International Mathematics 0607 Paper 41 Q10 May/June 2011

2. A cuboid has a square base of side x cm and a height of y cm.

Find, in terms of x and y,

a) the volume of the cuboid, [1]

b) the total surface area of the cuboid. [2]

Cambridge IGCSE International Mathematics 0607 Paper 2 Q5 October/November 2010

3.

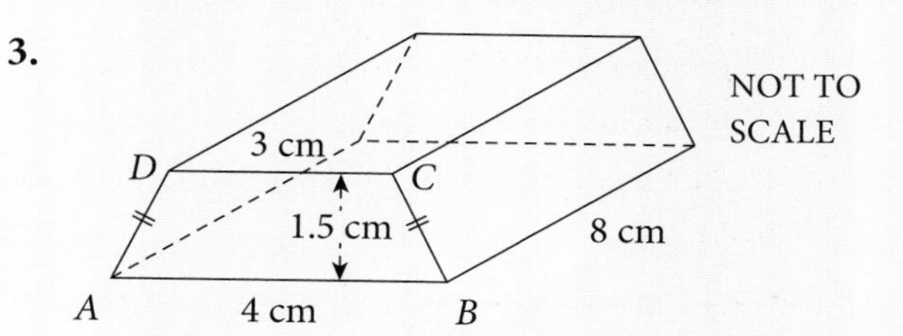

The diagram shows a gold bar of length 8 cm.
The cross-section of the bar, $ABCD$, is an isosceles trapezium.
$AB = 4$ cm, $DC = 3$ cm and these parallel edges are 1.5 cm apart.

a) Write down the mathematical name for this solid. [1]

b) i) Calculate the area of the trapezium. [2]

ii) One cubic centimetre of gold has a mass of 19.3 g.
Calculate the mass of the gold bar. [3]

iii) Calculate the **total** surface area of the gold bar. [4]

c) A box can hold a maximum of 20 kg.
Find the largest number of gold bars that can be put in the box. [3]

Cambridge IGCSE International Mathematics 0607 Paper 4 Q10 May/June 2010

5 Functions 1

René Descartes (1596–1650) was one of the greatest philosophers of his time. Strangely, his restless mind only found peace and quiet when he became a soldier and apparently he discovered the idea of 'Cartesian' geometry in a dream before the battle of Prague. The word 'Cartesian' is derived from his name and his work formed the link between geometry and algebra which inevitably led to the discovery of calculus. He finally settled in Holland for ten years, but later moved to Sweden where he died of pneumonia.

3.1 Notation
Domain and range (domain is $\mathbb{R}$ unless otherwise stated)
Mapping diagrams

3.2 Recognition of the following function types from the shape of their graphs (some of a, b, c or d may be 0):

linear	$f(x) = ax + b$
quadratic	$f(x) = ax^2 + bx + c$
cubic	$f(x) = ax^3 + bx^2 + cx + d$
reciprocal	$f(x) = a/x$
exponential	$f(x) = a^x$ with $0 < a < 1$ or $a > 1$ (compound interest)
absolute value	$f(x) = \|ax + b\|$
trigonometric	$f(x) = a\sin(bx)$; $a\cos(bx)$; $\tan x$

3.3 Determination of at most two of a, b, c or d in simple cases of 3·2

3.4 Finding the quadratic function given vertex and another point, x-intercepts and a point, vertex or x-intercepts with $a = 1$.
$y = a(x - h)^2 + k$ has a vertex of (h, k)

3.5 Understanding of the concept of asymptotes and graphical identification of examples e.g. $f(x) = \tan x$, asymptotes at 90°, 270° etc., excludes algebraic derivation of asymptotes, includes oblique asymptotes

3.9 Inverse function f^{-1}
Syllabus objectives 3.7, 3.8 and 3.10 are covered in Chapter 9.

5.1 Function notation

If A and B are non-empty sets then a **mapping** from A to B is a rule which associates an element of B with every element of A.
Here are two types of mapping:

- **Type 1**

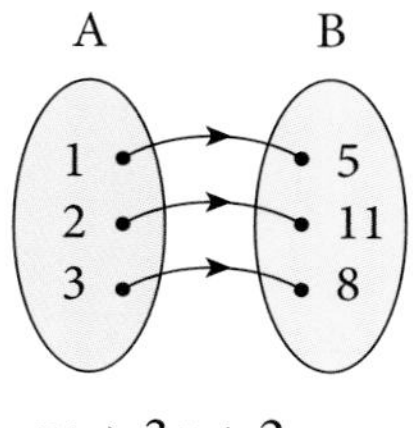

$x \mapsto 3x + 2$

This is called a **one-to-one** mapping.

We use 'an arrow with a tail' $\mapsto$ to avoid confusion with a simple arrow such as '$x \rightarrow 0$' which means '*x tends towards* 0'.

- **Type 2**

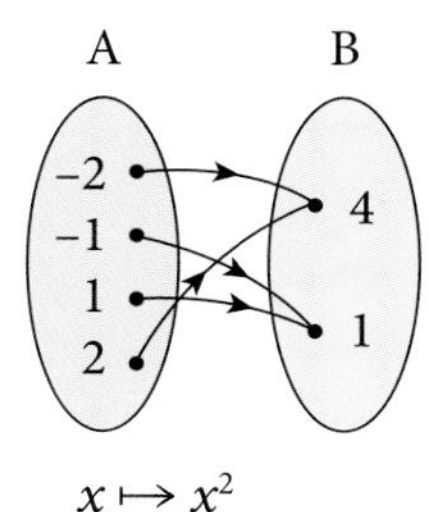

$x \mapsto x^2$

This is called a **two-to-one** mapping or a **many-to-one** mapping.

A **function** is a mapping which associates one and only one element of B with every element of A.
So a function is a one-to-one mapping or a many-to-one mapping.
The notations used for functions are either 'f(x) =' or 'f: $x \mapsto$'.

The graph of $y = 2x + 1$ is a one-to-one mapping

The graph of $y = \frac{1}{x}$ is a one-to-one mapping

The graph of $y = \sin x$ is a many-to-one mapping.

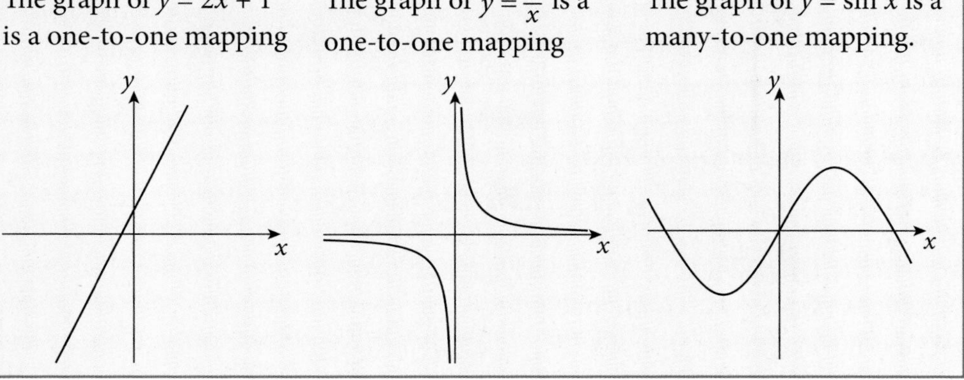

- The **domain** of a function is the set of all the points on which the function acts.
- The **range** of a function is the set of all the points to which the function maps.

In general the best way to find the range of a function is to sketch a graph of the function.

Example 1

Find the range of each function below.

a) $f(x) = x^2$, for $x \in \mathbb{R}$

The domain $[x \in \mathbb{R}]$ is the set of all real numbers.

The graph of $f(x) = x^2$ is shown.

The range of the function is $f(x) \geq 0$.

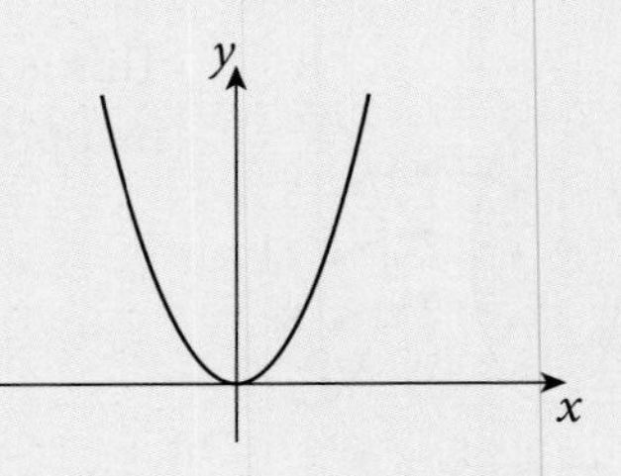

b) $f(x) = 2x + 1$, for $x \in \mathbb{R}, x > 0$

The domain is all values of x greater than 0.

From the graph of $f(x) = 2x + 1$, we see that the range is $f(x) > 1$.

Notice that since $x = 0$ is not in the domain, the range is $f(x) > 1$ and *not* $f(x) \geq 1$.

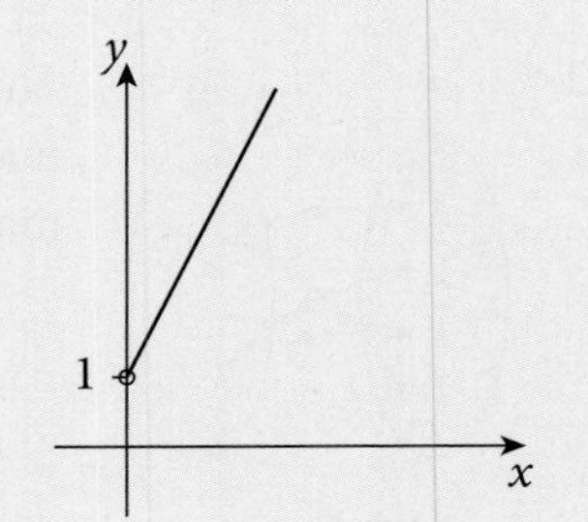

c) $f(x) = \sin x$, for $0° \leq x \leq 360°$

The range of the function is $-1 \leq f(x) \leq 1$.

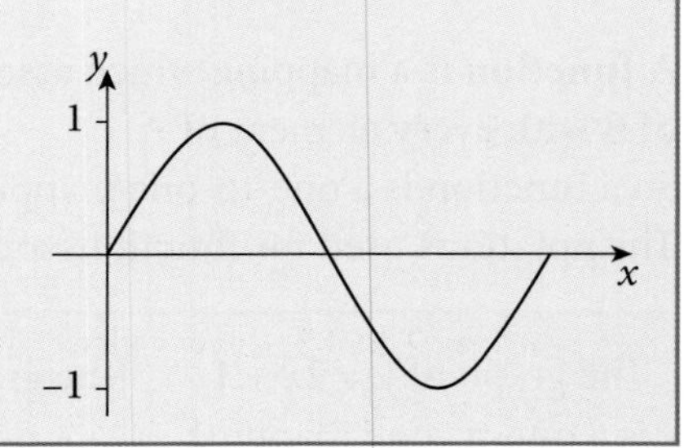

Example 2

If $f(x) = 3x - 1$, $g(x) = 1 - x^2$ and $h: x \mapsto \frac{10}{x}$ find:

a) $f(2)$ **b)** $f(-2)$ **c)** $h(2)$

d) $g(3)$ **e)** x if $f(x) = 1$

a) $f(2) = 5$ **b)** $f(-2) = -7$ **c)** $h(2) = 5$

d) $g(3) = -8$ **e)** If $f(x) = 1$

then $3x - 1 = 1$

$$3x = 2$$

$$x = \frac{2}{3}$$

Flow diagrams

The function f in Example 2 consisted of two simpler functions as illustrated by a flow diagram.

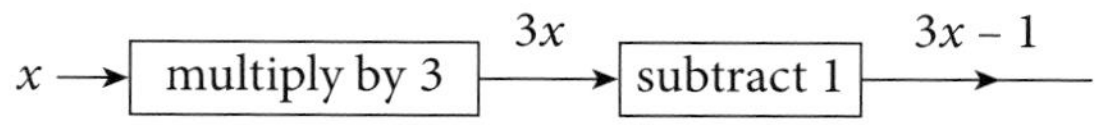

$x \rightarrow$ [multiply by 3] $\xrightarrow{3x}$ [subtract 1] $\xrightarrow{3x-1}$

It is obviously important to 'multiply by 3' and 'subtract 1' in the correct order.

Example 3

Draw flow diagrams for the following functions.

a) $f(x) = (2x + 5)^2$, **b)** $g(x) = \frac{5-7x}{3}$

a) $\xrightarrow{x}$ [multiply by 2] $\xrightarrow{2x}$ [add 5] $\xrightarrow{2x+5}$ [square] $\xrightarrow{(2x+5)^2}$

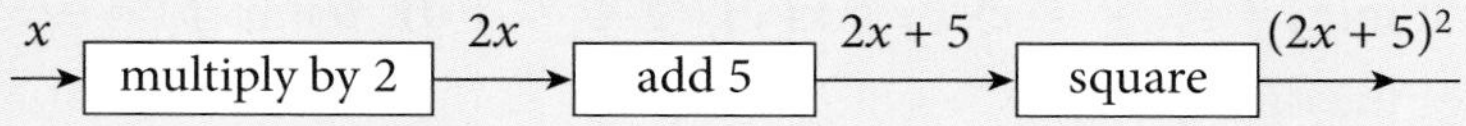

b) $\xrightarrow{x}$ [multiply by (−7)] $\xrightarrow{-7x}$ [add 5] $\xrightarrow{5-7x}$ [divide by 3] $\xrightarrow{\frac{5-7x}{3}}$

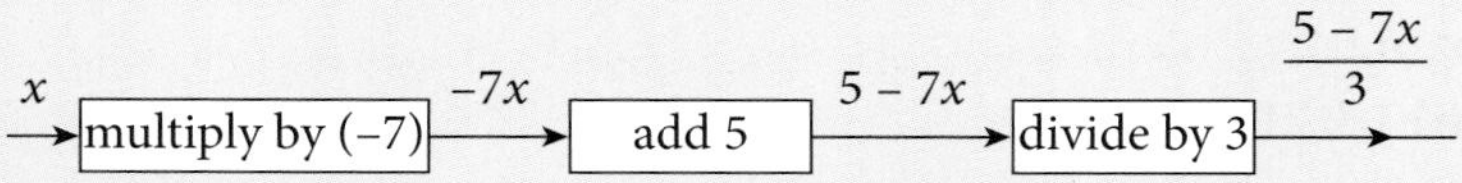

Exercise 5.1

1. Given the functions $h(x) = x^2 + 1$ and $g(x) = 10x + 1$, find:

a) h(2), h(−3), h(0) **b)** g(2), g(10), g(−3)

In questions **2** to **15**, draw a flow diagram for each function.

2. $f(x) = 5x + 4$

3. $f(x) = 3(x - 4)$

4. $f(x) = (2x + 7)^2$

5. $f(x) = \left(\frac{9+5x}{4}\right)$

6. $f(x) = \frac{4-3x}{5}$

7. $f(x) = 2x^2 + 1$

8. $f(x) = \frac{3x^2}{2} + 5$

9. $f(x) = \sqrt{(4x-5)}$

10. $f(x) = 4\sqrt{(x^2+10)}$

11. $f(x) = (7 - 3x)^2$

12. $f(x) = 4(3x + 1)^2 + 5$

13. $f(x) = 5 - x^2$

14. $f(x) = \frac{10\sqrt{(x^2+1)}+6}{4}$

15. $f(x) = \left(\frac{x^3}{4}+1\right)^2 - 6$

In question **14**, $\sqrt{\ }$ means 'the positive square root'

For questions **16**, **17** and **18**, the functions f, g and h are defined as follows:

$f(x) = 1 - 2x$ $\quad g(x) = \frac{x^3}{10}$ $\quad h(x) = \frac{12}{x}$

16. Find:

a) $f(5), f(-5), f\left(\frac{1}{4}\right)$ **b)** $g(2), g(-3), g\left(\frac{1}{2}\right)$ **c)** $h(3), h(10), h\left(\frac{1}{3}\right)$

17. Find:

a) x if $f(x) = 1$ **b)** x if $f(x) = -11$ **c)** x if $h(x) = 1$

18. Find:

a) y if $g(y) = 100$ **b)** z if $h(z) = 24$ **c)** w if $g(w) = 0{\cdot}8$

For questions **19** and **20**, the functions k, ℓ and m are defined as follows:

$$k(x) = \frac{2x^2}{3}$$

$$\ell(x) = \sqrt{[(x-1)(x-2)]}$$

$$m(x) = 10 - x^2$$

19. Find:

a) $k(3), k(6), k(-3)$ **b)** $\ell(2), \ell(0), \ell(4)$ **c)** $m(4), m(-2), m\left(\frac{1}{2}\right)$

20. Find:

a) x if $k(x) = 6$ **b)** x if $m(x) = 1$ **c)** y if $k(y) = 2\frac{2}{3}$ **d)** p if $m(p) = -26$

21. $f(x)$ is defined as the product of the digits of x,

e.g. $f(12) = 1 \times 2 = 2$

a) Find: **i)** $f(25)$ **ii)** $f(713)$

b) If x is an integer with three digits, find:

i) x such that $f(x) = 1$ **ii)** the largest x such that $f(x) = 4$

iii) the largest x such that $f(x) = 0$ **iv)** the smallest x such that $f(x) = 2$

22. $g(x)$ is defined as the sum of the prime factors of x,
e.g. $g(12) = 2 + 3 = 5$. Find:

a) $g(10)$ **b)** $g(21)$ **c)** $g(36)$ **d)** $g(99)$ **e)** $g(100)$ **f)** $g(1000)$

23. $h(x)$ is defined as the number of letters used to spell out
the number x, e.g. $h(1) = 3$. Find:

a) $h(2)$ **b)** $h(11)$ **c)** $h(18)$ **d)** the largest value of x for which $h(x) = 3$

24. If $f : x \mapsto$ 'next prime number greater than x', find:

a) $f(7)$ **b)** $f(14)$ **c)** $f[f(3)]$

25. If $g : x \mapsto 2^x + 1$, find:

a) $g(2)$ **b)** $g(4)$ **c)** $g(-1)$ **d)** the value of x if $g(x) = 9$

26. The function f is defined as $f : x \to ax + b$ where a and b are constants.

If $f(1) = 8$ and $f(4) = 17$, find the values of a and b.

27. The function g is defined as $g(x) = ax^2 + b$ where a and b are constants.

If $g(2) = 3$ and $g(-3) = 13$, find the values of a and b.

28. Functions h and k are defined as follows:

$h : x \mapsto x^2 + 1$, $k : x \mapsto ax + b$, where a and b are constants.

If $h(0) = k(0)$ and $k(2) = 15$, find the values of a and b.

Composite functions

The function $f(x) = 3x + 2$ is itself a composite function, consisting of two simpler functions: 'multiply by 3' and 'add 2'.

If $f(x) = 3x + 2$ and $g(x) = x^2$ then $f[g(x)]$ is a composite function where g is performed first and then f is performed on the result of g. $f[g(x)]$ is usually abbreviated to $fg(x)$.

The function fg may be found using a flow diagram.

Thus x → [square] → x^2 → [multiply by 3] → $3x^2$ → [add 2] → $3x^2 + 2$

(square: g; multiply by 3, add 2: f)

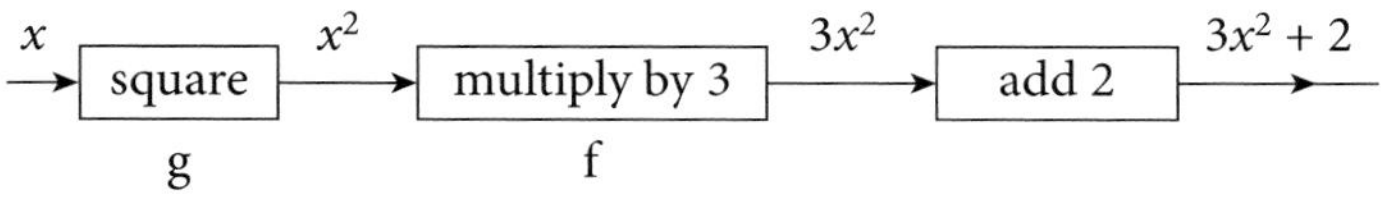

So $fg(x) = 3x^2 + 2$

5.2 Inverse function

If a function f maps a number n onto m, then the inverse function f^{-1} maps m onto n.
The inverse of a given function is found using a flow diagram.

Example 4

- **Method 1**

To find the inverse of f where $f: x \rightarrow \frac{5x-2}{3}$,

a) Draw a flow diagram for f:

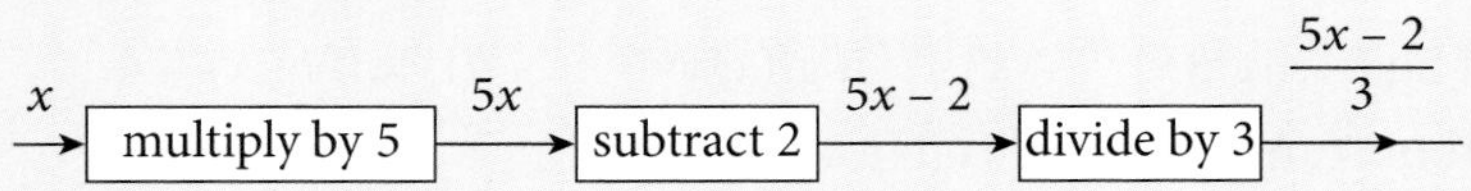

x → [multiply by 5] → $5x$ → [subtract 2] → $5x - 2$ → [divide by 3] → $\frac{5x-2}{3}$

b) Draw a new flow diagram with each operation replaced by its inverse. Start with x on the right:

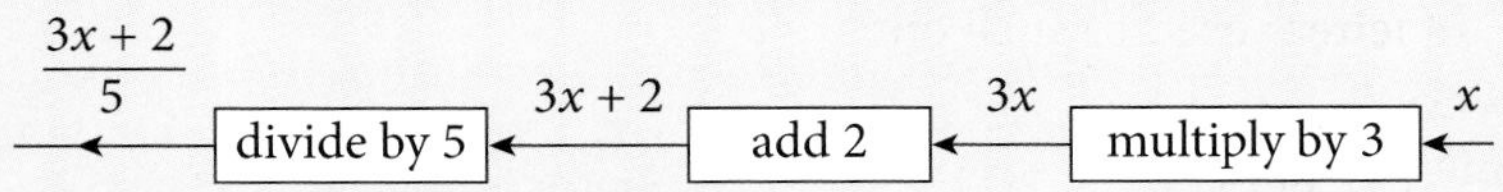

$\frac{3x+2}{5}$ ← [divide by 5] ← $3x + 2$ ← [add 2] ← $3x$ ← [multiply by 3] ← x

Thus the inverse of f is given by

$$f^{-1}: x \mapsto \frac{3x+2}{5} \text{ or } f^{-1}(x) = \frac{3x+2}{5}$$

- **Method 2**

Again, to find the inverse of f where $f: x \rightarrow \frac{5x-2}{3}$,

Let $y = \frac{5x-2}{3}$

Rearrange this equation to make x the subject:

$$3y = 5x - 2 \qquad 3y + 2 = 5x \qquad x = \frac{3y+2}{5}$$

For an inverse function we interchange x and y.

So the inverse function is $\frac{3x+2}{5}$.

Many people prefer this algebraic method. You should use the method which you find easier.

Exercise 5.2

For questions **1** and **2**, the functions f, g and h are as follows:

$f : x \mapsto 4x$

$g : x \mapsto x + 5$

$h : x \mapsto x^2$

1. Find the following in the form '$x \mapsto ...$'

a) fg **b)** gf **c)** hf **d)** fh **e)** gh **f)** fgh **g)** hfg

2. Find:

a) x if $hg(x) = h(x)$ **b)** x if $fh(x) = gh(x)$

For questions **3**, **4** and **5**, the functions f, g and h are as follows:

$f : x \mapsto 2x$

$g : x \mapsto x - 3$

$h : x \mapsto x^2$

3. Find the following in the form '$x \mapsto ...$'

a) fg **b)** gf **c)** gh **d)** hf **e)** ghf **f)** hgf

4. Evaluate:

a) fg(4) **b)** gf(7) **c)** gh(−3) **d)** fgf(2) **e)** ggg(10) **f)** hfh(−2)

5. Find:

a) x if $f(x) = g(x)$ **b)** x if $hg(x) = gh(x)$ **c)** x if $gf(x) = 0$ **d)** x if $fg(x) = 4$

For questions **6**, **7** and **8**, the functions ℓ, m and n are as follows:

$\ell : x \mapsto 2x + 1$

$m : x \mapsto 3x - 1$

$n : x \mapsto x^2$

6. Find the following in the form '$x \mapsto ...$'

a) ℓm **b)** mℓ **c)** ℓn **d)** nm **e)** ℓnm **f)** mℓn

7. Find:

a) ℓm(2) **b)** nℓ(1) **c)** mn(−2) **d)** mm(2) **e)** nℓn(2) **f)** $\ell\ell$m(0)

8. Find:

a) x if $\ell(x) = m(x)$ **b)** two values of x if $n\ell(x) = nm(x)$ **c)** x if $\ell n(x) = mn(x)$

In questions **9** to **22**, find the inverse of each function in the form '$x \mapsto ...$'

9. $f : x \mapsto 5x - 2$

10. $f : x \mapsto 5(x - 2)$

11. $f : x \mapsto 3(2x + 4)$

12. $g : x \mapsto \frac{2x + 1}{3}$

13. $f : x \mapsto \frac{3(x - 1)}{4}$

14. $g : x \mapsto 2(3x + 4) - 6$

15. $h : x \mapsto \frac{1}{2}(4 + 5x) + 10$

16. $k : x \mapsto -7x + 3$

17. $j : x \mapsto \frac{12 - 5x}{3}$

18. $\ell : x \mapsto \dfrac{4-x}{3}+2$

19. $\mathrm{m} : x \mapsto \dfrac{\left[\dfrac{(2x-1)}{4}-3\right]}{5}$

20. $\mathrm{f} : x \mapsto \dfrac{3(10-2x)}{7}$

21. $\mathrm{g} : x \mapsto \left[\dfrac{\frac{x}{4}+6}{5}\right]+7$

22. A calculator has the following function buttons:

$x \mapsto x^2$	$x \mapsto \sqrt{x}$	$x \mapsto \dfrac{1}{x}$	$x \mapsto \log x$	
$x \mapsto \ln x$	$x \mapsto \sin x$	$x \mapsto \cos x$	$x \mapsto \tan x$	$x \mapsto x!$

$x!$ is 'x factorial'
$4! = 4 \times 3 \times 2 \times 1$
$3! = 3 \times 2 \times 1$ etc

Find which button was used for the following input/outputs:

a) $1\,000\,000 \mapsto 1000$ **b)** $1000 \mapsto 3$ **c)** $3 \mapsto 6$ **d)** $0{\cdot}2 \mapsto 0{\cdot}04$

e) $10 \mapsto 0{\cdot}1$ **f)** $45 \mapsto 1$ **g)** $0{\cdot}5 \mapsto 2$ **h)** $64 \mapsto 8$

i) $60 \mapsto 0{\cdot}5$ **j)** $1 \mapsto 0$ **k)** $135 \mapsto -1$ **i)** $10 \mapsto 3\,628\,800$

m) $0 \mapsto 1$ **n)** $30 \mapsto 0{\cdot}5$ **o)** $90 \mapsto 0$ **p)** $0{\cdot}4 \mapsto 2{\cdot}5$

q) $4 \mapsto 24$ **r)** $1\,000\,000 \mapsto 6$

5.3 The absolute value function (modulus function)

The **absolute value** of x is written $|x|$ and means the positive value of x.
For example, $|-2| = 2$ and $|7| = 7$. This function is also called the **modulus function**.

Example 5

Sketch the graph of $y = |x-2|$

a) Sketch $y = x - 2$

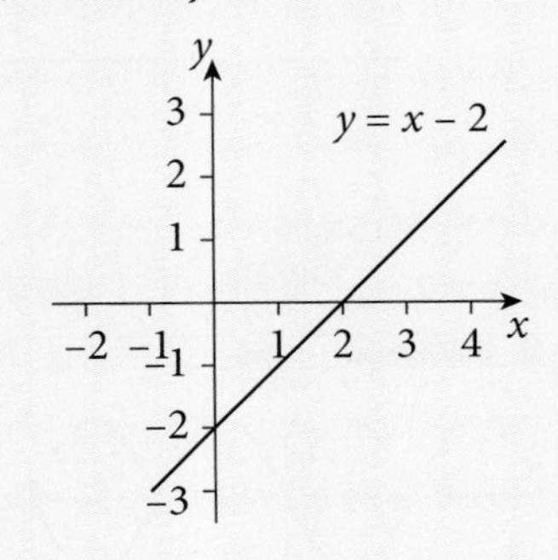

b) Where y is negative, draw the reflection of $y = x - 2$ in the x-axis.

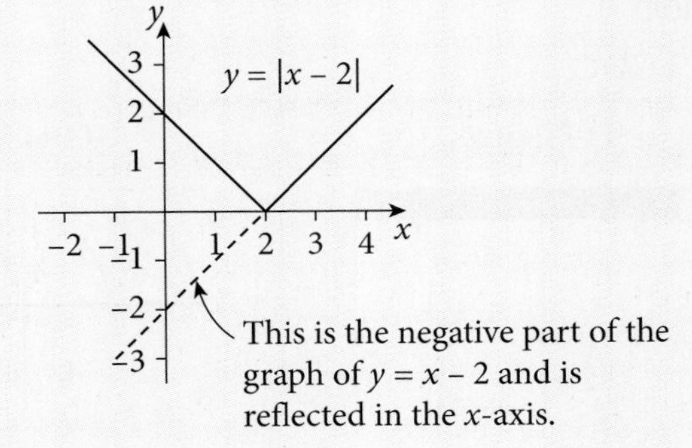

▶ Continued on next page

GDC help

Using a GDC

TI-Nspire	TI84-Plus	Casio
	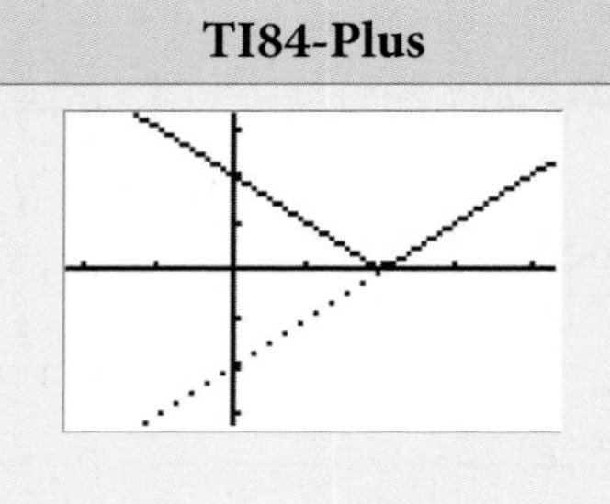	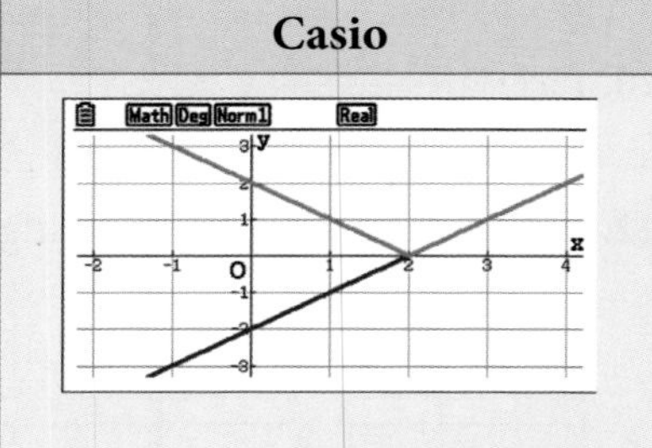

Example 6

Sketch the graph of $y = |\sin x|$ for $-360° \leq x \leq 360°$

a) Sketch $y = \sin x$

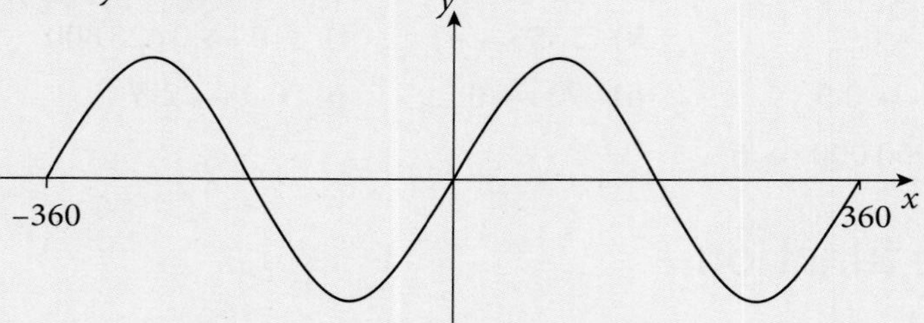

b) Where y is negative, reflect the curve in the x-axis.

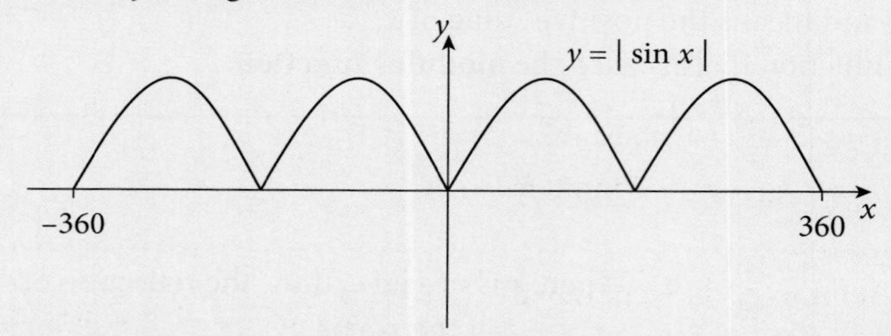

GDC help

Using a GDC

TI-Nspire	TI84-Plus	Casio
	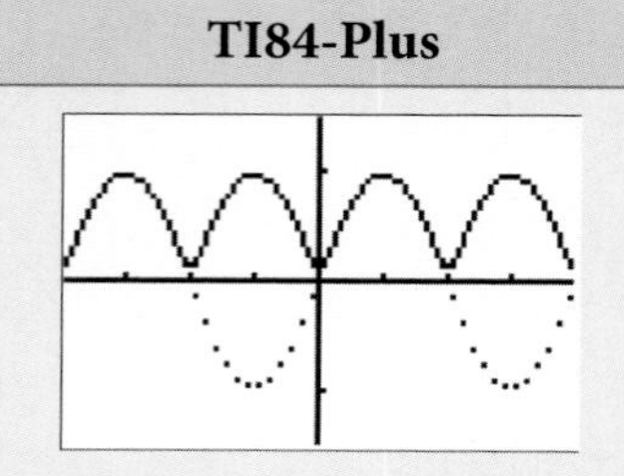	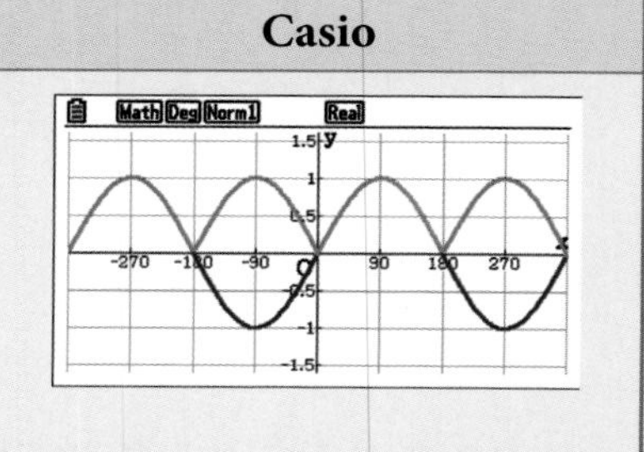

Exercise 5.3

1. Write down the value of the following.

a) $|6|$ **b)** $|-10|$ **c)** $|\sqrt{4}|$

d) $|x+3|$ when $x=-5$ **e)** $|-\sqrt{100}|$ **f)** $|x^2-20|$ when $x=3$

2. Sketch the graphs of the following, marking where the graphs cross the axes.

a) $y=|x-4|$ **b)** $y=|x+2|$ **c)** $y=|3x-1|$

d) $y=|x^2-2x|$ **e)** $y=|x^2-9|$ **f)** $y=|x^2-5x+4|$

g) $y=\left|\frac{1}{x}\right|$ **h)** $y=|x^3-x^2-6x|$

5.4 Sketch graphs

A **sketch** graph shows the general shape of a graph and you should learn the shapes of the following functions.

Asymptotes

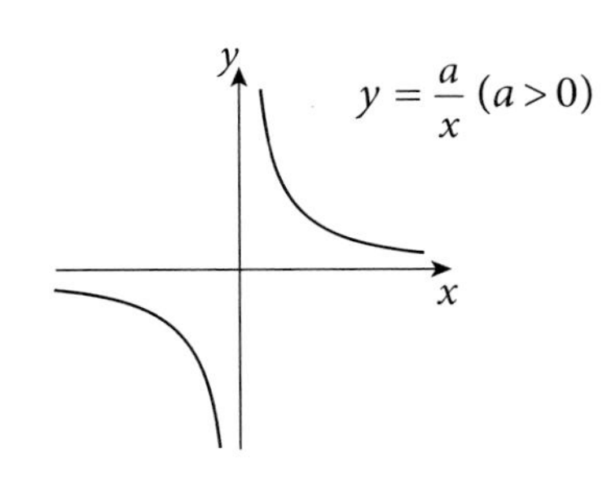

Consider the equation $y=\frac{1}{x}$

When $x=\frac{1}{2}$, $y=\frac{1}{\frac{1}{2}}=2$.

When $x=\frac{1}{100}$, $y=\frac{1}{\frac{1}{100}}=100$.

As the denominator of the fraction $\frac{1}{x}$ gets smaller, the answer gets larger. An 'infinitely small' denominator gives an 'infinitely large' answer.

We write $\frac{1}{0}\to\infty$ to mean '$\frac{1}{0}$ tends to an infinitely large number.'

When $x=1000$, $y=\frac{1}{1000}$

As $x\to\infty$, $y\to 0$

The curve has **asymptotes** at $x=0$ and $y=0$.
The curve gets very near but never actually touches the asymptotes.

Common curves

$y = ax^2 + bx + c$

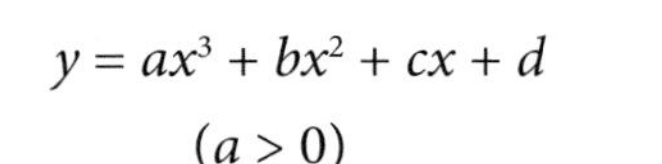

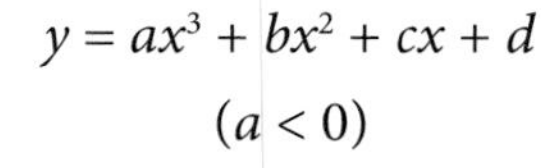

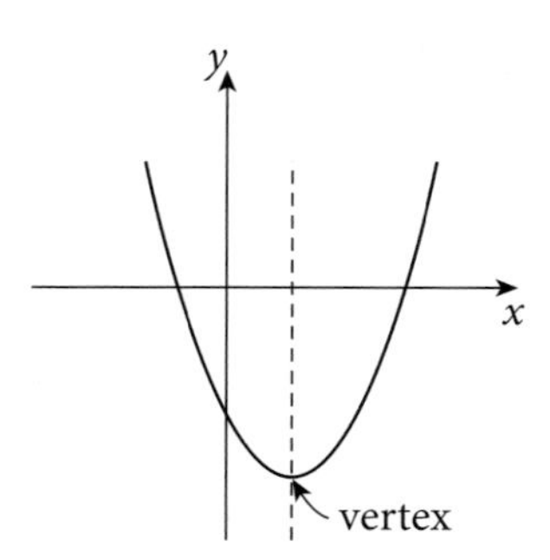

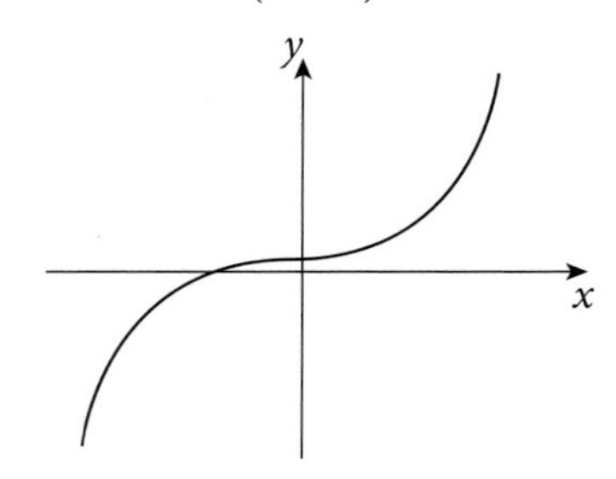

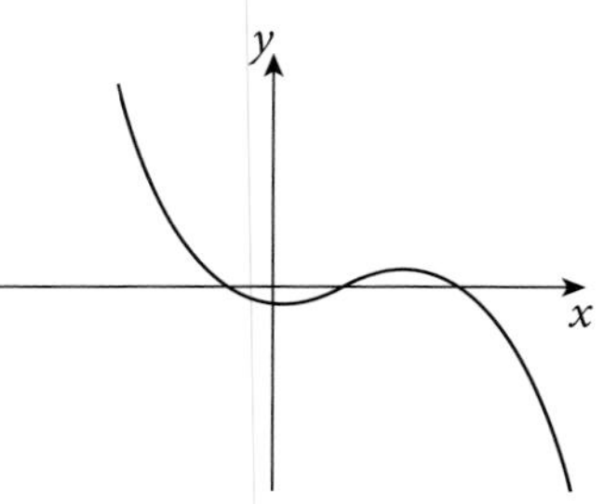

The curve is symmetrical about a line parallel to the y axis which passes through the lowest point on the curve (the vertex).

These two curves may cut or touch the x-axis at 1, 2 or 3 points.

$y = a^x$

$y = |ax + b|$

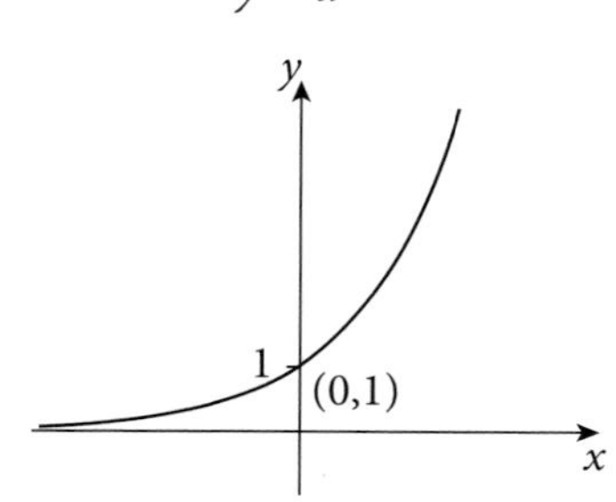

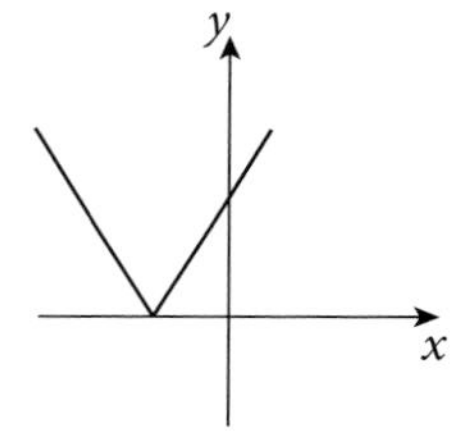

> The 'V' is symmetrical about a vertical axis.

$y = a^x$ is known as an **exponential** function.

$y = 0$ is an asymptote.

The curve cuts the y axis at (0, 1).

The absolute value function meets the x axis at $x = -\frac{b}{a}$

The graph is symmetrical about the line $x = -\frac{b}{a}$

Exercise 5.4

1. The sketches opposite have an equation of the form of one of the following:

$f(x) = ax + b$; $f(x) = ax^2 + bx + c$; $f(x) = ax^3 + bx^2 + cx + d$;

$f(x) = \frac{a}{x}$; $f(x) = a^x$; $f(x) = |ax + b|$

For each sketch state which equation matches the graph shown.

For the cubic functions state whether a is positive or negative.

a) 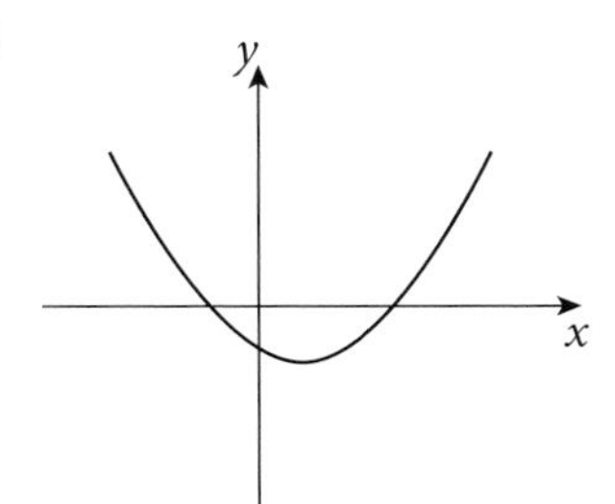

b) 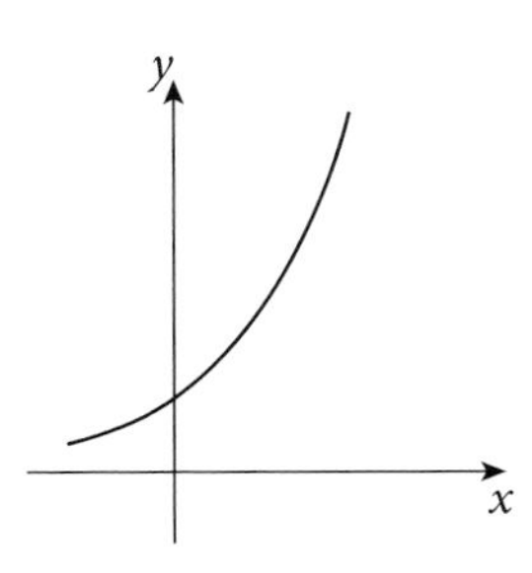

c) 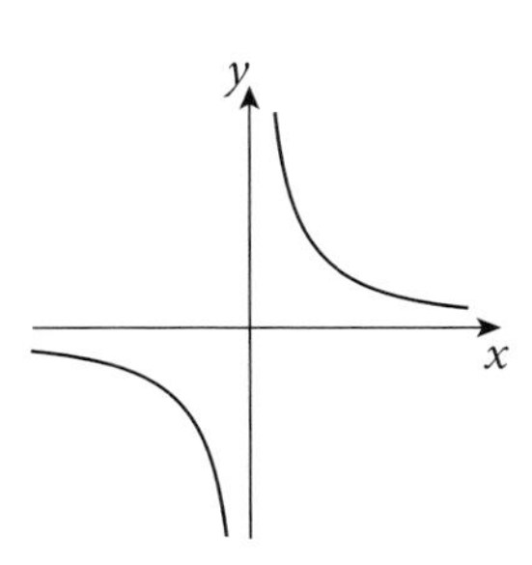

d) 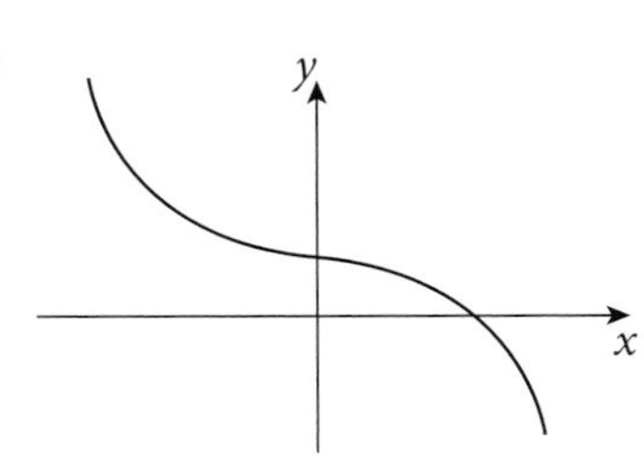

e) 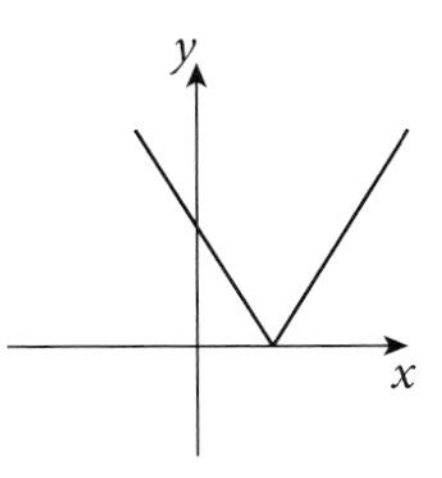

f) 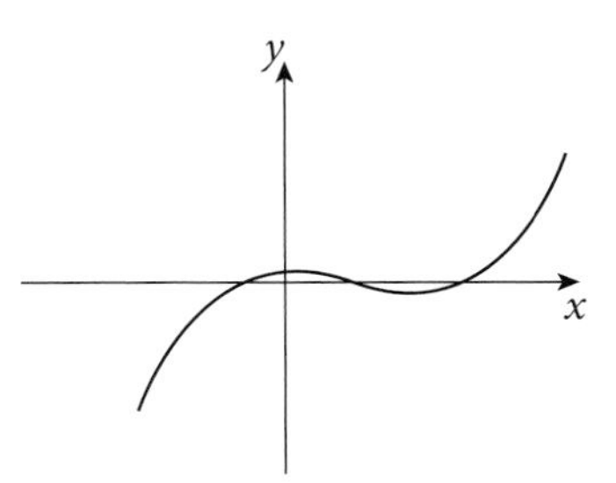

Exercise 5.5

Sketch the graphs of the following, first without a calculator and then check your answer with a calculator. Write down the equations of any asymptotes involved.

1. $y = x^2 + 4x - 1$

2. $y = x^3 + x + 2$

3. $y = \dfrac{15}{x}$

4. $y = 3^x$

5. $y = 8 + x - x^2$

6. $y = (x - 1)(x - 3)$

7. $y = 4 + x - x^3$

8. $y = \dfrac{12}{x-1}$

9. $y = |x - 3|$

10. $y = \left(\dfrac{1}{2}\right)^x$

11. $y = 2^x + 1$

12. $y = |3x - 6|$

13. $y = (x - 3)(x + 1)$

14. $y = x(2x - 7)$

15. $y = x(x - 2)(x - 5)$

16. $y = \dfrac{x}{2}(x+1)(x-4)$

17. $y = |(x-1)(x-3)|$

18. $y = \dfrac{10}{x^2}$

19. a) Display the curve $y = x + \dfrac{1}{x}$ on a calculator.

b) Copy and complete the sentence:
'As x get larger and larger, $\dfrac{1}{x}$ gets ________.'

c) The graph of $y = x + \dfrac{1}{x}$ has two asymptotes.
Write down the equations of the two asymptotes.

20. Sketch the curves

a) $y = \dfrac{10}{x+2}$

b) $y = \dfrac{12}{(x-1)(x+2)}$

c) $y = \dfrac{x}{(x-3)(x+3)}$

d) $y = \dfrac{(x-2)}{x(x-4)}$

e) $y = \dfrac{2x-1}{x^2+x}$

21. Here are five sketch graphs:

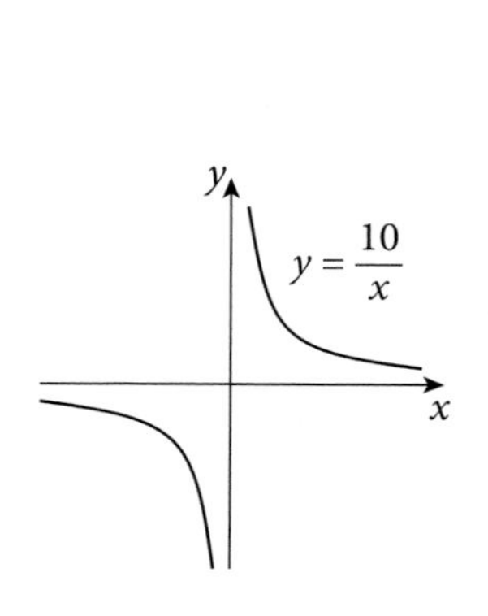

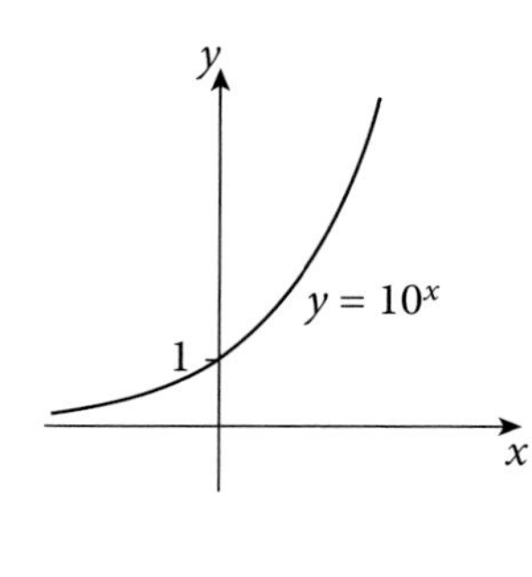

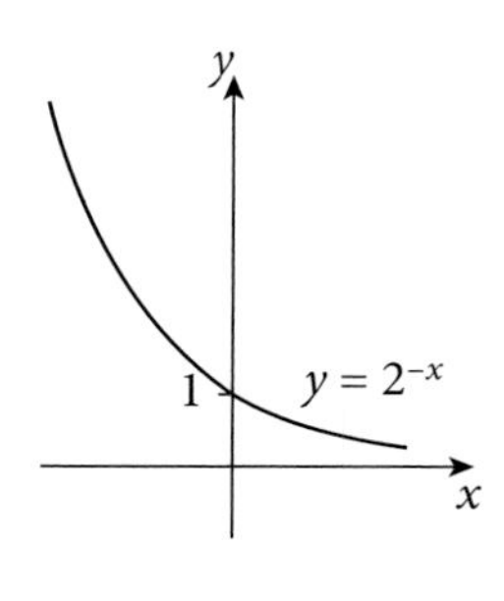

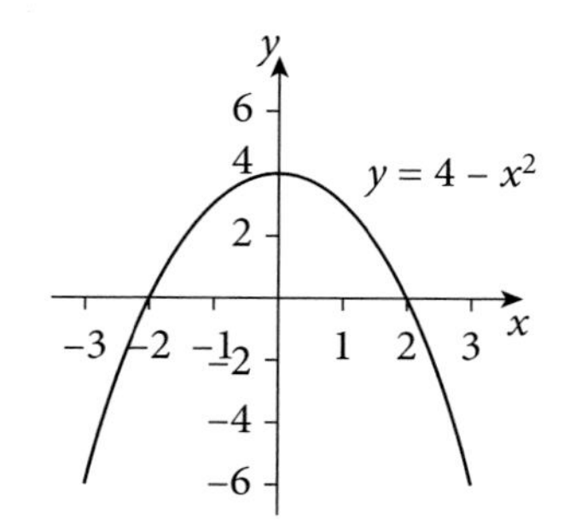

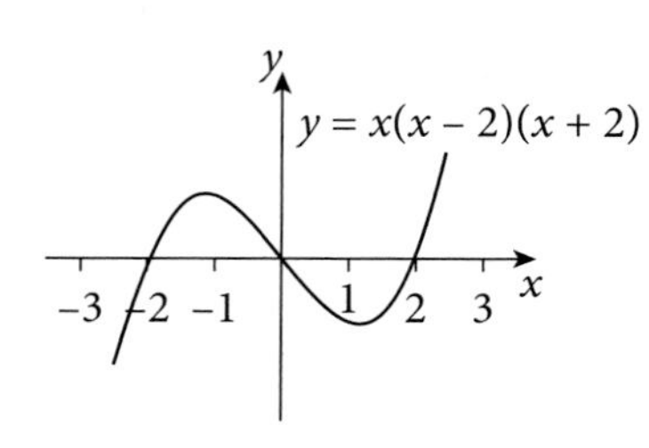

Make your own sketch graphs to find the *number of solutions* of the following equations:

a) $\frac{10}{x} = 10^x$ **b)** $4 - x^2 = 10^x$ **c)** $x(x-2)(x+2) = \frac{10}{x}$

d) $2^{-x} = 10^x$ **e)** $4 - x^2 = 2^{-x}$ **f)** $x(x - 2)(x + 2) = 0$

5.5 Interpreting graphs

Exercise 5.6

1. The graph shows how to convert miles into kilometres.

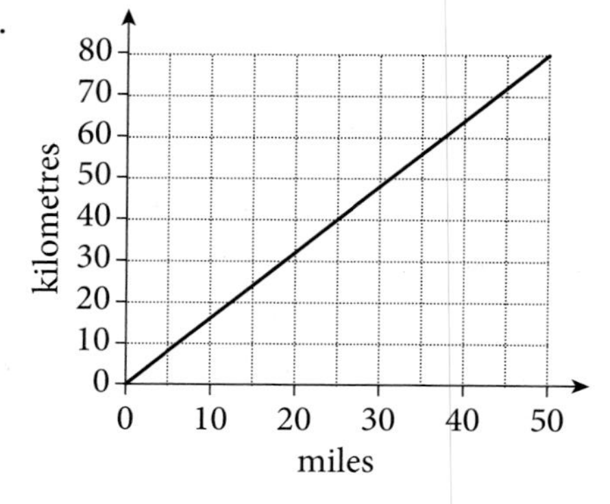

a) Use the graph to find approximately how many kilometres are the same as:

i) 25 miles **ii)** 15 miles

iii) 45 miles **iv)** 5 miles

b) Use the graph to find approximately how many miles are the same as:

i) 64 km **ii)** 56 km

iii) 16 km **iv)** 32 km

2. The graph shows how to convert pounds into euros.

   a) Use the graph to find approximately how many euros are the same as:

   i) £20 ii) £80 iii) £50

   b) Use the graph to find approximately how many pounds are the same as:

   i) €56 ii) €84 iii) €140

   c) Tim spends €154 on clothes in Paris. How many pounds has he spent?

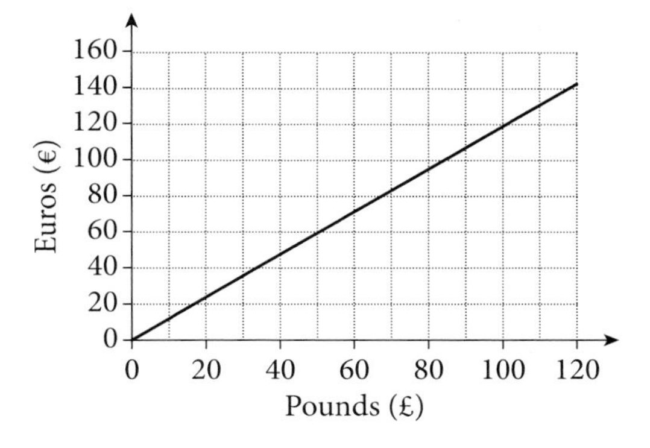

3. A company hires out vans at a basic charge of \$35 plus a charge of 20c per km travelled. Copy and complete the table where x is the number of km travelled and C is the total cost in dollars.

x	0	50	100	150	200	250	300
C	35			65			95

   Draw a graph of C against x, using scales of 2 cm for 50km on the x-axis and 1 cm for \$10 on the C-axis.

   a) Use the graph to find the number of miles travelled when the total cost was \$71.

   b) What is the formula connecting C and x?

4. A car travels along a motorway and the amount of petrol in its tank is monitored as shown on the graph.

   a) How much petrol was bought at the first stop?

   b) What was the petrol consumption in km per litre

   i) before the first stop

   ii) between the two stops?

   c) What was the average petrol consumption over the 200 km?

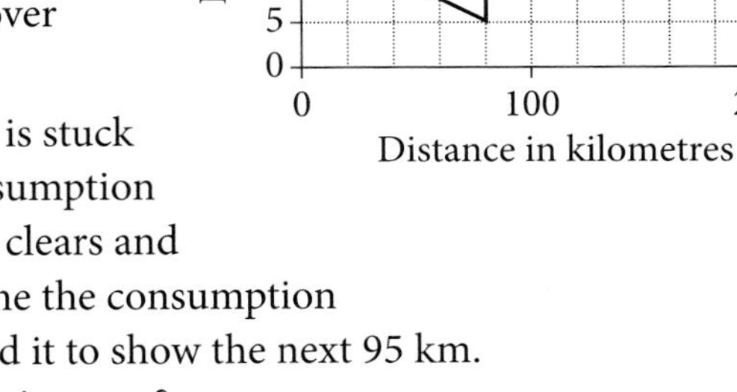

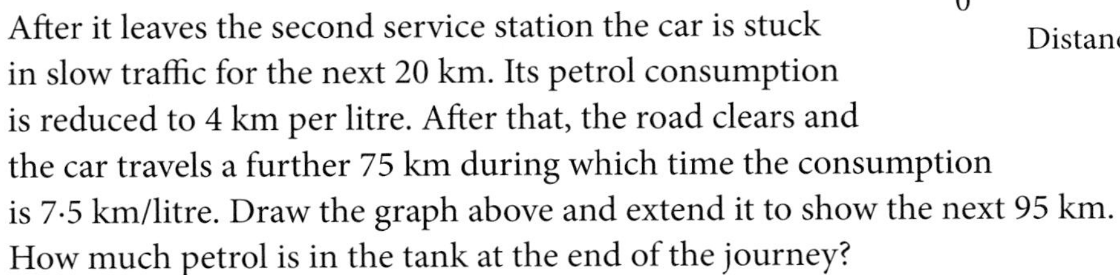

   After it leaves the second service station the car is stuck in slow traffic for the next 20 km. Its petrol consumption is reduced to 4 km per litre. After that, the road clears and the car travels a further 75 km during which time the consumption is 7·5 km/litre. Draw the graph above and extend it to show the next 95 km. How much petrol is in the tank at the end of the journey?

5. A firm makes a profit of P thousand dollars from producing x thousand tiles.
Corresponding values of P and x are given below

x	0	0·5	1·0	1·5	2·0	2·5	3·0
P	−1·0	0·75	2·0	2·75	3·0	2·75	2·0

Using a scale of 4 cm to one unit on each axis, draw the graph of P against x. Use your graph to find:

Plot x on the horizontal axis.

a) the number of tiles the firm should produce in order to make the maximum profit

b) the minimum number of tiles that should be produced to cover the cost of production

c) the range of values of x for which the profit is more than \$2850.

6. A small firm increases its monthly expenditure on advertising and records its monthly income from sales.

Month	1	2	3	4	5	6	7
Expenditure (\$)	100	200	300	400	500	600	700
Income (\$)	280	450	560	630	680	720	740

Draw a graph to display this information.

a) Is it wise to spend \$100 per month on advertising?

b) Is it wise to spend \$700 per month?

c) What is the most sensible level of expenditure on advertising?

5.6 The quadratic function

The general form of the quadratic function is $f(x) = ax^2 + bx + c$, where a, b and c are constants (numbers).

We can find the function if we know either

- the x-intercepts and a point,
- the vertex or x-intercepts with $a = 1$, or
- the vertex and another point.

Example 7

Find the equation of the quadratic curve with x-intercepts at $(-1, 0)$ and $(3, 0)$ and which passes through the point $(2, -6)$.

Since the curve has x-intercepts at $x = -1$ and $x = 3$, the quadratic is of the form $y = n\,(x + 1)(x - 3)$, where n is a number to be found.

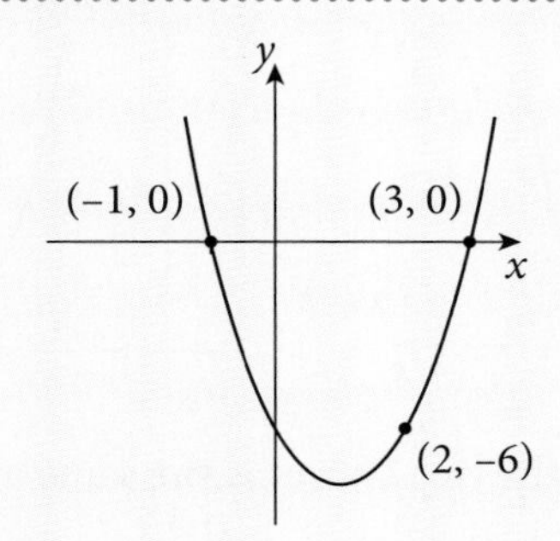

The point $(2, -6)$ lies on the curve

$\therefore\ -6 = n\,(2 + 1)(2 - 3)$

$-6 = n \times -3$

$n = 2$

So the equation of the curve is

$y = 2(x + 1)(x - 3)$

Example 8

Find the equation of the quadratic curve of the form $y = x^2 + bx + c$ with x-intercepts at $(-2, 0)$ and $(3, 0)$.

From the x-intercepts we have

$x^2 + bx + c = (x + 2)(x - 3)$

i.e. $x^2 + bx + c = x^2 - x - 6$

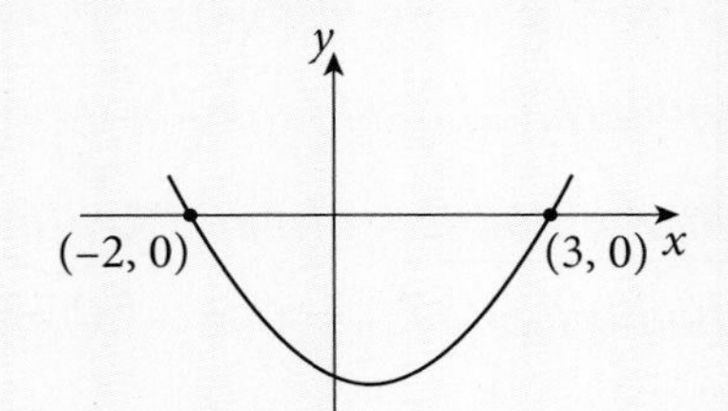

The equation of the curve is

$y = x^2 - x - 6$

Example 9

Find the quadratic function which has its vertex at $(-2, -27)$ and passes through the point $(-5, 0)$.

Quadratic functions are symmetrical about a vertical line through the vertex.

AB = 3 units, so BC = 3 units

C has coordinates $(1, 0)$

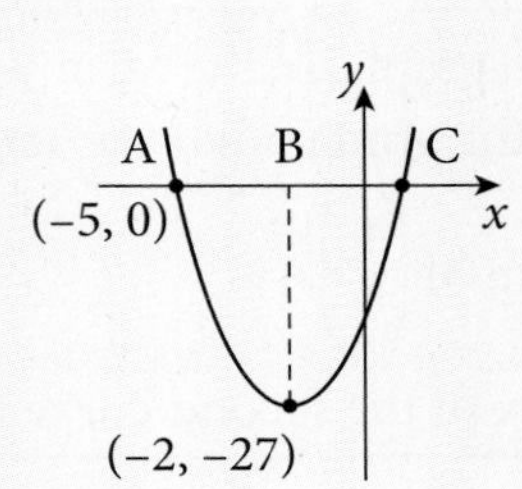

We now know that the function is of the form $n\,(x + 5)(x - 1)$, where n is a number to be found.

The curve passes through the point $(-2, -27)$

$\therefore\ -27 = n\,(-2 + 5)(-2 - 1)$

$-27 = n\,(-9)$

$3 = n$

The function is $y = 3(x + 5)(x - 1)$

Graphs of the form $y = a(x - h)^2 + k$

- Here are sketch graphs of $y = 2x^2$ and $y = 2(x - 3)^2 + 1$:

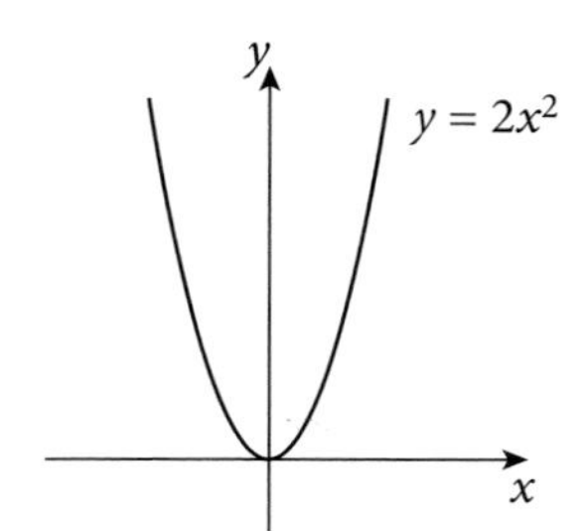

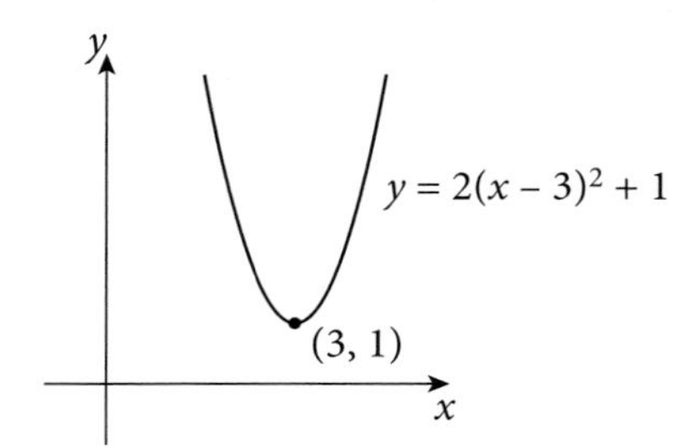

The shape of the two curves is the same and the second curve is obtained by the **translation** $\begin{pmatrix} 3 \\ 1 \end{pmatrix}$

- Here are sketch graphs of $y = \frac{1}{2}x^2$ and $y = \frac{1}{2}(x+2)^2 - 3$:

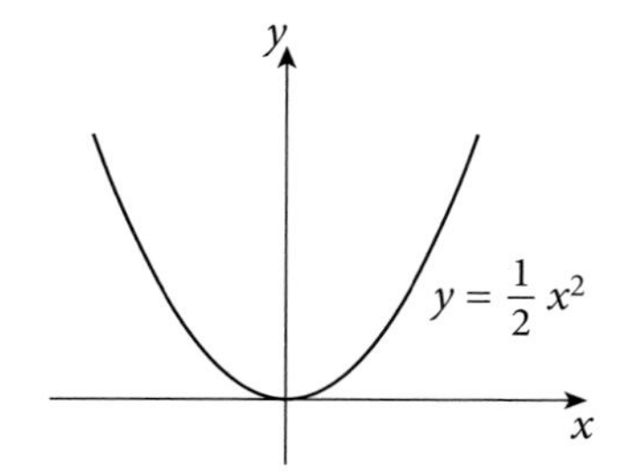

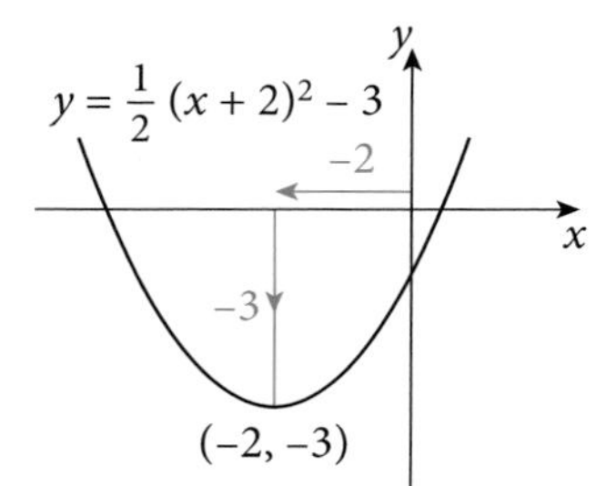

The shape of the two curves is the same, and the curve $y = \frac{1}{2}(x+2)^2 - 3$

is obtained by the translation of the curve $y = \frac{1}{2}x^2$ by $\begin{pmatrix} -2 \\ -3 \end{pmatrix}$.

> In general graphs of the form $y = a(x - h)^2 + k$ have the same shape as the graph of $y = ax^2$ and are obtained by a
>
> translation of $\begin{pmatrix} h \\ k \end{pmatrix}$.
>
> The vertex of the second curve has coordinates (h, k).

Example 10

Find the equation of the quadratic function with vertex (–1, –9) and which passes through the point (3, 7).

We know that the curve $y = a(x - h)^2 + k$ has a vertex at (h, k).

Here the vertex is at (–1, –9)

The curve may be written as

$$y = a[x - (-1)]^2 - 9$$

i.e. $y = a(x + 1)^2 - 9$ [*]

The curve who passes through the point (3, 7).

In [*], $7 = a(3 + 1)^2 - 9$

$$16 = a \times 16$$

$$a = 1$$

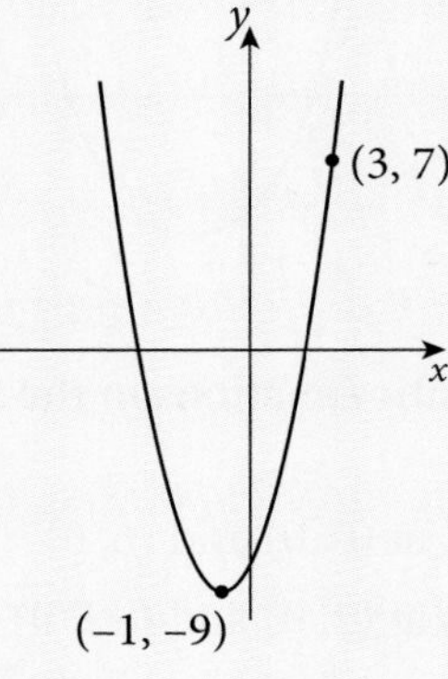

So the equation of the curve is $y = 1 \times (x + 1)^2 - 9$

or $y = x^2 + 2x + 1 - 9$

The function is $y = x^2 + 2x - 8$

Exercise 5.7

1. The graphs below have equation of the form $y = ax^2 + bx + c$, where $a = \pm 1$.

   Find the equation for each graph.

a)

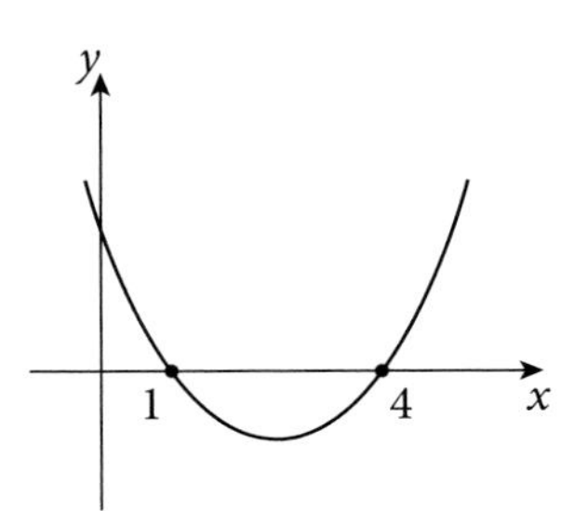

b)

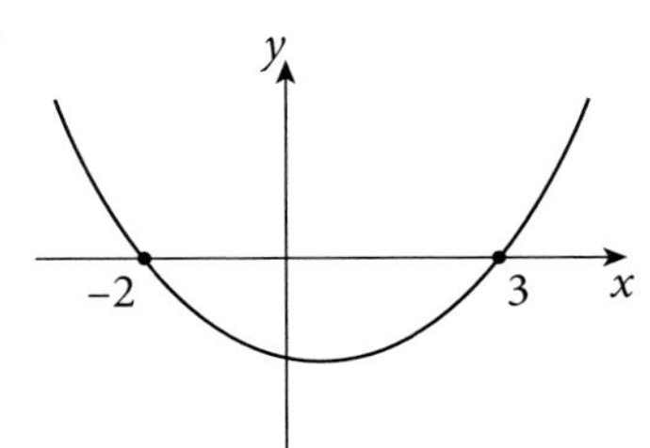

c)

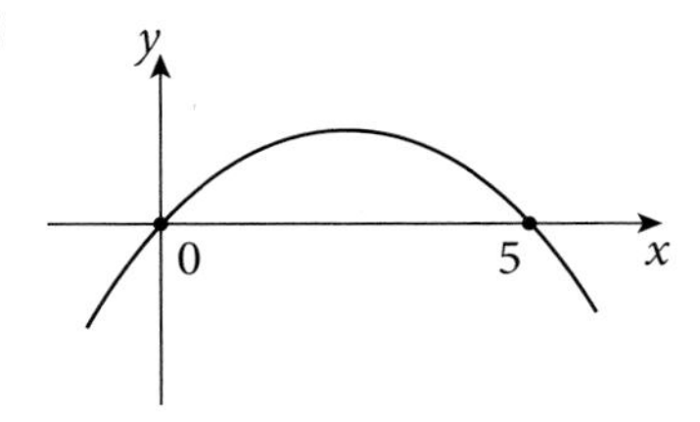

d)

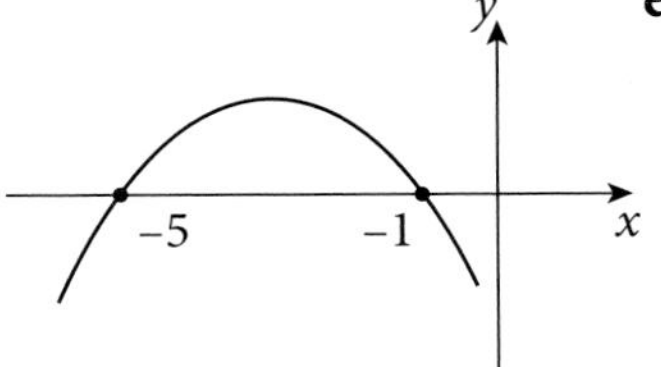

e)

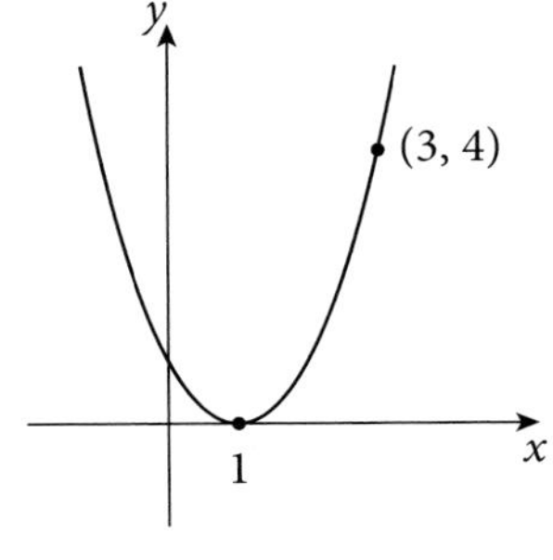

2. Find the equation of the quadratic functions below, giving your answer in the form $y = ax^2 + bx + c$.

a)

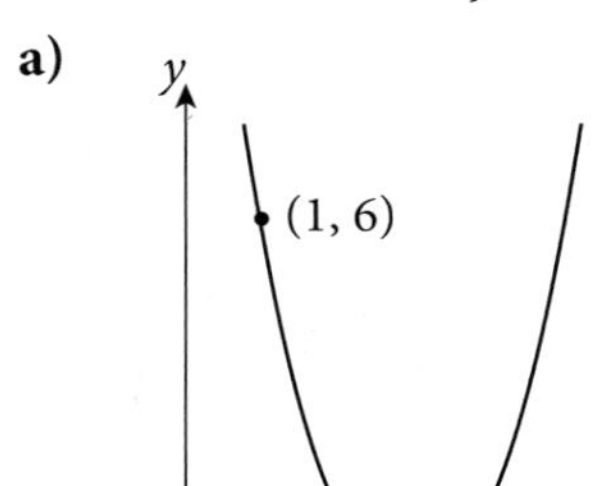

b)

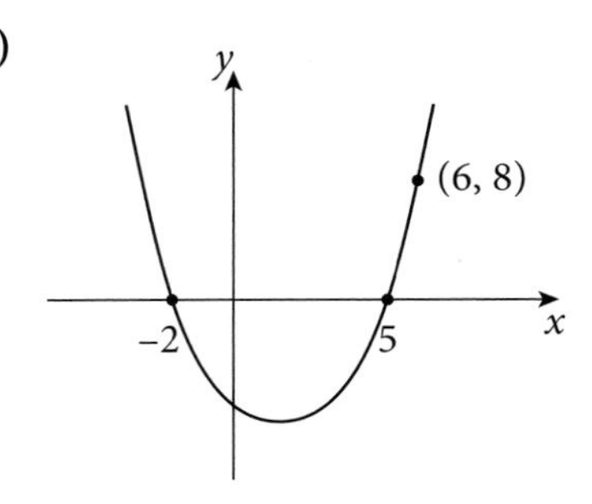

c)

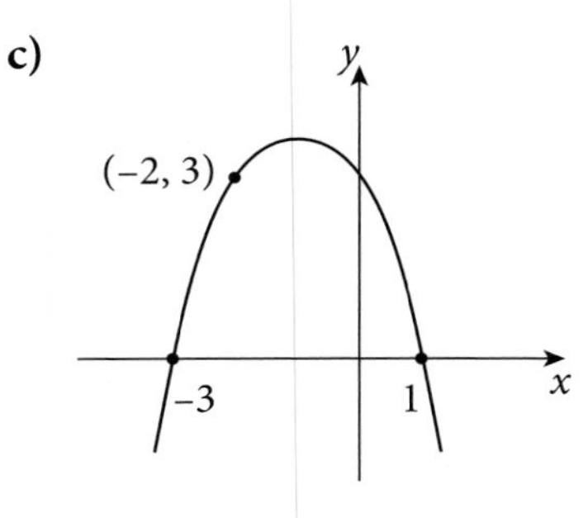

3. Find the equations of these quadratic functions in the form $f(x) = ax^2 + bx + c$.

   a) vertex at $(3, -27)$, x-intercepts at $(0, 0)$ and $(6, 0)$

   b) vertex at $(0, -12)$, x-intercepts at $(2, 0)$ and $(-2, 0)$

   c) vertex at $(-2, 5)$, x-intercepts at $(-7, 0)$ and $(3, 0)$.

4. Find the coordinates of the vertex of each quadratic function.

   a) $f(x) = 2(x - 3)^2 + 5$ **b)** $f(x) = 5(x - 1)^2 + 11$

   c) $f(x) = 5x^2 - 6$ **d)** $f(x) = (x + 2)^2$

Revision exercise 5A

1. **a)** If $f(x) = 3x + 4$ and $h(x) = \frac{x-2}{5}$ find

   i) $f^{-1}(x)$ **ii)** $h^{-1}(x)$

   b) Find the value of z if $f(z) = 20$

2. Given $f(x) = 2x - 3$ and $g(x) = x^2 - 1$, find

   a) $f(-1)$ **b)** $g(-1)$

   c) $fg(-1)$ **d)** $gf(3)$

3. Given that $f(x) = x - 5$, find

   a) the value of s such that $f(s) = -2$

   b) the values of t such that $t \times f(t) = 0$

4. The graphs **a)** to **d)** below show some of the functions (**A** to **F**).

   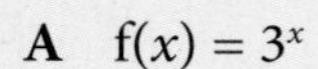

   A $f(x) = 3^x$ **B** $f(x) = x^3 + x^2 - 7$

   C $f(x) = 2x - x^2$ **D** $f(x) = |2x - 6|$

   E $f(x) = \frac{6}{x}$ **F** $f(x) = 2^{-x}$

a)

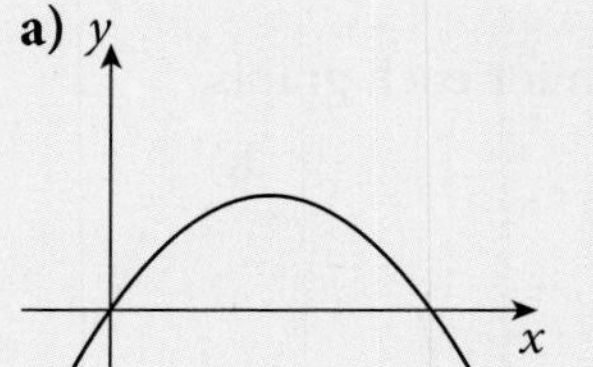

b)

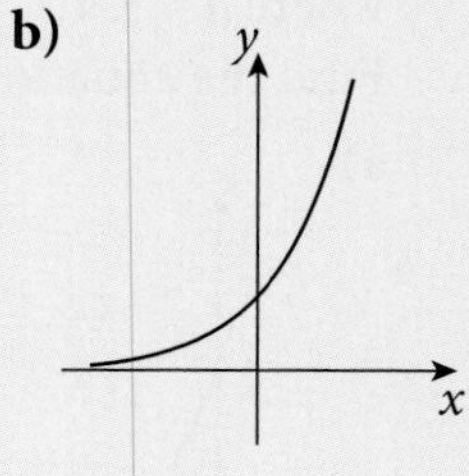

c)

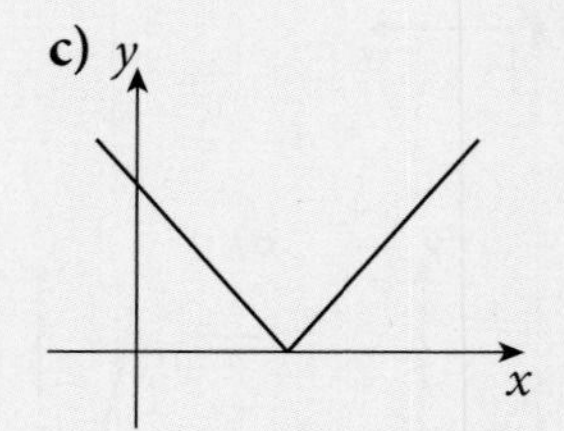

d)

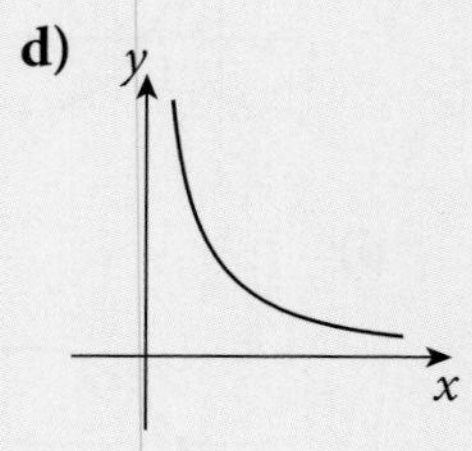

Match each graph with its correct function.

5. Write down the equations of the asymptotes to the graphs of

   a) $f(x) = \frac{10}{x}$ **b)** $f(x) = 2^x$

6. Sketch graphs of the following functions.

   a) $f(x) = 2^x$ for $-10 \le x \le 3$

   b) $f(x) = |x+1|$ for $-3 \le x \le 3$

   c) $f(x) = 2 \sin x$ for $0° \le x \le 360°$

Examination exercise 5B

1. The graphs **a)** to **f)** below show some of the following functions (**A** to **H**).

   A $f(x) = 4 - 2x$ **B** $f(x) = 2^x$

   C $f(x) = x^2 - 4x + 4$ **D** $f(x) = \cos x$

   E $f(x) = 2^{-x}$ **F** $f(x) = \frac{4}{x}$

   G $f(x) = |x-3|$ **H** $f(x) = \sin 2x$

   Match each graph with its correct function.

a)

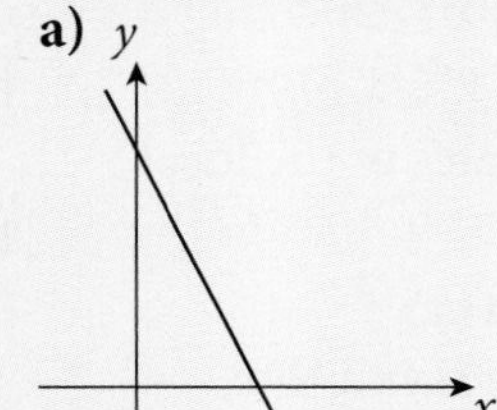

b)

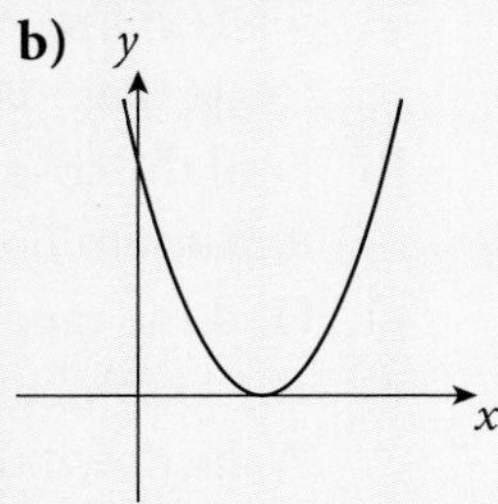

c) **d)**

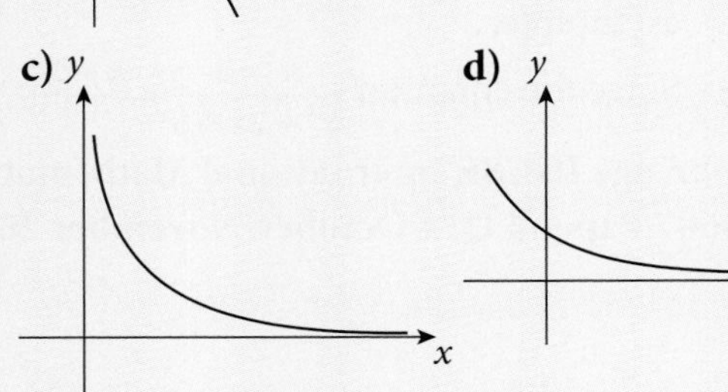

e) **f)**

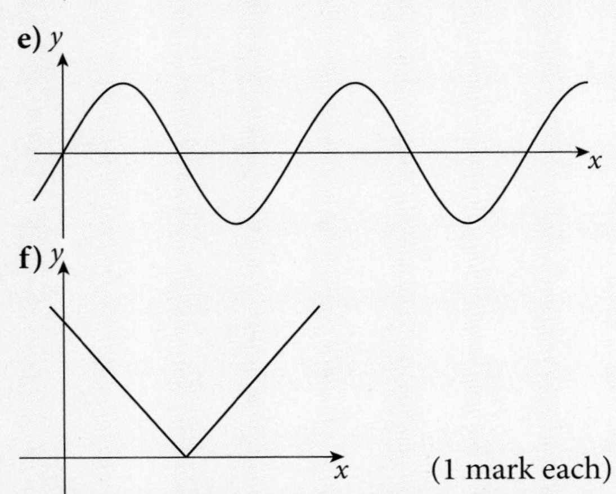

(1 mark each)

Cambridge IGCSE International Mathematics 0607 Specimen Paper 2 Q10 2010

2. $f(x) = 2x - 1$ $g(x) = 3x^2 + 1$

   Find

   a) $g(2)$, [1]

   b) $g(f(x))$, [2]

   c) the inverse function $f^{-1}(x)$. [2]

Cambridge IGCSE International Mathematics 0607 Paper 2 Q8 October/November 2010

3.

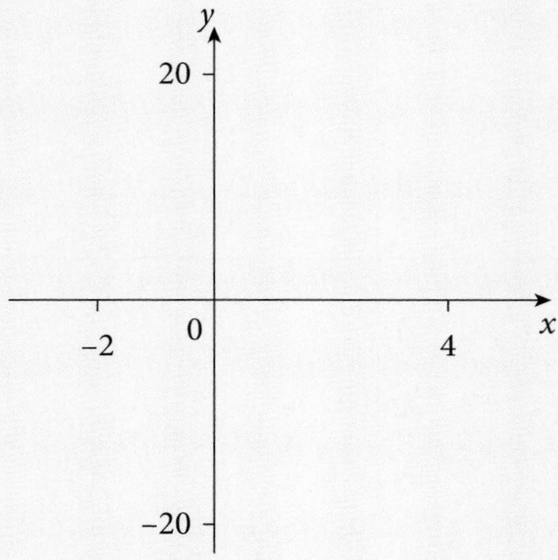

   a) On the axes, sketch the graph of $y = x^3 - 3x^2$. [3]

   b) Write down the zeros of $y = x^3 - 3x^2$. [1]

   c) Write down the co-ordinates of any local maximum or local minimum points. [2]

Cambridge IGCSE International Mathematics 0607 Paper 41 Q2 May/June 2011

4. **a)** Show clearly that

   $(x^2 - x + 1)(x + 1) = x^3 + 1$ [2]

   b) Show that $x^2 - x + 1 = 0$ has no solutions. [3]

   c) $f(x) = x^3 + 1$

   i) Find $f(2)$. [1]

   ii) Find $f(-1)$. [1]

   iii) Find $f^{-1}(x)$. [3]

   iv) Solve the equation $f^{-1}(x) = 3$. [1]

Cambridge IGCSE International Mathematics 0607 Paper 4 Q2 May/June 2010

5. $f(x)=\frac{5}{1-x}$.

a) Find f(−9). [1]

b) Solve $f(x) = 2$. [2]

c) Find $f^{-1}(x)$. [4]

Cambridge IGCSE International Mathematics 0607 Specimen Paper 4 Q2 2010

6.

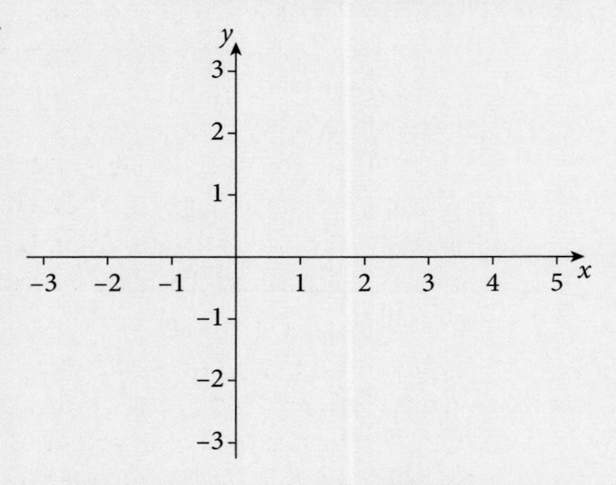

a) On the axes, sketch the graph of $y = f(x)$ where $f(x)=\frac{1}{(x^2-2x-3)}$. [3]

b) Write down the equations of the three asymptotes. [3]

c) Write down the co-ordinates of the local maximum point. [2]

d) Write down the domain and range of $f(x)$. [4]

e) How many solutions are there to these equations?

i) $f(x) = 0.5$ [1]

ii) $|f(x)| = 0.5$ [1]

Cambridge IGCSE International Mathematics 0607 Paper 4 Q5 May/June 2010

7.

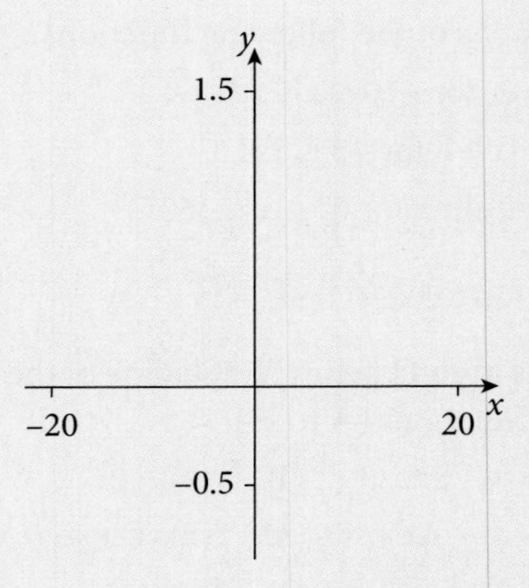

$f(x)=\frac{x^2+1}{x^2+2x+6}$

a) On the axes above, sketch the graph of $y = f(x)$ for $-20 \leqslant x \leqslant 20$. (Note that $-0.5 \leqslant y \leqslant 1.5$) [3]

b) Find the co-ordinates of the local maximum point. [2]

c) Find the range of $f(x)$. [3]

d) The graph has one asymptote. Write down the equation of this asymptote. [1]

e) Solve the equation $\frac{x^2+1}{x^2+2x+6}=\frac{x+5}{5}$. [2]

Cambridge IGCSE International Mathematics 0607 Paper 4 Q14 October/November 2010

6 Investigations and mathematical modelling

William Shockley (1910–89) Every time you use a calculator you are making use of integrated circuits which were developed from the first transistor. The transistor was invented by William Shockley, working with two scientists, in 1947. The three men shared the 1956 Nobel Prize for Physics. The story of the invention is a good example of how mathematics can be used to solve practical problems.

The first electronic computers did not make use of transistors or integrated circuits and were so big that they occupied whole rooms. A modern computer which can carry out the same functions (and do much more besides) can be carried around in your hand.

6.1 Investigations

There are a large number of possible starting points for these investigations so it may be possible for you to choose investigations which appeal to you. On other occasions, your teacher may set the same investigation for the whole class.

Here are a few guidelines for you:

- If the set problem is too complicated try an easier case.
- Draw your own diagrams.
- Make tables of your results and be systematic.
- Look for patterns.
- Is there a rule or formula to describe the results?
- Can you *predict* further results?
- Can you *prove* any rules which you may find?

1. Opposite corners

Here the numbers are arranged in 10 columns:

1	2	3	4	5	6	7	8	9	10
11	12	13	14	15	16	17	18	19	20
21	22	23	24	25	26	27	28	29	30
31	32	33	34	35	36	37	38	39	40
41	42	43	44	45	46	47	48	49	50
51	52	53	54	55	56	57	58	59	60
61	62	63	64	65	66	67	68	69	70
71	72	73	74	75	76	77	78	79	80
81	82	83	84	85	86	87	88	89	90
91	92	93	94	95	96	97	98	99	100

In the 2×2 square, multiply the numbers in opposite corners.

7	8
17	18

$7 \times 18 = 126$

$8 \times 17 = 136$

The difference between them is 10.

In the 3×3 square,

12	13	14
22	23	24
32	33	34

$12 \times 34 = 408$

$14 \times 32 = 448$

The difference between them is 40.

Investigate to see if you can find any rules or patterns connecting the size of square chosen and the difference.

If you find a rule, use it to *predict* the difference for larger squares.

Test your rule by looking at squares like 8×8 or 9×9.

Can you *generalise* the rule?

[What is the difference for a square of size $n \times n$?]

x		?
?		?

Can you prove the rule?

What happens if the numbers are arranged in six columns or seven columns?

1	2	3	4	5	6
7	8	9	10	11	12
13	14	15	16	17	18
19					

1	2	3	4	5	6	7
8	9	10	11	12	13	14
15	16	17	18	19	20	21
22						

2. How many dots?

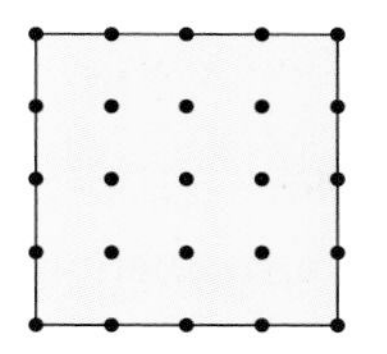

A The diagram shows a 5×5 square of dots.

There are 16 dots on the perimeter and 9 dots inside the square.

a) Draw a 3×3, a 4×4 and a 6×6 square of dots. For each diagram, count the number of dots on the perimeter and the number of dots inside the square.

b) For a 100×100 square of dots, how many dots are on the perimeter?

c) For a 57×57 square of dots, how many dots are inside the square?

B Rectangles are drawn so that the width is always 1 unit more than the height. The number of dots on the perimeter and the number of dots inside the rectangle are counted.

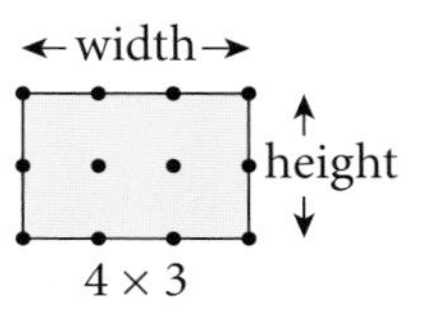

dots on perimeter = 10
dots inside = 2

a) How many dots are on the perimeter of a 100×100 rectangle?

b) How many dots are inside a 9×8 rectangle?

c) How many dots are inside a 52×51 rectangle? (Much harder)

3. Mystic rose

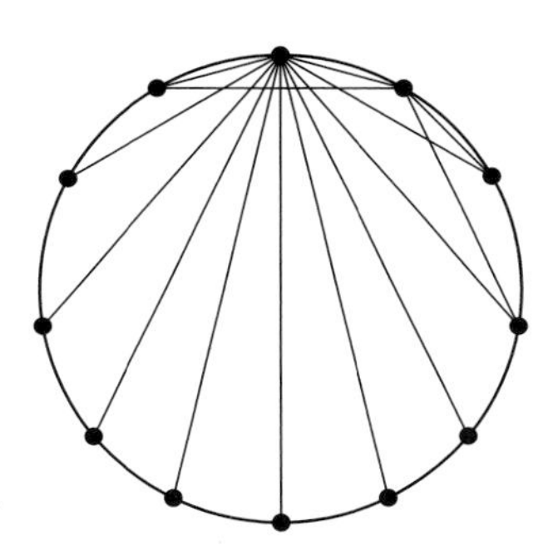

Straight lines are drawn between each of the 12 points on the circle. Every point is joined to every other point. How many straight lines are there?

Suppose we draw a mystic rose with 24 points on the circle.
How many straight lines are there?
How many straight lines would there be with n points on the circle?

4. Knockout competition

Eight teams reach the 'knockout' stage of the World Cup.

Round 1		Round 2		Round 3	
England	2	England	1	Brazil	1
Germany	1				
France	0	Brazil	3		
Brazil	1				
Argentina	3	Argentina	0	Morocco	2
Italy	0				
Scotland	2	Morocco	1		
Morocco	3				

How would you organise a knockout competition if there were 12 teams? Or 15?

How many matches are played up to and including the final, if there are

a) 8 teams **b)** 12 teams **c)** 15 teams

d) 23 teams **e)** n teams?

In a major tournament like the Wimbledon Tennis Championships, the better players are seeded from 1 to 16. Can you organise a tournament for 32 players so that, if they win all their games

a) seeds 1 and 2 can meet in the final

b) seeds 1, 2, 3 and 4 can meet in the semi-finals

c) seeds 1, 2, 3, 4, 5, 6, 7, 8 can meet in the quarter-finals?

5. Discs

a) You have five black discs and five white discs which are arranged in a line as shown.

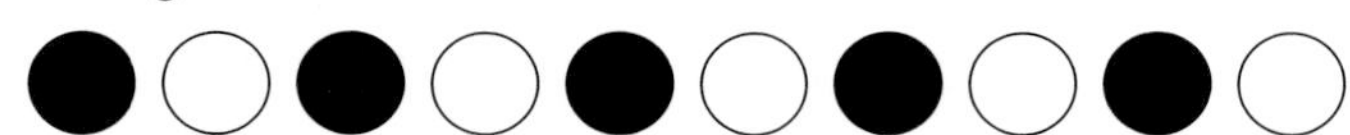

We want to get all the black discs to the right-hand end and all the white discs to the left-hand end.

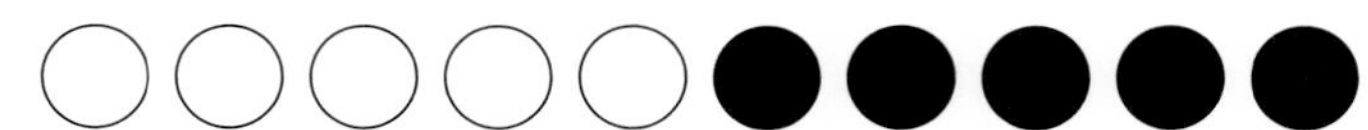

The only move allowed is to interchange two neighbouring discs.

becomes

How many moves does it take?
How many moves would it take if we had fifty black discs and fifty white discs arranged alternately?

b) Suppose the discs are arranged in pairs

... etc.

How many moves would it take if we had fifty black discs and fifty white discs arranged like this?

In both cases work with a smaller number of discs until you can see a pattern.

c) Now suppose you have three colours, black, white and green, arranged alternately.

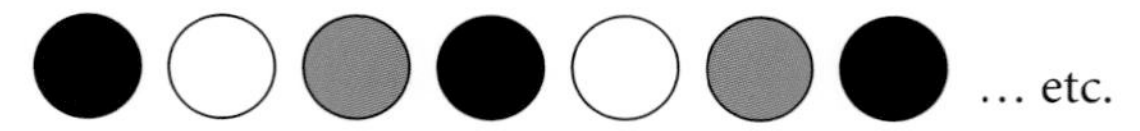

... etc.

You want to get all the black discs to the right, the green discs to the left and the white discs in the middle.
How many moves would it take if you have 30 discs of each colour?

6. Chess board

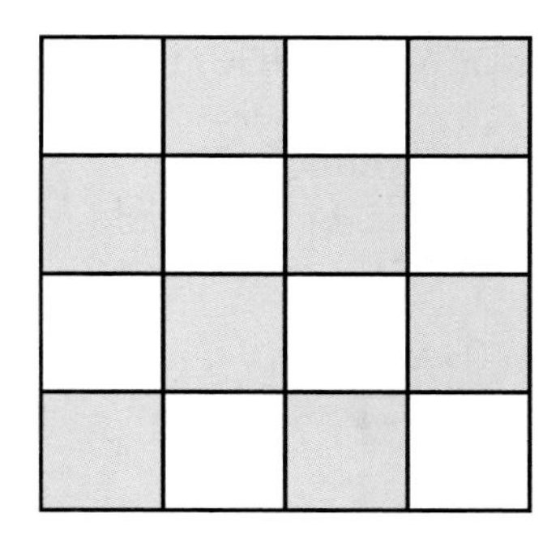

Start with a small board, just 4×4.

How many squares are there? [It is not just 16!]

How many squares are there on an 8×8 chess board?

How many squares are there on an $n \times n$ chess board?

7. Area and perimeter

This is about finding different shapes in which the area is numerically equal to the perimeter.

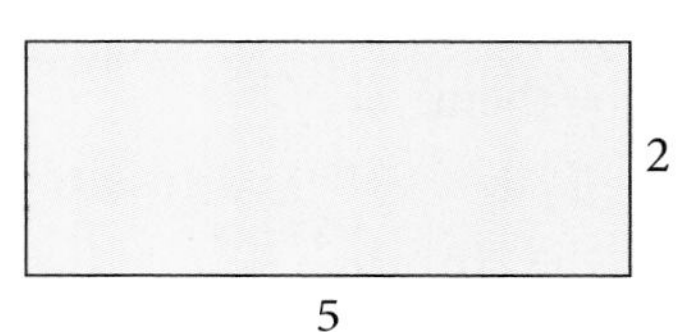

This rectangle has an area of 10 square units and a perimeter of 14 units, so we will have to try another one.
There are some suggestions below but you can investigate shapes of your own choice if you prefer.

a) Find rectangles with equal area and perimeter. After a while you can try adding on bits like this:

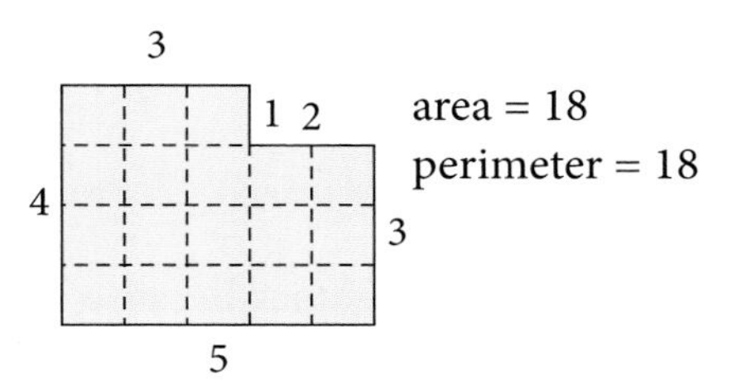

b) Suppose one dimension of the rectangle is fixed.
In this rectangle the length is 5 units.

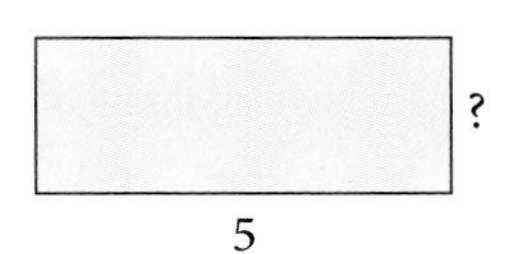

c) Try right-angled triangles and equilateral triangles.

d) Try circles, semi-circles and so on.

e) How about three-dimensional shapes? Now we are looking for cuboids, spheres, cylinders in which the volume is numerically equal to the surface area.

f) Can you find any connection between the square with equal area and perimeter and the circle with equal area and perimeter? How about the equilateral triangle with equal area and perimeter?

8. Happy numbers (and more)

a) Take the number 23.
Square the digits and add.

2 3

$2^2 + 3^2 = 1 \quad 3$

$1^2 + 3^2 = 1 \quad 0$

$1^2 + 0^2 = 1$

The sequence ends at 1 and so we call 23 a 'happy' number.
Investigate for other numbers.
Here are a few suggestions: 70, 85, 49, 44, 14, 15, 94.

b) Now change the rule. Instead of squaring the digits we will cube them.

2 1

$2^3 + 1^3 = 0 \quad 9$

$0^3 + 9^3 = 7 \quad 2 \quad 9$

$7^3 + 2^3 + 9^3 = 1 \quad 0 \quad 8 \quad 0$

$1^3 + 0^3 + 8^3 = 5 \quad 1 \quad 3$

$5^3 + 1^3 + 3^3 = 153$

And now we are stuck because 153 leads to 153 again.
Investigate for numbers of your own choice. Do any numbers lead to 1?

9. Prime numbers

Write all the numbers from 1 to 104 in eight columns and draw a ring around the prime numbers 2, 3, 5 and 7.

1	(2)	(3)	4	(5)	6	(7)	8
9	10	11	12	13	14	15	16
17	18	19	20	21	22	23	24
25...							

If we cross out all the multiples of 2, 3, 5 and 7, we will be left with all the prime numbers below 104. Can you see why this works?

Draw *four* lines to eliminate the multiples of 2.
Draw *six* lines to eliminate the multiples of 3.
Draw *two* lines to eliminate the multiples of 7.
Cross out all the numbers ending in 5.

Put a ring around all the prime numbers less than 104.
[Check that there are 27 numbers.]

Many prime numbers can be written as the sum of two squares.
For example $5 = 2^2 + 1^2$, $13 = 3^2 + 2^2$. Find all the prime numbers in your table which can be written as the sum of two squares.
Draw a red ring around them in the table.
What do you notice?
Check any 'gaps' you may have found.

Extend the table up to 200 and see if the pattern continues. In this case you will need to eliminate the multiples of 11 and 13 as well.

6.2 Mathematical modelling

1. Stopping distances

This diagram from the British Highway Code gives the overall stopping distance for cars travelling at various speeds.

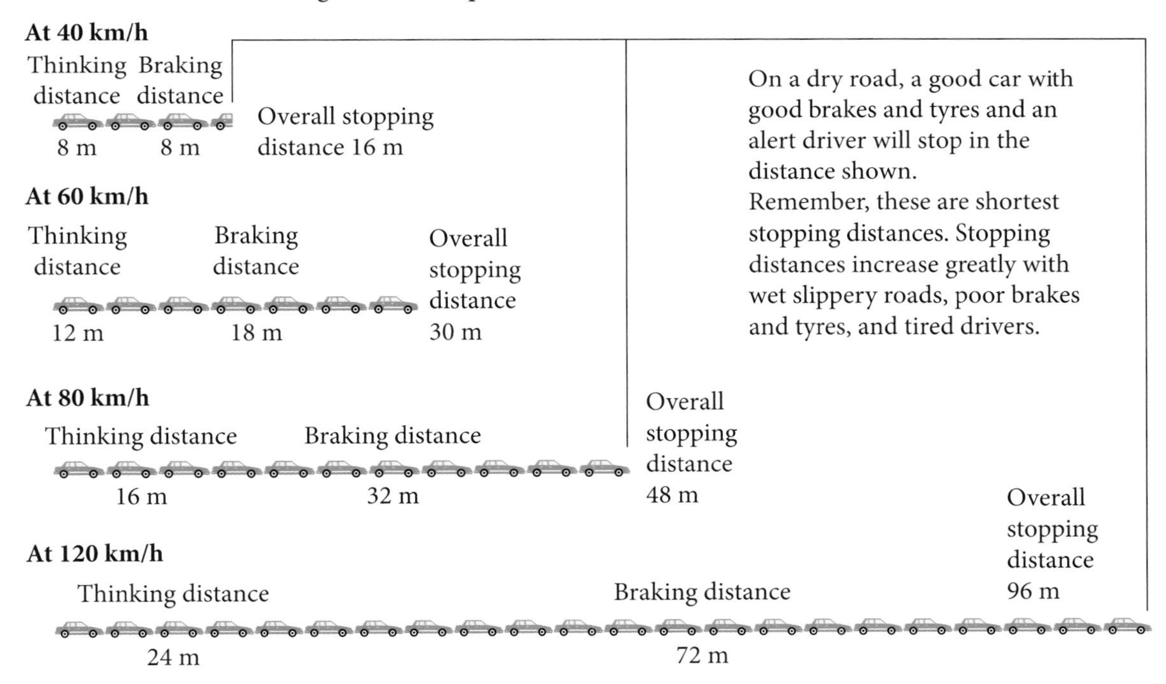

a) What is meant by 'thinking distance'?
Work out the thinking distance for a car travelling at a speed of 90 km/h. What is the formula which connects the speed of the car and the thinking distance?
Use S for speed and T for thinking distance.

b) Now we need to try to find a formula which connects the *braking* distance, B, with the speed of the car.
It is helpful to draw a graph of speed, S, (on the x-axis) against braking distance, B, (on the y-axis).

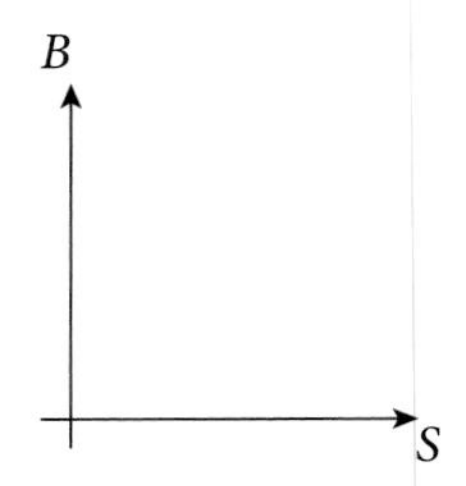

Try to work out the equation of the curve you obtain.
What curve are you reminded of?
You can use a graphic display calculator to help you.

c) When you think you have the equation connecting B and S, write down a formula which connects the speed of the car and the *overall* stopping distance (call it O).
Check that your formula gives the correct answer for the overall stopping distance at a speed of

i) 40 km/h

ii) 120 km/h

d) **Bad weather**

In bad weather, particularly on wet roads, the stopping distance will be greater.
Do you think the thinking distance will be increased or will it stay the same?
By how much do you think the braking distance will be increased? [E.g. Twice as long, three times as long – you decide].
Work out a new formula for the overall stopping distance, O, in bad weather.

2. Maximum box

a) You have a square sheet of card 24 cm by 24 cm.
You can make a box (without a lid) by cutting squares from the corners and folding up the sides.

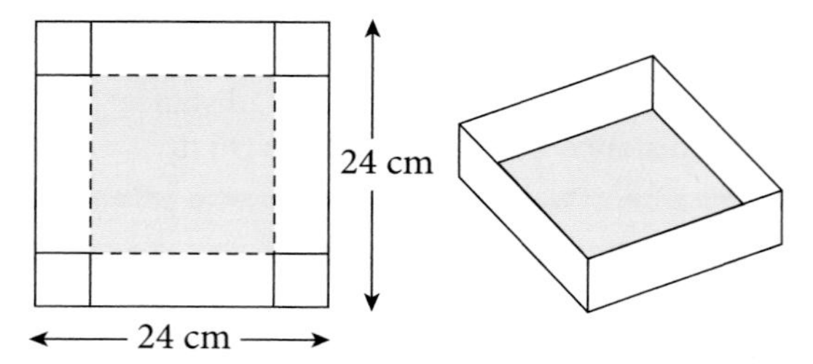

What size corners should you cut out so that the volume of the box is as large as possible?

Try different sizes for the corners and record the results in a table:

Length of the side of the corner square (cm)	Dimensions of the open box (cm)	Volume of the box (cm³)
1	$22 \times 22 \times 1$	484
2		
–		
–		

Now consider boxes made from different-sized cards:
15 cm × 15 cm and 20 cm by 20 cm.
What size corners should you cut out this time so that the volume of the box is as large as possible?
Is there a connection between the size of the corners cut out and the size of the square card?

b) A mathematical model

We will now try to produce a mathematical model for a box with any size corners cut from the original 24 × 24 cm square sheet.

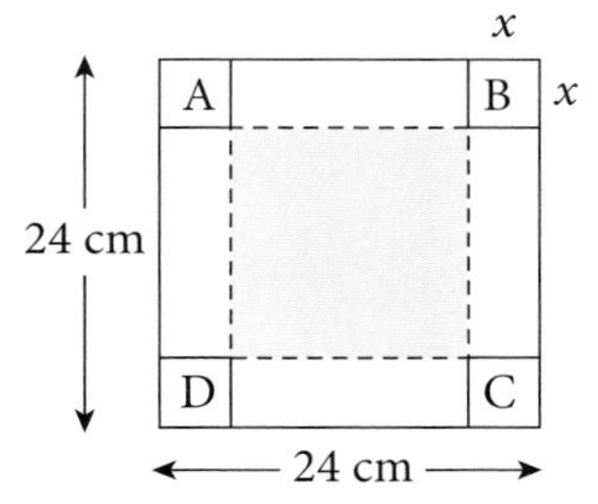

The base of the box is the square ABCD.
Find the length AB in terms of x.
Hence write a formula for the volume, V, of the box, in terms of x.
Check that your formula is correct: when $x = 1$, $V = 484$.
Use a graphic display calculator to find the value of x which makes the volume of the box its maximum.
Repeat the above calculations for a square cards of different sizes, say 18 cm × 18 cm and 27 cm × 27 cm.
State the connection between the size of the corners cut out and the size of the original square card.

c) Produce a mathematical model of the situation when the card is not square. Take rectangular cards where the length is twice the width (20 × 10, 12 × 6, 18 × 9 etc.)
For the maximum volume find the connection between the size of the corners cut out and the size of the original card.

7 Geometry

Pythagoras (569–500 BCE) was one of the first of the great mathematical names in Greek antiquity. He settled in southern Italy and formed a mysterious brotherhood with his students who were bound by an oath not to reveal the secrets of numbers and who exercised great influence. They laid the foundations of arithmetic through geometry but failed to resolve the concept of irrational numbers. The work of these and others was brought together by Euclid at Alexandria in a book called *The Elements* which was still studied in schools as recently as 1900.

4.1 Use and interpret the geometrical terms:
acute, obtuse, right angle, reflex, parallel, perpendicular, congruent, similar
Use and interpret vocabulary of triangles, quadrilaterals, polygons and simple solid figures
4.2 Line and rotational symmetry
4.3 Angle measurement in degrees
4.4 Angles round a point
Angles on a straight line and intersecting straight lines
Vertically opposite angles
Alternate and corresponding angles on parallel lines
Angle sum of a triangle, quadrilateral and polygons
Interior and exterior angles of a polygon
Angles of regular polygons
4.5 Similarity
Calculation of lengths of similar figures
Use of area and volume scale factors
4.6 Pythagoras' Theorem and its converse in two and three dimensions including:
chord length; distances of a chord from the centre of a circle; distance on a grid
4.7 Use and interpret vocabulary of circles
Properties of circles: tangent perpendicular to radius at the point of contact; tangents from a point; angle in a semicircle; angles at the centre and at the circumference on the same arc; cyclic quadrilateral

7.1 Fundamental results

You should already be familiar with the following results. They are used later in this section and are quoted here for reference.

- The angles on a straight line add up to 180°.

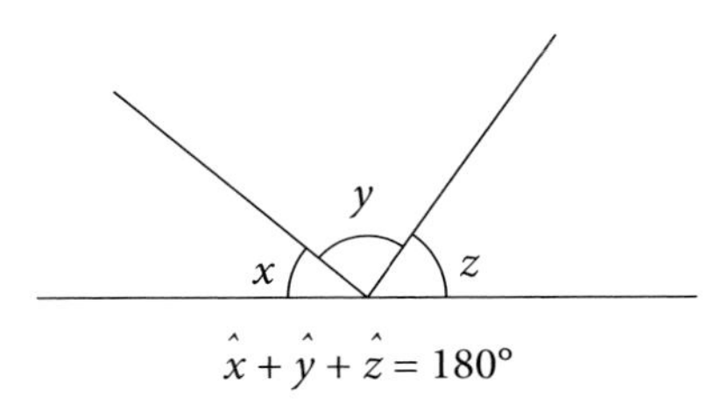

$\hat{x} + \hat{y} + \hat{z} = 180°$

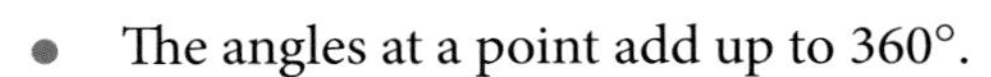

- The angles at a point add up to 360°.

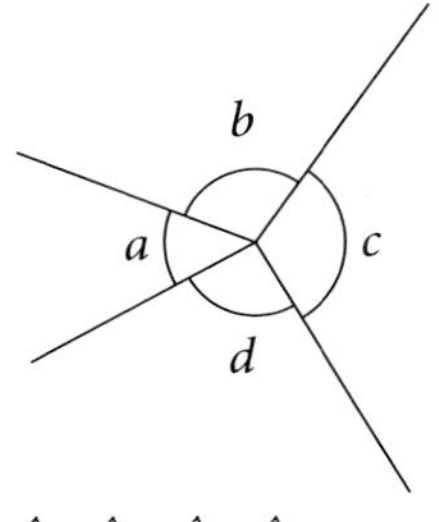

$\hat{a} + \hat{b} + \hat{c} + \hat{d} = 360°$

- Vertically opposite angles are equal.

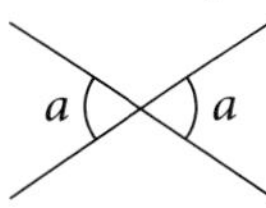

- The angle sum of a triangle is 180°.
- An isosceles triangle has 2 sides and 2 angles the same.

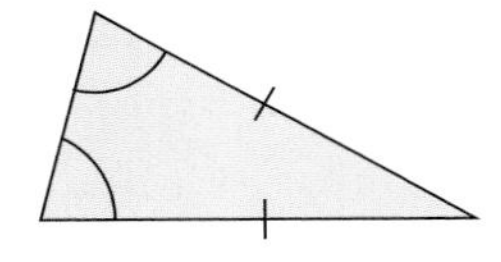

- The angle sum of a quadrilateral is 360°.
- An equilateral triangle has 3 sides and 3 angles the same.

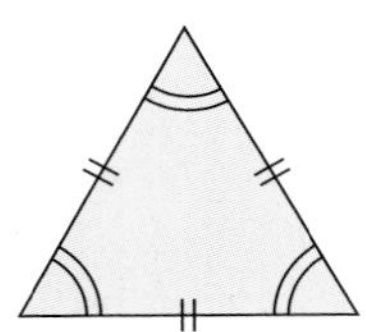

Exercise 7.1

Find the angles marked with letters. (AB is always a straight line.)

1.

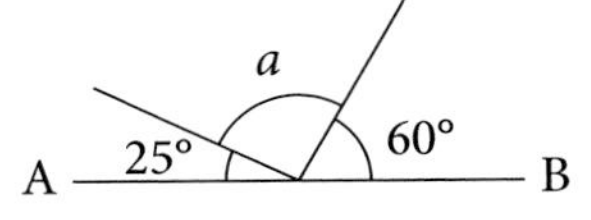

2.

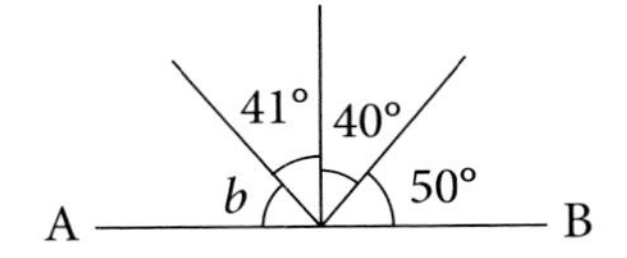

3.

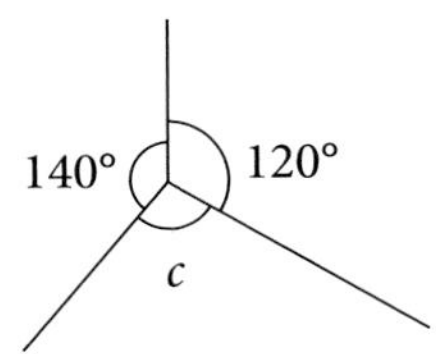

4.

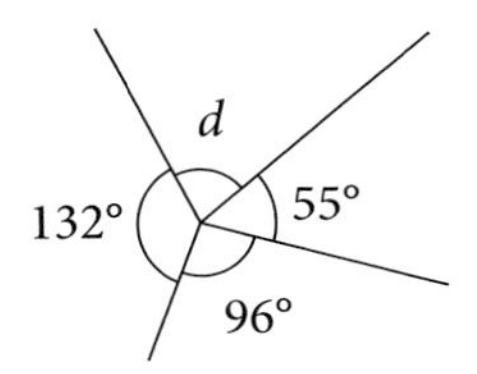

5.

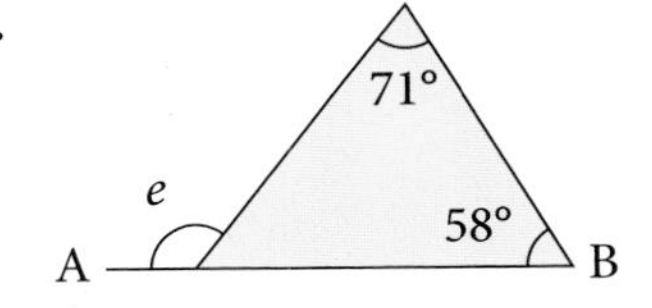

6.

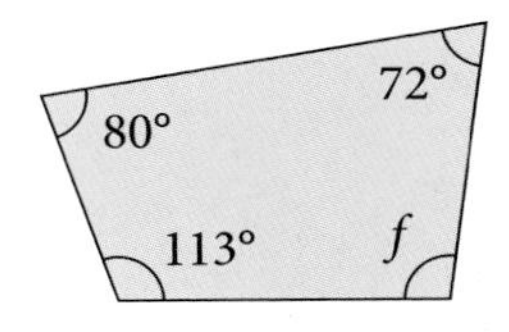

7.

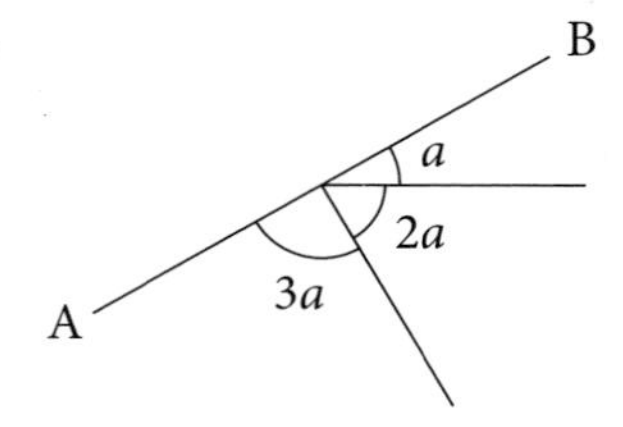

8.

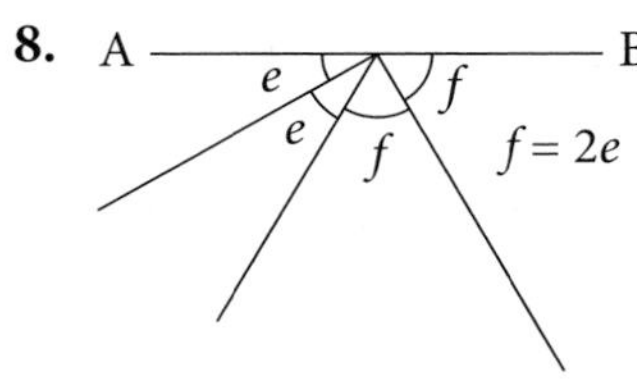

9.

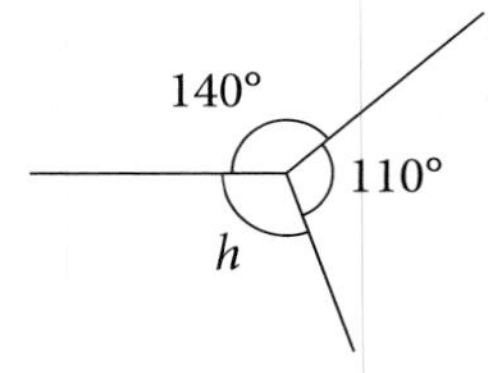

10.

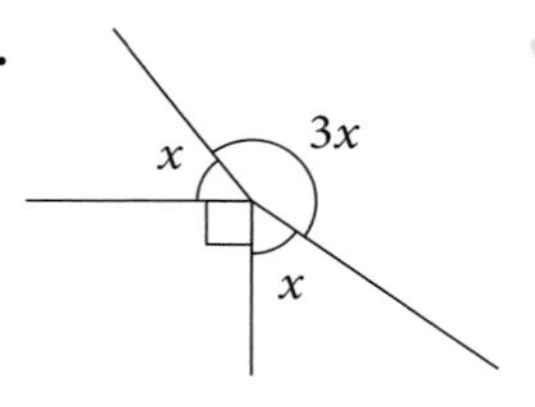

11.

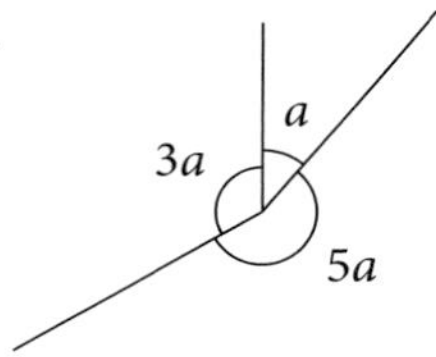

12.

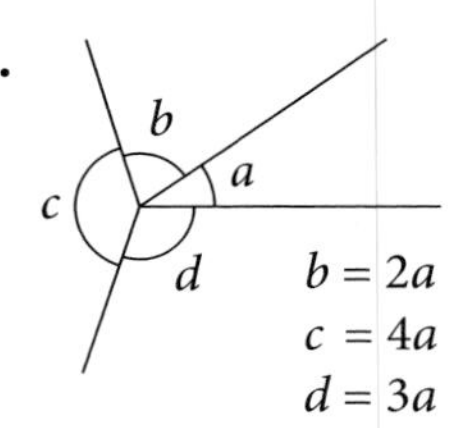

13.

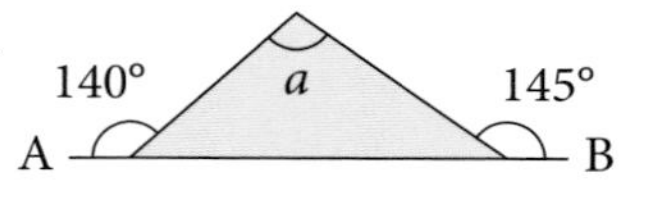

14.

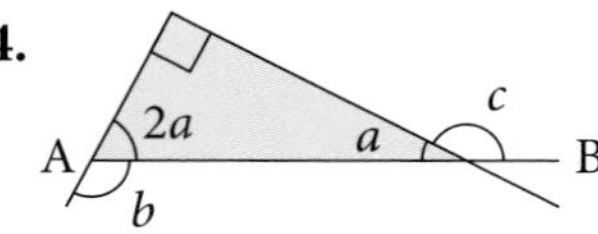

15.

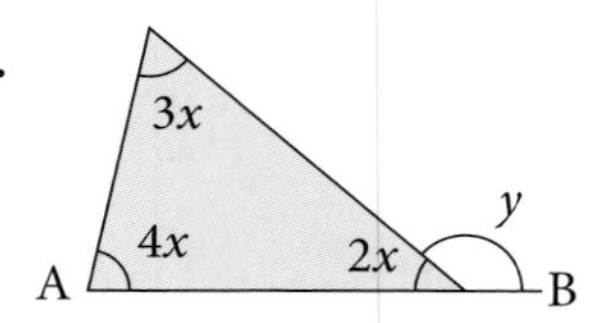

16.

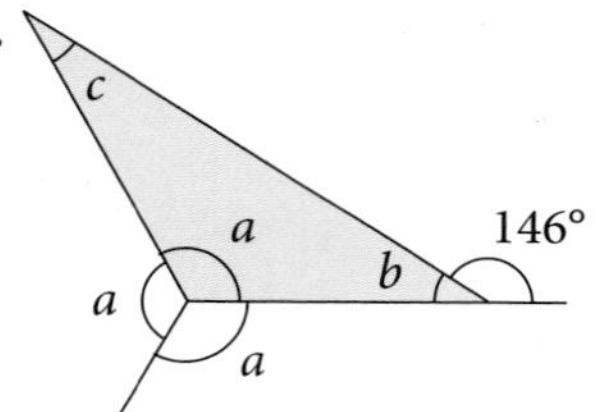

17.

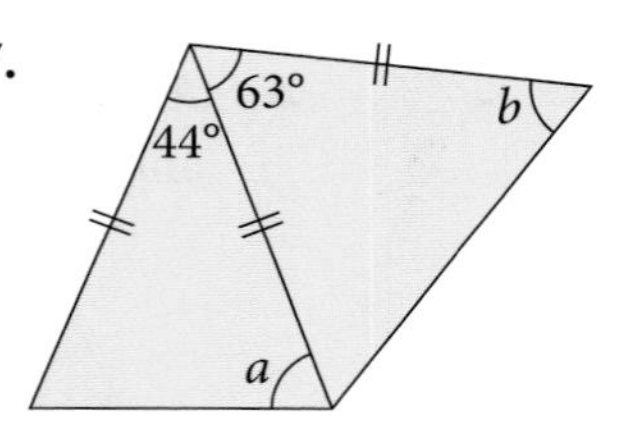

18.

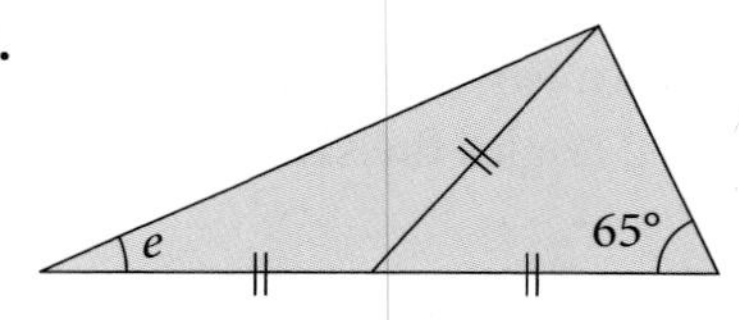

19.

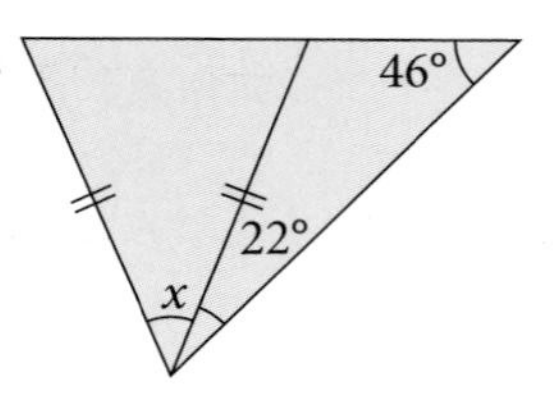

20.

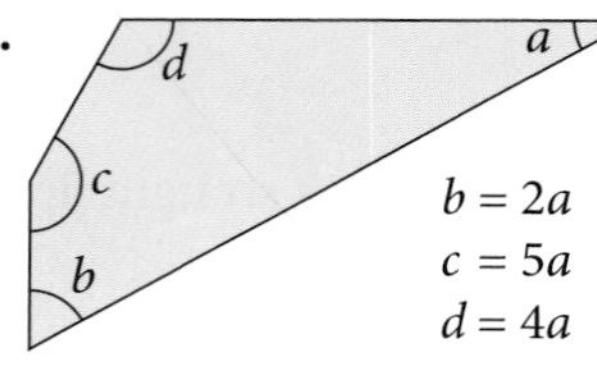

21.

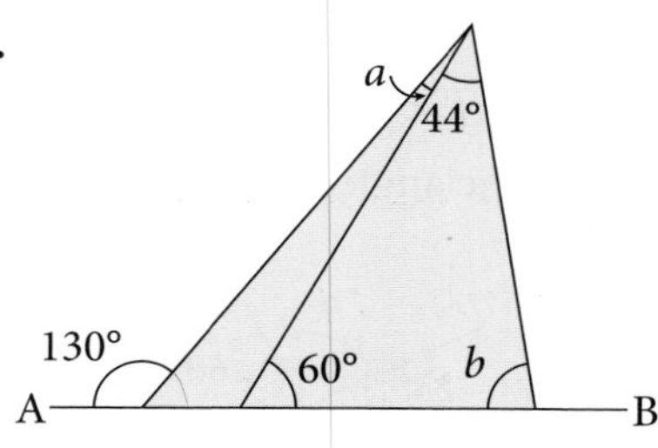

22.

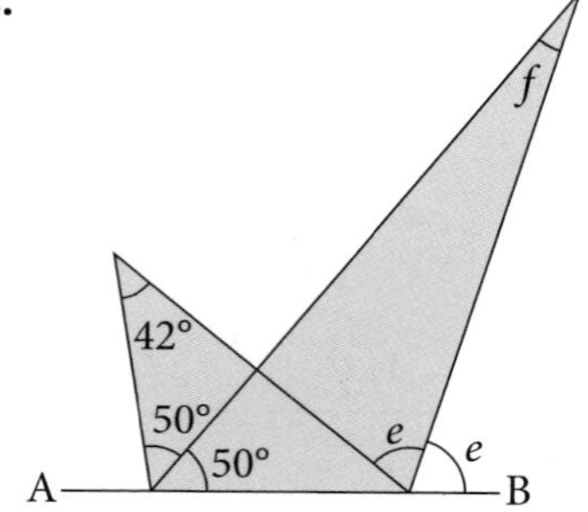

23. Calculate the largest angle of a triangle in which one angle is eight times each of the others.

24. In ΔABC, $\hat{A}$ is a right angle and D is a point on AC such that BD bisects $\hat{B}$. If $B\hat{D}C = 100°$, calculate $\hat{C}$.

25. WXYZ is a quadrilateral in which $\hat{W} = 108°$, $\hat{X} = 88°$, $\hat{Y} = 57°$ and $W\hat{X}Z = 31°$. Calculate $W\hat{Z}X$ and $X\hat{Z}Y$.

26. In quadrilateral ABCD, AB produced is perpendicular to DC produced. If $\hat{A} = 44°$ and $\hat{C} = 148°$, calculate $\hat{D}$ and $\hat{B}$.

AB **produced** means AB is extended

A B

27. Triangles ABD, CBD and ADC are all isosceles. Find the angle x.

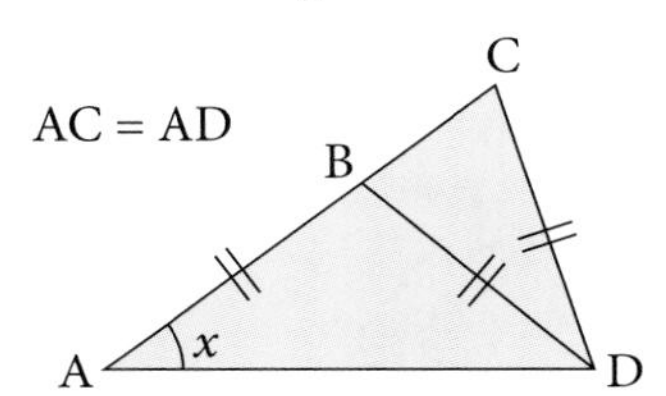

Example 1

Find the angles marked with letters.

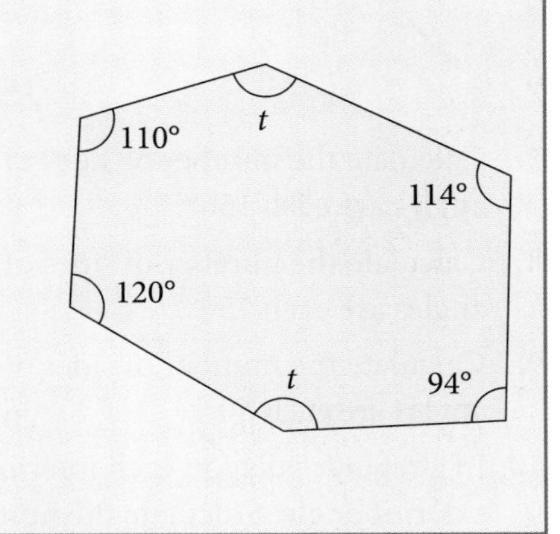

The sum of the interior angles = $(n - 2) \times 180°$ where n is the number of sides of the polygon.

In this case $n = 6$.

$$\therefore \quad 110 + 120 + 94 + 114 + 2t = 4 \times 180$$
$$438 + 2t = 720$$
$$2t = 282$$
$$t = 141°$$

Exercise 7.2

1. Find angles a and b for the regular pentagon.

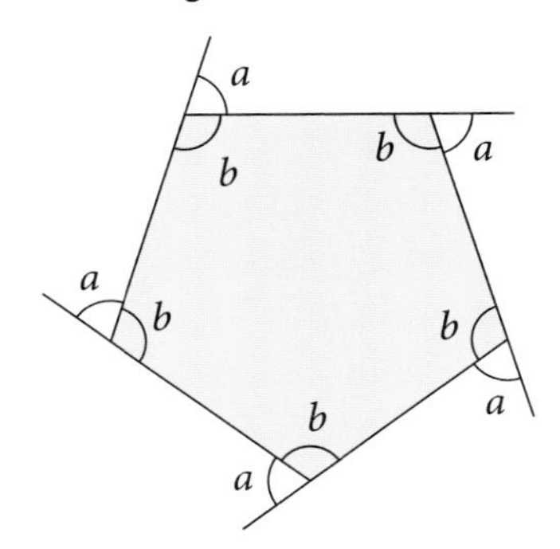

2. Find x and y.

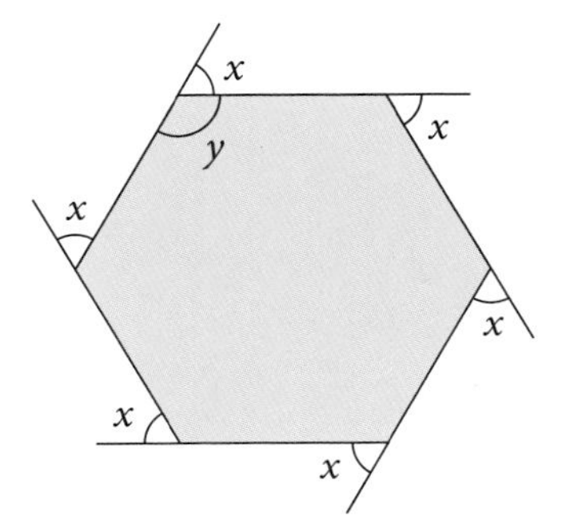

3. Consider the pentagon on the right which has been divided into three triangles.

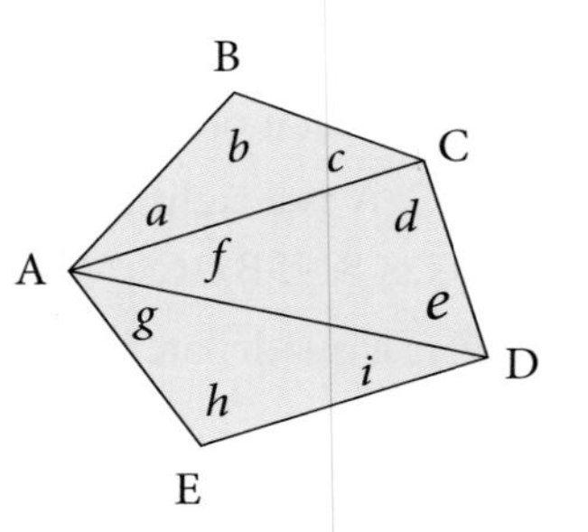

$\hat{A} = a + f + g$, $\hat{B} = b$, $\hat{C} = c + d$, $\hat{D} = e + i$, $\hat{E} = h$

Now $a + b + c = d + e + f = g + h + i = 180°$

$\therefore \hat{A} + \hat{B} + \hat{C} + \hat{D} + \hat{E} = a + b + c + d + e + f + g + h + i$

$= 3 \times 180°$

$= 6 \times 90°$

Draw further polygons and make a table of results.

Number of sides *n*	5	6	7	8...
Sum of interior angles	$3 \times 180°$			

What is the sum of the interior angles for a polygon with n sides?

4. Find a.

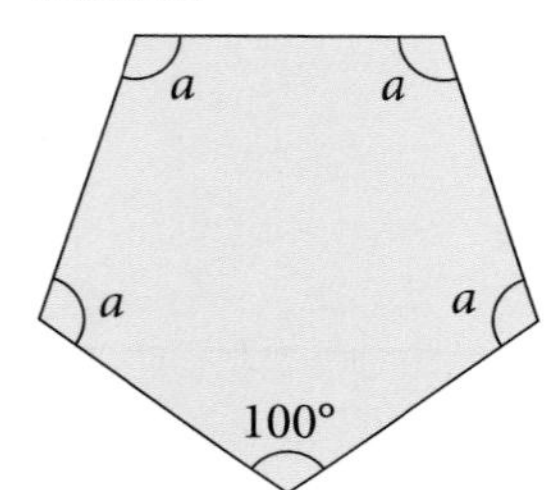

5. Find m.

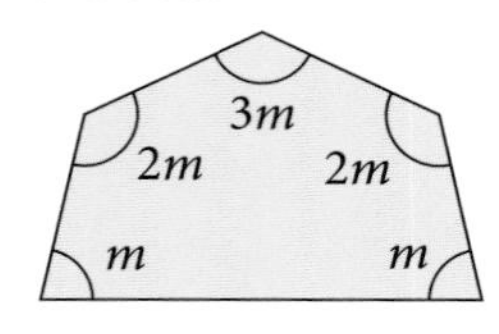

6. Find a.

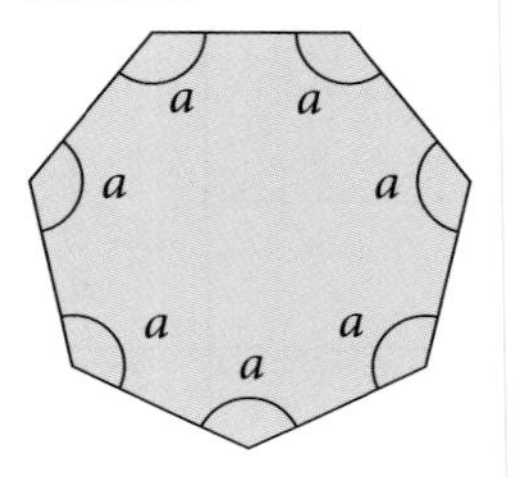

7. Calculate the number of sides of a regular polygon whose interior angles are each 156°.
8. Calculate the number of sides of a regular polygon whose interior angles are each 150°.
9. Calculate the number of sides of a regular polygon whose exterior angles are each 40°.
10. In a regular polygon each interior angle is 140° greater than each exterior angle. Calculate the number of sides of the polygon.
11. In a regular polygon each interior angle is 120° greater than each exterior angle. Calculate the number of sides of the polygon.
12. Two sides of a regular pentagon are produced to form angle x. What is x?

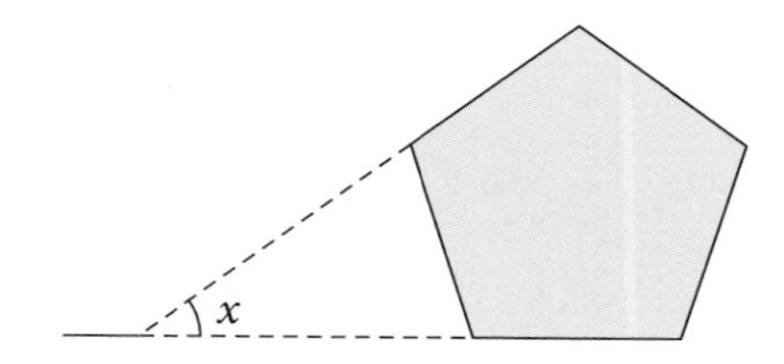

13.

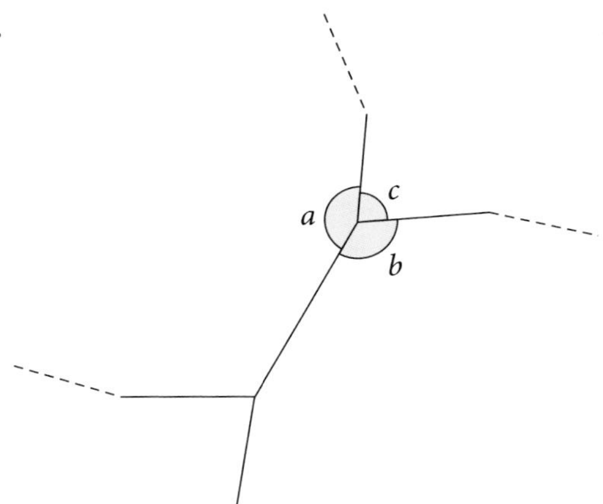

Angle a is the interior angle of a 20-sided regular polygon.

Angle b is the interior angle of a 15-sided regular polygon.

Calculate the value of angle c.

14. Satheesh measured all the interior angles in a polygon but unfortunately he missed out one angle. The sum of the angles he measured came to 1147°.
What is the size of the missing angle?

15. ABCDEFGH is a regular octagon.

Find **a)** angle EFG **b)** angle BGH **c)** angle BFC.

An octagon has 8 sides; a nonagon has 9 sides.

16. A nonagon has 8 equal angles x and one angle $4x$.
Find the value of x.

17.

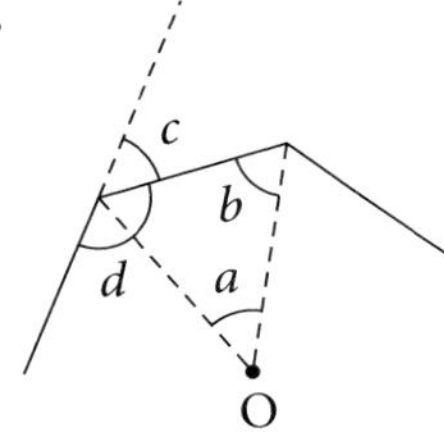

Find each angle marked with a letter in this 20-sided *regular* polygon. O is the centre of the polygon.

18. In a regular polygon each interior angle is 160° greater than each exterior angle. Calculate the number of sides of the polygon.

19. The diagram shown is formed by joining regular pentagons.
Find the angles x and y.

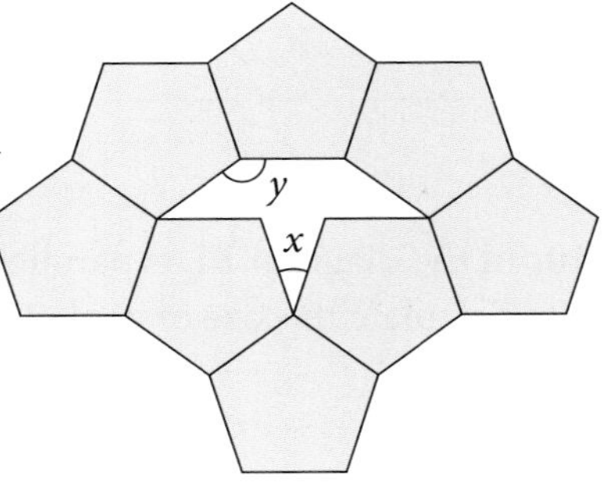

Parallel lines

Look at the diagram on the right. The following properties hold:

- $\hat{a} = \hat{c}$ (corresponding angles)
- $\hat{c} = \hat{d}$ (alternate angles)
- $\hat{b} + \hat{c} = 180°$ (allied angles)

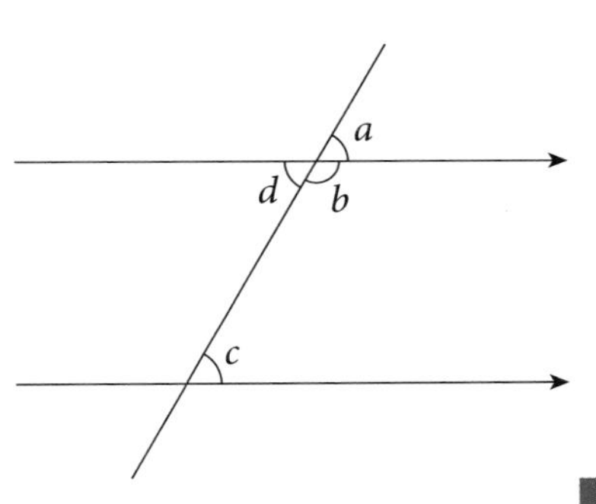

The acute angles (less than 90°) are the same and the obtuse angles (between 90° and 180°) are the same.

Exercise 7.3

In questions **1** to **9** find the angles marked with letters.

1.

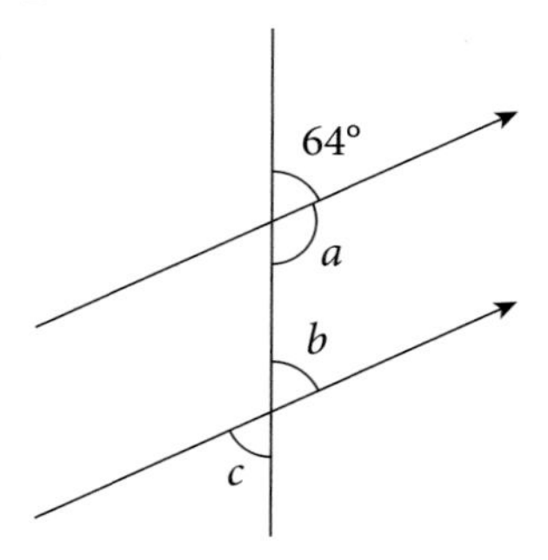

2.

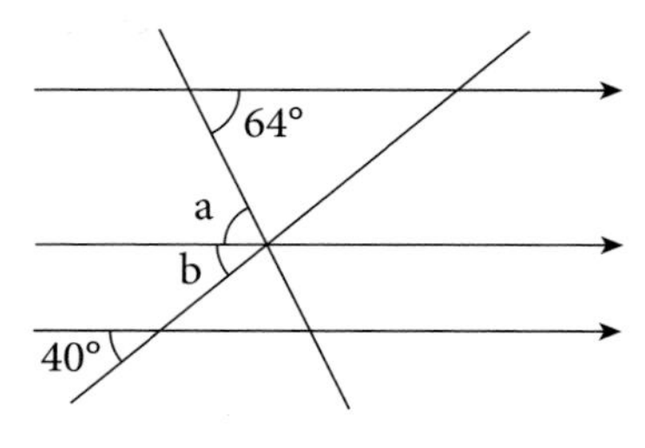

3.

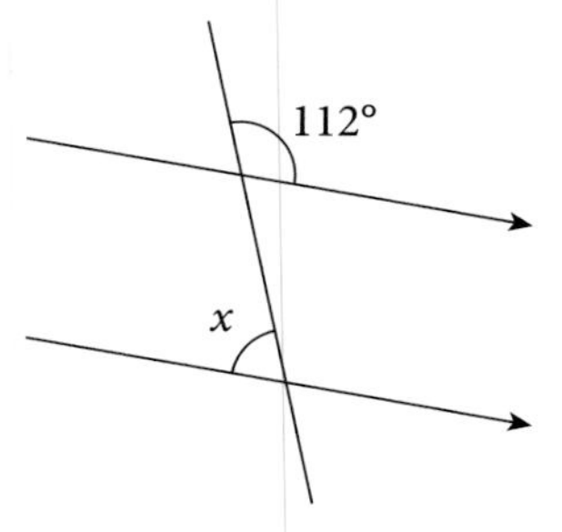

4.

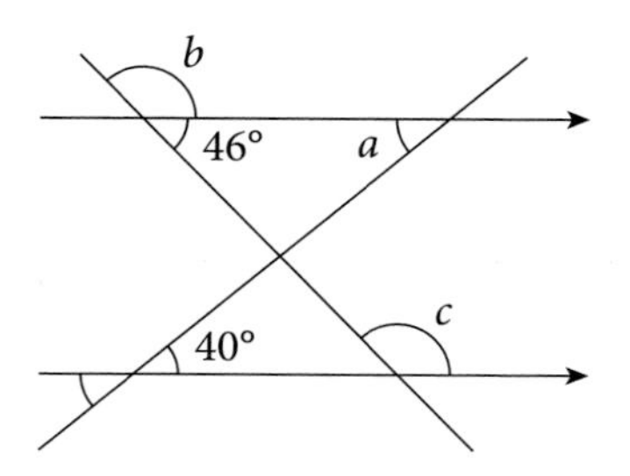

5.

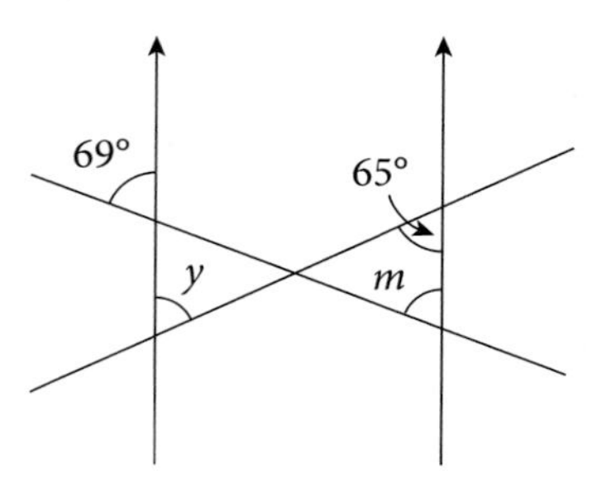

6.

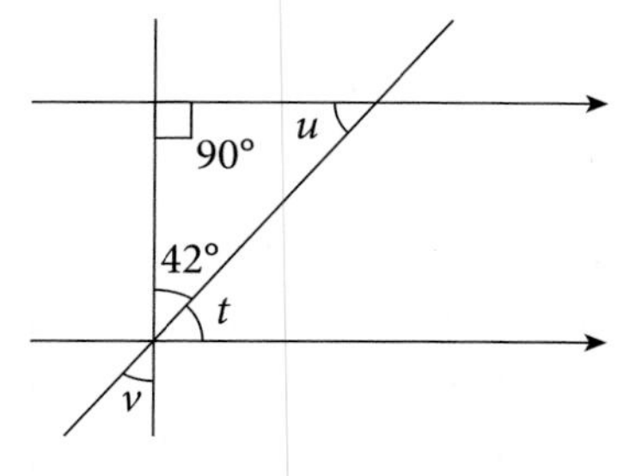

7.

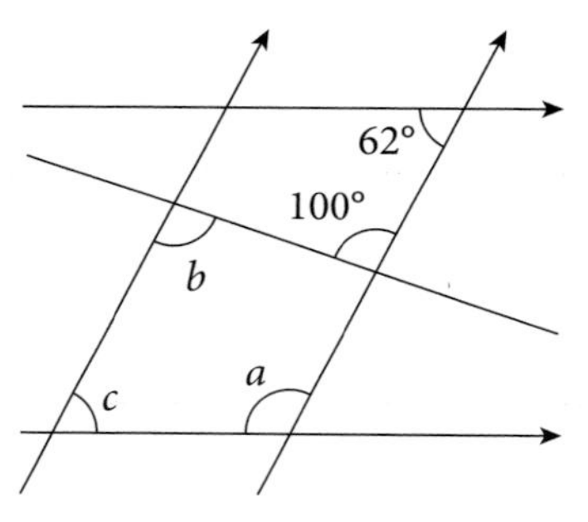

8.

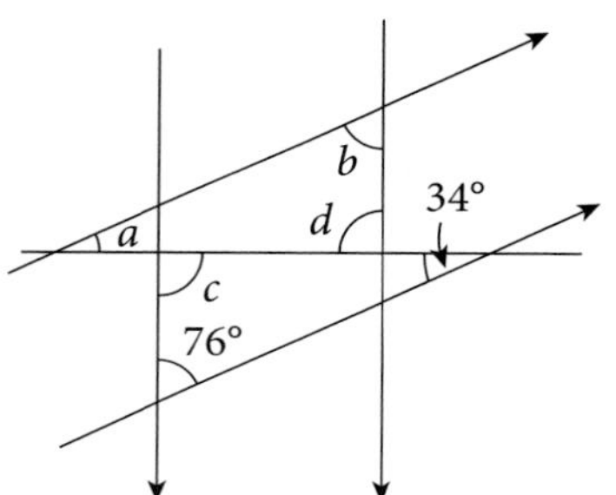

9.

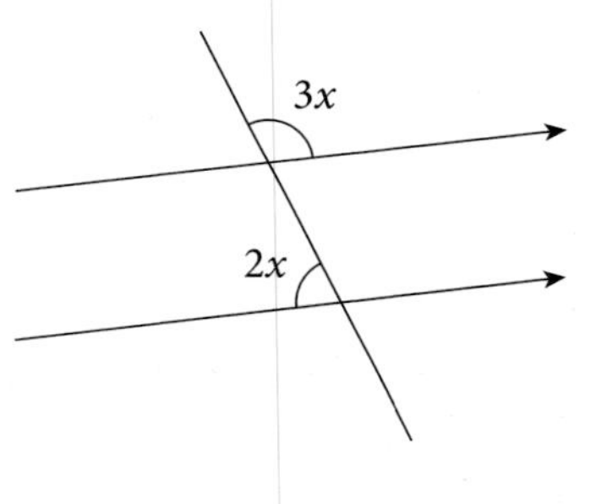

10. In the diagram KL is parallel to NM and LJ = LM.
Calculate the size of angle JLM.

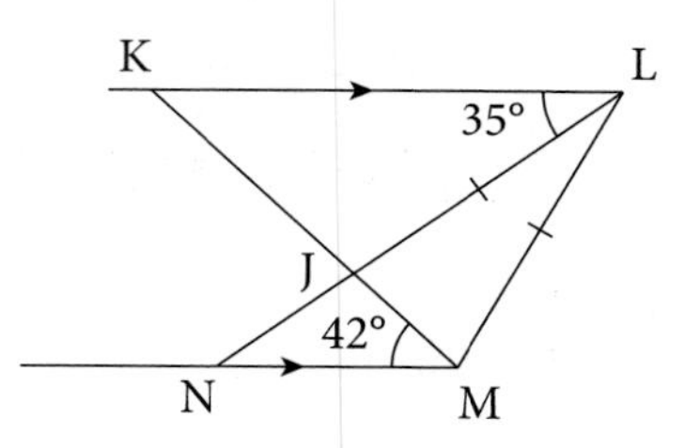

11. The diagram shows a series of isosceles triangles drawn between two lines.
Find the value of x.

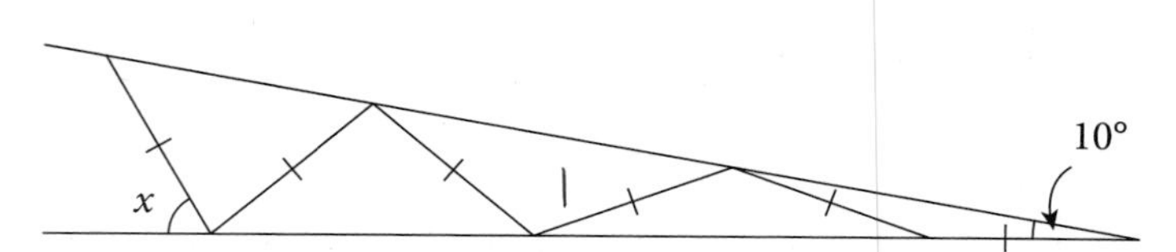

7.2 Pythagoras' theorem

In a right-angled triangle the square on the hypotenuse is equal to the sum of the squares on the other two sides.

$$a^2 + b^2 = c^2$$

This is the hypotenuse. It is the side opposite the right angle.

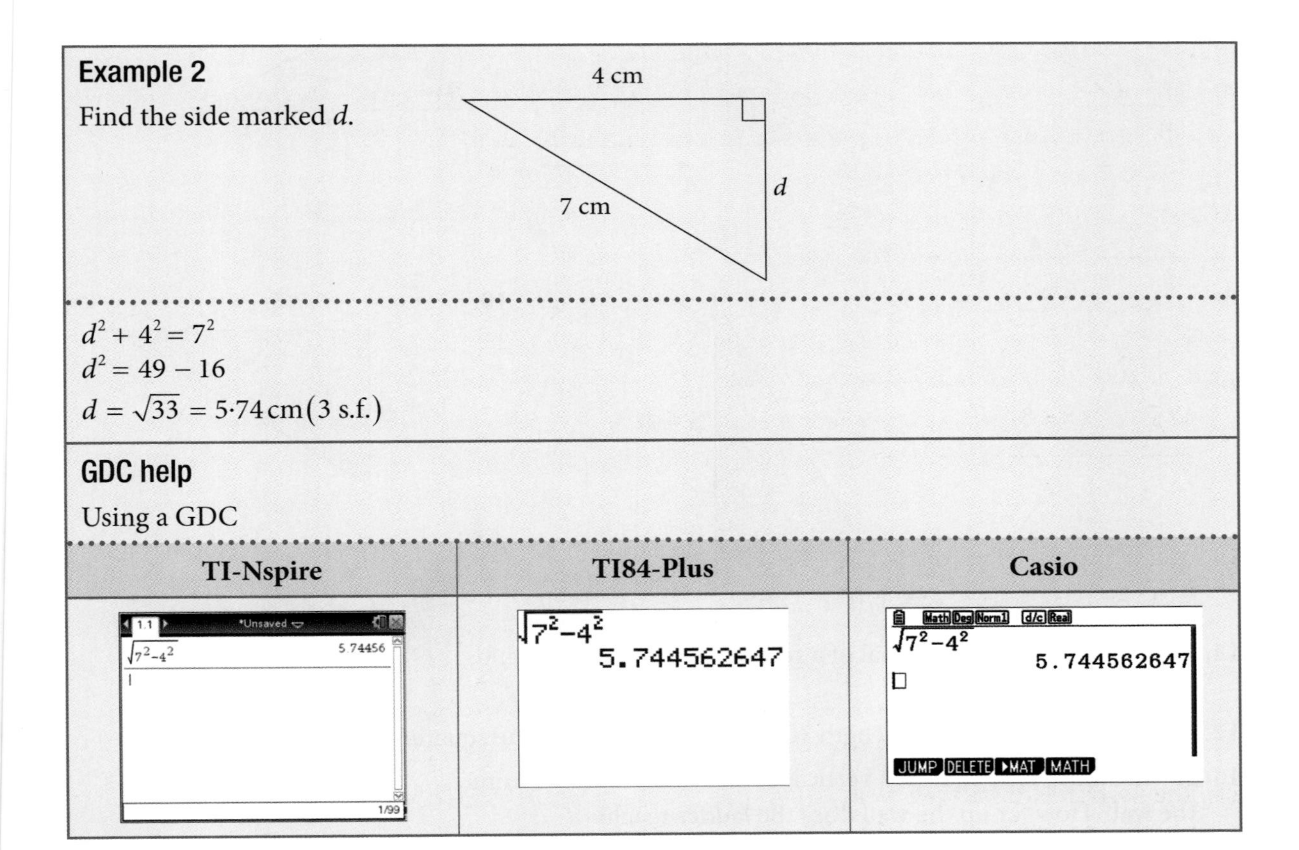

Example 2

Find the side marked d.

$d^2 + 4^2 = 7^2$

$d^2 = 49 - 16$

$d = \sqrt{33} = 5{\cdot}74\,\text{cm}\,(3\text{ s.f.})$

GDC help

Using a GDC

TI-Nspire	TI84-Plus	Casio
$\sqrt{7^2-4^2}$ 5.74456	$\sqrt{7^2-4^2}$ 5.744562647	$\sqrt{7^2-4^2}$ 5.744562647

The **converse** is also true:

If the square on one side of a triangle is equal to the sum of the squares on the other two sides, then the triangle is right-angled.

Exercise 7.4

In questions **1** to **10**, find x. All the lengths are in cm.

1.

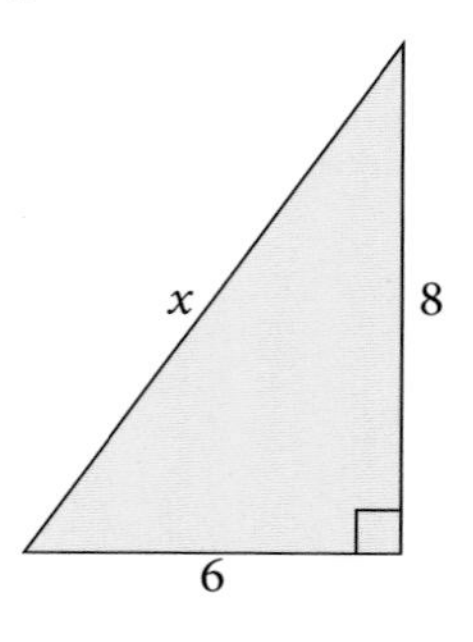

2.

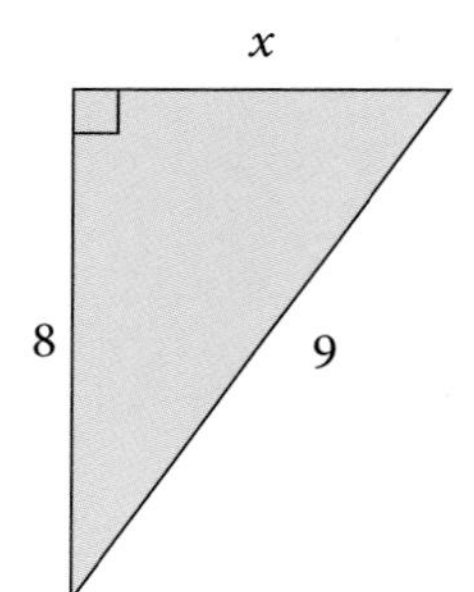

3.

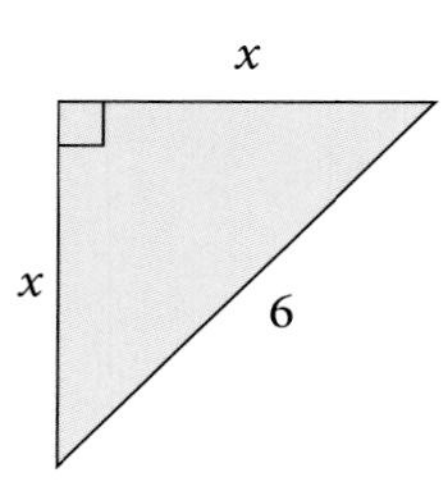

4.

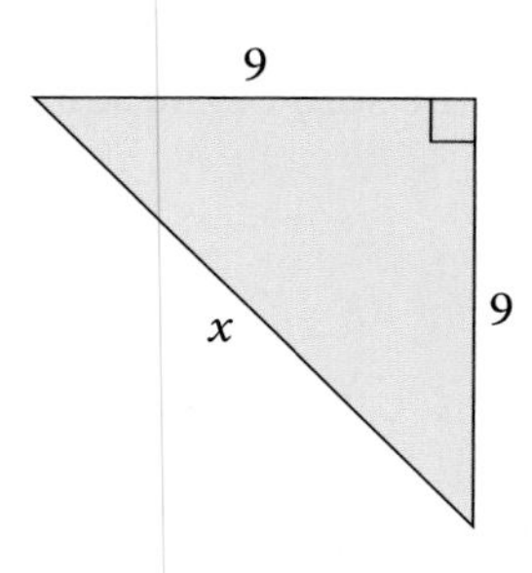

5.

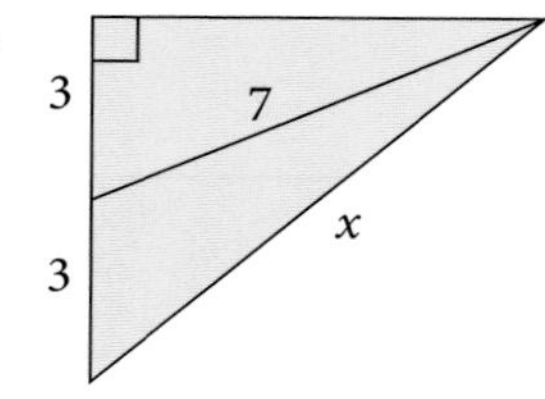

6.

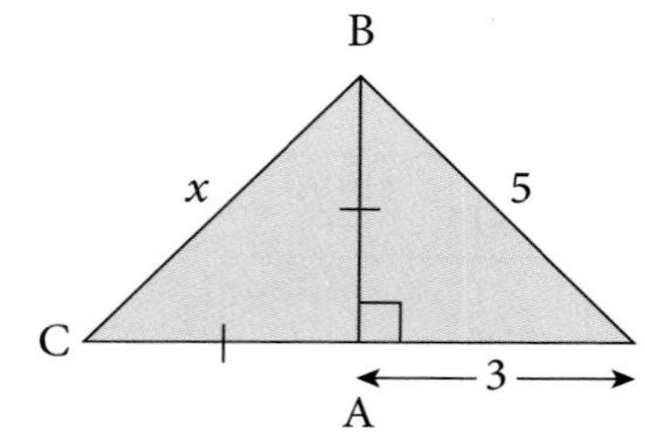

7.

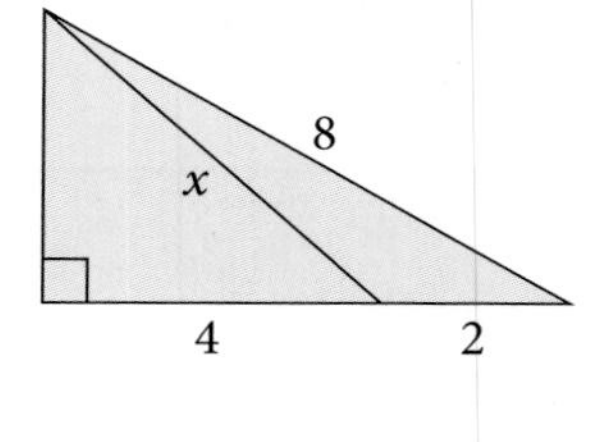

8.

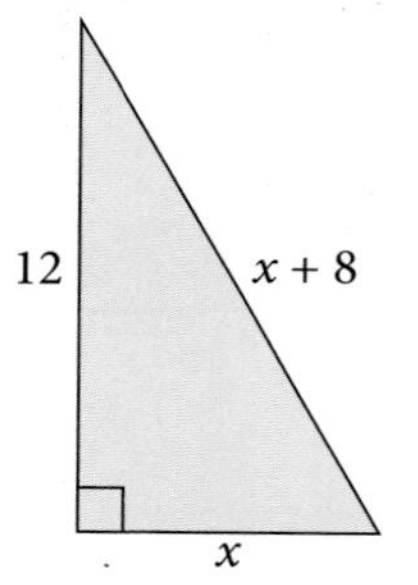

9.

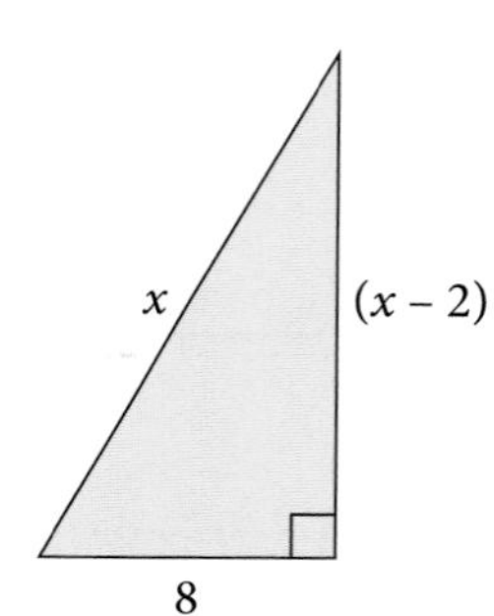

10.

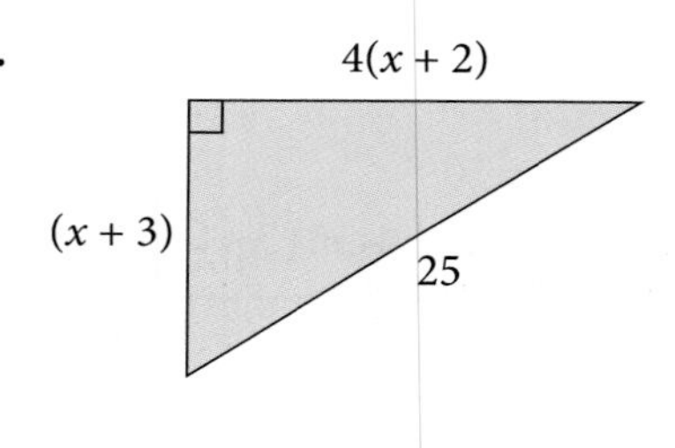

11. Find the length of a diagonal of a rectangle of length 9 cm and width 4 cm.

12. A square has diagonals of length 10 cm. Find the sides of the square.

13. A 4 m ladder rests against a vertical wall with its foot 2 m from the wall. How far up the wall does the ladder reach?

Example 3

A chord of length 10 cm is drawn in a circle of radius 7 cm. What is the distance of the chord from the centre of the circle?

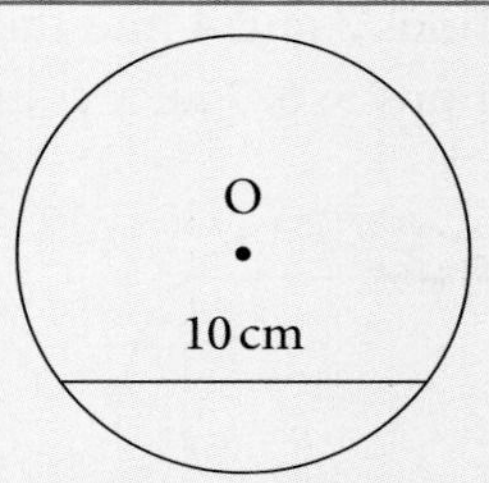

The line OP bisects the chord AB and angle OPB is 90°.

PB = 5 cm

By Pythagoras in triangle OPB,

$OP^2 + PB^2 = OB^2$

$OP^2 + 5^2 = 7^2$

$OP = \sqrt{7^2 - 5^2} = \sqrt{24}$ cm

$= 4{\cdot}90$ cm (to 2 d.p.)

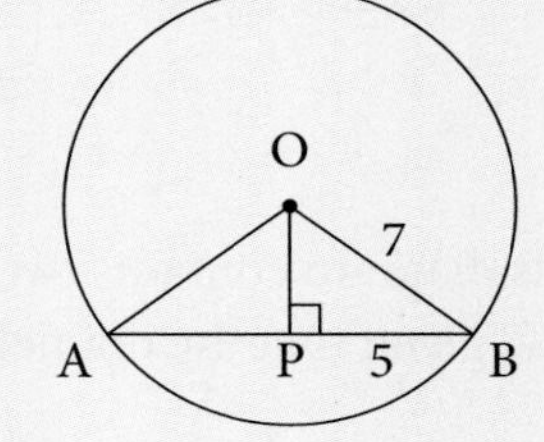

So the chord is 4·90 cm from the centre of the circle.

GDC help

Using a GDC

TI-Nspire	TI84-Plus	Casio
1.1 *Unsaved $\sqrt{7^2-5^2}$ 4.89898 1/99	$\sqrt{7^2-5^2}$ 4.898979486	Math Deg Norm1 d/c Real $\sqrt{7^2-5^2}$ 4.898979486 JUMP DELETE ▸MAT MATH

Exercise 7.5

1. A chord of length 12 cm is drawn in a circle of radius 8 cm. How far is the chord from the centre of the circle?
2. A chord of length 9 cm is 4·2 cm from the centre of the circle. What is the radius of the circle?

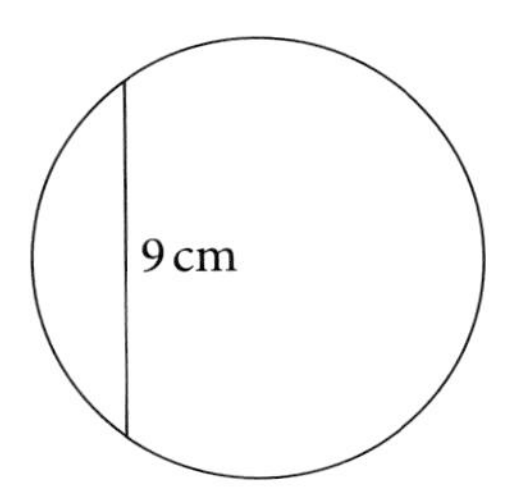

In questions **3** and **4** find the shaded area. Lengths are in cm.
In questions **5**, **6**, **7** all arcs are either semicircles or quarter-circles.

3.

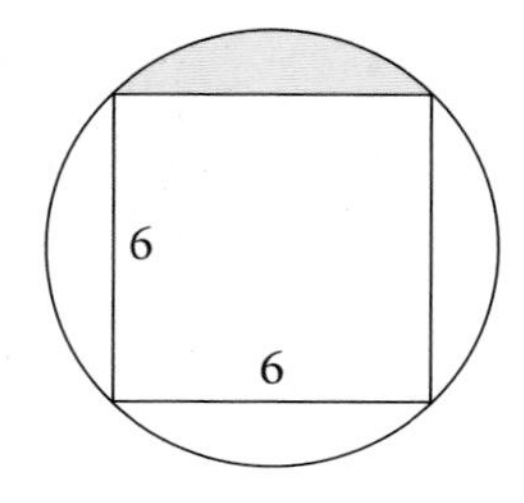

4.

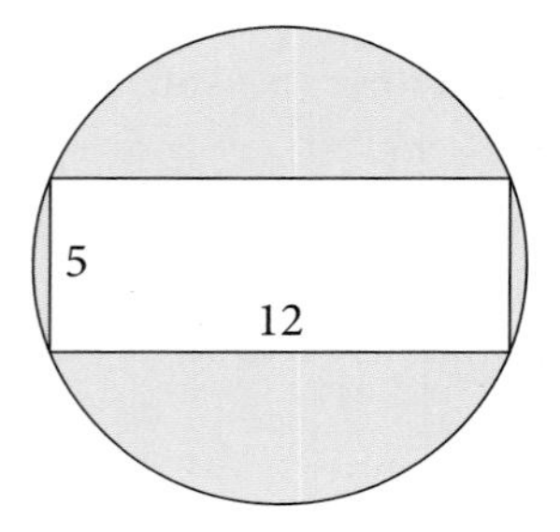

5.

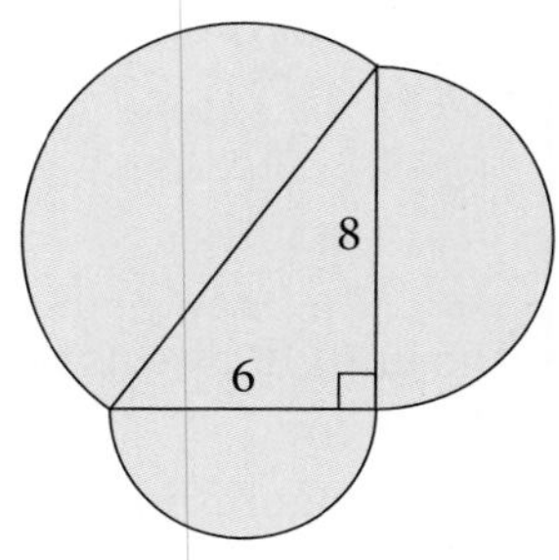

Find the total area.

6. This diagram contains two semicircles. Calculate the shaded area, given that the diameter of the larger semicircle is 12 cm.

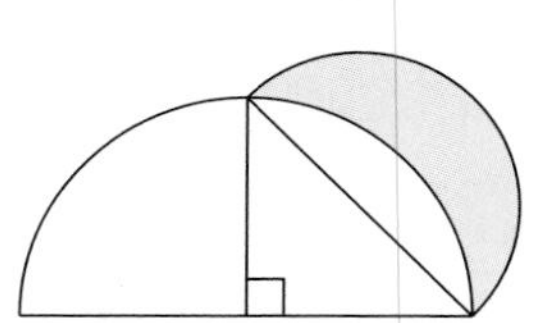

7. This diagram has one quarter-circle and two semicircles. Calculate the shaded area.

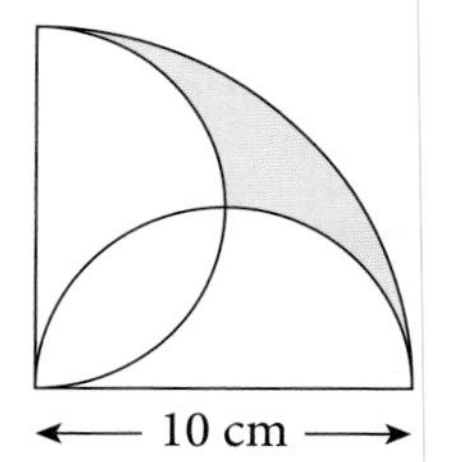

8. In the diagram, A is at (1, 2) and B is at (6, 4)
Work out the length AB.

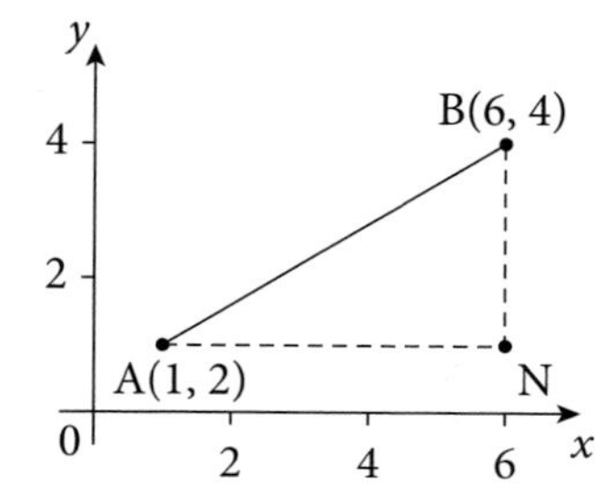

9. Use the diagram to work out the following lengths:

a) AC
b) AB
c) CD

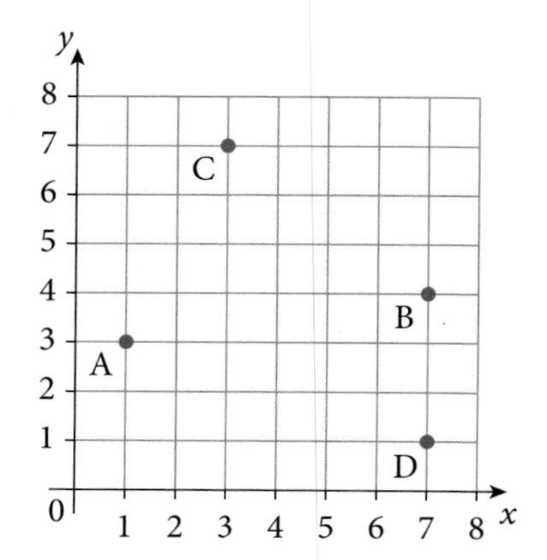

Exercise 7.6

1. A ship sails 20 km due North and then 35 km due East. How far is it from its starting point?

2. Find the length of a diagonal of a rectangular box of length 12 cm, width 5 cm and height 4 cm.

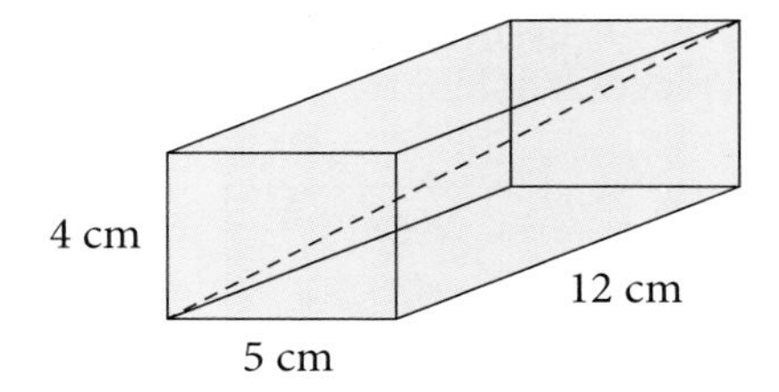

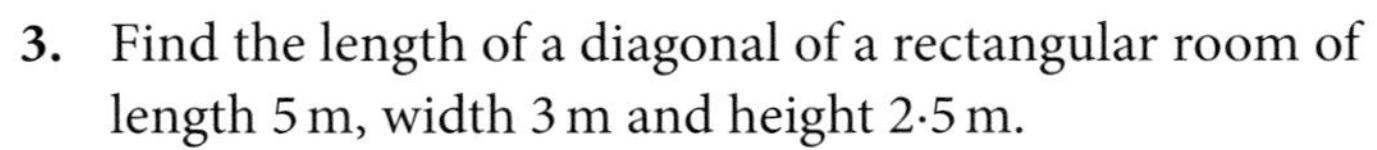

3. Find the length of a diagonal of a rectangular room of length 5 m, width 3 m and height 2·5 m.

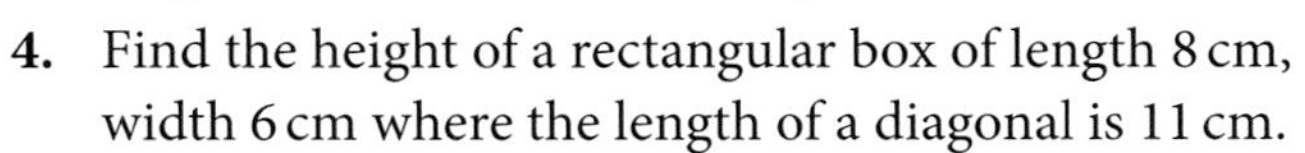

4. Find the height of a rectangular box of length 8 cm, width 6 cm where the length of a diagonal is 11 cm.

5. An aircraft flies equal distances South-East and then South-West to finish 120 km due South of its starting-point. How long is each part of its journey?

6. The diagonal of a rectangle exceeds the length by 2 cm. If the width of the rectangle is 10 cm, find the length.

7. A cone has base radius 5 cm and *slant* height 11 cm. Find its vertical height.

8. It is possible to find the sides of a right-angled triangle, with lengths which are whole numbers, by substituting different values of x into the expressions:

   a) $2x^2 + 2x + 1$ **b)** $2x^2 + 2x$ **c)** $2x + 1$

   [**a)** represents the hypotenuse, **b)** and **c)** the other two sides.]

   i) Find the sides of the triangles when $x = 1, 2, 3, 4$ and 5.

   ii) Confirm that $(2x+1)^2 + (2x^2+2x)^2 = (2x^2+2x+1)^2$

9. The diagram represents the starting position (AB) and the finishing position (CD) of a ladder as it slips. The ladder is leaning against a vertical wall.

   Given that: AC $= x$, OC $= 4$AC, BD $= 2$AC and OB $= 5$ m, form an equation in x, find x and hence find the length of the ladder.

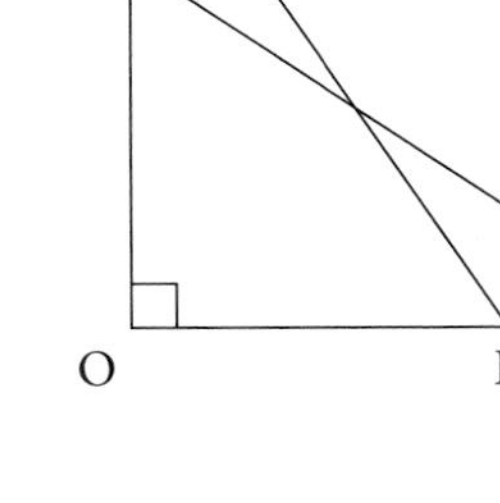

10. A thin wire of length 18 cm is bent into the shape shown. Calculate the length from A to B.

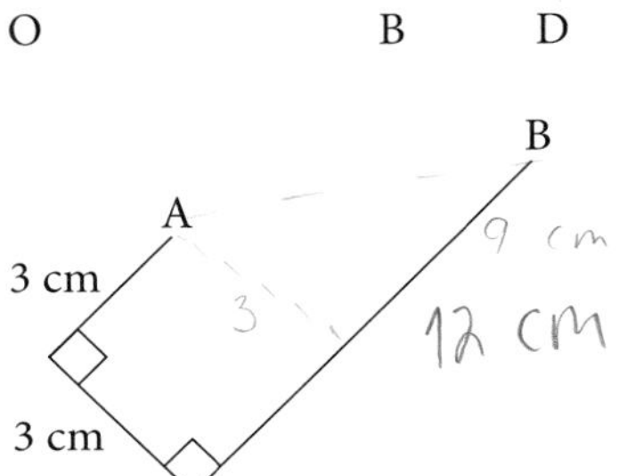

11. An aircraft is vertically above a point which is 10 km West and 15 km North of a control tower. If the aircraft is 4000 m above the ground, how far is it from the control tower?

In questions **12, 13, 14** find the length x, correct to 2 decimal places. All lengths are in cm.

12.

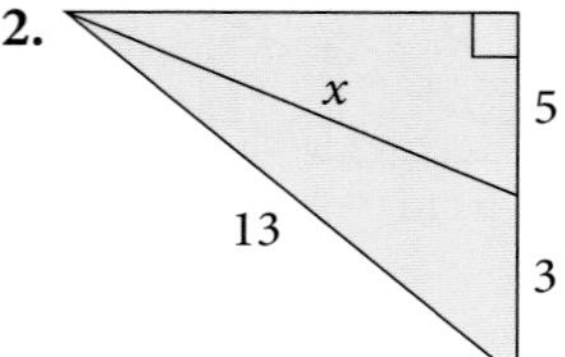

13.

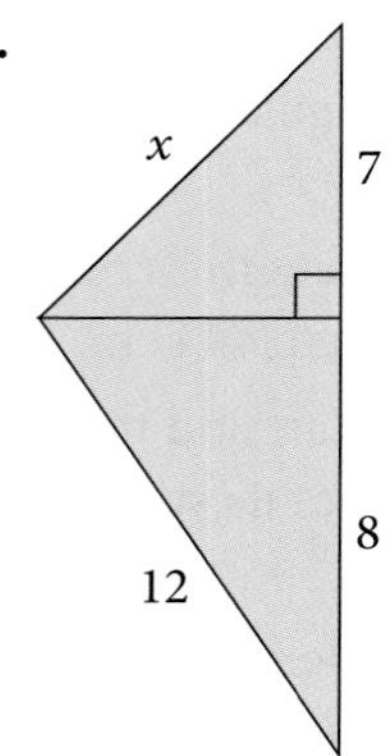

14.

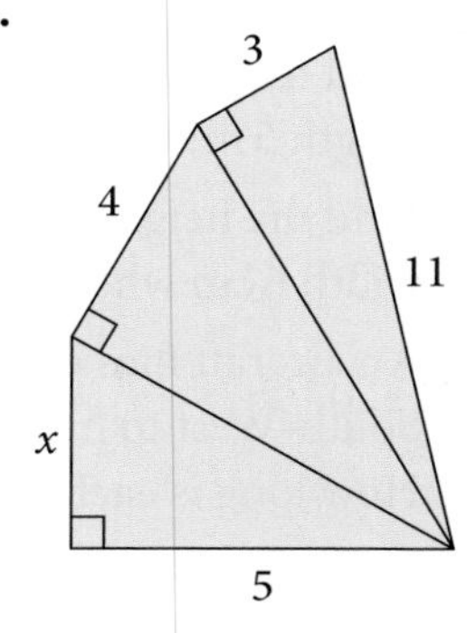

15. TC is a vertical pole whose base lies at a corner of the horizontal rectangle ABCD.
The top of the pole T is connected by straight wires to points A, B and D.

Calculate **a)** TC

b) TD

c) TA

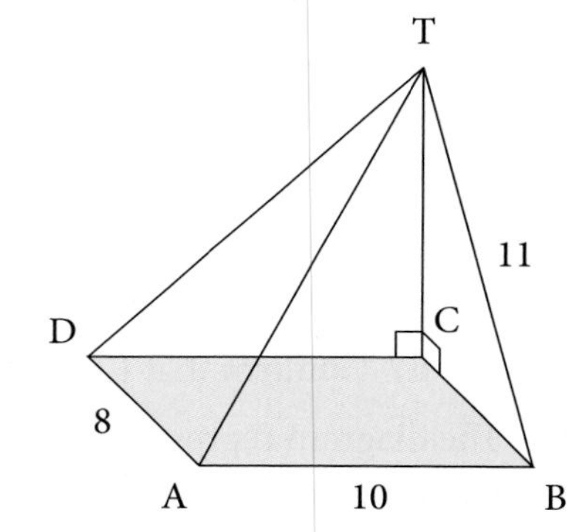

16. A ladder reaches H when held vertically against a wall.
When the base is 6 feet from the wall, the top of the ladder is 2 feet lower than H.
How long is the ladder?

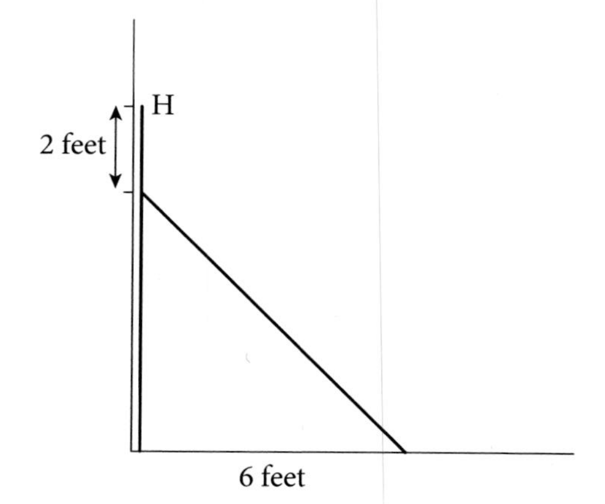

7.3 Symmetry

Line symmetry

The capital letter A has one line of symmetry, shown dotted.

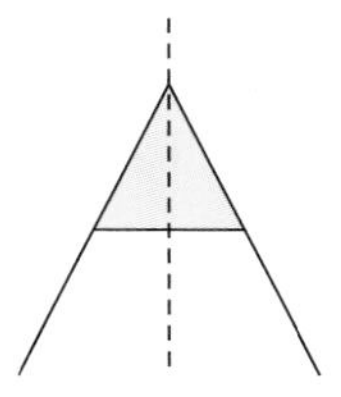

Rotational symmetry

The shape may be turned about O into three identical positions. It has rotational symmetry of order 3.

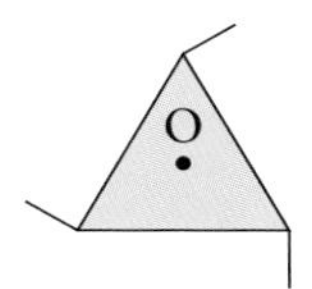

Quadrilaterals

- **Square**
  All sides are equal, all angles 90°, opposite sides parallel, diagonals bisect at right angles. Four lines of symmetry. Rotational symmetry of order of 4.

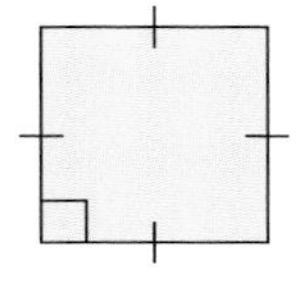

- **Rectangle**
  Opposite sides parallel and equal, all angles 90°, diagonals bisect each other. Two lines of symmetry. Rotational symmetry of order 2.

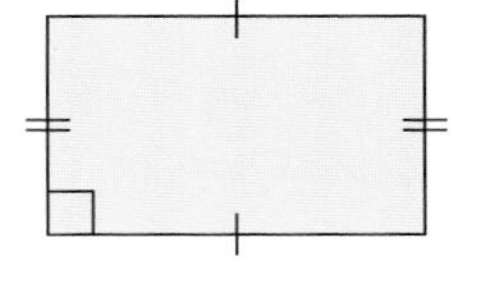

- **Parallelogram**
  Opposite sides parallel and equal, opposite angles equal, diagonals bisect each other (but not equal). No lines of symmetry. Rotational symmetry of order 2.

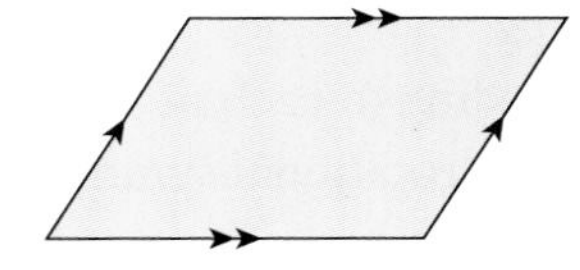

- **Rhombus**
  A parallelogram with all sides equal, diagonals bisect each other at right angles and bisect angles. Two lines of symmetry. Rotational symmetry of order 2.

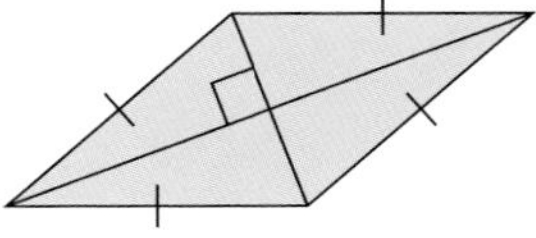
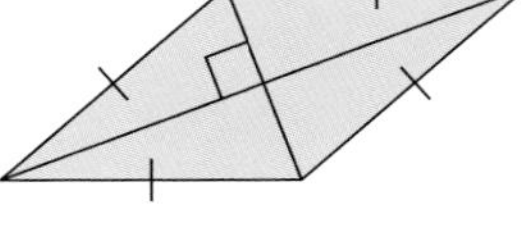

- **Trapezium**
  One pair of sides is parallel. No line or rotational symmetry.

- **Kite**
  Two pairs of adjacent sides equal, diagonals meet at right angles bisecting one of them. One line of symmetry. No rotational symmetry.

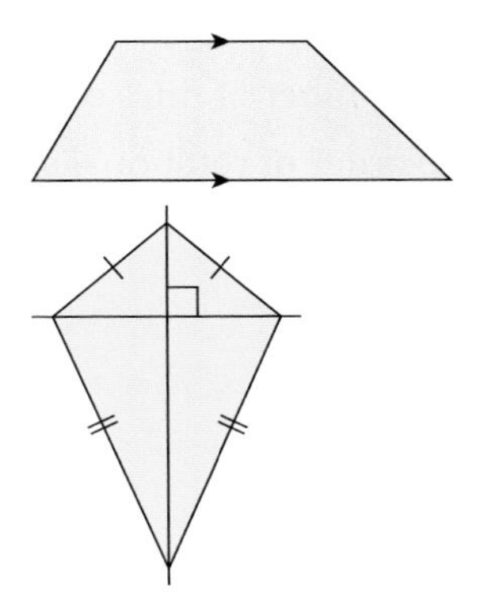

Exercise 7.7

1. For each shape state:

i) the number of lines of symmetry **ii)** the order of rotational symmetry.

a) 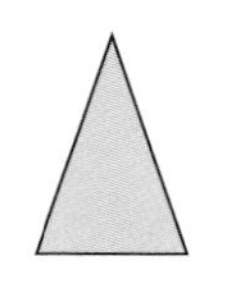

b)

c) 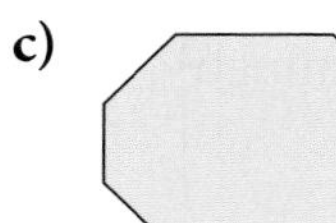

d) 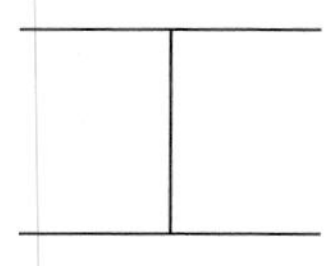

e) 

f) 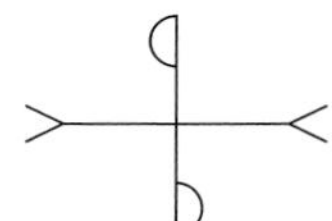

g) 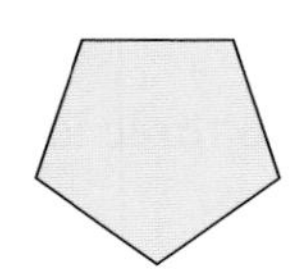

h) 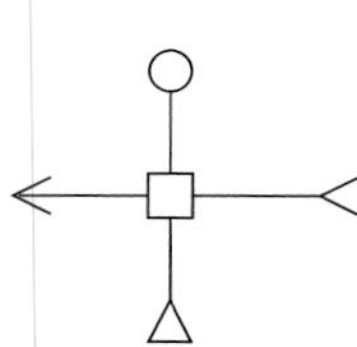

i) 

j) 

k) 

l) 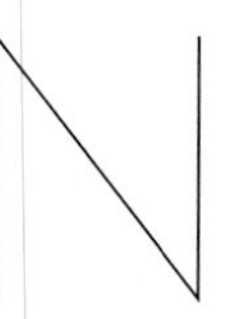

m) 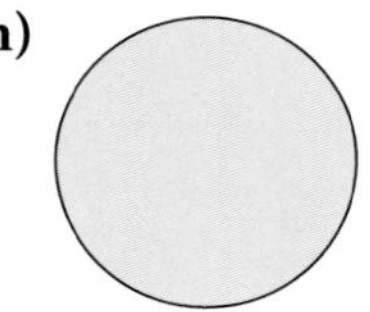

n) 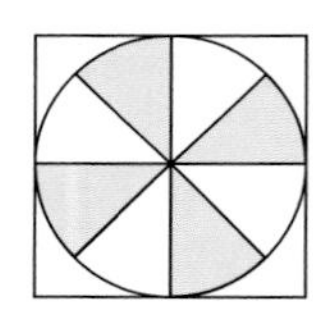

2. Add one line to each of the diagrams below so that the resulting figure has rotational symmetry but not line symmetry.

a) 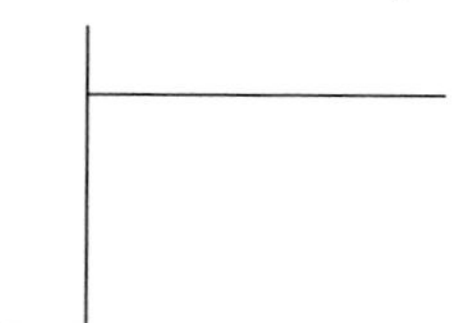

b) 

3. Draw a hexagon with just two lines of symmetry.

4. For each of the shapes listed, find:

a) the number of lines of symmetry

b) the order of rotational symmetry.

square	rectangle
parallelogram	rhombus
trapezium	kite
equilateral triangle	regular hexagon

In questions **5** to **15**, begin by drawing a diagram.

5. In a rectangle KLMN, $L\hat{N}M = 34°$. Calculate:

a) $K\hat{L}N$ **b)** $K\hat{M}L$

6. In a trapezium ABCD, $A\hat{B}D = 35°$, $B\hat{A}D = 110°$ and AB is parallel to DC. Calculate:

a) $A\hat{D}B$ **b)** $B\hat{D}C$

7. In a parallelogram WXYZ, $W\hat{X}Y = 72°$, $Z\hat{W}Y = 80°$. Calculate:

a) $W\hat{Z}Y$ **b)** $X\hat{W}Z$ **c)** $W\hat{Y}Z$

8. In a kite ABCD, AB = AD; BC = CD; $C\hat{A}D = 40°$ and $C\hat{B}D = 60°$. Calculate:

a) $B\hat{A}C$ **b)** $B\hat{C}A$ **c)** $A\hat{D}C$

9. In a rhombus ABCD, $A\hat{B}C = 64°$. Calculate:

a) $B\hat{C}D$ **b)** $A\hat{D}B$ **c)** $B\hat{A}C$

10. In a rectangle WXYZ, M is the mid-point of WX and $Z\hat{M}Y = 70°$. Calculate:

a) $M\hat{Z}Y$ **b)** $Y\hat{M}X$

11. In a trapezium ABCD, AB is parallel to DC, AB = AD, BD = DC and $B\hat{A}D = 128°$. Find:

a) $A\hat{B}D$ **b)** $B\hat{D}C$ **c)** $B\hat{C}D$

12. In a parallelogram KLMN, KL = KM and $K\hat{M}L = 64°$. Find:

a) $M\hat{K}L$ **b)** $K\hat{N}M$ **c)** $L\hat{M}N$

13. In a kite PQRS with PQ = PS and RQ = RS, $Q\hat{R}S = 40°$ and $Q\hat{P}S = 100°$. Find:

a) $Q\hat{S}R$ **b)** $P\hat{S}Q$ **c)** $P\hat{Q}R$

14. In a rhombus PQRS, $R\hat{P}Q = 54°$. Find:

a) $P\hat{R}Q$ **b)** $P\hat{S}R$ **c)** $R\hat{Q}S$

15. In a kite PQRS, $R\hat{P}S = 2P\hat{R}S$, PQ = QS = PS and QR = RS. Find:

a) $Q\hat{P}S$ **b)** $P\hat{R}S$ **c)** $Q\hat{S}R$ **d)** $P\hat{Q}R$

7.4 Similarity

Two triangles are similar if they have the same angles.
For other shapes, not only must corresponding angles be equal, but also corresponding sides must be in the same proportion.
The two rectangles A and B are *not* similar even though they have the same angles.

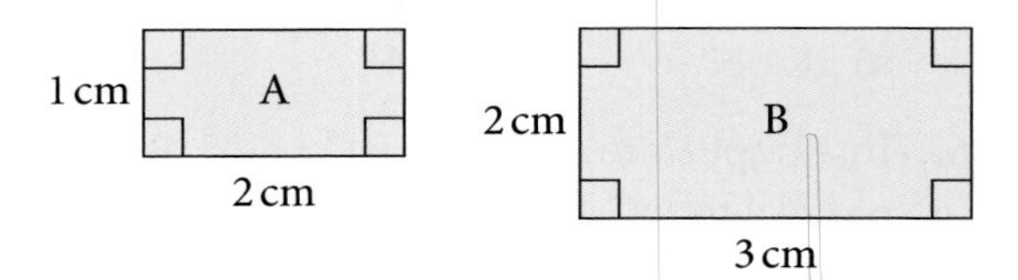

Example 4

In the triangles ABC and XYZ
$\hat{A} = \hat{X}$ and $\hat{B} = \hat{Y}$
so the triangles are similar.
($\hat{C}$ must be equal to $\hat{Z}$.)

We have $\dfrac{BC}{YZ} = \dfrac{AC}{XZ} = \dfrac{AB}{XY}$

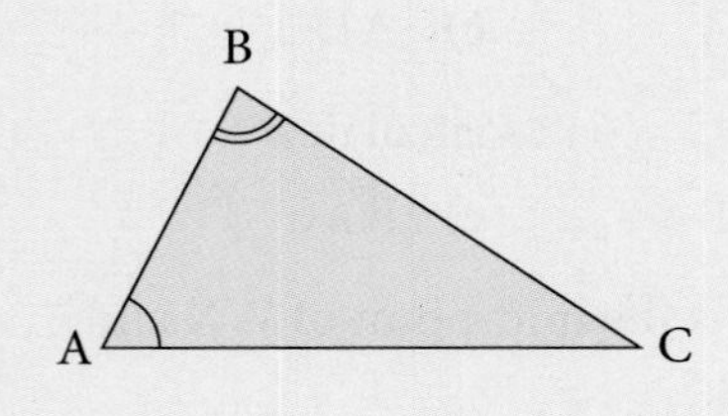

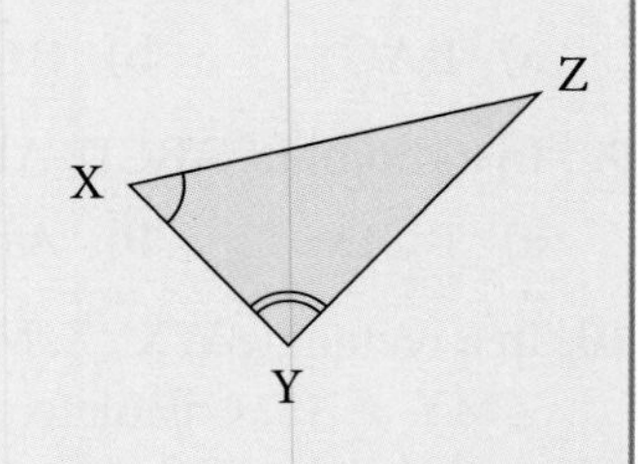

Exercise 7.8

In questions **1** to **11**, find the sides marked with letters. All lengths are given in centimetres.

1.

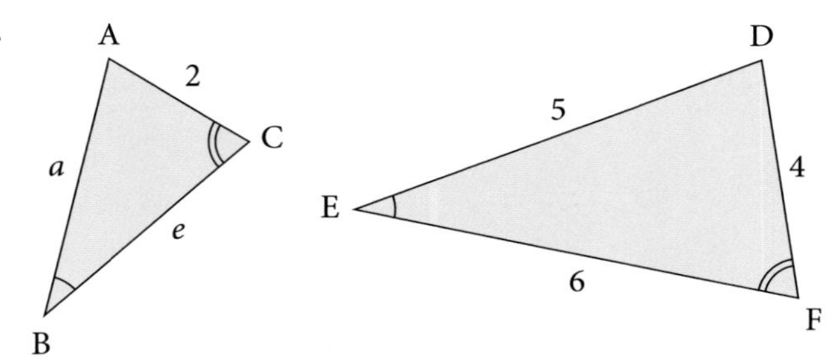

2.

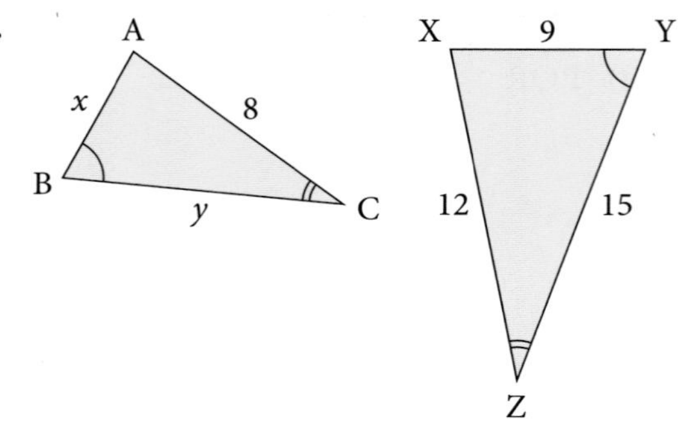

3.

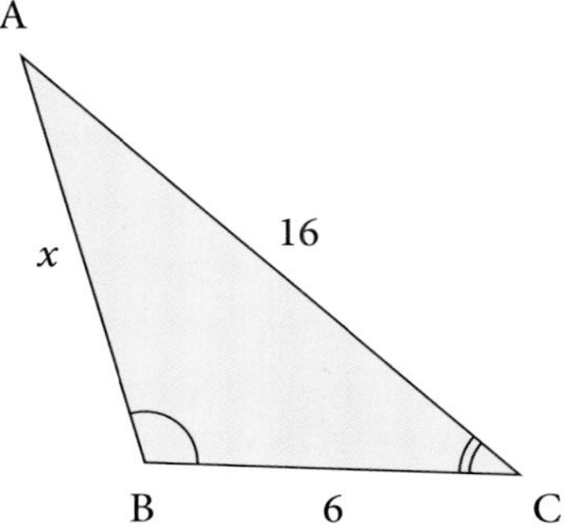

4.

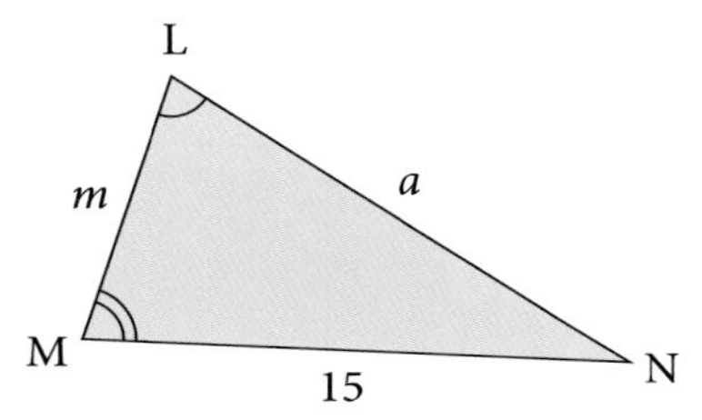

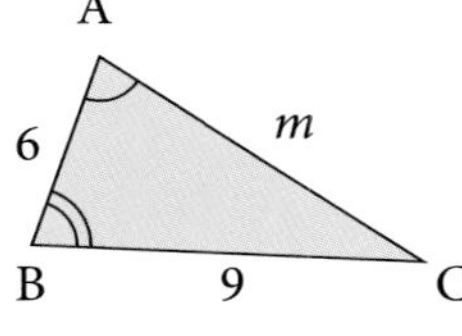

5.

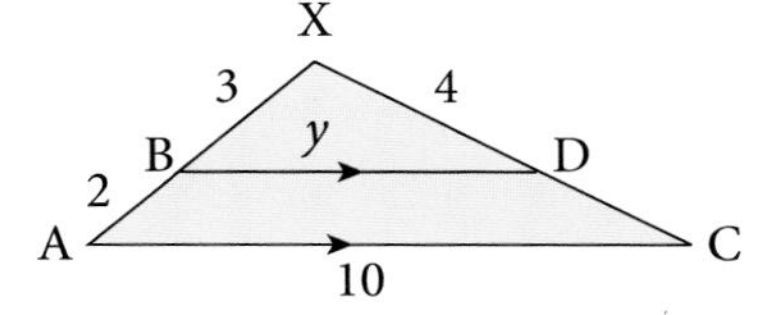

6.

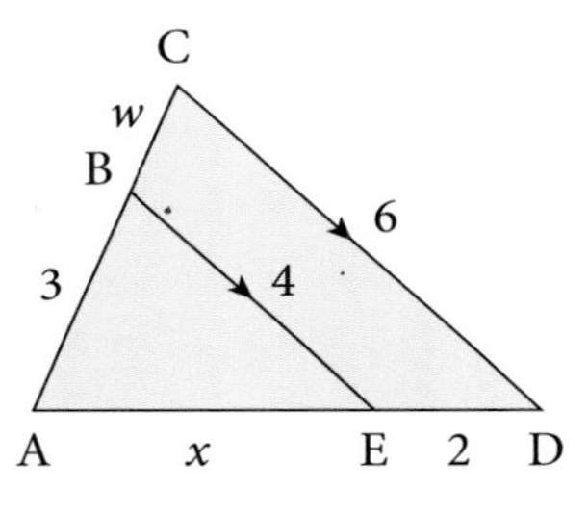

7.

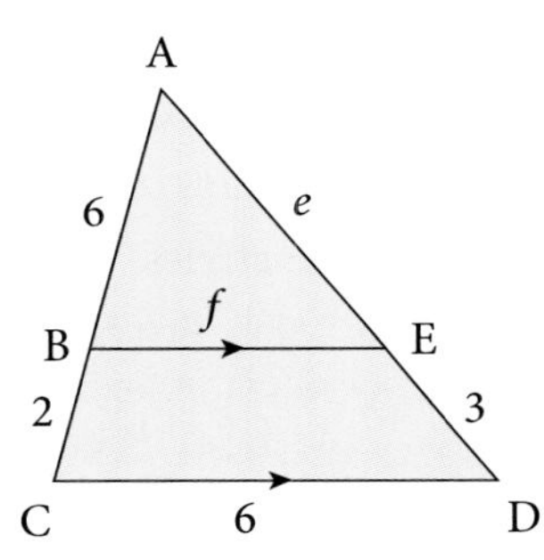

8.

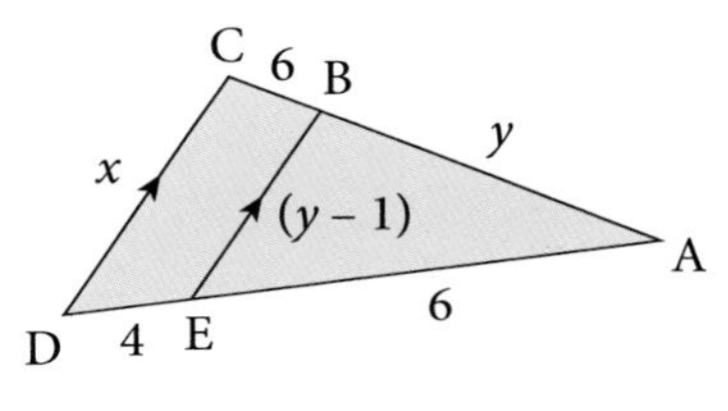

9. $B\hat{A}C = D\hat{B}C$

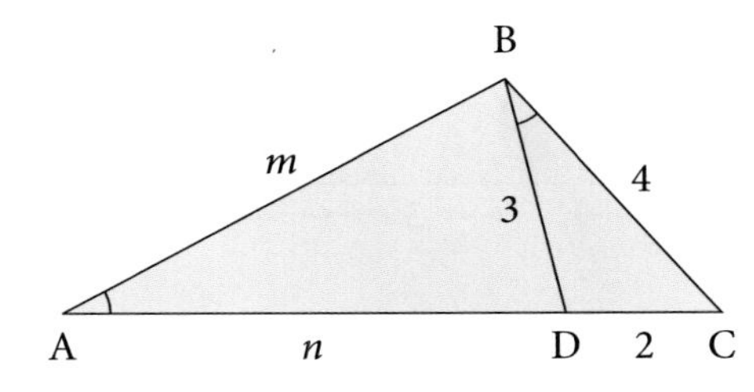

10.

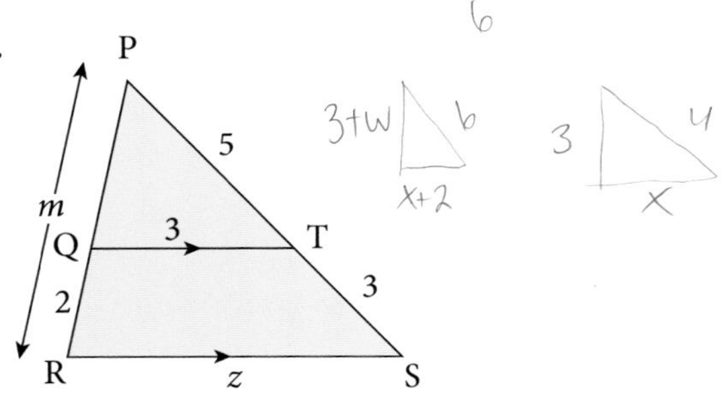

11.

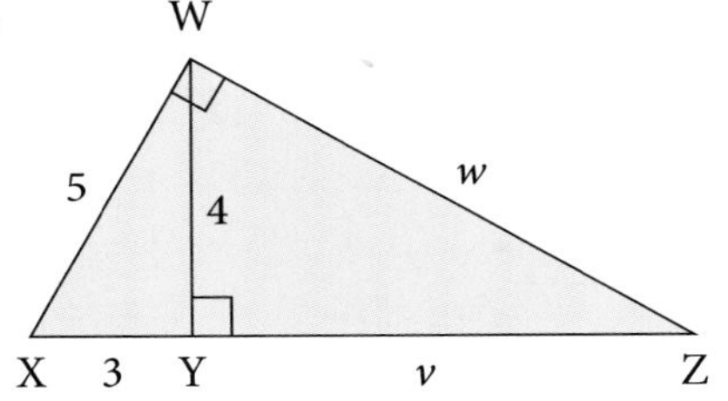

12. The drawing shows a rectangular photo 16 cm × 8 cm surrounded by a frame of width 4 cm. Are the two rectangles similar?

13. The diagonals of a trapezium ABCD intersect at O. AB is parallel to DC, AB = 3 cm and DC = 6 cm. If CO = 4 cm and OB = 3 cm, find AO and DO.

14. A tree of height 4 m casts a shadow of length 6·5 m. Find the height of a house casting a shadow 26 m long.

15. Which of the following *must* be similar to each other?

a) two equilateral triangles **b)** two rectangles

c) two isosceles triangles **d)** two squares

e) two regular pentagons **f)** two kites

g) two rhombuses **h)** two circles

Exercise 7.9

1. A simple musical instrument consists of 4 strings attached to a triangular wooden frame. If the shortest string P measures 10 cm, find the lengths of the other three strings.

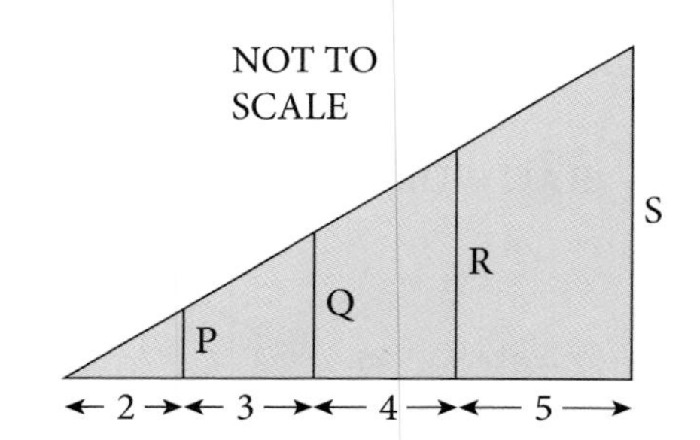

2. Using the diagram

a) name two similar triangles

b) calculate the length x.

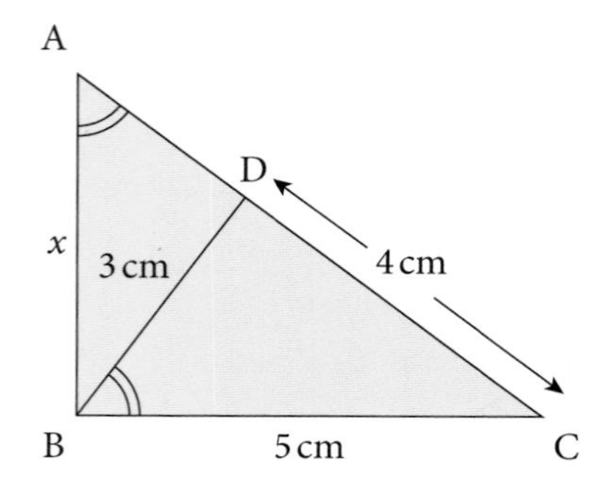

3. Photo B is an enlargement of photo A with scale factor 3.
If the area of photo A is 20 cm^2, find the area of photo B.

Photo A

Photo B

4. The sides of triangle ABC are each increased by 1 cm to form triangle DEF.
Are triangles ABC and DEF similar?

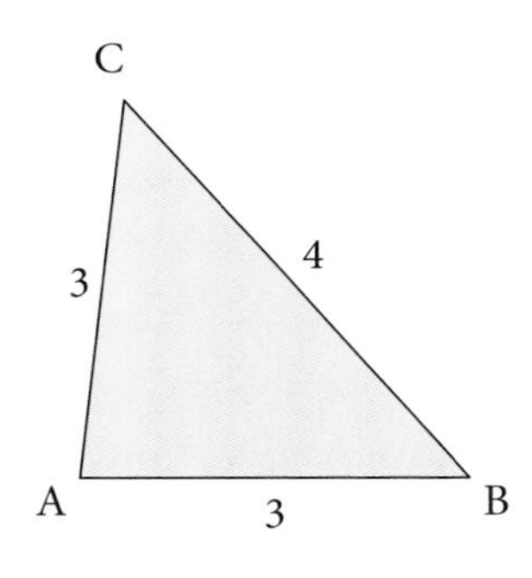

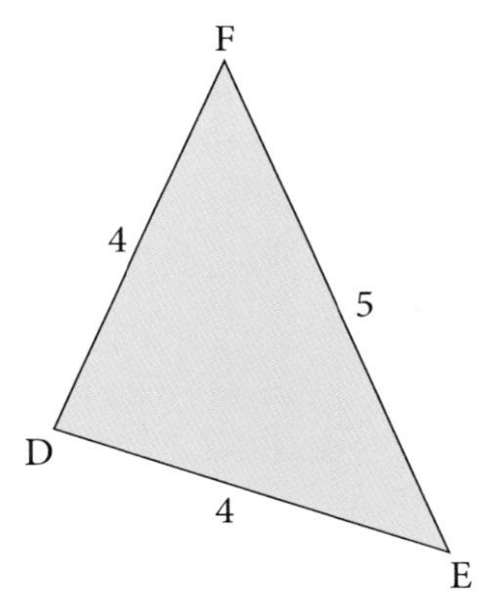

5. In the diagram $A\hat{B}E = A\hat{D}C$.
Find the length x.

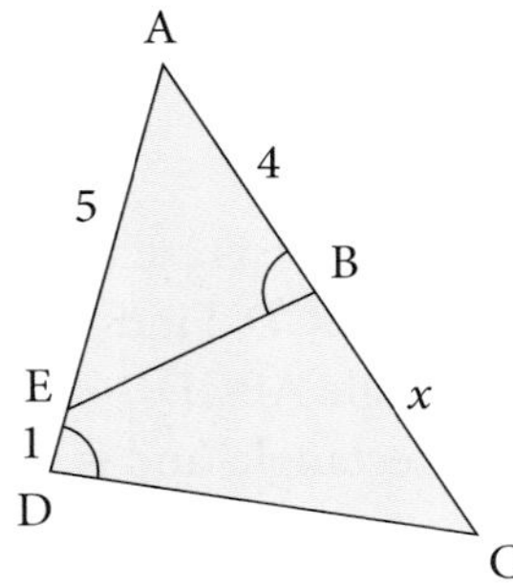

6. 

9 cm

14 cm

The photo shown is similar to a photo 3 cm by x cm.
Find the two possible values of x.

7. When the large rectangle is folded in half along PQ, each of the smaller rectangles is similar to the large rectangle.

Form an equation and use trial and improvement to find x, correct to 1 d.p.

P

$x - 1$

Q

x

8. In the diagram $A\hat{B}C = A\hat{D}B = 90°$, $AD = p$ and $DC = q$.

a) Use similar triangles to show that $x^2 = pz$.

b) Find a similar expression for y^2.

c) Add the expressions for x^2 and y^2 and hence prove Pythagoras' theorem.

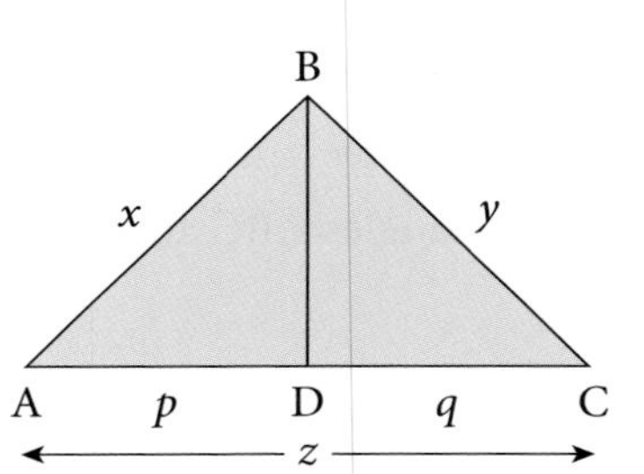

9. In a triangle ABC, a line is drawn parallel to BC to meet AB at D and AC at E. DC and BE meet at X.
Prove that:

a) the triangles ADE and ABC are similar

b) the triangles DXE and BXC are similar

c) $\dfrac{AD}{AB} = \dfrac{EX}{XB}$

10. From the rectangle ABCD a square is cut off to leave rectangle BCEF. Rectangle BCEF is similar to ABCD. Find x and hence state the ratio of the sides of rectangle ABCD.
ABCD is called the 'golden rectangle' and is an important shape in architecture.

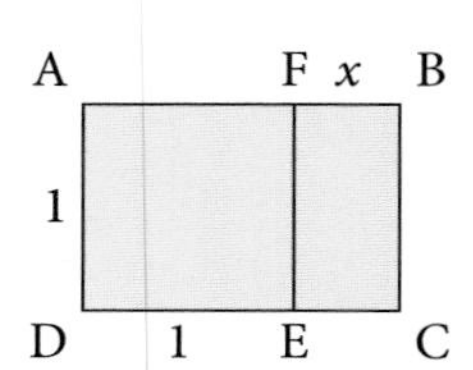

Congruence

Two plane figures are congruent if one fits exactly on the other. They must be the same size and the same shape.

Exercise 7.10

1. Identify pairs of congruent shapes below.

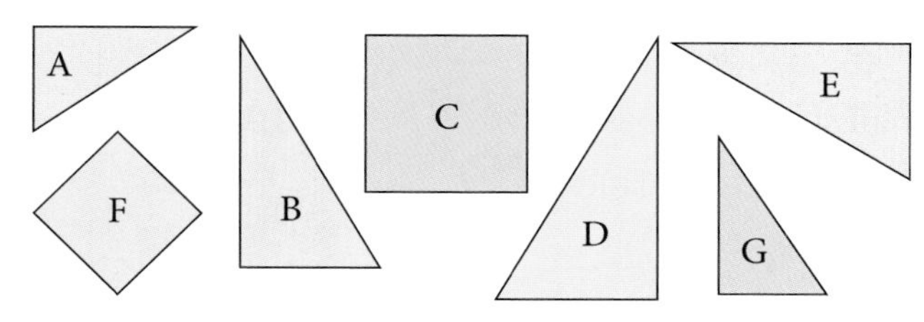

2. Triangle LMN is isosceles with LM = LN and X and Y are points on LM, LN respectively such that LX = LY. Prove that triangles LMY and LNX are congruent.

3. ABCD is a quadrilateral and a line through A parallel to BC meets DC at X. If $\hat{D} = \hat{C}$, prove that ΔADX is isosceles.

4. In the diagram, N lies on a side of the square ABCD, AM and LC are perpendicular to DN.
Prove that:

a) $A\hat{D}N = L\hat{C}D$ **b)** AM = LD

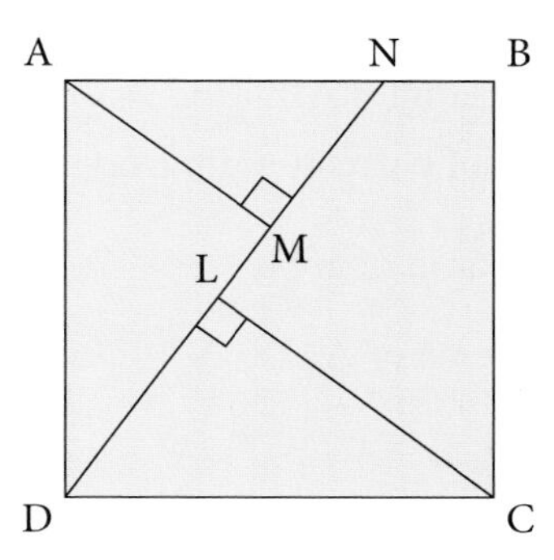

5. Points L and M on the side YZ of a triangle XYZ are drawn so that L is between Y and M. Given that XY = XZ and $Y\hat{X}L = M\hat{X}Z$, prove that YL = MZ.

6. Squares AMNB and AOPC are drawn on the sides of triangle ABC, so that they lie outside the triangle.
Prove that MC = OB.

7. In the diagram, $L\hat{M}N = O\hat{N}M = 90°$. P is the mid-point of MN, MN = 2ML and MN = NO.
Prove that:

a) the triangles MNL and NOP are congruent

b) $O\hat{P}N = L\hat{N}O$ **c)** $L\hat{Q}O = 90°$

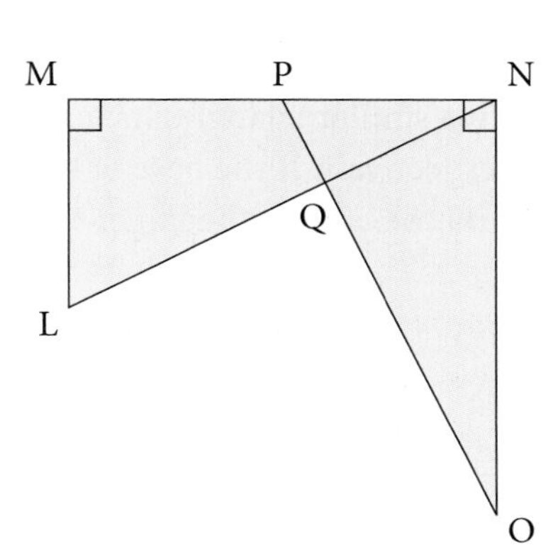

8. PQRS is a parallelogram in which the bisectors of the angles P and Q meet at X. Prove that the angle PXQ is a right angle.

Areas of similar shapes

The two rectangles shown are similar and the ratio of corresponding sides is k.

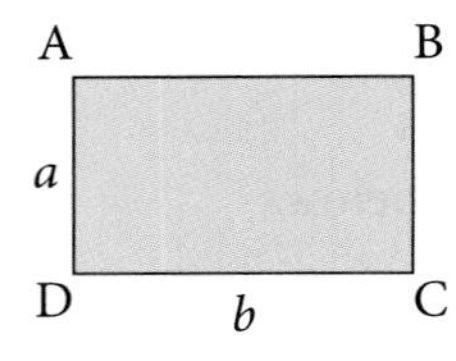

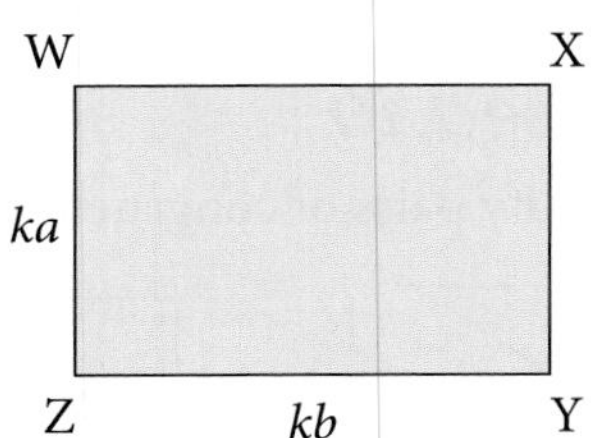

Area of ABCD $= ab$

Area of WXYZ $= ka \times kb = k^2ab$

$\therefore \quad \dfrac{\text{Area WXYZ}}{\text{Area ABCD}} = \dfrac{k^2ab}{ab} = k^2$

This illustrates an important general rule for all similar shapes:

- If two figures are similar and the ratio of corresponding sides is k, then the ratio of their areas is k^2.

This result also applies for the surface areas of similar three-dimensional objects.

k is sometimes called the linear scale factor.

Example 5

XY is parallel to BC.

$\dfrac{AB}{AX} = \dfrac{3}{2}$

If the area of $\Delta AXY = 4\,\text{cm}^2$, find the area of ΔABC.

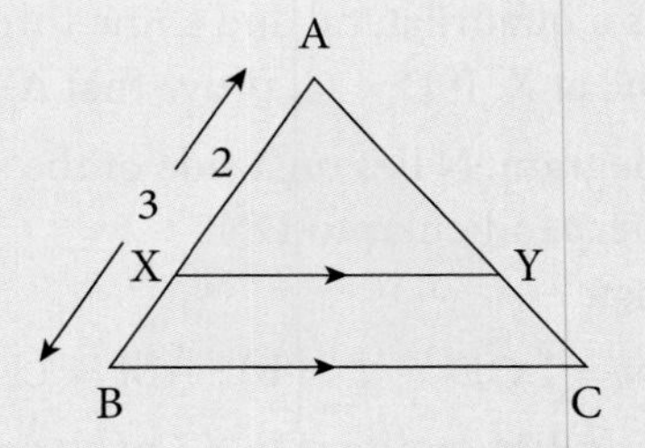

The triangles ABC and AXY are similar.

Ratio of corresponding sides $(k) = \dfrac{3}{2}$

$\therefore \quad$ Ratio of areas $\quad (k^2) = \dfrac{9}{4}$

$\therefore \quad$ Area of $\Delta ABC = \dfrac{9}{4} \times (\text{area of } \Delta AXY)$

$= \dfrac{9}{4} \times (4) = 9\,\text{cm}^2$

Example 6

Two similar triangles have areas of $18\,\text{cm}^2$ and $32\,\text{cm}^2$ respectively. If the base of the smaller triangle is 6 cm, find the base of the larger triangle.

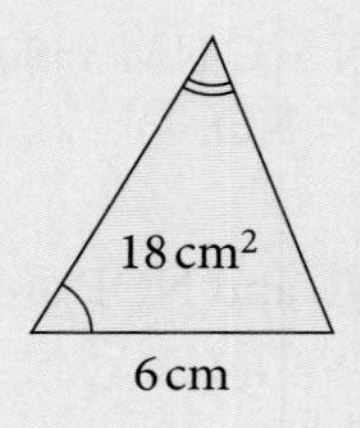

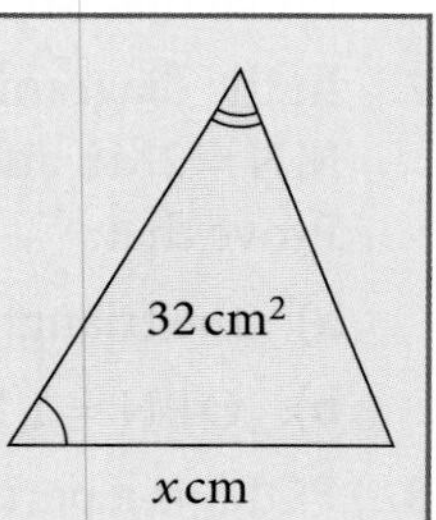

Ratio of areas $(k^2) = \dfrac{32}{18} = \dfrac{16}{9}$

$\therefore \quad$ Ratio of corresponding sides $(k) = \sqrt{\left(\dfrac{16}{9}\right)}$

$= \dfrac{4}{3}$

$\therefore \quad$ Base of larger triangle $= 6 \times \dfrac{4}{3} = 8$ cm

Exercise 7.11

In this exercise a number written inside a figure represents the area of the shape in cm². Numbers on the outside give linear dimensions in cm.

In questions **1** to **6** find the unknown area A. In each case the shapes are similar.

1.

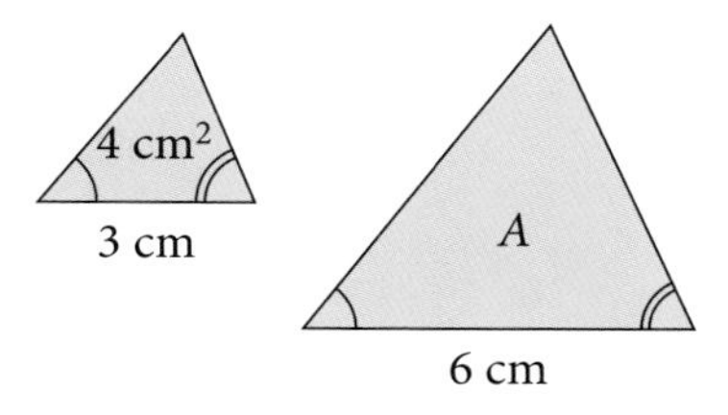

2.

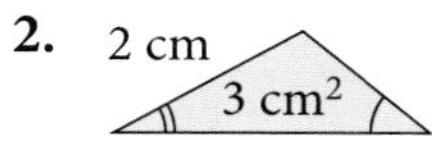

3.

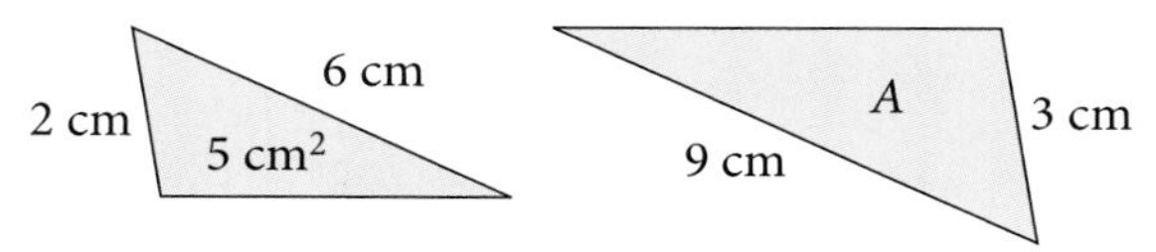

4.

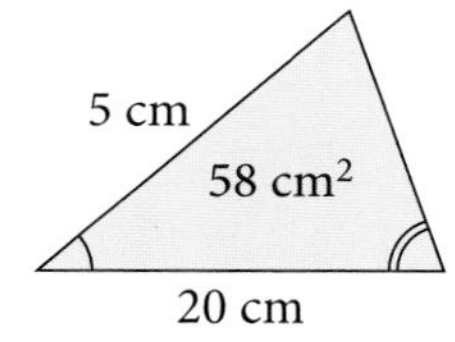

5.

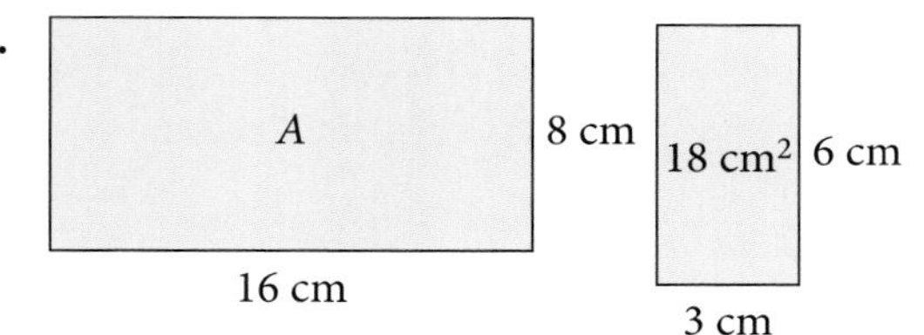

6.

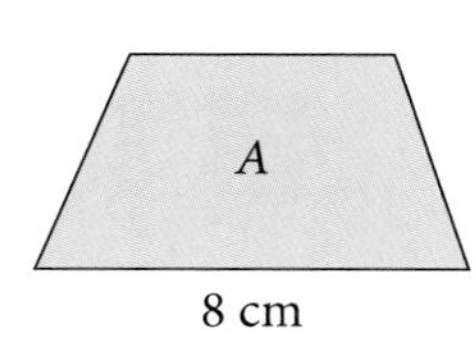

In questions **7** to **12**, find the lengths marked for each pair of similar shapes.

7.

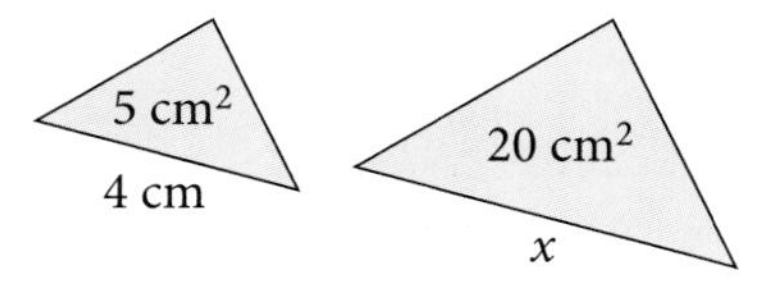

8.

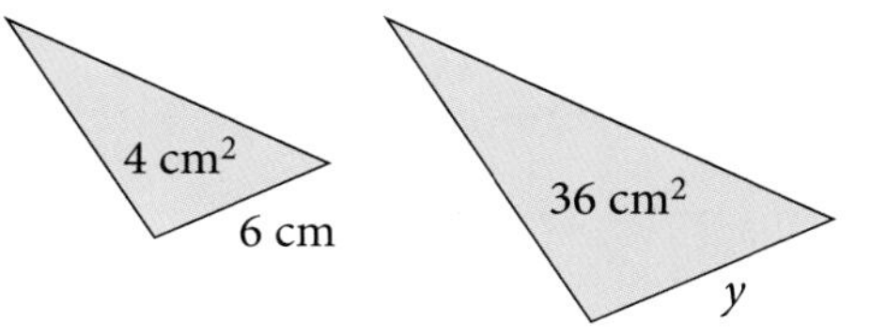

9.

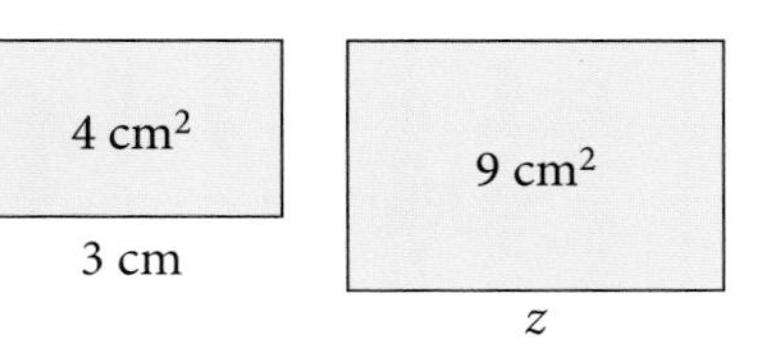

10.

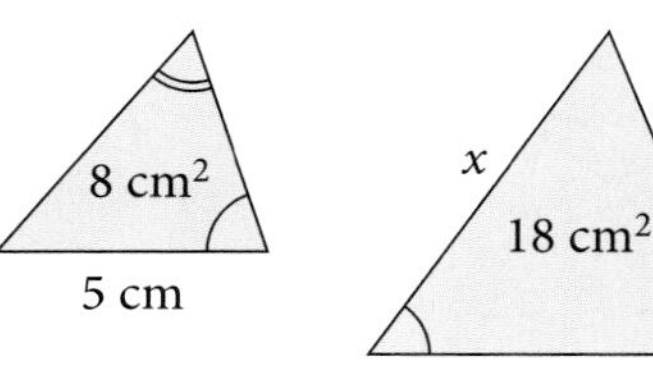

11.

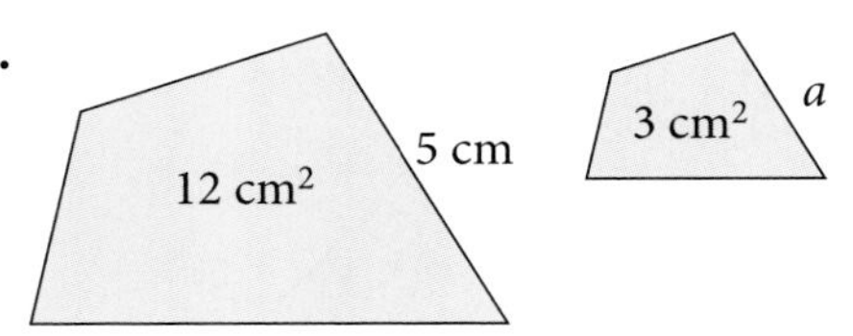

12.

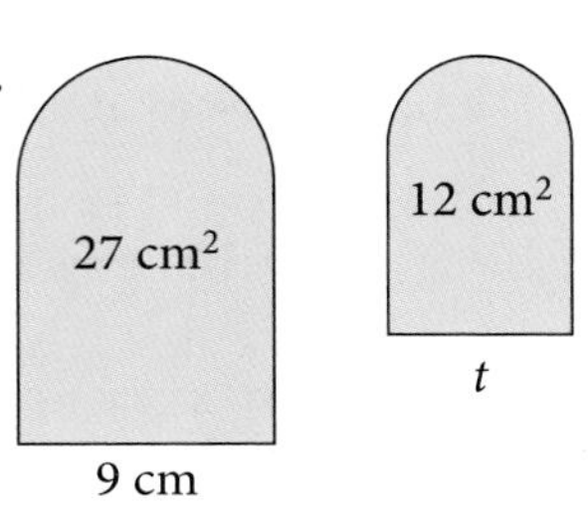

Exercise 7.12

In questions **1** to **13** you have a pair of similar three-dimensional objects. Find the surface area indicated.

1.

face area = 8 cm^2

3 cm

find surface area A

A

6 cm

2.

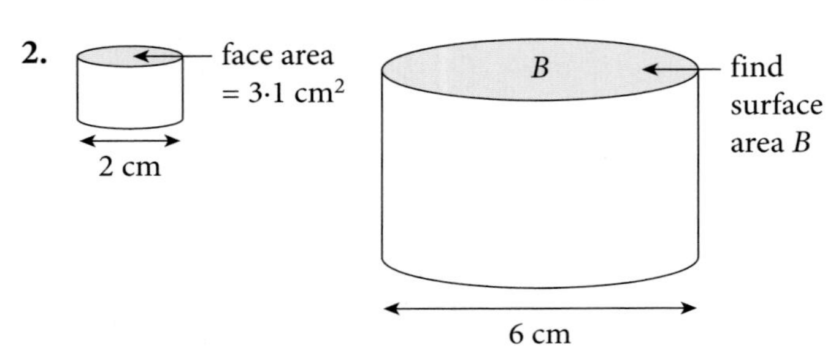

3. The radius of the large sphere is twice the radius of the small sphere.

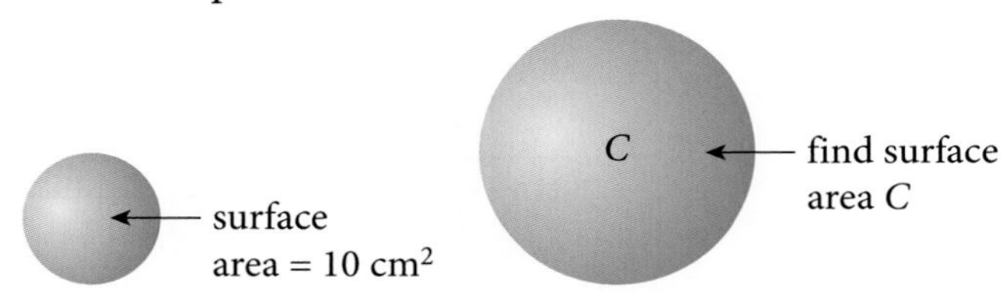

4. The length of large solid is 1·5 times the length of the small solid.

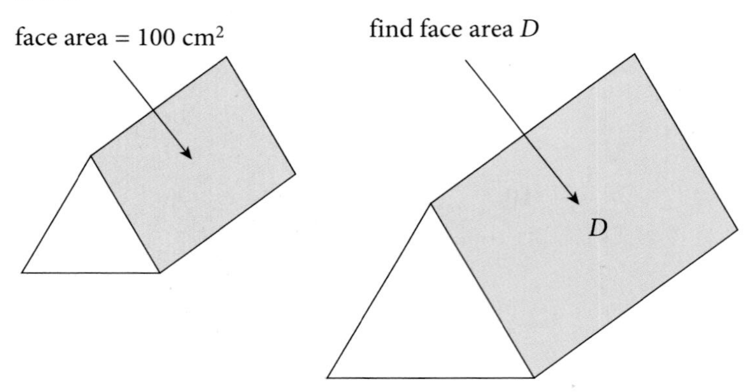

5. Given that AD = 3 cm, AB = 5 cm and area of ΔADE = 6 cm^2, find:

a) area of ΔABC **b)** area of DECB

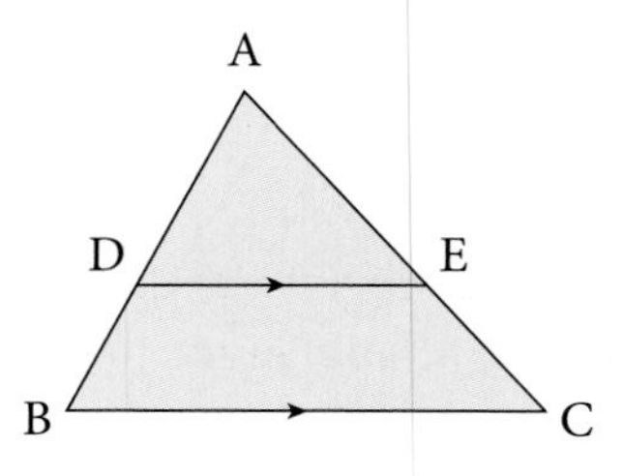

6. 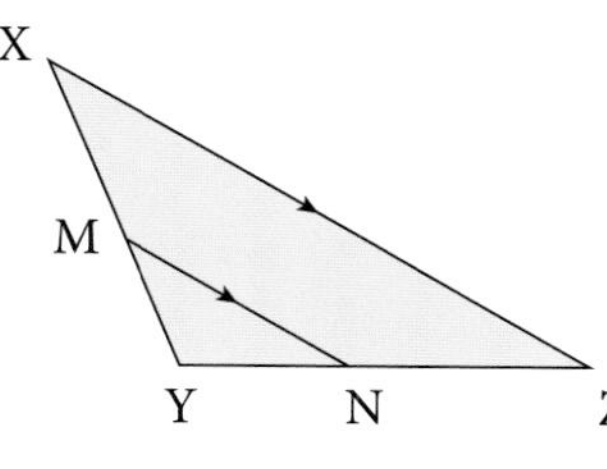

Given that XY = 5 cm, MY = 2 cm and area of ΔMYN = 4 cm^2, find:

a) the area of ΔXYZ

b) the area of MNZX

7. Given XY = 2 cm, BC = 3 cm and area of XYCB = 10 cm^2, find the area of ΔAXY.

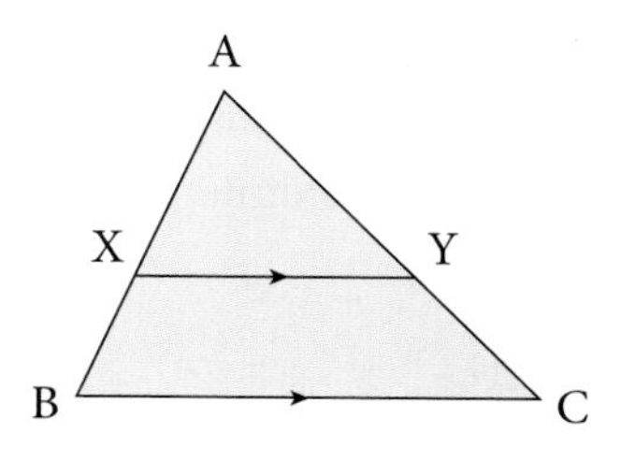

8. Given KP = 3 cm, area of ΔKOP = 2 cm^2 and area of OPML = 16 cm^2, find the length of PM.

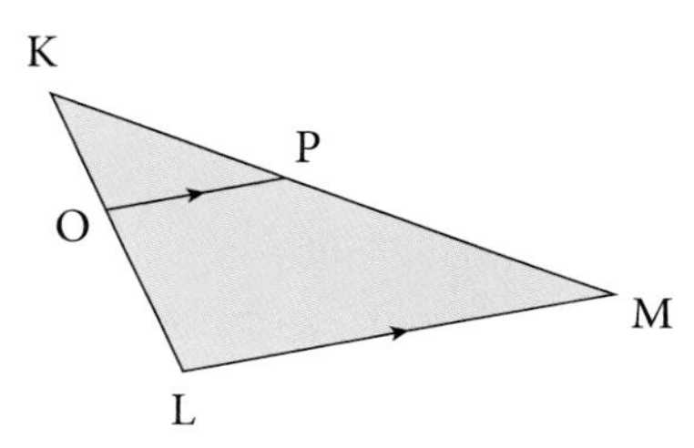

9. The triangles ABC and EBD are similar (AC and DE are *not* parallel). If AB = 8 cm, BE = 4 cm and the area of ΔDBE = 6 cm^2, find the area of ΔABC.

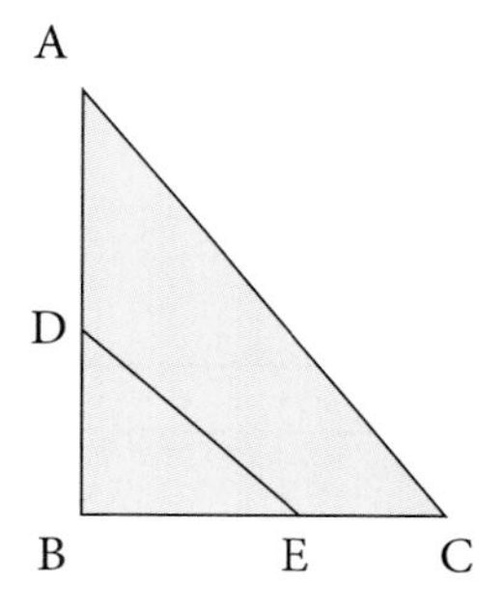

10. Given AZ = 3 cm, ZC = 2 cm, MC = 5 cm, BM = 3 cm, find:

a) XY
b) YZ
c) the ratio of areas AXY : AYZ
d) the ratio of areas AXY : ABM

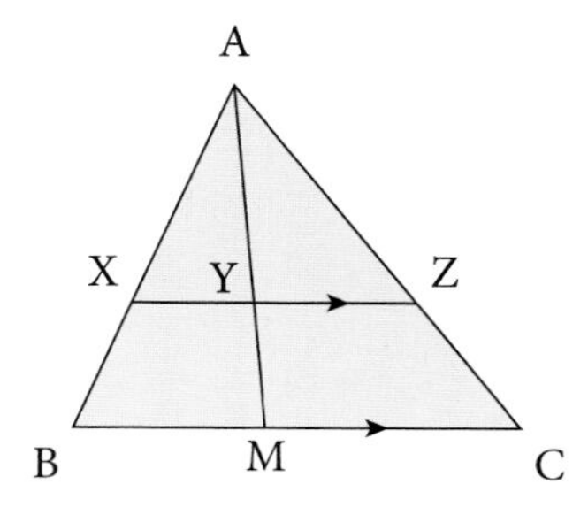

11. A floor is covered by 600 tiles which are 10 cm by 10 cm. How many 20 cm by 20 cm tiles are needed to cover the same floor?

12. A wall is covered by 160 tiles which are 15 cm by 15 cm. How many 10 cm by 10 cm tiles are needed to cover the same wall?

13. When potatoes are peeled do you lose more peel when you use big potatoes or small potatoes?

Volumes of similar objects

When solid objects are similar, one is an accurate enlargement of the other. If two objects are similar and the ratio of corresponding sides is k, then the ratio of their volumes is k^3.

A line has *one* dimension, and the scale factor is used *once.*

An area has *two* dimensions, and the scale factor is used *twice.*

A volume has *three* dimensions, and the scale factor is used *three* times.

Example 7

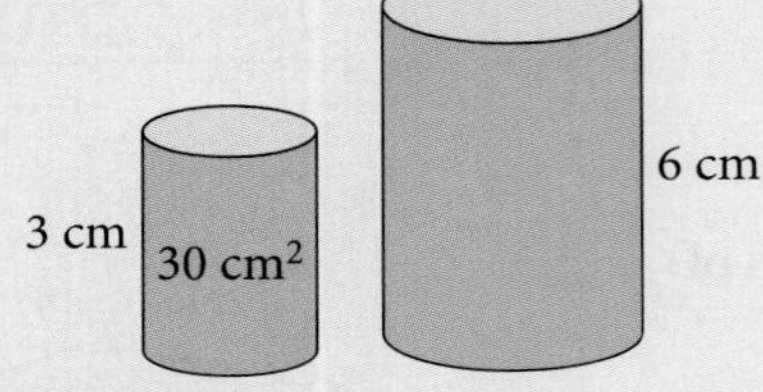

Two similar cylinders have heights of 3 cm and 6 cm respectively. If the volume of the smaller cylinder is $30\,\text{cm}^3$, find the volume of the larger cylinder.

If linear scale factor = k, then ratio of heights $(k) = \frac{6}{3} = 2$

$$\therefore \quad \text{Ratio of volumes } (k^3) = 2^3$$
$$= 8$$
$$\text{Volume of larger cylinder} = 8 \times 30$$
$$= 240\,\text{cm}^3$$

Example 8

Two similar solid spheres made of the same material have masses of 32 kg and 108 kg respectively. If the radius of the larger sphere is 9 cm, find the radius of the smaller sphere.

We may take the ratio of masses to be the same as the ratio of volumes.

$$\text{Ratio of volumes } (k^3) = \frac{32}{108}$$
$$= \frac{8}{27}$$
$$\text{Ratio of corresponding lengths } (k) = \sqrt[3]{\left(\frac{8}{27}\right)}$$
$$= \frac{2}{3}$$
$$\therefore \quad \text{Radius of smaller sphere} = \frac{2}{3} \times 9$$
$$= 6\,\text{cm}$$

Exercise 7.13

In this exercise, the objects are similar and a number written inside a figure represents the volume of the object in cm³. The numbers on the outside give linear dimensions in cm.

In questions **1** to **8**, find the unknown volume V.

1.

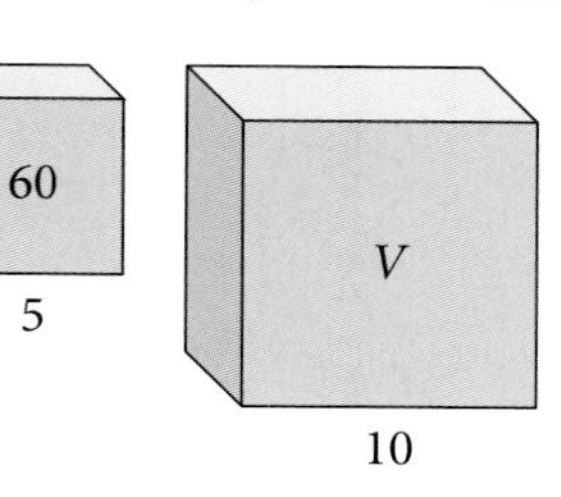

2.

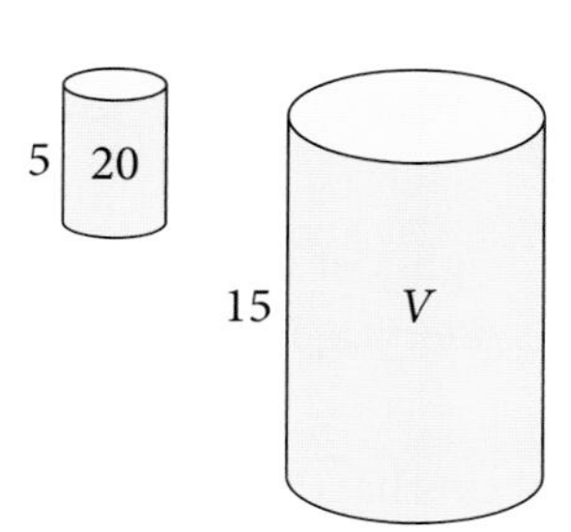

3.

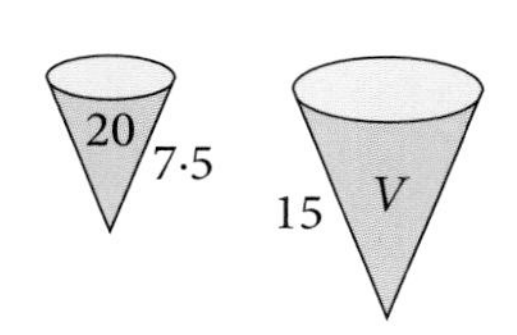

4.

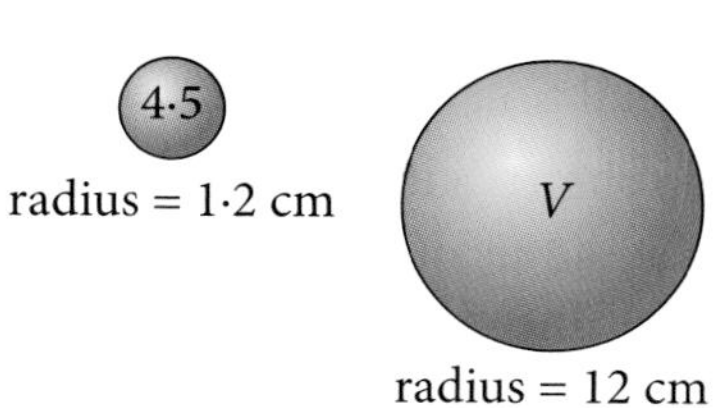

5.

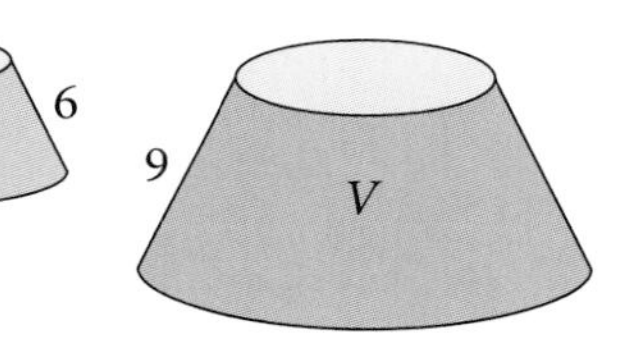

6.

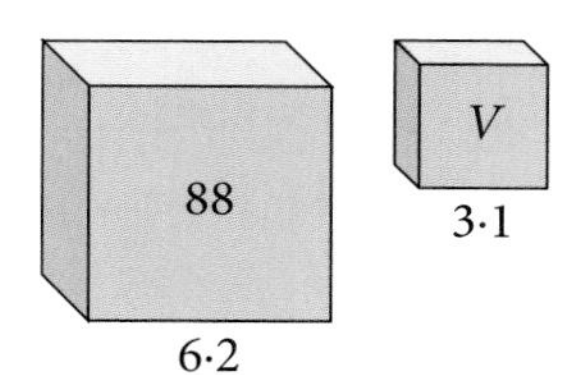

7.

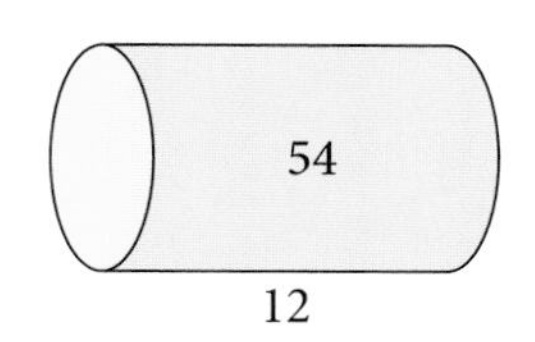

8.

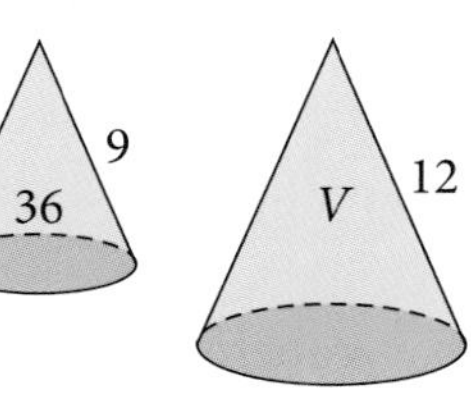

In questions **9** to **14**, find the lengths marked by a letter.

9.

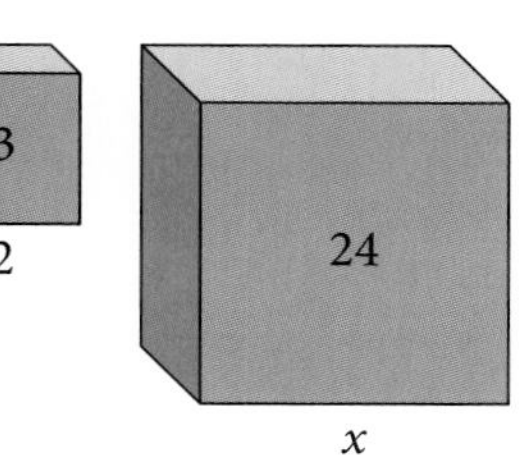

10.

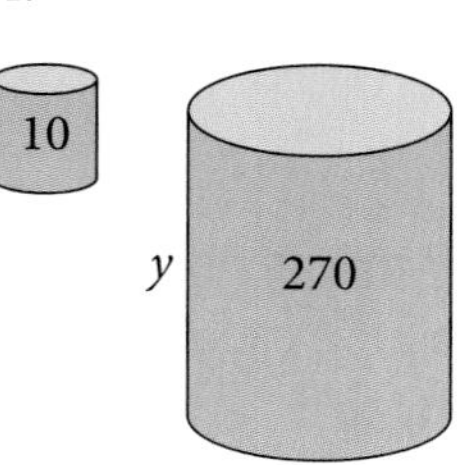

11.

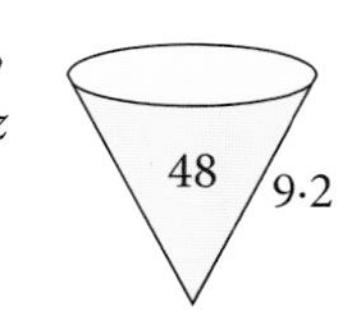

12.

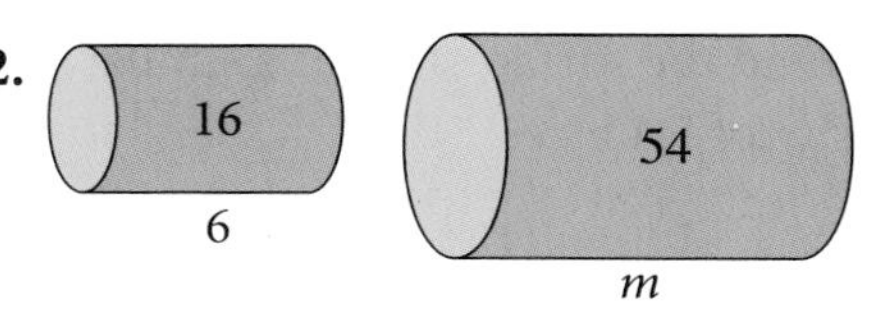

13.

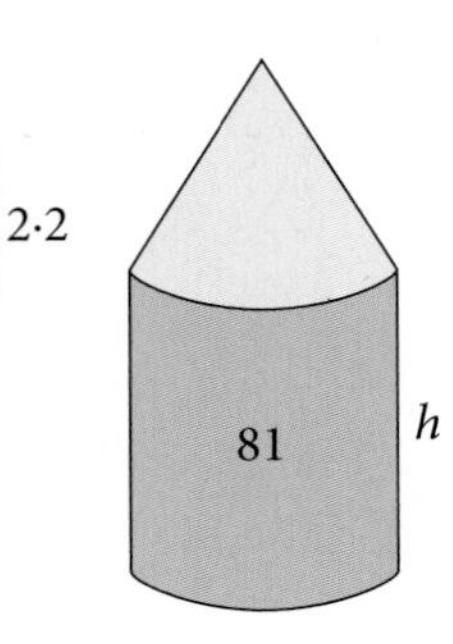

14. ←3→ ←d→

80 270

15. Two similar jugs have heights of 4 cm and 6 cm respectively. If the capacity of the smaller jug is 50 cm³, find the capacity of the larger jug.

16. Two similar cylindrical tins have base radii of 6 cm and 8 cm respectively. If the capacity of the larger tin is 252 cm³, find the capacity of the small tin.

17. Two solid metal spheres have masses of 5 kg and 135 kg respectively. If the radius of the smaller sphere is 4 cm, find the radius of the larger one.

18. Two similar cones have surface areas in the ratio 4 : 9. Find the ratio of

a) their lengths **b)** their volumes.

19. The area of the bases of two similar glasses are in the ratio 4 : 25. Find the ratio of their volumes.

20. Two similar solids have volumes V_1 and V_2 and corresponding sides of length x_1 and x_2. State the ratio $V_1 : V_2$ in terms of x_1 and x_2.

21. Two solid spheres have surface areas of 5 cm² and 45 cm² respectively and the mass of the smaller sphere is 2 kg. Find the mass of the larger sphere.

22. The masses of two similar icebergs are 24 tonnes and 81 tonnes respectively. If the surface area of the larger iceberg is 540 m², find the surface area of the smaller iceberg.

23. A cylindrical can has a circumference of 40 cm and a capacity of 4·8 litres. Find the capacity of a similar cylinder of circumference 50 cm.

24. A container has a surface area of 5000 cm² and a capacity of 12·8 litres. Find the surface area of a similar container which has a capacity of 5·4 litres.

7.5 Circle properties

- The angle subtended at the centre of a circle is twice the angle subtended at the circumference.

$A\hat{O}B = 2 \times A\hat{C}B$

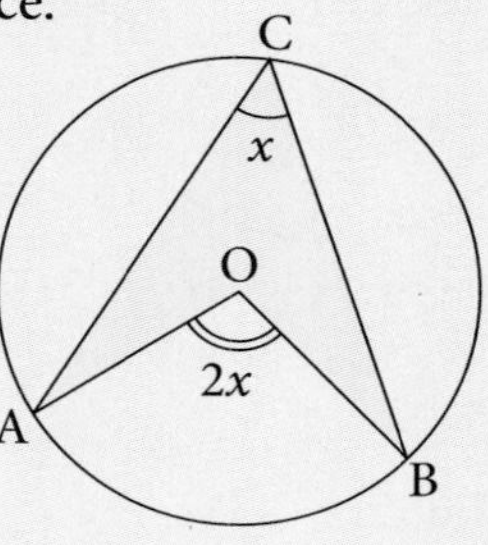

Proof:

Draw the straight line COD.

Let $A\hat{C}O = y$ and $B\hat{C}O = z$.

In triangle AOC,

	$AO = OC$	(radii)
$\therefore$	$O\hat{C}A = O\hat{A}C$	(isosceles triangle)
$\therefore$	$C\hat{O}A = 180 - 2y$	(angle sum of triangle)
$\therefore$	$A\hat{O}D = 2y$	(angles on a straight line)

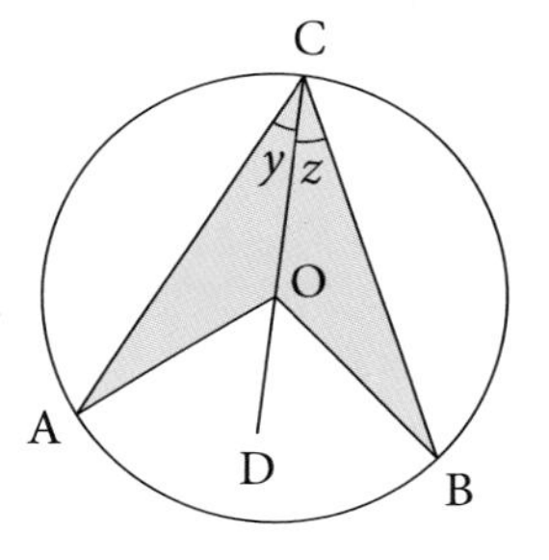

Similarly from triangle COB, we find

$D\hat{O}B = 2z$

Now $A\hat{C}B = y + z$

and $A\hat{O}B = 2y + 2z$

$\therefore$ $AOB = 2 \times A\hat{C}B$ as required.

- Angles subtended by an arc in the same segment of a circle are equal.

$A\hat{X}B = A\hat{Y}B = A\hat{Z}B$

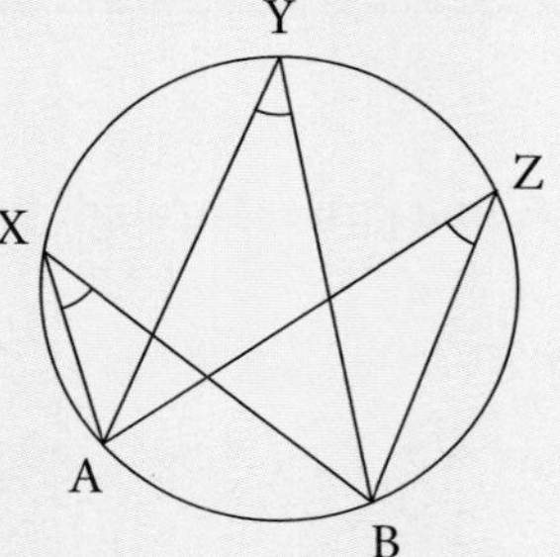

Example 9

Given $A\hat{B}O = 50°$, find $B\hat{C}A$.

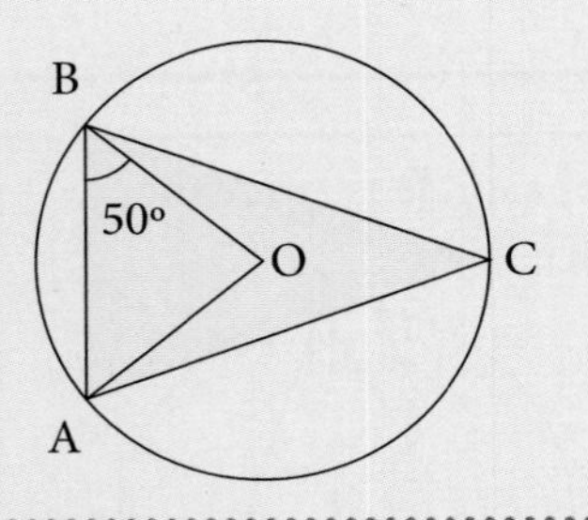

Triangle OBA is isosceles (OA = OB).

$\therefore \quad O\hat{A}B = 50°$

$\therefore \quad B\hat{O}A = 80°$ (angle sum of a triangle)

$\therefore \quad B\hat{C}A = 40°$ (angle at the circumference)

Example 10

Given $B\hat{D}C = 62°$ and $D\hat{C}A = 44°$, find $B\hat{A}C$ and $A\hat{B}D$.

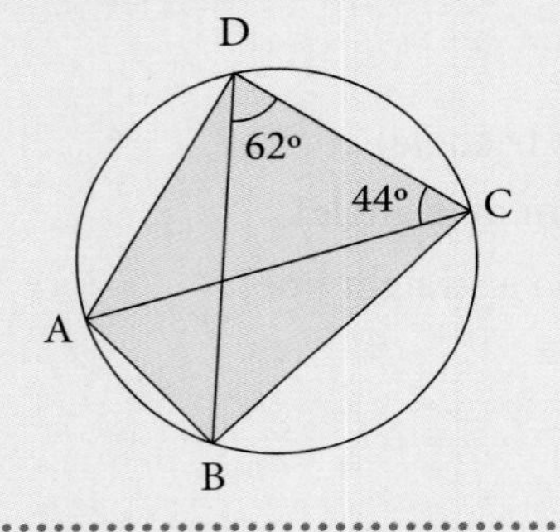

$B\hat{D}C = B\hat{A}C$ (both subtended by arc BC)

$\therefore \quad B\hat{A}C = 62°$

$D\hat{C}A = A\hat{B}D$ (both subtended by arc DA)

$\therefore \quad A\hat{B}D = 44°$

Exercise 7.14

Find the angles marked with letters. A line passes through the centre only when point O is shown.

1.

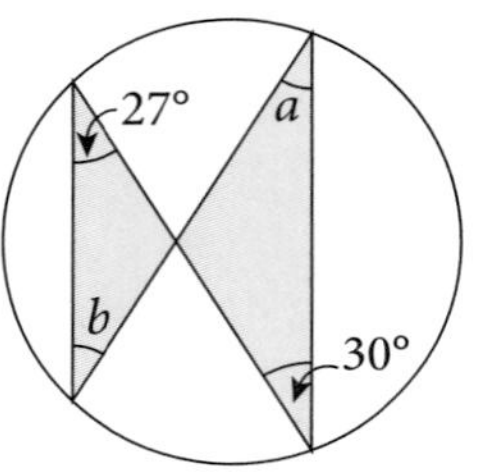

2.

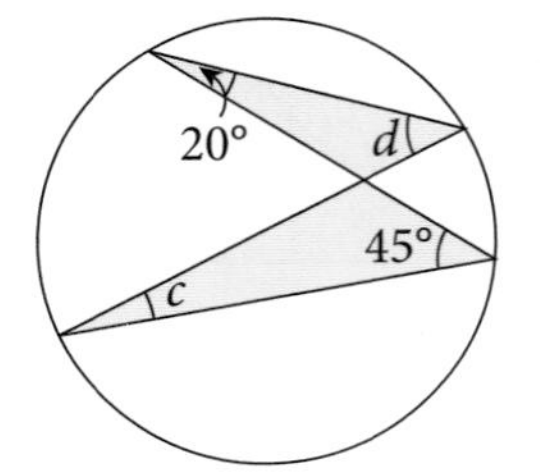

3.

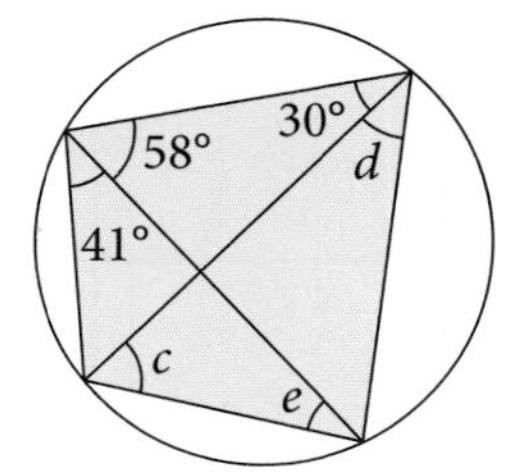

4.

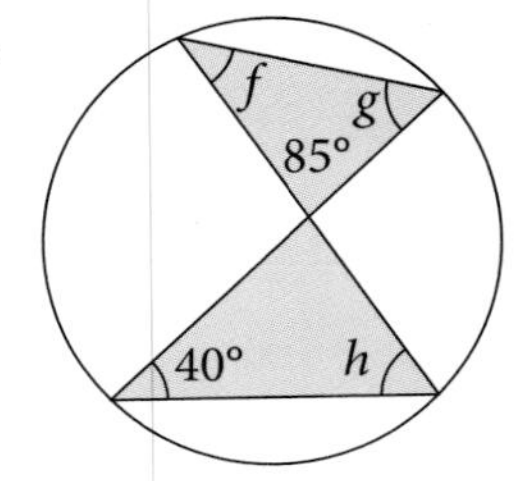

5.

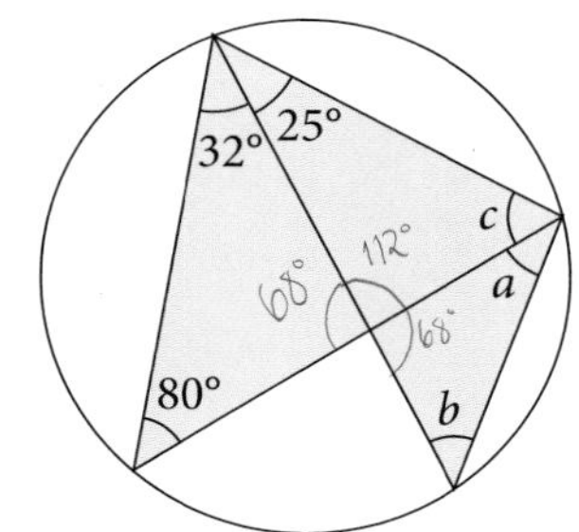

6.

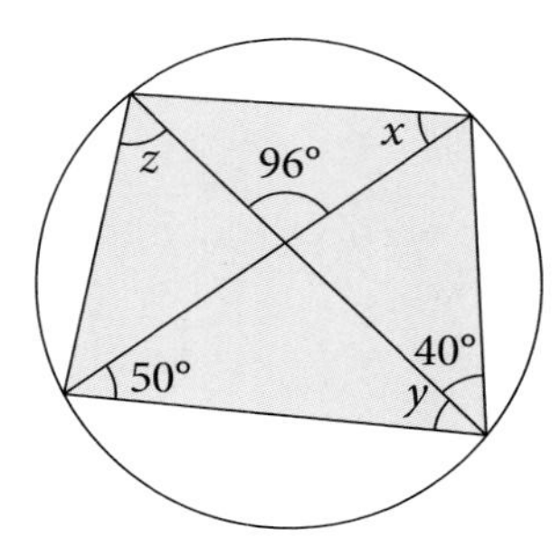

7.

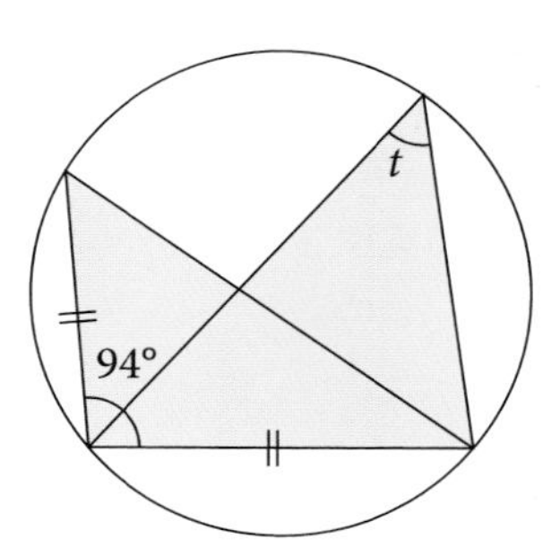

8.

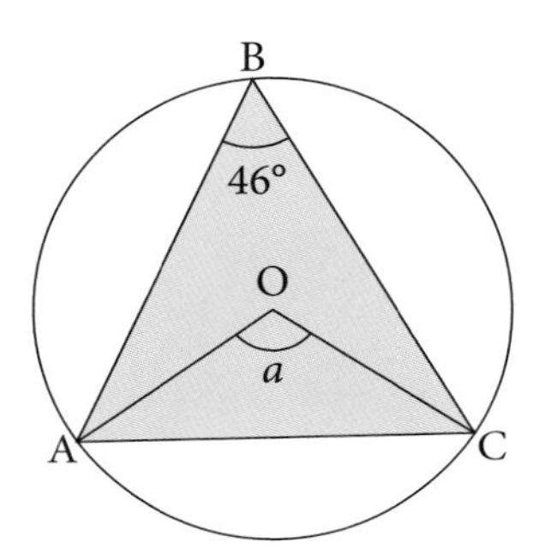

9.

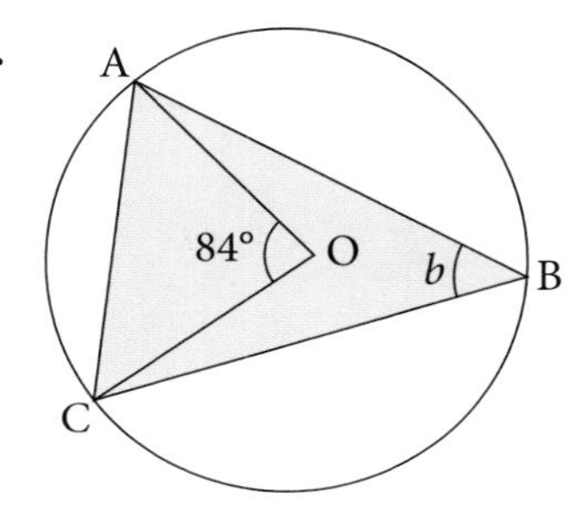

10.

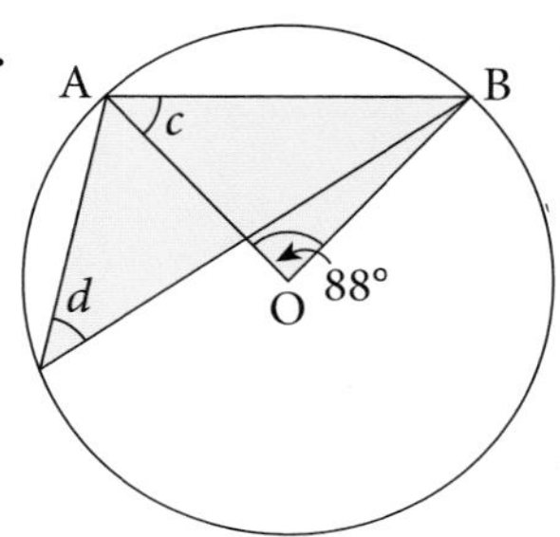

11.

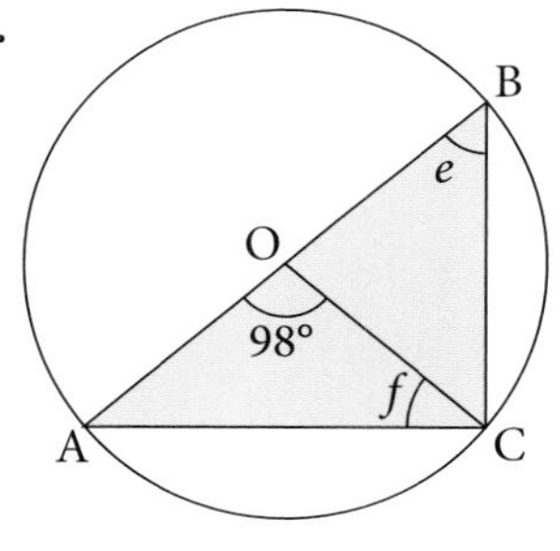

12.

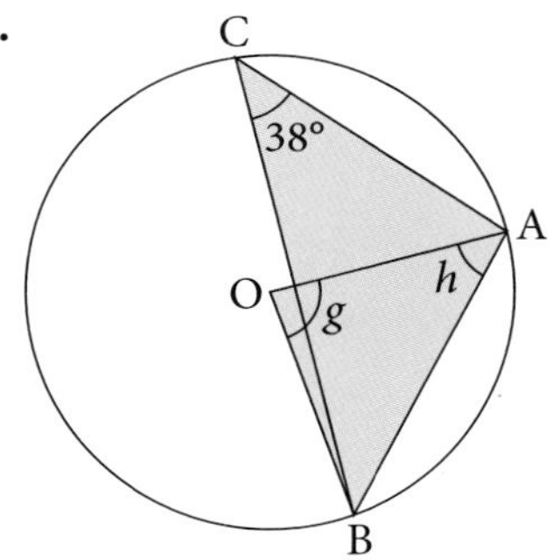

13.

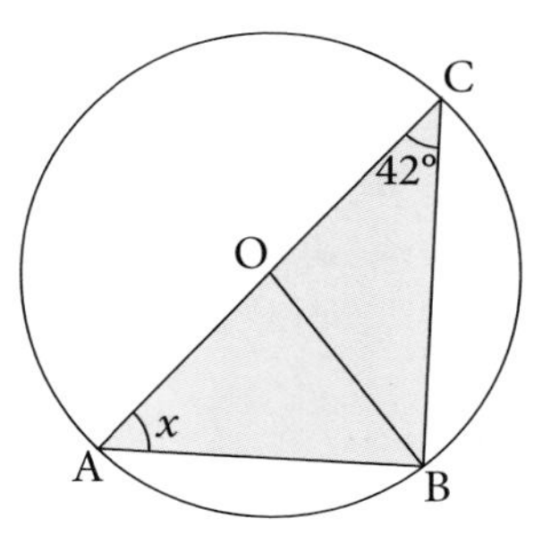

14.

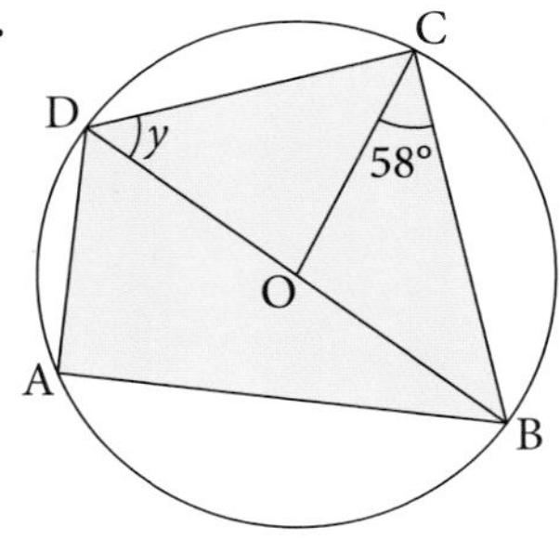

15.

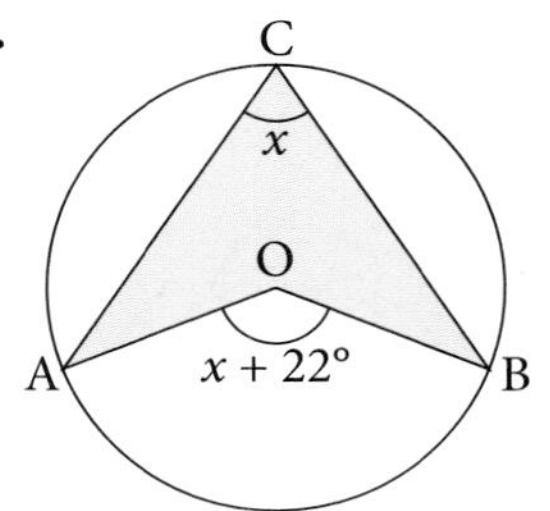

16.

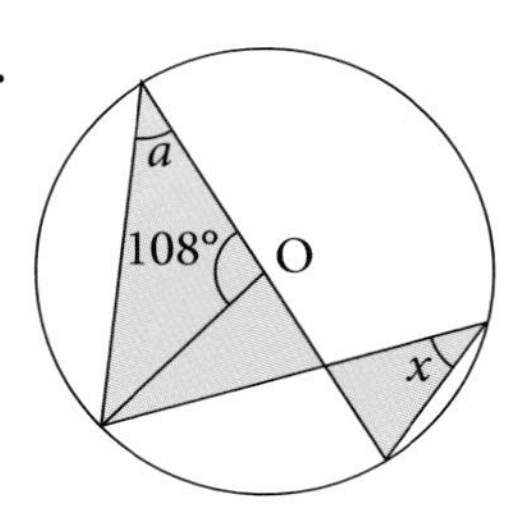

- The opposite angles in a cyclic quadrilateral add up to 180° (the angles are supplementary).

  ABCD is a cyclic quadrilateral. The vertices touch the circle.

  $\hat{A} + \hat{C} = 180°$

  $\hat{B} + \hat{D} = 180°$

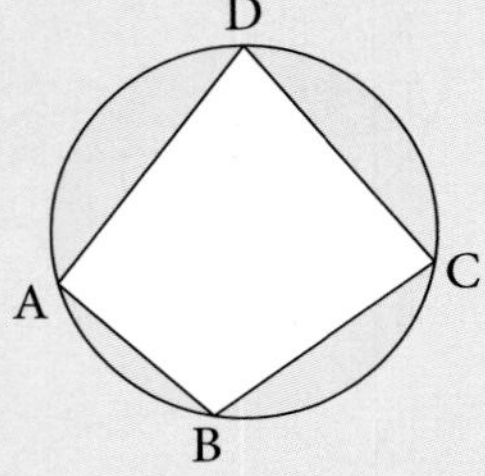

Proof:

Draw radii OA and OC.

Let $A\hat{D}C = x$ and $A\hat{B}C = y$.

$A\hat{O}C$ obtuse $= 2x$	(angle at the centre)
$A\hat{O}C$ reflex $= 2y$	(angle at the centre)
$\therefore \quad 2x + 2y = 360°$	(angles at a point)
$\therefore \quad x + y = 180°$	as required

- The angle in a semi-circle is a right angle.

  In the diagram, AB is a diameter.

  $A\hat{C}B = 90°$

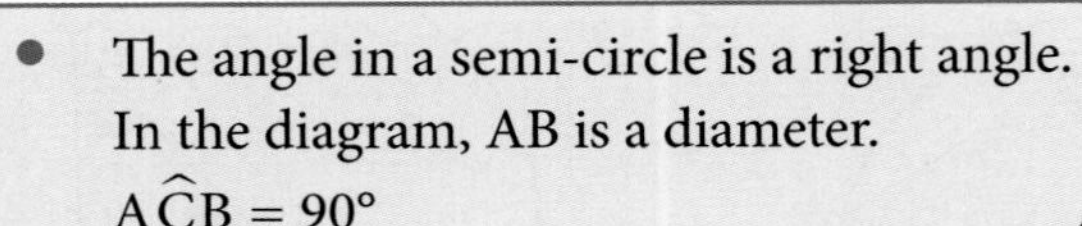

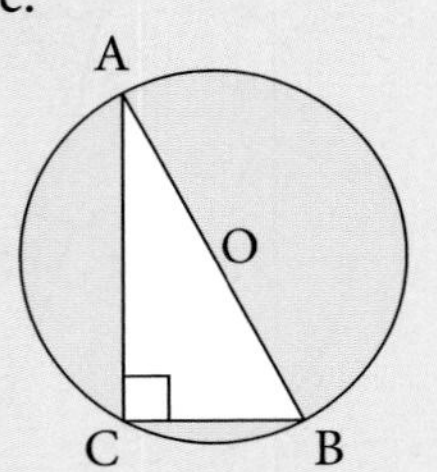

Example 11

Find a and x.

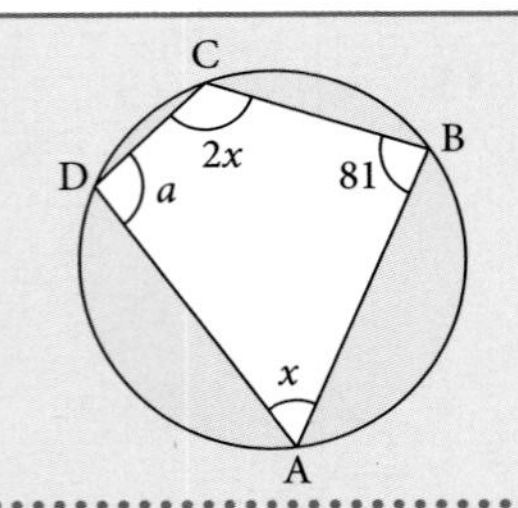

$a = 180° - 81°$ (opposite angles of a cyclic quadrilateral)

$\therefore \quad a = 99°$

$x + 2x = 180°$ (opposite angles of a cyclic quadrilateral)

$3x = 180°$

$x = 60°$

Example 12

Find b.

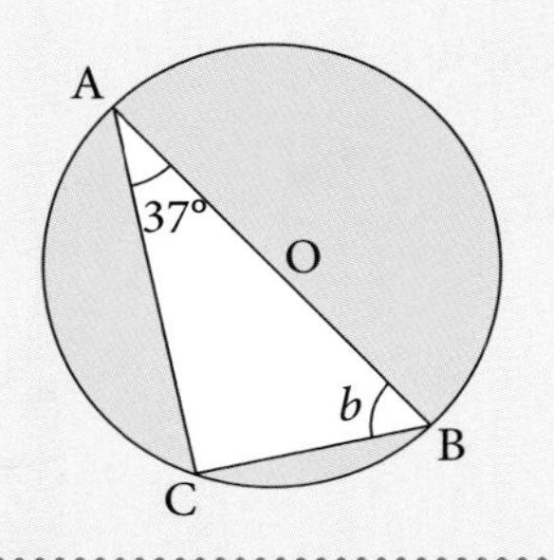

$A\hat{C}B = 90°$ (angle in a semi-circle)

$\therefore\ b = 180° - (90 + 37)°$

$= 53°$

Exercise 7.15

Find the angles marked with a letter. Point O is the centre of the circle.

1.

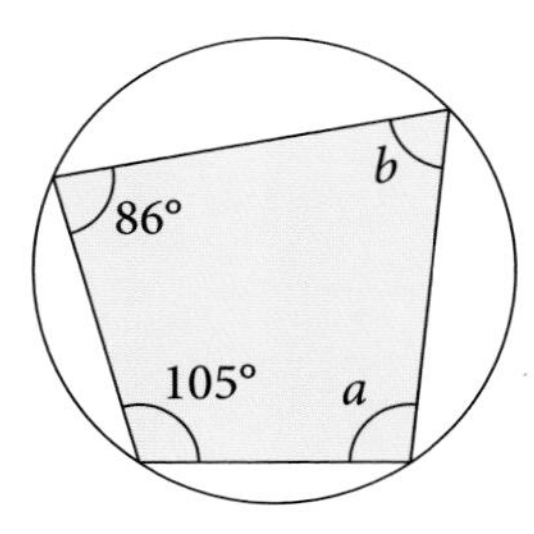

2.

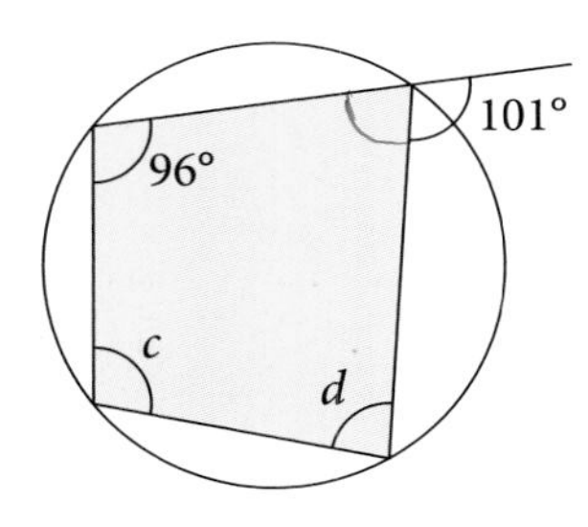

3.

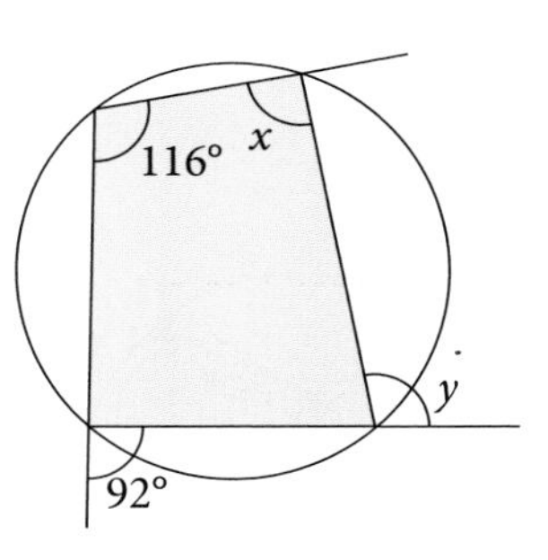

4.

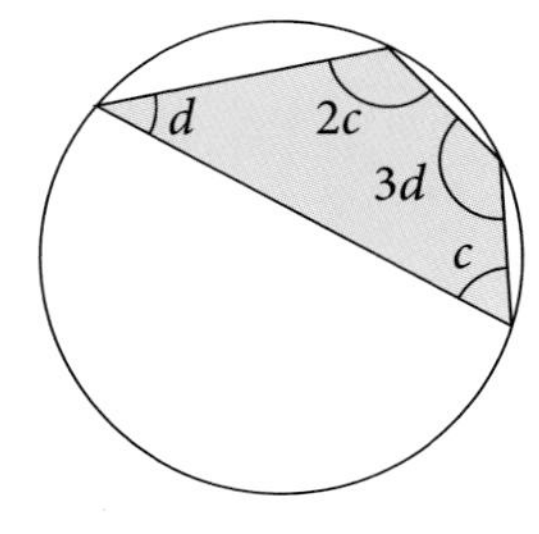

5.

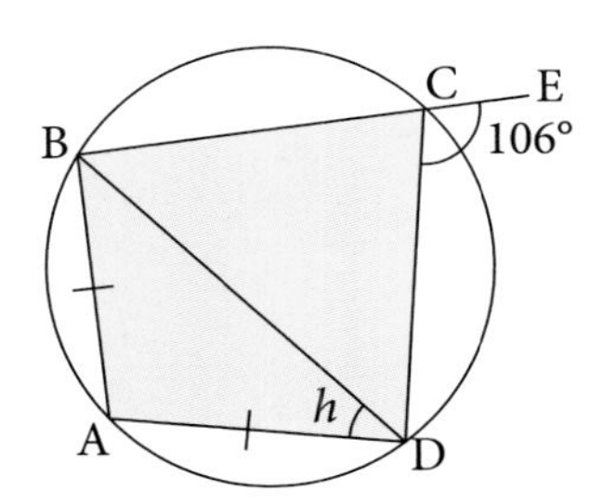

6.

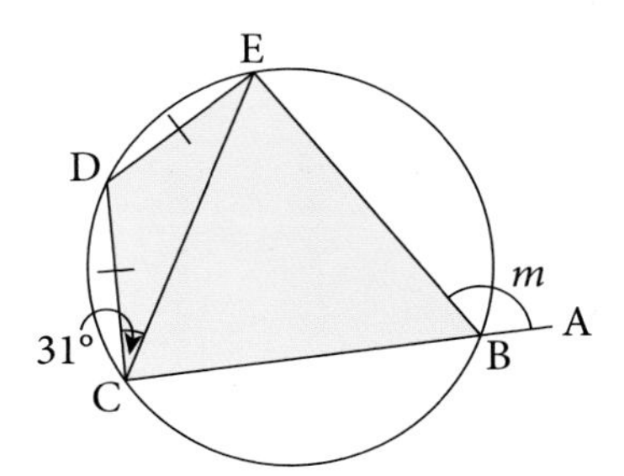

7.

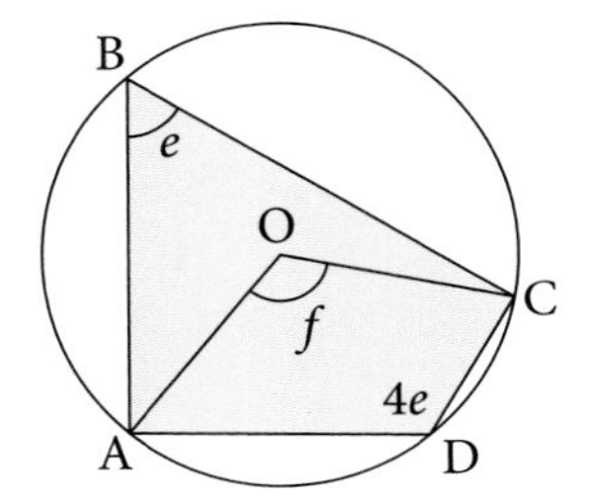

8.

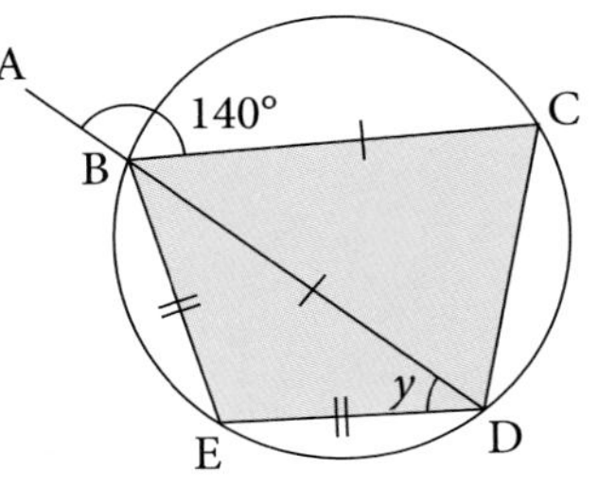

9.

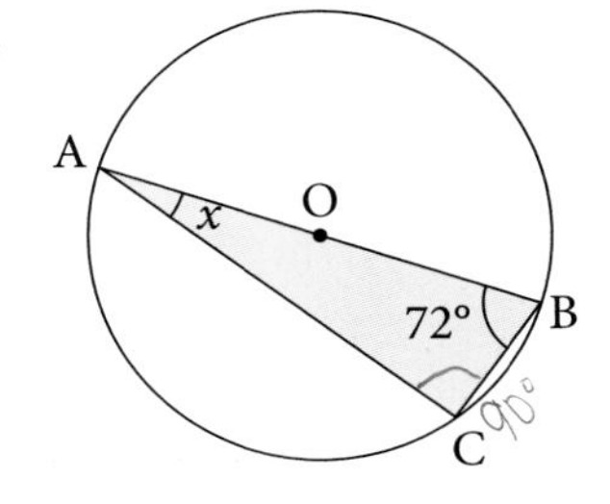

10.

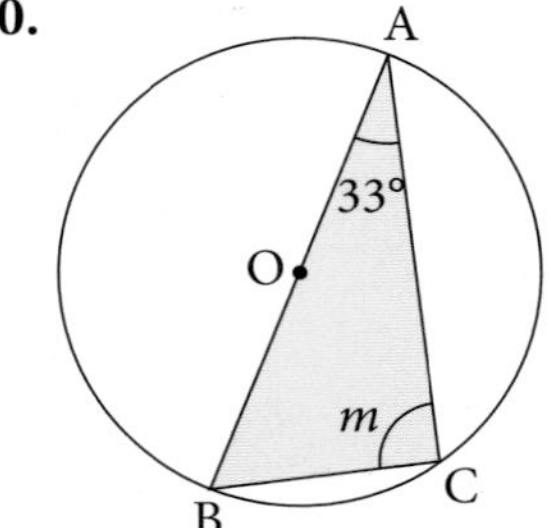

11.

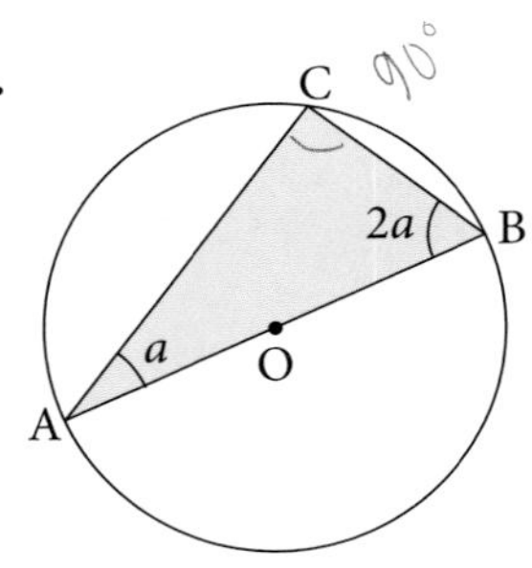

12.

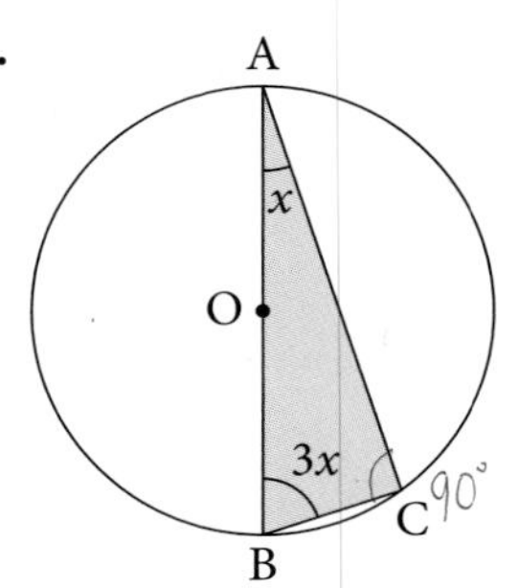

13.

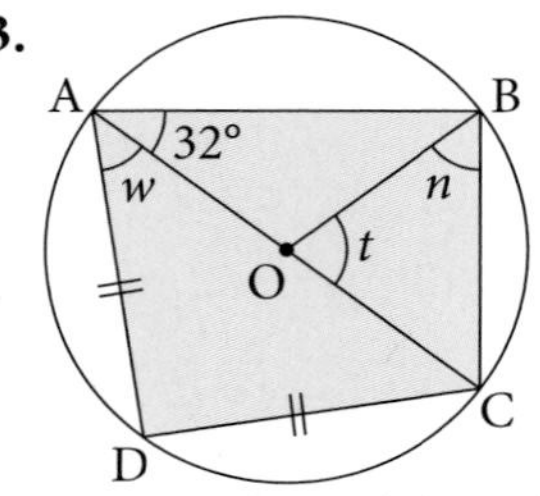

14.

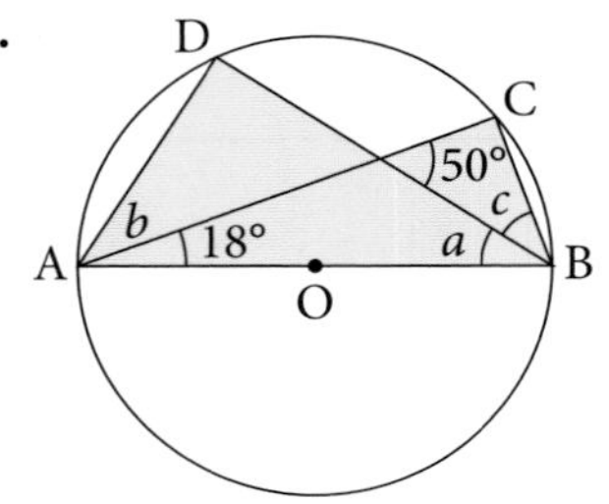

15.

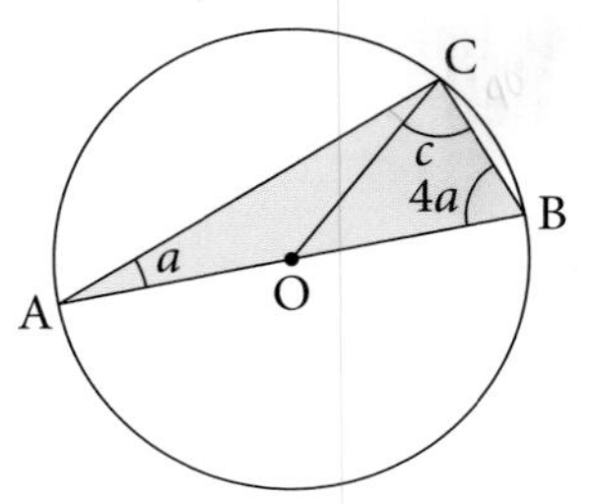

16.

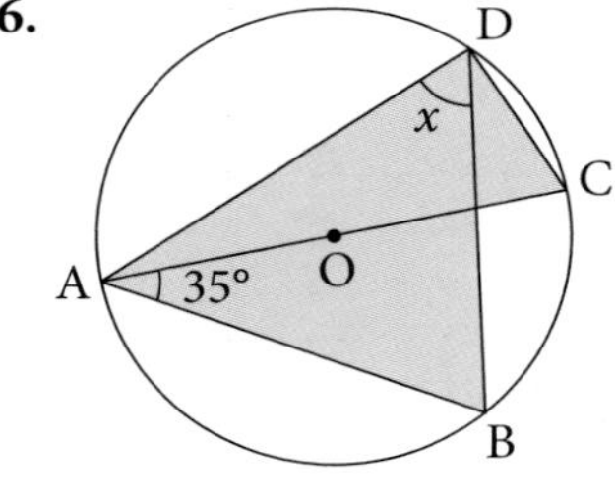

17.

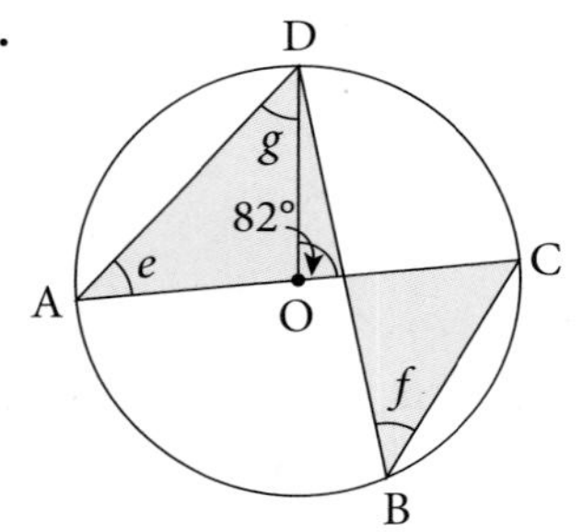

18.

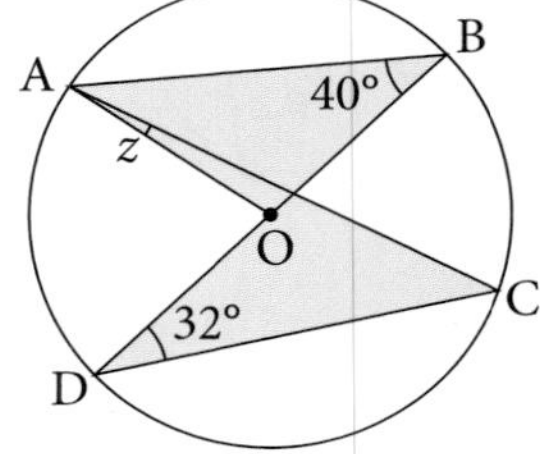

19.

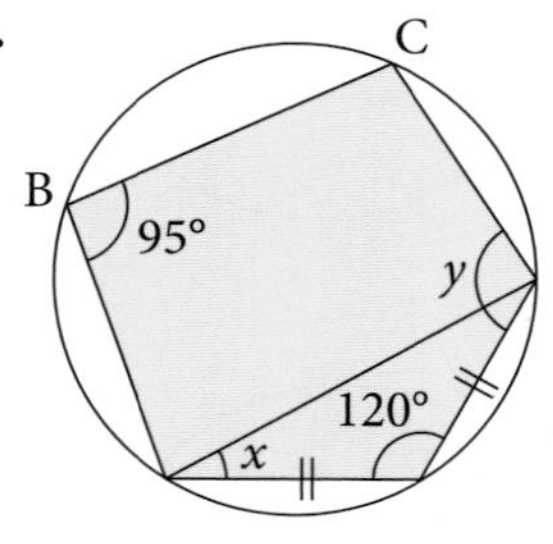

20.

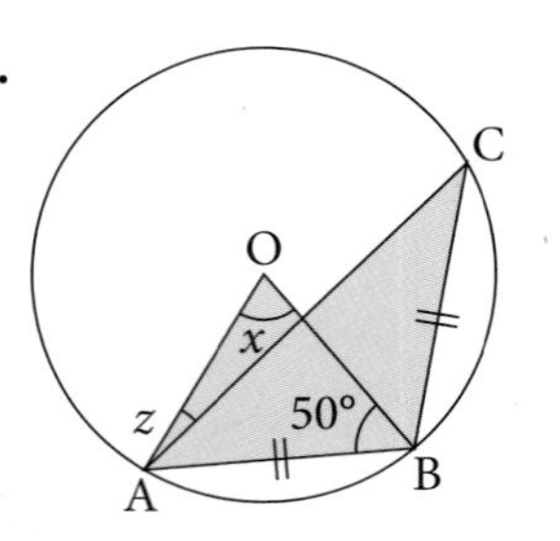

Tangents to circles

- The angle between a tangent and the radius drawn to the point of contact is 90°.

  $A\hat{B}O = 90°$

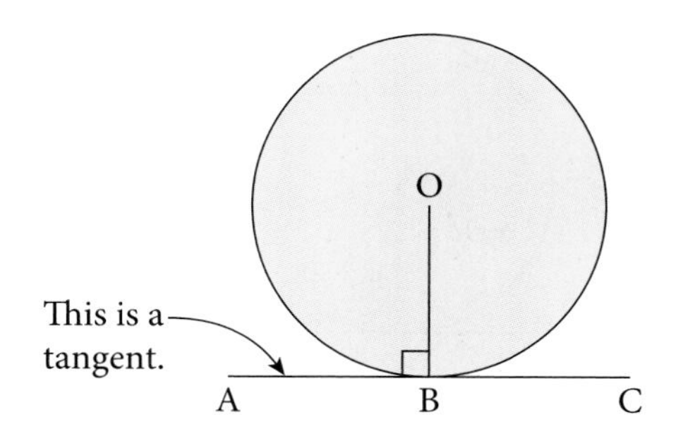

- From any point outside a circle just two tangents to the circle may be drawn and they are of equal length.

  TA = TB

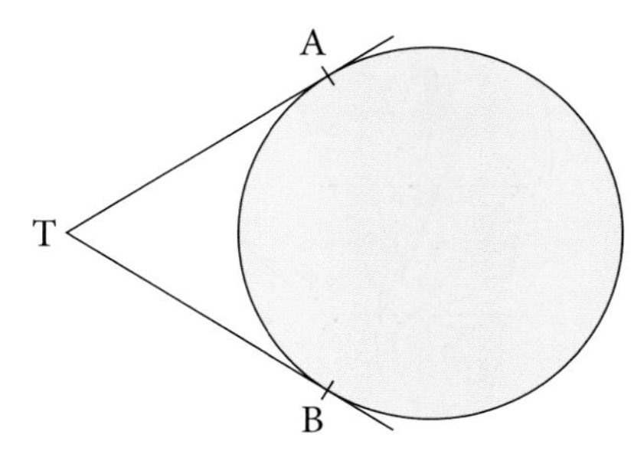

Example 13

TA and TB are tangents to the circle, centre O.
Given $A\hat{T}B = 50°$, find

a) $A\hat{B}T$

b) $O\hat{B}A$

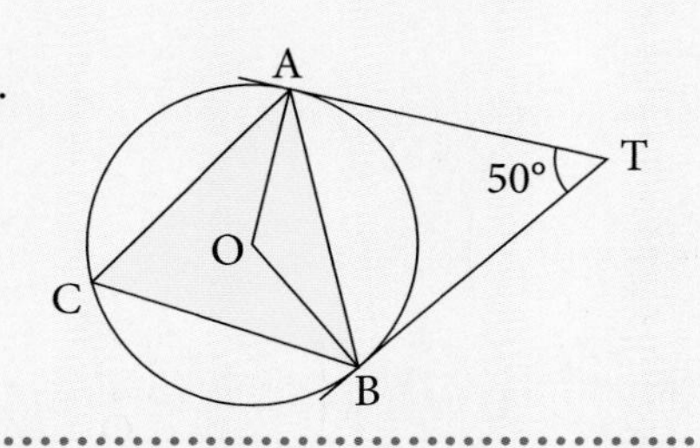

a) ΔTBA is isosceles (TA = TB)

$\therefore \quad A\hat{B}T = \frac{1}{2}(180 - 50) = 65°$

b) $O\hat{B}T = 90°$ (tangent and radius)

$\therefore \quad O\hat{B}A = 90° - 65°$
$= 25°$

Exercise 7.16

For questions **1** to **9**, find the angles marked with a letter.

The point O is the centre of the circle.

1.

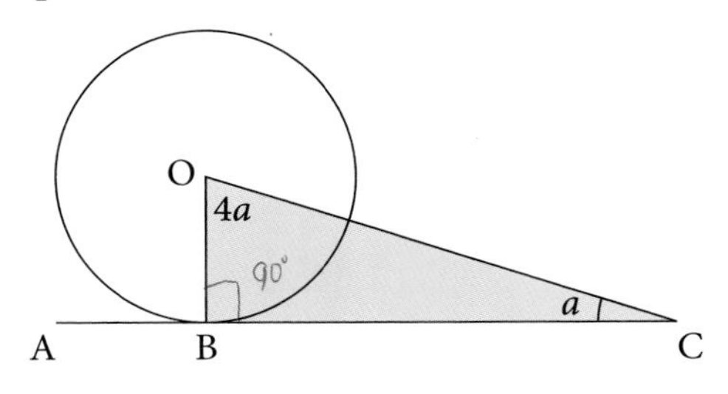

2.

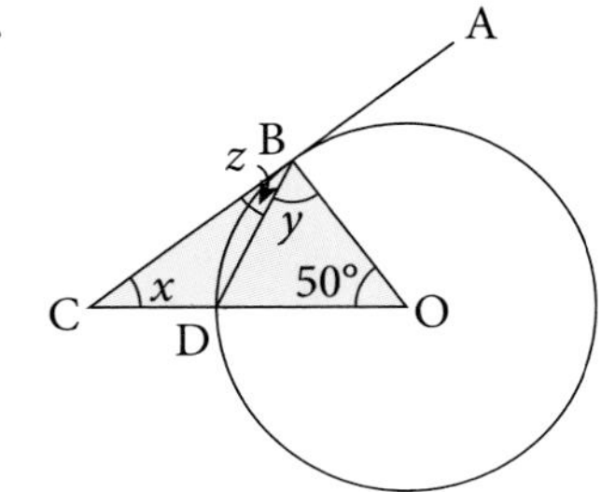

3.

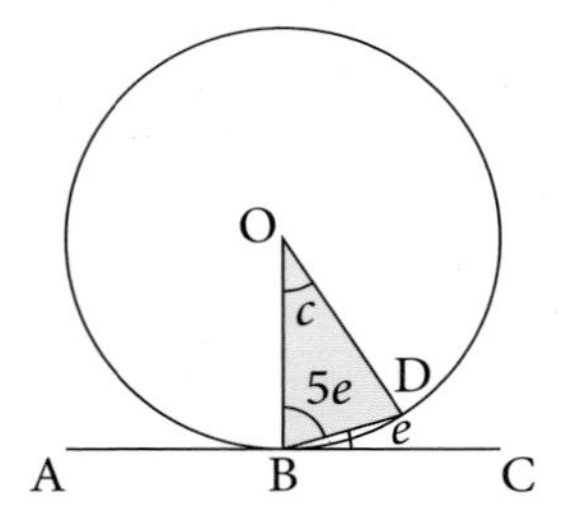

4.

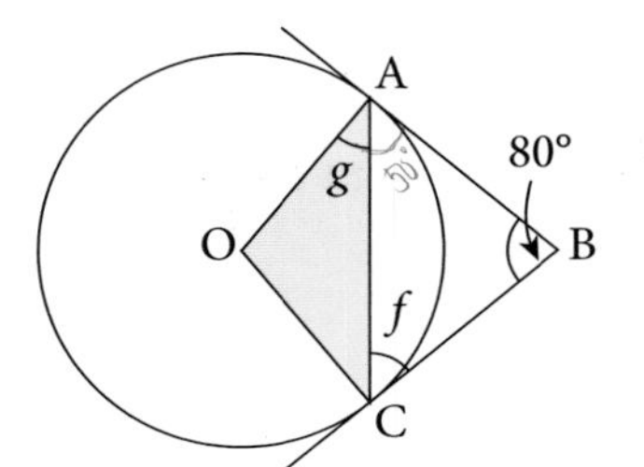

5.

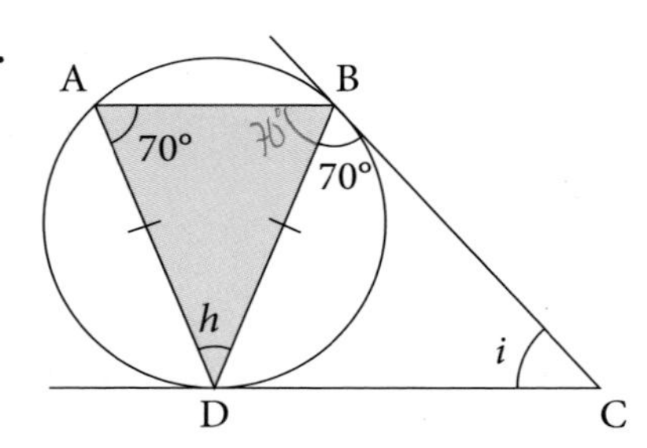

6.

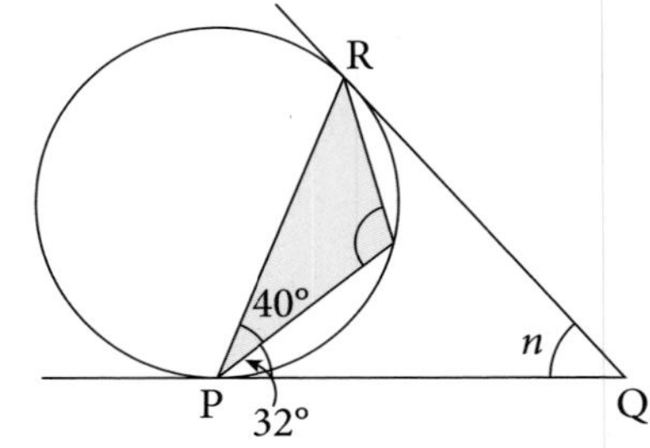

7.

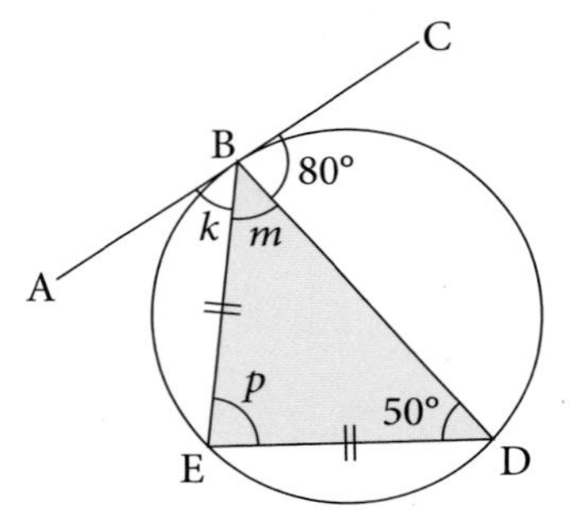

8.

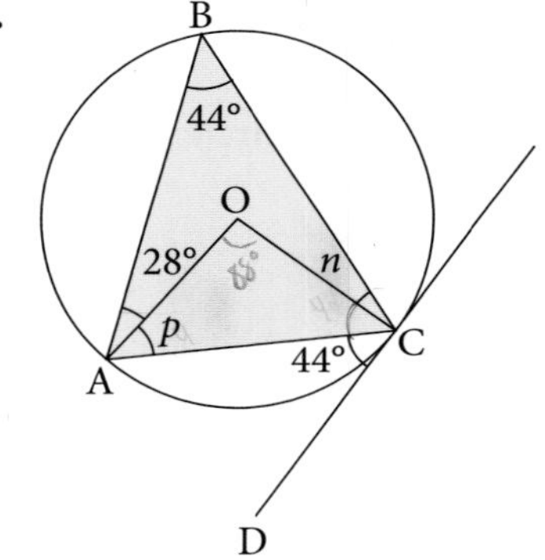

9. Find x, y and z.

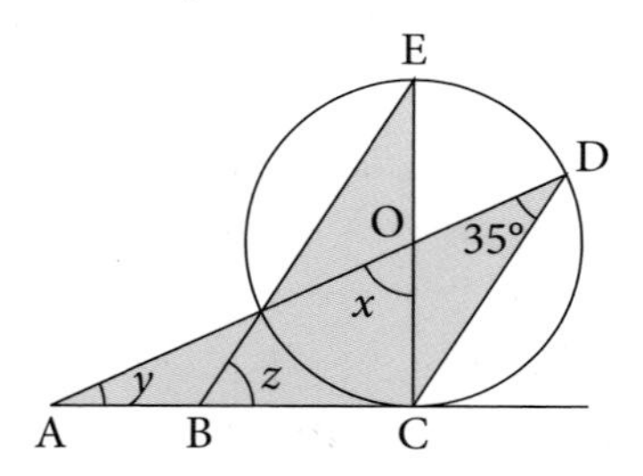

Revision exercise 7A

1. ABCD is a parallelogram and AE bisects angle A. Prove that DE = BC.

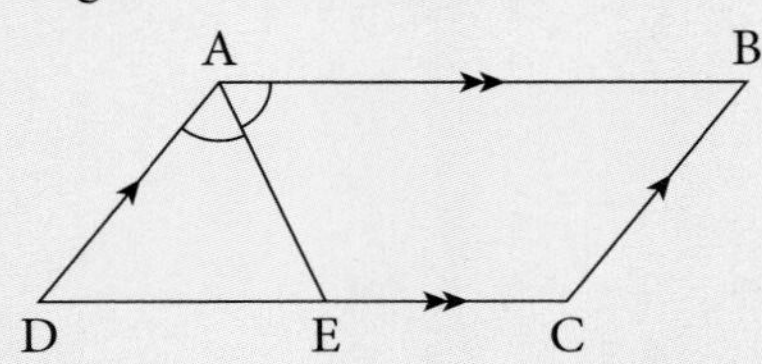

2. In a triangle PQR, $P\hat{Q}R = 50°$ and point X lies on PQ such QX = XR. Calculate $Q\hat{X}R$.

3. a) ABCDEF is a regular hexagon. Calculate $F\hat{D}E$.

b) ABCDEFGH is a regular eight-sided polygon. Calculate $A\hat{G}H$.

c) Each interior angle of a regular polygon measures 150°. How many sides has the polygon?

4.

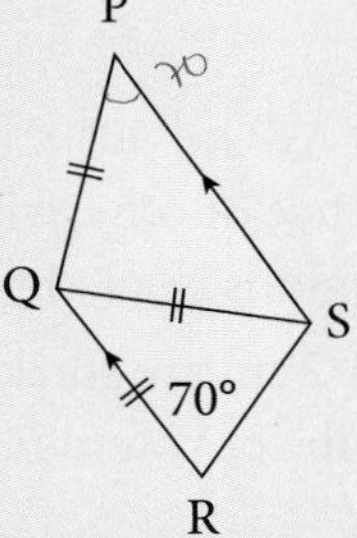

In the quadrilateral PQRS, PQ = QS = QR, PS is parallel to QR and $Q\hat{R}S = 70°$. Calculate

a) $R\hat{Q}S$ **b)** $P\hat{Q}S$

5. Find x.

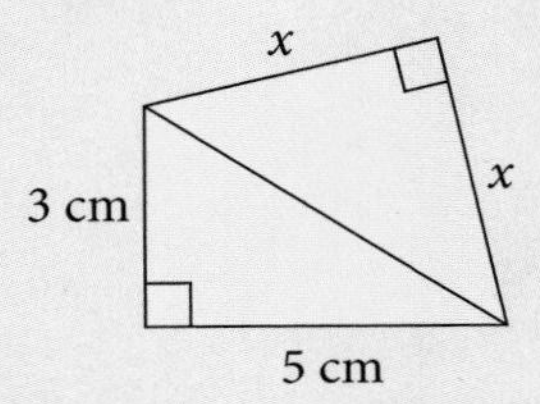

6. In the triangle ABC, AB = 7 cm, BC = 8 cm and $A\hat{B}C = 90°$. Point P lies inside the triangle such that BP = PC = 5 cm. Calculate

a) the perpendicular distance from P to BC

b) the length AP.

7. In triangle PQR the bisector of $P\hat{Q}R$ meets PR at S and the point T lies on PQ such that ST is parallel to RQ.

a) Prove that QT = TS.

b) Prove that the triangles PTS and PQR are similar.

c) Given that PT = 5 cm and TQ = 2 cm, calculate the length of QR.

8. In the quadrilateral ABCD, AB is parallel to DC and $D\hat{A}B = D\hat{B}C$.

a) Prove that the triangles ABD and DBC are similar.

b) If AB = 4 cm and DC = 9 cm, calculate the length of BD.

9. A rectangle 11 cm by 6 cm is similar to a rectangle 2 cm by x cm. Find the two possible values of x.

10.

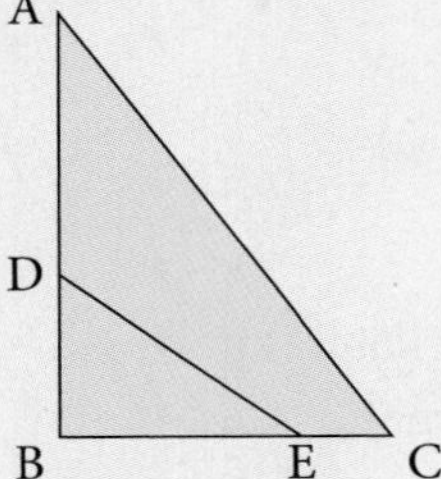

In the diagram, triangles ABC and EBD are similar but DE is *not* parallel to AC. Given that AD = 5 cm, DB = 3 cm and BE = 4 cm, calculate the length of BC.

11. The radii of two spheres are in the ratio 2 : 5. The volume of the smaller sphere is 16 cm³. Calculate the volume of the larger sphere.

12. The surface areas of two similar jugs are 50 cm² and 450 cm² respectively.

a) If the height of the larger jug is 10 cm, find the height of the smaller jug.

b) If the volume of the smaller jug is 60 cm³, find the volume of the larger jug.

13. A car is an enlargement of a model, the scale factor being 10.

  a) If the windscreen of the model has an area of $100\,\text{cm}^2$, find the area of the windscreen on the actual car. [Answer in m^2]

  b) If the capacity of the boot of the car is $1\ \text{m}^3$, find the capacity of the boot on the model. [Answer in cm^3]

14. Find the angles marked with letters. (O is the centre of the circle.)

  a)

  O a 40°

  b)

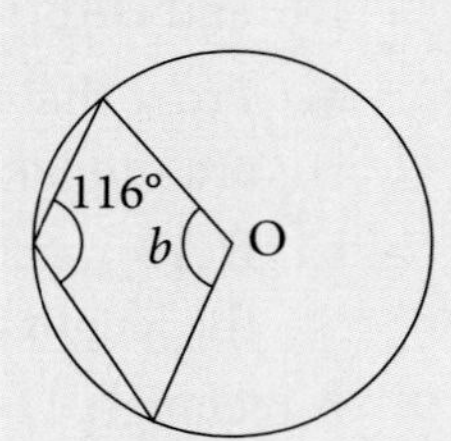

  c)

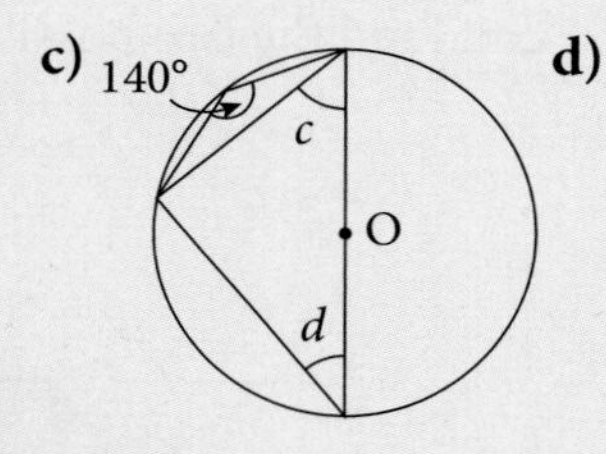

  d)

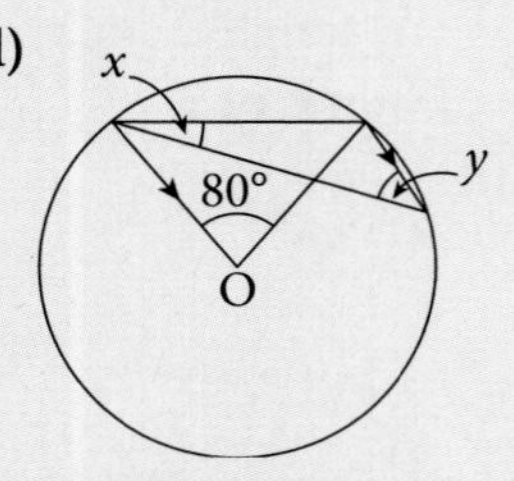

15. ABCD is a cyclic quadrilateral in which AB = BC and $A\hat{B}C = 70°$.

  AD produced meets BC produced at the point P, where $A\hat{P}B = 30°$.
  Calculate

  a) $A\hat{D}B$ b) $A\hat{B}D$

16.

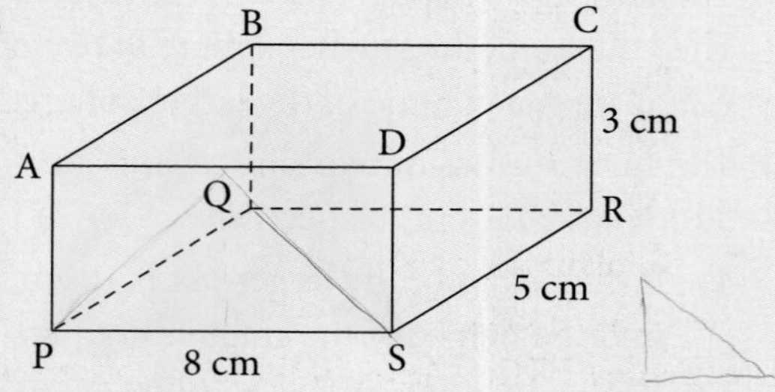

The diagram shows a cuboid.

  a) Use triangle PQS to find the length of QS.

  b) Use triangle BSQ to find the length of BS.

17. The diagram consists of a square and an equilateral triangle.

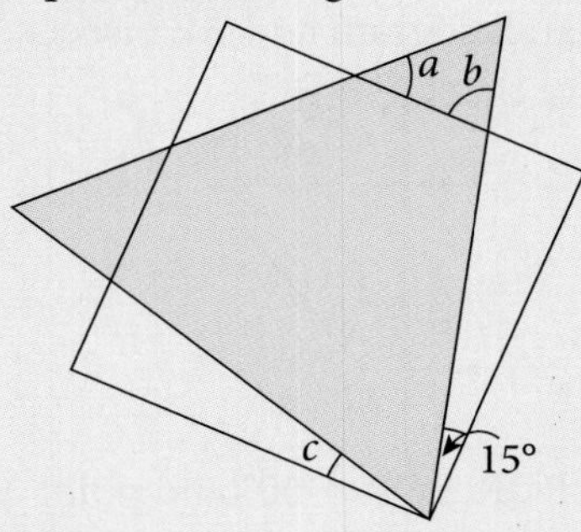

Find the angles a, b and c.

18. In the diagram triangle ABC is drawn inside a semicircle.

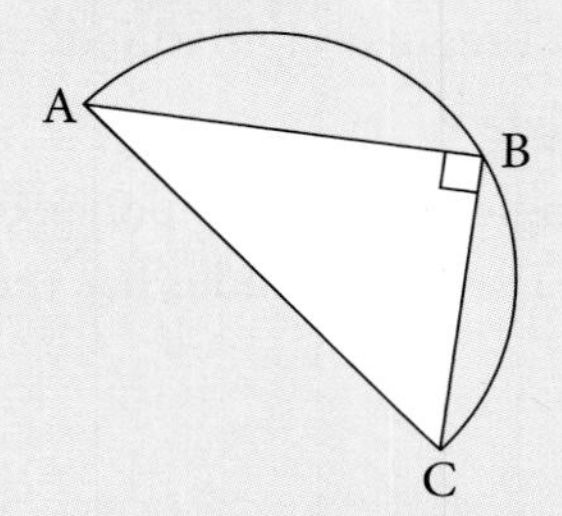

AB = 8 cm, BC = 6 cm and $A\hat{B}C = 90°$
Find the length AC and hence or otherwise find the total area of the regions shaded pink.

19. Many small cubes of side 1·2 cm are stuck together to make a large cube of volume $216\,\text{cm}^3$.
How many cubes are needed?

20. A paint tin is a cylinder of radius 12 cm and height 22 cm.

Leonardo, the painter, drops his stirring stick into the tin and it disappears.
Work out the maximum length of the stick.

21.

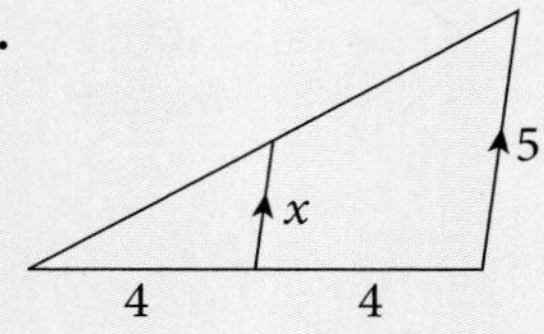

Use similar triangles to find x.

22.

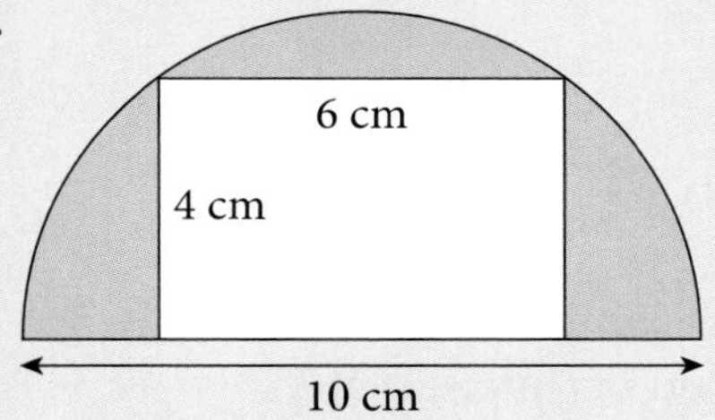

The diagram consists of a rectangle inside a semicircle.
Find the purple shaded area, correct to three significant figures.

23.

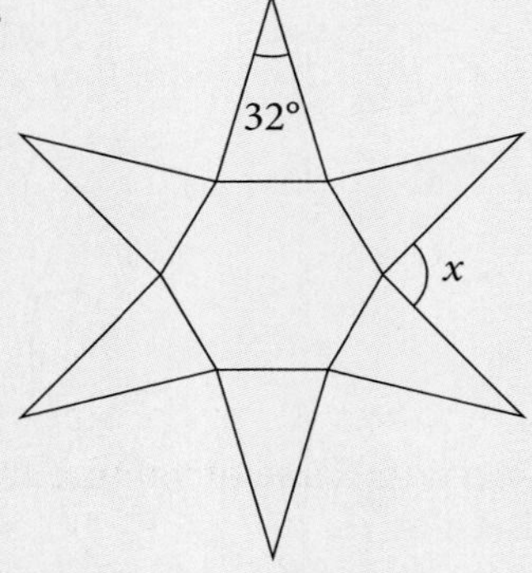

The diagram is a regular hexagon surrounded six congruent isosceles triangles. Find the size of angle x.

Examination exercise 7B

1.

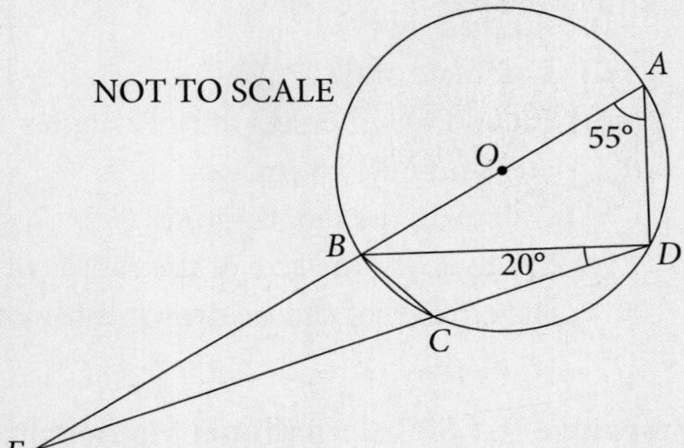

The points, A, B, C and D lie on a circle, centre O. AB is a diameter, angle $BAD = 55°$ and angle $BDC = 20°$. ABE and DCE are straight lines.
Find

a) angle ABD, [1]
b) angle BCD, [1]
c) angle AED. [1]

Cambridge IGCSE International Mathematics 0607 Paper 2 Q10 May/June 2010

2

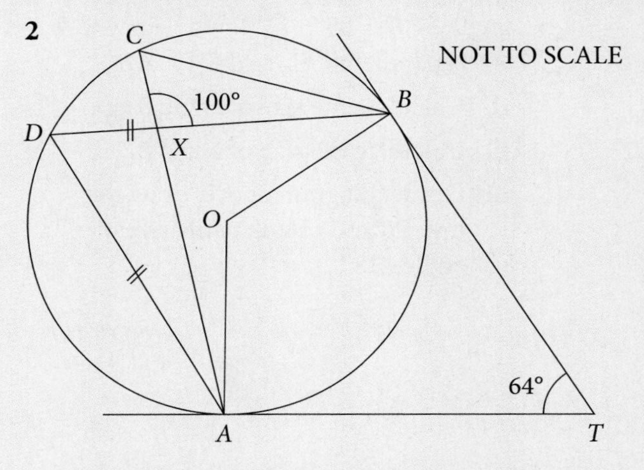

In the diagram, A, B, C and D lie on the circle, centre O.
TA and TB are tangents at A and B. The lines AC and BD cross at X.
$AD = BD$, angle $ATB = 64°$ and angle $CXB = 100°$.

a) Calculate
- **i)** angle AOB, [2]
- **ii)** angle OAB, [2]
- **iii)** angle BAD, [2]
- **iv)** angle CAO. [2]

b) Explain why $OATB$ is a cyclic quadrilateral. [1]

Cambridge IGCSE International Mathematics 0607 Specimen Paper 4 Q4 2010

3. a) NOT TO SCALE

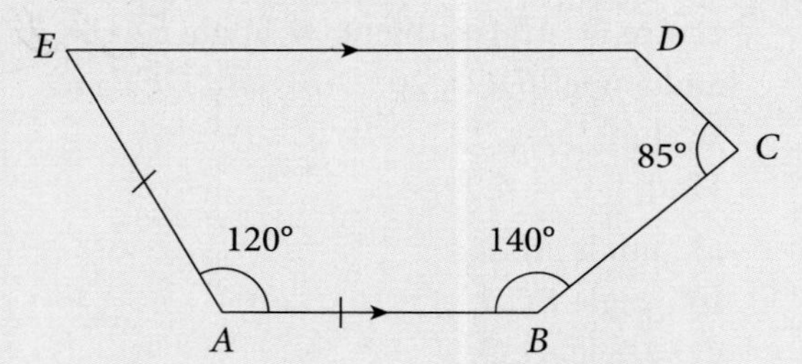

$ABCDE$ is a pentagon with angle $EAB = 120°$, angle $ABC = 140°$ and angle $BCD = 85°$.
AB is parallel to ED and $EA = AB$.

i) Calculate the size of angle AED. [1]

ii) Calculate the size of angle EDC. [2]

iii) On the diagram above, draw the line EB and calculate the size of angle EBC. [1]

b)

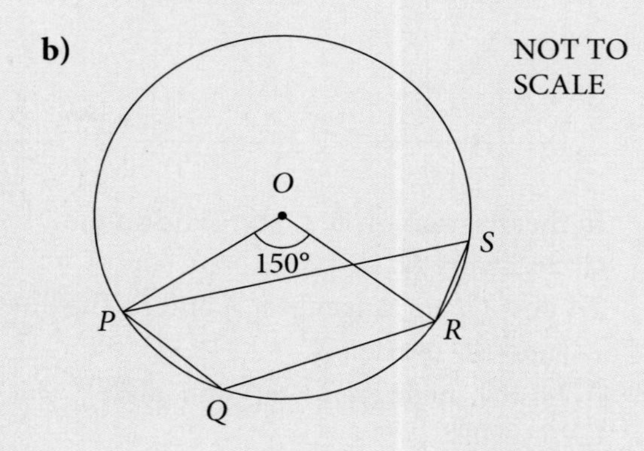

P, Q, R and S lie on a circle, centre O.
Angle $POR = 150°$.

Calculate the size of

i) angle PSR, [1]

ii) angle PQR. [1]

Cambridge IGCSE International Mathematics 0607 Paper 41 Q4 May/ June 2011

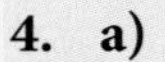

4. a)

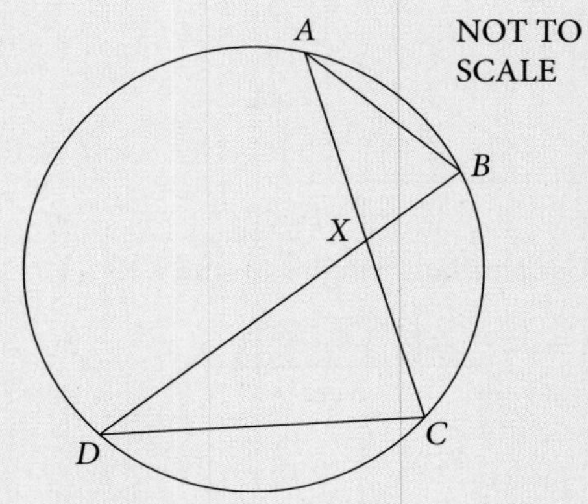

A, B, C and D lie on a circle.
AC and BD intersect at X.

i) Explain why triangles ABX and DCX are similar. [3]

ii) $BX = 2$ cm, $CX = 4$ cm and the area of triangle ABX is 4.5 cm^2. [2]
Calculate the area of triangle DCX.

b) NOT TO SCALE

$PQRS$ is a cyclic quadrilateral.
Angle $RSQ = 32°$ and angle $PRQ = 50°$.

i) Find angle PSQ. [1]

ii) Calculate angle PQR. [2]

iii) PR and QS intersect at right angles at Y and $QR = 8$ cm.
Calculate the length of RY. [2]

iv) Write down the size of the radius of the circle that can be drawn through Q, R and Y. [1]

Cambridge IGCSE International Mathematics 0607 Paper 4 Q10 October/November 2010

8 Algebra 2

Girolamo Cardan (1501–76) was a colourful character who became Professor of Mathematics at Milan. As well as being a distinguished academic, he was an astrologer, a physician, a gambler and a heretic–yet he received a pension from the Pope. His mathematical genius enabled him to open up the general theory of cubic and quartic equations, although a method for solving cubic equations which he claimed as his own was pirated from Niccolò Tartaglia.

2.1 Writing, showing and interpretation of inequalities, including those on the real number line

2.2 Solution of linear and quadratic inequalities
Solution of inequalities using a graphics calculator e.g. $2x^2 + 5x - 3 < 0$

2.4 Indices

2.5 Derivation, rearrangement and evaluation of formulae

2.9 Algebraic fractions:
- simplification, including use of factorisation
- addition or subtraction of fractions with linear denominators
- multiplication or division and simplification of two fractions

2.12 Continuation of a sequence of numbers or patterns
Determination of the nth term
Use of a difference method to find the formula for a linear sequence, a quadratic sequence or a cubic sequence
Identification of a simple geometric sequence and determination of its formula

2.13 Direct variation (proportion) $y \propto x$, $y \propto x^2$, $y \propto x^3$, $y \propto \sqrt{x}$
Inverse variation $y \propto \frac{1}{x}$, $y \propto \frac{1}{x^2}$, $y \propto \frac{1}{\sqrt{x}}$
Best variation model for given data

7.7 Linear inequalities on the Cartesian plane (shade unwanted regions)
Syllabus objectives 2.3, 2.6, 2.7, 2.8, 2.10 and 2.11 are covered in Chapter 11. Objectives 7.1, 7.2, 7.3, 7.4, 7.5 and 7.6 are covered in Chapter 9.

8.1 Inequalities

Notation

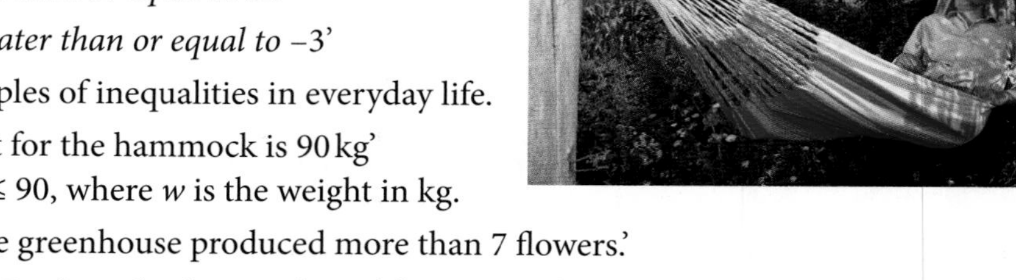

$x < 4$ means '*x is less than* 4'

$y > 7$ means '*y is greater than* 7'

$z \leq 10$ means '*z is less than or equal to* 10'

$t \geq -3$ means '*t is greater than or equal to* -3'

Here are some examples of inequalities in everyday life.

- 'The weight limit for the hammock is 90 kg'
  We can write $w \leq 90$, where w is the weight in kg.
- 'Each plant in the greenhouse produced more than 7 flowers.'
  We can write $f > 7$, where f is the number of flowers produced by each plant.

Solving inequalities

We follow the same procedure used for solving equations except that when we multiply or divide by a *negative* number the inequality is *reversed.*

e.g. $4 > -2$ but multiplying by -2 gives $-8 < 4$

Example 1

Solve the inequalities.

a) $2x - 1 > 5$

$$2x > 5 + 1$$
$$x > \frac{6}{2}$$
$$x > 3$$

b) $5 - 3x \leq 1$

$$5 \leq 1 + 3x$$
$$5 - 1 \leq 3x$$
$$\frac{4}{3} \leq x$$

Exercise 8.1

Introduce one of the symbols <, > or = between each pair of numbers.

1. $-2, 1$

2. $(-2)^2, 1$

3. $\frac{1}{4}, \frac{1}{5}$

4. $0{\cdot}2, \frac{1}{5}$

5. $10^2, 2^{10}$

6. $\frac{1}{4}, 0{\cdot}4$

7. $40\%, 0{\cdot}4$

8. $(-1)^2, \left(-\frac{1}{2}\right)^2$

9. $5^2, 2^5$

10. $3\frac{1}{3}, \sqrt{10}$

11. $\pi^2, 10$

12. $-\frac{1}{3}, -\frac{1}{2}$

13. $2^{-1}, 3^{-1}$

14. $50\%, \frac{1}{5}$

15. $1\%, 100^{-1}$

State whether the following are true or false.

16. $0{\cdot}7^2 > \frac{1}{2}$ **17.** $10^3 = 30$ **18.** $\frac{1}{8} > 12\%$

19. $(0{\cdot}1)^3 = 0.0001$ **20.** $\left(-\frac{1}{5}\right)^0 = -1$ **21.** $\frac{1}{5^2} > \frac{1}{2^5}$

22. $(0{\cdot}2)^3 < (0{\cdot}3)^2$ **23.** $\frac{6}{7} > \frac{7}{8}$ **24.** $0{\cdot}1^2 > 0{\cdot}1$

Solve the inequalities.

25. $x - 3 > 10$ **26.** $x + 1 < 0$ **27.** $5 > x - 7$

28. $2x + 1 \leq 6$ **29.** $3x - 4 > 5$ **30.** $10 \leq 2x - 6$

31. $5x < x + 1$ **32.** $2x \geq x - 3$ **33.** $4 + x < -4$

34. $3x + 1 < 2x + 5$ **35.** $2(x + 1) > x - 7$ **36.** $7 < 15 - x$

37. $9 > 12 - x$ **38.** $4 - 2x \leq 2$ **39.** $3(x - 1) < 2(1 - x)$

40. $7 - 3x < 0$

The number line

The inequality $x < 4$ is represented on the number line as:

$x \geq -2$ is shown as:

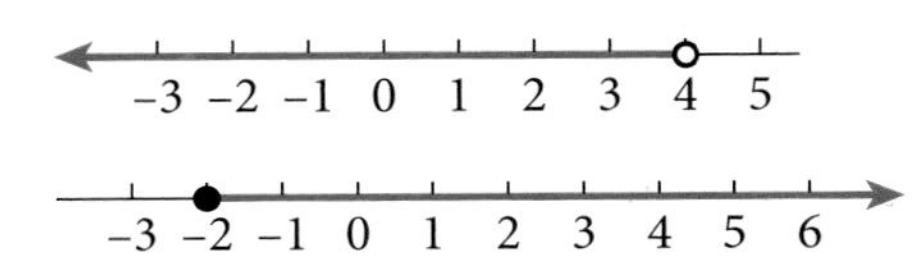

In the first case, 4 is *not* included so we have the empty circle ○.

In the second case, −2 *is* included so we have the filled circle ●.

$-1 \leq x < 3$ is shown as:

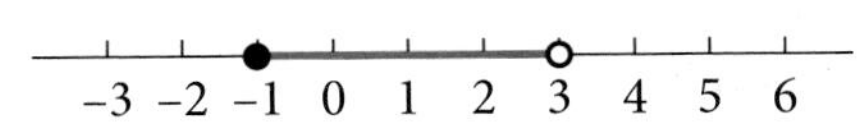

Exercise 8.2

For questions **1** to **25**, solve each inequality and show the result on a number line.

1. $2x + 1 > 11$ **2.** $3x - 4 \leq 5$ **3.** $2 < x - 4$

4. $6 \geq 10 - x$ **5.** $8 < 9 - x$ **6.** $8x - 1 < 5x - 10$

7. $2x > 0$ **8.** $1 < 3x - 11$ **9.** $4 - x > 6 - 2x$

10. $\frac{x}{3} < -1$ **11.** $1 < x < 4$ **12.** $-2 \leq x \leq 5$

13. $1 \leq x < 6$ **14.** $0 \leq 2x < 10$ **15.** $-3 \leq 3x \leq 21$

16. $1 < 5x < 10$ **17.** $\frac{x}{4} > 20$ **18.** $3x - 1 > x + 19$

19. $7(x + 2) < 3x + 4$ **20.** $1 < 2x + 1 < 9$ **21.** $10 \leq 2x \leq x + 9$

22. $x < 3x + 2 < 2x + 6$ **23.** $10 < 2x - 1 \leq x + 5$ **24.** $x < 3x - 1 < 2x + 7$

25. $x - 10 < 2(x - 1) < x$

> In questions **20** to **25**, solve the two inequalities separately.

For questions **26** to **35**, find the solutions, subject to the given condition.

26. $3a + 1 < 20$; a is a positive integer

27. $b - 1 \geq 6$; b is a prime number less than 20

28. $2e - 3 < 21$; e is positive even number

29. $1 < z < 50$; z is a square number

30. $0 < 3x < 40$; x is divisible by 5

31. $2x > -10$; x is a negative integer

32. $x + 1 < 2x < x + 13$; x is an integer

33. $x^2 < 100$; x is a positive square number

34. $0 \leq 2z - 3 \leq z + 8$; z is a prime number

35. $\frac{a}{2} + 10 > a$; a is a positive even number

Exercise 8.3

In questions **1** to **4** solve each pair of inequalities and then find the range of values of x for which both inequalities are true.

1. $x - 3 < 1$ and $2x + 1 > 0$

2. $\frac{x}{2} + 1 < 3$ and $3x > 0$

3. $1 - 5x > 6$ and $2x + 7 > 3$

4. $\frac{x}{3} - 3 > 0$ and $12 - x > 1$

5. The height of each photo below has to be greater than the width. Find the range of possible values of n in each case.

a)

$2(n + 1)$

$(n + 5)$

b)

$(n + 3)$

$2(n - 3)$

6. Given that $1 \leq x \leq 5$ and $-3 \leq y \leq 1$, find

a) the greatest possible value of $x - y$

b) the least possible value of $x^2 + y^2$

7. Given that $1 \leq a \leq 10$ and $-5 \leq b \leq 6$, find

a) the greatest possible value of $\frac{b}{a}$

b) the greatest possible value of $b^2 - a$

c) the greatest possible value of $a - b$

8. Given that $5x > 1$ and $x - 2 < 3$, list all the possible whole number values of x.

9. If $3^x > 1000$, what is the smallest whole number value of x?

10. Look at the series $1 + 2 + 4 + 8 + \ldots$

After how many terms is the sum of this series greater than 1000?

11. Find a simple fraction q such that $\frac{4}{9} < q < \frac{5}{9}$

12. Find an integer value of a such that $a - 3 \leq 11 \leq 2a + 10$

13. State the largest prime number z for which $3z < 66$

14. Find a simple fraction r such that $\frac{1}{3} < r < \frac{2}{3}$

15. Find the largest prime number p such that $p^2 < 400$

16. Illustrate on a number line the solution set of each pair of simultaneous inequalities:

a) $x < 6; -3 \leq x \leq 8$

b) $x > -2; -4 < x < 2$

c) $2x + 1 \leq 5; -12 \leq 3x - 3$

d) $3x - 2 < 19; 2x \geq -6$

17. Find the integer n such that $n < \sqrt{300} < n + 1$

Linear inequalities in the Cartesian plane

It is useful to represent inequalities on a graph, particularly where two variables (x and y) are involved.

You could be asked to draw a sketch graph and shade the area which represents the set of points that satisfy each of these inequalities.

a) $x < 2$

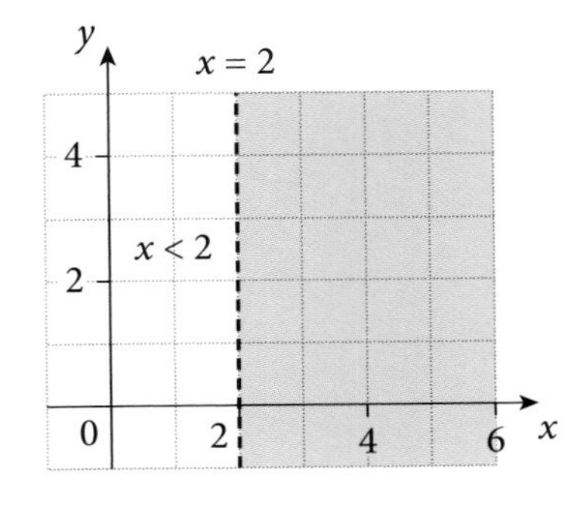

b) $y \leq 1, y \geq 5$

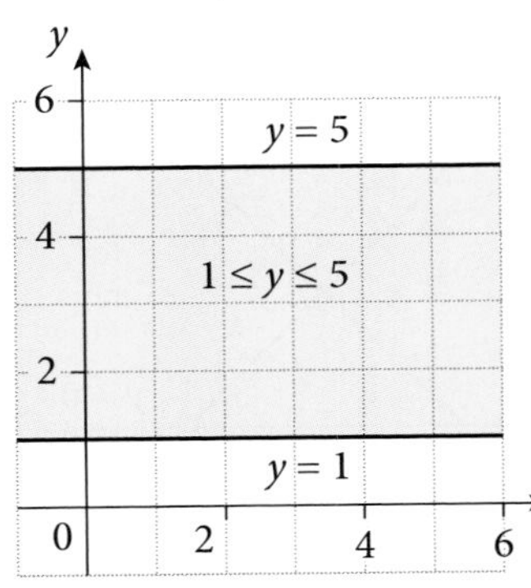

c) $x + y \geq 8$

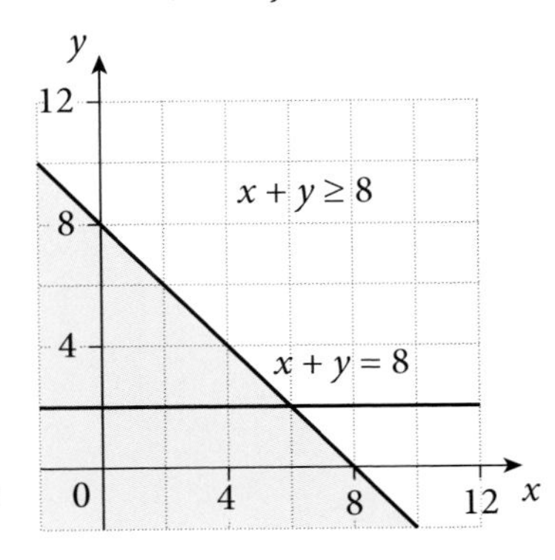

In each graph, the unwanted region is shaded.

In **a)**, the line $x = 2$ is shown as a broken line to indicate that the points on the line are not included.

In **b)** and **c)** points on the line are included 'in the region' and the lines are drawn unbroken.

To decide which side to shade when the line is sloping, we take a **trial point**. This can be any point which is not actually on the line.

In **c)** above, the trial point could be (1, 1).

Is (1, 1) in the region $x + y \geqslant 8$?

It does not satisfy $x + y \geqslant 8$ because $1 + 1 = 2$, which is less than 8.

So above the line is $x + y \geqslant 8$. We have shaded the unwanted region.

Exercise 8.4

In questions 1 to 3, describe the region shaded.

1.

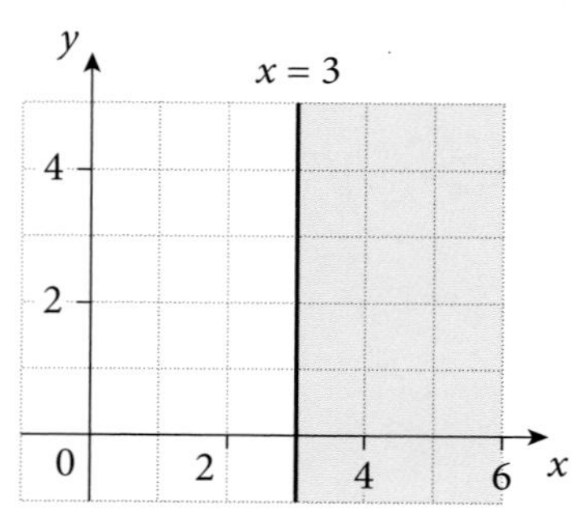

2.

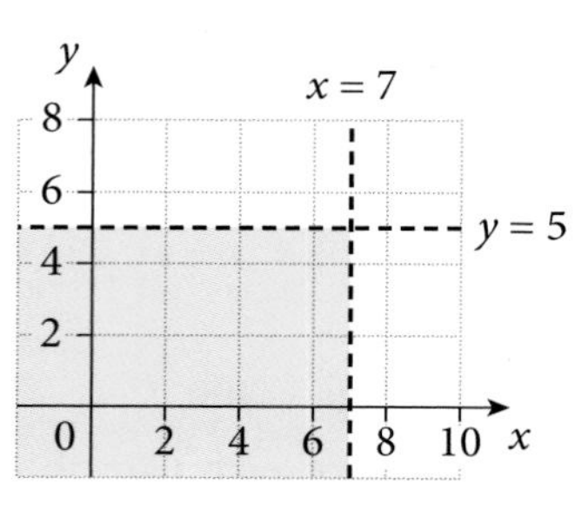

3.

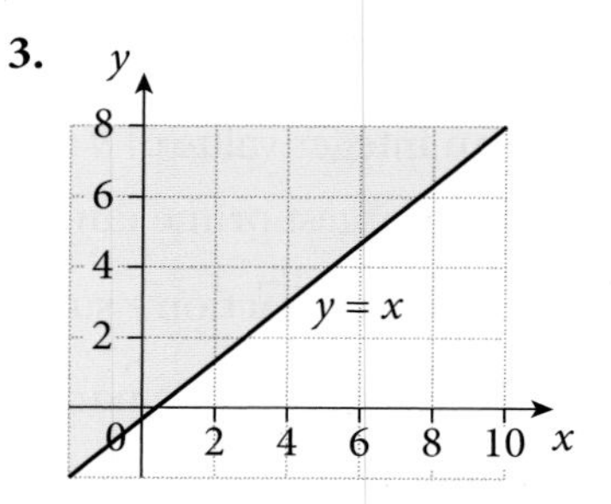

4. The point (3, 1), marked *, lies in the shaded triangle. Use this as a trial point to write down the three inequalities which describe the unshaded region.

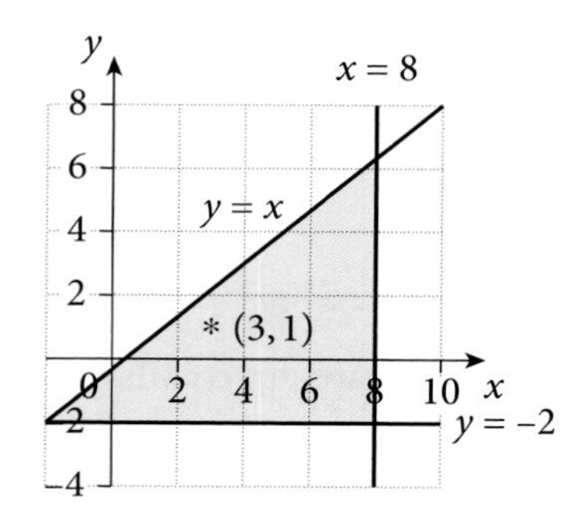

5. A trial point (1, 1) lies inside the shaded triangle. Write down the three inequalities which describe each shaded region.

a)

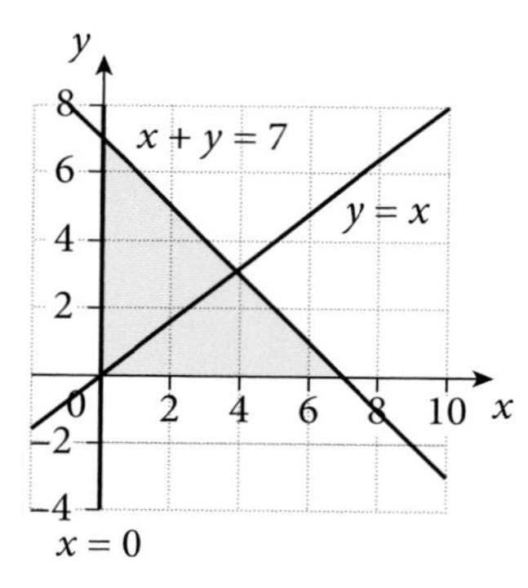

b)

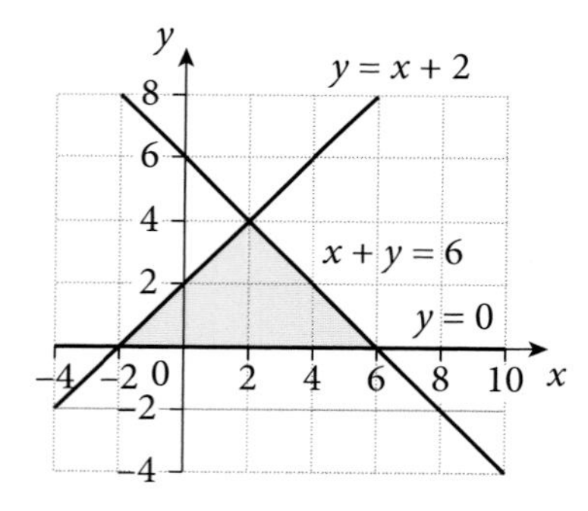

For Question **6** to **15**, draw a sketch graph similar to those above and indicate the set of points which satisfy the inequalities by shading the require region.

6. $2 < x < 7$

7. $0 < y < 3\frac{1}{2}$

8. $-2 < x < 2$

9. $x < 6$ and $y < 4$

10. $0 < x < 5$ and $y < 3$

11. $1 < x < 6$ and $2 < y < 8$

12. $-3 < x < 0$ and $-4 < y < 2$

13. $y < x$

14. $x + y < 5$

15. $y > x + 2$ and $y < 7$

Quadratic inequality

To solve a quadratic inequality sketch a graph of the quadratic function. The method is illustrated in the examples below.

Example 2

Solve $x^2 - 3x - 40 < 0$

We factorise to give $(x - 8)(x + 5) < 0$

The critical values are 8 and -5

Sketch the graph of $y = x^2 - 3x - 40$:

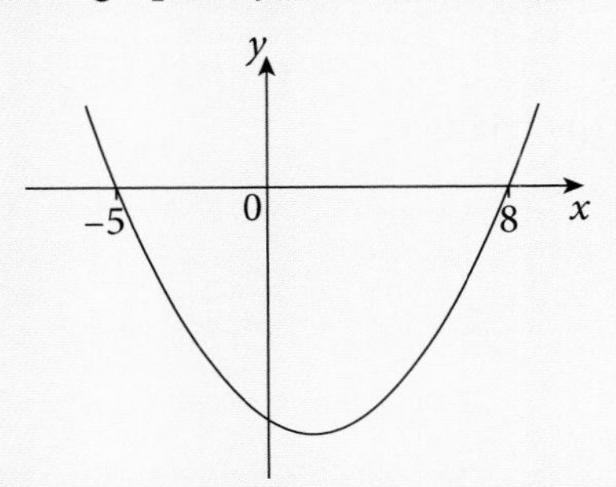

We are looking below the x-axis so we want *one* region, and hence *one* inequality. So the solution is $-5 < x < 8$

Example 3

Solve $x^2 - 10x + 21 \geq 0$

Factorise to give $(x - 3)(x - 7) \geq 0$

Critical values are 3 and 7

Sketch the graph of $y = x^2 - 10x + 21$

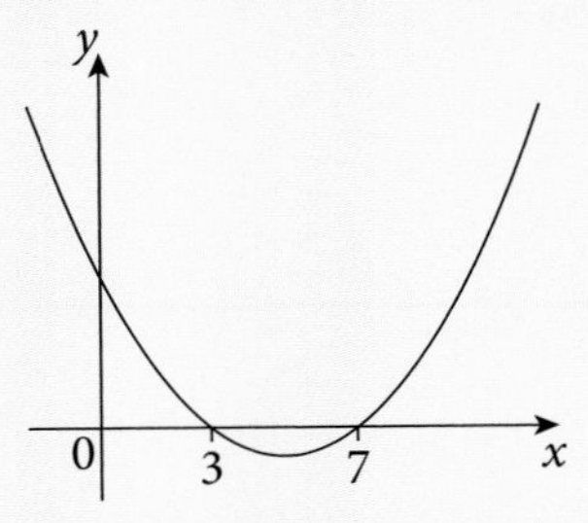

We are looking above x-axis so we want two regions, and hence *two* inequalities. So the solution is $x \leq 3$ or $x \geq 7$.

Exercise 8.5

1. Solve the following inequalities

   a) $x^2 < 1$ b) $x^2 > 16$ c) $2x^2 - 1 \leq 17$

   d) $3x^2 - 2 < 46$ e) $28 - x^2 < 3$ f) $15 - 2x^2 \leq -17$

2. Solve the following inequalities (by first factorising the quadratic).

   a) $x^2 - 4x - 12 \geq 0$ b) $x^2 + 6x + 8 < 0$ c) $x^2 - 5x - 6 \geq 0$

   d) $x^2 + 5x + 6 < 0$ e) $x^2 - 7x + 12 \geq 0$ f) $x^2 - 3x + 2 \leq 0$

   g) $2x^2 - 11x + 12 < 0$ h) $3x^2 - 13x - 8 < 2$ i) $4x^2 + 16x - 6 < 3$

   j) $3x^2 - 13x + 18 \geq 6$ k) $4x + 3 - 4x^2 \leq 0$ l) $4 + 7x - 3x^2 > -22$

   m) $9x - 2x^2 > 10$ n) $3 - 6x^2 < 8 - 17x$

3. The area of the rectangle must be greater than the area of the triangle. Find the range of possible values of x.

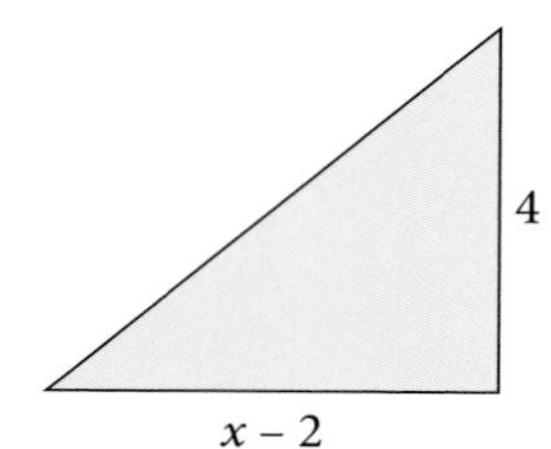

4. Find the range of values of x which satisfy both of the inequalities simultaneously.

   $x^2 - 3x - 4 \geq 0$ and $3x + 1 > 2x + 6$

5. Find the range of values of x which satisfy both of the inequalities simultaneously.

   $x^2 - x - 6 \geq 0$ and $5x + 1 < 2x - 8$

6. Find the complete set of values of p for which roots of the equation $2x^2 + px + 3p - 10 = 0$ exist.

8.2 Indices

Indices are used as a convenient way of writing products.

For example $4^3 = 4 \times 4 \times 4$ and $7^5 = 7 \times 7 \times 7 \times 7 \times 7$

For 7^5 we say '7 to the power 5' or just '7 to the 5'.

Rules of indices

1. $a^n \times a^m = a^{n+m}$	e.g. $7^2 \times 7^4 = 7^6$
2. $a^n \div a^m = a^{n-m}$	e.g. $6^6 \div 6^2 = 6^4$
3. $(a^n)^m = a^{nm}$	e.g. $(3^2)^5 = 3^{10}$
Also, $a^{-n} = \frac{1}{a^n}$	e.g. $5^{-2} = \frac{1}{5^2}$
$a^{\frac{1}{n}}$ means 'the nth root of a'	e.g. $9^{\frac{1}{2}} = \sqrt[2]{9}$
$a^{\frac{m}{n}}$ means 'the nth root of a raised to the power m'	e.g. $4^{\frac{3}{2}} = (\sqrt{4})^3 = 8$

Example 4

Simplify:

a) $x^7 \times x^{13}$ **b)** $x^3 \div x^7$ **c)** $(x^4)^3$

d) $(3x^2)^3$ **e)** $(2x^{-1})^2 \div x^{-5}$ **f)** $3y^2 \times 4y^3$

a) $x^7 \times x^{13} = x^{7+13} = x^{20}$

b) $x^3 \div x^7 = x^{3-7} = x^{-4} = \frac{1}{x^4}$

c) $(x^4)^3 = x^{12}$

d) $(3x^2)^3 = 3^3 \times (x^2)^3 = 27x^6$

e) $(2x^{-1})^2 \div x^{-5} = 4x^{-2} \div x^{-5}$
$= 4x^{(-2--5)}$
$= 4x^3$

f) $3y^2 \times 4y^3 = 12y^5$

Note that you cannot use the rules above if different numbers are raised to powers. So $3^4 \times 5^2$ or $7^5 \div 8^3$ cannot be found using the rules.

Exercise 8.6

In equation **1** to **21** write the answer in index form.

1. $2 \times 2 \times 2 \times 2$

2. $7 \times 7 \times 7 \times 7 \times 7 \times 7$

3. $3 \times 3 \times 2 \times 2 \times 2 \times 2 \times 2$

4. $a \times a \times a \times a \times a$

5. $p \times p \times p$

6. $5 \times 5 \times 5 \times 5 \times 8 \times 8$

7. $3^4 \times 3^6$

8. $4^2 \times 4^3$

9. $8^6 \times 8^2$

10. $7^4 \times 7^{40}$

11. $9^2 \times 9^{22}$

12. $n^3 \times n^2$

13. $a^7 \times a^3$

14. $n^5 \times n^5$

15. $y^7 \times y$

16. $n^7 \div n^2$

17. $a^{10} \div a^2$

18. $x^{11} \div x$

19. $(3^5 \times 3^2) \div 3^4$

20. $(2^3 \times 2^4) \div 2^2$

21. $(4^7 \times 4^2) \div 4^3$

In questions **22** to **30** copy and complete the equations.

22. $5^3 \times 5^2 = \square$

23. $7^5 \div \square = 7^2$

24. $\square \times 3^{10} = 3^{12}$

25. $\square \div 7^{10} = 7^3$

26. $n^3 \times n^3 = \square$

27. $3^{100} \div \square = 3^{20}$

28. $3^6 \div 3 = \square$

29. $10^{10} \div \square = 10$

30. Half of $2^5 = \square$

In questions **31** to **39** give the answer as an ordinary number.

31. $5^4 \div 5^2$

32. $2^5 \div 2^3$

33. $10^3 \div 10$

34. $3^5 \div 3^2$

35. $1^{11} \div 1^{10}$

36. $4^{12} \div 4^{10}$

37. 2×2^3

38. $10^6 \div 10^3$

39. $\left(\frac{1}{2}\right)^2 \times 4$

Exercise 8.7

Express in index form:

1. $3 \times 3 \times 3 \times 3$

2. $4 \times 4 \times 5 \times 5 \times 5$

3. $3 \times 7 \times 7 \times 7$

4. $2 \times 2 \times 2 \times 7$

5. $\frac{1}{10 \times 10 \times 10}$

6. $\frac{1}{2 \times 2 \times 3 \times 3 \times 3}$

7. $\sqrt{15}$

8. $\sqrt[3]{3}$

9. $\sqrt[5]{10}$

10. $(\sqrt{5})^3$

Simplify:

11. $x^3 \times x^4$

12. $y^6 \times y^7$

13. $z^7 \times z^3$

14. $z^{50} \times z^{50}$

15. $m^3 \div m^2$

16. $e^{-3} \times e^{-2}$

17. $y^{-2} \times y^4$

18. $w^4 \div w^{-2}$

19. $y^{\frac{1}{2}} \times y^{\frac{1}{2}}$

20. $(x^2)^5$

21. $x^{-2} \div x^{-2}$

22. $w^{-3} \times w^{-2}$

23. $w^{-7} \times w^2$

24. $x^3 \div x^{-4}$

25. $(a^2)^4$

26. $\left(k^{\frac{1}{2}}\right)^6$

27. $e^{-4} \times e^4$

28. $x^{-1} \times x^{30}$

29. $(y^4)^{\frac{1}{2}}$

30. $(x^{-3})^{-2}$

31. $z^2 \div z^{-2}$

32. $t^{-3} \div t$

33. $(2x^3)^2$

34. $(4y^5)^2$

35. $(x^2)^{\frac{1}{2}} \div \left(x^{\frac{1}{3}}\right)^3$

36. $7w^{-2} \times 3w^{-1}$

37. $(2n)^4 \div 8n^0$

38. $4x^{\frac{3}{2}} \div 2x^{\frac{1}{2}}$

Example 5

Evaluate:

a) $9^{\frac{1}{2}}$ **b)** 5^{-1} **c)** $4^{-\frac{1}{2}}$

d) $25^{\frac{3}{2}}$ **e)** $\left(5^{\frac{1}{2}}\right)^3 \times 5^{\frac{1}{2}}$ **f)** 7^0

a) $9^{\frac{1}{2}} = \sqrt{9} = 3$

b) $5^{-1} = \frac{1}{5}$

c) $4^{-\frac{1}{2}} = \frac{1}{4^{\frac{1}{2}}} = \frac{1}{\sqrt{4}} = \frac{1}{2}$

d) $25^{\frac{3}{2}} = (\sqrt{25})^3 = 5^3 = 125$

e) $\left(5^{\frac{1}{2}}\right)^3 \times 5^{\frac{1}{2}} = 5^{\frac{3}{2}} \times 5^{\frac{1}{2}} = 5^2 = 25$

f) $7^0 = 1$ [consider $\frac{7^3}{7^3} = 7^{3-3} = 7^0 = 1$]

Remember

$a^0 = 1$ for any non-zero value of a.

Exercise 8.8

Evaluate the following.

1. $3^2 \times 3$
2. 100^0
3. 3^{-2}
4. $(5^{-1})^{-2}$
5. $4^{\frac{1}{2}}$
6. $16^{\frac{1}{2}}$
7. $81^{\frac{1}{2}}$
8. $8^{\frac{1}{3}}$
9. $9^{\frac{3}{2}}$
10. $27^{\frac{1}{3}}$
11. $9^{-\frac{1}{2}}$
12. $8^{-\frac{1}{3}}$
13. $1^{\frac{5}{2}}$
14. $25^{-\frac{1}{2}}$
15. $1000^{\frac{1}{3}}$
16. $2^{-2} \times 2^5$
17. $2^4 \div 2^{-1}$
18. $8^{\frac{2}{3}}$
19. $27^{-\frac{2}{3}}$
20. $4^{-\frac{3}{2}}$
21. $36^{\frac{1}{2}} \times 27^{\frac{1}{3}}$
22. $10000^{\frac{1}{4}}$
23. $100^{\frac{3}{2}}$
24. $\left(100^{\frac{1}{2}}\right)^{-3}$
25. $\left(9^{\frac{1}{2}}\right)^{-2}$
26. $(-16{\cdot}371)^0$
27. $81^{\frac{1}{4}} \div 16^{\frac{1}{4}}$
28. $(5^{-4})^{\frac{1}{2}}$
29. $1000^{-\frac{1}{3}}$
30. $\left(4^{-\frac{1}{2}}\right)^2$
31. $8^{-\frac{2}{3}}$
32. $100^{\frac{5}{2}}$
33. $1^{\frac{4}{3}}$
34. 2^{-5}
35. $(0{\cdot}01)^{\frac{1}{2}}$
36. $(0{\cdot}04)^{\frac{1}{2}}$
37. $(2{\cdot}25)^{\frac{1}{2}}$
38. $(7{\cdot}63)^0$
39. $3^5 \times 3^{-3}$
40. $\left(3\frac{3}{8}\right)^{\frac{1}{3}}$
41. $\left(11\frac{1}{9}\right)^{-\frac{1}{2}}$
42. $\left(\frac{1}{8}\right)^{-2}$
43. $\left(\frac{1}{1000}\right)^{\frac{2}{3}}$
44. $\left(\frac{9}{25}\right)^{-\frac{1}{2}}$
45. $(10^{-6})^{\frac{1}{3}}$
46. $7^2 \div \left(7^{\frac{1}{2}}\right)^4$
47. $(0{\cdot}0001)^{-\frac{1}{2}}$
48. $\dfrac{9^{\frac{1}{2}}}{4^{-\frac{1}{2}}}$
49. $\dfrac{25^{\frac{3}{2}} \times 4^{\frac{1}{2}}}{9^{-\frac{1}{2}}}$
50. $\left(-\frac{1}{7}\right)^2 \div \left(-\frac{1}{7}\right)^3$
51. Find x in the following equations.

    a) $32^x = 1$ **b)** $32^x = 4$

Example 6

Simplify:

a) $(2a)^3 \div (9a^2)^{\frac{1}{2}}$ **b)** $(3ac^2)^3 \times 2a^{-2}$ **c)** $(2x)^2 \div 2x^2$

a) $(2a)^3 \div (9a^2)^{\frac{1}{2}} = 8a^3 \div 3a$ **b)** $(3ac^2)^3 \times 2a^{-2} = 27a^3c^6 \times 2a^{-2}$

$= \frac{8}{3}a^2$

$= 54ac^6$

c) $(2x)^2 \div 2x^2 = 4x^2 \div 2x^2$

$= 2$

Exercise 8.9

Rewrite without brackets:

1. $(5x^2)^2$ **2.** $(7y^3)^2$ **3.** $(10ab)^2$ **4.** $(2xy^2)^2$

5. $(4x^2)^{\frac{1}{2}}$ **6.** $(9y)^{-1}$ **7.** $(x^{-2})^{-1}$ **8.** $(2x^{-2})^{-1}$

9. $(5x^2y)^0$ **10.** $\left(\frac{1}{2}x\right)^{-1}$ **11.** $(3x)^2 \times (2x)^2$ **12.** $(5y)^2 \div y$

13. $\left(2x^{\frac{1}{2}}\right)^4$ **14.** $\left(3y^{\frac{1}{3}}\right)^3$ **15.** $(5x^0)^2$ **16.** $[(5x)^0]^2$

17. $(7y^0)^2$ **18.** $[(7y)^0]^2$ **19.** $(2x^2y)^3$ **20.** $(10xy^3)^2$

Simplify the following.

21. $(3x^{-1})^2 \div 6x^{-3}$ **22.** $(4x)^{\frac{1}{2}} \div x^{\frac{3}{2}}$ **23.** $x^2y^2 \times xy^3$ **24.** $4xy \times 3x^2y$

25. $10x^{-1}y^3 \times xy$ **26.** $(3x)^2 \times \left(\frac{1}{9}x^2\right)^{\frac{1}{2}}$ **27.** $z^3yx \times x^2yz$ **28.** $(2x)^{-2} \times 4x^3$

29. $(3y)^{-1} \div (9y^2)^{-1}$ **30.** $(xy)^0 \times (9x)^{\frac{3}{2}}$ **31.** $(x^2y)(2xy)(5y^3)$ **32.** $\left(4x^{\frac{1}{2}}\right) \times \left(8x^{\frac{3}{2}}\right)$

33. $5x^{-3} \div 2x^{-5}$ **34.** $[(3x^{-1})^{-2}]^{-1}$ **35.** $(2a)^{-2} \times 8a^4$ **36.** $(abc^2)^3$

37. Write in the form 2^p (e.g. $4 = 2^2$):

a) 32 **b)** 128 **c)** 64 **d)** 1

38. Write in the form 3^q:

a) $\frac{1}{27}$ **b)** $\frac{1}{81}$ **c)** $\frac{1}{3}$ **d)** $9 \times \frac{1}{81}$

In questions **39** to **50**, evaluate the equations, with $x = 16$ and $y = 8$.

39. $2x^{\frac{1}{2}} \times y^{\frac{1}{3}}$ **40.** $x^{\frac{1}{4}} \times y^{-1}$ **41.** $(y^2)^{\frac{1}{6}} \div (9x)^{\frac{1}{2}}$ **42.** $(x^2y^3)^0$

43. $x + y^{-1}$ **44.** $x^{-\frac{1}{2}} + y^{-1}$ **45.** $y^{\frac{1}{3}} \div x^{\frac{3}{4}}$ **46.** $(1000y)^{\frac{1}{3}} \times x^{-\frac{5}{2}}$

47. $\left(x^{\frac{1}{4}} + y^{-1}\right) \div x^{\frac{1}{4}}$ **48.** $x^{\frac{1}{2}} - y^{\frac{2}{3}}$ **49.** $\left(x^{\frac{3}{4}}y\right)^{-\frac{1}{3}}$ **50.** $\left(\frac{x}{y}\right)^{-2}$

In questions **51** to **64**, solve the equations for x.

51. $2^x = 8$ **52.** $3^x = 81$ **53.** $5^x = \frac{1}{5}$ **54.** $10^x = \frac{1}{100}$

55. $3^{-x} = \frac{1}{27}$ **56.** $4^x = 64$ **57.** $6^{-x} = \frac{1}{6}$ **58.** $100\,000^x = 10$

59. $12^x = 1$ **60.** $10^x = 0.0001$ **61.** $2^x + 3^x = 13$ **62.** $\left(\frac{1}{2}\right)^x = 32$

63. $5^{2x} = 25$ **64.** $1000\,000^{3x} = 10$

65. These two questions are more difficult. Use a calculator to find solutions correct to three significant figures.

a) $x^x = 100$ **b)** $x^x = 10\,000$

66. A white cell splits into two new cells every hour. So after one hour there are 2 cells, after two hours there are 4 cells and so on.

The number of cells, N, is given by the equation $N = 2^h$, where h is the number of hours after the start of the process.

a) How many cells are there after 10 hours?

b) After how many hours will there be more than one billion cells for the first time?

c) After how many hours will there be more than two billion cells?

Exercise 8.10

In questions **1** to **6** give your answer in index form.

1. The number of grains of sand in a bucket is 2^{20}.
The contents of the bucket are divided into two equal piles.
How many grains of sand are there in each pile?

2. There are 6^5 small cubes in a large model skyscraper and each cube has 6 faces. How many faces are there on all the cubes in the model?

3. The distance of the sun from the Earth is about 10^8 miles.
A spacecraft is about 10^5 miles from the Earth. How many times further from the Earth is the Sun than the spacecraft?

4. An imaginary cube of side 10^7 metres is drawn around the Moon.
Calculate the volume of the cube in cubic metres.

5. A scientist estimates that there are 10^{11} bacteria in a specimen dish.
After an antibiotic is added, the number is reduced to one millionth of the original. How many bacteria are left?

6. A spacecraft is moving at a steady speed of 10^4 m/s towards a planet which is 100 million km away. How long will it take the spacecraft to reach the planet?
Give your answer in seconds and state whether the time required is more or less than one year.

7. Fractional powers of numbers can be found using the (x^2) or the (^) button on a calculator.

   For example to work out $25^{\frac{1}{2}}$ press: (25) (^) (0.5) (=).

   Use a calculator to work out the following.

   a) $9^{\frac{1}{2}}$ **b)** $16^{\frac{1}{2}}$ **c)** $100^{\frac{1}{2}}$ **d)** $144^{\frac{1}{2}}$

   e) $8^{\frac{1}{3}}$ **f)** $27^{\frac{1}{3}}$ **g)** $1000^{\frac{1}{3}}$ **h)** $16^{\frac{1}{4}}$

8. Write in order of size, smallest first.

   4^{2^2} 2^5 2^{5^2} 5^2 2^{3^4} $3^{2^{2^2}}$

GDC help

Use your calculator to sketch the graph of $y = 4^x$ for values of x from −3 to 3. Use your graph to solve the equation $4^x = 20$.

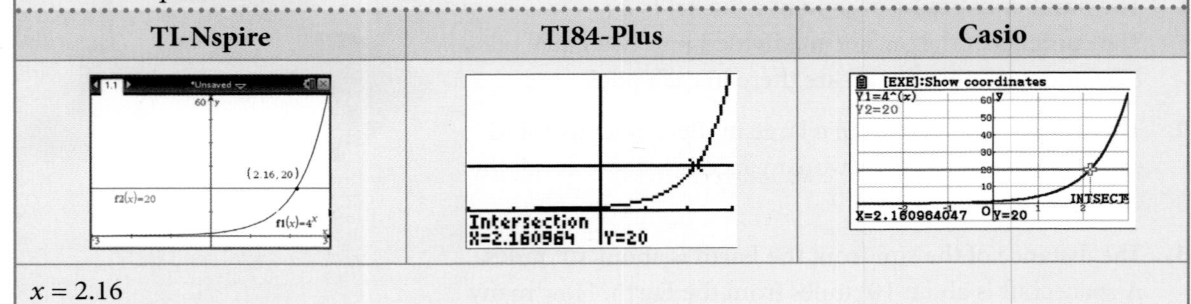

$x = 2.16$

Exercise 8.11

1. The graph shows the curve $y = 3^x$.
   Use the graph to give an estimate for the value of x which satisfies the equation $3^x = 15$.

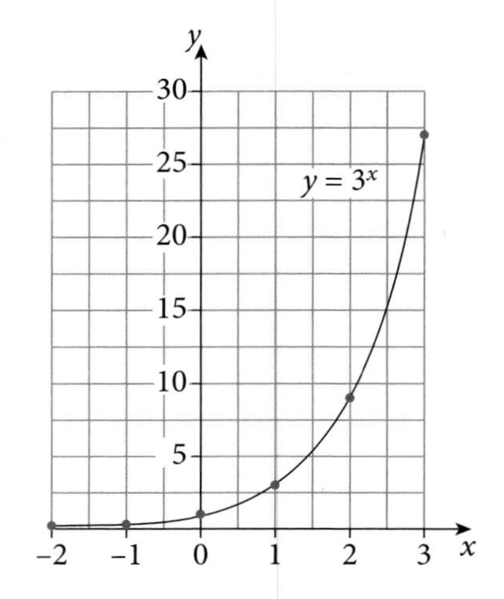

2. Draw the graph of $y = 2^x$ for values of x from −2 to + 4.
   Use a scale of 2 cm to 1 unit for x and 1 cm to 2 units for y.

3. Draw the graph of $y = \frac{x^2}{2^x}$ for values of x from 0 to 8.

Use a scale of 2 cm to 1 unit for x and 1 cm to 0·1 unit for y.

a) For what value of x is the value of y maximum?

b) Use your graph to give you two approximate solutions to the equation $\frac{x^2}{2^x} = 0{\cdot}8$, correct to 2 decimal places.

4. The cube root of 2 is written $\sqrt[3]{2}$.

The fourth root of 2 is written $\sqrt[4]{2}$.

The nth root of 2 is written $\sqrt[n]{2}$.

In index form:
$\sqrt[3]{2} = 2^{\frac{1}{3}}$
$\sqrt[4]{2} = 2^{\frac{1}{4}}$
$\sqrt[n]{2} = 2^{\frac{1}{n}}$

a) Use a calculator to work out $\sqrt[n]{2}$ for $n = 4, 5, 10, 100, 1000$.

Copy and complete the following sentence:
'As n becomes larger and larger, $\sqrt[n]{n}$ gets closer and closer to ______.'

b) Use a calculator to work out $\sqrt[n]{n}$ for $n = 4, 5, 10, 100, 1000$ and write a sentence similar to the one in part **a)**.

5. Solve the equations.

a) $2^x + 3^3 = 35$

b) $x^x = 2000$ (give your answer correct to 3 s.f.)

8.3 Rearrangement and evaluation of formulae

The operations involved in solving ordinary linear equations are exactly the same as the operations required in changing the subject of a formula.

Example 7

a) Solve the equation $3x + 1 = 12$.

b) Make x the subject of the formula $Mx + B = A$.

a) $3x + 1 = 12$

$3x = 12 - 1$

$x = \frac{12-1}{3} = \frac{11}{3}$

b) $Mx + B = A$

$Mx = A - B$

$x = \frac{A-B}{M}$

Example 8

a) Solve the equation $3(y - 2) = 5$.

b) Make y the subject of the formula $x(y - a) = e$.

a) $3(y - 2) = 5$

$3y - 6 = 5$

$3y = 11$

$y = \frac{11}{3}$

b) $x(y - a) = e$

$xy - xa = e$

$xy = e + xa$

$y = \frac{e+xa}{x}$

Exercise 8.12

Make x the subject of the following:

1. $2x = 5$
2. $7x = 21$
3. $Ax = B$
4. $Nx = T$
5. $Mx = K$
6. $xy = 4$
7. $Bx = C$
8. $4x = D$
9. $9x = T + N$
10. $Ax = B - R$
11. $Cx = R + T$
12. $Lx = N - R^2$
13. $R - S^2 = Nx$
14. $x + 5 = 7$
15. $x + 10 = 3$

Make y the subject of the following:

16. $L = y - B$
17. $N = y - T$
18. $3y + 1 = 7$
19. $2y - 4 = 5$
20. $Ay + C = N$
21. $By + D = L$
22. $Dy + E = F$
23. $Ny - F = H$
24. $Yy - Z = T$
25. $Ry - L = B$
26. $Vy + m = Q$
27. $ty - m = n + a$
28. $qy + n = s - t$
29. $ny - s^2 = t$
30. $V^2y + b = c$

Example 9

a) Solve the equation $\frac{3a+1}{2} = 4$.

b) Make a the subject of the formula $\frac{na+b}{m} = n$.

a) $\frac{3a+1}{2} = 4$

$3a + 1 = 8$

$3a = 7$

$a = \frac{7}{3}$

b) $\frac{na+b}{m} = n$

$na + b = mn$

$na = mn - b$

$a = \frac{mn-b}{n}$

Example 10

Make a the subject of the formula $x - na = y$

Make the 'a' term positive

$x = y + na$

$x - y = na$

$\frac{x-y}{n} = a \quad \therefore a = \frac{x-y}{n}$

Exercise 8.13

Make a the subject of the following formulae:

1. $\frac{a}{4} = 3$
2. $\frac{a}{5} = 2$
3. $\frac{a}{D} = B$
3. $\frac{a}{B} = T$
4. $\frac{a}{N} = R$
5. $b = \frac{a}{m}$
7. $\frac{a-2}{4} = 6$
8. $\frac{a-A}{B} = T$
9. $\frac{a-D}{N} = A$
10. $\frac{a+Q}{N} = B^2$
11. $g = \frac{a-r}{e}$
12. $\frac{2a+1}{5} = 2$
13. $\frac{Aa+B}{C} = D$
14. $\frac{na+m}{p} = q$
15. $\frac{ra-t}{S} = v$
16. $\frac{za-m}{q} = t$
17. $\frac{m+Aa}{b} = c$
18. $A = \frac{Ba+D}{E}$
19. $n = \frac{ea-f}{h}$
20. $q = \frac{ga+b}{r}$
21. $6 - a = 2$
22. $7 - a = 9$
23. $5 = 7 - a$
24. $A - a = B$
25. $r = v^2 - ra$
26. $t^2 = w - na$
27. $n - qa = 2$
28. $\frac{3-4a}{2} = 1$
29. $\frac{5-7a}{3} = 2$
30. $\frac{B-Aa}{D} = E$
31. $\frac{D-Ea}{N} = B$
32. $\frac{h-fa}{b} = x$
33. $\frac{v^2-ha}{C} = d$

Example 11

a) Solve the equation $\frac{4}{z} = 7$.

b) Make z the subject of the formula $\frac{n}{z} = k$.

a) $\frac{4}{z} = 7$

$4 = 7z$

$\frac{4}{7} = z$

b) $\frac{n}{z} = k$

$n = kz$

$\frac{n}{k} = z$

Example 12

Make t the subject of the formula $\frac{x}{t} + m = a$.

$\frac{x}{t} = a - m$

$x = (a - m)t$

$\frac{x}{(a-m)} = t \quad \therefore \frac{x}{(a-m)} = t$

Exercise 8.14

In questions **1** to **16**, make a the subject.

1. $\frac{7}{a} = 14$
2. $\frac{5}{a} = 3$
3. $\frac{B}{a} = C$
4. $\frac{T}{a} = X$
5. $\frac{M}{a} = B$
6. $m = \frac{n}{a}$
7. $t = \frac{v}{a}$
8. $\frac{n}{a} = \sin 20°$
9. $\frac{7}{a} = \cos 30°$
10. $\frac{B}{a} = x$
11. $\frac{5}{a} = \frac{3}{4}$
12. $\frac{N}{a} = \frac{B}{D}$
13. $\frac{H}{a} = \frac{N}{M}$
14. $\frac{t}{a} = \frac{b}{e}$
15. $\frac{v}{a} = \frac{m}{s}$
16. $\frac{t}{b} = \frac{m}{a}$

In questions **17** to **28**, make x the subject.

17. $\frac{2}{x} + 1 = 3$

18. $\frac{5}{x} - 2 = 4$

19. $\frac{A}{x} + B = C$

20. $\frac{V}{x} + G = H$

21. $\frac{r}{x} - t = n$

22. $q = \frac{b}{x} + d$

23. $t = \frac{m}{x} - n$

24. $h = d - \frac{b}{x}$

25. $C - \frac{d}{x} = e$

26. $r - \frac{m}{x} = e^2$

27. $t^2 = b - \frac{n}{x}$

28. $\frac{d}{x} + b = mn$

Example 13

Make x the subject of the following formulae.

a) $\sqrt{(x^2 + A)} = B$ **b)** $(Ax - B)^2 = M$ **c)** $\sqrt{(R - x)} = T$

a) $\sqrt{(x^2 + A)} = B$

$x^2 + A = B^2$ (square both sides)

$x^2 = B^2 - A$

$x = \pm\sqrt{(B^2 - A)}$

b) $(Ax - B)^2 = M$

$Ax - B = \pm\sqrt{M}$ (square root both sides)

$Ax = B \pm \sqrt{M}$

$x = \frac{B \pm \sqrt{M}}{A}$

c) $\sqrt{(R - x)} = T$

$R - x = T^2$

$R = T^2 + x$

$R - T^2 = x$

Exercise 8.15

In questions **1** to **28**, make x the subject.

1. $\sqrt{x} = 2$

2. $\sqrt{(x+1)} = 5$

3. $\sqrt{(x-2)} = 3$

4. $\sqrt{(x+a)} = B$

5. $\sqrt{(x+C)} = D$

6. $\sqrt{(x-E)} = H$

7. $\sqrt{(ax+b)} = c$

8. $\sqrt{(x-m)} = a$

9. $b = \sqrt{(gx-t)}$

10. $r = \sqrt{(b-x)}$

11. $\sqrt{(d-x)} = t$

12. $b = \sqrt{(x-d)}$

13. $c = \sqrt{(n-x)}$

14. $f = \sqrt{(b-x)}$

15. $g = \sqrt{(c-x)}$

16. $\sqrt{(M-Nx)} = P$

17. $\sqrt{(Ax+B)} = \sqrt{D}$

18. $\sqrt{(x-D)} = A^2$

19. $x^2 = g$

20. $x^2 + 1 = 17$

21. $x^2 = B$

22. $x^2 + A = B$

23. $x^2 - A = M$

24. $b = a + x^2$

25. $C - x^2 = m$

26. $n = d - x^2$

27. $mx^2 = n$

28. $b = ax^2$

In questions **29** to **50**, make k the subject.

29. $\frac{kz}{a} = t$

30. $ak^2 - t = m$

31. $n = a - k^2$

32. $\sqrt{(k^2 - 4)} = 6$

33. $\sqrt{(k^2 - A)} = B$

34. $\sqrt{(k^2 + y)} = x$

35. $t = \sqrt{(m + k^2)}$

36. $2\sqrt{(k + 1)} = 6$

37. $A\sqrt{(k + B)} = M$

38. $\sqrt{\left(\frac{M}{k}\right)} = N$

39. $\sqrt{\left(\frac{N}{k}\right)} = B$

40. $\sqrt{(a - k)} = b$

41. $\sqrt{(a^2 - k^2)} = t$

42. $\sqrt{(m - k^2)} = x$

43. $2\pi\sqrt{(k + t)} = 4$

44. $A\sqrt{(k + 1)} = B$

45. $\sqrt{(ak^2 - b)} = C$

46. $a\sqrt{(k^2 - x)} = b$

47. $k^2 + b = x^2$

48. $\frac{k^2}{a} + b = c$

49. $\sqrt{(c^2 - ak)} = b$

50. $\frac{m}{k^2} = a + b$

Example 14

Make x the subject of the following formulae.

a) $Ax - B = Cx + D$ **b)** $x + a = \frac{x + b}{c}$

a) $Ax - B = Cx + D$

$Ax - Cx = D + B$

$x(A - C) = D + B$ (factorise)

$x = \frac{D + B}{A - C}$

b) $x + a = \frac{x + b}{c}$

$c(x + a) = x + b$

$cx + ca = x + b$

$cx - x = b - ca$

$x(c - 1) = b - ca$ (factorise)

$x = \frac{b - ca}{c - 1}$

Exercise 8.16

Make y the subject of the following formulae.

1. $5(y - 1) = 2(y + 3)$

2. $7(y - 3) = 4(3 - y)$

3. $Ny + B = D - Ny$

4. $My - D = E - 2My$

5. $ay + b = 3b + by$

6. $my - c = e - ny$

7. $xy + 4 = 7 - ky$

8. $Ry + D = Ty + C$

9. $ay - x = z + by$

10. $m(y + a) = n(y + b)$

11. $x(y - b) = y + d$

12. $\frac{a - y}{a + y} = b$

13. $\frac{1 - y}{1 + y} = \frac{c}{d}$

14. $\frac{M - y}{M + y} = \frac{a}{b}$

15. $m(y + n) = n(n - y)$

16. $\frac{v - y}{v + y} = \frac{1}{2}$

17. $y(b - a) = a(y + b + c)$

18. $\sqrt{\left(\frac{y + x}{y - x}\right)} = 2$

19. $\sqrt{\left(\frac{z + y}{z - y}\right)} = \frac{1}{3}$

20. $\sqrt{\left[\frac{m(y + n)}{y}\right]} = p$

21. $n - y = \frac{4y - n}{m}$

Example 15

Make w the subject of the formula $\sqrt{\left(\frac{w}{w+a}\right)} = c$.

Squaring both sides, $\frac{w}{w+a} = c^2$

Multiplying by $(w + a)$, $w = c^2(w + a)$

$$w = c^2w + c^2a$$

$$w - c^2w = c^2a$$

$$w(1 - c^2) = c^2a$$

$$w = \frac{c^2}{1-c^2}$$

Exercise 8.17

Make the letter in square brackets the subject.

1. $ax + by + c = 0$ [x]

2. $\sqrt{a(y^2 - b)} = e$ [y]

3. $\frac{\sqrt{(k-m)}}{n} = \frac{1}{m}$ [k]

4. $a - bz = z + b$ [z]

5. $\frac{x+y}{x-y} = 2$ [x]

6. $\sqrt{\left(\frac{a}{z} - c\right)} = e$ [z]

7. $lm + mn + a = 0$ [n]

8. $t = 2\pi\sqrt{\left(\frac{d}{g}\right)}$ [d]

9. $t = 2\pi\sqrt{\left(\frac{d}{g}\right)}$ [g]

10. $\sqrt{(x^2 + a)} = 2x$ [x]

11. $\sqrt{\left\{\frac{b(m^2+a)}{e}\right\}} = t$ [m]

12. $\sqrt{\left(\frac{x+1}{x}\right)} = a$ [x]

13. $a + b - mx = 0$ [m]

14. $\sqrt{(a^2 + b^2)} = x^2$ [a]

15. $\frac{a}{k} + b = \frac{c}{k}$ [k]

16. $a - y = \frac{b+y}{a}$ [y]

17. $G = 4\pi\sqrt{(x^2 + T^2)}$ [x]

18. $M(ax + by + c) = 0$ [y]

Evaluation of a formula

Example 16

The formula for the volume of a cylinder, V, is $V = \pi r^2 h$, where r is the radius of the cylinder and h is the height.

Calculate the value of V when $r = 5$ and $h = 2$.

$V = \pi r^2 h$

Substitute $r = 5$ and $h = 2$:

$$V = \pi \times 5^2 \times 2$$

$$V = 50\pi$$

It is often convenient to leave π in the answer so that you can evaluate the answer to whatever degree of accuracy you require (e.g. 2 d.p.)

GDC help

Using a GDC

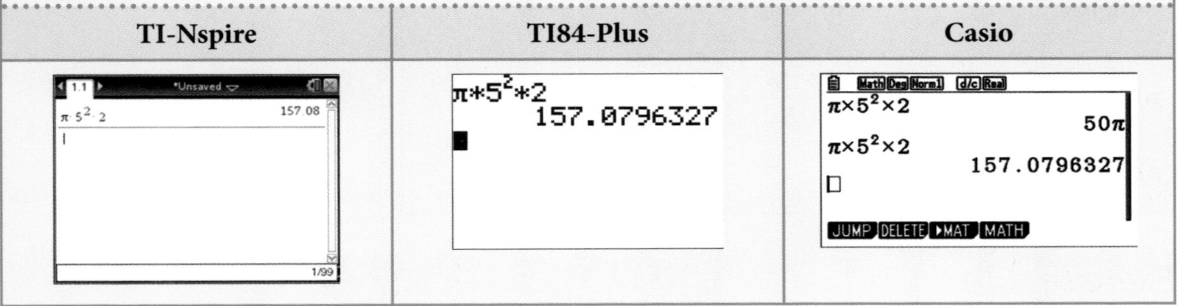

Exercise 8.18

1. Given that $s = t^2 + v$, find the value of s when $t = 7$ and $v = -2$.

2. Given that $q = \dfrac{n^2 - n}{6}$, find the value of q when $n = 9$.

3. When a stone is thrown vertically upwards with velocity u, its velocity v, after time t, is given by the formula $v = u + gt$.

   Calculate v when $g = -10$, $t = 20$ and $u = 640$.

4. Use the formula $x = \dfrac{-b \pm \sqrt{b^2 - 4ac}}{2a}$ to find x, when $a = 6$, $b = -7$ and $c = 2$.

5. The price £P, for one person staying in a hotel for t days is given by the formula $P = 215 + 75t$.
   a) Find the value of P when $t = 3$.
   b) The cost for one person of staying in the hotel was £815. For how many days did that person stay in the hotel?

6. A formula connecting x, y and P is $P = \frac{y \times 3^x}{(x^2 - 1)}$

   Find the value of P when $x = 5$ and $y = 1{\cdot}15$.

   (Give your answer correct to 1 d.p.)

8.4 Algebraic fractions

Simplifying fractions

Example 17

Simplify: **a)** $\frac{32}{56}$ **b)** $\frac{3a}{5a^2}$ **c)** $\frac{3y + y^2}{6y}$

a) $\frac{32}{56} = \frac{\cancel{8} \times 4}{\cancel{8} \times 7} = \frac{4}{7}$ **b)** $\frac{3a}{5a^2} = \frac{3 \times \cancel{a}}{5 \times a \times \cancel{a}} = \frac{3}{5a}$

c) $\frac{y(3 + y)}{6y} = \frac{3 + y}{6}$

Exercise 8.19

Simplify as far as possible, where you can.

1. $\frac{25}{35}$ **2.** $\frac{84}{96}$ **3.** $\frac{5y^2}{y}$ **4.** $\frac{y}{2y}$

5. $\frac{8x^2}{2x^2}$ **6.** $\frac{2x}{4y}$ **7.** $\frac{6y}{3y}$ **8.** $\frac{5ab}{10b}$

9. $\frac{8ab^2}{12ab}$ **10.** $\frac{7a^2b}{35ab^2}$ **11.** $\frac{(2a)^2}{4a}$ **12.** $\frac{7yx}{8xy}$

13. $\frac{5x + 2x^2}{3x}$ **14.** $\frac{9x + 3}{3x}$ **15.** $\frac{25 + 7}{25}$ **16.** $\frac{4a + 5a^2}{5a}$

17. $\frac{3x}{4x - x^2}$ **18.** $\frac{5ab}{15a + 10a^2}$ **19.** $\frac{5x + 4}{8x}$ **20.** $\frac{12x + 6}{6y}$

21. $\frac{5x + 10y}{15xy}$ **22.** $\frac{18a - 3ab}{6a^2}$ **23.** $\frac{4ab + 8a^2}{2ab}$ **24.** $\frac{(2x)^2 - 8x}{4x}$

Example 18

Simplify the fractions given.

a) $\dfrac{x^2+x-6}{x^2+2x-3}=\dfrac{(x-2)\cancel{(x+3)}}{\cancel{(x+3)}(x-1)}=\dfrac{x-2}{x-1}$

b) $\dfrac{x^2+3x-10}{x^2-4}=\dfrac{\cancel{(x-2)}(x+5)}{\cancel{(x-2)}(x+2)}=\dfrac{x+5}{x+2}$

c) $\dfrac{3x^2-9x}{x^2-4x+3}=\dfrac{3x\cancel{(x-3)}}{(x-1)\cancel{(x-3)}}=\dfrac{3x}{x-1}$

Exercise 8.20

Simplify as far as possible:

1. $\dfrac{x^2+2x}{x^2-3x}$ **2.** $\dfrac{x^2-3x}{x^2-2x-3}$ **3.** $\dfrac{x^2+4x}{2x^2-10x}$

4. $\dfrac{x^2+6x+5}{x^2-x-2}$ **5.** $\dfrac{x^2-4x-21}{x^2-5x-14}$ **6.** $\dfrac{x^2+7x+10}{x^2-4}$

7. $\dfrac{x^2+x-2}{x^2-x}$ **8.** $\dfrac{3x^2-6x}{x^2+3x-10}$ **9.** $\dfrac{6x^2-2x}{12x^2-4x}$

10. $\dfrac{3x^2+15x}{x^2-25}$ **11.** $\dfrac{12x^2-20x}{4x^2}$ **12.** $\dfrac{x^2+x-6}{x^2+2x-3}$

Addition and subtraction of algebraic fractions

Example 19

Write as a single fraction:

a) $\dfrac{2}{3}+\dfrac{3}{4}$ **b)** $\dfrac{2}{x}+\dfrac{3}{y}$

Compare these two workings line-for-line.

a) $\dfrac{2}{3}+\dfrac{3}{4}$; the LCM of 3 and 4 is 12.

$$\therefore \quad \frac{2}{3}+\frac{3}{4}=\frac{8}{12}+\frac{9}{12}$$

$$=\frac{17}{12}$$

b) $\dfrac{2}{x}+\dfrac{3}{y}$; the LCM of x and y is xy.

$$\therefore \quad \frac{2}{x}+\frac{3}{y}=\frac{2x}{xy}+\frac{3x}{xy}$$

$$=\frac{2x+3y}{xy}$$

Exercise 8.21

Simplify the following.

1. $\frac{2}{5} + \frac{1}{5}$ **2.** $\frac{2x}{5} + \frac{x}{5}$ **3.** $\frac{2}{x} + \frac{1}{x}$

4. $\frac{1}{7} + \frac{3}{7}$ **5.** $\frac{x}{7} + \frac{3x}{7}$ **6.** $\frac{1}{7x} + \frac{3}{7x}$

7. $\frac{5}{8} + \frac{1}{4}$ **8.** $\frac{5x}{8} + \frac{x}{4}$ **9.** $\frac{5}{8x} + \frac{1}{4x}$

10. $\frac{2}{3} + \frac{1}{6}$ **11.** $\frac{2x}{3} + \frac{x}{6}$ **12.** $\frac{2}{3x} + \frac{1}{6x}$

13. $\frac{3}{4} + \frac{2}{5}$ **14.** $\frac{3x}{4} + \frac{2x}{5}$ **15.** $\frac{3}{4x} + \frac{2}{5x}$

16. $\frac{3}{4} - \frac{2}{3}$ **17.** $\frac{3x}{4} - \frac{2x}{3}$ **18.** $\frac{3}{4x} - \frac{2}{3x}$

19. $\frac{x}{2} + \frac{x+1}{3}$ **20.** $\frac{x-1}{3} + \frac{x+2}{4}$ **21.** $\frac{2x-1}{5} + \frac{x+3}{2}$

22. $\frac{x+1}{3} - \frac{2x+1}{4}$ **23.** $\frac{x-3}{3} - \frac{x-2}{5}$ **24.** $\frac{2x+1}{7} - \frac{x+2}{2}$

25. $\frac{1}{x} + \frac{2}{x+1}$ **26.** $\frac{3}{x-2} + \frac{4}{x}$ **27.** $\frac{5}{x-2} + \frac{3}{x+3}$

28. $\frac{7}{x+1} - \frac{3}{x+2}$ **29.** $\frac{2}{x+3} - \frac{5}{x-1}$ **30.** $\frac{3}{x-2} + \frac{4}{x+1}$

Example 20

Rewrite the following as single fractions.

a) $\frac{x}{3} + \frac{x}{3}$ **b)** $\frac{5}{n} - \frac{3}{n}$

c) $\frac{3}{5} \times \frac{t}{4}$ **d)** $\frac{x}{5} \div \frac{x}{4}$

---

a) $\frac{x}{3} + \frac{x}{3} = \frac{2x}{3}$ **b)** $\frac{5}{n} - \frac{3}{n} = \frac{2}{n}$

c) $\frac{3}{5} \times \frac{t}{4} = \frac{3t}{20}$ **d)** $\frac{x}{5} \div \frac{x}{4} = \frac{x}{5} \times \frac{4}{x}$

$= \frac{4x}{5x} = \frac{4}{5}$

Exercise 8.22

Rewrite the following as single fractions.

1. $\frac{n}{5} + \frac{2n}{5}$ **2.** $\frac{3x}{5} - \frac{x}{5}$ **3.** $\frac{n}{2} + \frac{n}{4}$ **4.** $\frac{x}{5} - \frac{x}{10}$

5. $\frac{3}{4}x - \frac{1}{8}x$ **6.** $\frac{x}{3} - \frac{x}{4}$ **7.** $\frac{x}{5} + \frac{x}{4}$ **8.** $\frac{2}{5}x - \frac{1}{4}x$

9. $\frac{x}{4} \times \frac{x}{3}$

10. $\frac{2}{5}t \times \frac{1}{2}$

11. $\frac{1}{4}a \times \frac{1}{2}b$

12. $\left(\frac{3}{4}x\right)^2$

13. $\left(\frac{y}{3}\right)^2$

14. $\frac{4t}{5} \div \frac{1}{2}$

15. $\frac{5y}{3} \div \frac{y}{2}$

16. $2\frac{1}{2}x \times 1\frac{1}{2}x$

17. $\frac{3}{x} + \frac{5}{x}$

18. $\frac{9}{t} - \frac{3}{t}$

19. $\frac{11}{P} + \frac{10}{P}$

20. $\frac{4}{x} \times \frac{3}{x}$

Exercise 8.23

Write as a single fraction.

1. a) $\frac{3}{5} + \frac{1}{2}$ b) $\frac{3}{5}x + \frac{1}{2}x$ c) $\frac{3}{5} \times \frac{1}{2}$ d) $\frac{3}{5} \div \frac{1}{2}$
2. a) $\frac{1}{3} - \frac{1}{4}$ b) $\frac{m}{3} - \frac{m}{4}$ c) $\frac{m}{3} \times \frac{m}{4}$ d) $\frac{m}{3} \div \frac{m}{4}$
3. a) $\frac{3}{8} + \frac{1}{3}$ b) $\frac{3s}{8} + \frac{s}{3}$ c) $\frac{3s}{8} \times \frac{1}{3}$ d) $\frac{3s}{8} \div \frac{s}{8}$
4. a) $\frac{5}{7} - \frac{1}{2}$ b) $\frac{5t}{7} - \frac{t}{2}$ c) $\frac{5t}{7} \times \frac{t}{2}$ d) $\frac{5t}{7} \div \frac{t}{2}$
5. Here are some cards: $\frac{n}{2}$ $\frac{2}{n}$ $\left(\frac{n}{2}\right)^2$ $\frac{n}{2} + \frac{2}{n}$ $\frac{n}{2} - \frac{n}{4}$ $2 \div n$

   $n^2 \div 2$ $\frac{1}{2}n$ $\frac{n+2}{2n}$ $\frac{4}{n} - \frac{2}{n}$ $\frac{n}{2} \times \frac{n}{2}$

   a) Which cards will always be the same as $n \div 2$?

   b) Which cards will always be the same as $\frac{n^2}{4}$?

   c) Which cards will always be the same as $\frac{n}{4}$?

   d) Can you draw a new card which will always be the same as $\frac{n^2}{2} \div \frac{n}{4}$?

The next twelve questions are more difficult. Rewrite as a single fraction and simplify the final answer as far as possible.

6. $\frac{3}{x} + \frac{2}{y}$

7. $\frac{x}{t} \times \frac{3}{x}$

8. $\frac{4}{p} - \frac{5}{q}$

9. $\frac{\pi r^2}{h} \div \frac{\pi}{h}$

10. $\frac{p}{x} + \frac{q}{x}$

11. $\frac{5}{x} + \frac{1}{2x}$

12. $\frac{m}{n} \times \frac{n^2}{m^2}$

13. $\frac{3pq}{x} \div \frac{p^2}{x}$

14. $\frac{x}{a^2} \div \frac{ax}{y}$

15. $\frac{x}{2\frac{1}{2}} + \frac{x}{1\frac{1}{2}}$

16. $\frac{1}{2}$ of $a - \frac{1}{5}$ of b

17. $\frac{3}{4}$ of $t + \frac{1}{3}$ of z

18. a) Work out $\left(1+\frac{1}{2}\right)\left(1+\frac{1}{3}\right)\left(1+\frac{1}{4}\right)\ldots\left(1+\frac{1}{100}\right)$

    b) Simplify $\left(1+\frac{1}{2}\right)\left(1+\frac{1}{3}\right)\left(1+\frac{1}{4}\right)\ldots\left(1+\frac{1}{n}\right)$

Equations with fractions

Example 21

Solve each of the following.

a) $\frac{2x}{3} = 5$ **b)** $\frac{4}{x} = -2$ **c)** $\frac{x}{2} + 3 = 7$ **d)** $\frac{4}{x} - 1 = 14$

---

a) $\frac{2x}{3} = 5$

$2x = 15$ [Multiply by 3]

$x = \frac{15}{2}$ [Divide by 2]

$x = 7\frac{1}{2}$

b) $\frac{4}{x} = -2$

$4 = -2x$ [Multiply by x]

$\frac{4}{-2} = x$ [Divide by -2]

$-2 = x$

c) $\frac{x}{2} + 3 = 7$

$\frac{x}{2} = 4$ [Substract 3 from both sides.]

$\cancel{2} \times \frac{x}{\cancel{2}} = 4 \times 2$ [Multiply both sides by 2.]

$x = 8$

d) $\frac{4}{x} - 1 = 14$

$\frac{4}{x} = 15$ [Add 1 to both sides.]

$\cancel{x}\,\frac{4}{\cancel{x}} = 15x$ [Multiply both sides by x.]

$4 = 15x$

$\frac{4}{15} = x$

Exercise 8.24

Solve the following equations.

1. $\frac{x}{3} = 4$ **2.** $\frac{x}{5} = 2$ **3.** $5 = \frac{x}{4}$ **4.** $\frac{x}{7} = -2$

5. $\frac{x}{5} = -5$ **6.** $\frac{2x}{3} = 1$ **7.** $\frac{3x}{4} = 2$ **8.** $\frac{5x}{2} = 2$

9. $\frac{6}{x} = 7$ **10.** $\frac{4}{x} = 9$ **11.** $\frac{2}{x} = 1$ **12.** $\frac{3}{x} = \frac{1}{4}$

13. $3 = \frac{8}{x}$ **14.** $\frac{2}{3} = \frac{10}{x}$ **15.** $\frac{8}{x} = -11$ **16.** $-2 = \frac{100}{x}$

Exercise 8.25

Solve the following equations.

1. $\frac{x}{3} + 1 = 5$ **2.** $\frac{x}{2} - 1 = 8$ **3.** $\frac{x}{5} + 9 = 8$ **4.** $6 + \frac{x}{3} = 10$

5. $\frac{1}{2}x + 9 = 20$ **6.** $\frac{1}{3}x - 6 = 11$ **7.** $\frac{2}{3}x + 8 = 10$ **8.** $\frac{4}{5}x - 1 = 0$

Questions **9** to **17** are more difficult.

9. $\dfrac{3}{2x+1} = 2$

10. $\dfrac{3}{x} = \dfrac{2}{x-1}$

11. $\dfrac{9}{x+2} = \dfrac{7}{2x+1}$

12. $\dfrac{x-3}{5} = \dfrac{2x+1}{3}$

13. $\dfrac{1-x}{2} = \dfrac{1}{4}$

14. $\dfrac{3x}{5} = 2(x+1)$

15. $\dfrac{3}{x+1} + 3 = 0$

16. $\dfrac{8}{1-x} = 16$

17. $\dfrac{5(x-3)}{2} = \dfrac{2(x-1)}{5}$

8.5 Difference method, *n*th term of a sequence

In this section we will find the general term of linear, quadratic and cubic sequences using the difference method.

Linear sequence

A linear sequence has the general term $u_n = an + b$, where a and b are numbers. u_n is also called the nth term.

Here is the difference table for the sequence $u_n = 7n - 1$

n	1	2	3	4	5
u_n	6	13	20	27	34
First difference	7	7	7	7	

We see that the first difference is 7; this is the value of a in $u_n = an + b$

For the sequence 3, 8, 13, 18, . . . the first difference is 5.
We draw a table with a column for 5 times the term number (i.e. $5n$).

n	$5n$	term
1	5	3
2	10	8
3	15	13
4	20	18

We see that each term is 2 less than $5n$.

So, the 10th term is $(5 \times 10) - 2 = 48$

the 20th term is $(5 \times 20) - 2 = 98$

⋮

the nth term is $5 \times n - 2 = 5n - 2$

Exercise 8.26

1. Look at the sequence 5, 8, 11, 14, . . .
The difference between terms is 3.
Copy the table, which has a column for $3n$.
Copy and complete the following sentence:
'The nth term of the sequence is $3n + \square$.'

n	$3n$	Term
1	3	5
2	6	8
3	9	11
4	12	14

2. Look at the sequence and the table underneath. Find the nth term in each case.

a) Sequence 5, 9, 13, 17, . . .

n	$4n$	Term
1	4	5
2	8	9
3	12	13
4	16	17

nth term = ☐

b) Sequence 2, 8, 14, 20, . . .

n	$6n$	Term
1	6	2
2	12	8
3	18	14
4	24	20

nth term = ☐

3. In the sequence 6, 11, 16, 21, . . . the difference between terms is 5. Copy and complete the table and write an expression for the nth term of the sequence.

n	☐	Term
1	☐	6
2	☐	11
3	☐	16
4	☐	21

4. Look at the sequence 6, 10, 14, 18, . . .
Write down the difference between terms.
Make a table like the one in question **3** and use it to find and expression for the nth term.

5. Write down each sequence in a table and then find the nth term.
a) 5, 7, 9, 11, . . . **b)** 3, 7, 11, 15, . . . **c)** 2, 8, 14, 20, . . .

6. Make a table for each sequence and write the nth term.
a) 2, 10, 18, 26, . . . **b)** 7, 10, 13, 16, . . . **c)** 21, 30, 39, 48, . . .

7. Here is a sequence of triangles made from a number of toothpicks t.

$n = 1$, $t = 3$ $n = 2$, $t = 5$ $n = 3$, $t = 7$

n	t
1	3
2	5
3	7
4	

Draw the next diagram in the sequence and write the values for n and t in a table.
How many toothpicks are in the nth term?

8. Crosses are drawn on squared 'dotty' paper. The diagram number of the cross is recorded together with the total number of dots d on each cross.

n	d
1	5
2	9
3	13

$n = 1$ $n = 2$ $n = 3$

Find a formula connecting n and d.
Write it as '$d = \dots$'

9. Look at the tables below. In each case, find a formula connecting the two letters.

a)

n	h
2	10
3	13
4	16
5	19

write '$h = \ldots$'

b)

n	p
3	12
4	17
5	22
6	27

write '$p = \ldots$'

c)

n	s
2	4
3	$4\frac{1}{2}$
4	5
5	$5\frac{1}{2}$

write '$s = \ldots$'

Exercise 8.27

1. In each diagram below, a number of white squares w surrounds a rectangle of green squares. The length of each rectangle is one unit more than the height h.

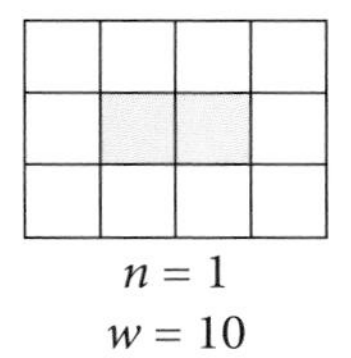

$n = 1$
$w = 10$

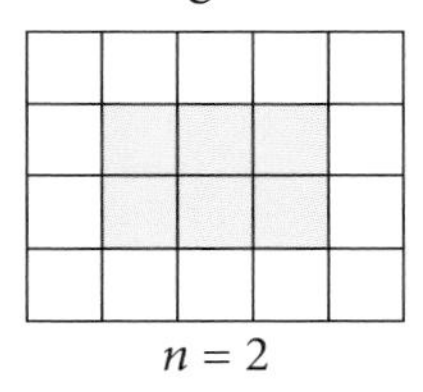

$n = 2$

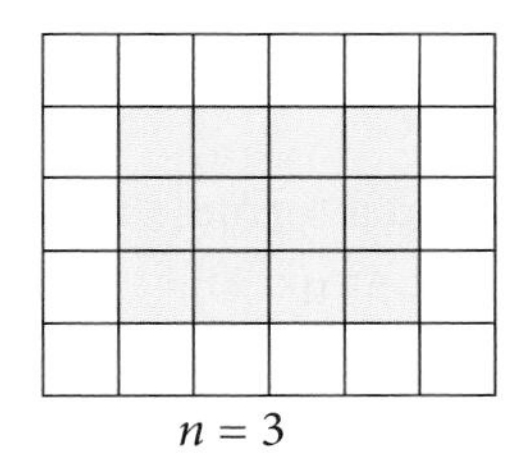

$n = 3$

n	w
1	10
2	
3	

Make a table of values of n and w and use it to find a formula connecting n and w.
Write it as '$w = \ldots$'

2. In a sequence of diagrams similar to the one in question **1**, white squares surround a rectangle but this time the length of the red rectangle is twice the height. The diagram with $n = 2$ is shown.
Draw the sequence of diagrams and make a table of values of n and w.
Write the formula connecting n and w in the form '$w = \ldots$'

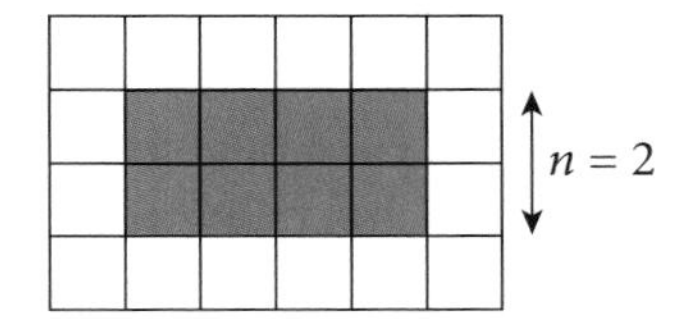

3. Open boxes are drawn on squared 'dotty' paper.

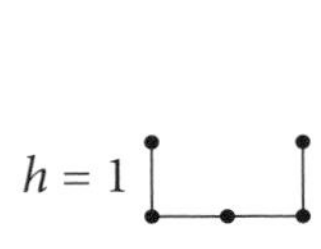

$h = 2$

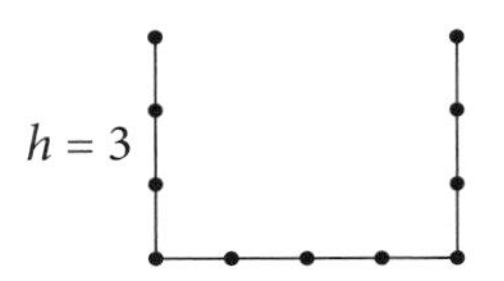

Count the number of dots d in each diagram and find a formula for d in terms of h (the height of the box).

4. Look at the tables below and in each case find a formula for z in terms of n.
Write the formula as '$z = \ldots$'
Notice that in part **a)** the values of n are not consecutive.

a)

n	z
2	7
3	13
5	25
6	31

b)

n	z
0	15
1	12
2	9
3	6

5. In these diagrams 4 triangles are joined either at a vertex or along a whole side. The number of common edges (shown dotted) is given by c and the perimeter of the shape is p.

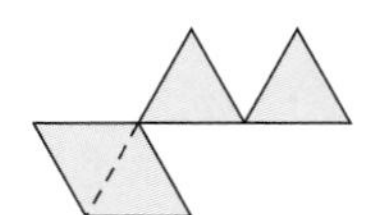
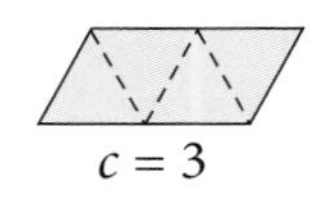

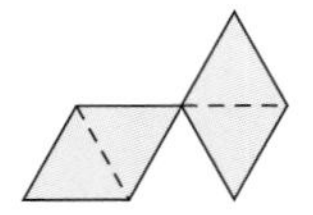

$c = 1, p = 10$

Make a table of values of p and c and find a formula for p in terms of c.

6. In each diagram below a 'V' is formed by shading squares inside a rectangle of height h and width w.

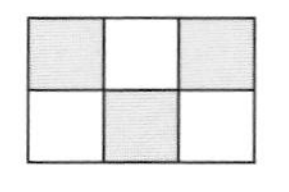
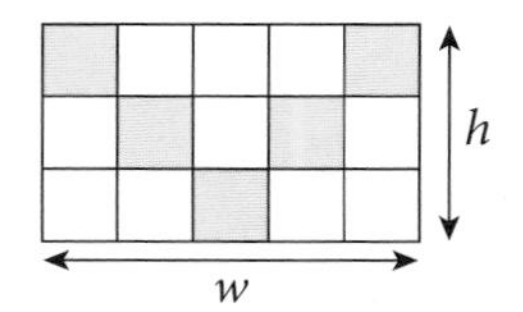

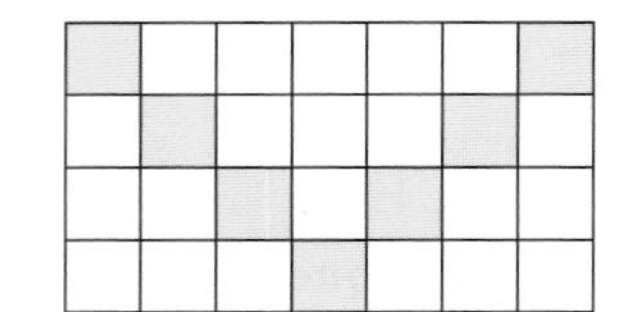

a) Record the number of shaded squares s and the height h of each rectangle. Find a formula for s in terms of h.

b) Find the width w for each value of h and hence find the number of *unshaded* squares in a rectangle with $h = 10$.

7. A chain of pentagons can be made from toothpicks as shown.

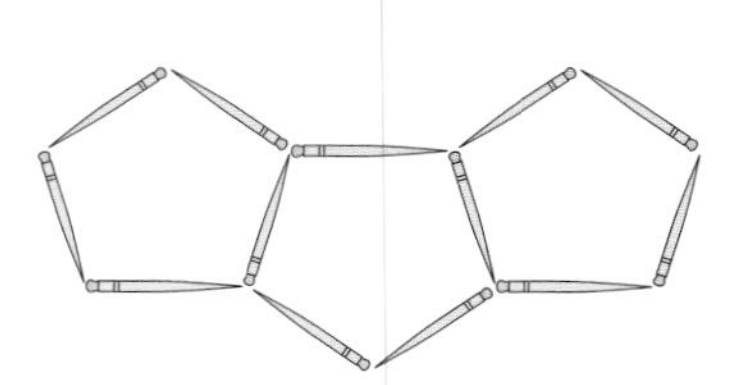

a) How many toothpicks are needed to make a chain of 5 pentagons?

b) How many toothpicks are needed to make a chain of n pentagons?

8. The nth term of a sequence is $\frac{3}{n^2 + 1}$.
The first term is $\frac{3}{2}$

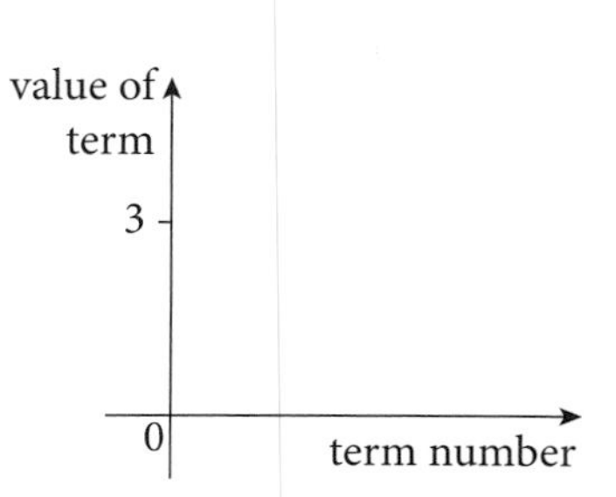

a) Write down the next three terms.

b) The sequences goes on and on for ever.
Draw a sketch graph to show how the sequence continues.

9. The lace edge of a tablecloth consists of a number of hexagons joined together as shown.

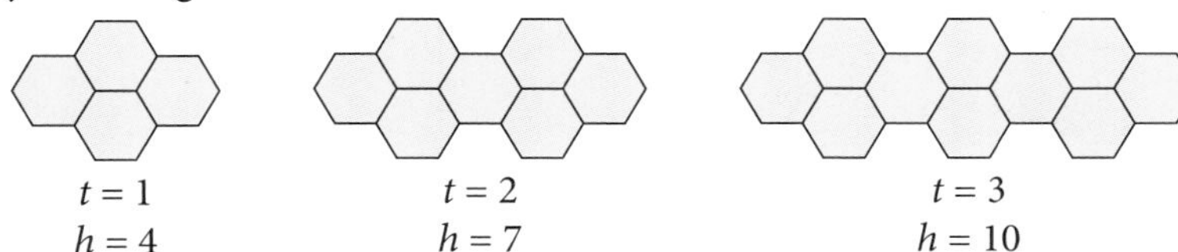

a) How many hexagons will be needed when $t = 7$?

b) How many hexagons will be needed when $t = 1000$?

c) How many hexagons will be needed for the nth pattern?

Quadratic sequences

Consider the quadratic sequence with general term $u_n = an^2 + bn + c$

n	1		2		3		4		5		6
u_n	$a + b + c$		$4a + 2b + c$		$9a + 3b + c$		$16a + 4b + c$		$25a + 5b + c$		$36a + 6b + c$
first difference		$3a + b$		$5a + b$		$7a + b$		$9a + b$		$11a + b$	
second difference			$2a$		$2a$		$2a$		$2a$		

We see that the second difference is $2a$

Example 22

The first and second differences are given for a sequence below. Find the nth term.

Sequence	11		21		33		47		63
First difference		10		12		14		16	
Second difference			2		2		2		

For *any* quadratic sequence, the nth term $= an^2 + bn + c$

The second difference above is 2, so we know that $a = 1$

So the nth term is now $n^2 + bn + c$

When $n = 1$, $1 + b + c = 11$

When $n = 2$, $4 + 2b + c = 21$

Solving the simultaneous equations, we obtain $b = 7$, $c = 3$

The nth term of the sequence is $n^2 + 7n + 3$

Exercise 8.28

1. Find an expression for the nth term of each sequence.

a) 4, 7, 10, 13, 16, ...

b) 9, 14, 19, 24, 29, ...

c) 8, 19, 30, 41, ...

In questions **2** to **9** use the difference method to find the nth term of each sequence.

2. 2, 7, 14, 23, 34 ...

3. 5, 11, 19, 29, 41 ...

4. 2, 3, 6, 11, 18 ...

5. 4, 11, 22, 37 ...

6. 7, 16, 31, 52 ...

7. 2, 10, 20, 32, 46 ...

8. 1, 3, 6, 10, 15...

9. 4, 5, 8, 13, 20...

10. In these diagrams 'steps' are made from sticks.

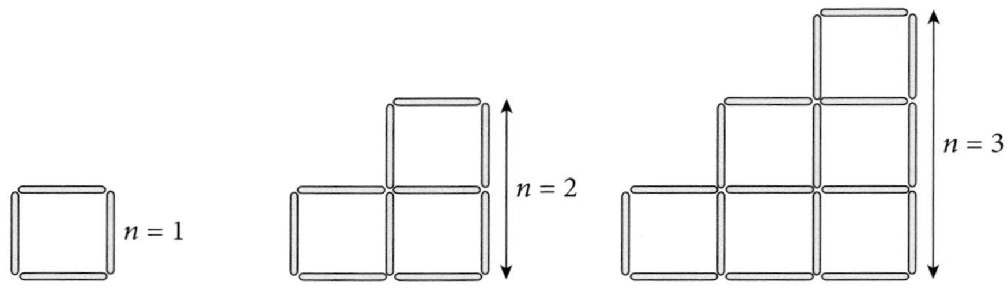

a) Draw the next diagram in the sequence.

b) Count the number of sticks s for each value of n, the height of the steps.

c) Find the number of sticks in the nth term of the sequence.

Cubic sequences

Consider the cubic sequence with nth term $u_n = an^3 + bn^2 + cn + d$.

Here is the difference table:

n	1	2	3	4	5
u_n	$a + b + c + d$	$8a + 4b + 2c + d$	$27a + 9b + 3c + d$	$64a + 16b + 4c + d$	$125a + 25b + 5c + d$
First difference	$7a + 3b + c$	$19a + 5b + c$	$37a + 7b + c$	$61a + 9b + c$	
Second difference	$12a + 2b$	$18a + 2b$	$24a + 2b$		
Third difference	$6a$	$6a$			

We see that the third difference for the cubic sequence above is $6a$.

Example 23

Find an expression for the nth term of the sequence 4, 16, 44, 94, 172, 284,…

Sequence	4		16		44		94		172		284
First difference		12		28		50		78		112	
Second difference			16		22		28		34		
Third difference				6		6		6			

We observe that the third difference is 6 so the sequence is a cubic with $a = 1$

The nth term is $n^3 + bn^2 + cn + d$

when $n = 1$, $1 + b + c + d = 4$ [A]

$n = 2$, $8 + 4b + 2c + d = 16$ [B]

$n = 3$, $27 + 9b + 3c + d = 44$ [C]

[B]–[A] $7 + 3b + c = 12$ [D]

[C]–[B] $19 + 5b + c = 28$ [E]

[E]–[D] $12 + 2b = 16$

So $b = 2$

From [D] $7 + (3 \times 2) + c = 12$

$c = -1$

From [A] $1 + 2 - 1 + d = 4$

$d = 2$

Finally the nth term of the sequence is $n^3 + 2n^2 - n + 2$

Exercise 8.29

Find an expression for the nth term of each sequence.

1. 3, 9, 25, 57, 111
2. 3, 11, 21, 33, 47
3. −5, −3, 11, 43, 99, 185
4. 8, 9, 12, 17, 24
5. −10, −3, 4, 11, 18
6. 4, 28, 88, 202, 388
7. 3, 11, 25, 45, 71
8. 2, 7, 18, 38, 70, 117
9. −5, −7, −5, 7, 35
10. Find a given that $(x - 2)$ is a factor of $x^3 + ax^2 + 3x + 2$
11. Find a given that $(x + 3)$ is a factor of $ax^3 - x^2 + 4x - 6$
12. Find a given that $(x + 4)$ is a factor of $f(x) = x^3 + 8x^2 + ax + 12$
    Hence factorise $f(x)$ completely.

Exercise 8.30

In questions **1** to **8** find an expression for the nth term of the sequence.

1. 5, 15, 37, 77, 141
2. 3, 11, 21, 33, 47
3. −2, 0, 14, 46, 102, 188
4. 1, 18, 61, 142, 273, 466
5. 4, 12, 32, 70, 132
6. 7, 25, 77, 181, 355

7. Look at this sequence of diagram made from sticks.

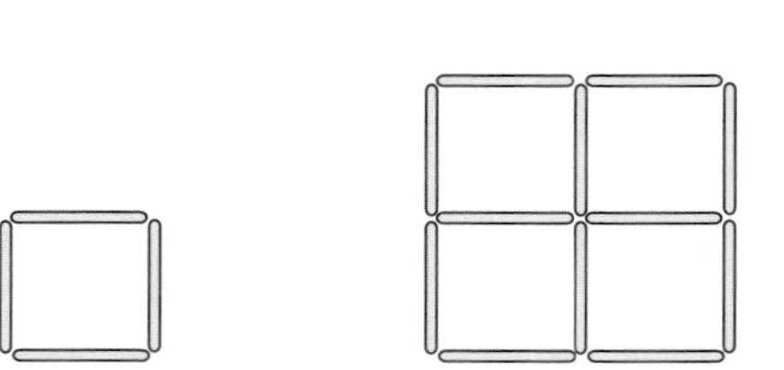
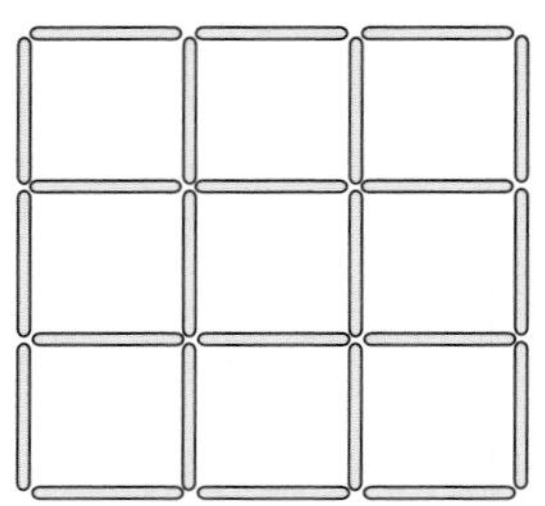

a) Find the number of sticks in the first 4 diagrams in the sequence.

b) Use your answers to find the number of sticks in the nth diagram in the sequence.

c) How many sticks are in the 8th diagram?

8. a) Find the value of a, given that

$(n - 2)(n^2 - 2n - 1) = n^3 + an^2 + 3n + 2$

b) Using the value of a obtained above, find the first four terms of the sequence with general term $n^3 + an^2 + 3n + 2$

9. The pyramid shown has

1 ball on layer 1

4 balls on layer 2

9 balls on layer 3, and so on.

The total number of balls in the first n layers is given by the formula

$$1^2 + 2^2 + 3^2 + \ldots + n^2 = \frac{n}{6}(n + 1)(2n + 1)$$

a) Show that this formula works with $n = 3$.

b) Use the formula to find the total number of balls in a pyramid with 20 layers.

10. The first five even numbers are 2, 4, 6, 8, 10.

a) Write down, in terms of n, an expression for the nth even number.

b) Write an expression for the next even number after the nth.

11. In the Fibonacci sequence below, each term is found by adding the two previous terms.

3, 4, 7, 11, 18, 29, …

Consider the Fibonacci sequence which starts a, b, $a + b$, …

a) Show that the 7th term of this sequence is $5a + 8b$.

b) Use algebra to show that the sum of the first six terms is four times the fifth term.

8.6 Geometric series

In a **geometric series** each term is obtained by multiplying the previous term by a constant r.

The number r is called the **common ratio**.

Geometric series are sometimes called **geometric progressions** ('GP' for short).

Here are three examples of geometric series:

1. $2 + 6 + 18 + 54 + \dots \quad (r = 3)$

2. $27 + 9 + 3 + 1 + \dots \quad \left(r = \frac{1}{3}\right)$

3. $4 - 8 + 16 - 32 + \dots \quad (r = -2)$

Consider the geometric series $3 + 6 + 12 + 24 + \dots$.

The terms may be written $3 + (3 \times 2) + (3 \times 2^2) + (3 \times 2^3) + \dots$

We see that the nth term is $3 \times 2^{n-1}$

A general series, with first term a and common ratio r, is

$$a + ar + ar^2 + ar^3 + \dots$$

The nth term is ar^{n-1}

The sum of a geometric series

The sum of the first n terms is

$$S = a + ar + ar^2 + \dots + ar^{n-1}$$

Multiplying by r, $rS = ar + ar^2 + \dots + ar^{n-1} + ar^n$

Subtracting, $S - rS = a - ar^n$

Factorising, $S(1 - r) = a(1 - r^n)$

Dividing by $(1 - r)$, $S = \frac{a(1-r^n)}{1-r}$

This result may also be written as $S = \frac{a(r^n-1)}{r-1}$

The sum of a geometric series is not on the Cambridge IGCSE International Mathematics syllabus. It is included here for you to try as extension material if you wish.

Example 24

Find an expression for the nth term of the geometric series

$3 + 12 + 48 + \ldots$

The first term, a is 3 and the common ratio, r, is 4.

The nth term is $3 \times 4^{n-1}$

Example 25

Find the 25th term of the geometric series

$5 + 10 + 20 + 40 + \ldots$

The first term is 5 and the common ratio is 2

The 25th term is $5 \times 2^{25-1}$

i.e. 5×2^{24}

GDC help

Using a GDC

TI-Nspire	TI84-Plus	Casio
$5 \cdot 2^{24}$ 83886080	$5*2^{24}$ 83886080	5×2^{24} 83886080

Example 26

Find the sum of the first 10 terms of the geometric series

$5 + 15 + 45 + 135 + \ldots$

The first term, a, is 5 and the common ratio, r, is 3

Using the formula $S = \dfrac{a(1-r^n)}{(1-r)}$,

we have $S = \dfrac{5(1-3^{10})}{(1-3)}$

$= 147\,620$

GDC help

Using a GDC

▶ Continued on next page

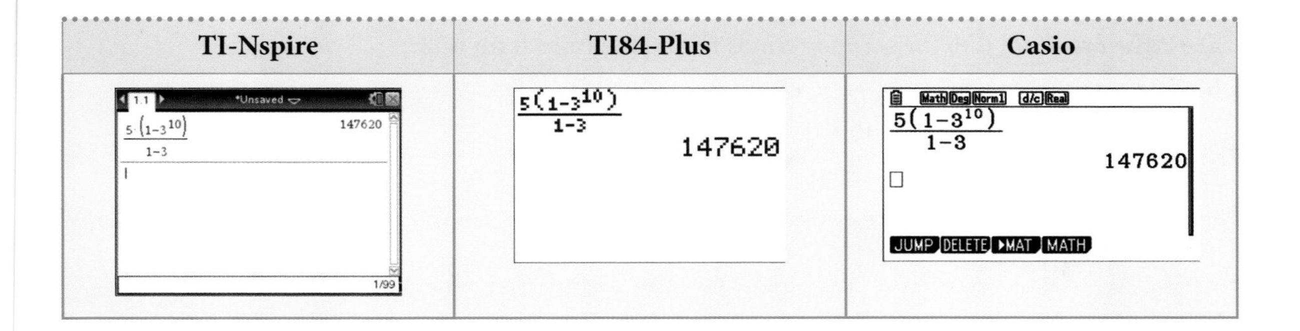

Example 27

Find the sum of the first 10 terms of the geometric series

$40 + 20 + 10 + 5 + \ldots$

The first term, a, is 40 and the common ratio, r, is $\frac{1}{2}$

Using the formula $S = \dfrac{a(1-r^n)}{(1-r)}$,

we have $S = \dfrac{40\left[1-\left(\frac{1}{2}\right)^{10}\right]}{1-\frac{1}{2}}$

$= 79{\cdot}9$ (to 3 s.f.)

GDC help

Using a GDC

TI-Nspire	TI84-Plus	Casio
$\dfrac{40\left(1-\left(\frac{1}{2}\right)^{10}\right)}{1-\frac{1}{2}}$ 79.9219	$\dfrac{40\left(1-\left(\frac{1}{2}\right)^{10}\right)}{1-\frac{1}{2}}$ 79.921875	$\dfrac{40\left(1-\left(\frac{1}{2}\right)^{10}\right)}{1-\frac{1}{2}}$ 79.921875

Example 28

The first term of a geometric series is 8 and its common ratio is r.

a) Write down the second and the third terms of the series, in terms of r.

b) Given that the sum of the first three terms is 38, find the two possible values of r.

▶ Continued on next page

a) If the first term is 8, the second term is $8r$ and the third term is $8r^2$.

b) We have $8 + 8r + 8r^2 = 38$

Rearrange: $8r^2 + 8r - 30 = 0$

Divide by 2: $4r^2 + 45 - 15 = 0$

Factorise: $(2r - 3)(2r + 5) = 0$

$r = \frac{3}{2}$ or $r = -\frac{5}{2}$

Exercise 8.31

1. Find an expression for the nth term of the following geometric sequences.
   a) 3, 6, 12, 24, ...
   b) 2, 10, 50, 250, ...
   c) 20, 10, 5, 2.5, ...
   d) 2000, 400, 80, 16, ...
   e) $\frac{1}{8}, \frac{1}{4}, \frac{1}{2}, 1, \ldots$
   f) 4, 12, 36, 108, ...
   g) 5, −15, 45, −135, ...
2. Find the following:
   a) the sum of the first ten terms of the geometric series $1 + 2 + 4 + 8 \ldots$
   b) the sum of the first seven terms of the geometric series $2 + 6 + 18 + 54 \ldots$
   c) the sum of the first eight terms of the geometric series $2 + 10 + 50 + 250 \ldots$
   d) the sum of the first five terms of the geometric series $3 - 12 + 48 - 192 \ldots$
3. Find the following (to 1 d.p.):
   a) the sum of the first ten terms of the geometric series $40 + 20 + 10 + 5 \ldots$
   b) the sum of the first twenty terms of the geometric series $40 + 20 + 10 + 5 \ldots$
   c) the sum of the first fifteen terms of the geometric series $405 + 135 + 45 + 15 \ldots$
   d) the sum of the first thirty terms of the geometric series $1280 + 320 + 80 + 20 \ldots$
   e) the sum of the first twelve terms of the geometric series $100 - 50 + 25 - 12{\cdot}5 \ldots$
4. The 5th term of a geometric series is 567 and the 6th term is 1701. Find the first term and the common ratio.
5. The 7th term of a geometric series is 70 and the 9th term is 280. Find the first term and the common ratio (given that it is positive).
6. A geometric series is such that the first term is 7 and its common ratio is r.
   a) Write down the second term of the series in terms of r.
   b) Given that the sum of the first 2 terms is 42, find r.

7. A geometric series is such that the first term is 5 and its common ratio is r.

   a) Write down the second and third terms of the sequence in terms of r.

   b) Given that the sum of the first 3 terms is 65, find the two possible values of r.

8. A man invested €1000 in a bank on 1st January 2012. It offered him
   5% interest per annum.

   a) By what factor does the amount in the account increase each year?

   b) How much would the money be worth on 1st January 2017?

9. a) Use the method from question **7** to find all the possible values of the common ratio of a geometric series in which the first term is 9 and the sum of its first three terms is 189.

   b) Find the first three terms of each series.

10. A geometric series is such that the sum of the first two terms is 28 and the sum of its fourth and fifth terms is 756.

    a) Write down two equations involving a and r.

    b) Factorise both equations.

    c) Divide one equation by the other to solve these to find the common ratio and the first term.

11. A geometric series has first term 3 and common ratio 2. If the sum of the first n terms of the sequence is 98 301.

    a) Show the value of 2^n is 32 768.

    b) Find the value of n, by trial and improvement.

12. A geometric series has first term 2 and common ratio -3. The sum of the first n terms of the series is 88 574.

    a) Find the value of $(-3)^n$.

    b) Explain why n must be odd.

    c) Find the value of n, by trial and improvement.

13. A geometric series has common ratio 3. If the sum of the first five terms is 605, find the first term.

14. a) Find an expression, in terms of n, for the sum of the first n terms of the geometric series $5 + 15 + 45 + \dots$

b) Find the sum of the first 12 terms of the series.

c) Find, by trial and improvement, the smallest number of terms whose total is more than 10^8.

15. A geometric series has common ratio 5. If the sum of the first ten terms is 17 089 842, find the first term.

16. Show that there are two possible geometric series in each of which the first term is 8 and the sum of the first three terms is 14. Find the second term in each series.

8.7 Variation, direct and inverse

Direct variation

There are several ways of expressing a relationship between two quantities x and y. Here are some examples:

- x varies as y
- x varies directly as y
- x is proportional to y

These three all mean the same and they are written in symbols like this:

$$x \propto y$$

The '$\propto$' sign can always be replaced by '$= k$', where k is a constant:

$$x = ky$$

Suppose $x = 3$ when $y = 12$.

Then $3 = k \times 12$

and $k = \frac{1}{4}$

We can then write $x = \frac{1}{4}y$, and this allows us to find the value of x for any value of y and vice versa.

Example 29

y varies as z, and $y = 2$ when $z = 5$.
Find

a) the value of y when $z = 6$

b) the value of z when $y = 5$

▶ Continued on next page

Because $y \propto z$, then $y = kz$, where k is a constant.

$y = 2$ when $z = 5$

$2 = k \times 5$

$k = \frac{2}{5}$

So $y = \frac{2}{5}z$

a) When $z = 6$, $y = \frac{2}{5} \times 6 = 2\frac{2}{5}$

b) When $y = 5$, $5 = \frac{2}{5}z$

$z = \frac{25}{2} = 12\frac{1}{2}$

Example 30

The value V of a diamond is proportional to the square of its mass M.

If a diamond with a mass of 10 grams is worth \$200, find

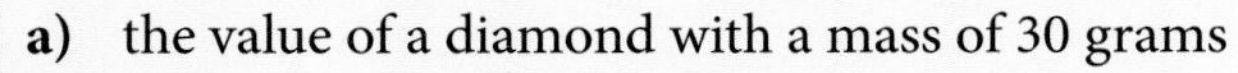

a) the value of a diamond with a mass of 30 grams

b) the mass of a diamond worth \$5000.

$V \propto M^2$ or $V = kM^2$ where k is a constant.

$V = 200$ when $M = 10$

$\therefore \quad 200 = k \times 10^2$

$k = 2$

So $\quad V = 2M^2$

a) When $M = 30$,

$V = 2 \times 30^2 = 2 \times 900$

$V = \$1800$

So a diamond with a mass of 30 grams is worth \$1800.

b) When $V = 5000$,

$5000 = 2 \times M^2$

$M^2 = \frac{5000}{2} = 2500$

$M = \sqrt{2500} = 50$

So a diamond of value \$5000 has a mass of 50 grams.

Exercise 8.32

1. Rewrite the statement connecting each pair of variables using a constant k instead of $\propto$

   a) $S \propto e$ b) $v \propto t$ c) $x \propto z^2$ d) $y \propto \sqrt{x}$

   e) $T \propto \sqrt{L}$ f) $C \propto r$ g) $A \propto r^2$ h) $V \propto r^2$

2. y varies as t. If $y = 6$ when $t = 4$, calculate:

   a) the value of y, when $t = 6$

   b) the value of t, when $y = 4$.

3. z is proportional to m. If $z = 20$, when $m = 4$, calculate:

   a) the value of z, when $m = 7$

   b) the value of m, when $z = 55$.

4. A varies directly as r^2. If $A = 12$, when $r = 2$, calculate:

   a) the value of A, when $r = 5$

   b) the value of r, when $A = 48$.

5. Given that $z \propto x$, copy and complete the table.

x	1	3		$5\frac{1}{2}$
z	4		16	

6. Given that $V \propto r^3$, copy and complete the table.

r	1	2		$1\frac{1}{2}$
V	4		256	

7. Given that $w \propto \sqrt{h}$, copy and complete the table.

h	4	9		$2\frac{1}{4}$
w	6		15	

8. s is proportional to $(v - 1)^2$. If $s = 8$, when $v = 3$, calculate:

   a) the value of s, when $v = 4$

   b) the value of v, when $s = 2$.

9. m varies as $(d + 3)$. If $m = 28$ when $d = 1$, calculate:

   a) the value of m, when $d = 3$

   b) the value of s, when $m = 49$.

10. The pressure of the water P at any point below the surface of the sea varies as the depth of the point below the surface d. If the pressure is 200 newtons/cm^2 at a depth of 3 m, calculate the pressure at a depth of 5 m.

11. The distance d through which a stone falls from rest is proportional to the square of the time taken t. If the stone falls 45 m in 3 seconds, how far will it fall in 6 seconds? How long will it take to fall 20 m?

12. The energy E stored in an elastic band varies as the square of the extension x. When the elastic is extended by 3 cm, the energy stored is 243 joules. What is the energy stored when the extension is 5 cm? What is the extension when the stored energy is 36 joules?

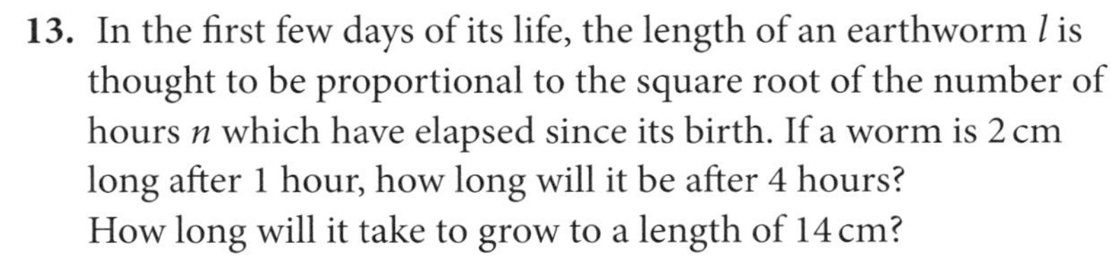

13. In the first few days of its life, the length of an earthworm l is thought to be proportional to the square root of the number of hours n which have elapsed since its birth. If a worm is 2 cm long after 1 hour, how long will it be after 4 hours? How long will it take to grow to a length of 14 cm?

14. The number of eggs which a goose lays in a week varies as the cube root of the average number of hours of sleep she has. When she has 8 hours sleep, she lays 4 eggs. How long does she sleep when she lays 5 eggs?

15. The resistance to motion of a car is proportional to the square of the speed of the car. If the resistance is 4000 newtons at a speed of 20 m/s, what is the resistance at a speed of 30 m/s? At what speed is the resistance 6250 newtons?

16. A road research organisation recently claimed that the damage to road surfaces was proportional to the fourth power of the axle load. The axle load of a 44-tonne HGV is about 15 times that of a car. Calculate the ratio of the damage to road surface made by a 44-tonne HGV and a car.

Inverse variation

There are several ways of expressing an inverse relationship between two variables:

- x varies inversely as y
- x is inversely proportional to y.

We write $x \propto \frac{1}{y}$ for both statements and proceed using the method outlined in the previous section.

Example 31

z is inversely proportional to t^2, and $z = 4$ when $t = 1$.
Calculate:

a) z when $t = 2$ **b)** t when $z = 16$.

We have $z \propto \frac{1}{t^2}$

or $z = k \times \frac{1}{t^2}$ (k is a constant)

$z = 4$ when $t = 1$

$\therefore\ 4 = k\left(\frac{1}{1^2}\right)$

$k = 4$

$\therefore\ z = 4 \times \frac{1}{t^2}$

a) when $t = 2$, $z = 4 \times \frac{1}{2^2} = 1$

b) when $z = 16$, $16 = 4 \times \frac{1}{t^2}$

$16t^2 = 4$

$t^2 = \frac{1}{4}$

$t = \pm\frac{1}{2}$

Exercise 8.33

1. Rewrite the statements connecting the variables using a constant of variation, k.

a) $x \propto \frac{1}{y}$ **b)** $s \propto \frac{1}{t^2}$ **c)** $t \propto \frac{1}{\sqrt{q}}$

d) m varies inversely as w

e) z is inversely proportional to t^2.

2. b varies inversely as e. If $b = 6$ when $e = 2$, calculate:

a) the value of b when $e = 12$

b) the value of e when $b = 3$.

3. q varies inversely as r. If $q = 5$ when $r = 2$, calculate:

a) the value of q when $r = 4$

b) the value of r when $q = 20$.

4. x is inversely proportional to y^2. If $x = 4$ when $y = 3$, calculate:

a) the value of x when $y = 1$

b) the value of y when $x = 2\frac{1}{4}$.

5. R varies inversely as v^2. If $R = 120$ when $v = 1$, calculate:

a) the value of R when $v = 10$

b) the value of v when $R = 30$.

6. T is inversely proportional to x^2. If $T = 36$ when $x = 2$, calculate:

a) the value of T when $x = 3$

b) the value of x when $T = 1{\cdot}44$.

7. p is inversely proportional to $\sqrt{y}$. If $p = 1{\cdot}2$ when $y = 100$, calculate:

a) the value of p when $y = 4$

b) the value of y when $p = 3$.

8. y varies inversely as z. If $y = \frac{1}{8}$ when $z = 4$, calculate:

a) the value of y when $z = 1$

b) the value of z when $y = 10$.

9. Given that $z \propto \frac{1}{y}$, copy and complete the table:

y	2	4		$\frac{1}{4}$
z	8		16	

10. Given that $v \propto \frac{1}{t^2}$, copy and complete the table:

t	2	5		10
v	25		$\frac{1}{4}$	

11. Given that $r \propto \frac{1}{\sqrt{x}}$, copy and complete the table:

x	1	4		
r	12		$\frac{3}{4}$	2

12. e varies inversely as $(y - 2)$. If $e = 12$ when $y = 4$, find:

a) e when $y = 6$ **b)** y when $e = \frac{1}{2}$.

13. M is inversely proportional to the square of l.
If $M = 9$ when $l = 2$, find:

a) M when $l = 10$ **b)** l when $M = 1$.

14. Given $z = \frac{k}{x^n}$, find k and n, then copy and complete the table.

x	1	2	4	
z	100	$12\frac{1}{2}$		$\frac{1}{10}$

15. Given $y = \frac{k}{\sqrt[n]{v}}$, find k and n, then copy and complete the table.

v	1	4	36	
y	12	6		$\frac{3}{25}$

16. The volume V of a given mass of gas varies inversely as the pressure P. When $V = 2\text{m}^3$, $P = 500\ \text{N/m}^2$. Find the volume when the pressure is $400\ \text{N/m}^2$. Find the pressure when the volume is $5\ \text{m}^3$.

17. The number of hours N required to dig a certain hole is inversely proportional to the number of men available x. When 6 men are digging, the hole takes 4 hours. Find the time taken when 8 men are available. If it takes $\frac{1}{2}$ hour to dig the hole, how many men are there?

18. The life expectancy L of a rat varies inversely as the square of the density d of poison distributed around his home. When the density of poison is $1\ \text{g/m}^2$ the life expectancy is 50 days. How long will he survive if the density of poison is

    a) $5\,\text{g/m}^2$?

    b) $\frac{1}{2}\,\text{g/m}^2$?

19. The force of attraction F between two magnets varies inversely as the square of the distance d between them. When the magnets are 2 cm apart, the force of attraction is 18 newtons. How far apart are they if the attractive force is 2 newtons?

Revision exercise 8A

1. Express the following as single fractions.

a) $\frac{x}{4}+\frac{x}{5}$ **b)** $\frac{1}{2x}+\frac{2}{3x}$

c) $\frac{x+2}{2}+\frac{x-4}{3}$ **d)** $\frac{7}{x-1}-\frac{2}{x+3}$

2. a) Factorise x^2-4

b) Simplify $\frac{3x-6}{x^2-4}$

3. Given that $s-3t=rt$, express:

a) s in terms of r and t

b) r in terms of s and t

c) t in terms of s and r.

4. a) Given that $x-z=5y$, express z in terms of x and y.

b) Given that $mk+3m=11$, express m in terms of k.

c) For the formula $T=C\sqrt{z}$, express z in terms of T and C.

5. It is given that $y=\frac{k}{x}$ and that $1\le x\le 10$.

a) If the smallest possible value of y is 5, find the value of the constant k.

b) Find the largest possible value of y.

6. Given that y varies as x^2 and that $y=36$ when $x=3$, find:

a) the value of y when $x=2$

b) the value of x when $y=64$.

7. a) Evaluate:

i) $9^{\frac{1}{2}}$ **ii)** $8^{\frac{2}{3}}$ **iii)** $16^{-\frac{1}{2}}$

b) Find x, given that

i) $3^x=81$ **ii)** $7^x=1$

8. List the integer values of x which satisfy

a) $2x-1<20<3x-5$

b) $5<3x+1<17$.

9. Given that $t=k\sqrt{(x+5)}$, express x in terms of t and k.

10. Given that $z=\frac{3y+2}{y-1}$, express y in terms of z.

11. Given that $y=\frac{k}{k+w}$,

a) find the value of y when $k=\frac{1}{2}$ and $w=\frac{1}{3}$

b) express w in terms of y and k.

12. On a suitable sketch graph, identify clearly the region A defined by $x\ge 0$, $x+y\le 8$ and $y\ge x$.

13. Without using a calculator, calculate the value of

a) $9^{-\frac{1}{2}}+\left(\frac{1}{8}\right)^{\frac{1}{3}}+(-3)^0$

b) $(1000)^{-\frac{1}{3}}-(0{\cdot}1)^2$

14. It is given that $10^x=3$ and $10^y=7$. What is the value of 10^{x+y}?

15. Make x the subject of the following formulae.

a) $x+a=\frac{2x-5}{a}$ **b)** $cz+ax+b=0$

c) $a=\sqrt{\left(\frac{x+1}{x-1}\right)}$

16. Write the following as single fractions.

a) $\frac{3}{x}+\frac{1}{2x}$ **b)** $\frac{3}{a-2}+\frac{1}{a^2-4}$

c) $\frac{3}{x(x+1)}-\frac{2}{x(x-2)}$

17. p varies jointly as the square of t and inversely as s. Given that $p=5$ when $t=1$ and $s=2$, find a formula for p in terms of t and s.

18. A positive integer r is such that $pr^2=168$, where p lies between 3 and 5. List the possible values of r.

19. The shaded region A is formed by the lines $y = 2$, $y = 3x$ and $x + y = 6$. Write down the three inequalities which define A.

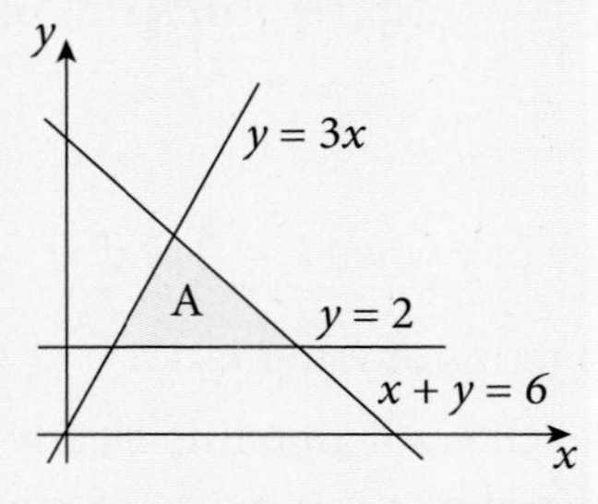

20. The shaded region B is formed by the lines $x = 0$, $y = 0$, $y = x - 2$ and $x + y = 7$.

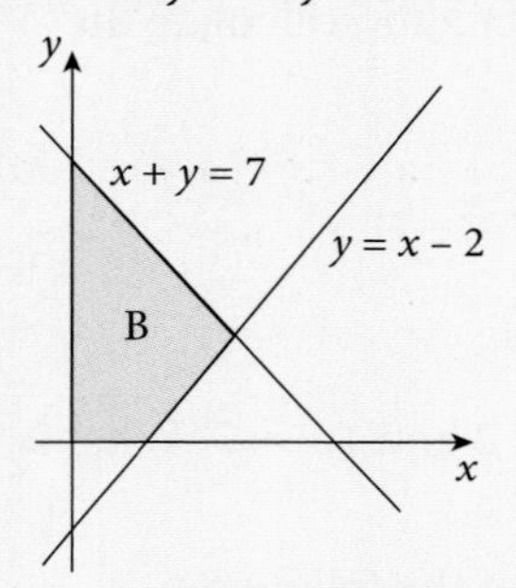

Write down the four inequalities which define B.

21. Find an expression for the nth term of each sequence below.

a) 10, 13, 16, 19 **b)** 6, 10, 16, 24, 34

c) 3, 13, 35, 75, 139

22. In a laboratory an experiment was started with 2 cells in a dish. The number of cells in the dish doubles every 30 minutes.

a) How many cells are in the dish after four hours?

b) After what time are there 2^{13} cells in the dish?

c) After $10\frac{1}{2}$ hours there are 2^{22} cells in the dish and an experimental fluid is added which eliminates half of the cells. How many cells are left?

23. Solve the inequalities:

a) $x^2 > 25$ **b)** $x^2 + 6x + 8 < 0$

24. Find an expressions for the nth term of the sequence

7, 15, 35, 73, 135

25. Find an expression for the nth term of the geometric sequences.

a) 5, 10, 20, 40, … **b)** 2, 6, 18, 54, …

c) 100, 50, 25, …

26. Find the sum o the first eight terms of the geometric series $3 + 6 + 12 + \ldots$

27. The seventh term of a geometric series is 160 and the ninth term is 640.
Find the first term and the common ratio (given that it is positive).

28. Find the common ratio of a geometric series in which the sum of the first two terms is 45 and the first term is 9.

29. Find all the possible values of the common ratio of a geometric series in which the sum of its first three terms is 105 and the first term is 5.

30. The first two terms of a geometric series are x and $x^2 + x$. Find, in terms of x,

a) the common ratio of the series

b) an expression for the one-hundredth term of the series.

Examination exercise 8B

1. Write down the value of

a) 7^{-2}, [1]

b) $64^{\frac{1}{3}}$. [1]

Cambridge IGCSE International Mathematics 0607 Specimen Paper 2 Q1 2010

2. $$\frac{d}{x-c} = \frac{x+c}{d}$$

Find x in terms of c and d. [3]

Cambridge IGCSE International Mathematics 0607 Paper 21 Q9 May/June 2011

3. a) Find the value of

i) 3^0, [1]

ii) $36^{\frac{1}{2}}$. [1]

b) $2^8 \div 2 = 2^x$

Find the value of x. [1]

Cambridge IGCSE International Mathematics 0607 Paper 2 Q2 May/June 2010

4. F varies inversely as the square of d.
When $F = 9$, $d = 2$.

a) Find F in terms of d. [2]

b) Find the value of F when d is 3. [1]

Cambridge IGCSE International Mathematics 0607 Paper 21 Q7 May/June 2011

5. For the sequence 2, 7, 14, 23, 34, 47, ...

a) find the next two terms, [2]

b) find a formula for the nth term. [2]

Cambridge IGCSE International Mathematics 0607 Paper 2 Q9 May/June 2010

6. a) Solve the equation $8x^2 = 7x + 3$, giving your answers correct to 2 decimal places. [4]

b) Solve the inequality $8x^2 < 7x + 3$ giving your answers correct to 2 decimal places. [1]

Cambridge IGCSE International Mathematics 0607 Specimen Paper 4 Q5 2010

7. The first four terms of a sequence are 0, 3, 8, 15.

a) Write down the next two terms of this sequence. [1]

b) Find an expression for the nth term of this sequence. [1]

Cambridge IGCSE International Mathematics 0607 Paper 21 Q2 May/June 2011

9 Functions 2

Pierre de Fermat (1601–65) worked as a lawyer in France and his is one of nearly 100 streets in Paris named after mathematicians. His famous last theorem stated that there are no integer solutions to the equation $x^n + y^n = z^n$ for $n > 2$. The theorem was eventually proved by Andrew Wiles in 1995. (See Chapter 15 of this book).

3.6 Use of a graphics calculator to:
sketch the graph of a function
produce a table of values (not mentioned explicitly in this syllabus)
find zeros, local maxima or minima
find the intersection of the graphs of functions (vertex of quadratic)

3.7 Simplify expressions such as $f(g(x))$ where $g(x)$ is a linear expression

3.8 Description and identification, using the language of tranformations, of the changes to the graph of $y = f(x)$ when
$y = f(x) + k$, $y = k\,f(x)$, $y = f(x + k)$, k an integer

3.10 Logarithmic function as the inverse of the exponential function
$y = a^x$ equivalent to $x = \log_a y$
Rules for logarithms corresponding to rules for exponents
Solution to $a^x = b$ as $x = \dfrac{\log b}{\log a}$.

7.1 Plotting of points and reading from a graph in the Cartesian plane

7.2 Distance between two points

7.3 Midpoint of a line segment

7.4 Gradient of a line segment

7.5 Gradient of parallel and perpendicular lines

7.6 Equation of a straight line as $y = mx + c$ and $ax + by = d$ (a, b and d integer)

7.8 Symmetry of diagram or graphs in the Cartesian plane

Syllabus objectives 3.1, 3.2, 3.3, 3.4, 3.5 and 3.9 are covered in Chapter 5.
Objective 7.7 is covered in Chapter 8.

9.1 Drawing and using graphs

Example 1

Draw the graph of $y = 2x - 3$ for values of x from -2 to $+4$.

The coordinates of points on the line are calculated in a table.

x	-2	-1	0	1	2	3	4
$2x$	-4	-2	0	2	4	6	8
-3	-3	-3	-3	-3	-3	-3	-3
y	-7	-5	-3	-1	1	3	5

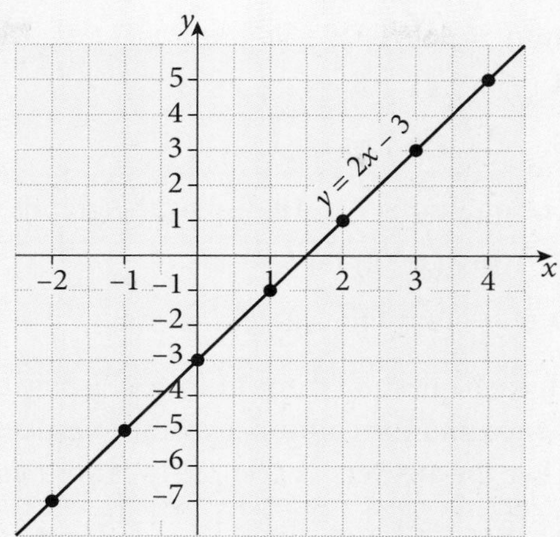

Draw and label axes using suitable scales.
Plot the points and draw a pencil line through them.
Label the line with its equation.

Using a GDC

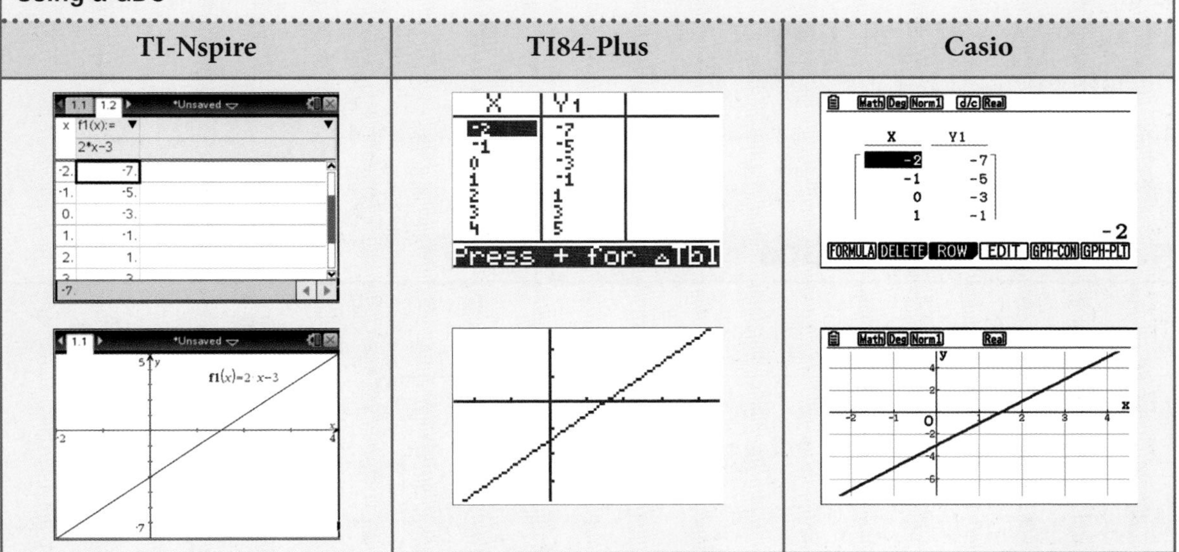

Exercise 9.1

For questions **1** to **6** draw the following graphs, using a scale of 2 cm to 1 unit on the x-axis and 1 cm to 1 unit on the y-axis.

1. $y = 2x + 1$ for $-3 \le x \le 3$

2. $y = 3x - 4$ for $-3 \le x \le 3$

3. $y = 2x - 1$ for $-3 \le x \le 3$

4. $y = 8 - x$ for $-2 \le x \le 4$

5. $y = 10 - 2x$ for $-2 \le x \le 4$

6. $y = \frac{x+5}{2}$ for $-3 \le x \le 3$

For questions **7** to **10** display the graphs on a calculator.

7. $y = 3(x - 2)$ for $-3 \le x \le 3$

8. $y = \frac{1}{2}x + 4$ for $-3 \le x \le 3$

9. $y = 2x - 3$ for $-2 \le x \le 4$

10. $y = 12 - 3x$ for $-2 \le x \le 4$

In each question from **11** to **15**, display the graphs on a calculator and hence find the coordinates of the vertices of the polygon formed. Give the answers as accurately as your calculator will allow.

11. a) $y = x$ **b)** $y = 8 - 4x$ **c)** $y = 4x$

Take $-1 \le x \le 3$ and $-4 \le y \le 14$.

12. a) $y = 2x + 1$ **b)** $y = 4x - 8$ **c)** $y = 1$

Take $0 \le x \le 5$ and $-8 \le y \le 12$.

13. a) $y = -x$ **b)** $y = 3x + 6$ **c)** $y = 8$ **d)** $x = 3\frac{1}{2}$

Take $-2 \le x \le 5$ and $-6 \le y \le 10$.

14. a) $y = \frac{1}{2}(x - 8)$ **b)** $2x + y = 6$ **c)** $y = 4(x + 1)$

Take $-3 \le x \le 4$ and $-7 \le y \le 7$.

15. a) $y = 2x + 7$ **b)** $3x + y = 10$ **c)** $y = x$ **d)** $2y + x = 4$

Take $-2 \le x \le 3$ and $0 \le y \le 13$.

9.2 Gradient of a line

The gradient of a straight line is a measure of how steep it is.

Example 2

Find the gradient of the line joining the points A(1,2) and B(6,5).

It is possible to use the formula

$$\text{gradient} = \frac{\text{increase in } y}{\text{increase in } x}$$

$$\text{gradient of AB} = \frac{5-2}{6-1} = \frac{3}{5}$$

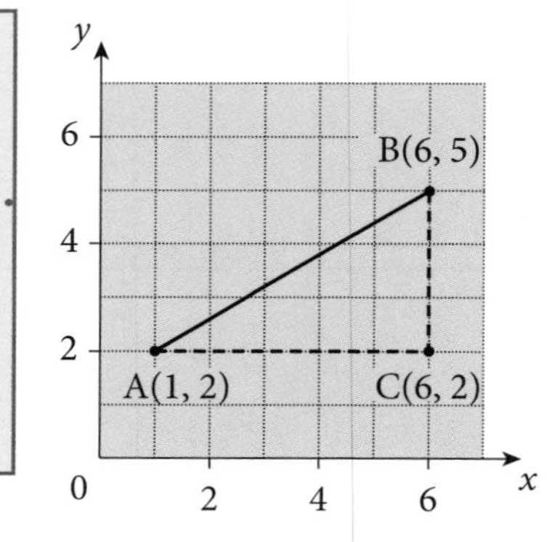

Example 3

Find the gradient of the line joining the points D(1, 5) and E(5, 2).

gradient of DE $= \frac{5-2}{1-5} = \frac{3}{-4} = -\frac{3}{4}$

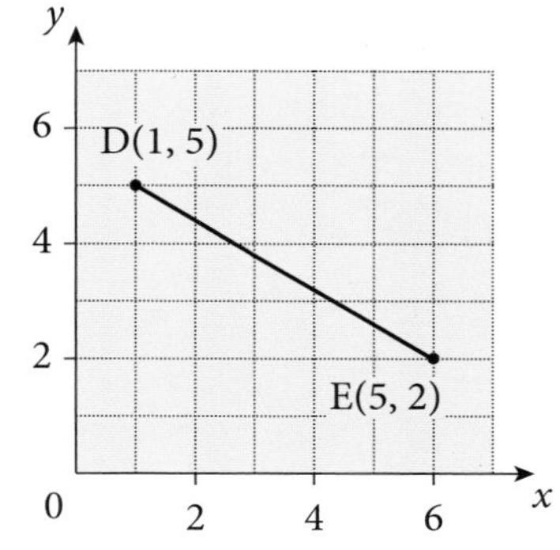

Note:

- Lines which slope upward to the right have a *positive* gradient.
- Lines which slope downward to the right have a *negative* gradient.

Hint:
Think of **N** for negative.

Example 4

a) Find the length of the line segment drawn from P(2, 1) to Q(6, 4).

b) Find the coordinates of the mid-point of the line segment PQ.

a) Draw triangle PQR. PR = 4 units and QR = 3 units. By Pythagoras' theorem $PQ^2 = 4^2 + 3^2 = 25$

$PQ = 5$ units

b) Add the x-coordinates of P and Q and then divide by 2. Add the y-coordinates of P and Q and then divide by 2.

The mid-point has coordinates $\left[\frac{(2+6)}{2}, \frac{(1+4)}{2}\right]$

i. e. $\left(4, 2\frac{1}{2}\right)$

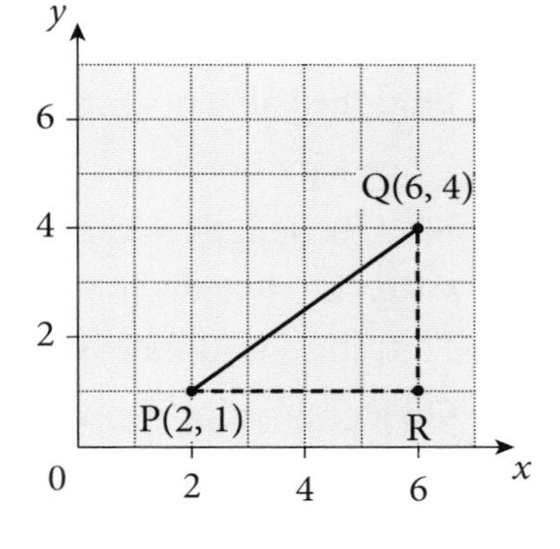

Exercise 9.2

Calculate the gradient of the line joining the following pairs of points.

1. $(3, 1)(5, 4)$
2. $(1, 1)(3, 5)$
3. $(3, 0)(4, 3)$
4. $(-1, 3)(1, 6)$
5. $(-2, -1)(0, 0)$
6. $(7, 5)(1, 6)$
7. $(2, -3)(1, 4)$
8. $(0, -2)(-2, 0)$
9. $\left(\frac{1}{2}, 1\right)\left(\frac{3}{4}, 2\right)$
10. $\left(-\frac{1}{2}, 1\right)(0, -1)$
11. $(3{\cdot}1, 2)(3{\cdot}2, 2{\cdot}5)$
12. $(-7, 10)(0, 0)$
13. $\left(\frac{1}{3}, 1\right)\left(\frac{1}{2}, 2\right)$
14. $(3, 4)(-2, 4)$
15. $(2, 5)(1{\cdot}3, 5)$
16. $(2, 3)(2, 7)$
17. $(-1, 4)(-1, 7{\cdot}2)$
18. $(2{\cdot}3, -2{\cdot}2)(1{\cdot}8, 1{\cdot}8)$
19. $(0{\cdot}75, 0)(0{\cdot}375, -2)$
20. $(17{\cdot}6, 1)(1{\cdot}4, 1)$
21. $(a, b)(c, d)$
22. $(m, n)(a, -b)$
23. $(2a, f)(a, -f)$
24. $(2k, -k)(k, 3k)$
25. $(m, 3n)(-3m, 3n)$
26. $\left(\frac{c}{2}, -d\right)\left(\frac{c}{4}, \frac{d}{2}\right)$

In questions **27** and **28**, find the gradient of each straight line.

27.

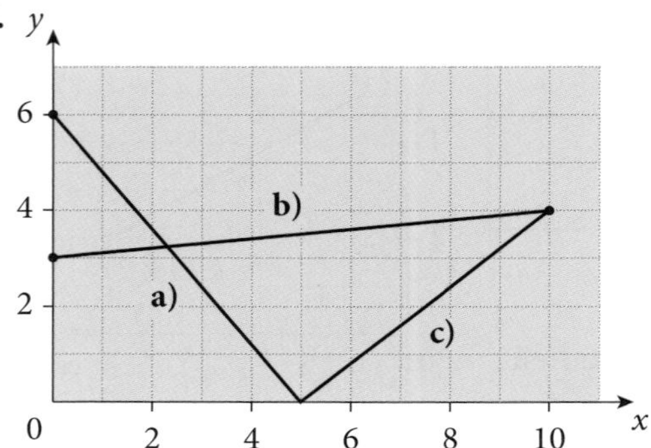

28.

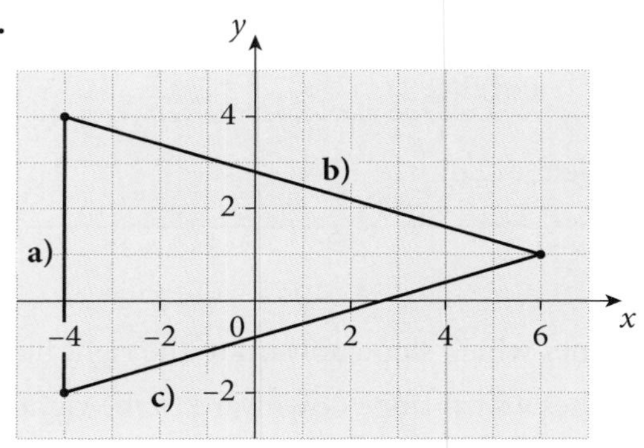

29. Find the value of a if the line joining the points $(3a, 4)$ and $(a, -3)$ has a gradient of 1.

30. a) Write down the gradient of the line joining the points $(2m, n)$ and $(3, -4)$,

b) Find the value of n if the line is parallel to the x-axis,

c) Find the value of m if the line is parallel to the y-axis.

31. a) Draw a pair of x and y axes and plot the points $A(1, 2)$ and $B(7, 6)$.

b) Find the length of the line segment AB. Give your answer correct to one decimal place.

c) Find the coordinates of the mid-point of AB.

32. a) On a graph plot the points $P(1, 4)$, $Q(4, 8)$, $R(5, 1)$.

b) Determine whether or not the triangle PQR is isosceles.

c) Find the coordinates of the mid-point of PR.

9.3 The forms $y = mx + c$ and $ax + by = d$

When the equation of a straight line is written in the form $y = mx + c$, the gradient of the line is m and the intercept on the y-axis is c.

Example 5

Draw the line $y = 2x + 3$ on a *sketch* graph.

The word 'sketch' implies that we do not plot a series of points but simply show the position and slope of the line.

The line $y = 2x + 3$ has a gradient of 2 and cuts the y-axis at $(0, 3)$.

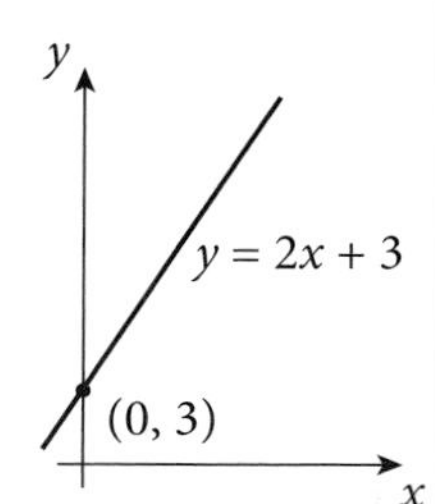

Example 6

Draw the line $x + 2y - 6 = 0$ on a sketch graph.

Rearrange the equation in the form $ax + by = d$

$x + 2y - 6 = 0$

$x + 2y = 6$

when $x = 0$, $y = 3$ and when $y = 0$, $x = 6$

The line has a gradient of $-\frac{1}{2}$ and cuts the y-axis at $(0, 3)$.

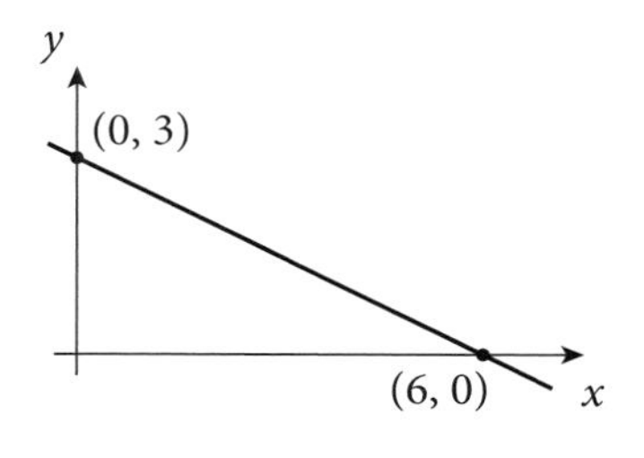

Exercise 9.3

In questions **1** to **20**, find the gradient of the line and the intercept on the y-axis. Hence draw a small sketch graph of each line.

1. $y = x + 3$ **2.** $y = x - 2$ **3.** $y = 2x + 1$

4. $y = 2x - 5$ **5.** $y = 3x + 4$ **6.** $y = \frac{1}{2}x + 6$

7. $y = 3x - 2$ **8.** $y = 2x$ **9.** $y = \frac{1}{4}x - 4$

10. $y = -x + 3$ **11.** $y = 6 - 2x$ **12.** $y = 2 - x$

13. $y + 2x = 3$ **14.** $3x + y + 4 = 0$ **15.** $2y - x = 6$

16. $3y + x - 9 = 0$ **17.** $4x - y = 5$ **18.** $3x - 2y = 8$

19. $10x - y = 0$ **20.** $y - 4 = 0$

Finding the equation of a line

Example 7

Find the equation of the straight line which passes through $(1, 3)$ and $(3, 7)$.

Let the equation of the line take the form $y = mx + c$

The gradient, $m = \frac{7-3}{3-1} = 2$

so we may write the equation as

$$y = 2x + c \qquad [1]$$

Since the line passes through $(1, 3)$, substitute 3 for y and 1 for x in [1]

$\therefore \quad 3 = 2 \times 1 + c$

$1 = c$

The equation of the line is $y = 2x + 1$

Exercise 9.4

In questions **1** to **11** find the equation of the line with the properties indicated.

1. Passes through $(0, 7)$ at a gradient of 3

2. Passes through $(0, -9)$ at a gradient of 2

3. Passes through $(0, 5)$ at a gradient of -1

4. Passes through $(2, 3)$ at a gradient of 2

5. Passes through $(2, 11)$ at a gradient of 3

6. Passes through $(4, 3)$ at a gradient of -1

7. Passes through $(6, 0)$ at a gradient of $\frac{1}{2}$

8. Passes through $(2, 1)$ and $(4, 5)$

9. Passes through $(5, 4)$ and $(6, 7)$

10. Passes through $(0, 5)$ and $(3, 2)$

11. Passes through $(3, -3)$ and $(9, -1)$

Exercise 9.5

1. Find the equations of the lines A and B.

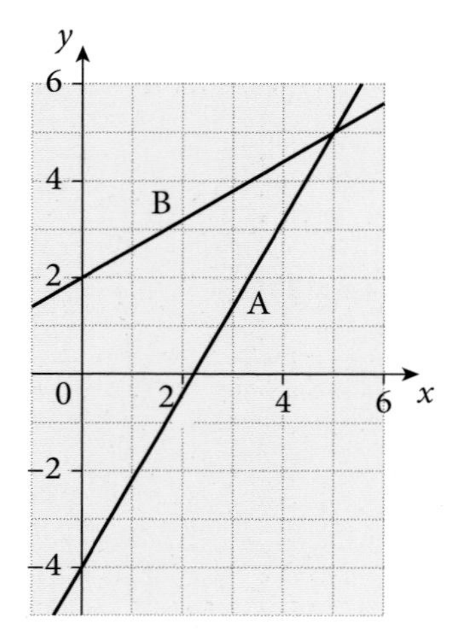

2. Find the equations of the lines C and D.

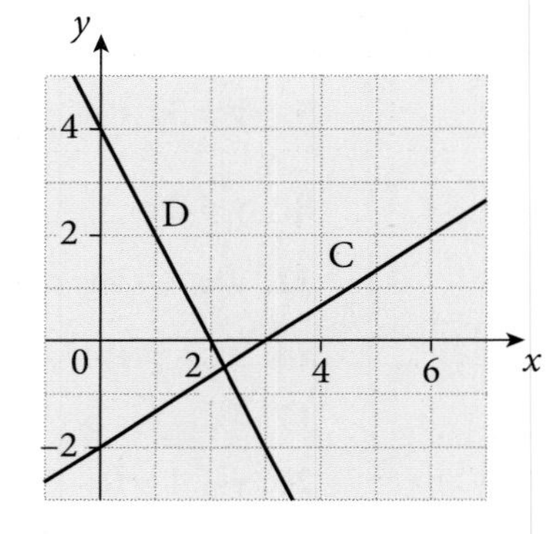

3. Look at the graph on the right.

a) Find the equation of the line which is parallel to line A and which passes through the point $(0, 5)$.

b) Find the equation of the line which is parallel to line B and which passes through the point $(0, 3)$.

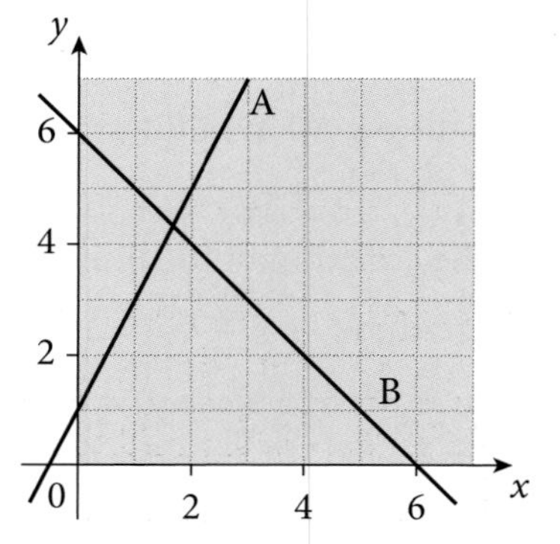

4. Look at the graph in Question **1**.

a) Find the equation of the line which is parallel to line A and which passes through the point $(0, 1)$.

b) Find the equation of the line which is parallel to line B and which passes through the point $(0, -2)$.

5. The sketch shows the graph of $2y + x = 8$.

   Find

   a) the coordinates of A

   b) the equation of the line which is the reflection of the line AB is the y axis.

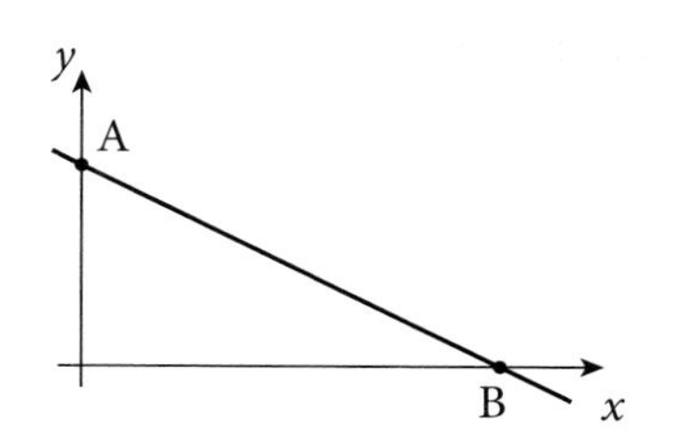

6. LMN is a right-angled triangle with vertices at L(1, 3), M(3, 5) and N(6, n). Given that angle LMN is 90°, find n.

9.4 Curved graphs

Example 8

Draw the graph of the function $y = 2x^2 + x - 6$, for $-3 \le x \le 3$.

x	−3	−2	−1	0	1	2	3
$2x^2$	18	8	2	0	2	8	18
x	−3	−2	−1	0	1	2	3
−6	−6	−6	−6	−6	−6	−6	−6
y	9	0	−5	−6	−3	−4	15

Draw and label axes using suitable scales.

Plot the points and draw a smooth curve through them with a pencil.

Check any points which interrupt the smoothness of the curve.

Label the curve with its equation.

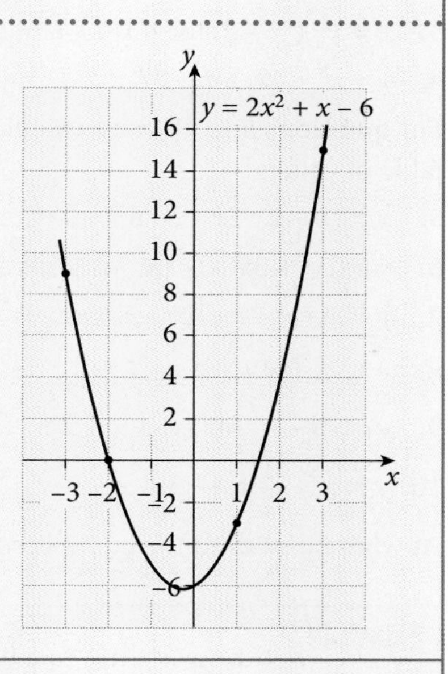

Using a GDC

TI-Nspire	TI84-Plus	Casio
f1(x):= 2*x^2+x−6; −3. 9.; −2. 0.; −1. −5.; 0. −6.; 1. −3.	X Y1; −3 9; −2 0; −1 −5; 0 −6; 1 −3; 2 4; 3 15; X= −3	X Y1; −3 9; −2 0; −1 −5; 0 −6; −3

▶ Continued on next page

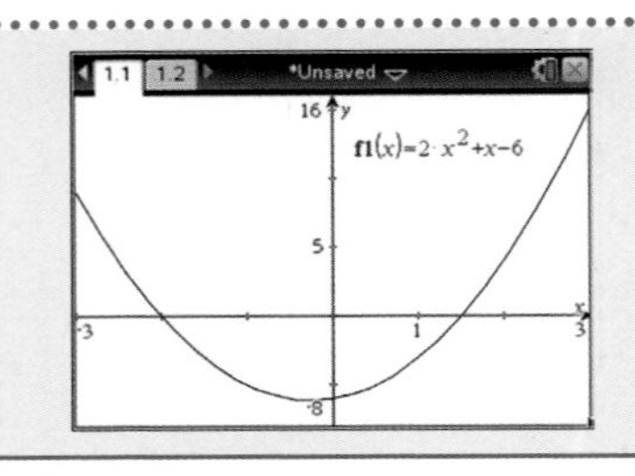

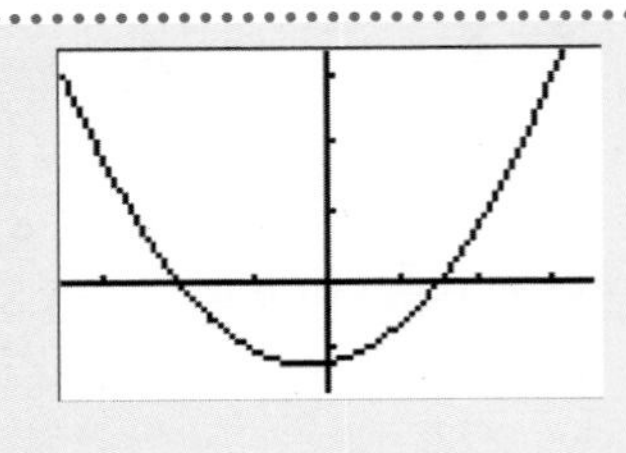
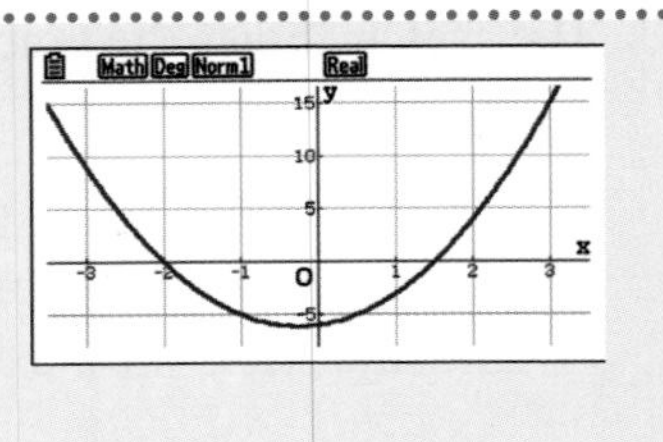

Exercise 9.6

For questions **1** to **4**, draw the graphs of the functions given, using a scale of 2 cm for 1 unit on the x-axis and 1 cm for 1 unit on the y-axis.

1. $y = x^2 + 2x$, for $-3 \leq x \leq 3$

2. $y = x^2 - 3x$, for $-3 \leq x \leq 3$

3. $y = x^2 + x - 2$, for $-3 \leq x \leq 3$

4. $y = x^3 - 4x$, for $-4 \leq x \leq 4$

For questions **5** to **10** use a calculator to display the curve and to produce a table of values.

5. $y = (x + 1)(2x - 5)$ for $-3 \leq x \leq 3$

6. $y = 2x^2 + 3x - 6$, for $-4 \leq x \leq 2$

7. $y = 2 + x - x^2$, for $-3 \leq x \leq 3$

8. $y = 2^x$, for $-2 \leq x \leq 5$

9. $y = 10 - 3^x$, for $-3 \leq x \leq 2{\cdot}5$

10. $y = x^2 - \frac{1}{x}$ for $-3 \leq x \leq 3$

At what point does the curve cross the x-axis.?

Example 9

Draw the graph of, $y = \frac{12}{x} + x - 6$, for $1 \leq x \leq 8$.

Use the graph to find approximate values for:

a) the minimum value of $\frac{12}{x} + x - 6$

b) the value of $\frac{12}{x} + x - 6$, when $x = 2{\cdot}25$

c) the gradient of the tangent to the curve drawn at the point where $x = 5$.

d) Display the graph on a calculator and find the, gradient of a tangent to the curve at $x = 5$.

▶ Continued on next page

Here is the table of values:

x	1	2	3	4	5	6	7	8	1·5
$\frac{12}{x}$	12	6	4	3	2·4	2	1·71	1·5	8
x	1	2	3	4	5	6	7	8	1·5
−6	−6	−6	−6	−6	−6	−6	−6	−6	−6
y	7	2	1	1	1·4	2	2·71	3·5	3·5

Notice that an 'extra' value of y has been calculated at $x = 1·5$ because of the large difference between the y-values at $x = 1$ and $x = 2$.

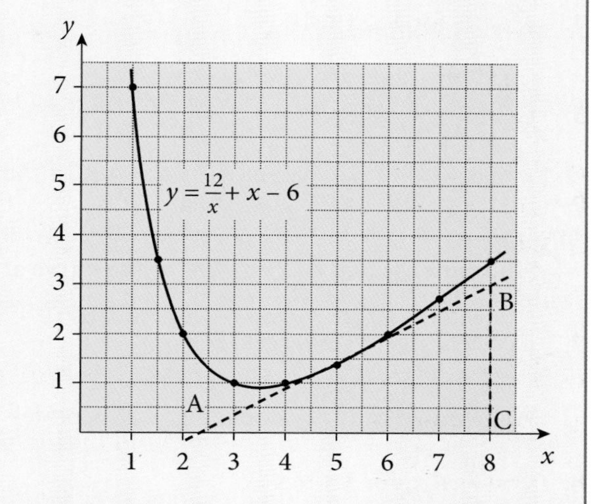

a) From the graph, the minimum value of $\frac{12}{x} + x - 6$ (i.e. y) is approximately 0·9.

b) At $x = 2·25$, y is approximately 1·6.

c) The tangent AB is drawn to touch the curve at $x = 5$

The gradient of AB $= \frac{BC}{AC}$

gradient $= \frac{3}{8-2·4} = \frac{3}{5·6} \approx 0·54$

d) The graph is displayed above. Try this on your calculator.

Exercise 9.7

Display the following curves on a calculator and answer the questions using the calculator.

1. $y = x^2$, for $0 \le x \le 6$

Find:

a) the gradient of the tangent to the curve at $x = 2$

b) the gradient of the tangent to the curve at $x = 4$

c) the y-value at $x = 3·25$.

2. $y = x^2 - 3x$, for $-2 \le x \le 5$

Find:

a) the gradient of the tangent to the curve at $x = 3$

b) the gradient of the tangent to the curve at $x = -1$

c) the value of x where the gradient of the curve is zero.

3. $y = 5 + 3x - x^2$, for $-2 \le x \le 5$

Find:

a) the maximum value of the function $5 + 3x - x^2$

b) the gradient of the tangent to the curve at $x = 2{\cdot}5$

c) the two values of x for which $y = 2$

4. $y = \frac{12}{x}$, for $1 \le x \le 10$

5. $y = \frac{12}{x+1}$, for $0 \le x \le 8$

6. $y = \frac{x}{x+4}$, for $-3{\cdot}5 \le x \le 4$

Find the gradient of a tangent to the curve at $x = 0$.

7. $y = \frac{x+8}{x+1}$, for $0 \le x \le 8$.

Write down the equations of the asymptotes to the curve.

8. $y = \frac{10}{x} + x$, for $1 \le x \le 7$.

Write down the equations of the asymptotes to the curve.

9. $y = \frac{15}{x} + x - 7$, for $1 \le x \le 7$.

Find:

a) the minimum value of y

b) the y value when $x = 5{\cdot}5$

10. $y = \frac{1}{10}(x^3 + 2x + 20)$, for $-3 \le x \le 3$.

Find:

a) the x value where $x^3 + 2x + 20 = 0$

b) the gradient of the tangent to the curve at $x = 2$

9.5 Graphical solution of equations

Using graphs enables us to find approximate solutions to a wide range of equations, many of which are impossible to solve exactly by 'conventional' methods.

Example 10

Draw the graph of the function

$y = 2x^2 - x - 3$ for $-2 \leq x \leq 3$

Use the graph to find approximate solutions to the following equations.

a) $2x^2 - x - 3 = 6$ **b)** $2x^2 - x = x + 5$

The table of values for $y = 2x^2 - x - 3$ is found. Note the 'extra' value at $x = \frac{1}{2}$

x	-2	-1	0	1	2	3	$\frac{1}{2}$
$2x^2$	8	2	0	2	8	18	$\frac{1}{2}$
$-x$	2	1	0	-1	-2	-3	$-\frac{1}{2}$
-3	-3	-3	-3	-3	-3	-3	-3
y	7	0	-3	-2	3	12	-3

The graph drawn from this table is opposite.

a) To solve the equation $2x^2 - x - 3 = 6$, the line $y = 6$ is drawn. At the points of intersection (A and B), y simultaneously equals both 6 and $(2x^2 - x - 3)$.
So we may write $2x^2 - x - 3 = 6$
The solutions are the x-values of the points A and B,
i.e. $x = -1{\cdot}9$ and $x = 2{\cdot}4$ approx.

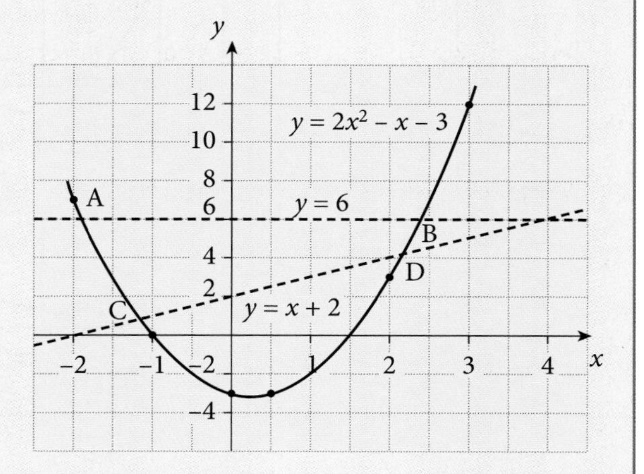

b) To solve the equation $2x^2 - x = x + 5$, we rearrange the equation to obtain the function $(2x^2 - x - 3)$ on the left-hand side.
In this case, subtract 3 from both sides.

$2x^2 - x - 3 = x + 5 - 3$

$2x^2 - x - 3 = x + 2$

If we now draw the line $y = x + 2$, the solutions of the equation are given by the x-values of C and D, the points of intersection.
i.e. $x = -1{\cdot}2$ and $x = 2{\cdot}2$ approx.
It is important to rearrange the equation to be solved so that the function already plotted is on one side.

▶ Continued on next page

Using a GDC

a)

TI-Nspire	TI84-Plus	Casio
(-1.89, 6) (2.39, 6) f2(x)=6 f1(x)=2·x²−x−3	Intersection X=-1.886001 Y=6	[EXE]:Show coordinates Y1=2x²−x−3 Y2=6 INTSECT X=-1.886000936 Y=6
	Intersection X=2.3860009 Y=6	[EXE]:Show coordinates Y1=2x²−x−3 Y2=6 INTSECT X=2.386000936 Y=6

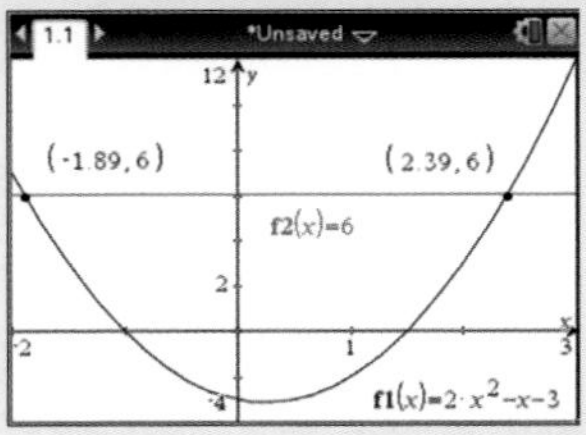

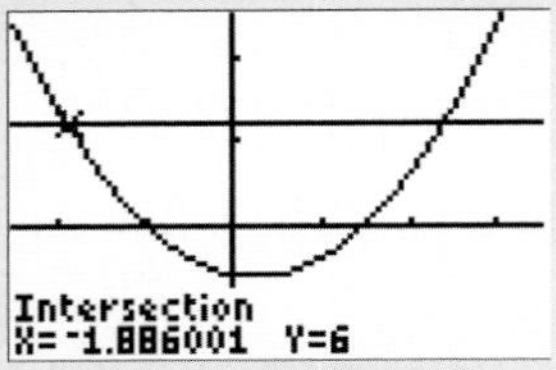

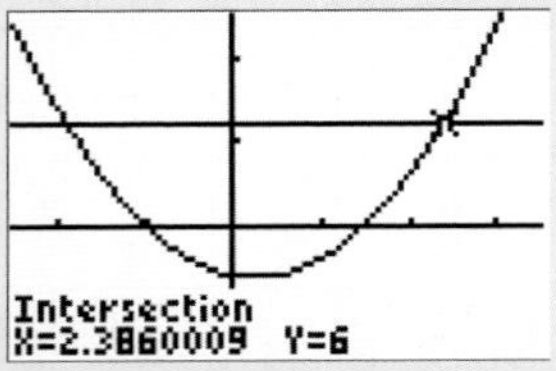

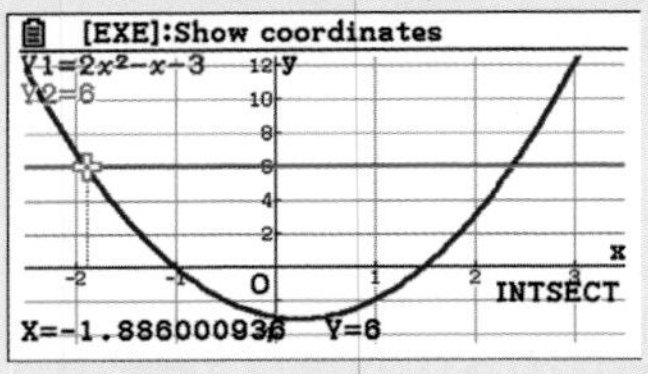

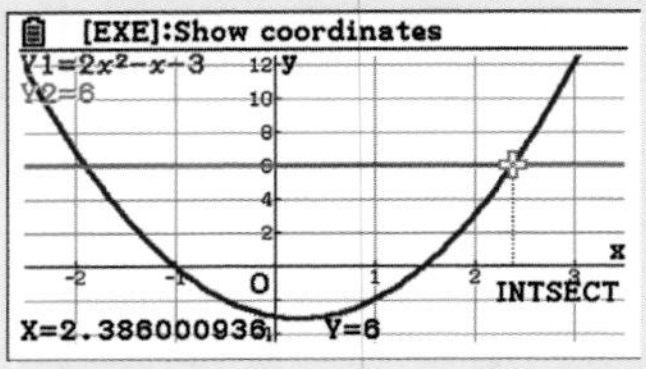

b)

TI-Nspire	TI84-Plus	Casio
(-1.16, 0.842) (2.16, 4.16) f2(x)=x+2 f1(x)=2·x²−x−3	Intersection X=-1.158312 Y=.8416876	[EXE]:Show coordinates Y1=2x²−x−3 Y2=x+2 INTSECT X=-1.158312395 Y=0.8416876048
	Intersection X=2.1583124 Y=4.1583124	[EXE]:Show coordinates Y1=2x²−x−3 Y2=x+2 INTSECT X=2.158312395 Y=4.158312395

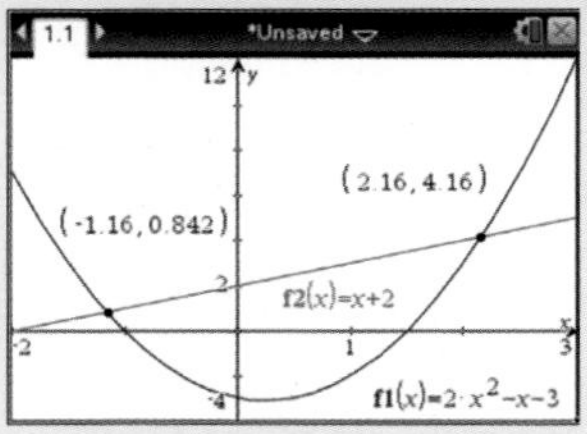

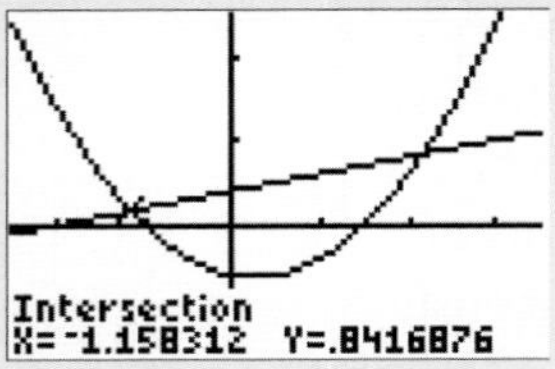

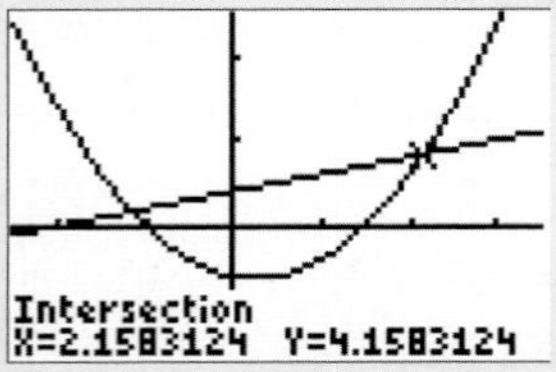

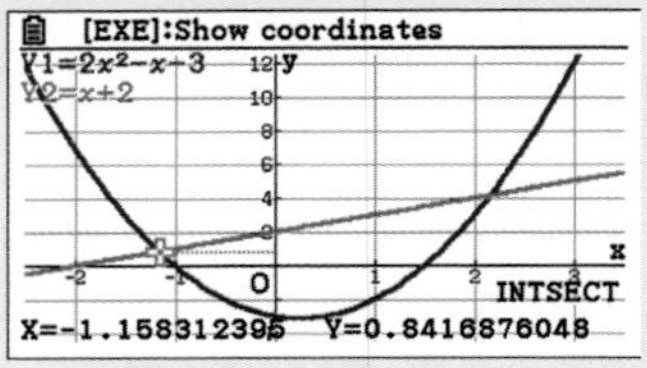

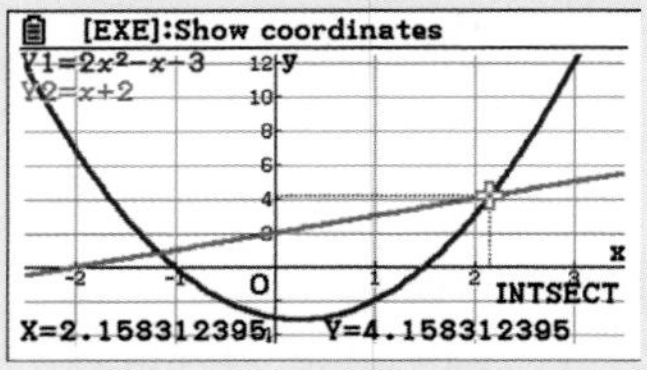

Example 11

Assuming that the graph of $y = x^2 - 3x + 1$ has been drawn, find the equation of the line which should be drawn to solve the equation

$x^2 - 4x + 3 = 0$

Rearrange $x^2 - 4x + 3 = 0$ in order to obtain $(x^2 - 3x + 1)$ on the left-hand side.

$$x^2 - 4x + 3 = 0$$

add x: $\quad x^2 - 3x + 3 = x$

subtract: $\quad 2x^2 - 3x + 1 = x - 2$

Therefore the line $y = x - 2$ should be drawn to solve the equation.

Exercise 9.8

1. In the diagram, the graphs of $y = x^2 - 2x - 3$, $y = -2$ and $y = x$ have been drawn.

Use the graphs to find approximate solutions to the following equations.

a) $x^2 - 2x - 3 = -2$ **b)** $x^2 - 2x - 3 = x$

c) $x^2 - 2x - 3 = 0$ **d)** $x^2 - 2x - 1 = 0$

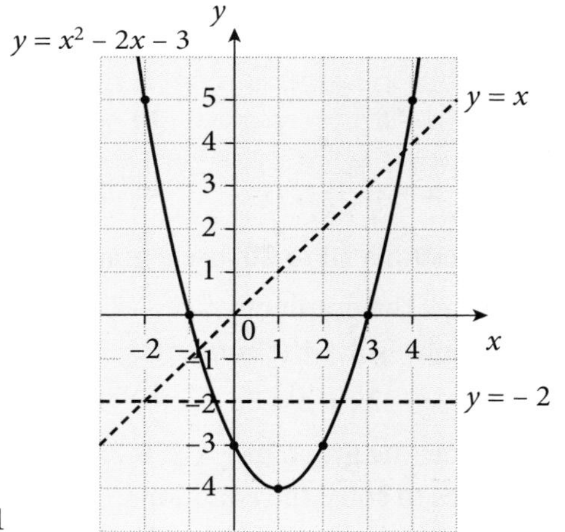

For questions **2** to **4**, use a graphic display calculator.

2. Draw the graphs of the functions $y = x^2 - 2x$ and $y = x + 1$ for $-1 \le x \le 4$. Hence find approximate solutions of the equation $x^2 - 2x = x + 1$

3. Draw the graphs of the functions $y = x^2 - 3x + 5$ and $y = x + 3$ for $-1 \le x \le 5$. Hence find approximate solutions of the equation $x^2 - 3x + 5 = x + 3$

4. Draw the graphs of the functions $y = 6x - x^2$ and $y = 2x + 1$ for $0 \le x \le 5$. Hence find approximate solutions of the equation $6x - x^2 = 2x + 1$

In questions **5** to **9**, do *not* draw any graphs.

5. Assuming the graph of $y = x^2 - 5x$ has been drawn, find the equation of the line which should be drawn to solve the equations.

a) $x^2 - 5x = 3$ **b)** $x^2 - 5x = -2$ **c)** $x^2 - 5x = x + 4$

d) $x^2 - 6x = 0$ **e)** $x^2 - 5x - 6 = 0$

6. Assuming the graph of $y = x^2 + x + 1$ has been drawn, find the equation of the line which should be drawn to solve the equations.

a) $x^2 + x + 1 = 6$ **b)** $x^2 + x + 1 = 0$ **c)** $x^2 + x - 3 = 0$

d) $x^2 - x + 1 = 0$ **e)** $x^2 - x - 3 = 0$

7. Assuming the graph of $y = 6x - x^2$ has been drawn, find the equation of the line which should be drawn to solve the equations.

a) $4 + 6x - x^2 = 0$ **b)** $4x - x^2 = 0$ **c)** $2 + 5x - x^2 = 0$

d) $x^2 - 6x = 3$ **e)** $x^2 - 6x = -2$

8. Assuming the graph of $y = x + \frac{4}{x}$ has been drawn, find the equation of the line which should be drawn to solve the equations.

a) $x + \frac{4}{x} - 5 = 0$ **b)** $\frac{4}{x} - x = 0$ **c)** $x + \frac{4}{x} = 0{\cdot}2$

d) $2x + \frac{4}{x} - 3 = 0$ **e)** $x^2 + 4 = 3x$.

9. Assuming the graph of $y = x^2 - 8x - 7$ has been drawn, find the equation of the line which should be drawn to solve the equations.

a) $x = 8 + \frac{7}{x}$ **b)** $2x^2 = 16x + 9$ **c)** $x^2 = 7$

d) $x = \frac{4}{x-8}$ **e)** $2x - 5 = \frac{14}{x}$.

For questions **10** to **20**, use a graphic display calculator.

10. Draw the graph of $y = x^2 - 2x + 2$ for $-2 \le x \le 4$. By drawing other graphs, solve the equations.

a) $x^2 - 2x + 2 = 8$ **b)** $x^2 - 2x + 2 = 5 - x$ **c)** $x^2 - 2x - 5 = 0$

11. Draw the graph of $y = x^2 - 7x$ for $0 \le x \le 7$. Draw suitable straight lines to solve the equations.

a) $x^2 - 7x + 9 = 0$ **b)** $x^2 - 5x + 1 = 0$

12. Draw the graph of $y = x^2 + 4x + 5$ for $-6 \le x \le 1$. Draw suitable straight lines to find approximate solutions of the equations.

a) $x^2 + 3x - 1 = 0$ **b)** $x^2 + 5x + 2 = 0$

13. Draw the graph of $y = 2x^2 + 3x - 9$ for $-3 \le x \le 2$. Draw suitable straight lines to find approximate solutions of the equations.

a) $2x^2 + 3x - 4 = 0$ **b)** $2x^2 + 2x - 9 = 1$

14. Draw the graph of $y = 2 + 3x - 2x^2$ for $-2 \le x \le 4$.

a) Draw suitable straight lines to find approximate solutions of the equations.

i) $2 + 4x - 2x^2 = 0$ **ii)** $2x^2 - 3x - 2 = 0$

b) Find the range of values of x for which $2 + 3x - 2x^2 \ge -5$

15. Draw the graph of $y = \frac{18}{x}$ for $1 \le x \le 10$. Use the graph to solve approximately:

a) $\frac{18}{x} = x + 2$ b) $\frac{18}{x} + x = 10$ c) $x^2 = 18$

16. Draw the graph of $y = \frac{1}{2}x^2 - 6$ for $-4 \le x \le 4$.

a) Use your graph to solve approximately the equation
$\frac{1}{2}x^2 - 6 = 1$.

b) Using tables or a calculator confirm that your solutions are approximately $\pm\sqrt{14}$ and explain why this is so.

c) Use your graph to find the square roots of 8.

17. Draw the graph of $y = 6 - 2x - \frac{1}{2}x^3$ for $-2 \le x \le 2$.
Use your graph to find approximate solutions of the equations:

a) $\frac{1}{2}x^3 + 2x - 6 = 0$ b) $x - \frac{1}{2}x^3 = 0$

Using tables confirm that two of the solutions to the equation in part **b**) are $\pm\sqrt{2}$ and explain why this is so.

18. Draw the graph of $y = x + \frac{12}{x} - 5$ for $1 \le x \le 8$.

a) From your graph find the range of values of x for which $x + \frac{12}{x} \le 9$

b) Find an approximate solution of the equation $2x - \frac{12}{x} - 12 = 0$.

19. Draw the graph of $y = 2^x$ for $-4 \le x \le 4$.
Find approximate solutions to the equations:

a) $2^x = 6$ b) $2^x = 3x$ c) $x2^x = 1$

Find also the approximate value of $2^{2 \cdot 5}$.

20. Draw the graph of $y = \frac{1}{x}$ for $-4 \le x \le 4$.

Find approximate solutions to the equations:

a) $\frac{1}{x} = x + 1$ b) $2x^2 - x - 1 = 0$

Exponential curves have a term involving a^x, where a is a constant.

For example, the graphs of $y = 3x$ and $y = \left(\frac{1}{2}\right)^x$ are exponential curves.

The x-axis is an asymptote to the curve.

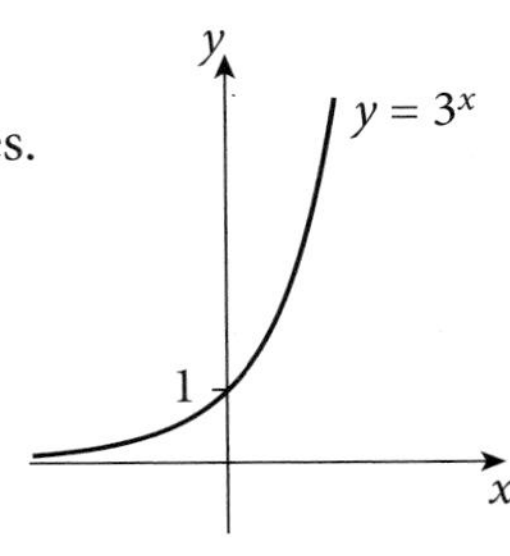

Example 12

Sketch the graph of $y = x^2 - x - 2$.

When we *sketch* a graph we show the general shape of the curve and show where it cuts the axes.

The curve cuts the x-axis when $y = 0$

i.e $\quad x^2 - x - 2 = 0$

$(x - 2)(x + 1) = 0$

$x = 2, -1$

Mark 2 and –1 on the x-axis.

The x^2 term is positive so the curve is '∪-shaped'.

Notice that when $x = 0$, $y = -2$.

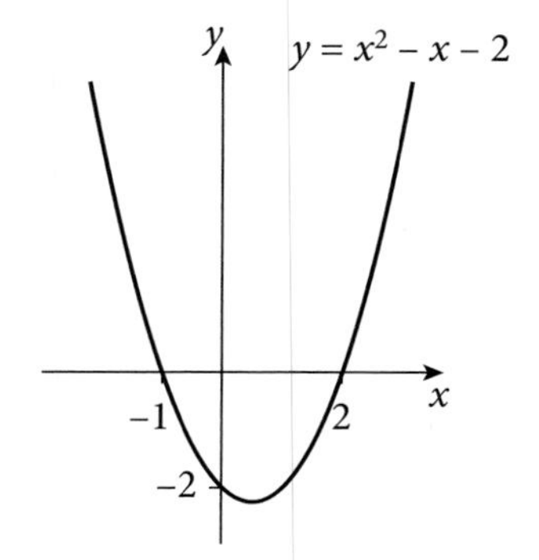

Example 13

Sketch the graph of $y = (x + 3)(x + 1)(x - 5)$.

The curve cuts the x-axis where $y = 0$

$(x + 3)(x + 1)(x - 5) = 0$

$x = -3, -1, 5$

Mark these points on the x-axis.

The x^3 term is positive so the curve may be sketched.

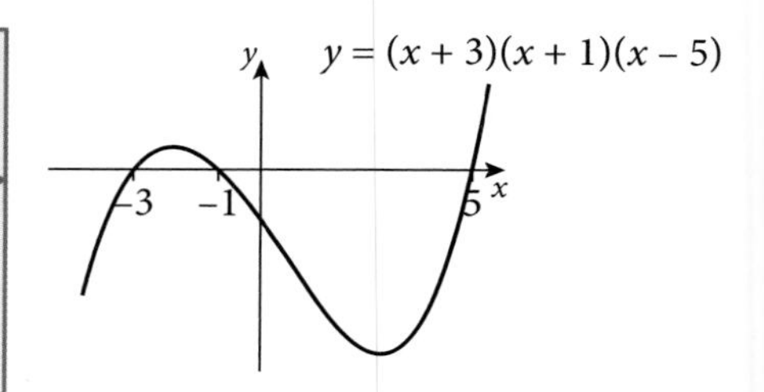

Exercise 9.9

1. What sort of curves are these? State as much information as you can.

a)

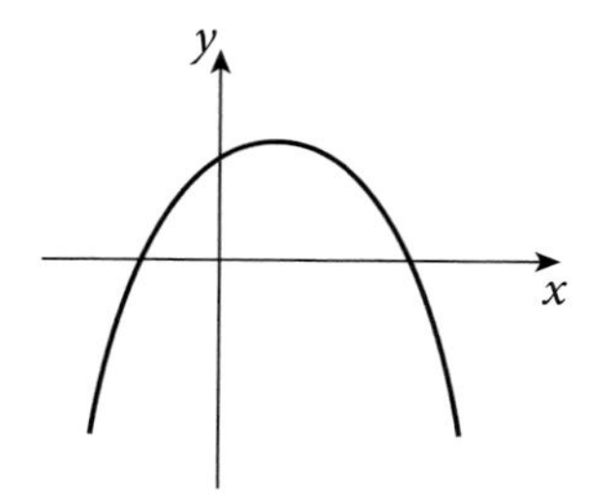

b)

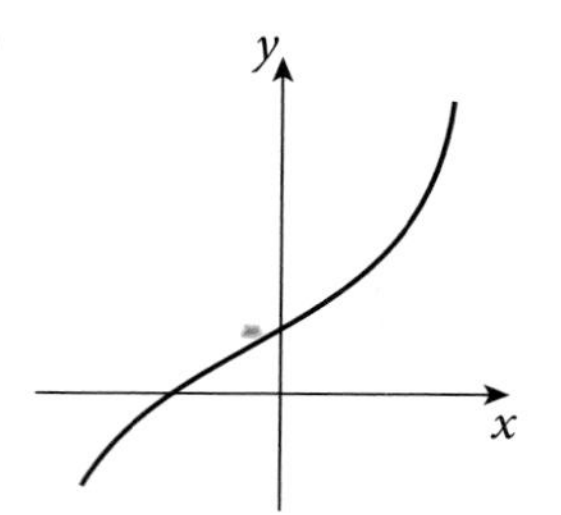

c)

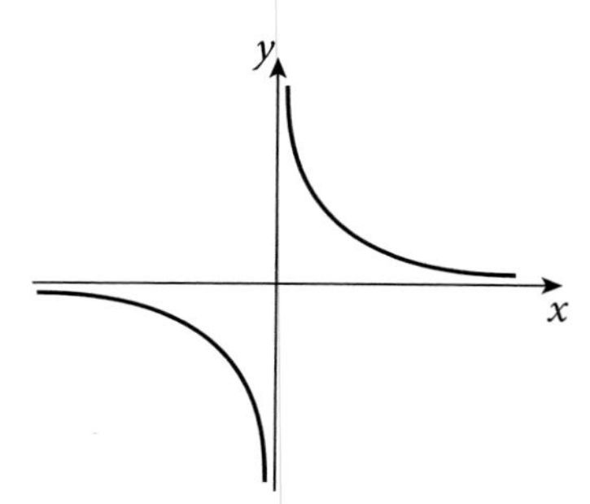

d)

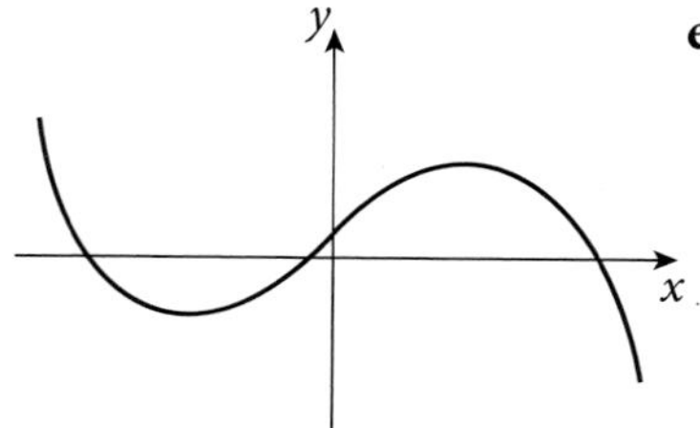

e)

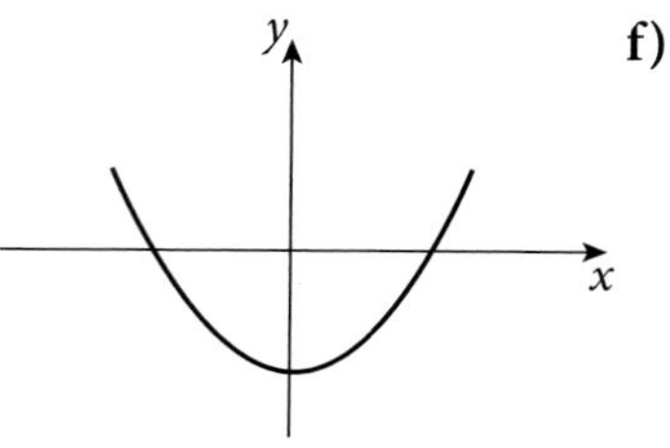

f)

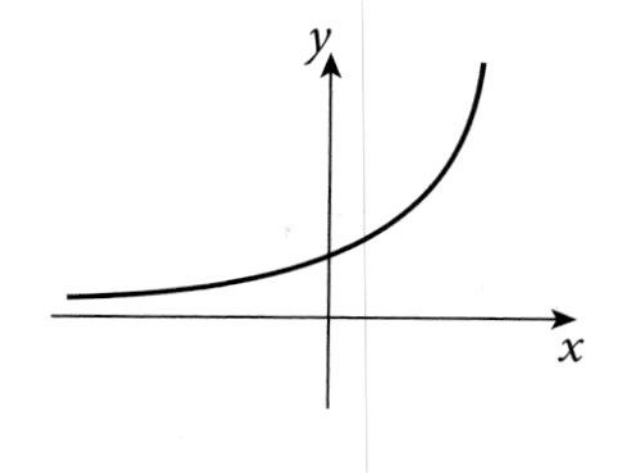

2. Draw the general shape of the following curves.

a) $y = 3x^2 - 7x + 11$ b) $y = 2^x$ c) $y = \frac{100}{x}$

d) $y = 8x - x^2$ e) $y = 10x^3 + 7x - 2$ f) $y = \frac{1}{x^2}$

3. Here are the equations of the six curves in Question **1**, but not in the correct order.

i) $y = \frac{8}{x}$ ii) $y = 2x^3 + x + 2$ iii) $y = 5 + 3x - x^2$

iv) $y = x^2 - 6$ v) $y = 5^x$ vi) $y = 12 + 11x - 2x^2 - x^3$

Decide which equations fit the curves **a** to **f**.

4. Sketch the two curves given and state the number of times the curves intersect.

a) $y = x^3$, $y = 10 - x$ b) $y = x^2$, $y = 10 - x^2$ c) $y = 3^x$, $y = 2 - x$ d) $y = x^3$, $y = x$

5. Sketch the graph of $y = (x - 1)(x + 3)$
Start by finding the two points where the graph cuts the x-axis (i.e. where $y = 0$)

6. Sketch the graphs of

a) $y = (x + 4)(2x - 1)$ b) $y = x(x - 4)$ c) $y = x(5 - x)$

7. Sketch the graph of the cubic equation $y = (x + 2)(x - 1)(x - 5)$

8. Sketch the graphs of

a) $y = x(x - 3)(x - 6)$ b) $y = (x + 2)(x + 1)(x - 4)$ c) $y = x(2x - 5)(x - 10)$

9.6 Transformations of the graph of $y = f(x)$

The notation $f(x)$ means 'function of x'. A function of x is an expression which (usually) varies, depending on the value of x. Examples of functions are:

$$f(x) = x^2 + 3; \quad f(x) = \frac{1}{x} + 7; \quad f(x) = \sin x.$$

We can imagine a box which performs the function f on any input.

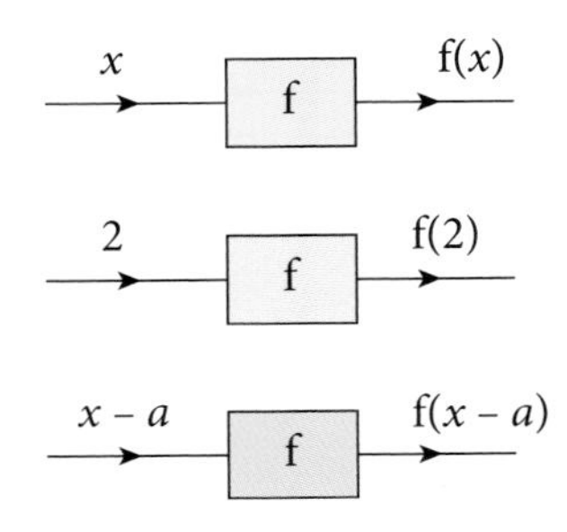

For example,

If $f(x) = x^2 + 7x + 2$, $f(3) = 3^2 + 7 \times 3 + 2 = 32$

Here are three ways in which any function f(x) can be tranformed, which you can investigate below:

a) f(x) + a **b)** f(x – a) **c)** af(x)

a) y = f(x) + a

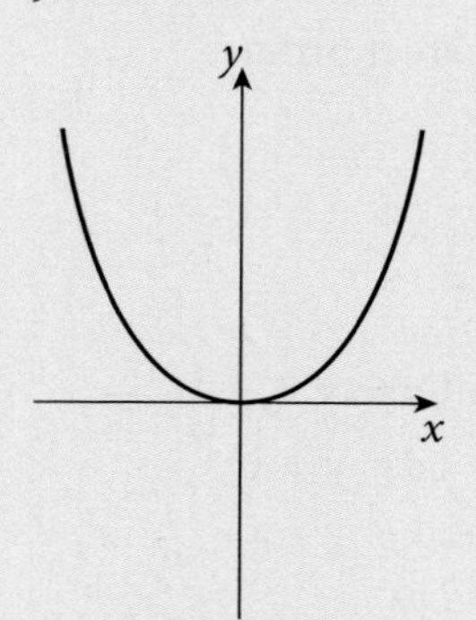

① y = f(x)

2

② y = f(x) + 2

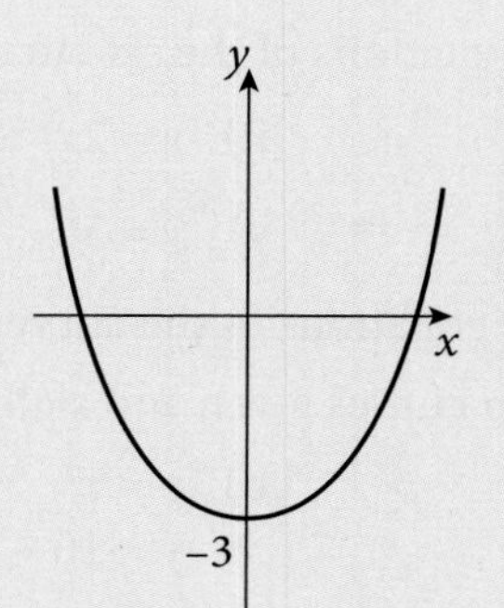

③ y = f(x) – 3

- Describe the transformation from ① to ②
- Describe the transformation from ① to ③

b) y = f (x – a)

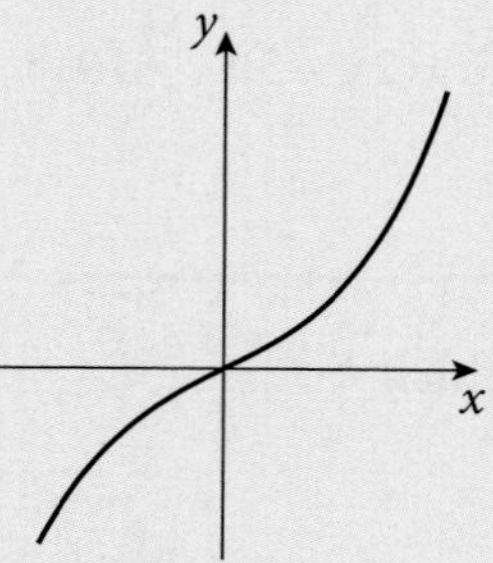

① y = f(x)

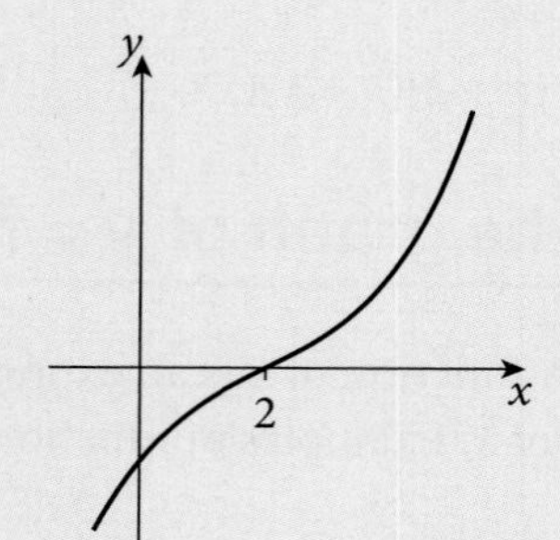

② y = f(x – 2)

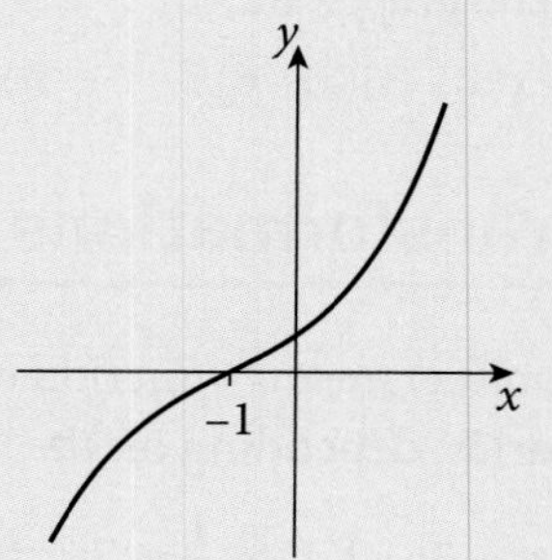

③ y = f(x + 1)

- Describe the transformation from ① to ②
- Describe the transformation from ① to ③

c) y = af(x)

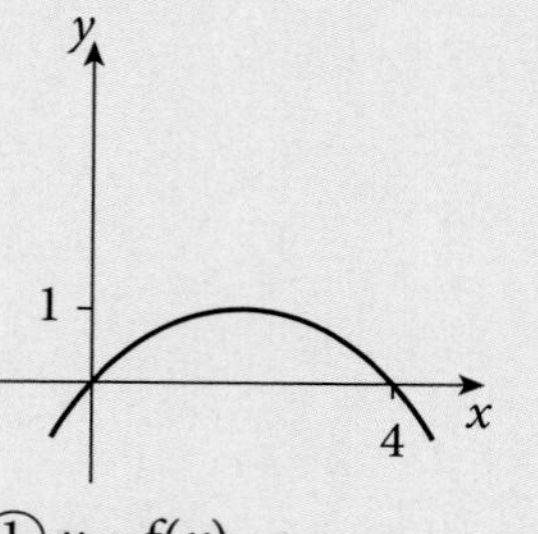

① y = f(x)

2

4

② y = 2f(x)

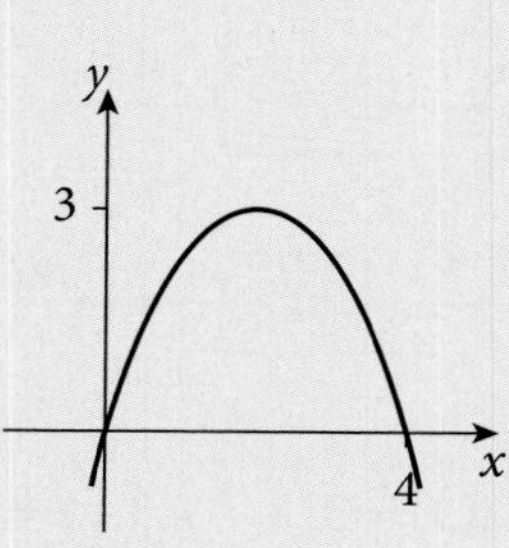

③ y = 3f(x)

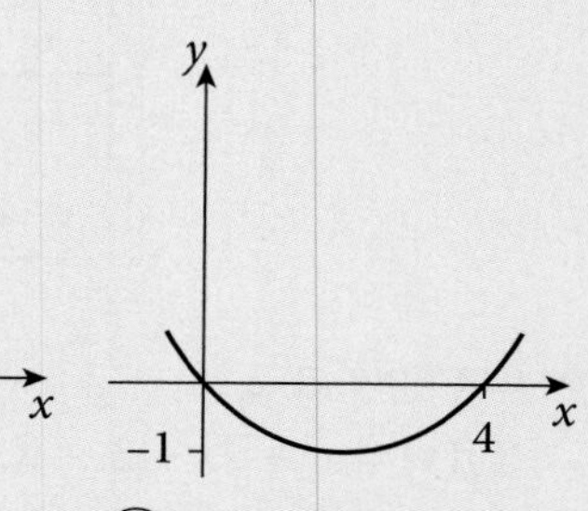

④ y = –f(x)

- Describe the transformation: from ① to ② from ① to ③ from ① to ④

Summary of rules for curve transformations

- $y = f(x) + a$: translation by a units parallel to the y-axis.
- $y = f(x - a)$: translation by a units parallel to the x-axis (note the negative sign).
- $y = af(x)$: stretch parallel to the y-axis by a scale factor a.

Example 14

The diagram show a graph of $y = f(x)$

Sketch the graphs of

a) $y = f(x) + 1$ **b)** $y = f(x + 1)$ **c)** $y = 2f(x)$

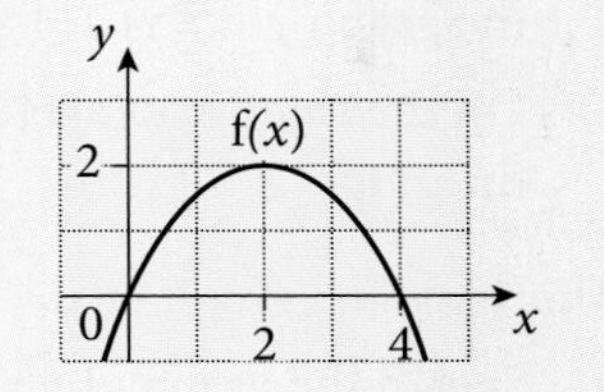

a) Translation 1 unit ↑

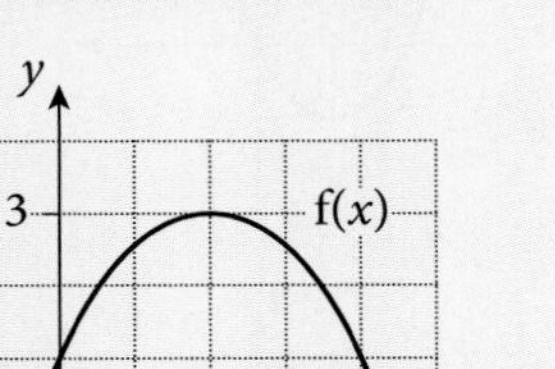

b) Translation 1 unit ←

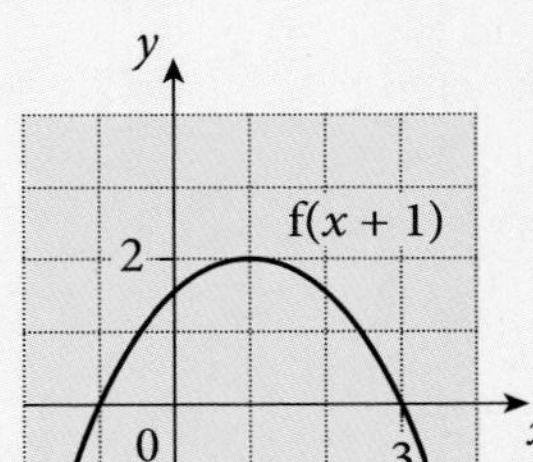

c) Stretch ↕, scale factor 2

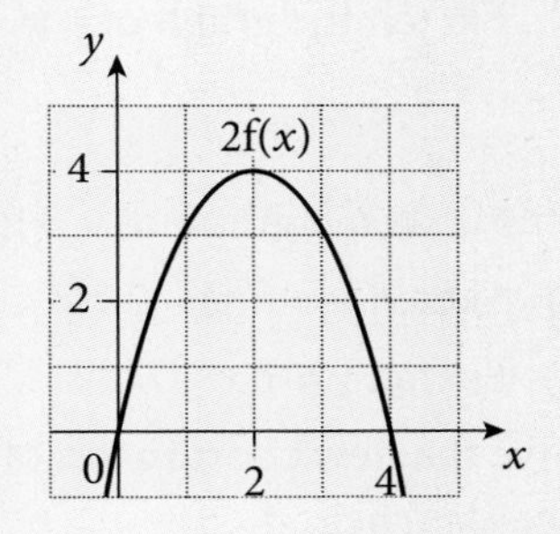

Example 15

From the sketch of $y = f(x)$, draw $y = f(x + 3)$

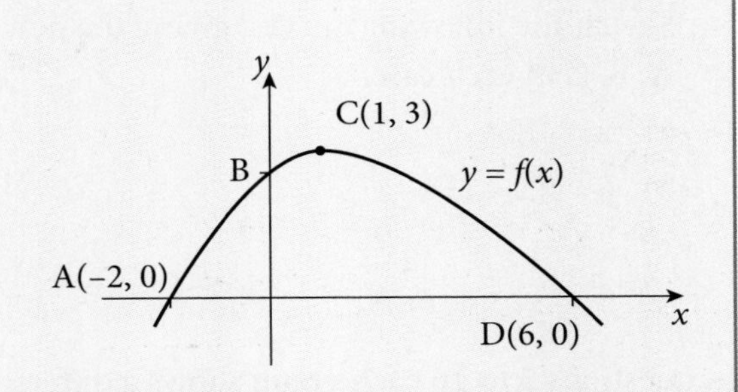

This is a translation of −3 units parallel to the x-axis.

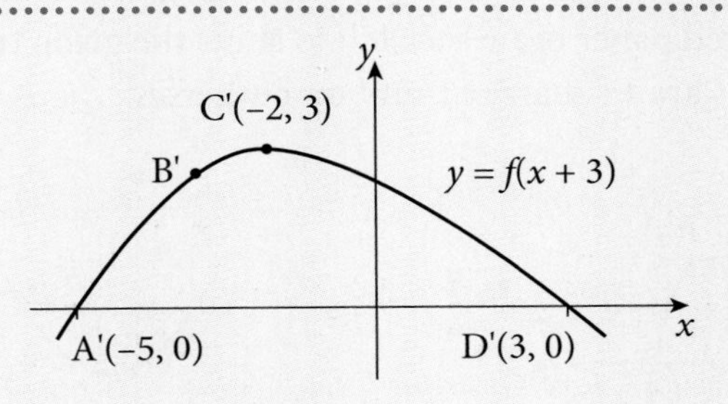

Exercise 9.10

A graphic display calculator or computer can be used effectively in this exercise.

1. a) Draw a sketch graph of $y = x^2$ for values of x from -3 to 3. Label the graph $y = f(x)$

   b) On the same axes draw a sketch graph of $y = x^2 + 4$ and label the graph $y = f(x) + 4$

   c) On the same axes sketch the graph of $y = (x - 1)^2$ and label the graph $y = f(x - 1)$

2. This is sketch graph of $y = f(x)$

   a) Sketch the graph of $y = f(x) + 3$

   b) Sketch the graph of $y = f(x + 1)$

   c) Sketch the graph of $y = -f(x)$

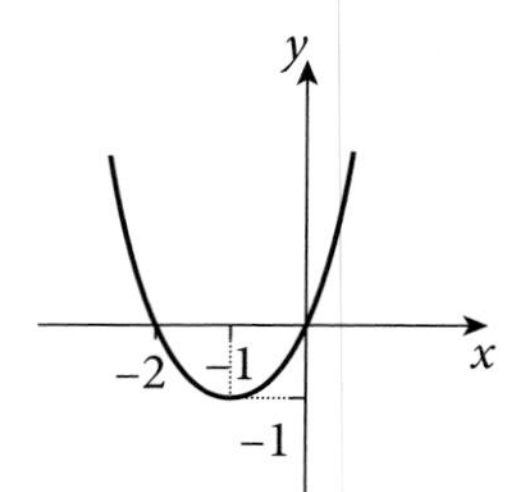

3. This is sketch graph of $y = f(x)$

   a) Sketch $y = f(x) - 2$

   b) Sketch $y = f(x - 7)$

   Give the new coordinates of the point A on the two sketches.

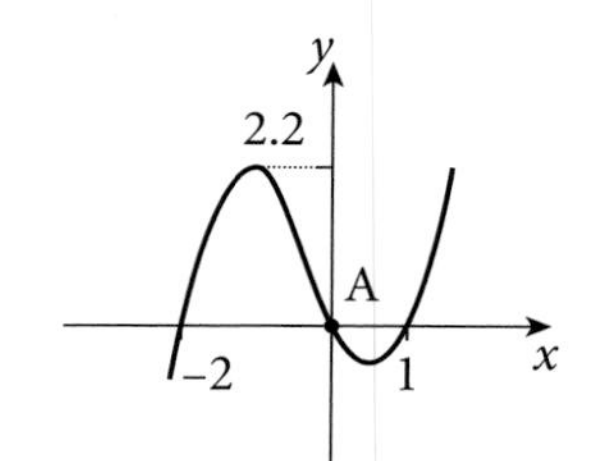

4. This is the sketch of $y = f(x)$ which passes through A, B, C. Sketch the following curves, giving the new coordinates of A, B, C in each case.

   a) $y = -f(x)$

   b) $y = f(x - 2)$

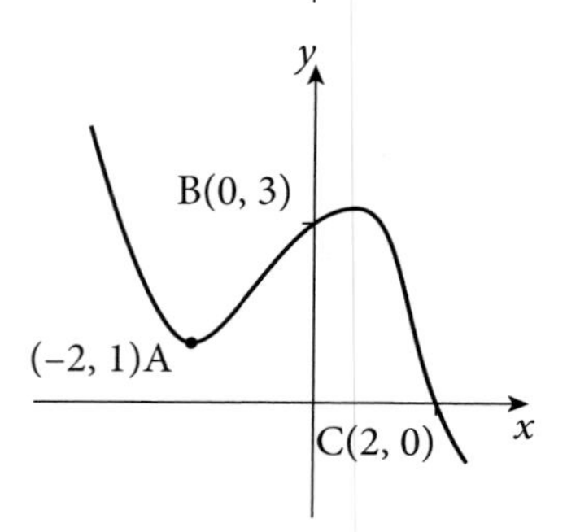

In questions **5** to **16** each graph shows a different function $f(x)$. On squared paper draw a sketch to show the given transformation. The scales are 1 square = 1 unit on both axes.

5.

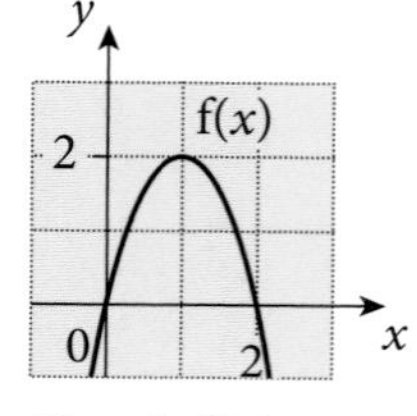

Sketch $f(x) + 3$

6.

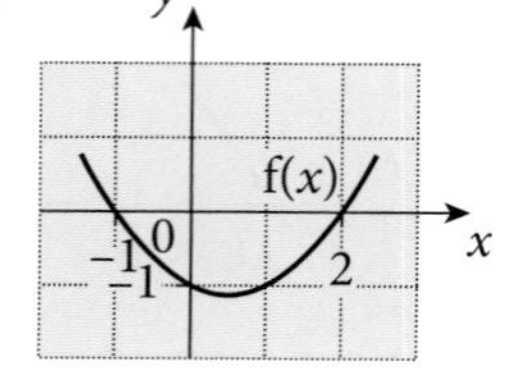

Sketch $f(x - 1)$

7.

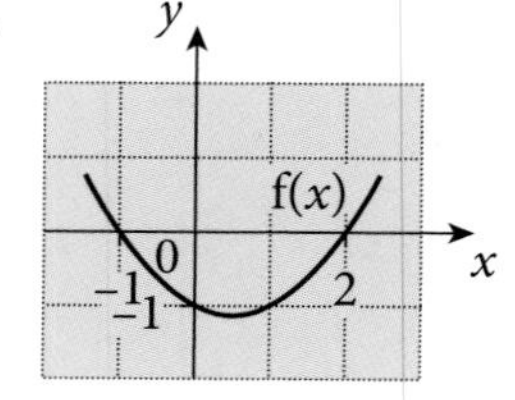

Sketch $2f(x)$

8.

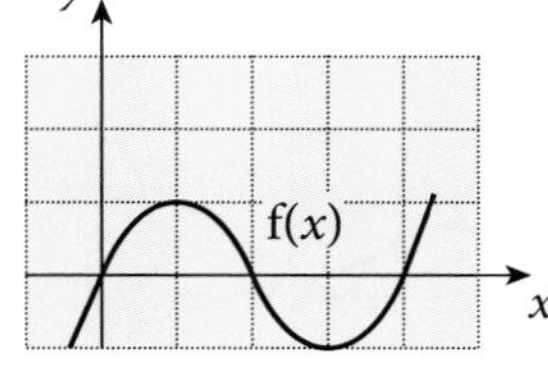

Sketch f$(x + 2)$

9.

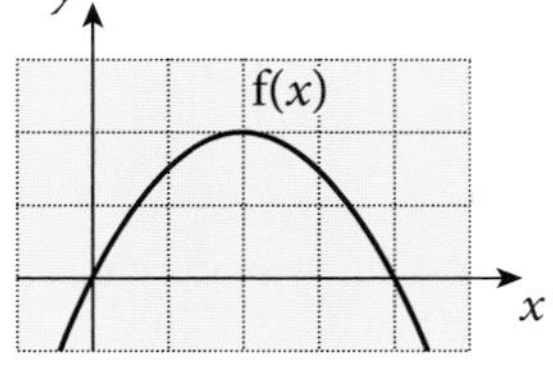

Sketch f$(2x)$

10.

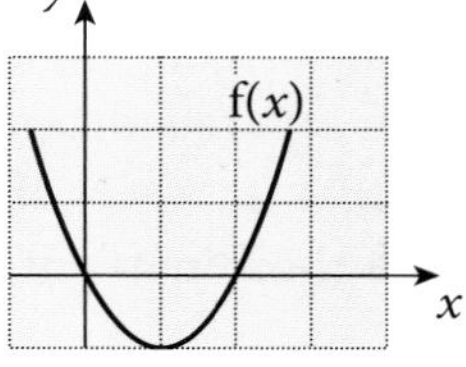

Sketch f$\left(\frac{x}{2}\right)$

11.

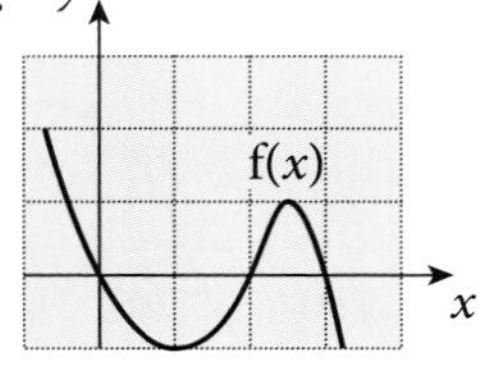

Sketch 3f(x)

12.

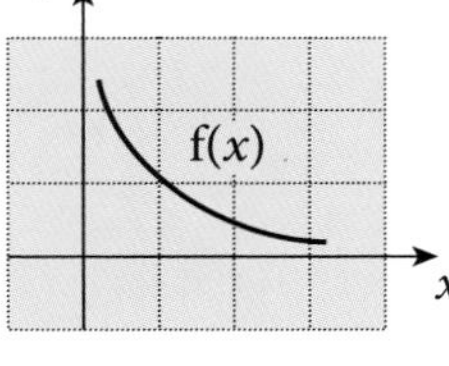

Sketch f(x) −2

13.

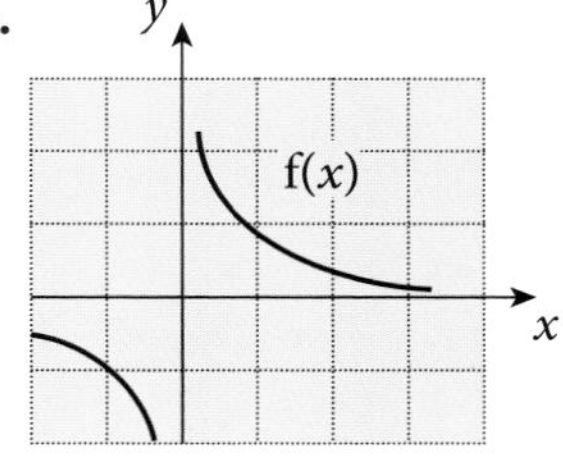

Sketch f$(x - 2)$

14.

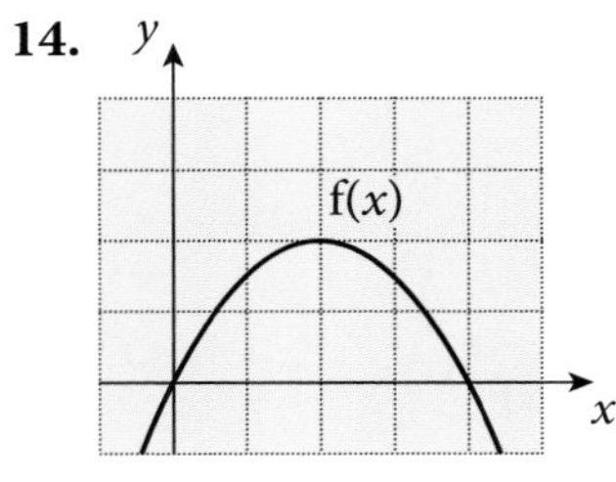

Sketch f$(4x)$

15.

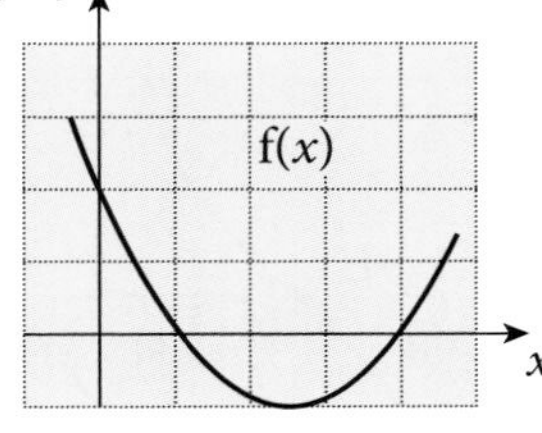

Sketch 2f(x)

16.

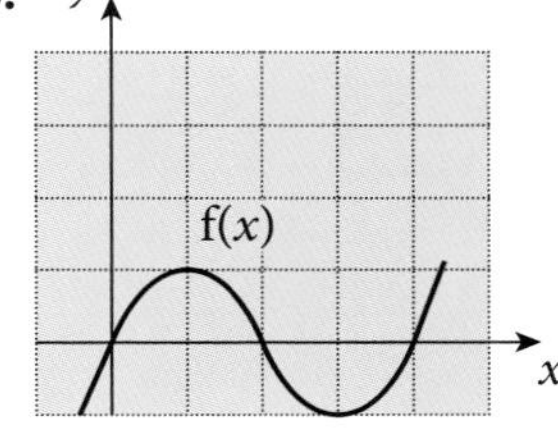

Sketch −f(x)

Exercise 9.11

1. On the same axes, sketch and label the graphs of:

a) $y = x^2$ **b)** $y = x^2 - 4$ **c)** $y = x^2 + 2$

2. On the same axes, sketch and label:

a) $y = x^2$ **b)** $y = (x - 3)^2$ **c)** $y = (x + 2)^2$

3. On the same axes, sketch and label:

a) $y = x^2$ **b)** $y = 4x^2$ **c)** $y = \frac{1}{2}x^2$

4. On the same axes, sketch and label:

a) $y = x^3$ **b)** $y = (x - 1)^3$ **c)** $y = (x - 1)^3 + 4$

5. On the same axes, sketch and label:

a) $y = \frac{1}{x}$ **b)** $y = \frac{1}{x - 2}$

6. The diagram shows the graph of $y = f(x)$. The curve passes through the origin, and has a maximum point at (1, 1). Sketch, on separate diagrams, the graphs of

a) $y = f(x) + 2$ **b)** $y = f(x + 2)$

giving the coordinates of the maximum point in each case.

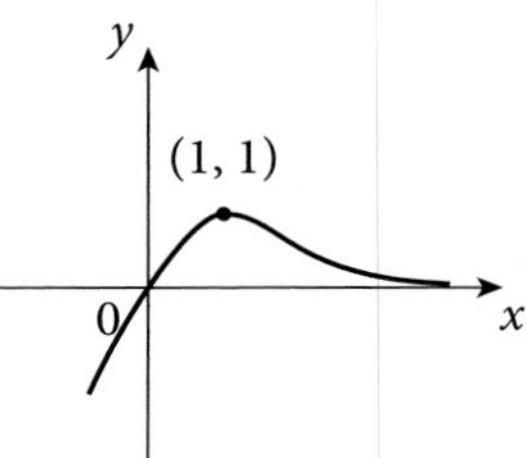

7. On squared paper copy the sketch of $y = \cos x$.

Using different colours, sketch:

a) $y = \cos(x + 90°)$ **b)** $y = 2\cos x$

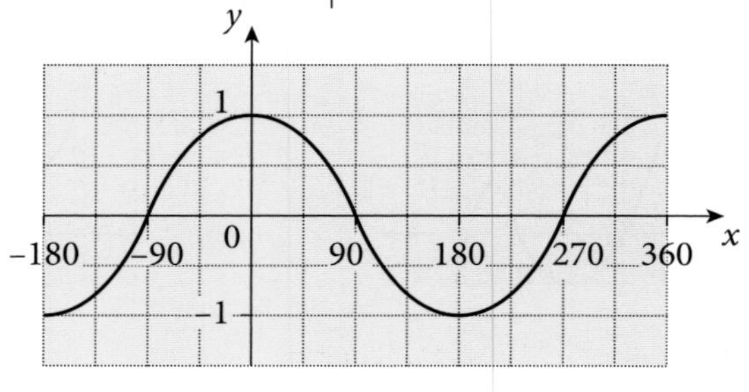

8. On squared paper copy the sketch graphs of $y = \sin x$ and $y = \tan x$.

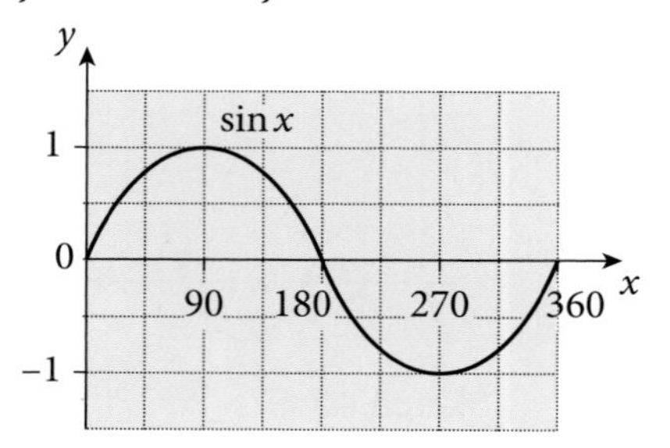

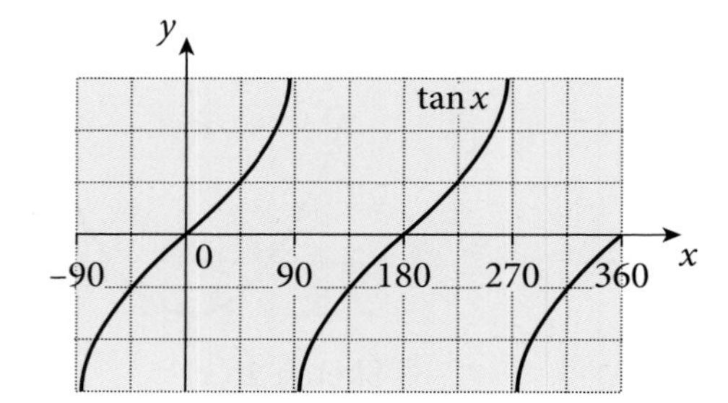

Using different colours, sketch the graphs of:

a) $y = \sin x + 1$ **b)** $y = \tan(x - 90°)$ **c)** $y = 2\sin x$

9.7 Logarithm function

A logarithm is an index.

The statement $10^2 = 100$ can be written in logarithm form,

$$10^2 = 100 \Leftrightarrow \log_{10} 100 = 2$$

Similarly $10^3 = 1000 \Leftrightarrow \log_{10} 1000 = 3$

$$3^4 = 81 \Leftrightarrow \log_3 81 = 4$$

We use 'log' for short and '$\log_{10} 100$' is 'log base 10 of 100'.

In general $a^n = b \Leftrightarrow \log_a b = n$

Consider $y = 2^x$. The inverse is $x = \log_2 y$

See page 161 in Chapter 5 for the inverse function.

The graph of $y = \log_2 x$ is the reflection of $y = 2^x$ in the line $y = x$.
The line $x = 0$ is an asymptote.
When converting from index form to log form, many people find it helpful to recall that '$\log_{10} 100 = 2$'.

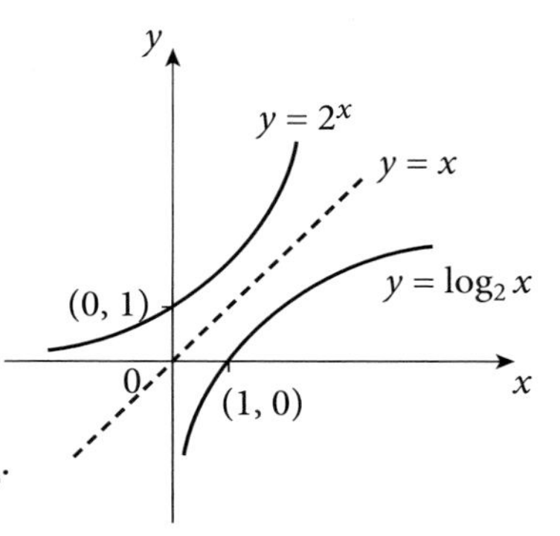

Most calculators have two 'log' keys.

log is $\log_{10}$ and ln is $\log_e$

The important constant e is not on the syllabus at this level.

Because $\log_{10}$ is used so often, mathematicians write 'log' to mean $\log_{10}$.
If any other base is used, the base must be written.

Here are two results which you should remember:

- Since $a^1 = a \Leftrightarrow \log_a a = 1$ (for $a > 0$)
- Since $a^0 = 1 \Leftrightarrow \log_a 1 = 0$ (for $a > 0$)

Laws of logarithms

We have three laws of indices:

A	$a^m \times a^n = a^{m+n}$
B	$a^m \div a^n = a^{m-n}$
C	$(a^m)^n = a^{mn}$

If $a^m = b$ and $a^n = c$ then from the above result $m = \log_a b$ and $n = \log_n c$.
From **A** we have $bc = a^{m+n}$

and so $\log_a(bc) = m + n$

Hence $\log_a(bc) = \log_a b + \log_a c$

From **B** we have $\frac{b}{c} = a^{m-n}$

and so $\log_a\left(\frac{b}{c}\right) = m - n$

Hence $\log_a\left(\frac{b}{c}\right) = \log_a b - \log_a c$

From **C** we have $b^n = (a^m)^n = a^{mn}$

and so $\log_a(b^n) = mn$

and so $n \log b = \log_a(b^n)$

Hence n $\log_a b = \log_a(b^n)$

Each law can be illustrated by an example below:

- **A** $\log 5 + \log 3 = \log 15$
- **B** $\log 35 - \log 7 = \log\left(\frac{35}{7}\right) = \log 5$
- **C** $\log 5^3 = 3 \log 5$

Example 16

Evaluate **a)** $\log_3 81$ **b)** $\log_2\left(\frac{1}{4}\right)$

a) Look for the connection between 3 and 81, ie $3^4 = 81$.

So $\log_3 81 = \log_3 3^4$
$= 4\log_3 3$
$= 4$

b) Look for the connection between 2 and $\frac{1}{4}$, ie $2^{-2} = \frac{1}{4}$

So $\log_2\left(\frac{1}{4}\right) = \log_2 2^{-2}$
$= -2\log_2 2$
$= -2$

Remember

$\log_a a = 1$

Example 17

Write x as a logarithm:

a) $3^x = 7$ **b)** $9^x = 2$ **c)** $10^x = 5{\cdot}1$

a) $3^x = 7 \Leftrightarrow \log_3 7 = x$ **b)** $9^x = 2 \Leftrightarrow \log_9 2 = x$ **c)** $10^x = 5{\cdot}1 \Leftrightarrow \log_{10} 5{\cdot}1 = x$

Example 18

Write as a single logarithm:

a) $\log 2 + \log 5$ **b)** $\log 48 - \log 6$ **c)** $2\log 5 + 3\log 2$

a) $\log 2 + \log 5 = \log 10$

b) $\log 48 - \log 6 = \log\left(\frac{48}{6}\right)$
$= \log 8$

c) $2\log 5 + 3\log 2 = \log 5^2 + \log 2^3$
$= \log 25 + \log 8$
$= \log 400$

Example 19

Express in terms of $\log a$ and $\log b$:

a) $\log a^2 b$ **b)** $\log\left(\frac{b^3}{a}\right)$ **c)** $\log\left(\frac{a}{\sqrt{b}}\right)$

a) $\log a^2 b = \log a^2 + \log b$
$= 2\log a + \log b$

b) $\log\left(\frac{b^3}{a}\right) = \log b^3 - \log a$
$= 3\log b - \log a$

c) $\log\left(\frac{a}{\sqrt{b}}\right) = \log a - \log\sqrt{b}$
$= \log a - \log b^{\frac{1}{2}}$
$= \log a - \frac{1}{2}\log b$

Example 20

Solve the equations for x correct to 3 significant figures.

a) $3^x = 4$ **b)** $5^{x+2} = 101$

a) $3^x = 4$

Take logarithm base 10 of both sides. [Remember '$\log_{10}$' is 'log']

$$\log 3^x = \log 4$$
$$x \log 3 = \log 4$$
$$x = \frac{\log 4}{\log 3} \approx \frac{0{\cdot}6020599}{0{\cdot}4771212}$$
$$= 1{\cdot}26 \text{(to 3 s.f.)}$$

Be careful here:

$\frac{\log 4}{\log 3}$ is *not* $\log \frac{4}{3}$!

b) $5^{x+2} = 101$

$$\log 5^{x+2} = \log 101$$
$$(x+2)\log 5 = \log 101$$
$$x + 2 = \frac{\log 101}{\log 5} \approx 2{\cdot}86757$$
$$x = 0{\cdot}868 \text{ (to 3 s.f.)}$$

Using a GDC

TI-Nspire	TI84-Plus	Casio
$\frac{\log_{10}(4)}{\log_{10}(3)}$ 1.26186 $\frac{\log_{10}(101)}{\log_{10}(5)}-2$ 0.867536	$\frac{\log(4)}{\log(3)}$ 1.261859507 $\frac{\log(101)}{\log(5)}-2$ .8675356043	$\frac{\log 4}{\log 3}$ 1.261859507 $\frac{\log 101}{\log 5}-2$ 0.8675356043

Exercise 9.12

1. Write in log form. [For example $3^4 = 81 \Leftrightarrow \log_3 81 = 4$]

a) $5^3 = 125$ **b)** $2^5 = 32$ **c)** $10^4 = 10000$

d) $2^3 = 8$ **e)** $49 = 7^2$ **f)** $x^n = 10$

2. Evaluate the following.

a) $\log_7 49$ **b)** $\log_2 8$ **c)** $\log_5 25$

d) $\log_2 32$ **e)** $\log_4 1$ **f)** $\log_2\left(\frac{1}{2}\right)$

g) $\log_{10} 0{\cdot}01$ **h)** $\log_3 \sqrt{3}$ **i)** $\log_{36}\left(\frac{1}{6}\right)$

j) $\log_{10} 1000$ **k)** $\log_4 8$ **l)** $\log_e e^2$

3. Write x as a logarithm in the following.

a) $10^x = 7$ b) $5^x = 3$ c) $8^x = 2$

4. Express the following as a single logarithm.

a) $\log 3 + \log 7$ b) $\log 4 + \log 5$

c) $3\log 4$ d) $2\log 3 + 3\log 2$

e) $2\log 11 + \log 1$ f) $\log 6 - \log 3$

g) $2\log 8 - 3\log 2$ h) $3\log 5 + 4\log 6 - 3\log 10$

5. Express the following in terms of $\log a$, $\log b$ and $\log c$.

a) $\log a^3 b$ b) $\log bc^2$ c) $\log \sqrt{a}$

d) $\log c^5$ e) $\log \frac{ab^2}{c^3}$ f) $\log \frac{a^4 b^3}{\sqrt{c}}$

6. Find x in the following equations. Try not to use a calculator.

a) $2^x = 32$ b) $5^x = 625$ c) $7^x = \frac{1}{49}$

d) $13^x = 1$ e) $3^{2x+1} = 243$ f) $2^{3x-5} = 16$

7. Solve the following equations to 3 s.f.

a) $10^x = 17$ b) $10^x = 251$ c) $4^x = 9$

8. Find x (to 3 s.f. if necessary) in the following equations.

a) $2^x = 17$ b) $3^x = 308$

c) $5^x = 27$ d) $7^x = 251$

e) $4^{x+1} = 13$ f) $5^{3x+1} = 350$

g) $7^{5x-4} = 80$ h) $2^{2x-3} = 65$

i) $2^{x+1} = 3^{3x-2}$ j) $5^{2x-3} = 7^{3x-1}$

9. Sketch the curves $y = \log x$ and $y = \log 2x$ on one graph.

Example 21

Solve the equation $\log_4(x + 3) - \log_4 x = 3$

$\log_4(x + 3) - \log_4 x = 3$

$$\log_4\left(\frac{x+3}{x}\right) = 3$$

$$\frac{x+3}{x} = 4^3 = 64$$

$$x + 3 = 64x$$

$$3 = 63x$$

$$x = \frac{1}{21}$$

Example 22

If $\log_2 y = 15$, find the value of $\log_2 8y$

$$\begin{aligned}\log_2 8y &= \log_2 8 + \log_2 y \\ &= \log_2 2^3 + \log_2 y \\ &= 3\log_2 2 + \log_2 y \\ &= 3 + \log_2 y\end{aligned}$$

Now $\log_2 y = 15$ (given).

So $\log_2 8y = 3 + 15$

$= 18$

Example 23

Solve the equation $5^{2x} - 6(5^x) = 0$

Now $5^{2x} = (5x)^2$ so we may write

$(5^x)^2 - 6(5^x) = 0$

$5^x(5^x - 6) = 0$

$5^x = 6 \quad (5^x \neq 0)$

$\log_{10} 5^x = \log_{10} 6$

$x\log_{10} 5 = \log_{10} 6$

$x = \dfrac{\log_{10} 6}{\log_{10} 5} = 1{\cdot}11$ (to 3 s.f.)

Exercise 9.13

The questions in this exercise are more difficult.

1. Find the following:

   a) $\log_2 32$ b) $\log_3 9$ c) $\log_2 \sqrt{2}$

   d) $\log_7 1$ e) $\log_{36} 6$ f) $\log_9 \sqrt{3}$

   g) $\log_3 \dfrac{1}{27}$ h) $\log_{49} \dfrac{1}{7}$ i) $\log_5 (5\sqrt{5})$

2. Write the following as single logarithms.

   a) $\log_{10}(x + 4) - \log_{10}(x)$ b) $\log_{10}(x + 1) - \log_{10}(x + 2)$

   c) $\log_3(x + 3) + \log_3(x + 5) - \log_3(x + 1)$ d) $\log_7(3x + 2) + \log_7(x - 3)$

   e) $\log_2(x^2 - 1) + \log_2(x + 2) - \log(x + 1)$

3. a) Express $\log_3(x + 4) - \log_3 x$ as a single logarithm.

   b) Hence solve $\log_3(x + 4) - \log_2 x = 2$.

4. Solve the following equations.

   a) $\log_7(x + 4) - \log_7(x) = 2$ b) $\log_2(2x + 9) - \log_2(x + 1) = 3$

   c) $\log_5(3x + 95) = 2 + \log_5(x + 3)$

5. If $a \log_2 x = 12$ then find the following.

a) $\log_2 x^3$ b) $\log_2 16x$ c) $\log_2 \sqrt{x}$

6. Solve the following equations.

a) $3^{x+1} = 9$ b) $4^{2x-1} = 65$ c) $\log_{10}(5x) = 6$

d) $\log_{10}(3x + 1) = 2$ e) $\log_7(x - 3) = 8$ f) $\log_x 5 = 4$

g) $\log_x 15 = 3$ h) $\log_3 9 = \log_{10} x$ i) $\log_5 x = \log_4 2$

7. Solve the equation $3^{2x} - 8(3^x) = 0$.

8. If $7^y = 3^x$, show that $y = kx$ for some constant k which is to be determined.

9. If $8^y = 5^x$, show that $y = kx$ for some constant k and find the value of k.

10. a) If $\log x^3 + \log xy - \log 3x = 0$, find an equation involving $\log x$ and $\log y$.

b) Hence find y in terms of x.

11. a) If $\log x^2 - \log xy + \log 5x^2 = 0$, find an equation involving $\log x$ and $\log y$.

b) Hence find y in terms of x.

12. Find the value of x for which $2^{3x+1} = 3^{x+2}$, giving three significant figures in your answer.

Revision exercise 9A

1. Find the equation of the straight line satisfied by the points given in the following tables.

a)

x	2	7	10
y	–5	0	3

b)

x	1	2	3
y	7	9	11

c)

x	1	2	3
y	8	6	4

d)

x	3	4	5
y	2	$2\frac{1}{2}$	3

2. Find the gradient of the line joining each pair of points.

a) (3, 3)(5, 7) b) (3, –1)(7, 3)

c) (–1, 4)(1, –3) d) (2, 4)(–3, 4)

e) (0·5, –3)(0·4, –4)

3. Find the gradient and the intercept on the y-axis for the following lines. Draw a *sketch* graph of each line.

a) $y = 2x - 7$ b) $y = 5 - 4x$

c) $2y = x + 8$ d) $2y = 10 - x$

e) $y + 2x = 12$ f) $2x + 3y = 24$

4. In the diagram, the equations of the lines are $y = 3x$, $y = 6$, $y = 10 - x$ and $y = \frac{1}{2}x - 3$.

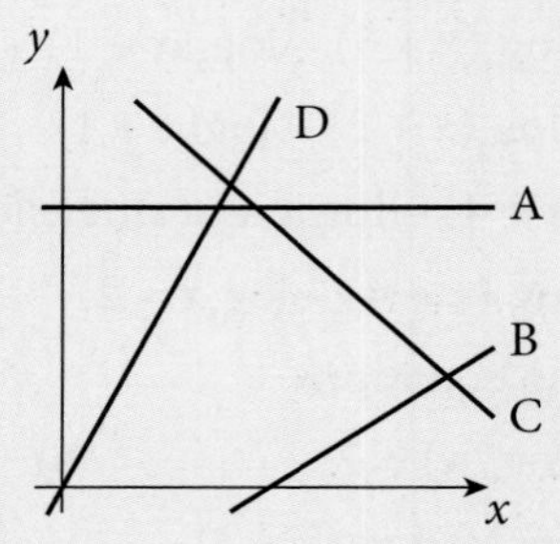

Find the equation corresponding to each line.

5. In the diagram, the equations of the lines are $2y = x - 8$, $2y + x = 8$, $4y = 3x - 16$ and $4y + 3x = 16$.

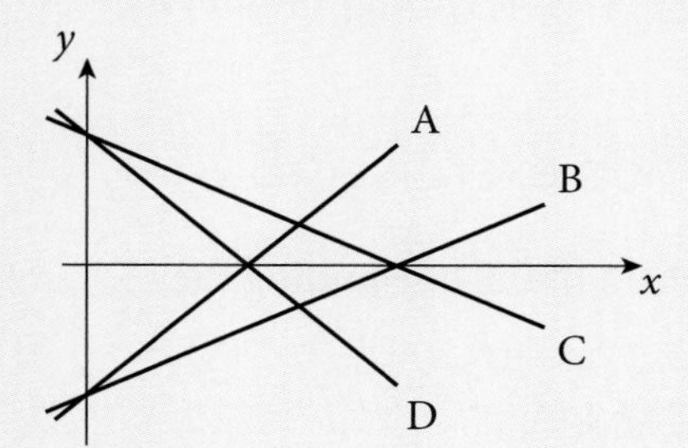

Find the equation corresponding to each line.

6. Find the equations of the lines which pass through the following pairs of points.

a) (2, 1)(4, 5) **b)** (0, 4)(−1, 1)

c) (2, 8)(−2, 12) **d)** (0, 7)(−3, 7)

7. The sketch represents a section of the curve $y = x^2 - 2x - 8$.

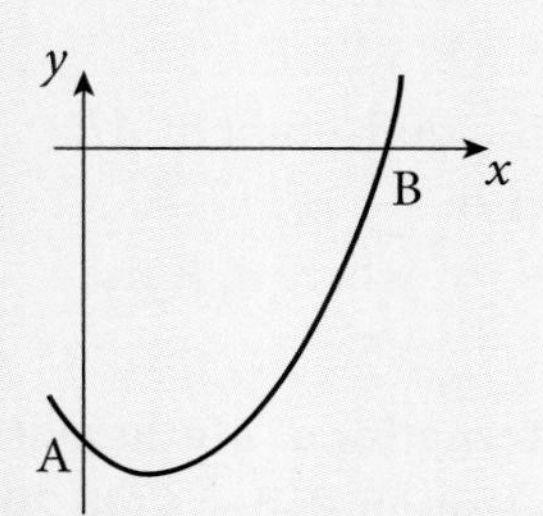

Calculate:

a) the coordinates of A and of B

b) the gradient of the line AB

c) the equation of the straight line AB.

8. Find the area of the triangle formed by the intersection of the lines $y = x$, $x + y = 10$ and $x = 0$.

9. Use a calculator to draw the graph of $y = 7 - 3x - 2x^2$ for $-4 \le x \le 2$. Find the gradient of the tangent to the curve at the point where the curve cuts the y-axis.

10. Use a calculator to draw the graph of $y = \frac{4000}{x} + 3x$ for $10 \le x \le 80$. Find the minimum value of y.

11.

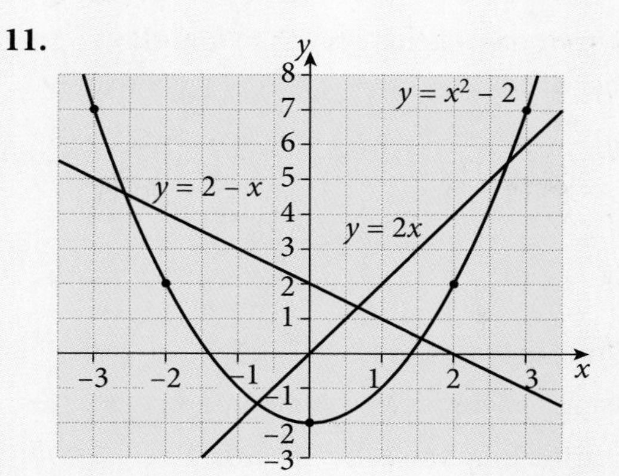

The graphs of $y = x^2 - 2$, $y = 2x$ and $y = 2 - x$ are shown. Use the graphs to find approximate solutions to the following equations.

a) $x^2 - 2 = 2 - x$ **b)** $x^2 - 2 = 2x$

c) $x^2 - 2 = 2$ **d)** $2 - x = 2x$

12. Draw the graph of $y = x^2 - 2x + 2$ for $-2 \le x \le 4$. By drawing other graphs, solve the equations:

a) $x^2 - 2x + 2 = 8$

b) $x^2 - 2x + 2 = 5 - x$ **c)** $x^2 - 2x - 5 = 0$

13. Here are five sketch graphs.

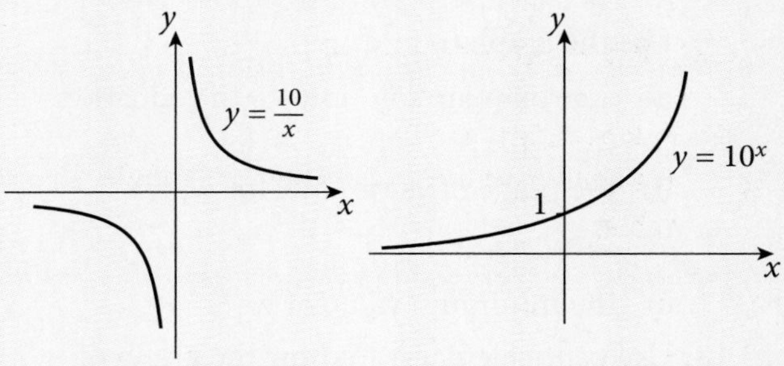

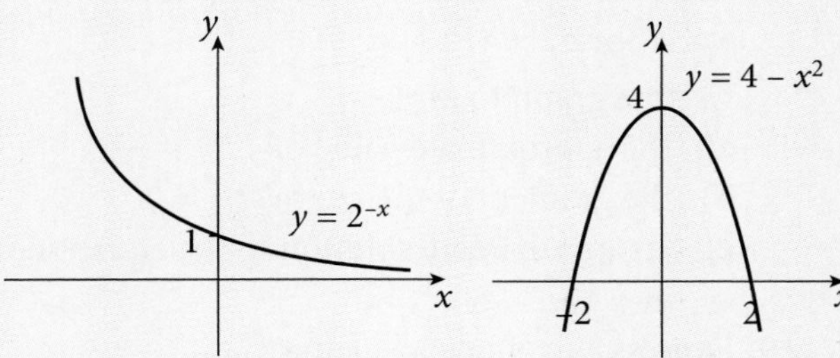

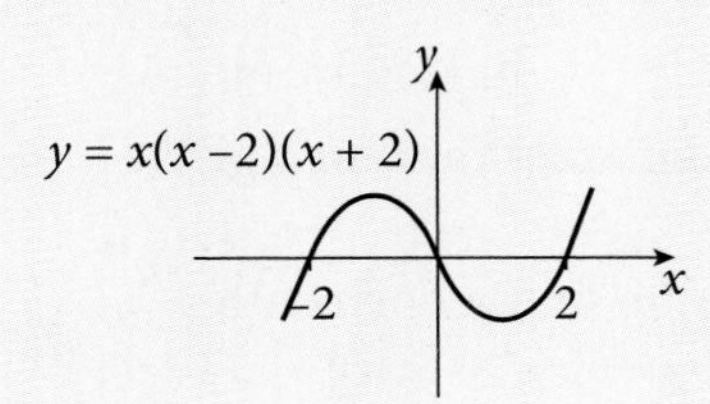

Make your own sketch graphs to find the number of solutions of the following equations.

a) $\frac{10}{x} = 10^x$ b) $4 - x^2 = 10^x$

c) $x(x-2)(x+2) = \frac{10}{x}$ d) $2^{-x} = 10^x$

e) $4 - x^2 = 2^{-x}$ f) $x(x-2)(x+2) = 0$

14. Draw the graph of $y = \frac{18}{x}$ for $1 \le x \le 10$, using scales of 1cm to one unit on both axes. Use the graph to solve approximately:

a) $\frac{18}{x} = x + 2$ b) $\frac{18}{x} + x = 10$ c) $x^2 = 18$

15. Draw the graph of $y = 3x^2 - x^3$ for $-2 < x < 3$. Use your graph to find the range of values of k for which the equation $3x^2 - x^3 = k$ has three solutions.

16. Show that the line $y = 2x$ is a tangent to curve $y = x^2 + 1$. Find the coordinates of the point of contact.

17. Use a calculator to draw the graph of $y = \frac{5}{x} + 2x - 3$, for $\frac{1}{2} \le x \le 7$

Use the graph to find:

a) approximate solutions to the equation $2x^2 - 10x + 7 = 0$

b) the range of values of x for which $\frac{5}{x} + 2x - 3 < 6$

c) the minimum value of y.

18. Use your calculator to draw the graph of $y = 4^x$ for $-2 \le x \le 2$

Use the graph to find:

a) the approximate value of $4^{1.6}$

b) the gradient of the curve at $x = 0$

c) an approximate solution to the equation $4^x = 10$.

19. Express as a single logarithm:

a) $\log 2 + 2\log 8$ b) $\log 36 - \log 4$

20. a) Write down the exact value of x given that $4^x = 8$.

b) Use logarithm to find y, correct to 3 decimal places, when $5^y = 10$.

21. a) Simplify $\frac{x^2 + 4x + 3}{x^2 + x}$

b) Find the value of x for which $\log_2(x^2 + 4x + 3) - \log_2(x^2 + x) = 4$.

Examination exercise 9B

1. The gradient of the line joining the points (2, 1) and (6, a) is $\frac{3}{2}$. [3]

Find the value of a.

Cambridge IGCSE International Mathematics 0607 Paper 2 Q3 October/November 2010

2. A is the point (0, 2) and B is the point (3, 8).

a) Find the equation of the straight line which passes through A and B. [3]

b) Find the equation of the line perpendicular to AB, which passes through the mid-point of AB. Give your answer in the form $ax + by = d$ where a, b and d are integers. [5]

Cambridge IGCSE International Mathematics 0607 Specimen Paper 4 Q6 2010

3.

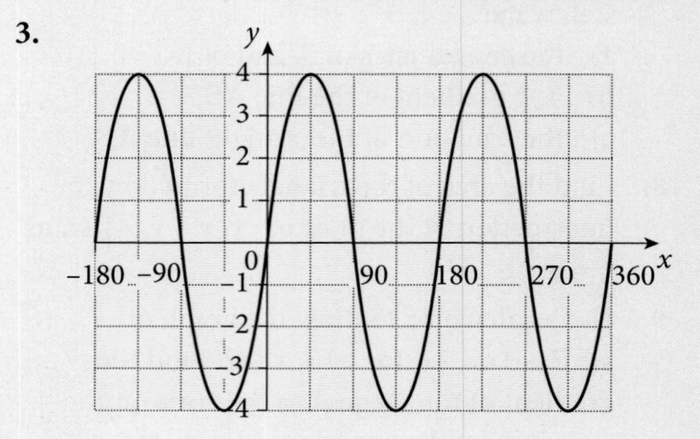

The diagram shows the graph of $y = f(x)$, where $f(x) = a\sin(bx)$.

Find the values of a and b. [2]

Cambridge IGCSE International Mathematics 0607 Paper 2 Q4 May/June 2010

4. The graphs shown are translations of the graph of $y = x^2$.
Write down their equations.

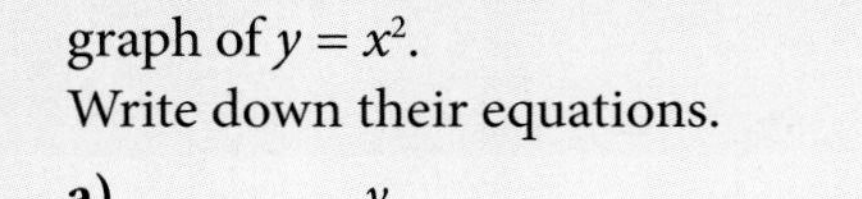

a)

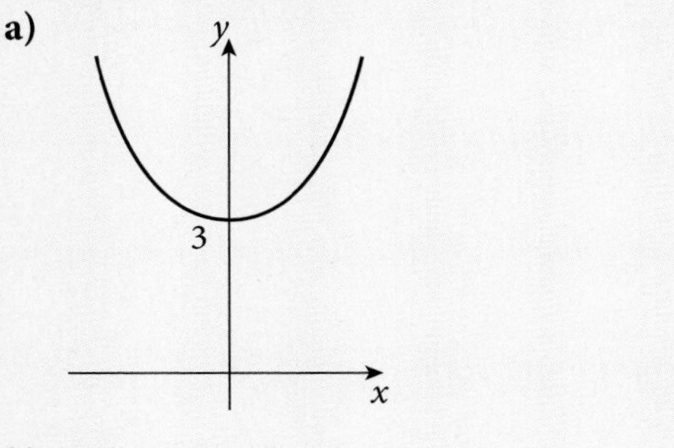

[1]

b)

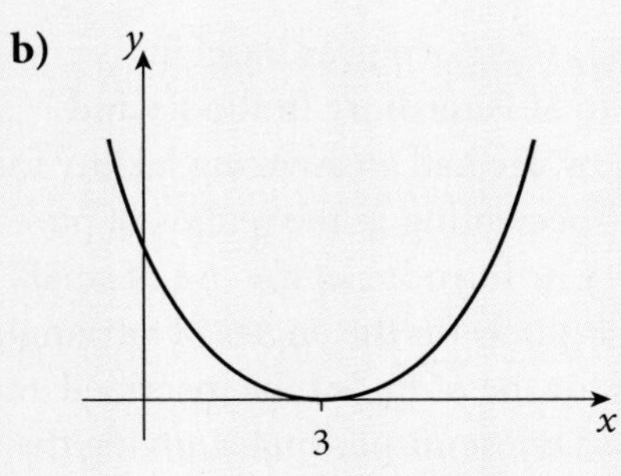

[1]

Cambridge International IGCSE Mathematics
0607 Specimen Paper 2 Q2 2010

5. $f(x) = \sin x°$ $g(x) = 2\sin x°$
$h(x) = 3\sin(4x)°$ $k(x) = \sin(x + 60)°$

a) Write down the domain of $f(x)$. [1]

b) Write down the amplitude and period of $h(x)$. [2]

c) Describe fully a **single** transformation that maps the graph of $y = f(x)$ onto the graph of

i) $y = g(x)$, [3]

ii) $y = k(x)$. [2]

Cambridge IGCSE International Mathematics
0607 Paper 4 Q7 October/November 2010

6. a) $3\log 2 + 2\log 3 = \log k$ [2]

Find the value of k. [1]

b) Find the value of $\frac{\log 25}{\log 5}$.

Cambridge IGCSE International Mathematics
0607 Paper 2 Q6 May/June 2010

7.

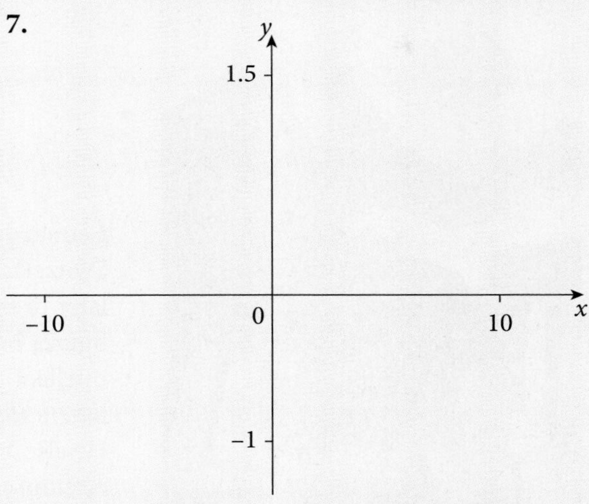

a) i) On the axes, sketch the graph of $y = f(x)$, where $f(x) = \log x$ for $0 < x \leqslant 10$. [2]

ii) Write down the co-ordinates of the point where the graph of $y = f(x)$ crosses the x-axis. [1]

iii) Write down the equation of the asymptote of the graph of $y = f(x)$. [1]

iv) Find the range of $f(x)$ for the domain $0 < x \leqslant 10$. [2]

b) Solve the equation $\log x = \frac{10 - x}{20}$. [2]

c) $g(x) = \frac{10 - x}{20}$ [2]

Find the range of $g(x)$ for the domain $-10 \leqslant x \leqslant 10$.

d) On the axes, sketch the graph of $y = f(x + 1)$. [2]

Cambridge IGCSE International Mathematics
0607 Paper 4 Q9 May/June 2011

10 Trigonometry

Leonard Euler (1707–83) was born near Basel in Switzerland but moved to St Petersburg in Russia and later to Berlin in Germany. He had an amazing facility for figures but delighted in speculating in the realms of pure intellect. In trigonometry he introduced the use of small letters for the sides and capitals for the angles of a triangle. He also wrote r, R and s for the radius of the inscribed and circumscribed circles and the semi-perimeter, giving the beautiful formula $4rRs = abc$.

8.1 Right-angled triangle trigonometry
8.2 Exact values for the trigonometric ratios 0°, 30°, 45°, 60°, 90°
8.3 Extension to the four quadrants i.e. 0°–360°
8.4 Sine Rule, formula given, ASA, SSA (ambiguous case)
8.5 Cosine Rule (formula given, SAS, SSS)
8.6 Area of a triangle (formula given)
8.7 Applications:
three-figure bearings and North, East, South, West problems
in two and three dimensions

10.1 Right-angled triangles

Look at the diagram.
The side opposite the right angle is called the **hypotenuse** (we will use H). It is the longest side.
The side opposite the marked angle of 35° is called the **opposite** (we will use O).
The other side is called the **adjacent** (we will use A).
Consider two triangles, one of which is an enlargement of the other.

It is clear that the *ratio* $\frac{O}{H}$ will be the same in both triangles.

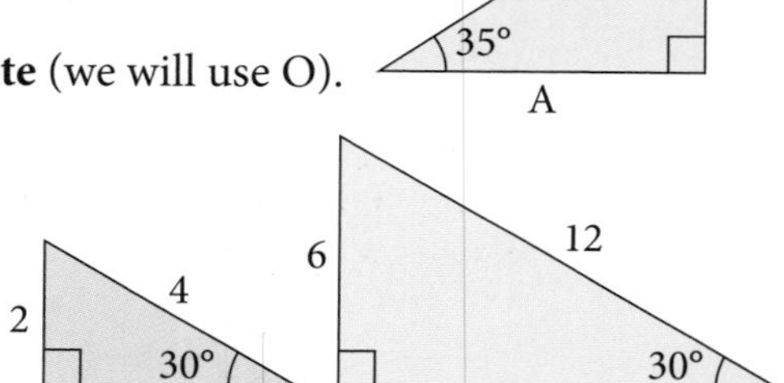

Sine, cosine and tangent

Three important functions are defined as follows:

- The **sine**, $\sin x = \frac{O}{H}$
- The **cosine**, $\cos x = \frac{A}{H}$
- The **tangent**, $\tan x = \frac{O}{A}$

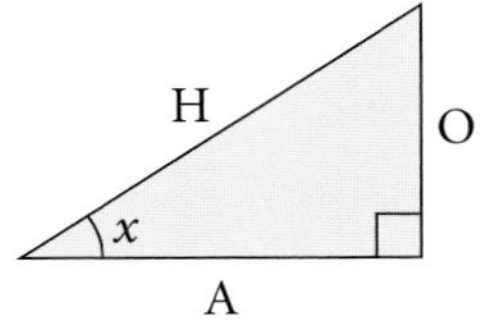

It is important to get the letters in the right order. Some people find a simple sentence helpful when the first letters of each word describe sine, cosine or tangent and Hypotenuse, Opposite and Adjacent. An example is:

Silly Old Harry Caught A Herring Trawling Off Australia.

e.g. S O H means $\sin = \frac{O}{H}$

For any angle x the values for $\sin x$, $\cos x$ and $\tan x$ can be found using a calculator.

Exercise 10.1

1. Draw a circle of radius 10 cm and construct a tangent to touch the circle at T.

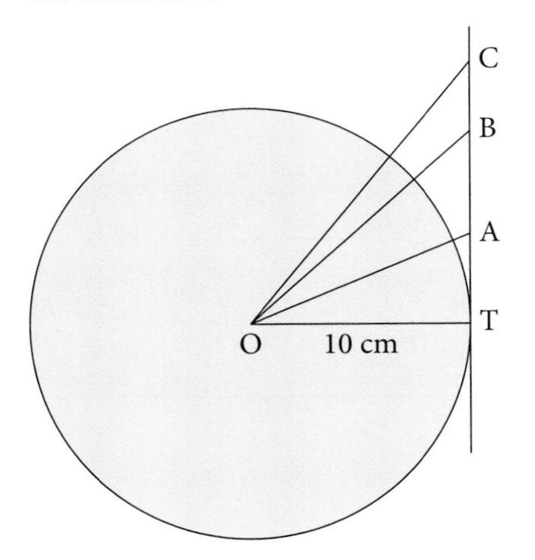

Draw OA, OB and OC where $A\hat{O}T = 20°$

$B\hat{O}T = 40°$

$C\hat{O}T = 50°$

Measure the length AT and compare it with the value for tan 20° given on a calculator. Repeat for BT, CT and for other angles of your own choice.

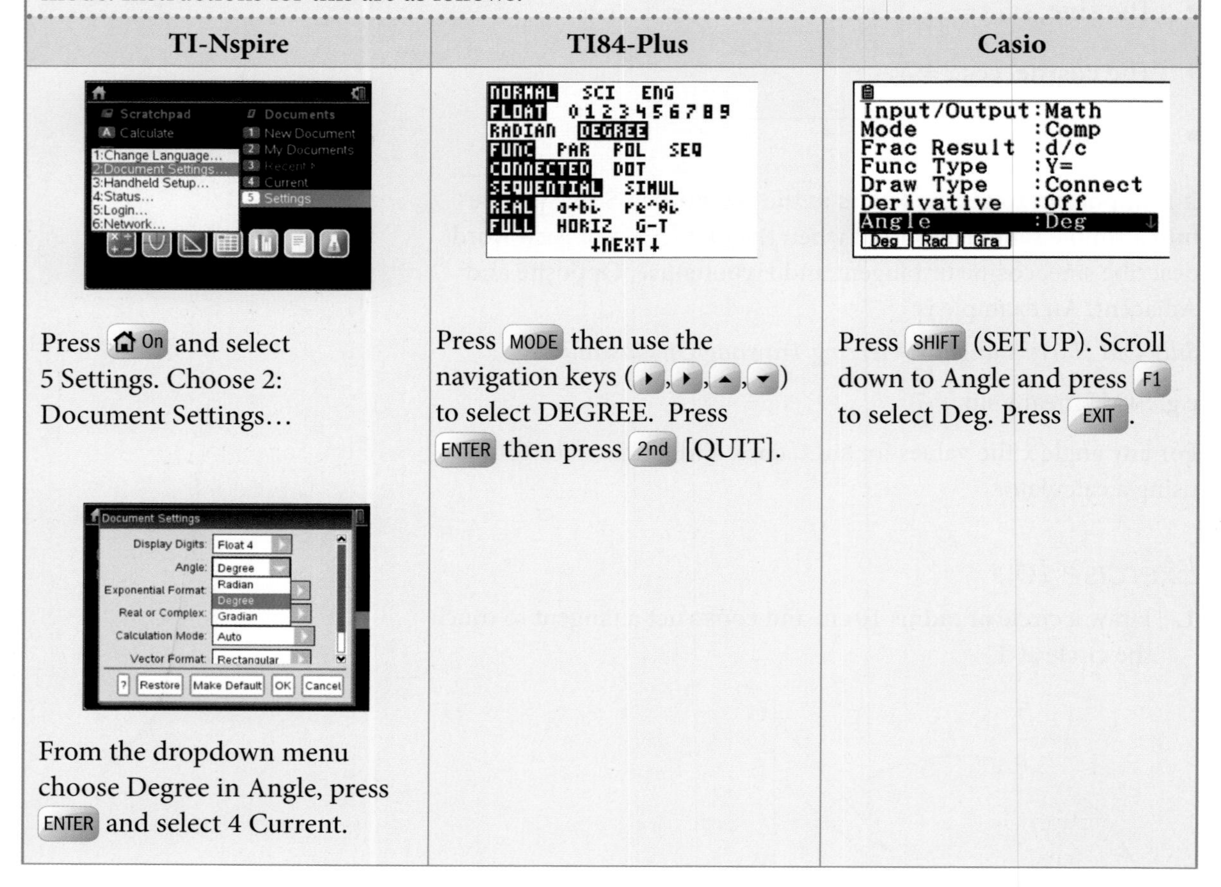

GDC help

When using your GDC in trigonometry calculations, you must ensure that it is set in 'Degree' mode. Instructions for this are as follows.

TI-Nspire	TI84-Plus	Casio
Press [On] and select 5 Settings. Choose 2: Document Settings…	Press [MODE] then use the navigation keys ([▶], [▶], [▲], [▼]) to select DEGREE. Press [ENTER] then press [2nd] [QUIT].	Press [SHIFT] (SET UP). Scroll down to Angle and press [F1] to select Deg. Press [EXIT].
From the dropdown menu choose Degree in Angle, press [ENTER] and select 4 Current.		

Finding the length of a side

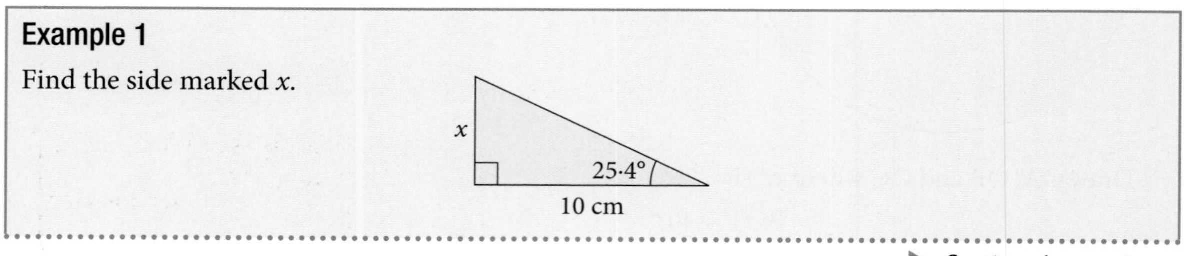

Example 1

Find the side marked x.

▶ Continued on next page

a) Label the sides of the triangle H, O, A (in brackets).

$\tan 25{\cdot}4° = \frac{O}{A} = \frac{x}{10}$

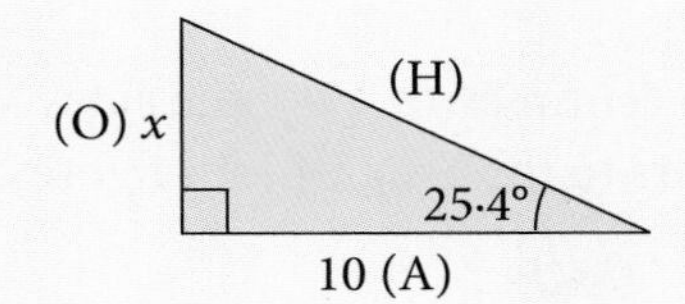

b) In this example, we know nothing about H so we need the function involving O and A.

c) Find tan 25·4° from a calculator.

$\tan 25{\cdot}4° = \frac{x}{10}$

d) Solve for x.

$x = 10 \times \tan 25{\cdot}4$

$x = 4{\cdot}75\,\text{cm}$ (3 s.f.)

Using GDC

TI-Nspire	TI84-Plus	Casio
10·tan(25.4) 4.74835	10*tan(25.4) 4.74834911	10×tan 25.4 4.74834911

Example 2

Find the side marked z.

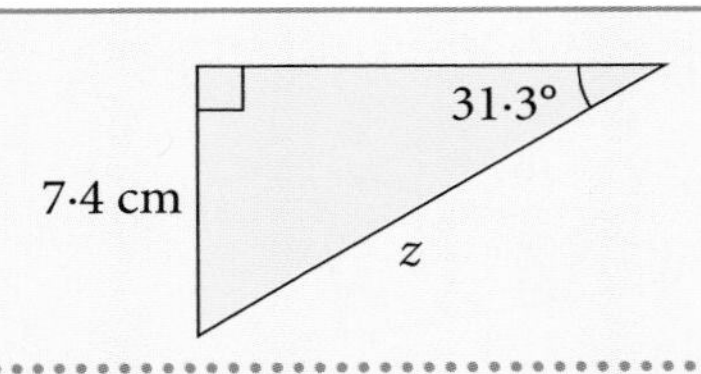

a) Label H, O, A.

b) $\sin 31{\cdot}3° = \frac{O}{H} = \frac{7{\cdot}4}{z}$

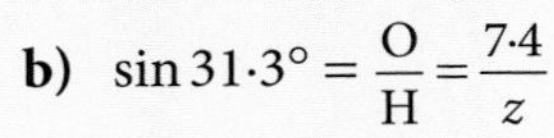

c) Multiply by z.

$z \times (\sin 31{\cdot}3°) = 7{\cdot}4$

$z = \frac{7.4}{\sin 31.3}$

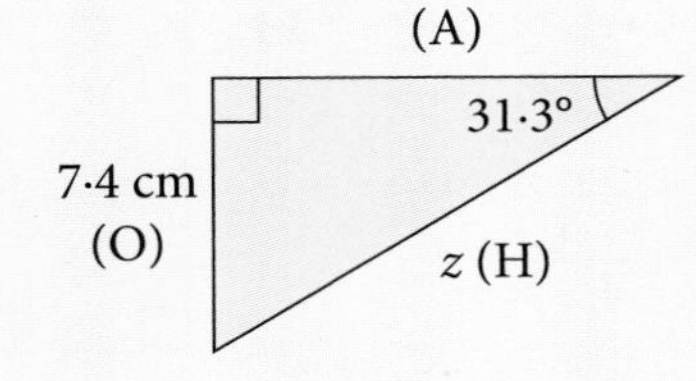

d) Use a GDC to solve

TI-Nspire	TI84-Plus	Casio
$\frac{7.4}{\sin(31.3)}$ 14.2439	7.4/sin(31.3) 14.24394181	$\frac{7.4}{\sin 31.3}$ 14.24394181

You can also use the fraction template.

Exercise 10.2

In questions **1** to **22** all lengths are in centimetres. Find the sides marked with letters. Give your answers to three significant figures.

1.

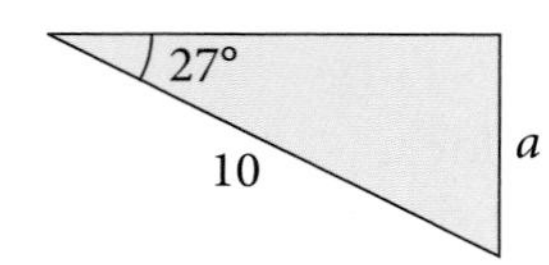

2.

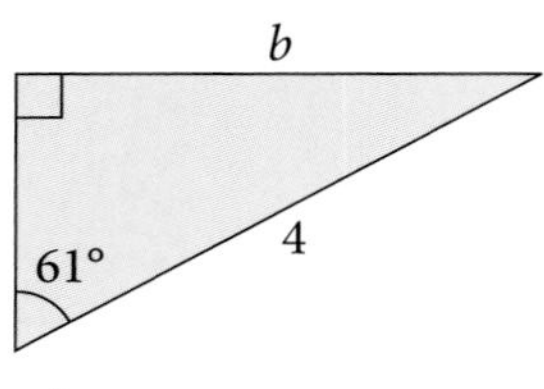

3.

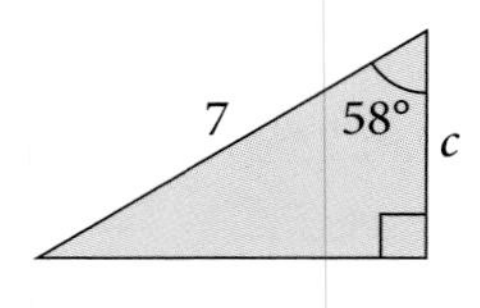

4

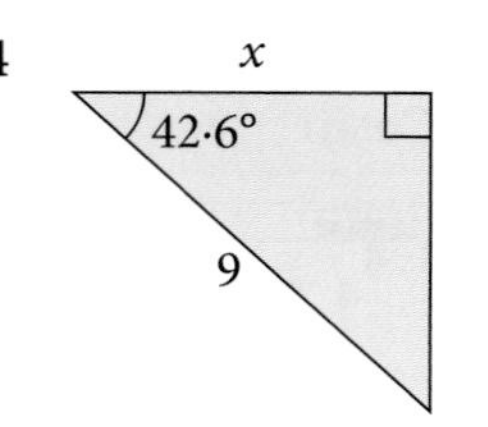

5.

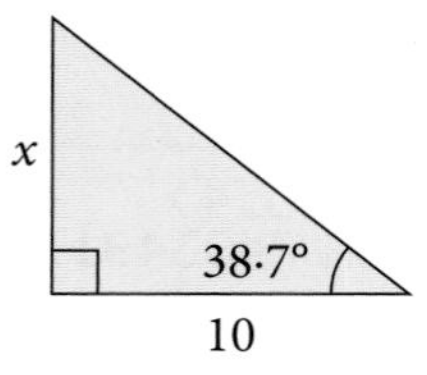

6.

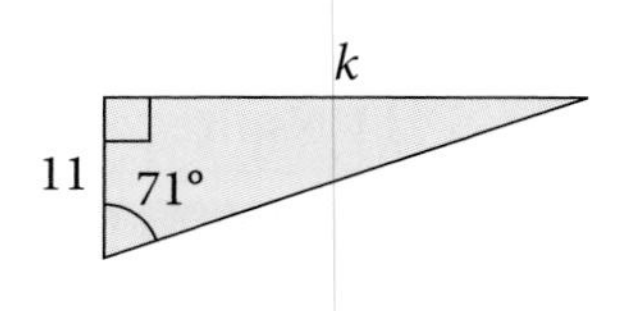

7.

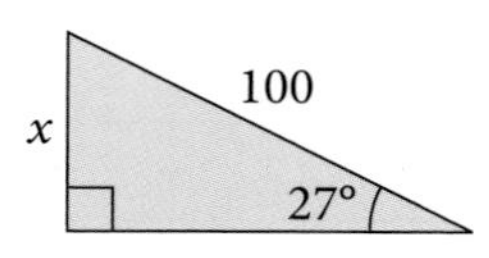

8.

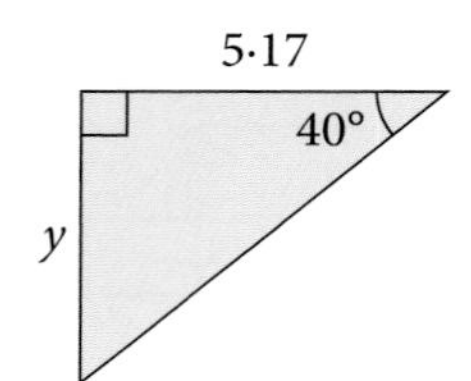

9.

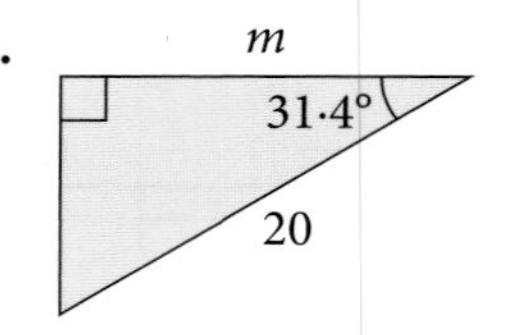

10.

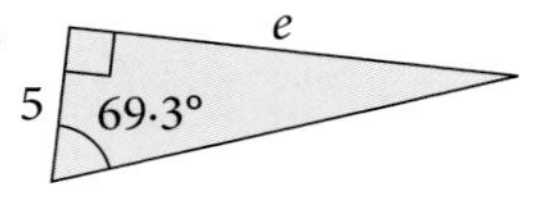

11.

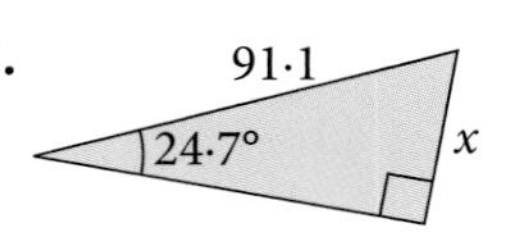

12.

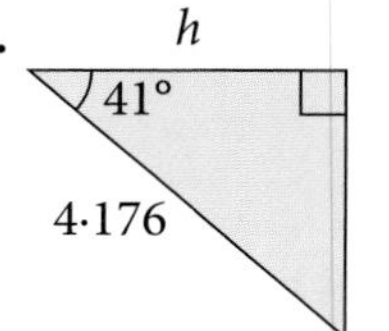

13.

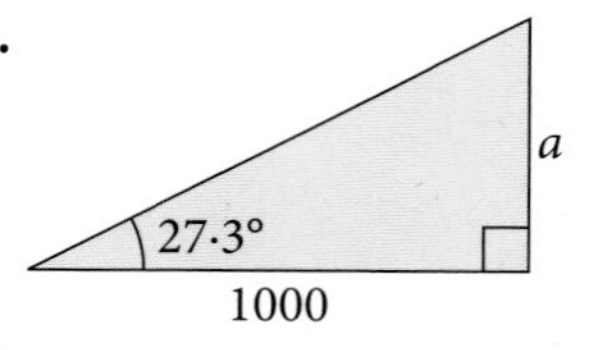

14.

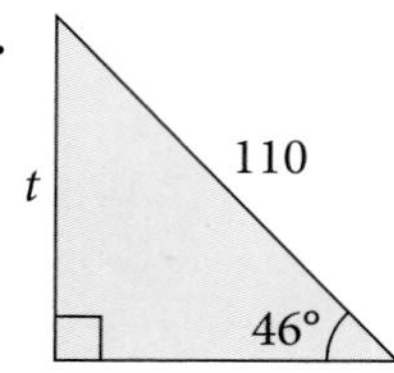

15.

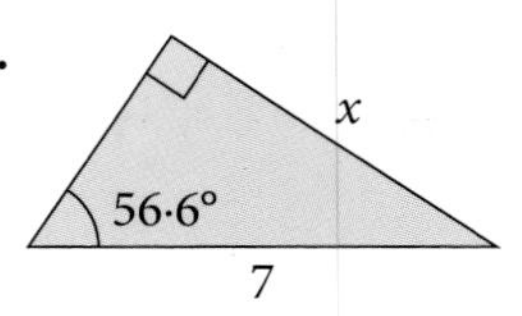

16.

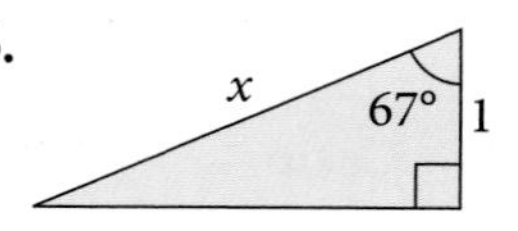

17.

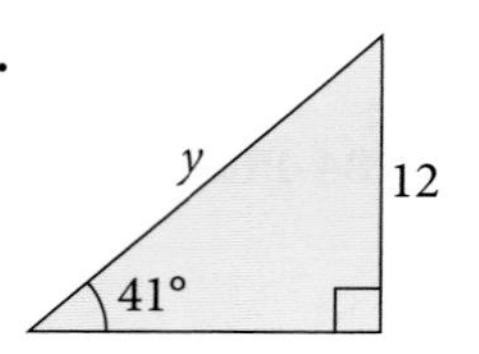

18.

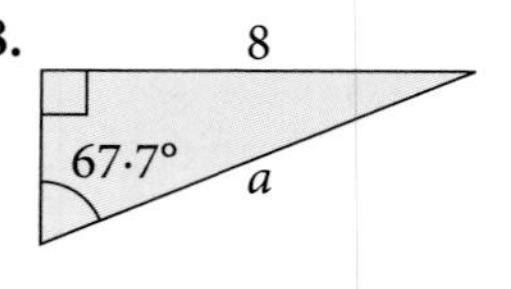

19.

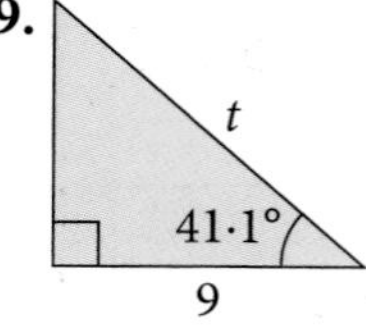

20.

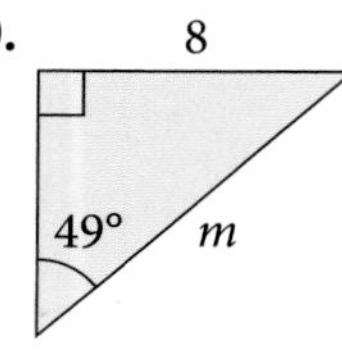

21.

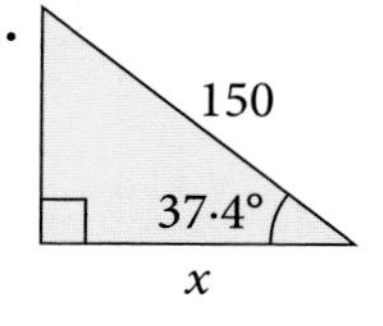

22.

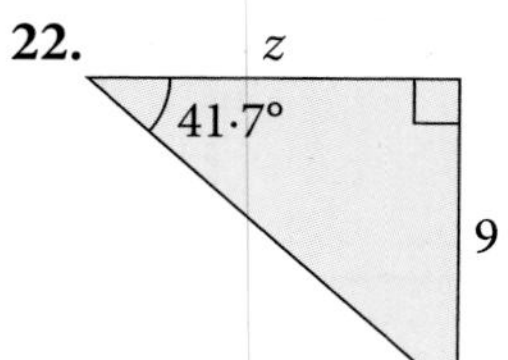

In questions **23** to **34**, the triangle has a right angle at the middle letter.

23. In ΔABC, $\widehat{C} = 40°$, BC = 4 cm. Find AB.

24. In ΔDEF, $\widehat{F} = 35{\cdot}3°$, DF = 7 cm. Find ED.

25. In ΔGHI, $\widehat{I} = 70°$, GI = 12 m. Find HI.

26. In ΔJKL, $\widehat{L} = 55°$, KL = 8·21 m. Find JK.

27. In ΔMNO, $\widehat{M} = 42{\cdot}6°$, MO = 14 cm. Find ON.

28. In ΔPQR, $\widehat{P} = 28°$, PQ = 5·071 m. Find PR.

29. In ΔSTU, $\widehat{S} = 39°$, TU = 6 cm. Find SU.

30. In ΔVWX, $\widehat{X} = 17°$, WV = 30·7 m. Find WX.

31. In ΔABC, $\widehat{A} = 14{\cdot}3°$, BC = 14 m. Find AC.

32. In ΔKLM, $\widehat{K} = 72{\cdot}8°$, KL = 5·04 cm. Find LM.

33. In ΔPQR, $\widehat{R} = 31{\cdot}7°$, QR = 0·81 cm. Find PR.

34. In ΔXYZ, $\widehat{X} = 81{\cdot}07°$, YZ = 52·6 m. Find XY.

Example 3

Find the length marked x.

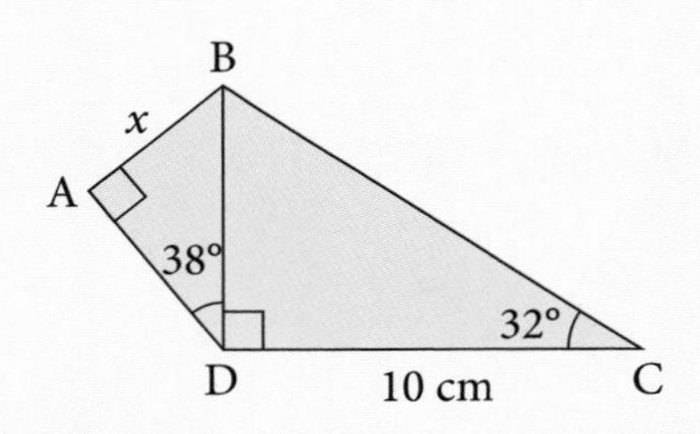

a) Find BD from triangle BDC.

$$\tan 32° = \frac{BD}{10}$$

$\therefore$ BD = 10 × tan 32° [1]

b) Now find x from triangle ABD.

$$\sin 38° = \frac{x}{BD}$$

$\therefore$ x = BD × sin 38°

x = 10 × tan 32° × sin 38° (from [1])

x = 3·85 cm (to 3 s.f.)

Notice that BD was *not* calculated explicitly in [1]. It is better to do all the multiplications at the same time.

Using a GDC

TI-Nspire	TI84-Plus	Casio
10·tan(32)·sin(38) 3.84708	10*tan(32)*sin(3.847079871	10×tan 32×sin 38 3.847079871

Exercise 10.3

In questions **1** to **10**, find each side marked with a letter.
All lengths are in centimetres.

1.

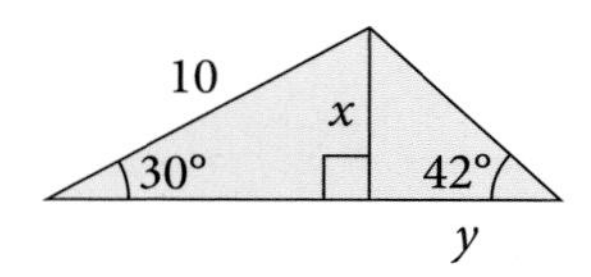

2.

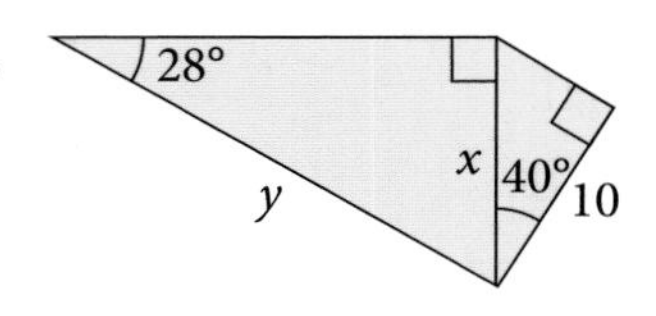

3.

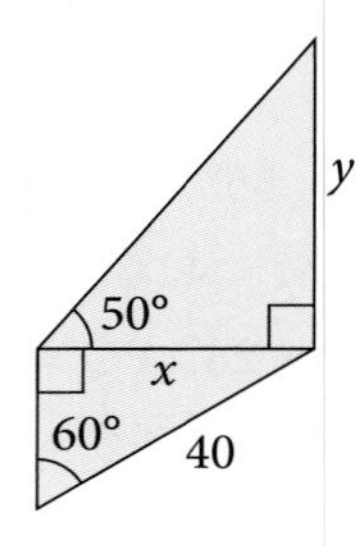

4.

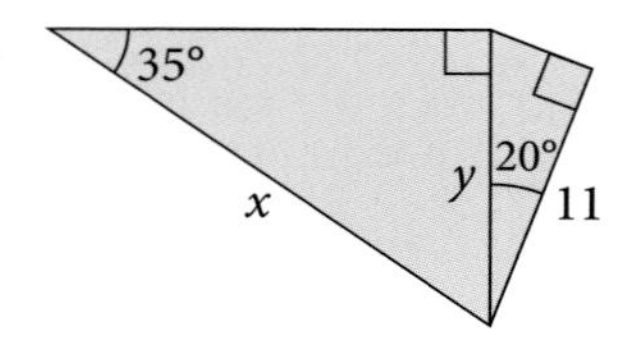

5.

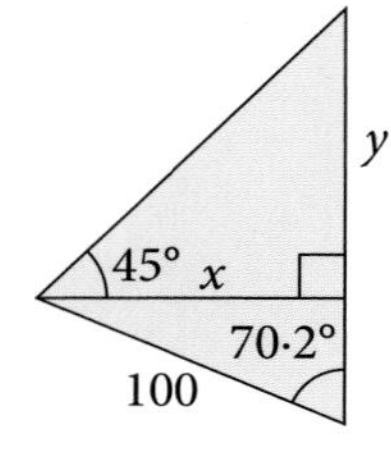

6.

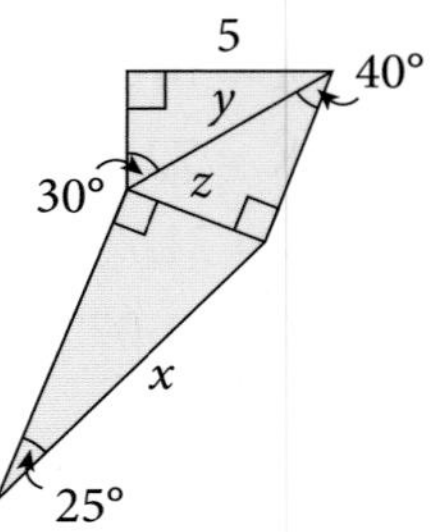

7.

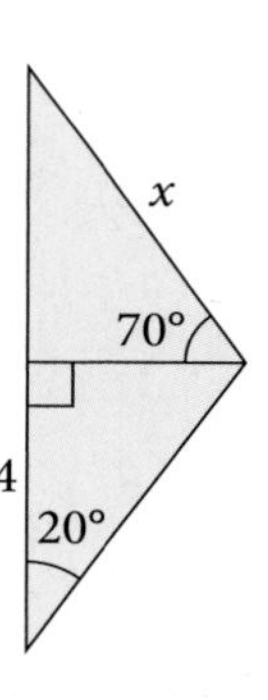

8.

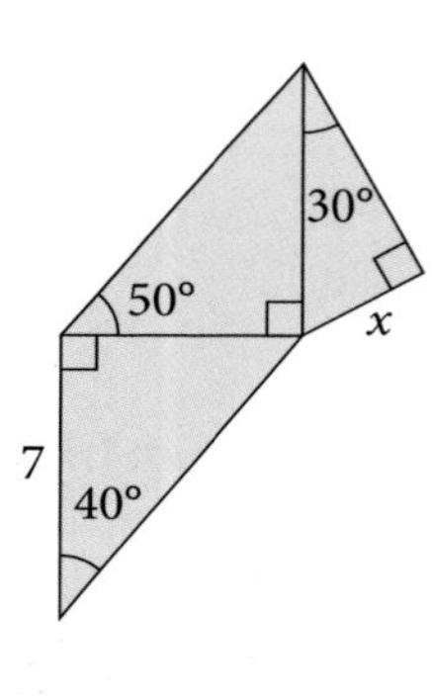

9.

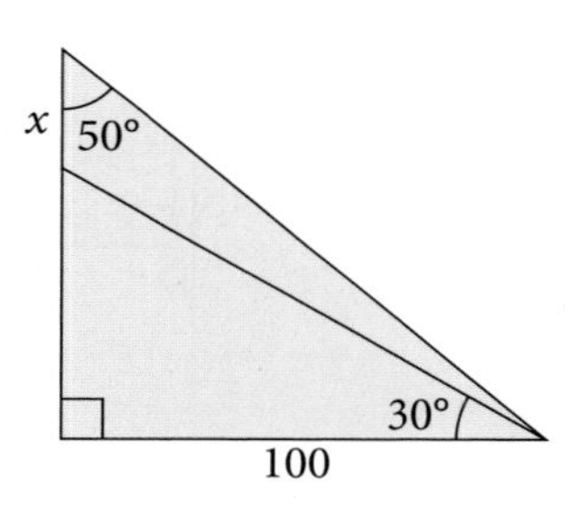

10.

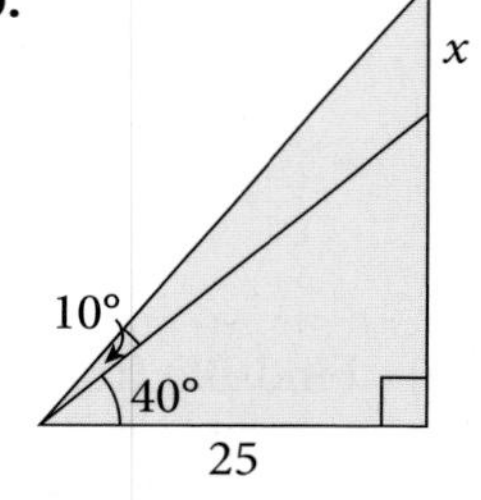

11. $B\hat{A}D = A\hat{C}D = 90°$
$C\hat{A}D = 35°$
$B\hat{D}A = 41°$
AD = 20 cm
Calculate:

a) AB **b)** DC **c)** BD

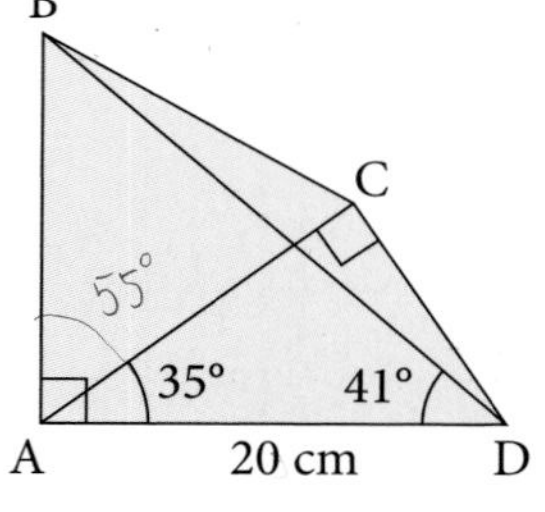

12. $A\hat{B}D = A\hat{D}C = 90°$
$C\hat{A}D = 31°$
$B\hat{D}A = 43°$
AD = 10 cm
Calculate:

a) AB **b)** CD **c)** DB

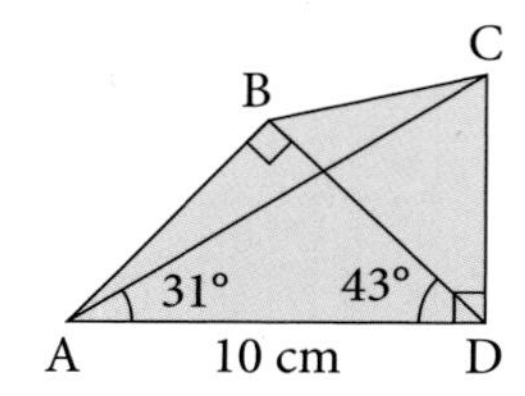

Finding an unknown angle

Example 4

Find the angle marked m.

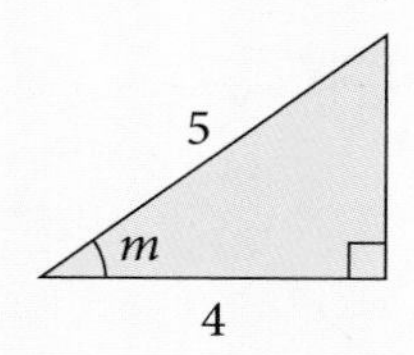

a) Label the sides of the triangle H, O, A in relation to angle m.

b) In this example, we do not know O so we need the cosine.

$\cos m = \left(\frac{A}{H}\right) = \frac{4}{5}$

c) Change $\frac{4}{5}$ to a decimal: $\frac{4}{5} = 0{\cdot}8$

d) $\cos m = 0{\cdot}8$

Find angle m from your calculator: $m = 36{\cdot}9°$

Note: On a calculator, angles can be found as follows:

Press Cos⁻¹ ($\frac{4}{5}$) (using the fraction template)

This will give the angle as 36·86989765.

We require the angle to 1 d.p. so $m = 36{\cdot}9°$

Using a GDC

TI-Nspire	TI84-Plus	Casio

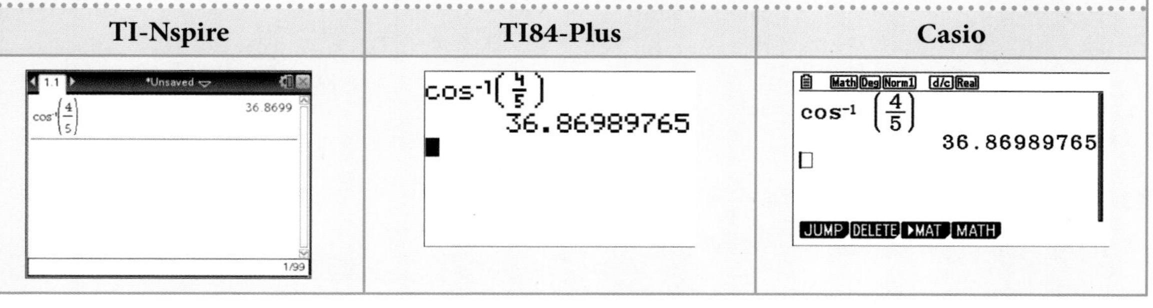

Exercise 10.4

In questions **1** to **15**, find the angle marked with a letter.
All lengths are in cm.

1.

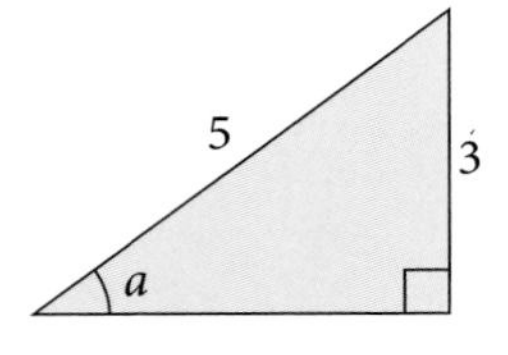

2.

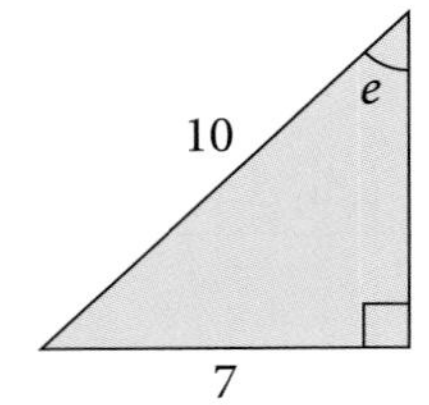

3.

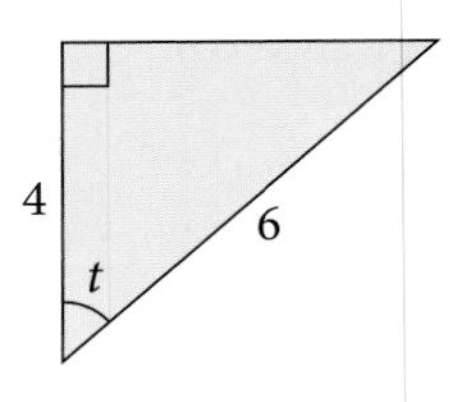

4.

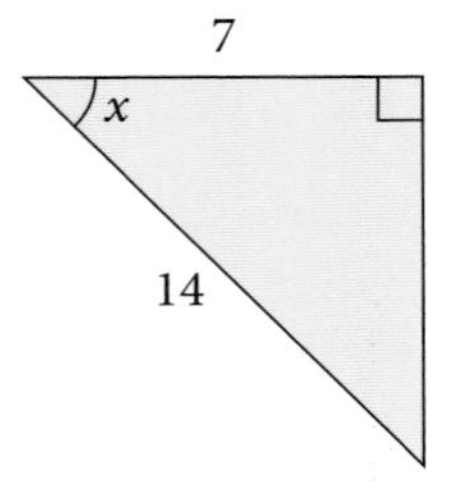

5.

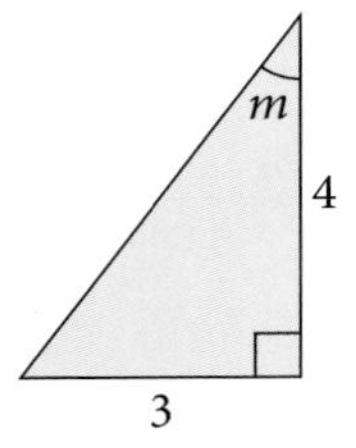

6.

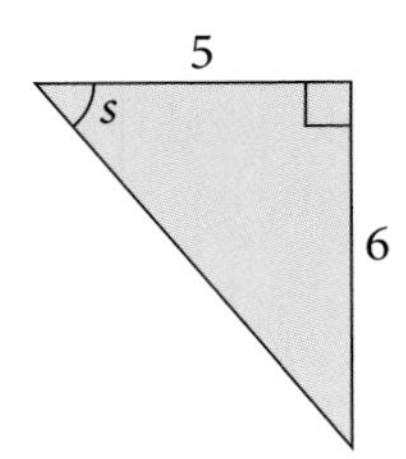

7.

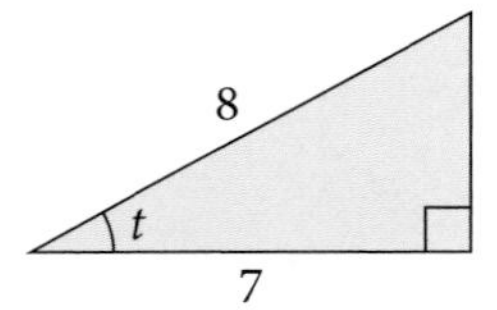

8.

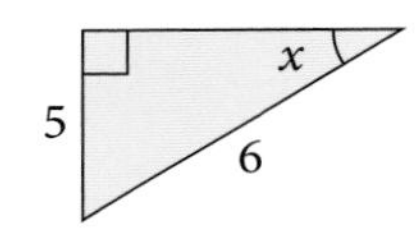

9.

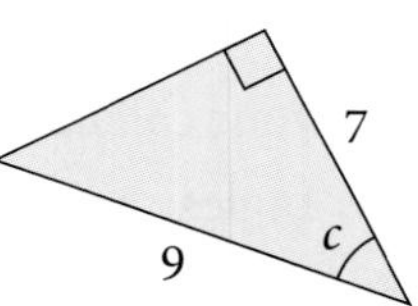

10.

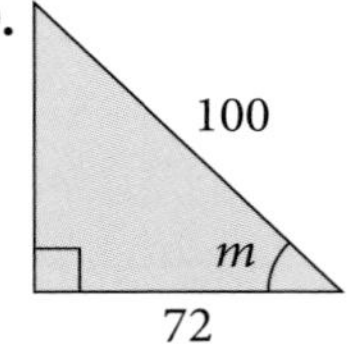

11.

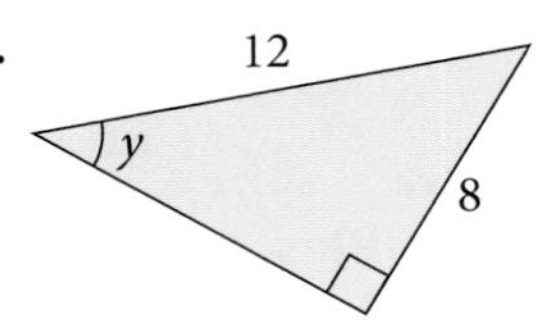

12.

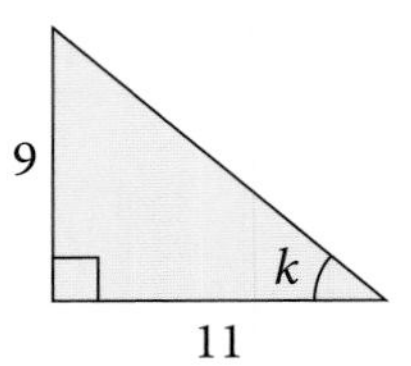

13.

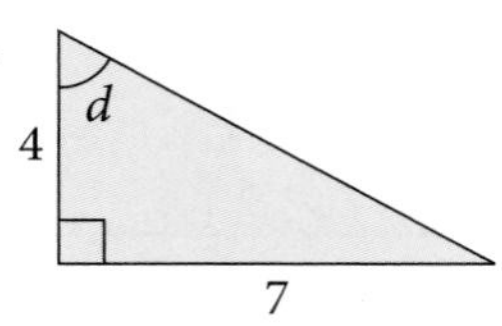

14.

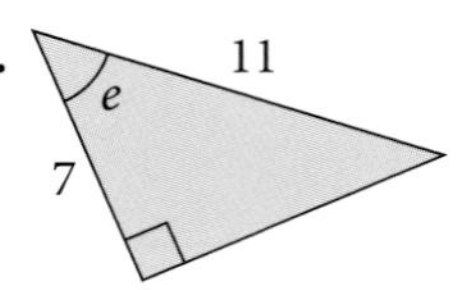

15.

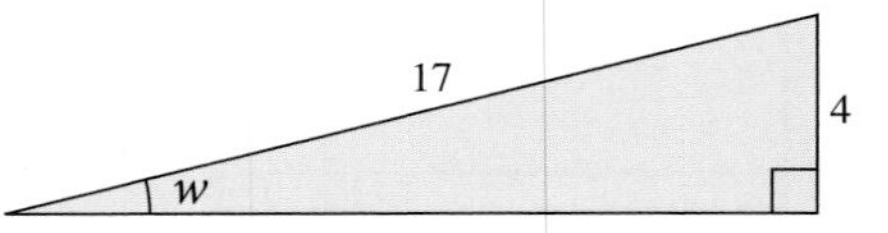

In questions **16** to **20**, the triangle has a right angle at the middle letter.

16. In ΔABC, BC = 4, AC = 7. Find $\hat{A}$.

17. In ΔDEF, EF = 5, DF = 10. Find $\hat{F}$.

18. In ΔGHI, GH = 9, HI = 10. Find $\hat{I}$.

19. In ΔJKL, JL = 5, KL = 3. Find $\hat{J}$.

20. In ΔMNO, MN = 4, NO = 5. Find $\widehat{M}$.

In questions **21** to **26**, find the angle x.

21.

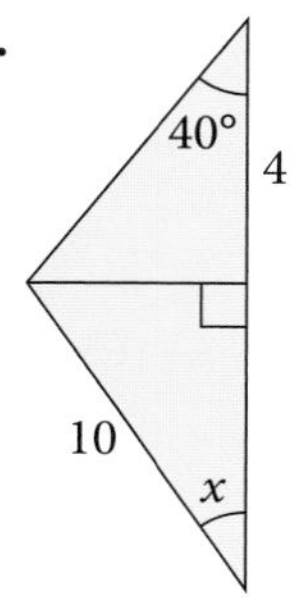

22.

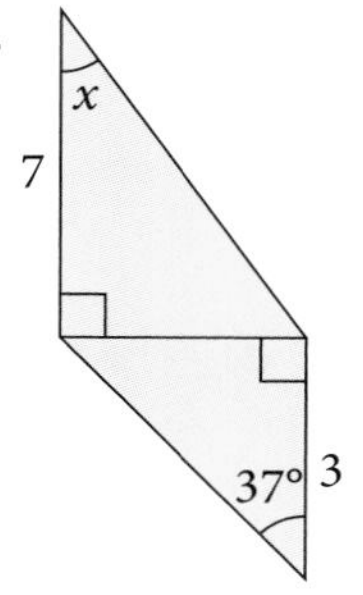

23.

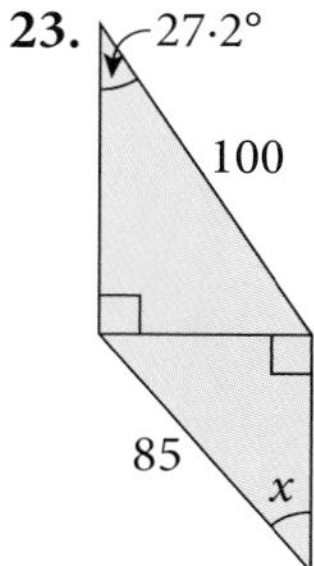

24.

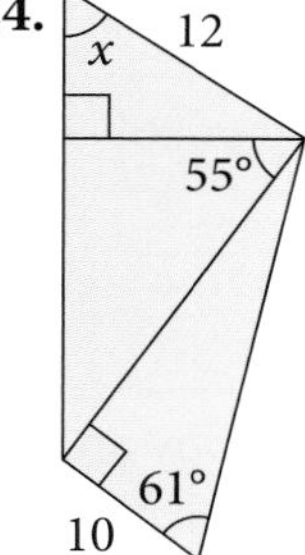

25.

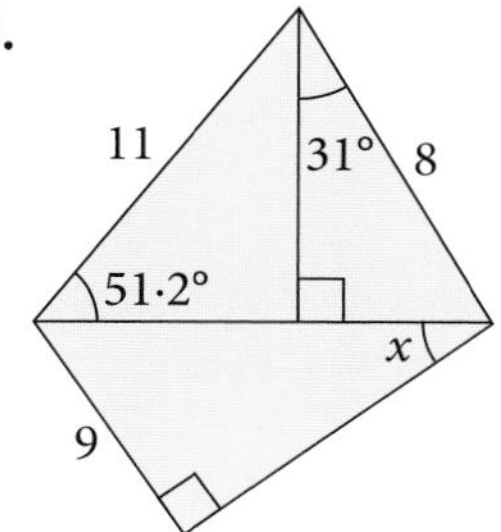

26.

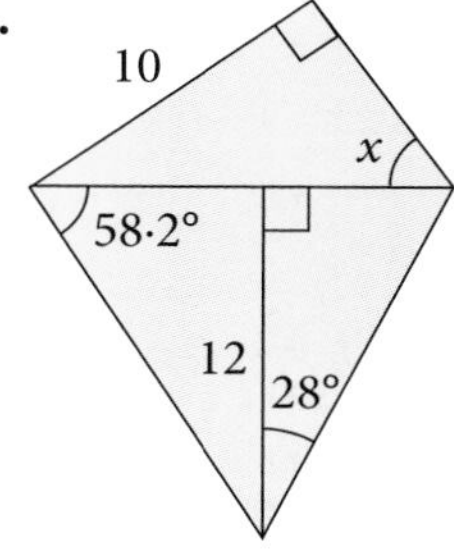

Bearings

A bearing is an angle measured clockwise from North.
It is given using three digits.
In the diagram:

the bearing of B from A is 052°
the bearing of A from B is 232°.

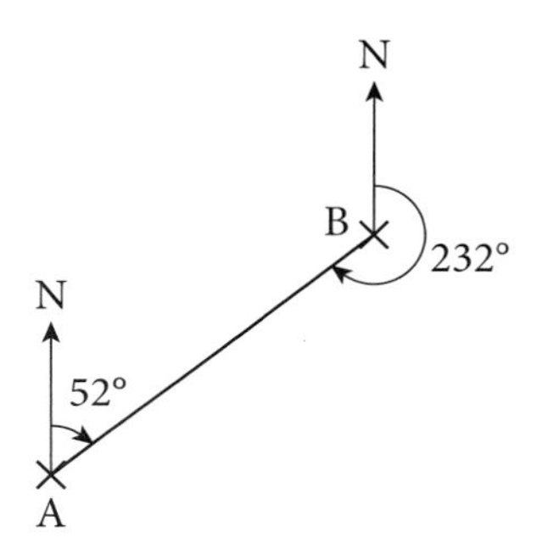

Example 5

A ship sails 22 km from A on a bearing of 042°, and a further 30 km on a bearing of 090° to arrive at B. What is the distance and bearing of B from A?

a) Draw a clear diagram and label extra points as shown.

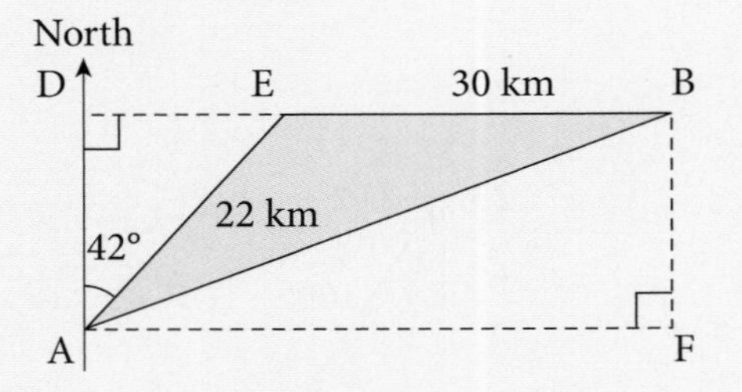

b) Find DE and AD.

i) $\sin 42° = \dfrac{DE}{22}$

$\therefore \quad DE = 22 \times \sin 42° = 14{\cdot}72\text{ km}$

ii) $\cos 42° = \dfrac{AD}{22}$

$\therefore \quad AD = 22 \times \cos 42° = 16{\cdot}35\text{ km}$

c) Using triangle ABF,

$AB^2 = AF^2 + BF^2$ (Pythagoras' theorem)

and $AF = DE + EB$

$AF = 14{\cdot}72 + 30 = 44{\cdot}72\text{ km}$

and $BF = AD = 16{\cdot}35\text{ km}$

$\therefore \quad AB^2 = 44{\cdot}72^2 + 16{\cdot}35^2$

$= 2267{\cdot}2$

$AB = 47{\cdot}6\text{ km}$ (to 3 s.f.)

d) The bearing of B from A is given by the angle DAB.

But $D\hat{A}B = A\hat{B}F$.

$\tan A\hat{B}F = \dfrac{AF}{BF} = \dfrac{44{\cdot}72}{16{\cdot}35}$

$= 2{\cdot}7352$

$\therefore \quad A\hat{B}F = 69{\cdot}9°$

B is 47·6 km from A on a bearing of 069·9°

Using a GDC

TI-Nspire	TI84-Plus	Casio
Using copy and paste	Using copy and paste	**Note**: there is no copy and paste on Casio calculators.
1.1 *Unsaved 22·sin(42) 14.7209 22·cos(42) 16.3492 14.720873339895+30 44.7209 $\sqrt{(44.720873339895)^2+(16.349186160503)}$▸ 47.6157 $\tan^{-1}\left(\frac{44.720873339895}{16.349186160503}\right)$ 69.9185 1/5	22sin(42) 14.72087334 22cos(42) 16.34918616 14.72087334+30 44.72087334 $\sqrt{44.72087334^2+16}$▸	Math Deg Norm1 d/c Real 22sin 42→A 14.72087334 22cos 42→B 16.34918616 A+30→C 44.72087334 $\sqrt{C^2+B^2}$ JUMP DELETE ▸MAT MATH
	$\sqrt{44.72087334^2+16}$▸ 47.61567389 $\tan^{-1}\left(\frac{44.72087334}{16.34918616}\right)$ 69.91845246	Math Deg Norm1 d/c Real $\sqrt{C^2+B^2}$ 47.61567389 $\tan^{-1}\left(\frac{C}{B}\right)$ 69.91845246 JUMP DELETE ▸MAT MATH

Exercise 10.5

In questions **1** to **8**, start by drawing a clear diagram.

1. A ladder of length 6 m leans against a vertical wall so that the base of the ladder is 2 m from the wall. Calculate the angle between the ladder and the wall.

2. A ladder of length 8 m rests against a wall so that the angle between the ladder and the wall is 31°. How far is the base of the ladder from the wall?

3. A ship sails 35 km on a bearing of 042°

a) How far north has it travelled?

b) How far east has it travelled?

4. A ship sails 200 km on a bearing of 243·7°

a) How far south has it travelled?

b) How far west has it travelled?

5. Find TR if PR = 10 m and QT = 7 m.

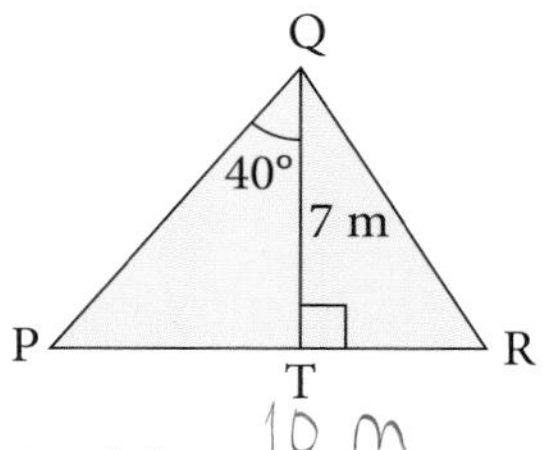

6. Find d.

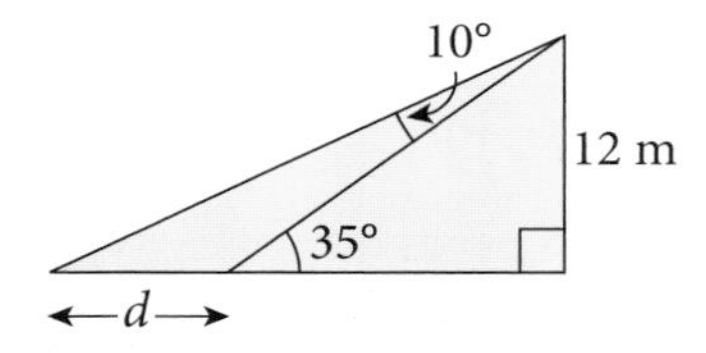

7. An aircraft flies 400 km from a point O on a bearing of 025° and then 700 km on a bearing of 080° to arrive at B.

a) How far north of O is B?

b) How far east of O is B?

c) Find the distance and bearing of B from O.

8. An aircraft flies 500 km on a bearing of 100° and then 600 km on a bearing of 160°.
Find the distance and bearing of the finishing point from the starting point.

For questions **9** to **12**, plot the points for each question on a sketch graph with x- and y-axes drawn to the same scale.

9. For the points A(5, 0) and B(7, 3), calculate the angle between AB and the x-axis.

10. For the points C(0, 2) and D(5, 9), calculate the angle between CD and the y-axis.

11. For the points A(3, 0), B(5, 2) and C(7, −2), calculate the angle BAC.

12. For the points P(2, 5), Q(5, 1) and R(0, −3), calculate the angle PQR.

13. From the top of a tower of height 75 m, a man sees two goats, both due west of him. If the angles of depression of the two goats are 10° and 17°, calculate the distance between them.

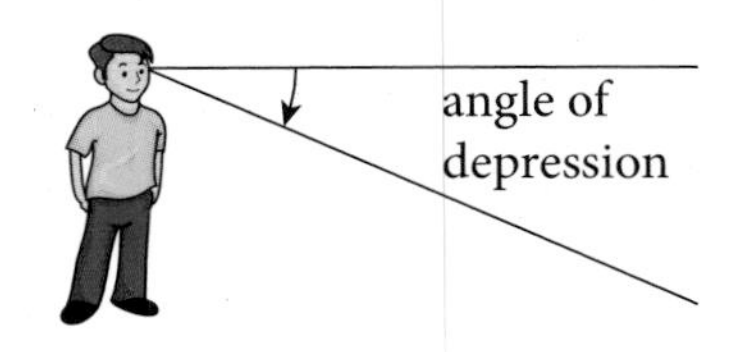

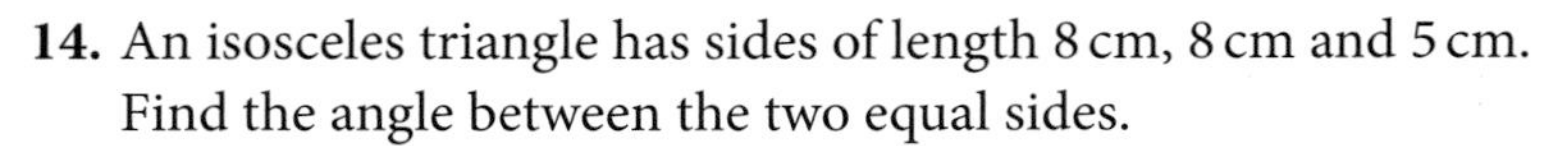

14. An isosceles triangle has sides of length 8 cm, 8 cm and 5 cm. Find the angle between the two equal sides.

15. The angles of an isosceles triangle are 66°, 66° and 48°. If the shortest side of the triangle is 8·4 cm, find the length of one of the two equal sides.

Exercise 10.6

1. A chord of length 12 cm subtends an angle of 78·2° at the centre of a circle. Find the radius of the circle.

2. Find the acute angle between the diagonals of a rectangle whose sides are 5 cm and 7 cm.

3. Barbed wire of length 76 m is fixed to opposite corners of a square field. How long are the sides of the field?

4. A kite flying at a height of 55 m is attached to a string which makes an angle of 55° with the horizontal. What is the length of the string?

5. A boy is flying a kite from a string of length 150 m. If the string is taut and makes an angle of 67° with the horizontal, what is the height of the kite?

6. A rocket flies 10 km vertically, then 20 km at an angle of 15° to the vertical and finally 60 km at an angle of 26° to the vertical. Calculate the vertical height of the rocket at the end of the third stage.

7. Find x, given that AD = BC = 6 m.

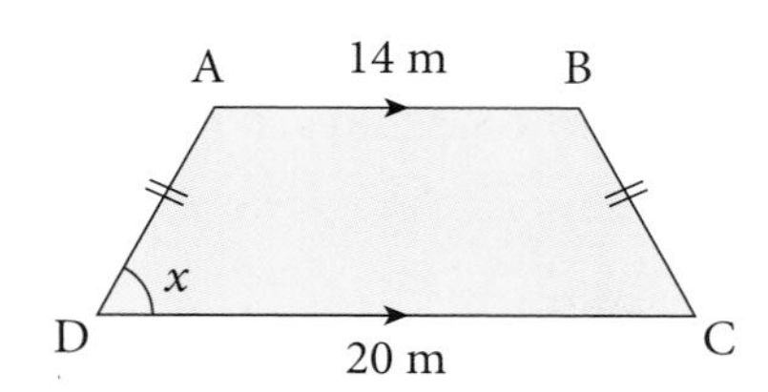

8. 

The shadow of a sunflower in a field is 60 cm long.
The angle of elevation of the sun is 76°.
How tall is the sunflower, to the nearest cm?

Angle of elevation = angle formed between base of flower and the sun.

9. Find x.

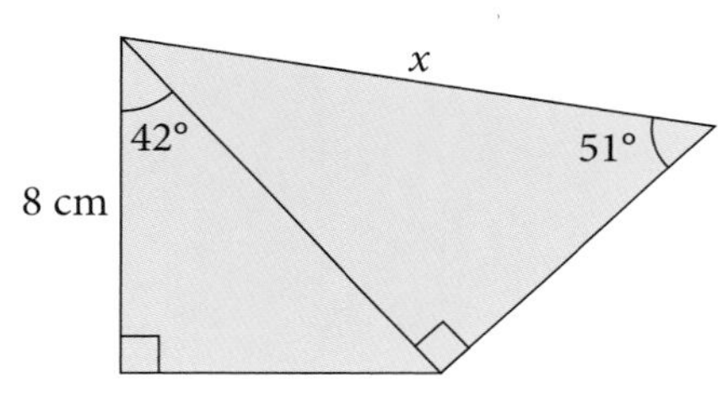

10. Diagram A shows a kitchen floor tiled with regular octogons and squares.

Diagram B shows an enlarged view of one octagon.

The smaller green squares have sides of length 12 cm.

Find the width of the octagons.

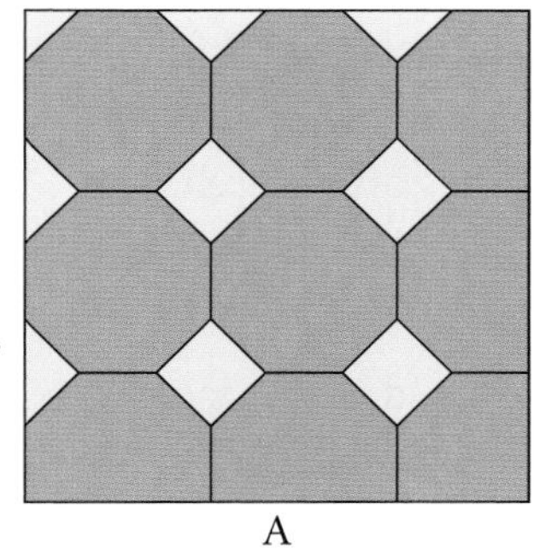

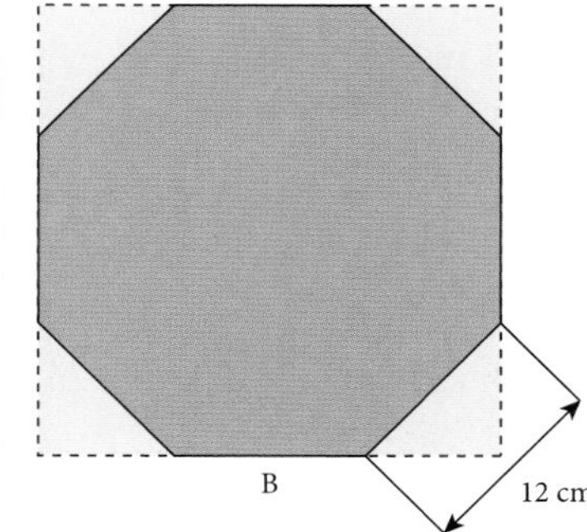

11. Ants can hear each other up to a range of 2 m. An ant at A, 1 m from a wall, sees her friend at B about to be eaten by a spider. If the angle of elevation of B from A is 62°, will the spider have a meal or not? (Assume B escapes if he hears A calling.)

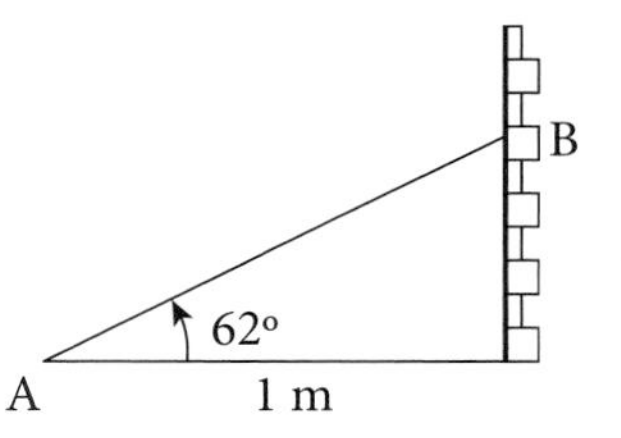

12. A hedgehog wishes to cross a road without being run over. He observes the angle of elevation of a lamp post on the other side of the road to be 27° from the edge of the road and 15° from a point 10 m back from the road. How wide is the road? If he can run at 1 m/s, how long will he take to cross? If cars are travelling at 20 m/s, how far apart must they be if he is to survive?

13. The tips of the blades of a pair of scissors are 8·6 cm apart and the angle between the blades is 38°. How long is each blade to the point where they meet?

14. From a point 10 m from a vertical wall, the angles of elevation of the bottom and the top of a statue of Sir Isaac Newton, set in the wall, are 40° and 52°. Calculate the height of the statue.

15. A farm gate is 1·3 m high. The diagonal crossbar has a width of 15 cm and makes an angle of 24° with the horizontal. Find the length of x and y in the diagram. Hence find the width of the gate.

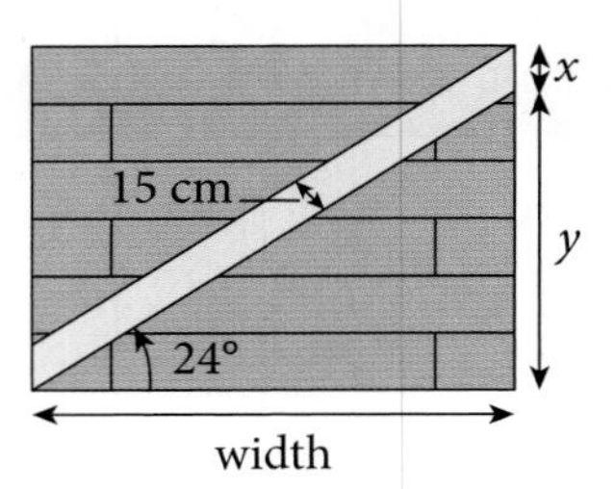

16. A rectangular paving stone 3 m by 1 m rests against a vertical wall as shown.

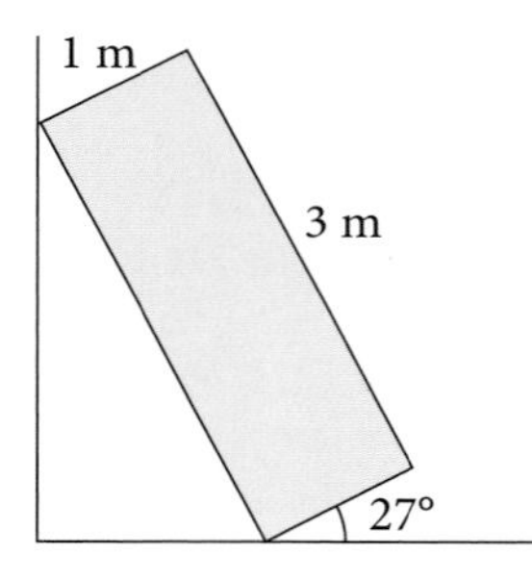

What is the height of the highest point of the stone above the ground?

17. A rectangular piece of paper 30 cm by 21 cm is folded so that opposite corners coincide. How long is the crease?

18. The diagram shows the cross section of a rectangular fish tank. When AB is inclined at 40°, the water just comes up to A.

The tank is then lowered so that BC is horizontal. What is now the depth of water in the tank?

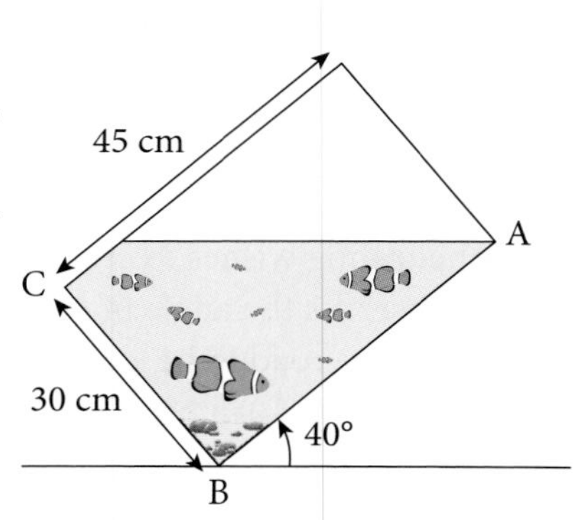

10.2 Sine, cosine and tangent for 0°, 30°, 45°, 60°, 90°

Here are three right angled triangles:

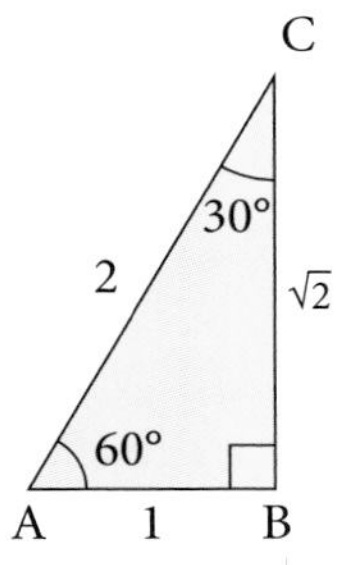

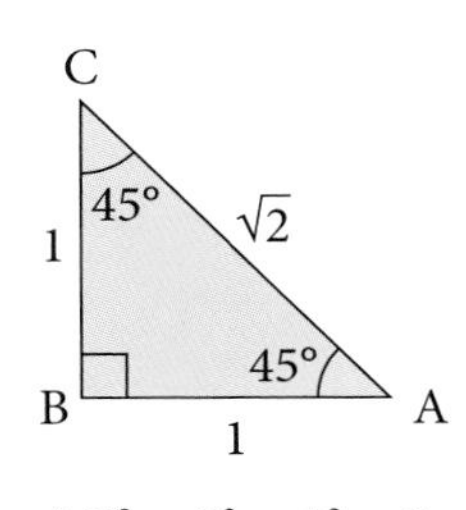

This triangle is 'imaginary'.

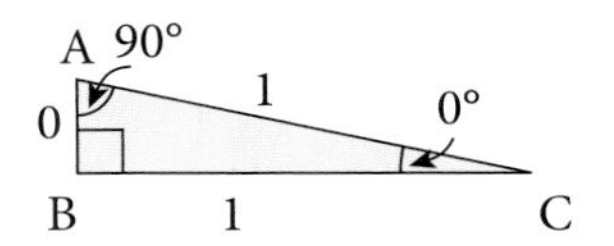

By Pythagoras: $BC^2 = 2^2 - 1^2$

$BC = \sqrt{3}$

$AC^2 = 1^2 + 1^2 = 2$

$AC = \sqrt{2}$

	30°	**60°**	**45°**	**0°**	**90°**
sin	$\frac{1}{2}$	$\frac{\sqrt{3}}{2}$	$\frac{1}{\sqrt{2}}$	$\frac{0}{1}=0$	$\frac{1}{1}=1$
cos	$\frac{\sqrt{3}}{2}$	$\frac{1}{2}$	$\frac{1}{\sqrt{2}}$	$\frac{1}{1}=1$	$\frac{0}{1}=0$
tan	$\frac{1}{\sqrt{3}}$	$\sqrt{3}$	1	$\frac{0}{1}=0$	$\frac{1}{0}$ undefined (infinitely large)

Exercise 10.7

1. Write down the exact values of

   a) sin 30° b) tan 60° c) cos 60° d) sin 60°

   e) tan 45° f) cos 0° g) tan 30° h) cos 30°

2. Using $x = 30°$, show that $\sin 2x = 2 \sin x \cos x$
3. Using $x = 45°$, show that $\sin^2 x + \cos^2 x = 1$
4. Using $x = 30°$, show that $\cos 2x = 2\cos^2 x - 1$
5. Find the exact value of the side marked with a letter.
   All lengths are in cm

It is customary to write $(\sin x)^2$ as $\sin^2 x$. Always use $(\sin x)^2$ on your calculator.

a)

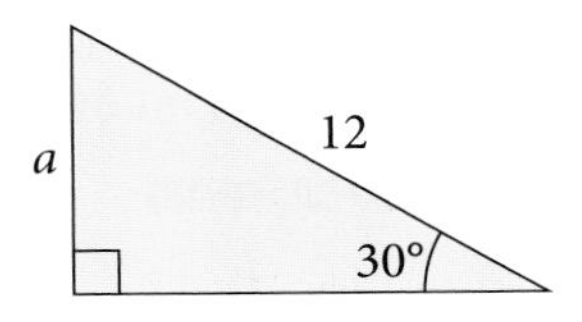

b)

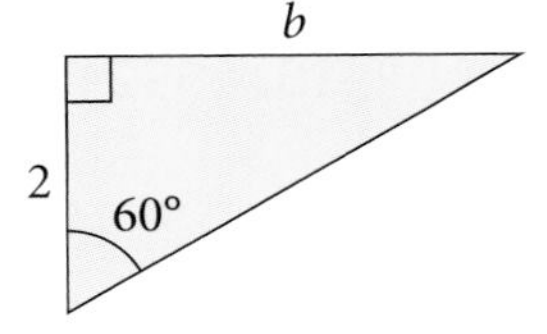

c)

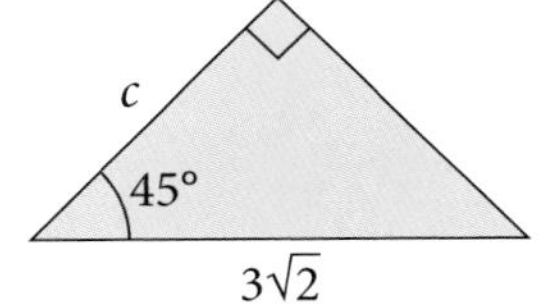

d) 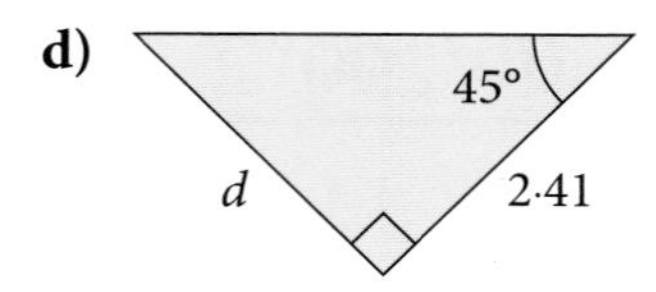

e) 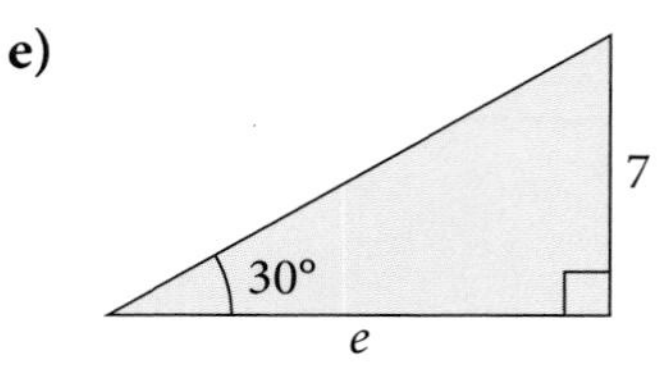

f) 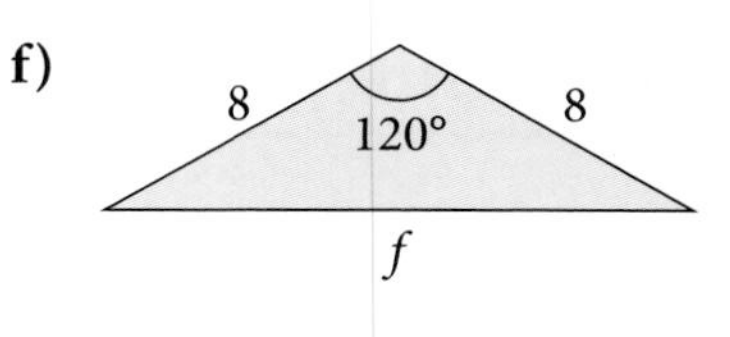

6. Substitute $x = 30°$ to show that $\tan 2x = \dfrac{2\tan x}{1-\tan^2 x}$
7. Using A = 60° and B = 30°, show that
   $\sin(A - B) = \sin A \cos B - \sin B \cos A$

10.3 Three-dimensional problems

Always draw a large, clear diagram. It is often helpful to redraw the triangle which contains the length or angle to be found.

Example 6

A rectangular box with top WXYZ and base ABCD has AB = 6 cm, BC = 8 cm and WA = 3 cm.

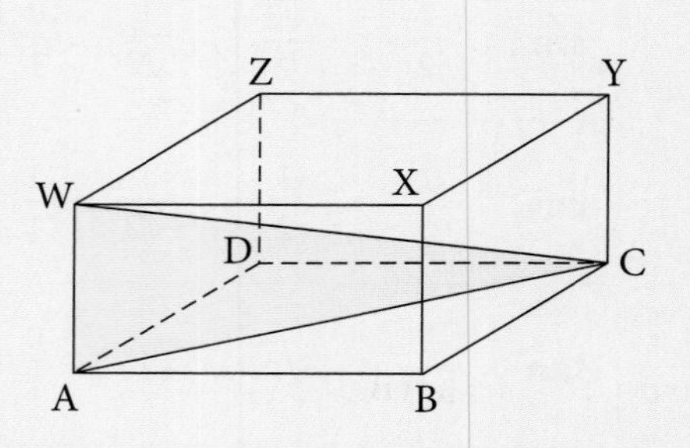

Calculate:

a) the length of AC

b) the angle between WC and AC.

a) Redraw triangle ABC.

$AC^2 = 6^2 + 8^2 = 100$

AC = 10 cm

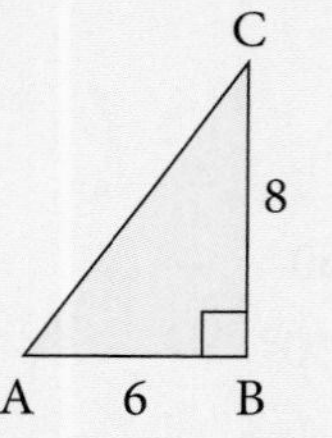

b) Redraw triangle WAC.

Let $W\hat{C}A = \theta$

$\tan\theta = \dfrac{3}{10}$

$\theta = 16{\cdot}7°$

The angle between WC and AC is 16·7°

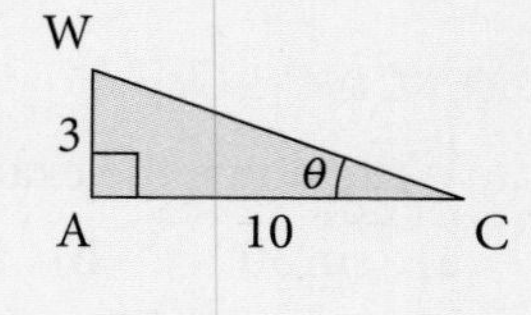

Using a GDC

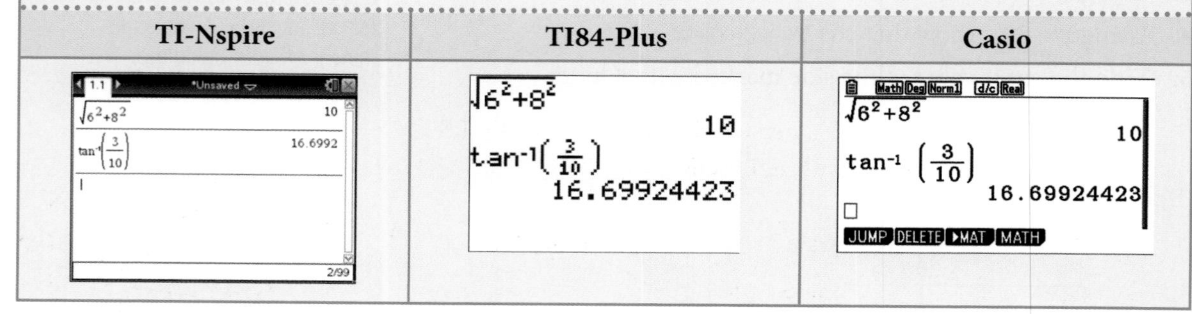

Exercise 10.8

1. In the rectangular box shown, find:

   a) AC

   b) AR

   c) the angle between AC and AR.

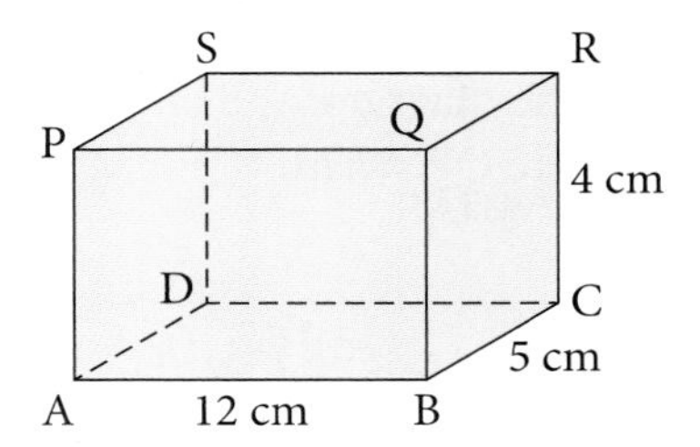

2. A vertical pole BP stands at one corner of a horizontal rectangular field as shown.
   If AB = 10 m, AD = 5 m and the angle of elevation of P from A is 22°, calculate:

   a) the height of the pole

   b) the angle of elevation of P from C

   c) the length of a diagonal of the rectangle ABCD

   d) the angle of elevation of P from D.

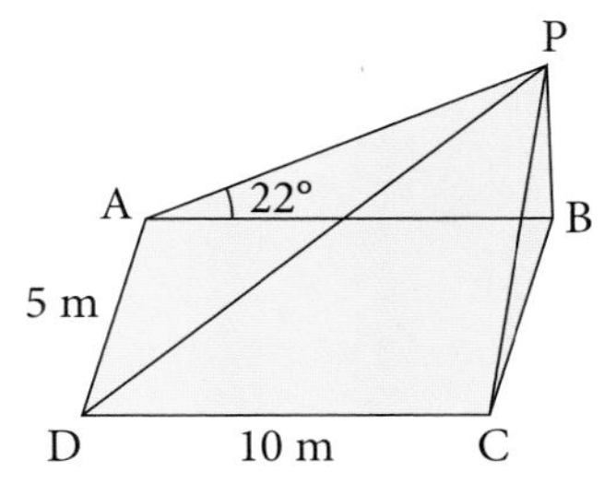

3. In the cube shown, find:

   a) BD

   b) AS

   c) BS

   d) the angle SBD

   e) the angle ASB.

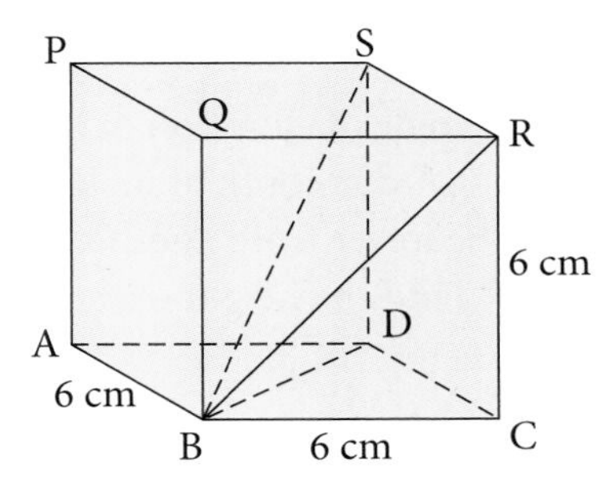

4. In the square-based pyramid, V is vertically above the middle of the base, AB = 10 cm and VC = 20 cm. Find:

   a) AC

   b) the height of the pyramid

   c) the angle between VC and the base ABCD

   d) the angle AVB

   e) the angle AVC.

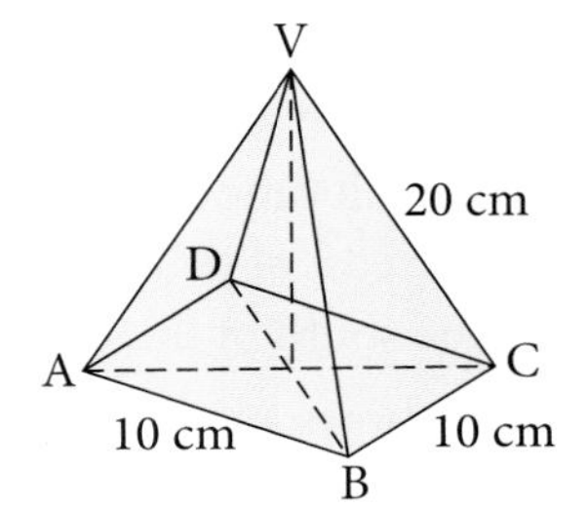

5. In the wedge shown, PQRS is perpendicular to ABRQ. PQRS and ABRQ are rectangles with AB = QR = 6 m, BR = 4 m, RS = 2m. Find:

   a) BS
   b) AS
   c) angle BSR
   d) angle ASR
   e) angle PAS.

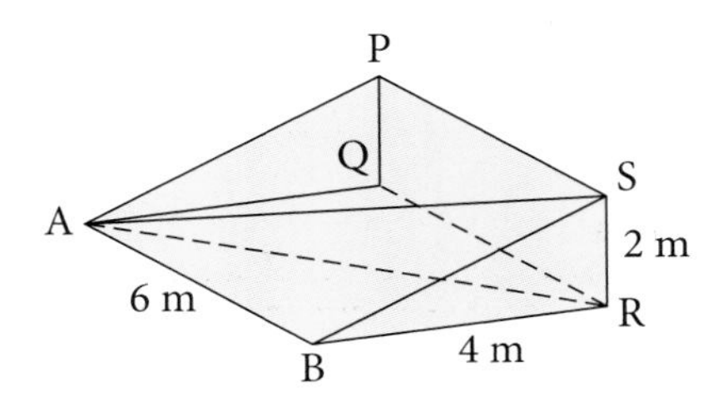

6. The edges of a box are 4 cm, 6 cm and 8 cm. Find the length of a diagonal and the angle it makes with the diagonal on the largest face.

7. In the diagram A, B and O are points in a horizontal plane and P is vertically above O, where OP = h m.

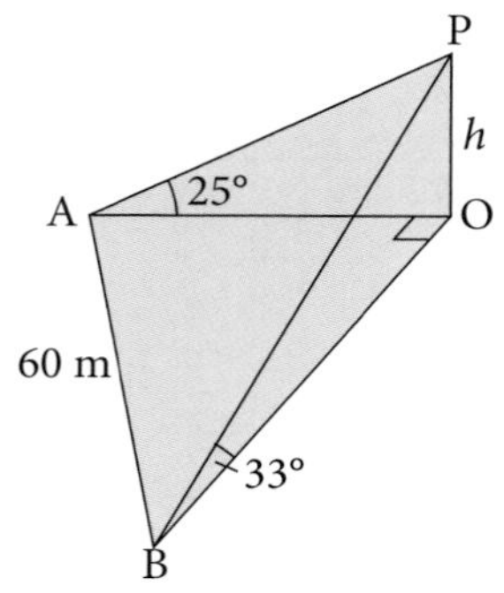

A is due West of O, B is due South of O and AB = 60 m. The angle of elevation of P from A is 25° and the angle of elevation of P from B is 33°.

a) Find the length AO in terms of h.

b) Find the length BO in terms of h.

c) Find the value of h.

8. The angle of elevation of the top of a tower is 38° from a point A due south of it. The angle of elevation of the top of the tower from another point B, due east of the tower is 29°. Find the height of the tower if the distance AB is 50 m.

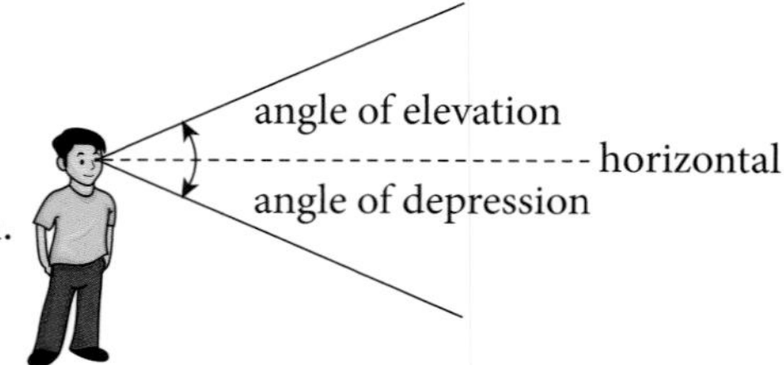

9. An observer at the top of a tower of height 15 m sees a man due west of him at an angle of depression 31°. He sees another man due south at an angle of depression 17°. Find the distance between the men.

10. The angle of elevation of the top of a tower is 27° from a point A due east of it. The angle of elevation of the top of the tower is 11° from another point B due south of the tower. Find the height of the tower if the distance AB is 40 m.

11. The figure shows a triangular pyramid on a horizontal base ABC. V is vertically above B where VB = 10 cm, $\hat{ABC} = 90°$ and AB = BC = 15 cm. Point M is the mid-point of AC. Calculate the size of angle VMB.

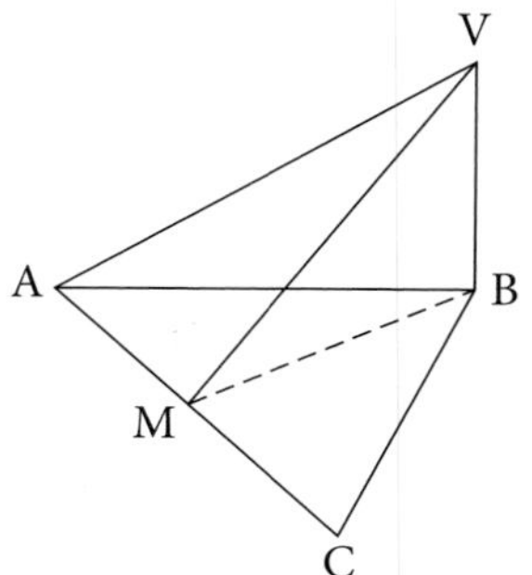

10.4 Sine, cosine, tangent for any angle

So far we have used sine, cosine and tangent only in right-angled triangles. For angles greater than 90°, we will see that there is a close connection between trigonometric ratios and circles.

The circle on the right is of radius 1 unit with centre (0, 0).
A point P with coordinates (x, y) moves round the circumference of the circle. The angle that OP makes with the positive x-axis as it turns in an anticlockwise direction is θ.

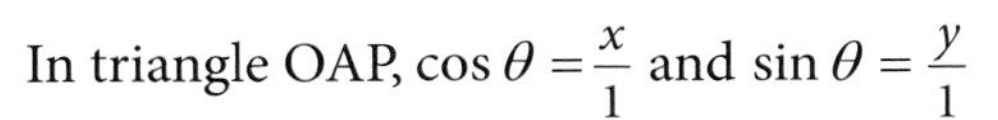

In triangle OAP, $\cos\theta = \frac{x}{1}$ and $\sin\theta = \frac{y}{1}$

The x-coordinate of P is $\cos\theta$.
The y-coordinate of P is $\sin\theta$.
This idea is used to define the cosine and the sine of any angle, including angles greater than 90°.
Here is an angle greater than 90°:

$\cos 120° = -0{\cdot}5$
$\sin 120° = 0{\cdot}866$

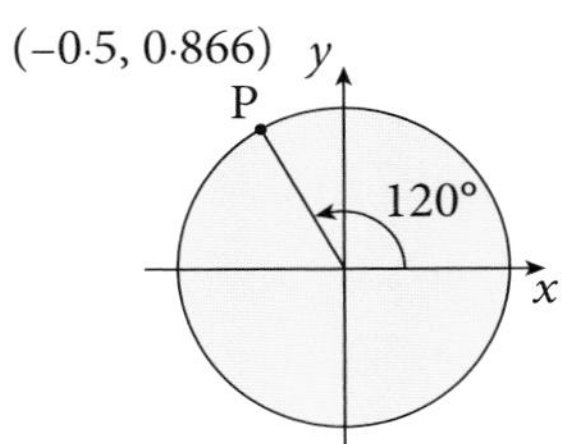

A graphic display calculator can be used to show the graph of $y = \sin x$ for any range of angles. The graph on the right shows $y = \sin x$ for x from 0° to 360°.

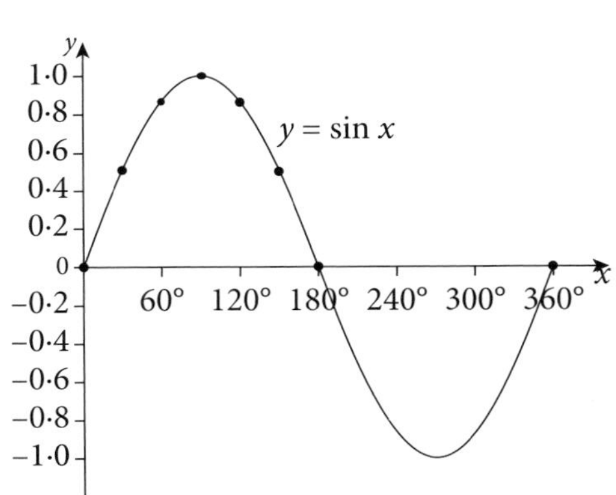

Note:

$\sin 150° = \sin 30°$ and $\cos 150° = -\cos 30°$
$\sin 110° = \sin 70°$ $\cos 110° = -\cos 70°$
$\sin 163° = \sin 17°$ $\cos 163° = -\cos 17°$
or $\sin x = \sin(180° - x)$
or $\cos x = -\cos(180° - x)$

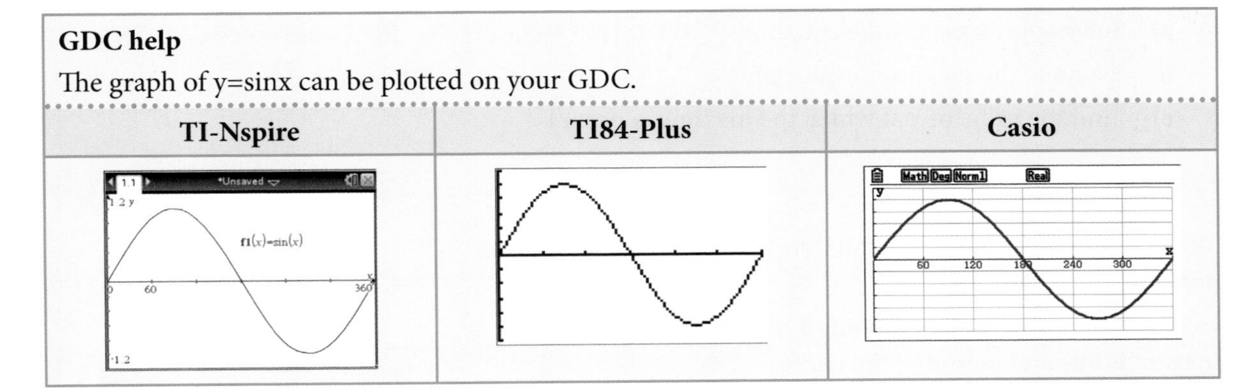

GDC help

The graph of y=sinx can be plotted on your GDC.

TI-Nspire	TI84-Plus	Casio

Exercise 10.9

1. Use a graphic display calculator to display the graph of $y = \cos x$ for values of x from 0° to 360°. Sketch the graph.

2. a) Display the graphs of i) $y = \sin x$ and ii) $y = \tan x$ for values of x from 0° to 360°.
   b) Sketch the two graphs.

In questions **3** to **9** do not use a calculator. Use the symmetry of the graphs $y = \sin x$ and $y = \cos x$.

Angles are given to the nearest degree.

3. If $\sin 18° = 0{\cdot}309$, give another angle whose sine is 0·309.

4. If $\sin 27° = 0{\cdot}454$, give another angle whose sine is 0·454.

5. Give another angle which has the same sine as:

   a) 40° b) 70° c) 130°

6. If $\cos 70° = 0{\cdot}342$, give an angle whose cosine is −0·342.

7. If $\cos 45° = 0{\cdot}707$, give an angle whose cosine is −0·707.

8. If $\sin 20° = 0{\cdot}342$, what other angle has a sine of 0·342?

9. If $\sin 98° = 0{\cdot}990$, give another angle whose sine is 0·990.

10. Find two values for x, between 0° and 180°, if $\sin x = 0{\cdot}848$. Give each angle to the nearest degree.

11. If $\sin x = 0{\cdot}35$, find two solutions for x between 0° and 180°.

12. Find *two* solutions of the equation $(\cos x)^2 = \frac{1}{4}$ for x between 0° and 180°.

For questions **13** and **14** use a graphics calculator.

13. Draw the graph of $y = 2\sin x + 1$, for $0 \le x \le 180°$

    Find approximate solutions to the equations:

    a) $2\sin x + 1 = 2{\cdot}3$ b) $\dfrac{1}{(2\sin x + 1)} = 0{\cdot}5$

14. Draw the graph of $y = 2\sin x + \cos x$ for $0 \le x \le 180°$.

    a) Solve approximately the equations: i) $2\sin x + \cos x = 1{\cdot}5$ ii) $2\sin x + \cos x = 0$
    b) Estimate the maximum value of y.
    c) Find the value of x at which the maximum occurs.

Amplitude and period

The functions $\sin x$, $\cos x$ and $\tan x$ are all **periodic** functions.

Here is part of a periodic function showing the amplitude and period of the curve.

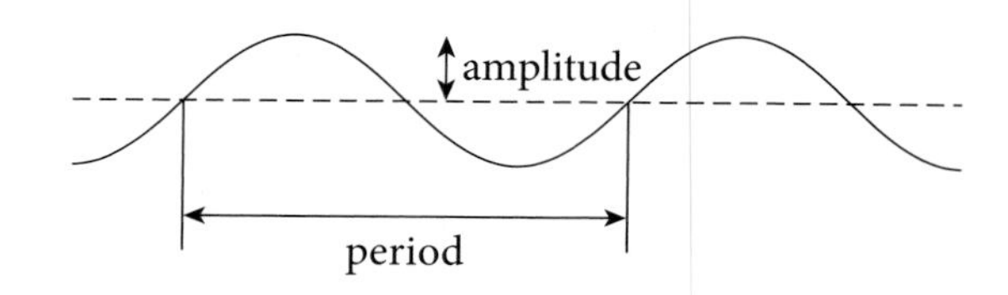

Exercise 10.10

1. Here are the graphs of **a)** $y = \sin 2x$ and **b)** $y = \sin 3x$

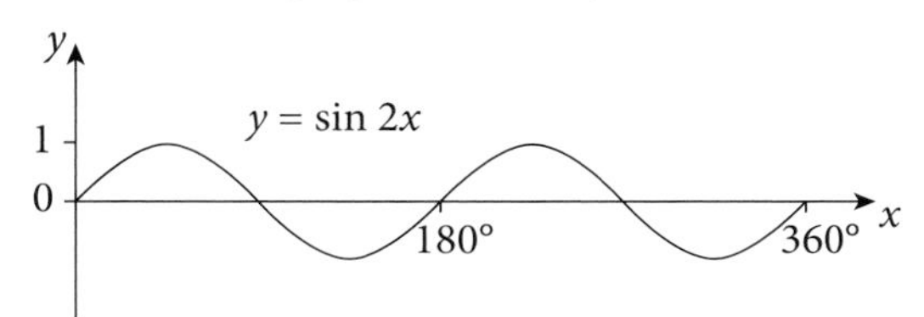

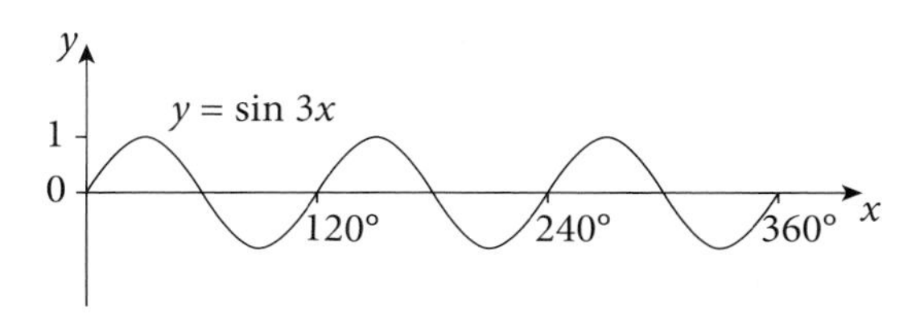

For each graph write down the amplitude and the period of the curve.

2. On a graphics calculator display the graph of $y = \tan x$ for $0 < x < 360°$
   a) Write down the period of the curve.
   b) What can you write about the amplitude of the curve?
3. **a)** On a graphics calculator display the following:
   i) $y = \sin x$ **ii)** $y = 2\sin x$ **iii)** $y = 3\sin x$
   b) For the graph of $y = a\sin x$, what does the value of a tell you?
4. Write down the amplitude of the following:
   a) $y = 2\cos x$ **b)** $y = \frac{1}{2}\sin x$ **c)** $y = 3\sin 2x$
5. On the same graph *sketch* the graphs of $y = \sin x$ and $y = \tan x$.
6. Match each equation with one of the sketch graphs which follow.
   A $y = \sin 2x$ **B** $y = \cos(x - 90°)$ **C** $y = \sin(x - 90°)$
   D $y = \sin x + 1$ **E** $y = \cos x - 1$ **F** $y = 2\sin x$

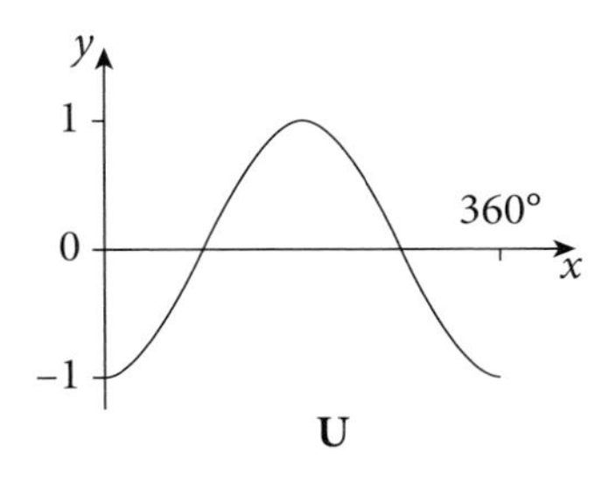

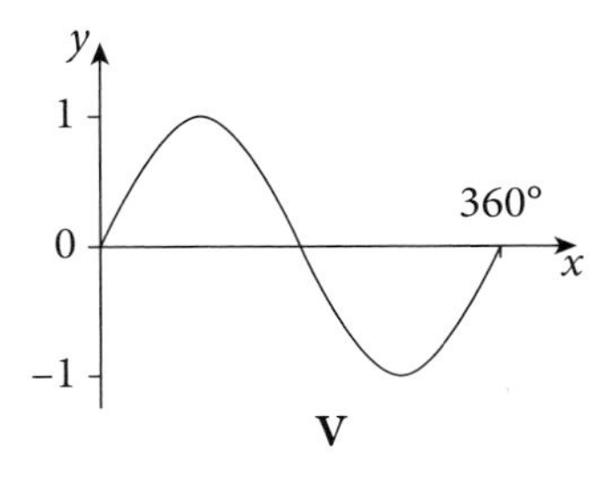

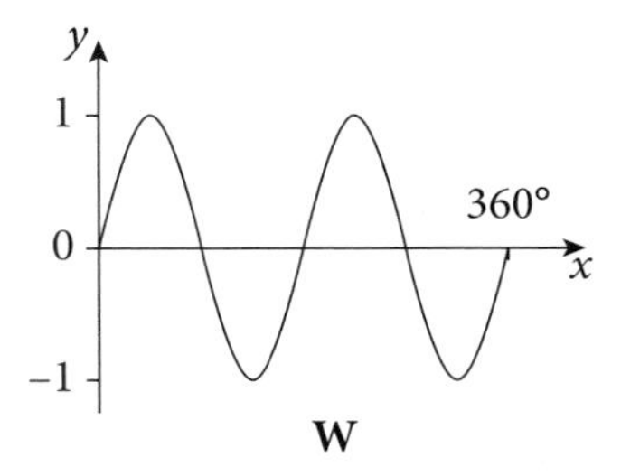

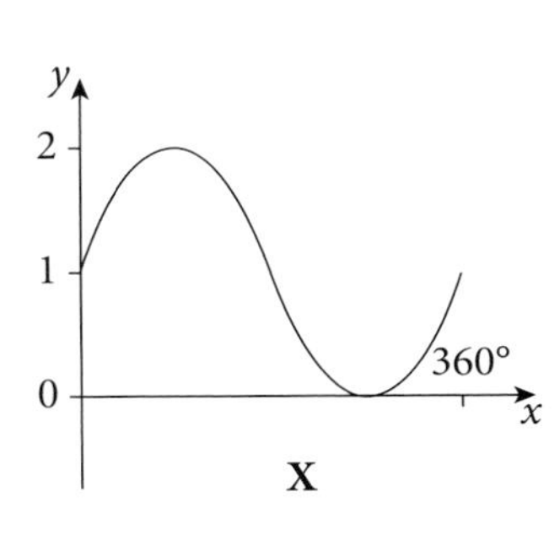

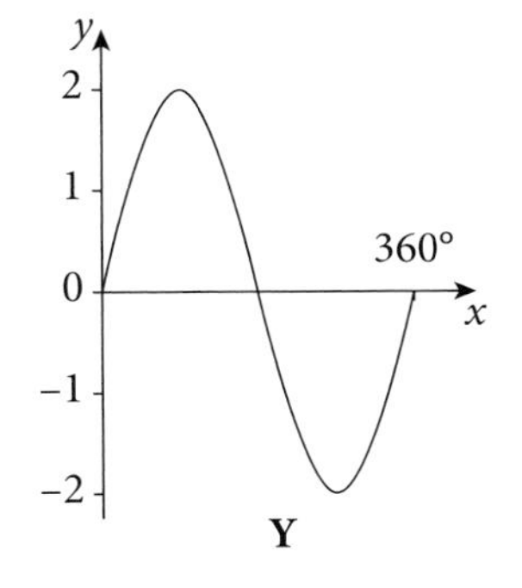

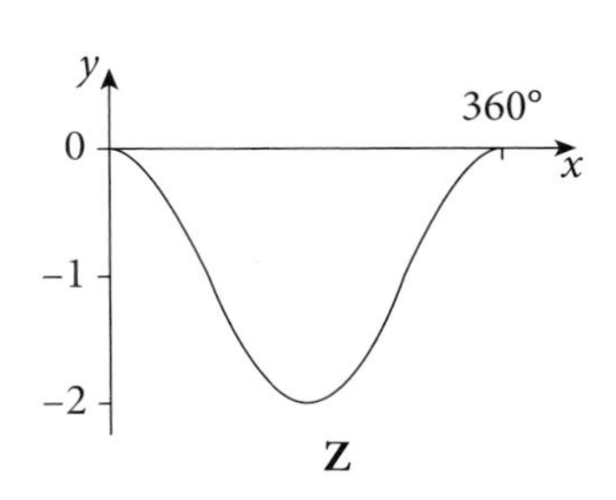

10.5 The sine rule

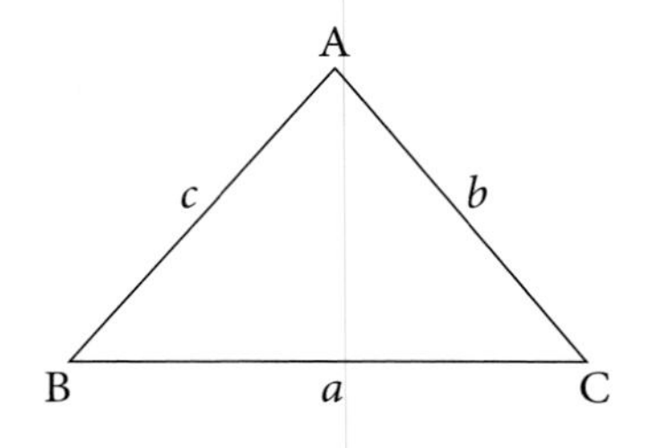

The sine rule enables us to calculate sides and angles in some triangles where there is not a right angle.

In ΔABC, we use the convention that

a is the side opposite $\hat{A}$

b is the side opposite $\hat{B}$, etc.

The sine rule states that either $\dfrac{a}{\sin A} = \dfrac{b}{\sin B} = \dfrac{c}{\sin C}$ [1]

or $\dfrac{\sin A}{a} = \dfrac{\sin B}{b} = \dfrac{\sin C}{c}$ [2]

Use [1] when finding a *side*, and [2] when finding an *angle*.

Example 7

Find c.

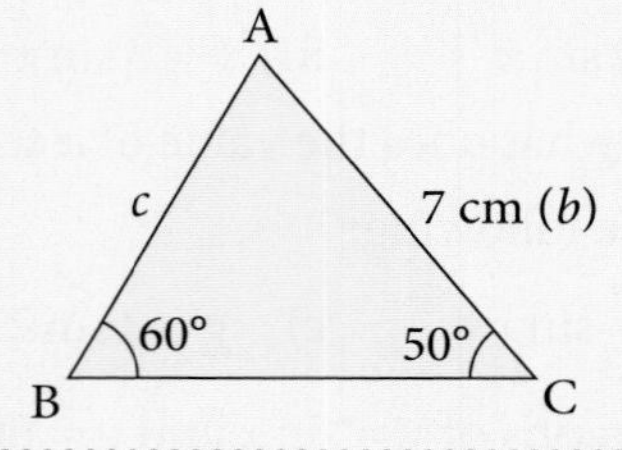

$$\frac{c}{\sin C} = \frac{b}{\sin B}$$

$$\frac{c}{\sin 50°} = \frac{7}{\sin 60°}$$

$$c = \frac{7 \times \sin 50°}{\sin 60°} = 6.19\text{cm (3 s.f.)}$$

Using a GDC

TI-Nspire	TI84-Plus	Casio
1.1 *Unsaved; 7·sin(50) / sin(60) 6.19186; 1/99	7sin(50) / sin(60) 6.191863516	Math Deg Norm1 d/c Real; 7sin 50 / sin 60 6.191863516; JUMP DELETE ▸MAT MATH

Although we cannot have an angle of more than 90° in a right-angled triangle, it is still useful to define sine, cosine and tangent for these angles.

For an obtuse angle x, we have

- $\sin x = \sin(180 - x)$

Examples: $\sin 130° = \sin 50°$
$\sin 170° = \sin 10°$
$\sin 116° = \sin 64°$

Most people simply use a calculator when finding the sine of an obtuse angle.

Example 8

Find $\hat{B}$

$\frac{\sin B}{b} = \frac{\sin A}{a}$

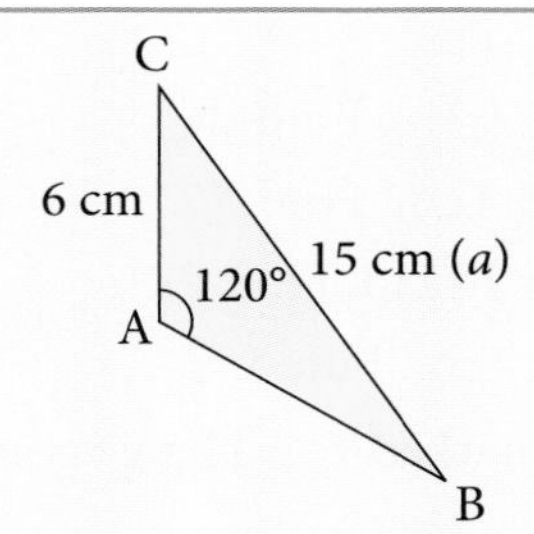

$$\frac{\sin B}{b} = \frac{\sin A}{a}$$

$$\frac{\sin B}{6} = \frac{\sin 120°}{15}$$

$$\sin B = \frac{6 \times \sin 120°}{15}$$

$$\sin B = 0{\cdot}346$$

$\hat{B} = 20{\cdot}3°$, correct to 1 d.p.

Using a GDC

TI-Nspire	TI84-Plus	Casio

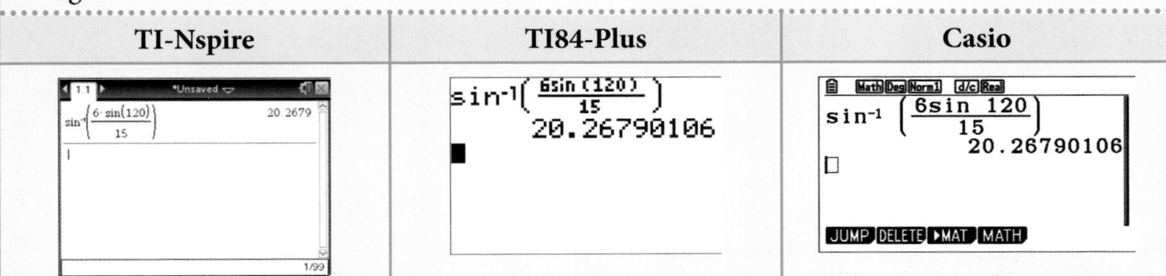

Exercise 10.11

For questions **1** to **6**, find each side marked with a letter.
Give answers to 3 s.f.

1.

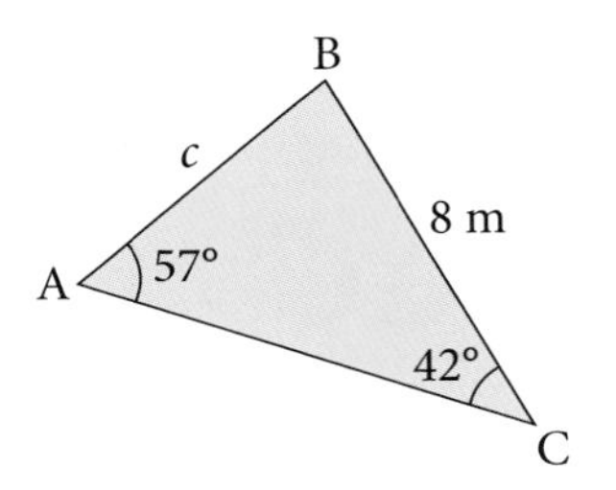

2.

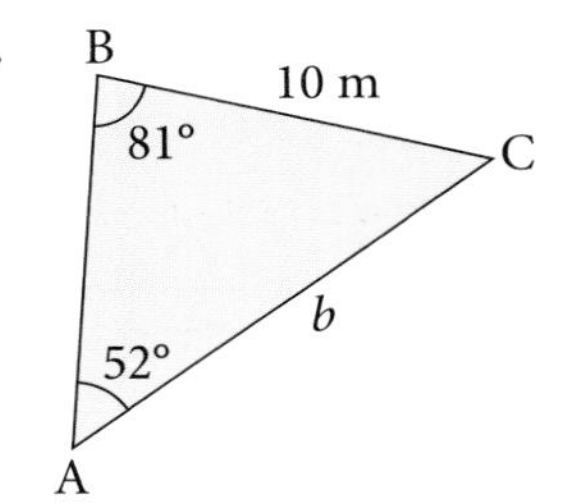

3.

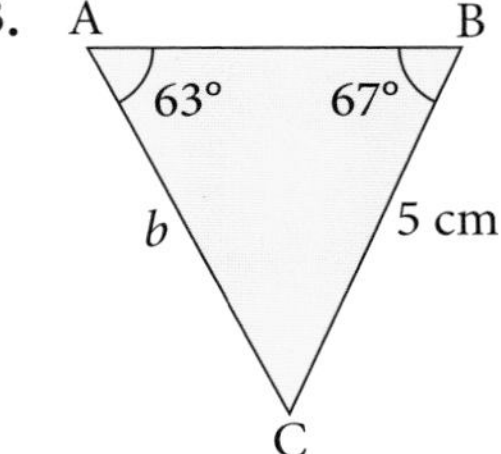

4. 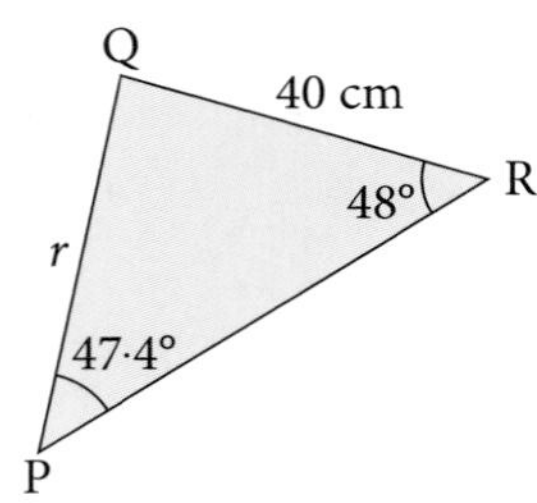

5. 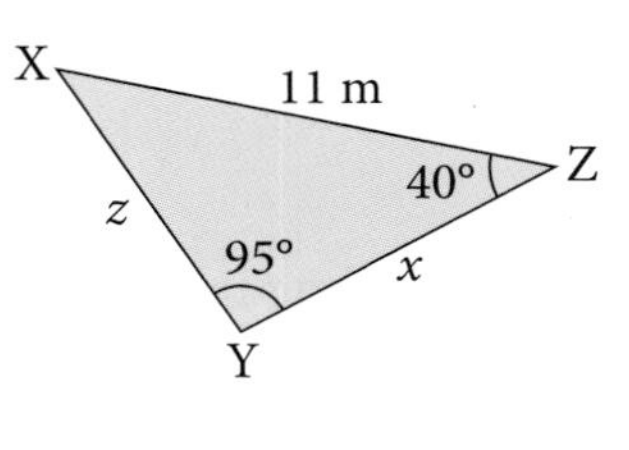

6. 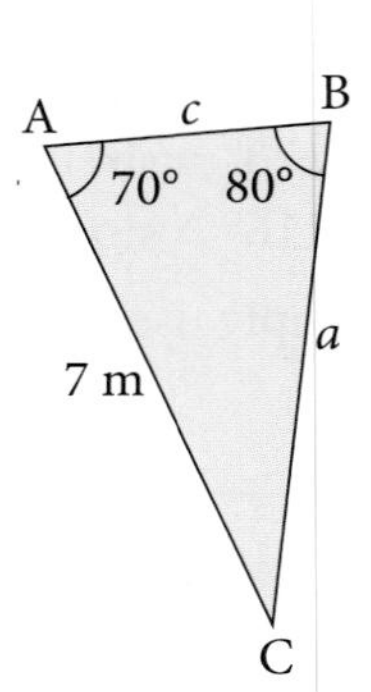

7. In ΔABC, $\widehat{A} = 61°$, $\widehat{B} = 47°$, AC = 7·2 cm. Find BC.

8. In ΔXYZ, $\widehat{Z} = 32°$, $\widehat{Y} = 78°$, XY = 5·4 cm. Find XZ.

9. In ΔPQR, $\widehat{Q} = 100°$, $\widehat{R} = 21°$, PQ = 3·1 cm. Find PR.

10. In ΔLMN, $\widehat{L} = 21°$, $\widehat{N} = 30°$, MN = 7 cm. Find LN.

In questions **11** to **18**, find each angle marked $*$. All lengths are in centimetres.

11. 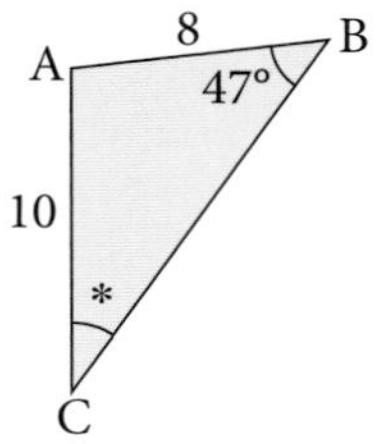

12. 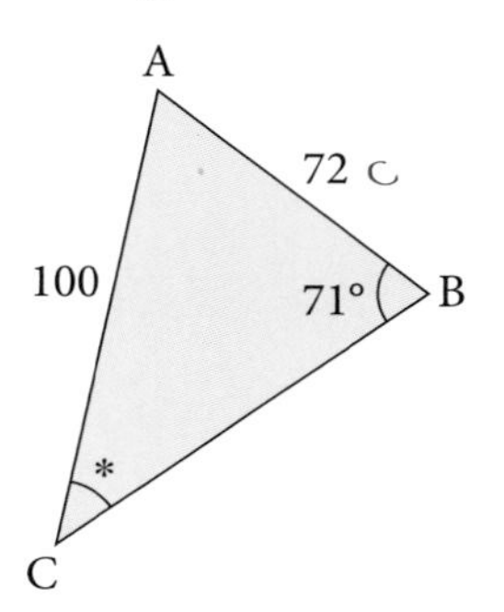

13. 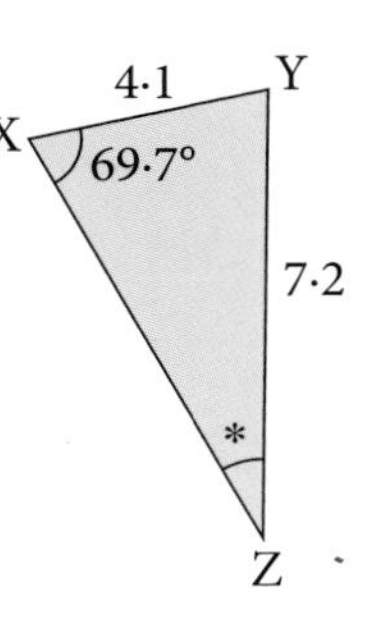

14. 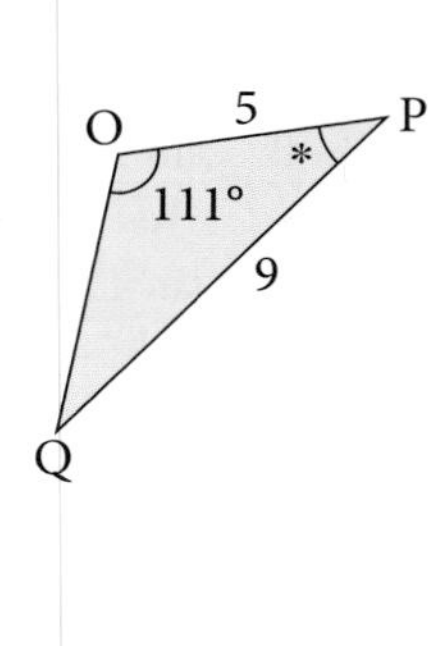

15. 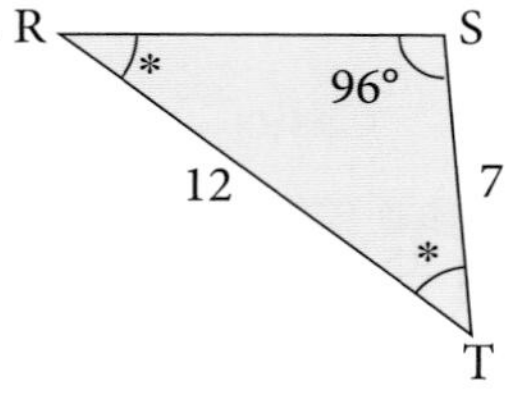

16. 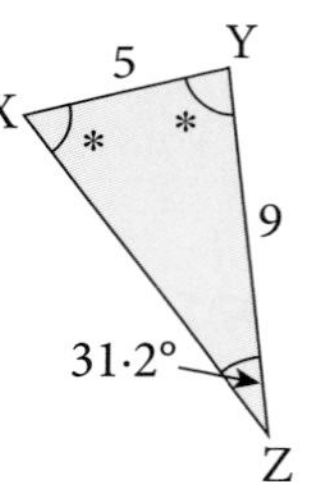

17. 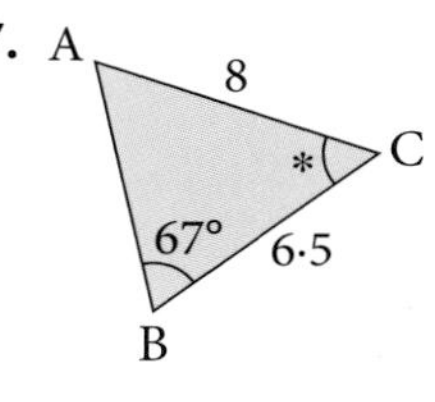

18. 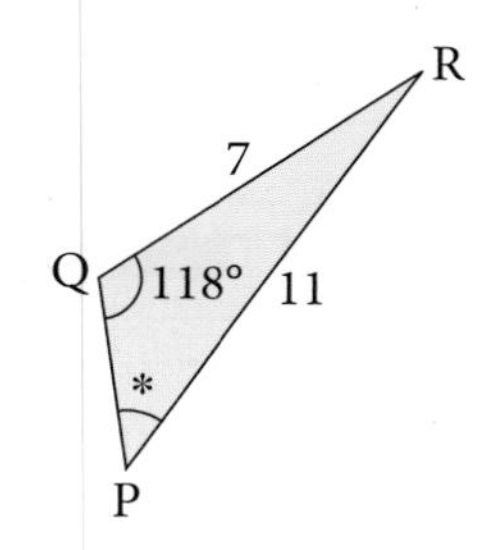

19. In ΔABC, $\widehat{A} = 62°$, BC = 8, AB = 7. Find $\widehat{C}$.

20. In ΔXYZ, $\widehat{Y} = 97·3°$, XZ = 22, XY = 14. Find $\widehat{Z}$.

21. In ΔDEF, $\widehat{D} = 58°$, EF = 7·2, DE = 5·4. Find $\widehat{F}$.

22. In ΔLMN, $\widehat{M} = 127·1°$, LN = 11·2, LM = 7·3. Find $\widehat{L}$.

Area of a triangle

In triangle BCD, $\sin C = \frac{h}{a}$

$\therefore \quad h = a \sin C$

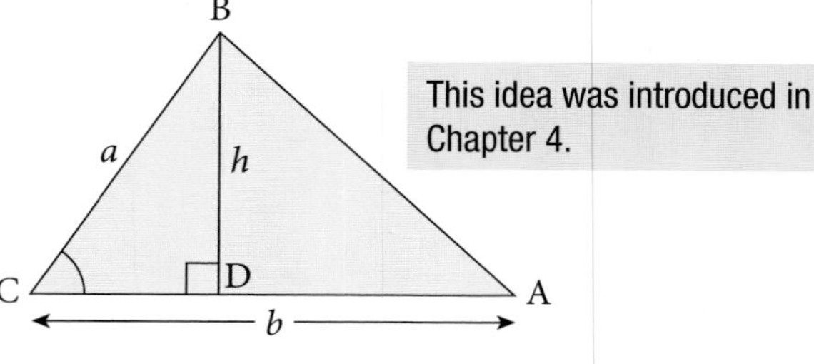

This idea was introduced in Chapter 4.

$$\therefore \quad \text{area of triangle ABC} = \frac{1}{2} \times b \times a \sin C$$

This formula is useful when *two sides* and the *included angle* are known.

Area of a triangle $= \frac{1}{2} \times b \times h$

10.6 The cosine rule

We use the cosine rule when we have either

- two sides and the included angle, or
- all three sides.

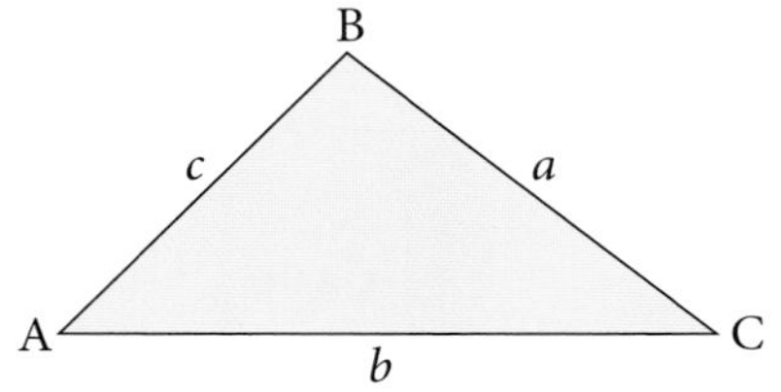

There are two forms.

1. To find the length of a side.

$$a^2 = b^2 + c^2 - (2bc \cos A)$$

or $b^2 = c^2 + a^2 - (2ac \cos B)$

or $c^2 = a^2 + b^2 - (2ab \cos C)$

2. To find an angle when given all three sides.

$$\cos A = \frac{b^2 + c^2 - a^2}{2bc}$$

or $\cos B = \dfrac{a^2 + c^2 - b^2}{2ac}$

or $\cos C = \dfrac{a^2 + b^2 - c^2}{2ab}$

For an obtuse angle x we have

- $\cos x = -\cos(180 - x)$

Examples: $\cos 120° = -\cos 60°$
$\cos 142° = -\cos 38°$

Example 9

Find b.

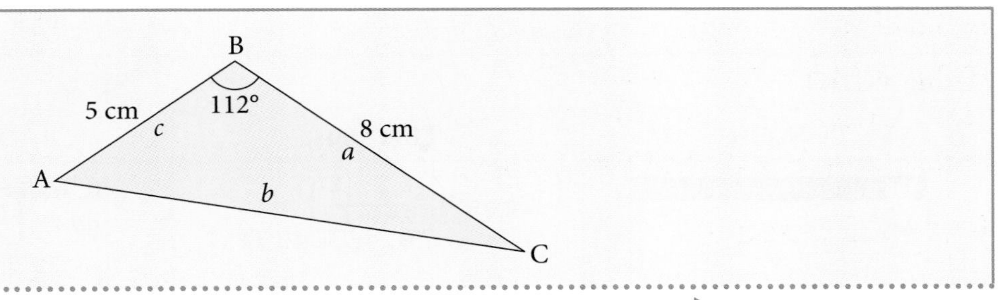

▶ Continued on next page

$b^2 = a^2 + c^2 - (2ac \cos B)$

$b^2 = 8^2 + 5^2 - (2 \times 8 \times 5 \times \cos 112°)$

$b^2 = 64 + 25 - [80 \times (-0{\cdot}3746)]$

$b^2 = 64 + 25 + 29{\cdot}968$

(Notice the change of sign for the obtuse angle.)

$b = \sqrt{118{\cdot}968} = 10{\cdot}9$ cm (to 3 s.f.)

Using a GDC

TI-Nspire	TI84-Plus	Casio
$\sqrt{8^2+5^2-2\cdot 8\cdot 5\cdot \cos(112)}$ 10.9073	$\sqrt{8^2+5^2-2*8*5*\cos}$ 10.90726948	$\sqrt{8^2+5^2-2\times 8\times 5\times \cos 112}$ 10.90726948

Example 10

Find angle C.

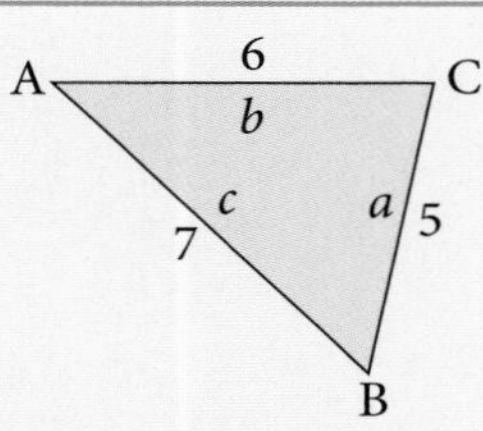

$$\cos C = \frac{a^2+b^2-c^2}{2ab}$$

$$\cos C = \frac{5^2+6^2-7^2}{2\times 5\times 6} = \frac{12}{60} = 0{\cdot}200$$

$$\hat{C} = 78{\cdot}5°$$

Using a GDC

TI-Nspire	TI84-Plus	Casio
$\cos^{-1}\left(\frac{5^2+6^2-7^2}{2\cdot 5\cdot 6}\right)$ 78.463	$\cos^{-1}\left(\frac{5^2+6^2-7^2}{2*5*6}\right)$ 78.46304097	$\cos^{-1}\left(\frac{5^2+6^2-7^2}{2\times 5\times 6}\right)$ 78.46304097

Exercise 10.12

Find the sides marked $*$. All lengths are in centimetres.

1.

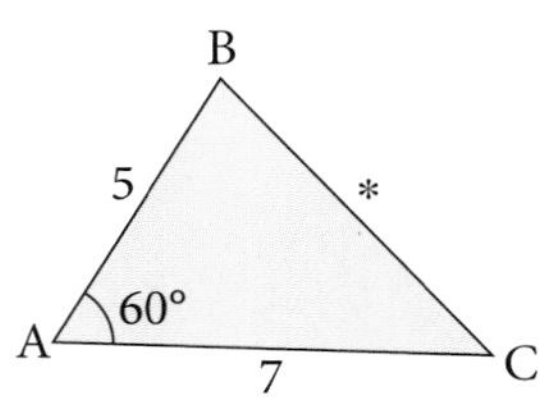

2.

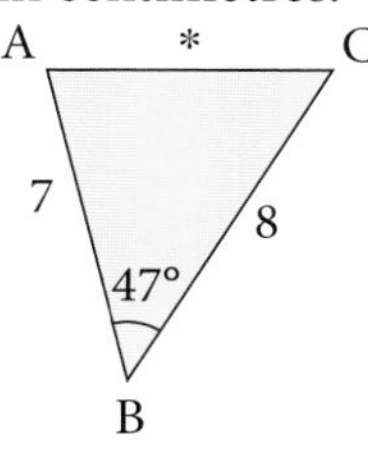

3.

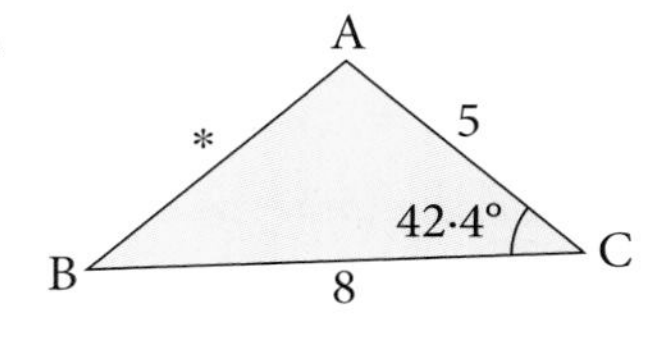

4.

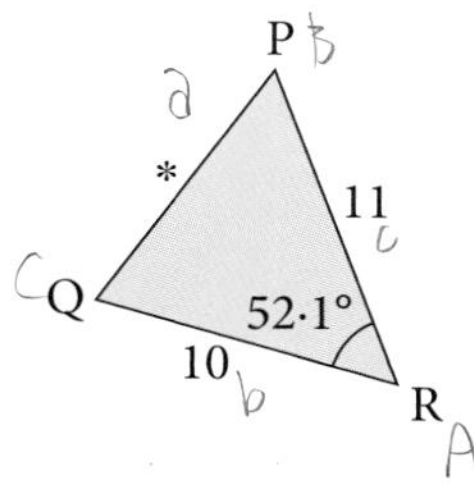

5.

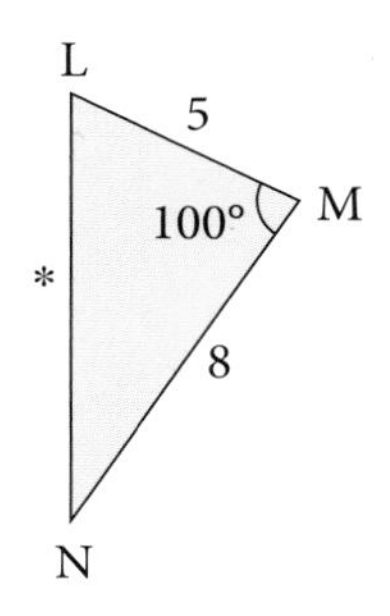

6.

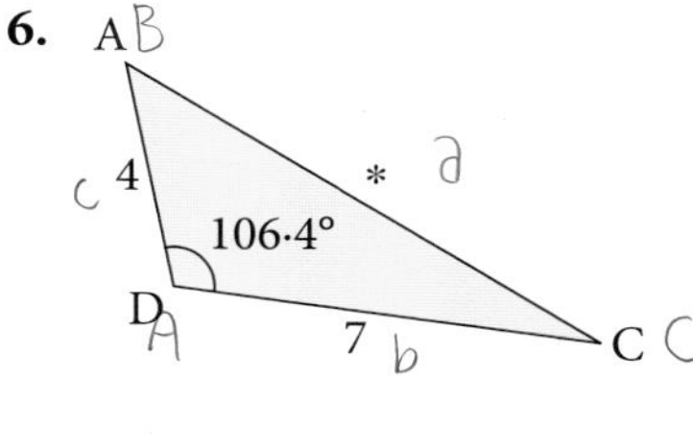

7. In ΔABC, AB = 4 cm, AC = 7 cm, $\widehat{A} = 57°$. Find BC.

8. In ΔXYZ, XY = 3 cm, YZ = 3 cm, $\widehat{Y} = 90°$. Find XZ.

9. In ΔLMN, LM = 5·3 cm, MN = 7·9 cm, $\widehat{M} = 127°$. Find LN.

10. In ΔPQR, $\widehat{Q} = 117°$, PQ = 80 cm, QR = 100 cm. Find PR.

In questions **11** to **16**, find each angle marked $*$.

11.

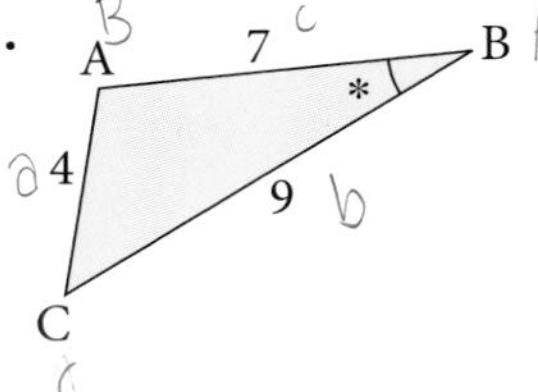

12.

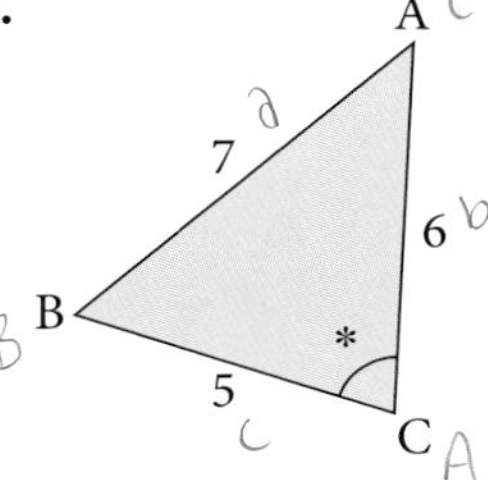

13.

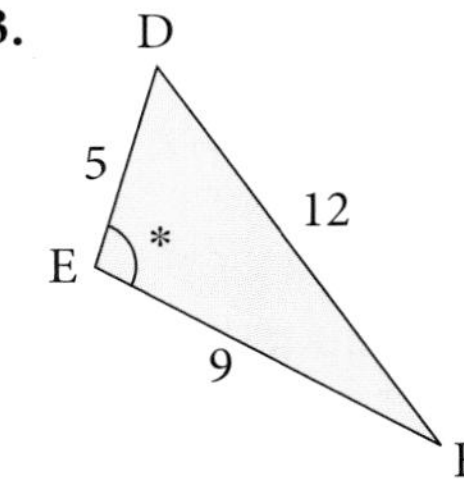

14.

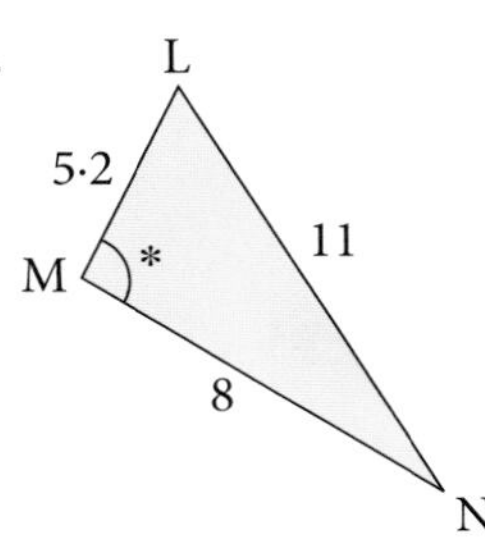

15.

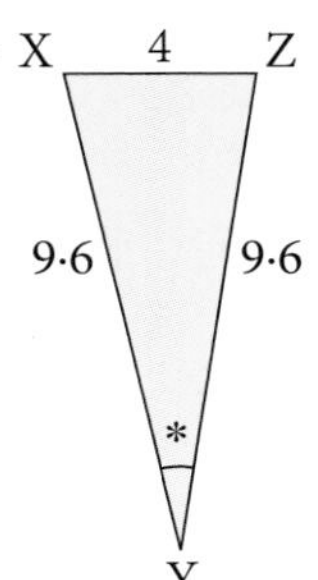

16.

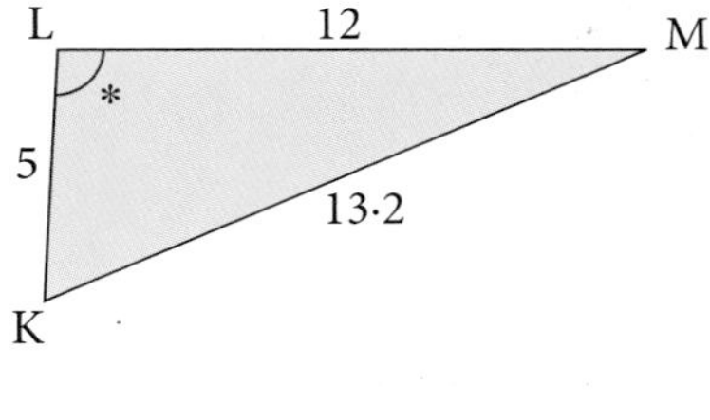

17. In ΔABC, $a = 4{\cdot}3$, $b = 7{\cdot}2$, $c = 9$. Find $\widehat{\mathrm{C}}$.

18. In ΔDEF, $d = 30$, $e = 50$, $f = 70$. Find $\hat{\mathrm{E}}$

19. In ΔPQR, $P = 8$, $q = 14$, $r = 7$. Find $\widehat{\mathrm{Q}}$

20. In ΔLMN, $l = 7$, $m = 5$, $n = 4$. Find $\widehat{\mathrm{N}}$

21. In ΔXYZ, $x = 5{\cdot}3$, $y = 6{\cdot}7$, $z = 6{\cdot}14$. Find $\hat{\mathrm{Z}}$

22. In ΔABC, $a = 4{\cdot}1$, $c = 6{\cdot}3$, $\hat{\mathrm{B}} = 112{\cdot}2°$.
Find b.

23. In ΔPQR, $r = 0{\cdot}72$, $p = 1{\cdot}14$, $\widehat{\mathrm{Q}} = 94{\cdot}6°$.
Find q.

24. In ΔLMN, $n = 7{\cdot}206$, $l = 6{\cdot}3$, $\hat{\mathrm{L}} = 51{\cdot}2°$, $\widehat{\mathrm{N}} = 63°$.
Find m.

Example 11

A ship sails from a port P a distance of 7 km on a bearing of 306° and then a further 11 km on a bearing of 070° to arrive at X.
Calculate the distance from P to X.

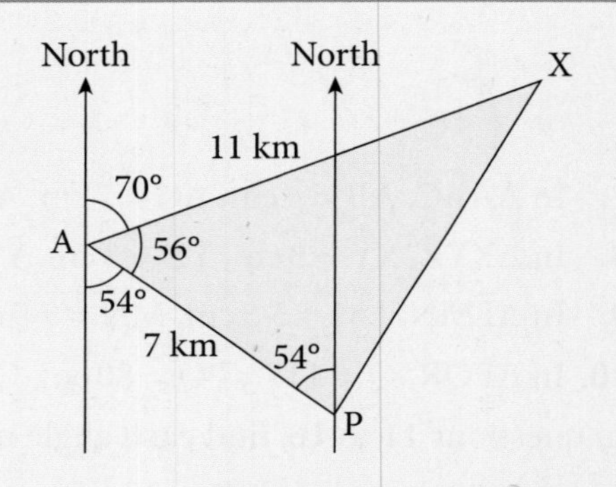

$\mathrm{PX}^2 = 7^2 + 11^2 - (2 \times 7 \times 11 \times \cos 56°)$
$= 49 + 121 - (86{\cdot}12)$
$\mathrm{PX}^2 = 83{\cdot}88$
$\mathrm{PX} = 9{\cdot}16$ km (to 3 s.f.)
The distance from P to X is 9·16 km.

Using a GDC

TI-Nspire	TI84-Plus	Casio
1.1 *Unsaved $\sqrt{7^2+11^2-2\cdot 7\cdot 11\cdot \cos(56)}$ 9.15884 1/99	$\sqrt{7^2+11^2-2*7*11*c}$ 9.158836873	Math Deg Norm1 d/c Real $\sqrt{7^2+11^2-2\times 7\times 11\times \cos 56}$ 9.158836873 JUMP DELETE ▸MAT MATH

Exercise 10.13

Start each question by drawing a large, clear diagram.

1. In triangle PQR, $\widehat{Q} = 72°$, $\widehat{R} = 32°$ and PR = 12 cm. Find PQ.
2. In triangle LMN, $\widehat{M} = 84°$, LM = 7 m and MN = 9 m. Find LN.
3. A destroyer D and a cruiser C leave port P at the same time. The destroyer sails 25 km on a bearing 040° and the cruiser sails 30 km on a bearing of 320°. How far apart are the ships?
4. Two honeybees A and B leave the hive H at the same time. A flies 27 m due South and B flies 9 m on a bearing of 111°. How far apart are they?
5. Find all the angles of a triangle in which the sides are in the ratio 5 : 6 : 8.
6. A golfer hits his ball B a distance of 170 m towards a hole H which measures 195 m from the tee T to the green. If his shot is directed 10° away from the true line to the hole, find the distance between his ball and the hole.
7. From A, B lies 11 km away on a bearing of 041° and C lies 8 km away on a bearing of 341°. Find:
   - **a)** the distance between B and C
   - **b)** the bearing of B from C.
8. From a lighthouse L an aircraft carrier A is 15 km away on a bearing of 112° and a submarine S is 26 km away on a bearing of 200°. Find:
   - **a)** the distance between A and S
   - **b)** the bearing of A from S.
9. If the line BCD is horizontal find:
   - **a)** AE
   - **b)** $E\widehat{A}C$
   - **c)** the angle of elevation of E from A.

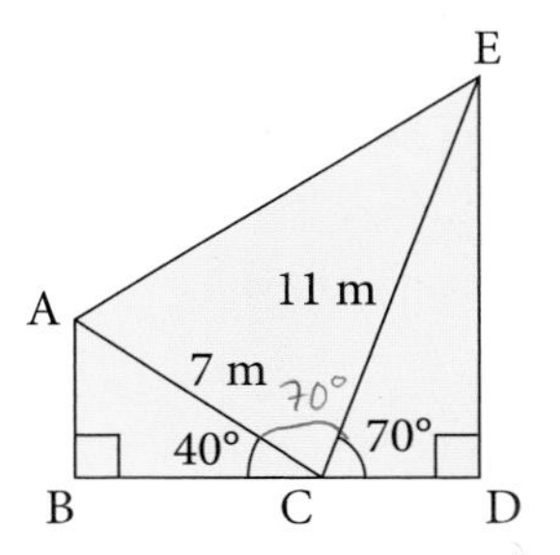

10. An aircraft flies from its base 200 km on a bearing 162°, then 350 km on a bearing 260°, and then returns directly to base. Calculate the length and bearing of the return journey.
11. Shop Y is 9 km due North of shop Z. Shop X is 8 km from Y, 5 km from Z and somewhere to the west of the line YZ.
    - **a)** Draw triangle XYZ and find angle YZX.
    - **b)** To improve sales, shop X is moved due South until it is due West of Z. Find how far it has moved.

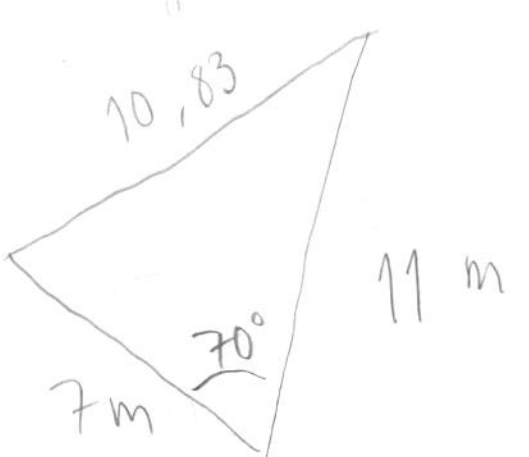

12. Calculate WX, given YZ = 15 m.

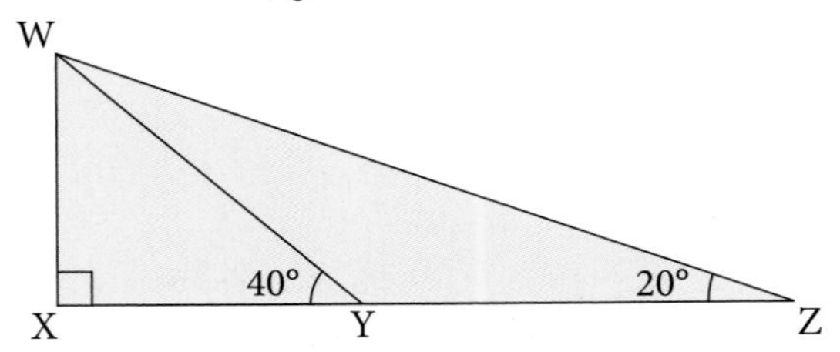

13. A golfer hits her ball a distance of 127 m so that it finishes 31 m from the hole. If the length of the hole is 150 m, calculate the angle between the line of her shot and the direct line to the hole.

Revision exercise 10A

1. Calculate the side or angle marked with a letter.

a)

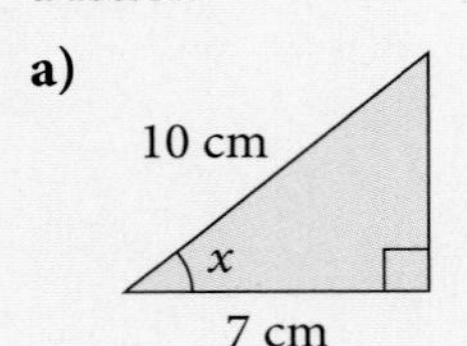

b)

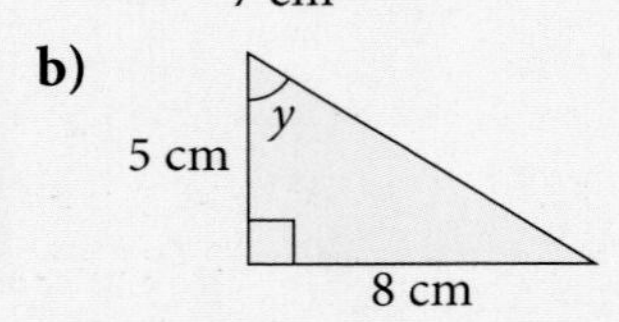

c)

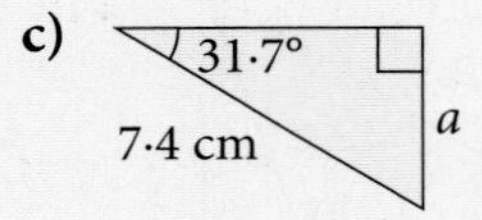

d)

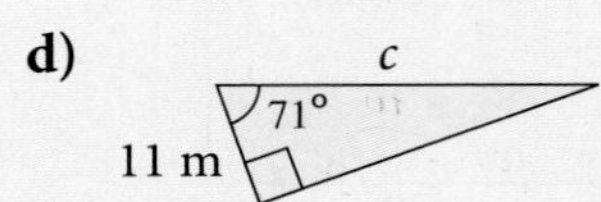

2. Given that x is an acute angle and that $3\tan x - 2 = 4\cos 35{\cdot}3°$ calculate:

a) $\tan x$

b) the value of x in degrees correct to 1 d.p.

3. In the triangle XYZ, XY = 14 cm, XZ = 17 cm and angle YXZ = 25°. A is the foot of the perpendicular from Y to XZ. Calculate:

a) the length XA **b)** the length YA

c) the angle ZYA.

4. Calculate the length of AB.

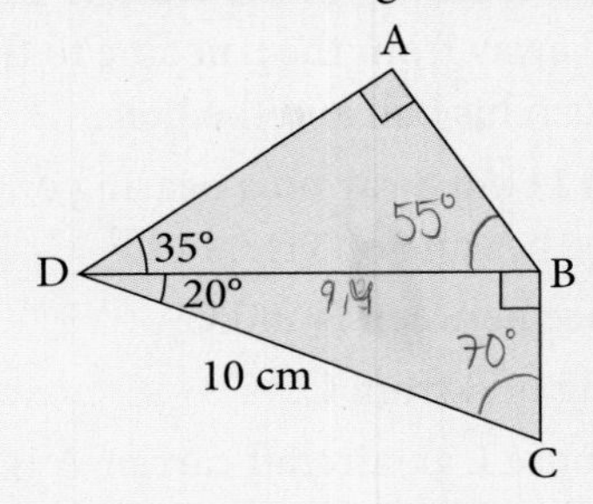

5. a) A lies on a bearing of 040° from B. Calculate the bearing of B from A.

b) The bearing of X from Y is 115°. Calculate the bearing of Y from X.

$B\hat{A}D = A\hat{C}D = 90°$

6. Given BD = 1 m, calculate the length AC.

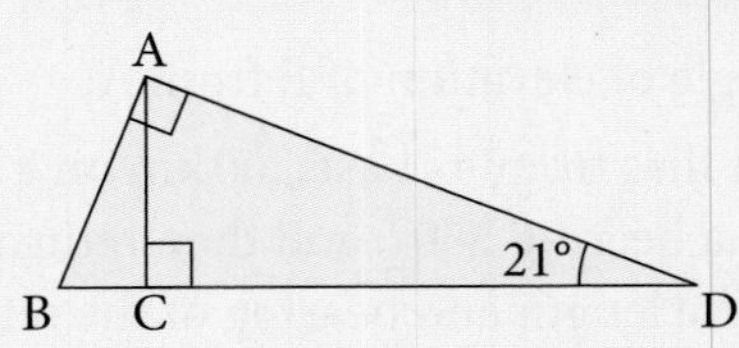

7. In the triangle PQR, angle PQR = 90° and angle RPQ = 31°. The length of PQ is 11 cm. Calculate:

a) the length of QR

b) the length of PR

c) the length of the perpendicular from Q to PR.

8. $B\hat{A}D = D\hat{C}A = 90°$, $C\hat{A}D = 32.4°$, $B\hat{D}A = 41°$ and AD = 100 cm.

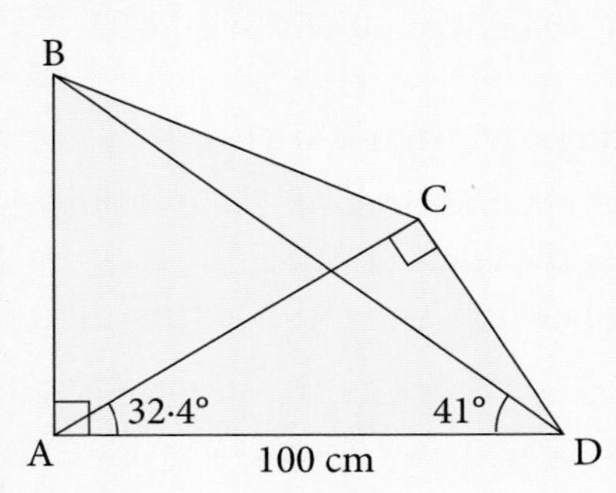

Calculate:

a) the length of AB

b) the length of DC

c) the length of BD.

9. Calculate the area of rectangle ABCD.

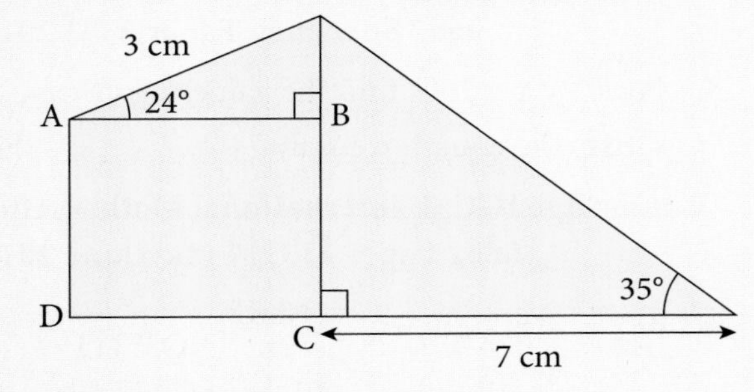

10. The figure shows a cube of side 10 cm.

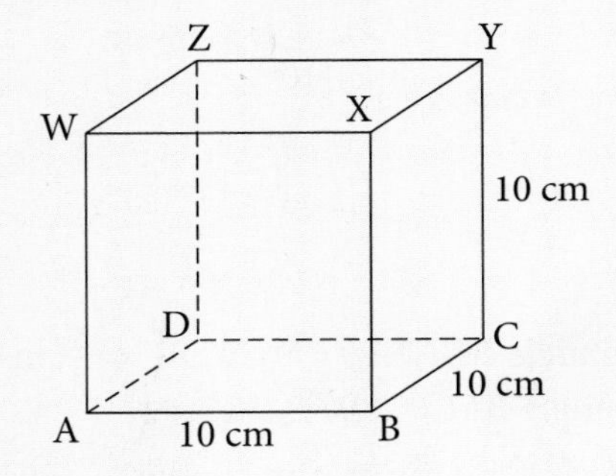

Calculate:

a) the length of AC

b) the angle YAC

c) the angle ZBD.

11.

The diagram shows a rectangular block.

AY = 12 cm, AB = 8 cm, BC = 6 cm.

Calculate:

a) the length YC **b)** the angle $Y\hat{A}Z$.

12. VABCD is a pyramid in which the base ABCD is a square of side 8 cm; V is vertically above the centre of the square and VA = VB = VC = VD = 10 cm.

Calculate:

a) the length AC

b) the height of V above the base

c) the angle $V\hat{C}A$.

13. Two lighthouses A and B are 25 km apart and A is due West of B. A submarine S is on a bearing of 137° from A and on a bearing of 170° from B. Find the distance of S from A and the distance of S from B.

14. In triangle PQR, PQ = 7 cm, PR = 8 cm and QR = 9 cm. Find angle QPR.

15. In triangle XYZ, XY = 8 m, $\hat{X} = 57°$ and $\hat{Z} = 50°$. Find the lengths YZ and XZ.

16. In triangle ABC, $\hat{A} = 22°$ and $\hat{C} = 44°$.

Find the ratio $\frac{BC}{AB}$.

17. Given $\cos A\hat{C}B = 0{\cdot}6$, AC = 4 cm, BC = 5 cm and CD = 7 cm, find the length of AB and AD.

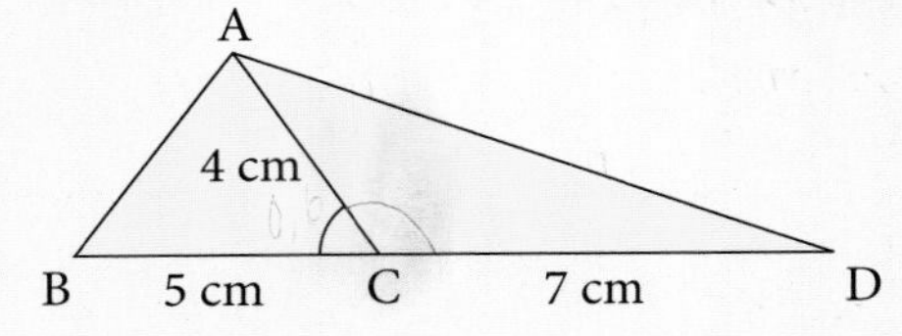

18. Find the smallest angle in a triangle whose sides are of length $3x$, $4x$ and $6x$.

19. Find the perimeter of the rhombus.

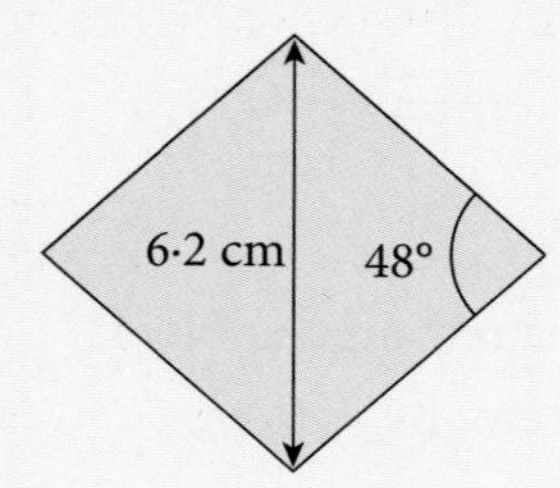

20. A control tower observes the flight of an aircraft. At 09:23 the aircraft is 580 km away on a bearing of 043°. At 09:43 the aircraft is 360 km away on a bearing of 016°. What is the speed of the aircraft?

21. A regular pentagon is inscribed in a circle of radius 7 cm.

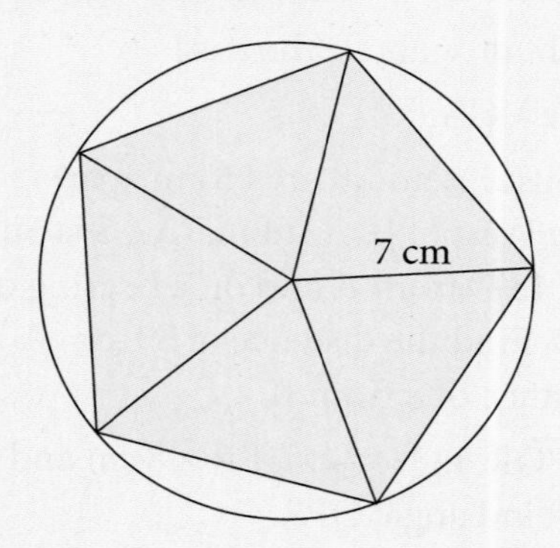

Find the length of a side of the pentagon.

22. Write down the exact values of the following.

a) $\sin 30°$ **b)** $\tan 60°$

c) $\cos 45°$ **d)** $2 - \cos 60°$

e) $\cos^2 30°$ **f)** $2 \tan 45°$

g) $\sin^2 45° + \cos^2 45°$

Examination exercise 10B

1. The distance between towns A and B is 50 km.
The bearing of A from B is 210°.

a) Sketch the positions of A and B showing clearly the angle of 210°. [1]

b) Calculate how far west A is from B. [2]

Cambridge IGCSE International Mathematics 0607 Paper 2 Q6 October/November 2010

2. Solve $2 \sin x° = 1$ for $0 \leqslant x \leqslant 360$. [2]

Cambridge IGCSE International Mathematics 0607 Specimen Paper 2 Q3 2010

3. For $0° < x < 360°$ find the values of x that satisfy the equation $\cos x = -\frac{1}{2}$. [2]

Cambridge IGCSE International Mathematics 0602 Paper 21 Q11 May/June 2011

4. a)

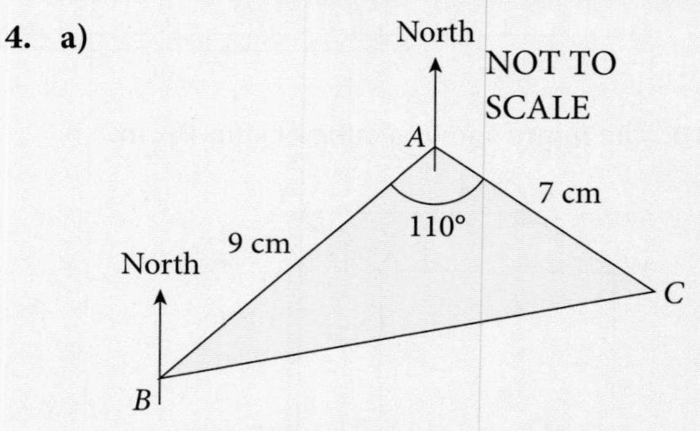

In triangle ABC, $AB = 9$ cm, $AC = 7$ cm and angle $BAC = 110°$.

i) Calculate the area of triangle ABC. [2]

ii) Calculate the length of BC. [3]

iii) The bearing of A from B is 050°. Find the bearing of C from A. [2]

b) NOT TO SCALE

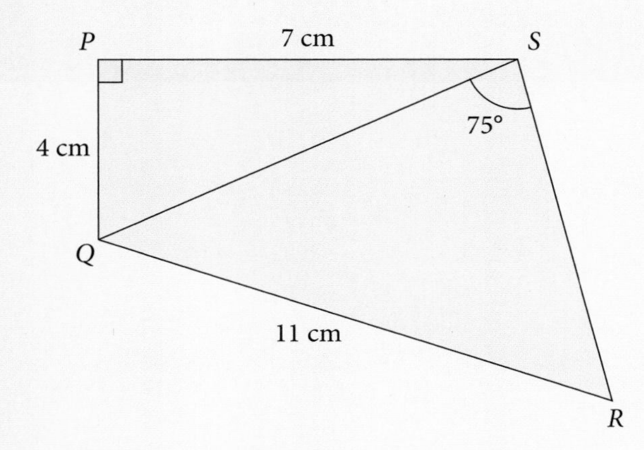

In the quadrilateral $PQRS$, $PQ = 4$ cm, $PS = 7$ cm and $QR = 11$ cm.

Angle $QPS = 90°$ and angle $QSR = 75°$.

Calculate the size of angle QRS. [5]

Cambridge IGCSE International Mathematics 0607 Paper 41 Q6 May/June 2011

5. NOT TO SCALE

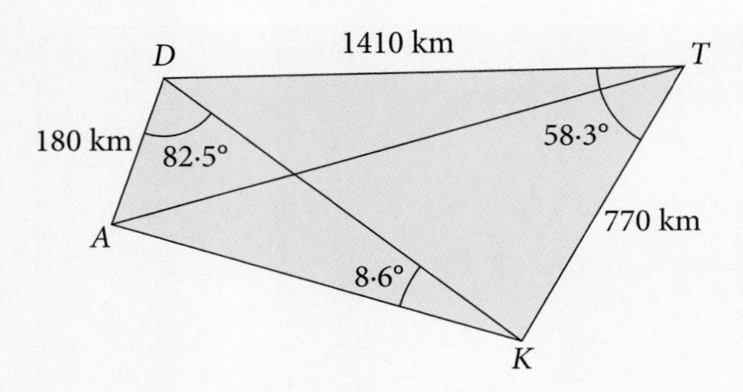

The diagram shows the cities of Amman (A), Damascus (D), Tehran (T) and Kuwait (K).

$AD = 180$ km, $DT = 1410$ km and $TK = 770$ km.

Angle $ADK = 82.5°$, angle $AKD = 8.6°$ and angle $DTK = 58.3°$.

a) Use the sine rule in triangle ADK to calculate the distance AK. [3]

b) Use the cosine rule in triangle DKT to calculate the distance DK. [3]

c) Calculate the area of the quadrilateral $ADTK$. [3]

d) Calculate the distance AT. [5]

e) A map is drawn to a scale of 1 : 5 000 000.

Calculate the length of DT on the map, in centimetres. [2]

Cambridge IGCSE International Mathematics 0607 Paper 4 Q9 May/June 2010

6. In triangle ABC, $AB = 10$ cm, $BC = 6$ cm and angle $BAC = 30°$.

a) Calculate the **sine** of angle ACB. Give your answer correct to 4 decimal places. [3]

b) To draw triangle ABC accurately, the line AB and an angle 30° have been drawn.

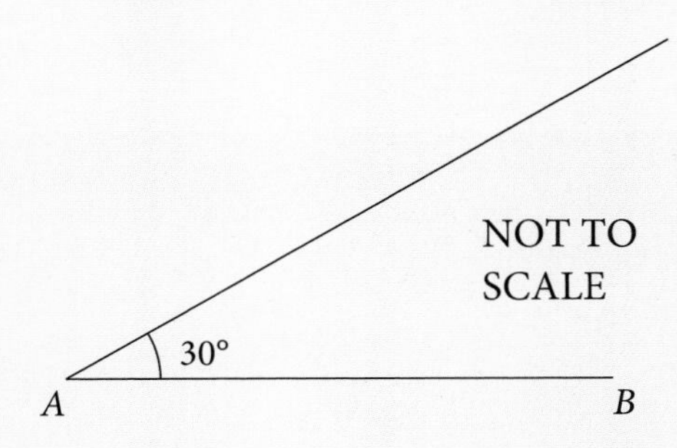

i) On the diagram, mark the two possible positions of C, so that $BC = 6$ cm.
Label them C_1 and C_2. [2]

ii) **Use your answer to part a)** to calculate the sizes of angle AC_1B and angle AC_2B. Give your answers to 1 decimal place. [2]

iii) Calculate the size of angle C_1BC_2. [1]

Cambridge IGCSE International Mathematics 0607 Paper 4 Q12 October/November/2010

11 Vectors and transformations

Stephen Hawking (1942–) was born on the 300th anniversary of the death of Galileo. At the age of 21 he was diagnosed with a crippling disease in which the nerves that controlled his muscles were failing. He is able to communicate by using a speaking device that he directs by head or eye movement. He has produced a large number of scientific papers as well as *A Brief History of Time* which has sold over 10 million copies.

5.1 Notation:
Vector **a**; directed line segment $\overrightarrow{AB}$;
Component form $\begin{pmatrix} x \\ y \end{pmatrix}$

5.2 Addition and subtraction of vectors
Negative of a vector
Multiplication of a vector by a scalar

5.3 Magnitude |**a**|

5.4 Transformations on the Cartesian plane:
translation, reflection, rotation, enlargement (reduction), stretch
Description of a translation using the notation in 5.1

5.5 Inverse of a transformation

5.6 Combined transformations

11.1 Vectors

A **vector** quantity has both magnitude and direction. Problems involving forces, velocities and displacements are often made easier when vectors are used.

In most books and examination papers, vectors are denoted by bold lower case letters as on this page.

Addition of vectors

Vectors **a** and **b** represented by the line segments in the diagram can be added using the parallelogram rule or the 'nose-to-tail' method.

The 'tail' of vector **b** is joined to the 'nose' of vector **a**.

Alternatively the 'tail' of **a** can be joined to the 'nose' of vector **b**.

In both cases the vector $\overrightarrow{XY}$ has the same length and direction and therefore

$$\mathbf{a} + \mathbf{b} = \mathbf{b} + \mathbf{a}$$

Multiplication by a scalar

A **scalar** quantity has a magnitude but no direction (e.g. mass, volume, temperature). Ordinary numbers are scalars.

When vector **x** is multiplied by 2, the result is 2**x**.

When **x** is multiplied by –3 the result is –3**x**.

Note:

1. The negative sign reverses the direction of the vector.
2. The result of **a** – **b** is **a** + –**b**.
   i.e. Subtracting **b** is equivalent to adding the negative of **b**.

Example 1

The diagram shows vectors **a** and **b**.

Find $\overrightarrow{OP}$ and $\overrightarrow{OQ}$ such that

$\overrightarrow{OP} = 3\mathbf{a} + \mathbf{b}$

$\overrightarrow{OQ} = -2\mathbf{a} - 3\mathbf{b}$

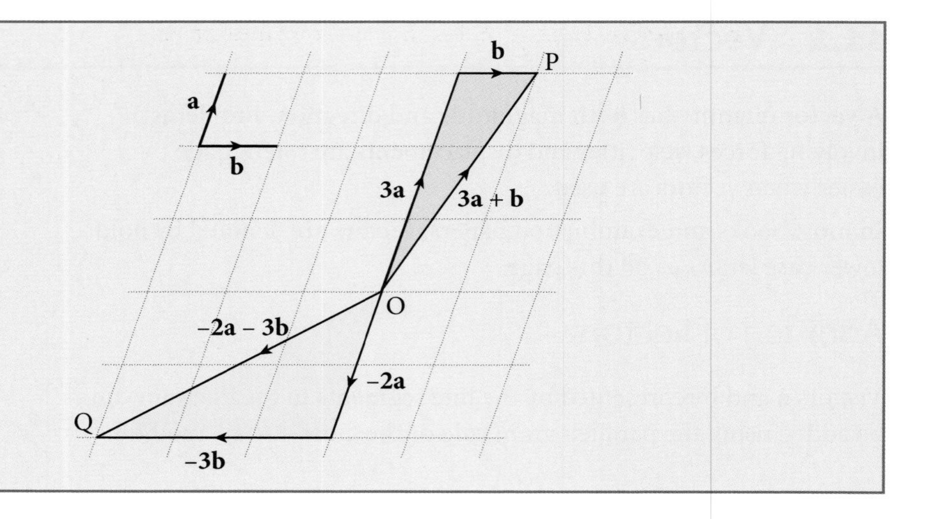

Exercise 11.1

In questions **1** to **26**, use the diagram below to describe the vectors given in terms of **c** and **d** where $\mathbf{c} = \overrightarrow{QN}$ and $\mathbf{d} = \overrightarrow{QR}$.

e.g. $\overrightarrow{QS} = 2\mathbf{d}$, $\overrightarrow{TD} = \mathbf{c} + \mathbf{d}$.

1. $\overrightarrow{AB}$ **2.** $\overrightarrow{SG}$ **3.** $\overrightarrow{VK}$

4. $\overrightarrow{KH}$ **5.** $\overrightarrow{OT}$ **6.** $\overrightarrow{WJ}$

7. $\overrightarrow{FH}$ **8.** $\overrightarrow{FT}$ **9.** $\overrightarrow{KV}$

10. $\overrightarrow{NQ}$ **11.** $\overrightarrow{OM}$ **12.** $\overrightarrow{SD}$

13. $\overrightarrow{PI}$ **14.** $\overrightarrow{YG}$ **15.** $\overrightarrow{OI}$

16. $\overrightarrow{RE}$ **17.** $\overrightarrow{XM}$ **18.** $\overrightarrow{ZH}$

19. $\overrightarrow{MR}$ **20.** $\overrightarrow{KA}$ **21.** $\overrightarrow{RZ}$

22. $\overrightarrow{CR}$ **23.** $\overrightarrow{NV}$ **24.** $\overrightarrow{EV}$

25. $\overrightarrow{JS}$ **26.** $\overrightarrow{LE}$

In questions **27** to **38**, use the same diagram as above to find vectors for the following in terms of the capital letters, starting from **Q** each time.

e.g. $3\mathbf{d} = \overrightarrow{QT}$, $\mathbf{c} + \mathbf{d} = \overrightarrow{QA}$

27. $2\mathbf{c}$ **28.** $4\mathbf{d}$ **29.** $2\mathbf{c} + \mathbf{d}$

30. $2\mathbf{d} + \mathbf{c}$ **31.** $3\mathbf{d} + 2\mathbf{c}$ **32.** $2\mathbf{c} - \mathbf{d}$

33. $-\mathbf{c} + 2\mathbf{d}$ **34.** $\mathbf{c} - 2\mathbf{d}$ **35.** $2\mathbf{c} + 4\mathbf{d}$

36. $-\mathbf{c}$ **37.** $-\mathbf{c} - \mathbf{d}$ **38.** $2\mathbf{c} - 2\mathbf{d}$

In questions **39** to **42**, write each vector in terms of **a** and/or **b**.

39. a) $\overrightarrow{BA}$ **b)** $\overrightarrow{AC}$

c) $\overrightarrow{DB}$ **d)** $\overrightarrow{AD}$

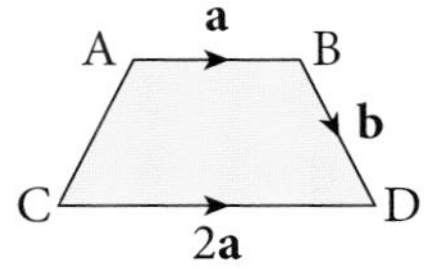

40. a) $\overrightarrow{ZX}$ **b)** $\overrightarrow{YW}$

c) $\overrightarrow{XY}$ **d)** $\overrightarrow{XZ}$

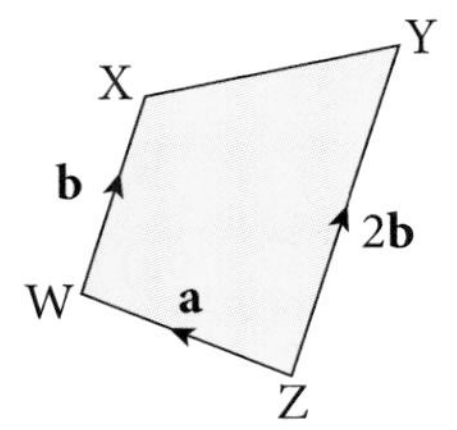

41. a) $\overrightarrow{MK}$ **b)** $\overrightarrow{NL}$

c) $\overrightarrow{NK}$ **d)** $\overrightarrow{KN}$

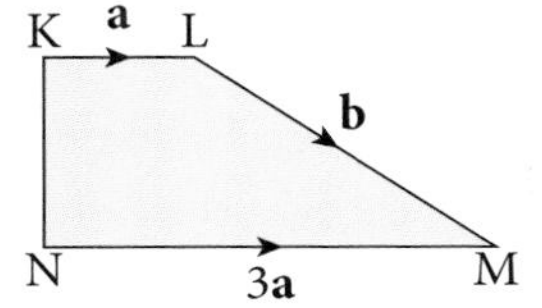

42. a) $\overrightarrow{FE}$ **b)** $\overrightarrow{BC}$

c) $\overrightarrow{FC}$ **d)** $\overrightarrow{DA}$

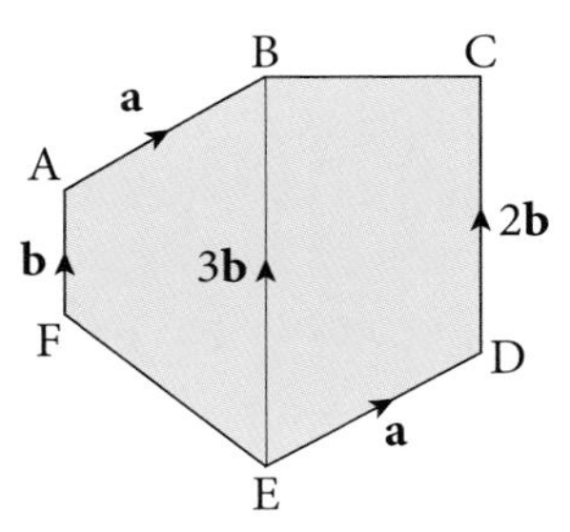

43. The points A, B and C lie on a straight line and the vector $\overrightarrow{AB}$ is **a** + 2**b**.

Which of the following vectors is possible for $\overrightarrow{AC}$?

a) 3**a** + 6**b** **b)** 4**a** + 4**b**

c) **a** – 2**b** **d)** 5**a** + 10**b**

In questions **44** to **46**, write each vector in terms of **a**, **b** and **c**.

44. a) $\overrightarrow{FC}$ **b)** $\overrightarrow{GB}$ **c)** $\overrightarrow{AB}$

d) $\overrightarrow{HE}$ **e)** $\overrightarrow{CA}$

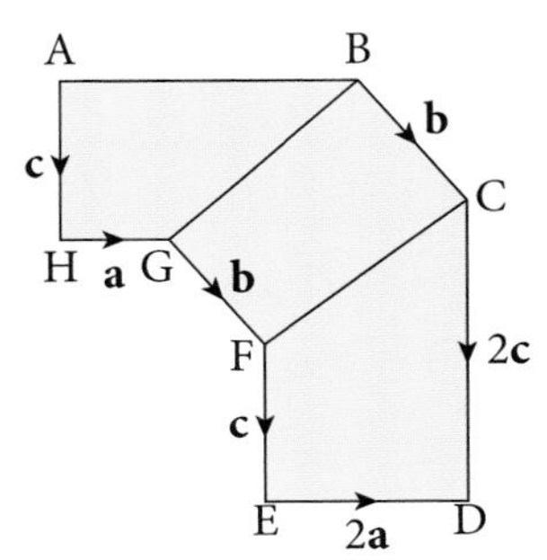

45. **a)** $\overrightarrow{OF}$ **b)** $\overrightarrow{OC}$ **c)** $\overrightarrow{BC}$
d) $\overrightarrow{EB}$ **e)** $\overrightarrow{FB}$

46. **a)** $\overrightarrow{GD}$ **b)** $\overrightarrow{GE}$ **c)** $\overrightarrow{AD}$
d) $\overrightarrow{AF}$ **e)** $\overrightarrow{FE}$

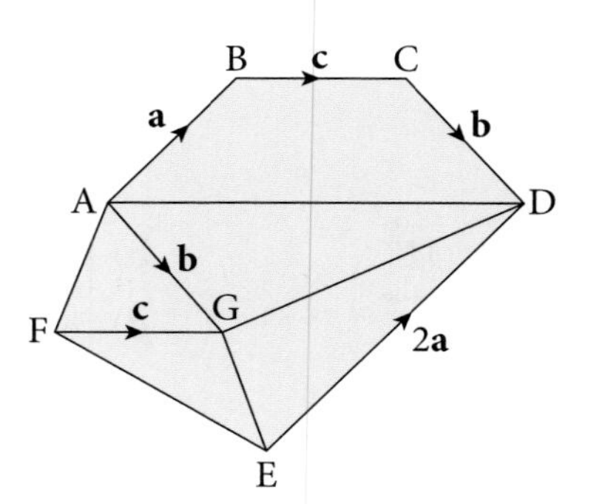

Example 2

Using the diagram, express each of the following vectors in terms of **a** and/or **b**.

a) $\overrightarrow{AP}$ **b)** $\overrightarrow{AB}$ **c)** $\overrightarrow{OQ}$ **d)** $\overrightarrow{PO}$
e) $\overrightarrow{PQ}$ **f)** $\overrightarrow{PN}$ **g)** $\overrightarrow{ON}$ **h)** $\overrightarrow{AN}$
i) $\overrightarrow{BP}$ **j)** $\overrightarrow{QA}$

OA = AP
BQ = 3OB
N is the mid-point of PQ
$\overrightarrow{OA} = \mathbf{a}$, $\overrightarrow{OB} = \mathbf{b}$

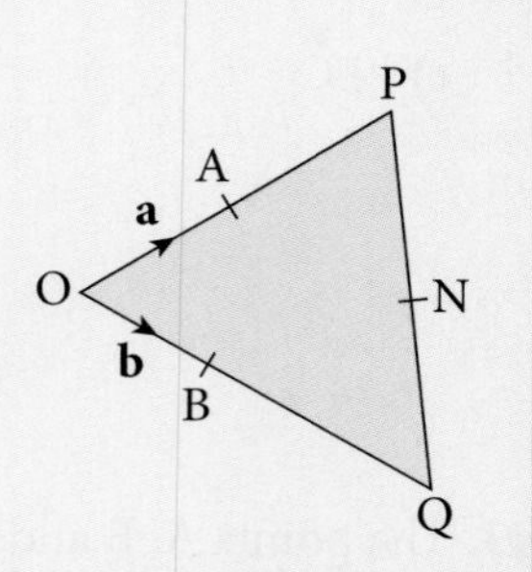

a) $\overrightarrow{AP} = \mathbf{a}$

b) $\overrightarrow{AB} = -\mathbf{a} + \mathbf{b}$

c) $\overrightarrow{OQ} = 4\mathbf{b}$

d) $\overrightarrow{PO} = -2\mathbf{a}$

e) $\overrightarrow{PQ} = \overrightarrow{PO} = \overrightarrow{OQ}$
$= -2\mathbf{a} + 4\mathbf{b}$

f) $\overrightarrow{PN} = \frac{1}{2}\overrightarrow{PQ}$
$= -\mathbf{a} + 2\mathbf{b}$

g) $\overrightarrow{ON} = \overrightarrow{OP} + \overrightarrow{PN}$
$= 2\mathbf{a} + (-\mathbf{a} + 2\mathbf{b})$
$= \mathbf{a} + 2\mathbf{b}$

h) $\overrightarrow{AN} = \overrightarrow{AP} + \overrightarrow{PN}$
$= \mathbf{a} + (-\mathbf{a} + 2\mathbf{b})$
$= 2\mathbf{b}$

i) $\overrightarrow{BP} = \overrightarrow{BO} + \overrightarrow{OP}$
$= -\mathbf{b} + 2\mathbf{a}$

j) $\overrightarrow{QA} = \overrightarrow{QO} + \overrightarrow{OA}$
$= -4\mathbf{b} + \mathbf{a}$

Exercise 11.2

In questions **1** to **6**, $\overrightarrow{OA} = \mathbf{a}$ and $\overrightarrow{OB} = \mathbf{b}$. Copy each diagram and use the information given to express the following vectors in terms of **a** and/or **b**.

a) $\overrightarrow{AP}$ **b)** $\overrightarrow{AB}$ **c)** $\overrightarrow{OQ}$ **d)** $\overrightarrow{PO}$ **e)** $\overrightarrow{PQ}$

f) $\overrightarrow{PN}$ **g)** $\overrightarrow{ON}$ **h)** $\overrightarrow{AN}$ **i)** $\overrightarrow{BP}$ **j)** $\overrightarrow{QA}$

1. A, B and N are mid-points of OP, OQ and PQ respectively.

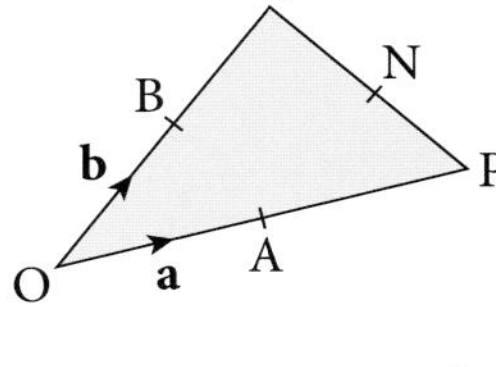

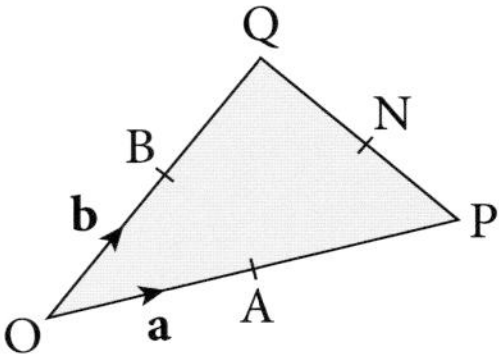

2. A and N are mid-points of OP and PQ and BQ = 2OB.

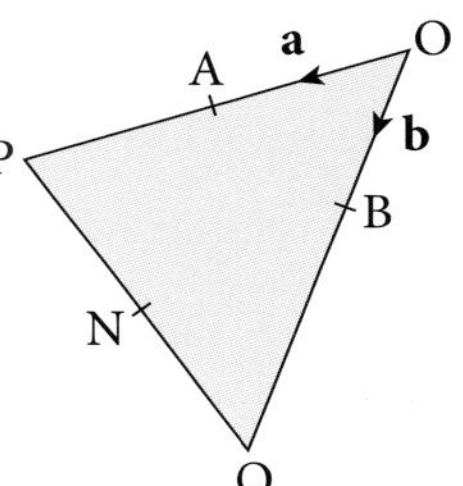

3. AP = 2OA, BQ = OB, PN = NQ.

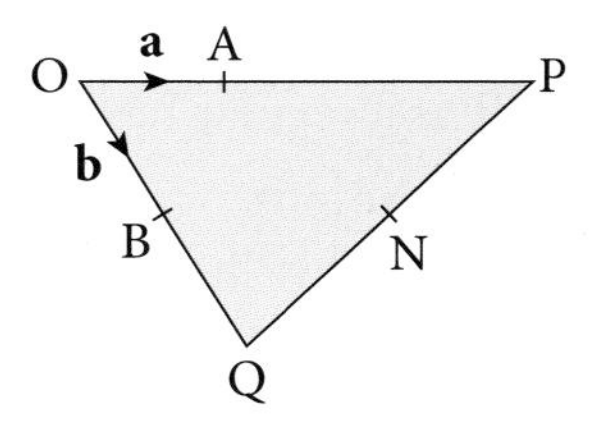

4. OA = 2AP, BQ = 3OB, PN = 2NQ.

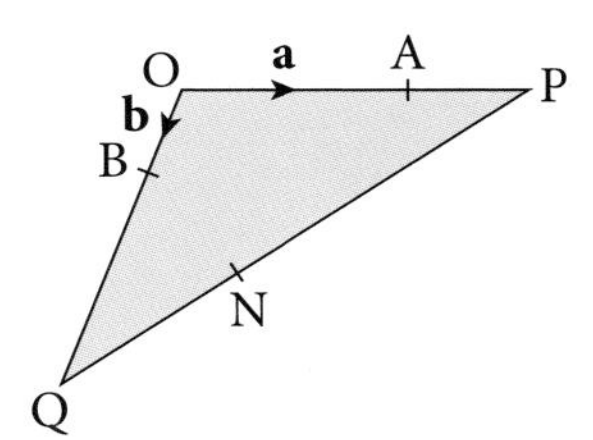

5. AP = 5OA, OB = 2BQ, NP = 2QN.

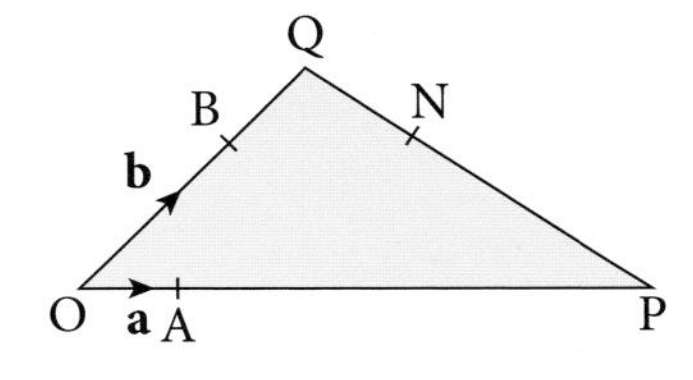

6. $OA = \frac{1}{5}OP$, OQ = 3OB, N is $\frac{1}{4}$ the way along PQ.

7. In ΔXYZ, the mid-point of YZ is M.
If $\overrightarrow{XY} = \mathbf{s}$ and $\overrightarrow{ZX} = \mathbf{t}$, find $\overrightarrow{XM}$ in terms of **s** and **t**.

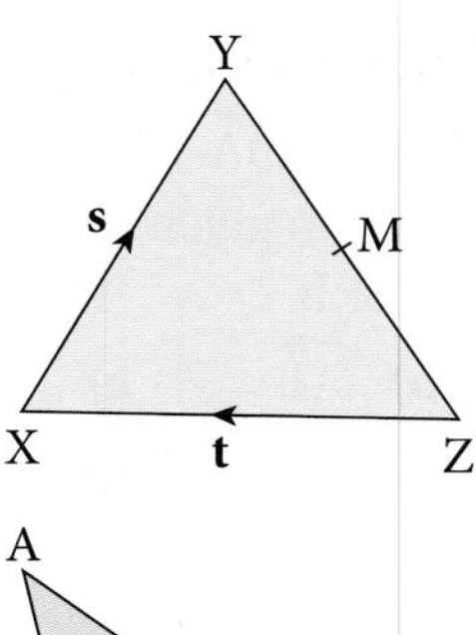

8. In ΔAOB, AM:MB= 2:1. If $\overrightarrow{OA} = \mathbf{a}$ and $\overrightarrow{OB} = \mathbf{b}$, find $\overrightarrow{OM}$ in terms of **a** and **b**.

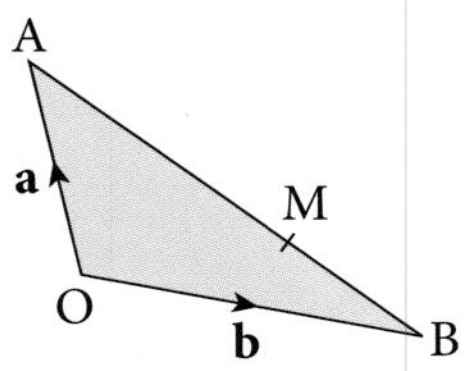

9. O is any point in the plane of the square ABCD.
The vectors $\overrightarrow{OA}$, $\overrightarrow{OB}$ and $\overrightarrow{OC}$ are **a**, **b** and **c** respectively.
Find the vector $\overrightarrow{OD}$ in terms of **a**, **b** and **c**.

10. ABCDEF is a regular hexagon with $\overrightarrow{AB}$ representing the vector **m** and $\overrightarrow{AF}$ representing the vector **n**. Find the vector representing $\overrightarrow{AD}$.

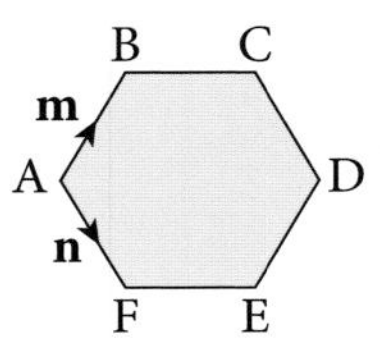

11. ABCDEF is a regular hexagon with centre O.
$\overrightarrow{FA} = \mathbf{a}$ and $\overrightarrow{FB} = \mathbf{b}$.
Express the following vectors in terms of **a** and/or **b**.

a) $\overrightarrow{AB}$ **b)** $\overrightarrow{FO}$ **c)** $\overrightarrow{FC}$
d) $\overrightarrow{BC}$ **e)** $\overrightarrow{AO}$ **f)** $\overrightarrow{FD}$

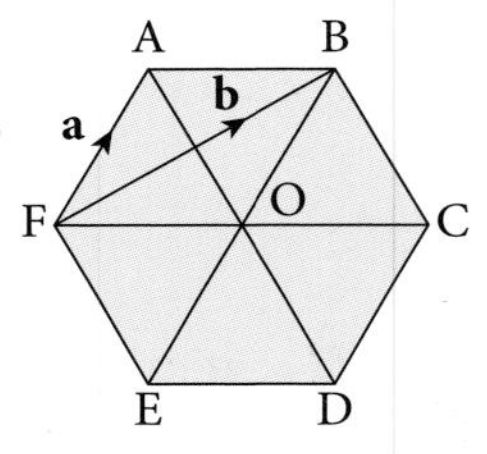

12. In the diagram, **M** is the mid-point of CD, BP:PM = 2:1,
$\overrightarrow{AB} = \mathbf{x}$, $\overrightarrow{AC} = \mathbf{y}$ and $\overrightarrow{AD} = \mathbf{z}$.
Express the following vectors in terms of **x**, **y** and **z**.

a) $\overrightarrow{DC}$ **b)** $\overrightarrow{DM}$ **c)** $\overrightarrow{AM}$
d) $\overrightarrow{BM}$ **e)** $\overrightarrow{BP}$ **f)** $\overrightarrow{AP}$

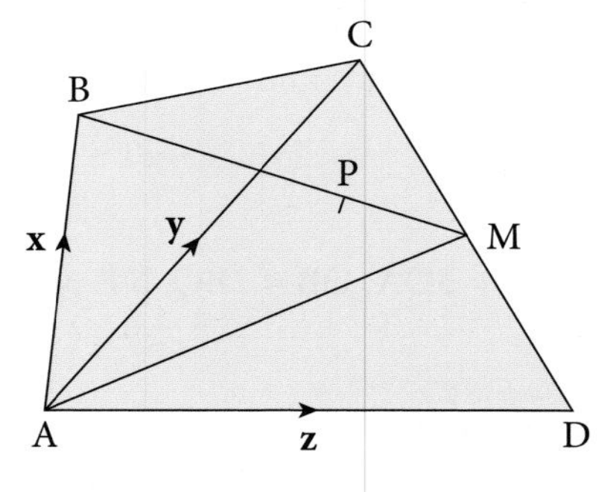

11.2 Column vectors

The vector $\overrightarrow{AB}$ may be written as a **column vector**.

$$AB = \begin{pmatrix} 5 \\ 3 \end{pmatrix}$$

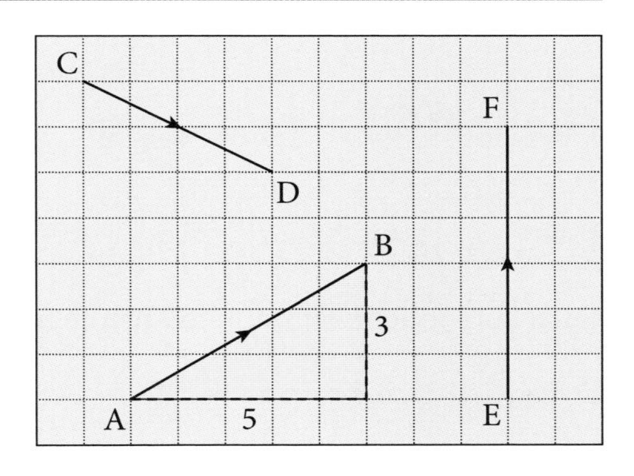

The top number is the horizontal component of $\overrightarrow{AB}$ (i.e. 5) and the bottom number is the vertical component (i.e. 3).

Similarly $\overrightarrow{CD} = \begin{pmatrix} 4 \\ -2 \end{pmatrix}$ and $\overrightarrow{EF} = \begin{pmatrix} 0 \\ 6 \end{pmatrix}$

Addition of vectors

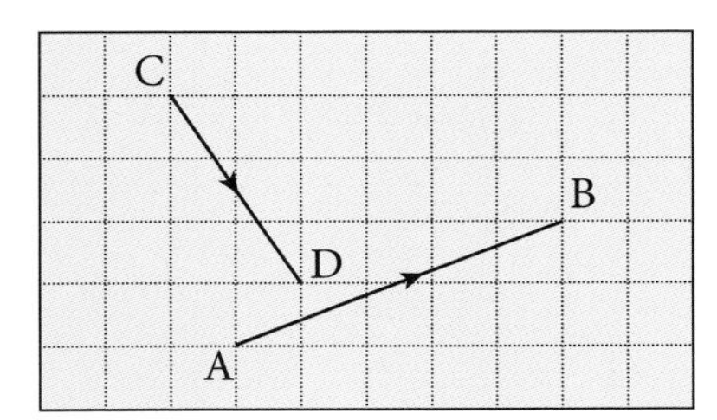

Suppose we wish to add vectors $\overrightarrow{AB}$ and $\overrightarrow{CD}$.

First move $\overrightarrow{CD}$ so that $\overrightarrow{AB}$ and $\overrightarrow{CD}$ join 'nose to tail'. Remember that changing the *position* of a vector does not change the vector. A vector is described by its length and direction.

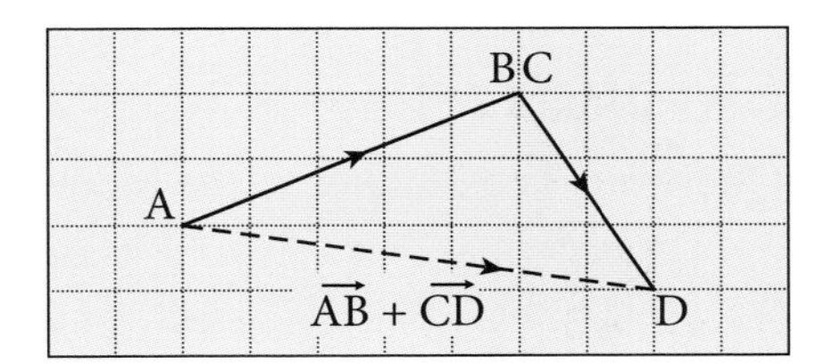

The broken line shows the result of adding $\overrightarrow{AB}$ and $\overrightarrow{CD}$.

In column vectors,

$$\overrightarrow{AB} + \overrightarrow{CD} = \begin{pmatrix} 5 \\ 2 \end{pmatrix} + \begin{pmatrix} 2 \\ -3 \end{pmatrix}$$

We see that the column vector for the broken line is $\begin{pmatrix} 7 \\ -1 \end{pmatrix}$

So we perform addition with vectors by adding together the corresponding components of the vectors.

Subtraction of vectors

The diagram shows $\overrightarrow{AB} - \overrightarrow{CD}$.

To subtract vector $\overrightarrow{CD}$ from $\overrightarrow{AB}$ we add the *negative* of $\overrightarrow{CD}$ to $\overrightarrow{AB}$.

So $\overrightarrow{AB} - \overrightarrow{CD} = \overrightarrow{AB} + (-\overrightarrow{CD})$

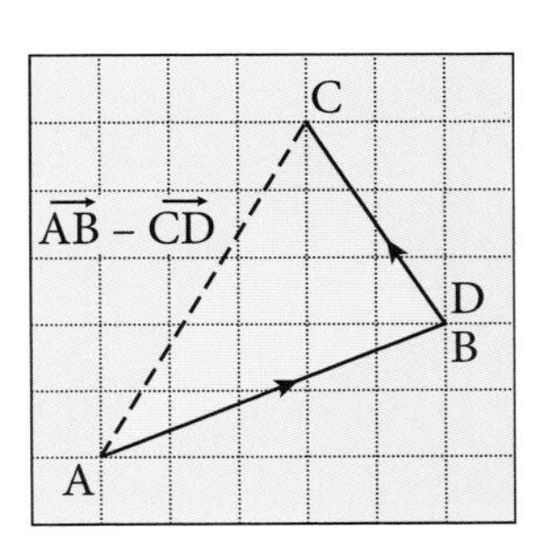

In column vectors,

$$\overrightarrow{AB} + (-\overrightarrow{CD}) = \begin{pmatrix} 5 \\ 2 \end{pmatrix} + \begin{pmatrix} -2 \\ 3 \end{pmatrix} = \begin{pmatrix} 3 \\ 5 \end{pmatrix}$$

Multiplication by a scalar

If $\mathbf{a} = \begin{pmatrix} 3 \\ -4 \end{pmatrix}$ then $2\mathbf{a} = 2\begin{pmatrix} 3 \\ -4 \end{pmatrix} = \begin{pmatrix} 6 \\ -8 \end{pmatrix}$

Each component is multiplied by the number 2.

Parallel vectors

Vectors are parallel if they have the same direction. Both components of one vector must be in the same ratio to the corresponding components of the parallel vector.

e.g. $\begin{pmatrix} 3 \\ -5 \end{pmatrix}$ is parallel to $\begin{pmatrix} 6 \\ -10 \end{pmatrix}$,

because $\begin{pmatrix} 6 \\ -10 \end{pmatrix}$ may be written $2\begin{pmatrix} 3 \\ -5 \end{pmatrix}$.

- In general the vector $k\begin{pmatrix} a \\ b \end{pmatrix}$ is parallel to $\begin{pmatrix} a \\ b \end{pmatrix}$.

Exercise 11.3

Questions **1** to **36** refer to the following vectors.

$\mathbf{a} = \begin{pmatrix} 3 \\ 4 \end{pmatrix}$ $\mathbf{b} = \begin{pmatrix} 1 \\ 4 \end{pmatrix}$ $\mathbf{c} = \begin{pmatrix} 4 \\ -3 \end{pmatrix}$ $\mathbf{d} = \begin{pmatrix} -1 \\ 1 \end{pmatrix}$

$\mathbf{e} = \begin{pmatrix} 5 \\ 12 \end{pmatrix}$ $\mathbf{f} = \begin{pmatrix} 3 \\ -2 \end{pmatrix}$ $\mathbf{g} = \begin{pmatrix} -4 \\ -2 \end{pmatrix}$ $\mathbf{h} = \begin{pmatrix} -12 \\ 5 \end{pmatrix}$

Draw and label the following vectors on graph paper (take 1 cm to 1 unit).

1. $\mathbf{c}$ **2.** $\mathbf{f}$ **3.** $2\mathbf{b}$ **4.** $-\mathbf{a}$

5. $-\mathbf{g}$ **6.** $3\mathbf{a}$ **7.** $\frac{1}{2}\mathbf{e}$ **8.** $5\mathbf{d}$

9. $-\frac{1}{2}\mathbf{h}$ **10.** $\frac{3}{2}\mathbf{g}$ **11.** $\frac{1}{5}\mathbf{h}$ **12.** $-3\mathbf{b}$

Find the following vectors in component form.

13. $\mathbf{b} + \mathbf{h}$ **14.** $\mathbf{f} + \mathbf{g}$ **15.** $\mathbf{e} - \mathbf{b}$

16. $\mathbf{a} - \mathbf{d}$ **17.** $\mathbf{g} - \mathbf{h}$ **18.** $2\mathbf{a} + 3\mathbf{c}$

19. $3\mathbf{f} + 2\mathbf{d}$ **20.** $4\mathbf{g} - 2\mathbf{b}$ **21.** $5\mathbf{a} + \frac{1}{2}\mathbf{g}$

22. $\mathbf{a} + \mathbf{b} + \mathbf{c}$ **23.** $3\mathbf{f} - \mathbf{a} + \mathbf{c}$ **24.** $\mathbf{c} + 2\mathbf{d} + 3\mathbf{e}$

In each of the following, find $\mathbf{x}$ in component form.

25. $\mathbf{x} + \mathbf{b} = \mathbf{e}$ **26.** $\mathbf{x} + \mathbf{d} = \mathbf{a}$ **27.** $\mathbf{c} + \mathbf{x} = \mathbf{f}$

28. $\mathbf{x} - \mathbf{g} = \mathbf{h}$ **29.** $2\mathbf{x} + \mathbf{b} = \mathbf{g}$ **30.** $2\mathbf{x} - 3\mathbf{d} = \mathbf{g}$

31. $2\mathbf{b} = \mathbf{d} - \mathbf{x}$ **32.** $\mathbf{f} - \mathbf{g} = \mathbf{e} - \mathbf{x}$ **33.** $2\mathbf{x} + \mathbf{b} = \mathbf{x} + \mathbf{e}$

34. $3\mathbf{x} - \mathbf{b} = \mathbf{x} + \mathbf{h}$ **35.** $\mathbf{a} + \mathbf{b} + \mathbf{x} = \mathbf{b} + \mathbf{a}$ **36.** $2\mathbf{x} + \mathbf{e} = 0$ (zero vector)

37. a) Draw and label each of the following vectors on graph paper.

$\mathbf{l} = \begin{pmatrix} -3 \\ -3 \end{pmatrix}$ $\mathbf{m} = \begin{pmatrix} 2 \\ 0 \end{pmatrix}$ $\mathbf{n} = \begin{pmatrix} 3 \\ 2 \end{pmatrix}$ $\mathbf{p} = \begin{pmatrix} 1 \\ -2 \end{pmatrix}$ $\mathbf{q} = \begin{pmatrix} 3 \\ 0 \end{pmatrix}$

$\mathbf{r} = \begin{pmatrix} 6 \\ 4 \end{pmatrix}$ $\mathbf{s} = \begin{pmatrix} 2 \\ 2 \end{pmatrix}$ $\mathbf{t} = \begin{pmatrix} 2 \\ -4 \end{pmatrix}$ $\mathbf{u} = \begin{pmatrix} -1 \\ -3 \end{pmatrix}$ $\mathbf{v} = \begin{pmatrix} 0 \\ 3 \end{pmatrix}$

b) Find four pairs of parallel vectors amongst the ten vectors.

38. Decide whether the statements are 'true' or 'false'.

a) $\begin{pmatrix} 3 \\ -1 \end{pmatrix}$ is parallel to $\begin{pmatrix} 9 \\ -3 \end{pmatrix}$ **b)** $\begin{pmatrix} -2 \\ 0 \end{pmatrix}$ is parallel to $\begin{pmatrix} 4 \\ 0 \end{pmatrix}$

c) $\begin{pmatrix} -1 \\ 1 \end{pmatrix}$ is parallel to $\begin{pmatrix} -1 \\ 1 \end{pmatrix}$ **d)** $\begin{pmatrix} 5 \\ -15 \end{pmatrix} = 5\begin{pmatrix} 1 \\ -3 \end{pmatrix}$

e) $\begin{pmatrix} 4 \\ 0 \end{pmatrix}$ is parallel to $\begin{pmatrix} 0 \\ 6 \end{pmatrix}$ **f)** $\begin{pmatrix} 3 \\ -1 \end{pmatrix} + \begin{pmatrix} -4 \\ -2 \end{pmatrix} = \begin{pmatrix} -1 \\ 1 \end{pmatrix}$

39. a) Draw a diagram to illustrate the vector addition $\overrightarrow{AB} + \overrightarrow{CD}$.

b) Draw a diagram to illustrate $\overrightarrow{AB} - \overrightarrow{CD}$.

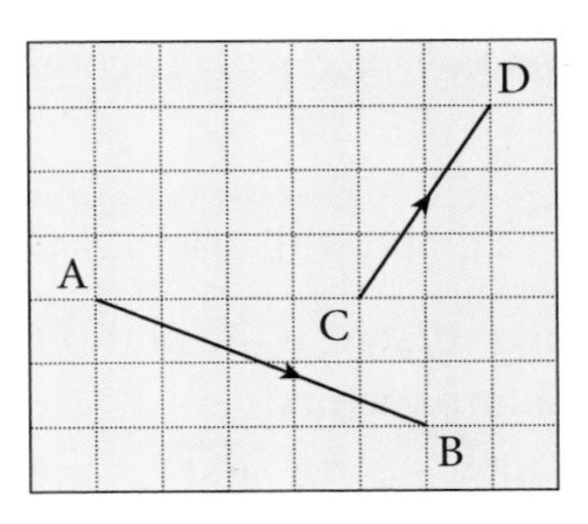

40. Draw separate diagrams to illustrate the following.

a) $\overrightarrow{FE} + \overrightarrow{JI}$

b) $\overrightarrow{HG} + \overrightarrow{FE}$

c) $\overrightarrow{JI} - \overrightarrow{FE}$

d) $\overrightarrow{HG} + \overrightarrow{JI}$

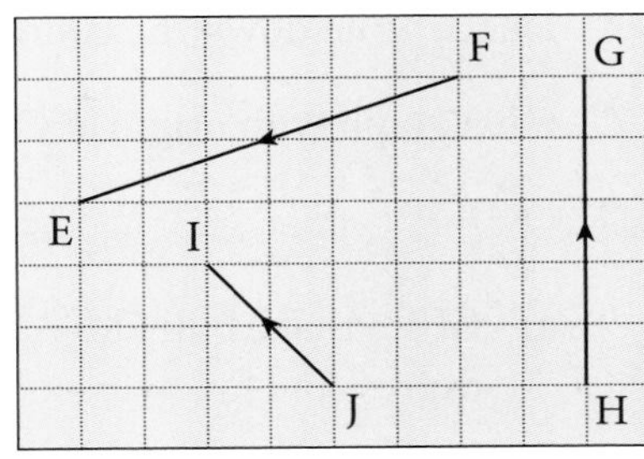

Exercise 11.4

1. If D has coordinates (7, 2) and E has coordinates (9, 0), find the column vector for $\overrightarrow{DE}$.

2. Find the column vector $\overrightarrow{XY}$ where X and Y have coordinates (−1, 4) and (5, 2) respectively.

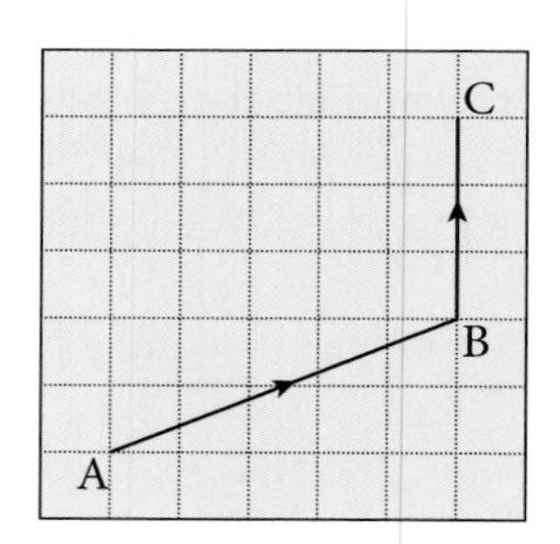

3. In the diagram $\overrightarrow{AB}$ represents the vector $\begin{pmatrix} 5 \\ 2 \end{pmatrix}$ and $\overrightarrow{BC}$ represents the vector $\begin{pmatrix} 0 \\ 3 \end{pmatrix}$.
   a) Copy the diagram and mark point D such that ABCD is a parallelogram.
   b) Write $\overrightarrow{AD}$ and $\overrightarrow{CA}$ as column vectors.

4. **a)** On squared paper draw $\overrightarrow{AB} = \begin{pmatrix} 3 \\ -2 \end{pmatrix}$ and $\overrightarrow{BC} = \begin{pmatrix} 4 \\ 2 \end{pmatrix}$ and mark point D such that ABCD is a parallelogram.
   b) Write $\overrightarrow{AD}$ and $\overrightarrow{CA}$ as column vectors.

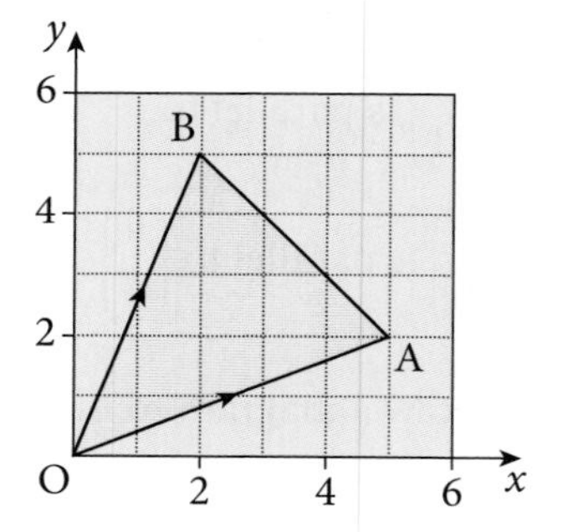

5. Copy the diagram in which $\overrightarrow{OA} = \begin{pmatrix} 5 \\ 2 \end{pmatrix}$ and $\overrightarrow{OB} = \begin{pmatrix} 2 \\ 5 \end{pmatrix}$.
   M is the mid-point of AB. Express the following as column vectors:
   a) $\overrightarrow{BA}$ **b)** $\overrightarrow{BM}$ **c)** $\overrightarrow{OM}$ (use $\overrightarrow{OM} = \overrightarrow{OB} + \overrightarrow{BM}$)
   Hence write down the coordinates of M.

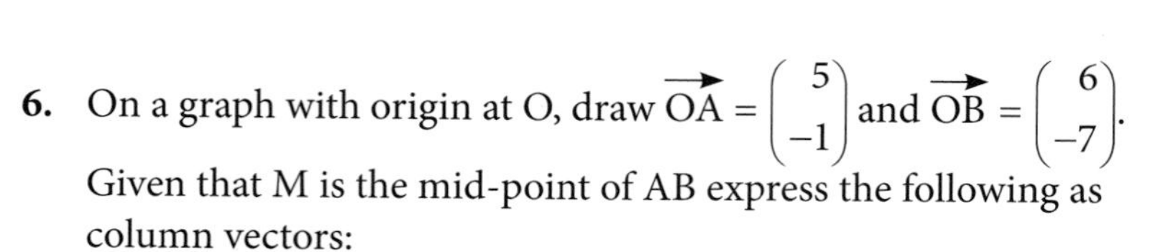

6. On a graph with origin at O, draw $\overrightarrow{OA} = \begin{pmatrix} 5 \\ -1 \end{pmatrix}$ and $\overrightarrow{OB} = \begin{pmatrix} 6 \\ -7 \end{pmatrix}$.
   Given that M is the mid-point of AB express the following as column vectors:
   a) $\overrightarrow{BA}$ **b)** $\overrightarrow{BM}$ **c)** $\overrightarrow{OM}$
   Hence write down the coordinates of M.

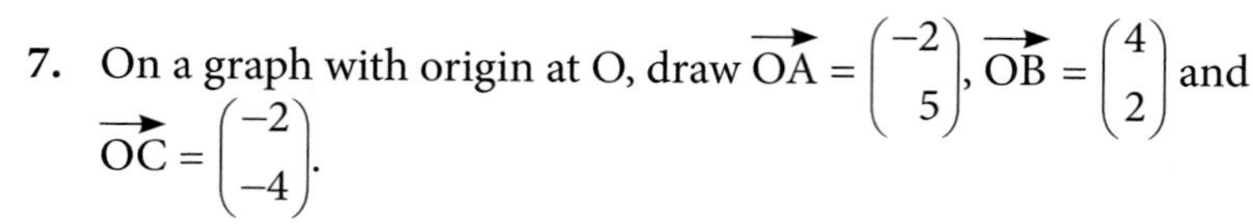

7. On a graph with origin at O, draw $\overrightarrow{OA} = \begin{pmatrix} -2 \\ 5 \end{pmatrix}$, $\overrightarrow{OB} = \begin{pmatrix} 4 \\ 2 \end{pmatrix}$ and $\overrightarrow{OC} = \begin{pmatrix} -2 \\ -4 \end{pmatrix}$.
   a) Given that M divides AB such that AM:MB = 2:1, express the following as column vectors:
      i) $\overrightarrow{BA}$ **ii)** $\overrightarrow{BM}$ **iii)** $\overrightarrow{OM}$
   b) Given that N divides AC such that AN:NC = 1:2, express the following as column vectors:
      i) $\overrightarrow{AC}$ **ii)** $\overrightarrow{AN}$ **iii)** $\overrightarrow{ON}$

8. In square ABCD, side AB has column vector $\begin{pmatrix} 2 \\ 1 \end{pmatrix}$. Find two possible column vectors for $\overrightarrow{BC}$.

9. Rectangle KLMN has an area of 10 square units and $\overrightarrow{KL}$ has column vector $\begin{pmatrix} 5 \\ 0 \end{pmatrix}$. Find two possible column vectors for $\overrightarrow{LM}$.

10. In the diagram, ABCD is a trapezium in which $\overrightarrow{DC} = 2\overrightarrow{AB}$. If $\overrightarrow{AB} = \mathbf{p}$ and $\overrightarrow{AD} = \mathbf{q}$, express in terms of **p** and **q**:

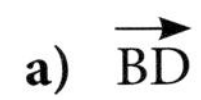

a) $\overrightarrow{BD}$ **b)** $\overrightarrow{AC}$ **c)** $\overrightarrow{BC}$

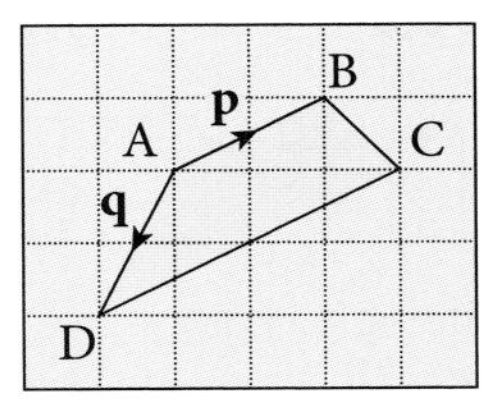

11. Find the image of the vector $\begin{pmatrix} 1 \\ 3 \end{pmatrix}$ after reflection in the following lines:

a) $y = 0$ **b)** $x = 0$ **c)** $y = x$ **d)** $y = -x$

Magnitude of a vector

The **magnitude** of a vector **a** is written $|\mathbf{a}|$ and represents the length (or magnitude) of the vector.

In the diagram above, $\mathbf{a} = \begin{pmatrix} 5 \\ 3 \end{pmatrix}$

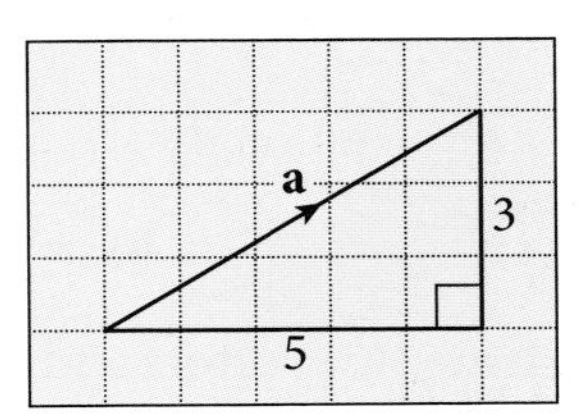

By Pythagoras' theorem, $|\mathbf{a}| = \sqrt{(5^2 + 3^2)}$

$|\mathbf{a}| = \sqrt{34}$ units

- In general if $\mathbf{x} = \begin{pmatrix} m \\ n \end{pmatrix}$, $|\mathbf{x}| = \sqrt{(m^2 + n^2)}$

Exercise 11.5

Questions **1** to **12** refer to the following vectors:

$\mathbf{a} = \begin{pmatrix} 3 \\ 4 \end{pmatrix}$ $\mathbf{b} = \begin{pmatrix} 4 \\ 1 \end{pmatrix}$ $\mathbf{c} = \begin{pmatrix} 5 \\ 12 \end{pmatrix}$

$\mathbf{d} = \begin{pmatrix} -3 \\ 0 \end{pmatrix}$ $\mathbf{e} = \begin{pmatrix} -4 \\ -3 \end{pmatrix}$ $\mathbf{f} = \begin{pmatrix} -3 \\ 6 \end{pmatrix}$

Find the following, leaving the answer in square root form where necessary.

1. $|\mathbf{a}|$ **2.** $|\mathbf{b}|$ **3.** $|\mathbf{c}|$ **4.** $|\mathbf{d}|$

5. $|\mathbf{e}|$ **6.** $|\mathbf{f}|$ **7.** $|\mathbf{a} + \mathbf{b}|$ **8.** $|\mathbf{c} - \mathbf{d}|$

9. $|2\mathbf{e}|$ **10.** $|\mathbf{f} + 2\mathbf{b}|$

11. a) Find $|\mathbf{a} + \mathbf{c}|$ **b)** Is $|\mathbf{a} + \mathbf{c}|$ equal to $|\mathbf{a}| + |\mathbf{c}|$?

12. a) Find $|\mathbf{c} + \mathbf{d}|$ **b)** Is $|\mathbf{c} + \mathbf{d}|$ equal to $|\mathbf{c}| + |\mathbf{d}|$?

13. If $\overrightarrow{AB} = \begin{pmatrix} 3 \\ -1 \end{pmatrix}$ and $\overrightarrow{BC} = \begin{pmatrix} 2 \\ 3 \end{pmatrix}$, find $|\overrightarrow{AC}|$

14. If $\overrightarrow{PQ} = \begin{pmatrix} 5 \\ -2 \end{pmatrix}$ and $\overrightarrow{QR} = \begin{pmatrix} 0 \\ 1 \end{pmatrix}$, find $|\overrightarrow{PR}|$

15. If $\overrightarrow{WX} = \begin{pmatrix} 1 \\ 3 \end{pmatrix}$, $\overrightarrow{XY} = \begin{pmatrix} -2 \\ 1 \end{pmatrix}$ and $\overrightarrow{YZ} = \begin{pmatrix} 2 \\ -1 \end{pmatrix}$, find $|\overrightarrow{WZ}|$

16. Given that $\overrightarrow{OP} = \begin{pmatrix} 0 \\ 5 \end{pmatrix}$ and $\overrightarrow{OQ} = \begin{pmatrix} n \\ 3 \end{pmatrix}$, find:

a) $|\overrightarrow{OP}|$ **b)** a value for n if $|\overrightarrow{OP}| = |\overrightarrow{OQ}|$

17. Given that $\overrightarrow{OA} = \begin{pmatrix} 5 \\ 12 \end{pmatrix}$ and $\overrightarrow{OB} = \begin{pmatrix} 0 \\ m \end{pmatrix}$, find:

a) $|\overrightarrow{OA}|$ **b)** a value for m if $|\overrightarrow{OA}| = |\overrightarrow{OB}|$

18. Given that $\overrightarrow{LM} = \begin{pmatrix} -3 \\ 4 \end{pmatrix}$ and $\overrightarrow{MN} = \begin{pmatrix} -15 \\ p \end{pmatrix}$, find:

a) $|\overrightarrow{LM}|$ **b)** a value for p if $|\overrightarrow{MN}| = 3|\overrightarrow{LM}|$

19. $\mathbf{a}$ and $\mathbf{b}$ are two vectors and $|\mathbf{a}| = 3$.
Find the value of $|\mathbf{a} + \mathbf{b}|$ when:

a) $\mathbf{b} = 2\mathbf{a}$

b) $\mathbf{b} = -3\mathbf{a}$

c) $\mathbf{b}$ is perpendicular to $\mathbf{a}$ and $|\mathbf{b}| = 4$

20. $\mathbf{r}$ and $\mathbf{s}$ are two vectors and $|\mathbf{r}| = 5$.

Find the value of $|\mathbf{r} + \mathbf{s}|$ when:

a) $\mathbf{s} = 5\mathbf{r}$

b) $\mathbf{s} = -2\mathbf{r}$

c) $\mathbf{r}$ is perpendicular to $\mathbf{s}$ and $|\mathbf{s}| = 5$

d) $\mathbf{s}$ is perpendicular to $(\mathbf{r} + \mathbf{s})$ and $|\mathbf{s}| = 3$

11.3 Simple transformations

Reflection

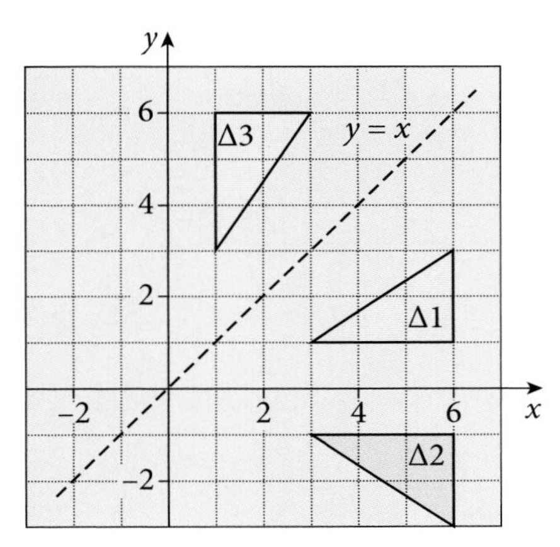

You must describe a mirror line fully.

Δ2 is the image of Δ1 after reflection in the x-axis.

Δ3 is the image of Δ1 after reflection in the line $y = x$.

Exercise 11.6

In questions **1** to **6** draw the object and its image after reflection in the broken line.

1.

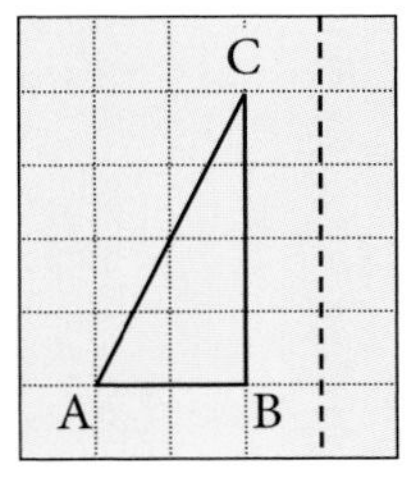

2.

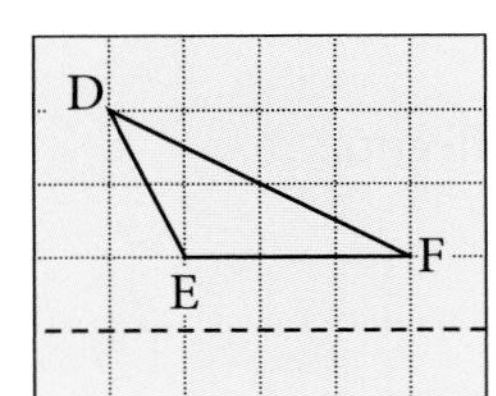

3.

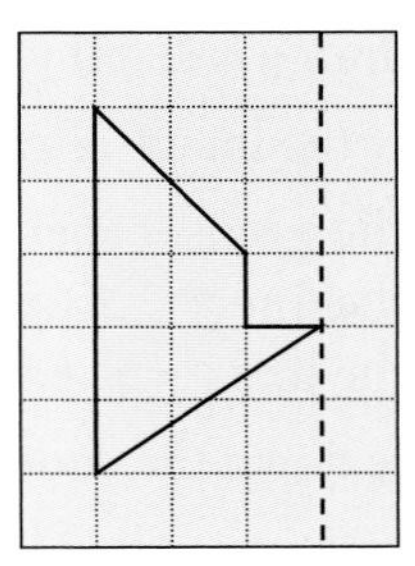

4.

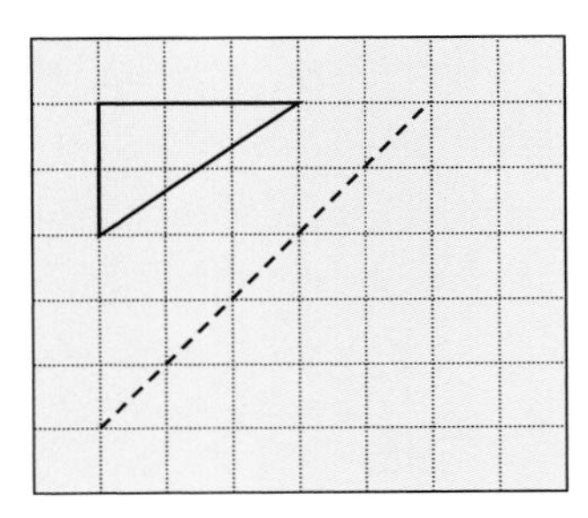

5.

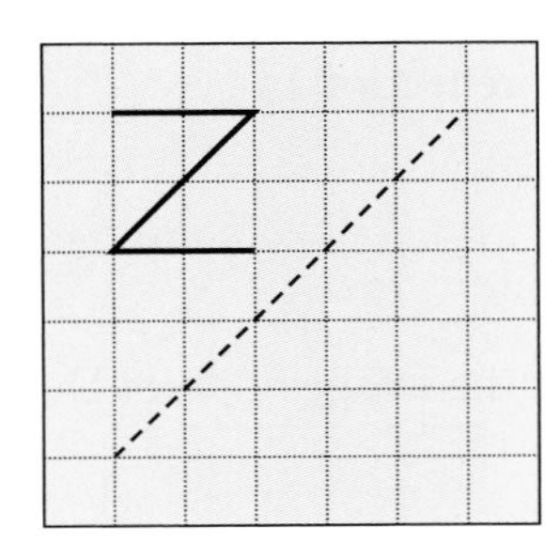

6.

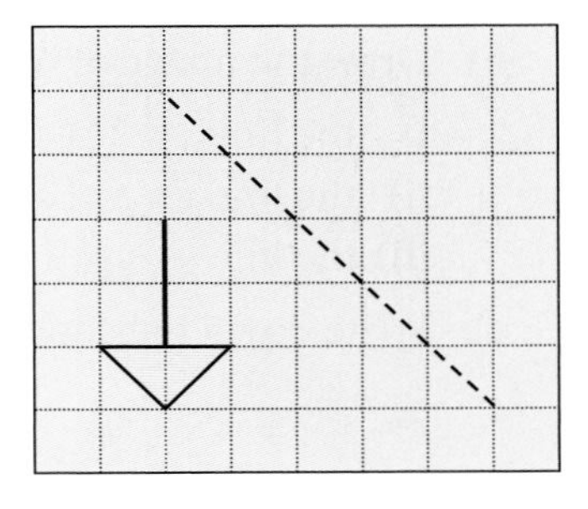

In questions **7**, **8**, **9** draw the image of the given shape after reflection in line M_1 and then reflect this new shape in line M_2.

7.

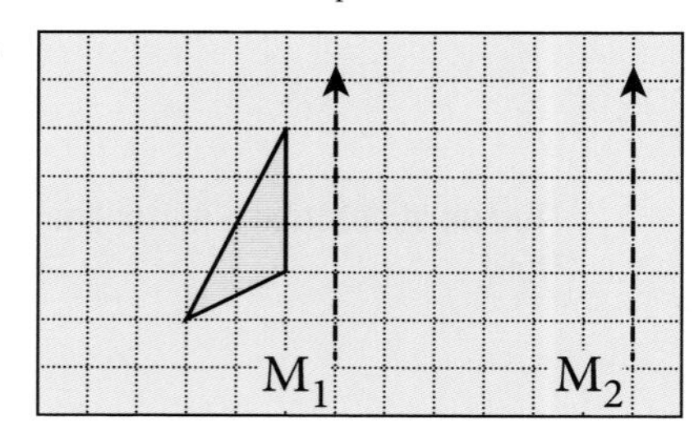

8.

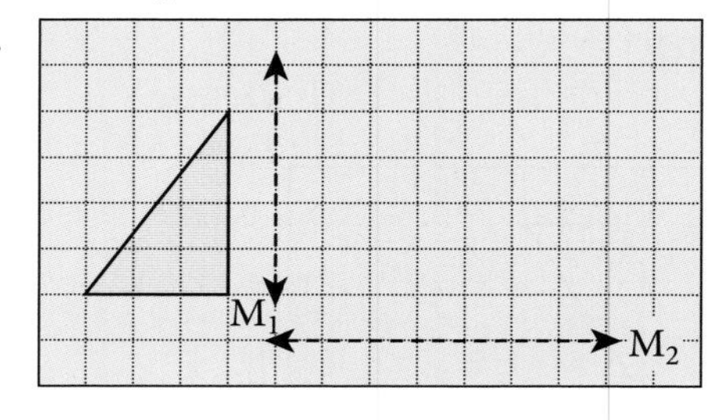

9.

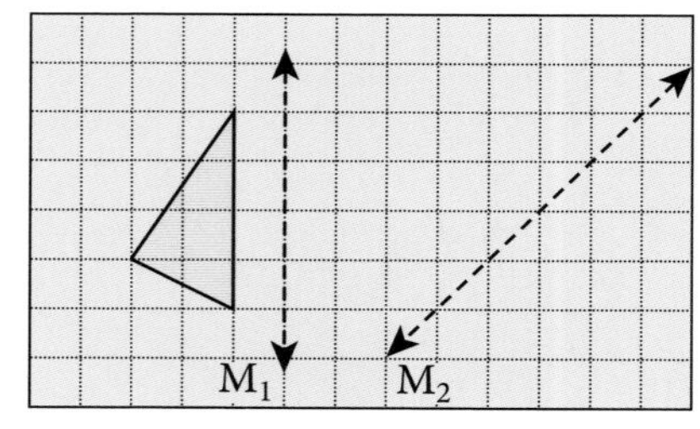

Exercise 11.7

For each questions draw x- and y-axes with values from –8 to 8.

1. **a)** Draw the triangle ABC at A(6, 8), B(2, 8), C(2, 6). Draw the lines $y = 2$ and $y = x$.

b) Draw the image of ΔABC after reflection in

i) the y-axis. Label it Δ1.
ii) the line $y = 2$. Label it Δ2.
iii) the line $y = x$. Label it Δ3.

c) Write down the coordinates of the image of point A in each case.

2. **a)** Draw the triangle DEF at D(–6, 8), E(–2, 8), F(–2, 6).

Draw the lines $x = 1$, $y = x$, $y = -x$.

b) Draw the image of ΔDEF after reflection in

i) the line $x = 1$. Label it Δ1.
ii) the line $y = x$. Label it Δ2.
iii) the line $y = -x$. Label it Δ3.

c) Write down the coordinates of the image of point D in each case.

3. **a)** Draw the triangle ABC at A(5, 1), B(8, 1), C(8, 3).
Draw the lines $x + y = 4$, $y = x - 3$, $x = 2$.

b) Draw the image of ΔABC after reflection in

i) the line $x + y = 4$. Label it $\Delta 1$.
ii) the line $y = x - 3$. Label it $\Delta 2$.
iii) the line $x = 2$. Label it $\Delta 3$.

c) Write down the coordinates of the image of point A in each case.

4. **a)** Draw and label the following triangles.

i) $\Delta 1 : (3, 3), (3, 6), (1, 6)$

$\Delta 2 : (3, -1), (3, -4), (1, -4)$

$\Delta 3 : (3, 3), (6, 3), (6, 1)$

$\Delta 4 : (-6, -1), (-6, -3), (-3, -3)$

$\Delta 5 : (-6, 5), (-6, 7), (-3, 7)$

b) Find the equation of the mirror line for the reflection:

i) $\Delta 1$ onto $\Delta 2$ **ii)** $\Delta 1$ onto $\Delta 3$
iii) $\Delta 1$ onto $\Delta 4$ **iv)** $\Delta 4$ onto $\Delta 5$

5. **a)** Draw $\Delta 1$ at (3, 1), (7, 1), (7, 3).

b) Reflect $\Delta 1$ in the line $y = x$ onto $\Delta 2$.

c) Reflect $\Delta 2$ in the x-axis onto $\Delta 3$.

d) Reflect $\Delta 3$ in the line $y = -x$ onto $\Delta 4$.

e) Reflect $\Delta 4$ in the line $x = 2$ onto $\Delta 5$.

f) Write down the coordinates of $\Delta 5$.

6. **a)** Draw $\Delta 1$ at (2, 6), (2, 8), (6, 6).

b) Reflect $\Delta 1$ in the line $x + y = 6$ onto $\Delta 2$.

c) Reflect $\Delta 2$ in the line $x = 3$ onto $\Delta 3$.

d) Reflect $\Delta 3$ in the line $x + y = 6$ onto $\Delta 4$.

e) Reflect $\Delta 4$ in the line $y = x - 8$ onto $\Delta 5$.

f) Write down the coordinates of $\Delta 5$.

Rotation

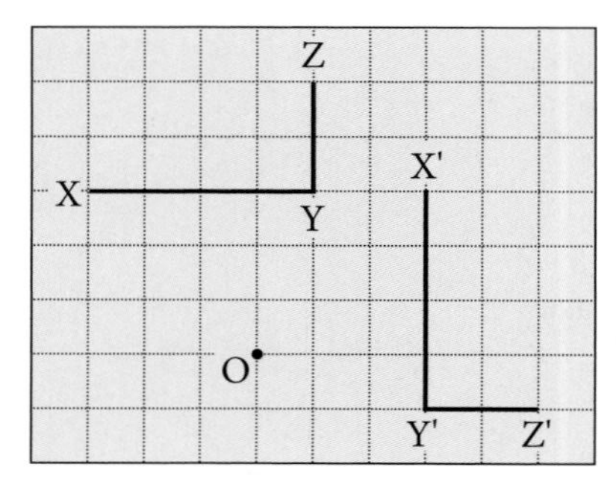

The letter L has been rotated through 90° clockwise about the centre O. The angle, direction, and centre are needed to fully describe a rotation.

We say that the object **maps** onto the image. Here,

X maps onto X′
Y maps onto Y′
Z maps onto Z′

In this work, a clockwise rotation is *negative* and an anticlockwise rotation is *positive*: in this example, the letter L has been rotated through −90°. The angle, the direction, and the centre of rotation can be found using tracing paper and a sharp pencil placed where you think the centre of rotation is.

For more accurate work, draw the perpendicular bisector of the line joining two corresponding points, e.g. Y and Y′. Repeat for another pair of corresponding points. The centre of rotation is at the intersection of the two perpendicular bisectors.

Exercise 11.8

In question **1** to **4** draw the object and its image under the rotation given. Take O as the centre of rotation in each case.

1.

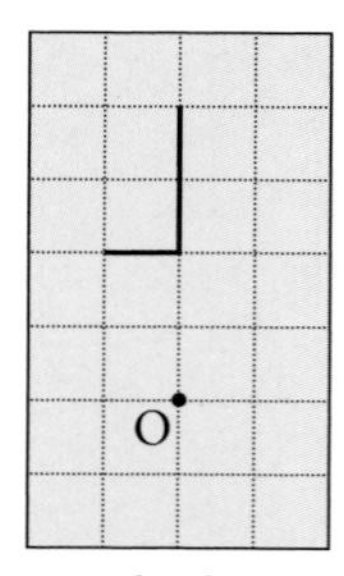

90° clockwise

2.

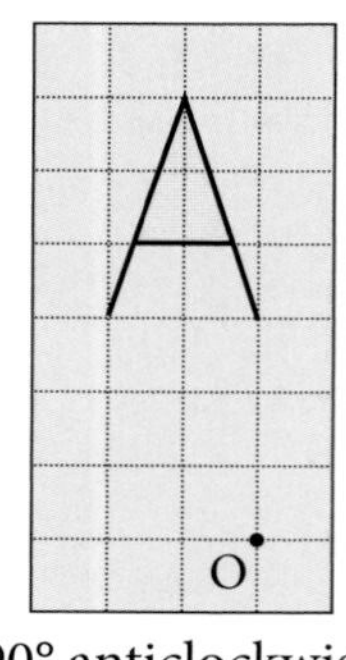

90° anticlockwise

3.

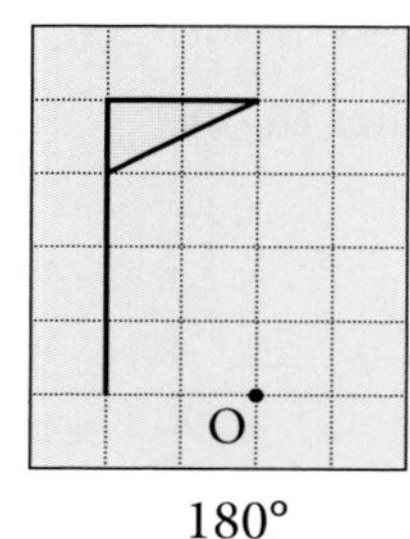

180°

4.

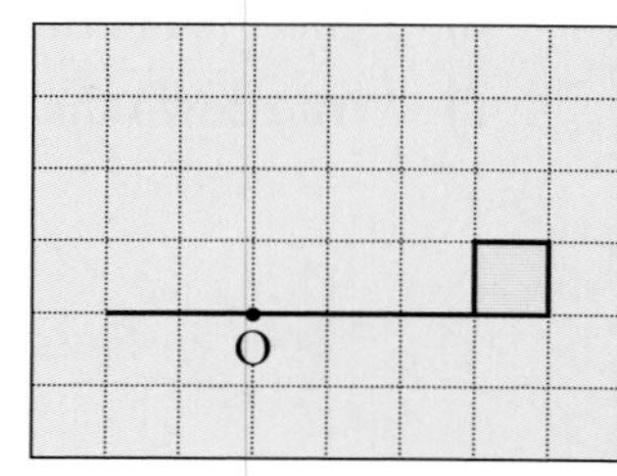

90° clockwise

In question **5** to **8**, copy the diagram on squared paper and find the angle, the direction, and the centre of the rotation.

5

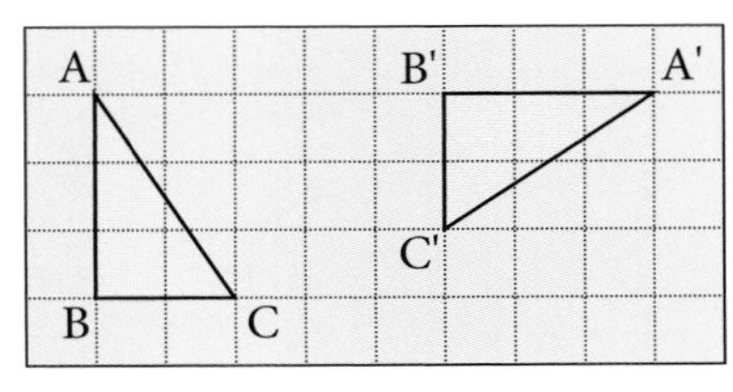

6

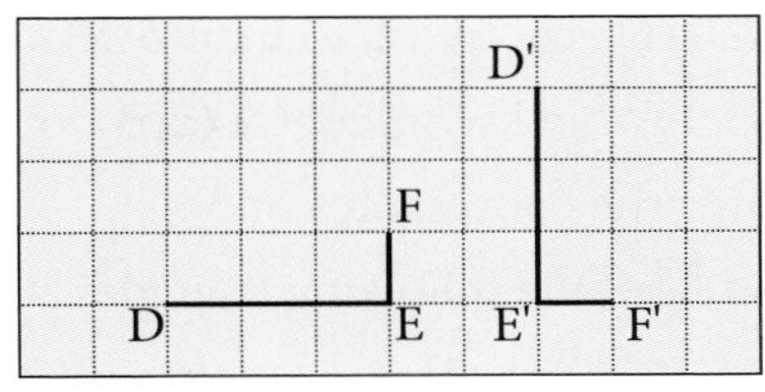

7

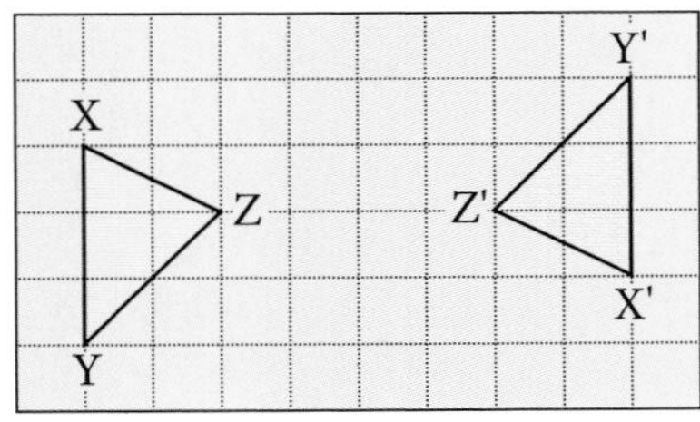

8

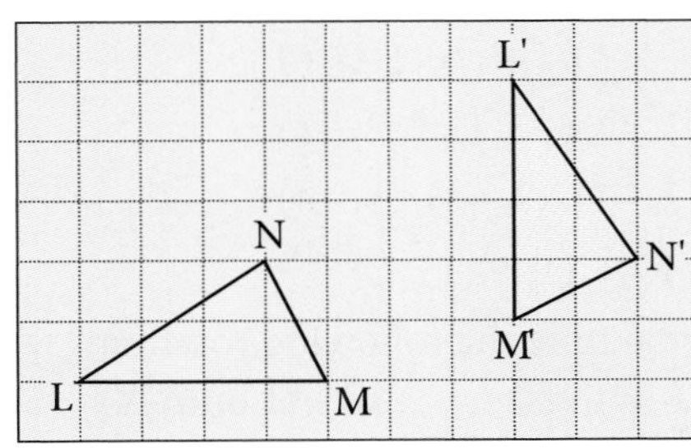

Exercise 11.9

For all questions draw x- and y-axes for values from −8 to +8.

1. a) Draw the object triangle ABC at A(1, 3), B(1, 6), C(3, 6), rotate ABC through 90° clockwise about (0, 0), mark A′ B′ C′.

b) Draw the object triangle DEF at D(3, 3), E(6, 3), F(6, 1), rotate DEF through 90° clockwise about (0, 0), mark D′ E′ F′.

c) Draw the object triangle PQR at P(−4, 7), Q(−4, 5), R(−1, 5), rotate PQR through 90° anticlockwise about (0, 0), mark P′ Q′ R′.

2. a) Draw Δ1 at (1, 4), (1, 7), (3, 7).

b) Find the image of PQR under the following rotations.

i) 90° clockwise, centre (0,0). Label it Δ2.
ii) 180°, centre (0, 0). Label it Δ3.
iii) 90° anticlockwise, centre (0, 0). Label it Δ4.

3. a) Draw triangle PQR at P(1, 2), Q(3, 5), R(6, 2).

b) Find the image of PQR under the following rotations.

i) 90° anticlockwise, centre (0, 0). Label the iamge P′ Q′ R′.
ii) 90° clockwise, centre (−2, 2). Label the iamge P″ Q″ R″.
iii) 180°, centre (1, 0). Label the iamge P* Q* R*.

c) Write down the coordiantes of P′, P″, P*.

4. **a)** Draw Δ1 at (1, 2), (1, 6), (3, 5).
   b) Rotate Δ1 90° clockwise, centre (1, 2), onto Δ2.
   c) Rotate Δ2 180°, centre (2, –1), onto Δ3.
   d) Rotate Δ3 90° clockwise, centre (2, 3), onto Δ4.
   e) Write down the coordiantes of Δ4.
5. **a)** Draw and label the following triangles.
   Δ1 : (3, 1), (6, 1), (6, 3)
   Δ2 : (–1, 3), (–1, 6), (–3, 6)
   Δ3 : (1, 1), (–2, 1), (–2, –1)
   Δ4 : (3, –1), (3, –4), (5, –4)
   Δ5 : (4, 4), (1, 4), (1, 2)
   b) Describe fully the following rotations.
   i) Δ1 onto Δ2 **ii)** Δ1 onto Δ3
   iii) Δ1 onto Δ4 **iv)** Δ1 onto Δ5
   v) Δ5 onto Δ4 **vi)** Δ3 onto Δ2
6. **a)** Draw Δ1 at (4, 7), (8, 5), (8, 7).
   b) Rotate Δ1 90° clockwise, centre (4, 3) onto Δ2.
   c) Rotate Δ2 180°, centre (5, –1) onto Δ3.
   d) Rotate Δ3 90° anticlockwise, centre (0, –8) onto Δ4.
   e) Describe fully the following rotations.
   i) Δ4 onto Δ1 **ii)** Δ4 onto Δ2

Translation

The triangle ABC below has been transformed onto the triangle A′ B′ C′ by a **translation**.

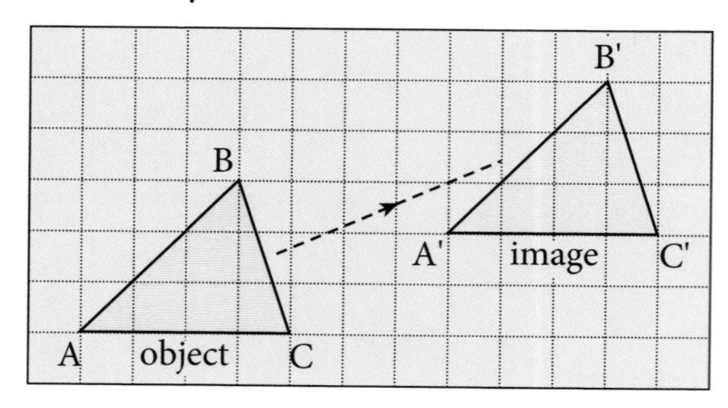

Here the translation is 7 squares to the right and 2 squares up the page.

The translation can be described by a column vector.

In the case the translation is $\begin{pmatrix} 7 \\ 2 \end{pmatrix}$.

The **modulus** (or **magnitude**) of the vector $\begin{pmatrix} 7 \\ 2 \end{pmatrix}$ is $\sqrt{7^2 + 2^2} = \sqrt{53}$.

Exercise 11.10

1. Make a copy of the diagram below and write down the column vector for each of the following translations.

   a) D onto A **b)** B onto F **c)** E onto A
   d) A onto C **e)** E onto C **f)** C onto B
   g) F onto E **h)** B onto C.

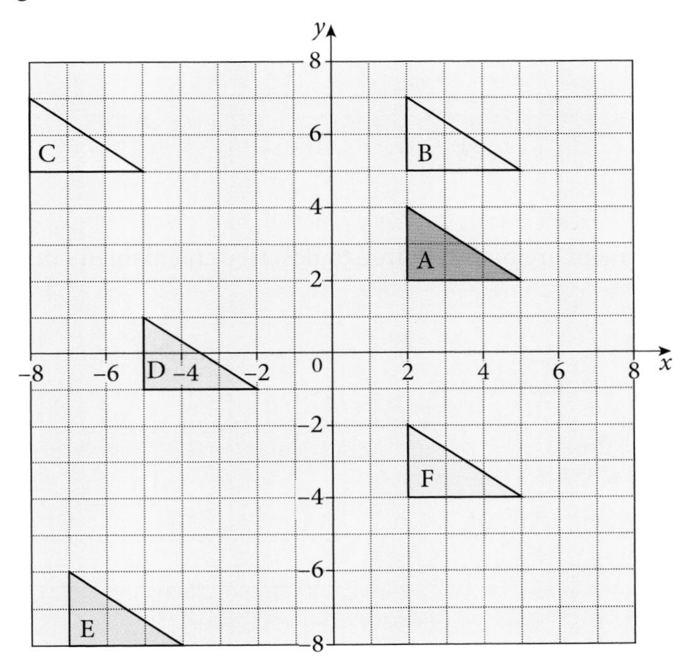

For questions **2** to **11** draw x and y axes with values from −8 to 8. Draw triangle ABC at A(−4, −1), B(−4, 1), C(−1, −1) and shade it.

This is the object. Draw the image of ABC under the translations described by the vectors below. For each question, write down the new coordinates of point C.

2. $\begin{pmatrix} 6 \\ 3 \end{pmatrix}$ **3.** $\begin{pmatrix} 6 \\ 7 \end{pmatrix}$ **4.** $\begin{pmatrix} 9 \\ -4 \end{pmatrix}$ **5.** $\begin{pmatrix} 1 \\ 7 \end{pmatrix}$

6. $\begin{pmatrix} 5 \\ -6 \end{pmatrix}$ **7.** $\begin{pmatrix} -2 \\ 5 \end{pmatrix}$ **8.** $\begin{pmatrix} -2 \\ -4 \end{pmatrix}$ **9.** $\begin{pmatrix} 0 \\ -7 \end{pmatrix}$

10. $\begin{pmatrix} 3 \\ 1 \end{pmatrix}$ followed by $\begin{pmatrix} 3 \\ 2 \end{pmatrix}$

11. $\begin{pmatrix} -2 \\ 0 \end{pmatrix}$ followed by $\begin{pmatrix} 0 \\ 3 \end{pmatrix}$ followed by $\begin{pmatrix} 1 \\ -1 \end{pmatrix}$

Enlargement

In the diagram below, the letter T has been enlarged by a scale factor of 2 using the point O as the centre of the enlargement.

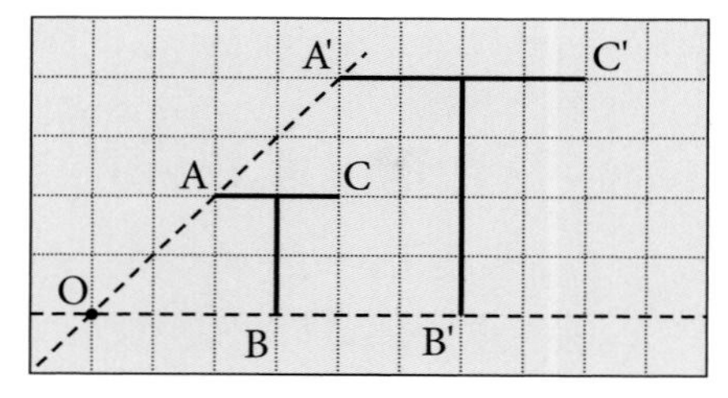

Notice that $OA' = 2 \times OA$

$OB' = 2 \times OB$

The scale factor and the centre of the enlargement are both required to describe an enlargement.

Example 3

Draw the image of triangle ABC under an enlargement scale factor of $\frac{1}{2}$ using O as centre of enlargement.

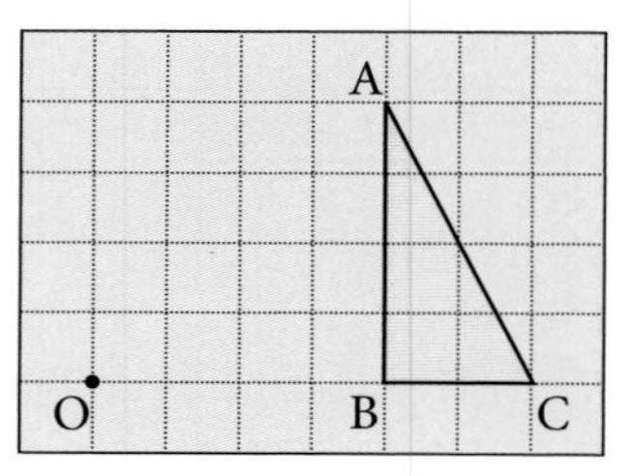

a) Draw lines through OA, OB and OC.

b) Mark A′ so that $OA' = \frac{1}{2}OA$

Mark B′ so that $OB' = \frac{1}{2}OB$

Mark C′ so that $OC' = \frac{1}{2}OC$.

c) Join A′B′C as shown.

Remember always to measure the lengths from O, not from A, B or C.

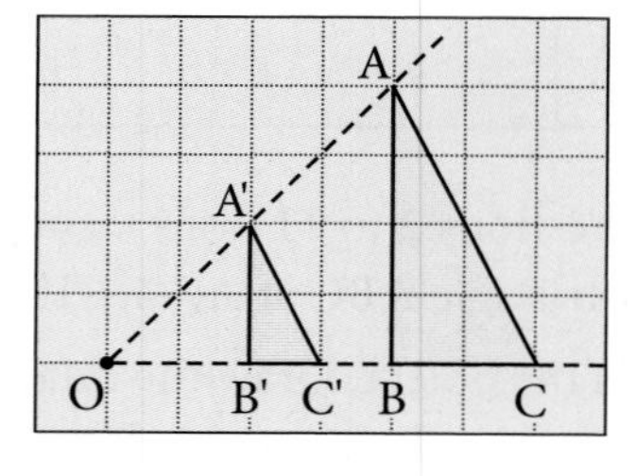

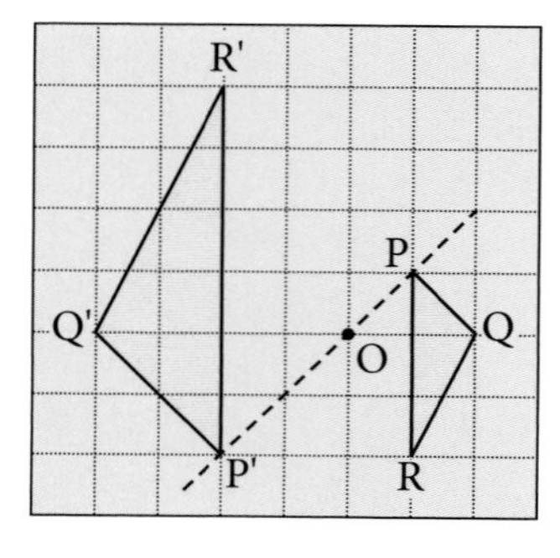

In the diagram, P′Q′R′ is the image of PQR after enlargement with scale factor −2 and centre O. Notice that P′ and P are on opposite sides of point O. Similarly for Q′ and Q, R′ and R.

Exercise 11.11

In questions **1** to **6** copy the diagram and draw an enlargement using the centre O and the scale factor given.

1. Scale factor 2

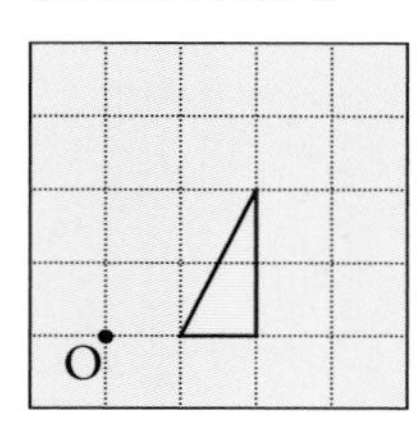

2. Scale factor 3

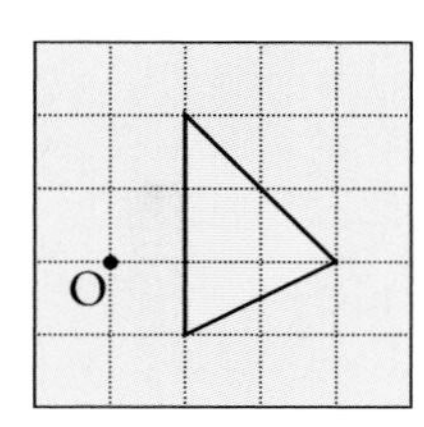

3. Scale factor 3

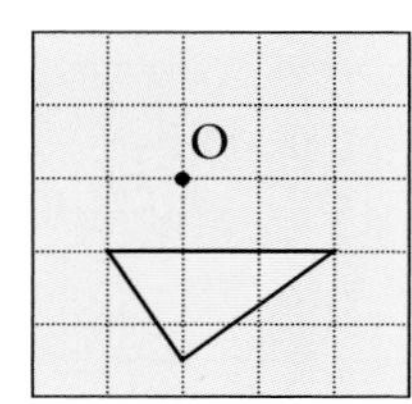

4. Scale factor −2

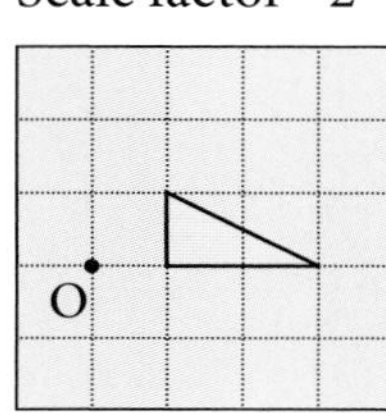

5. Scale factor −3

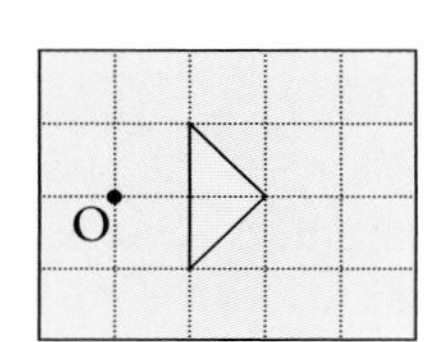

6. Scale factor $1\frac{1}{2}$

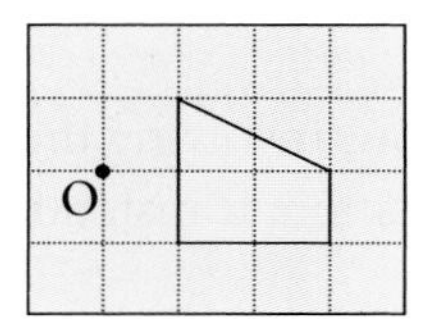

Answer questions **7** to **19** on graph paper, drawing the x- and y- axes from 0 to 15.

The vertices of the object are given in coordinate from.

In questions **7** to **10**, enlarge the object with the centre of enlargement and scale factor indicated.

	Object	Centre	Scale factor
7.	(2, 4) (4, 2) (5, 5)	(0, 0)	+2
8.	(2, 4) (4, 2) (5, 5)	(1, 2)	+2
9.	(1, 1) (4, 2) (2, 3)	(1, 1)	+3
10.	(4, 4) (7, 6) (9, 3)	(7, 4)	+2

In questions **11** to **14** plot the object and image and find the centre of enlargement and the scale factor.

11. Object A(2, 1), B(5, 1), C(3, 3)
Image A′(2, 1), B′(11, 1), C′(5, 7)

12. Object A(2, 5), B(9, 3), C(5, 9)
Image $A'\left(6\frac{1}{2}, 7\right)$, B′(10, 6), C′(8, 9)

13. Object A(2, 2), B(4, 4), C(2, 6)
Image A′(11, 8), B′(7, 4), C′(11, 0)

14. Object A(0, 6), B(4, 6), C(3, 0)
Image A′(12, 6), B′(8, 6), C′(9, 12)

In questions **15** to **19** enlarge the object using the centre of enlargement and scale factor indicated.

	Object	Centre	Scale factor
15.	(1, 2), (13, 2), (1, 10)	(0, 0)	$+\frac{1}{2}$
16.	(5, 10), (5, 7), (11, 7)	(2, 1)	$+\frac{1}{3}$
17.	(7, 3), (9, 3), (7, 8)	(5, 5)	−1
18.	(1, 1), (3, 1), (3, 2)	(4, 3)	−2
19.	(9, 2), (14, 2), (14, 6)	(7, 4)	$-\frac{1}{2}$

Exercise 11.12 contains questions involving the basic transformations: reflection, rotation, translation, enlargement.

Exercise 11.12

1. a) Copy the diagram below.

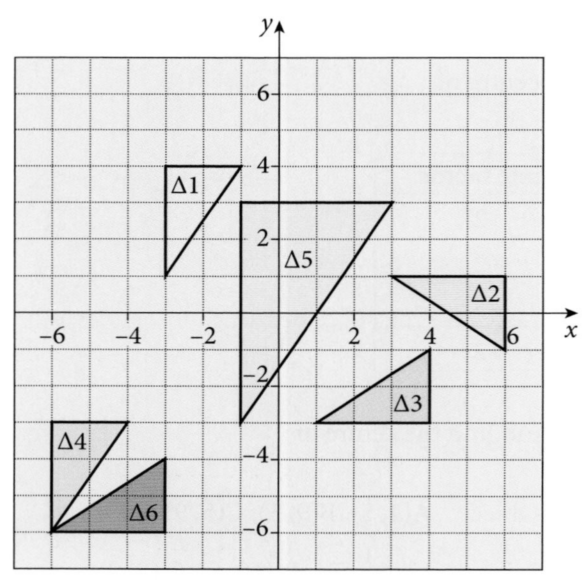

b) Describe fully the following transformations.

i) Δ1 → Δ2 **ii)** Δ1 → Δ3

iii) Δ4 → Δ1 **iv)** Δ1 → Δ5

v) Δ3 → Δ6 **vi)** Δ6 → Δ4

2. Plot and label the following triangles.

   $\Delta 1$: (−5, −5), (−1, −5), (−1, −3)

   $\Delta 2$: (1, 7), (1, 3), (3, 3)

   $\Delta 3$: (3, −3), (7, −3), (7, −1)

   $\Delta 4$: (−5, −5), (−5, −1), (−3, −1)

   $\Delta 5$: (1, −6), (3, −6), (3, −5)

   $\Delta 6$: (−3, 3), (−3, 7), (−5, 7)

   Describe fully the following transformations.

   a) $\Delta 1 \rightarrow \Delta 2$ **b)** $\Delta 1 \rightarrow \Delta 3$

   c) $\Delta 1 \rightarrow \Delta 4$ **d)** $\Delta 1 \rightarrow \Delta 5$

   e) $\Delta 1 \rightarrow \Delta 6$ **f)** $\Delta 5 \rightarrow \Delta 3$

   g) $\Delta 2 \rightarrow \Delta 3$

3. Plot and label the following triangles.

   $\Delta 1$: (−3, −6), (−3, −2), (−5, −2)

   $\Delta 2$: (−5, −1), (−5, −7), (−8, −1)

   $\Delta 3$: (−2, −1), (2, −1), (2, 1)

   $\Delta 4$: (6, 3), (2, 3), (2, 5)

   $\Delta 5$: (8, 4), (8, 8), (6, 8)

   $\Delta 6$: (−3, 1), (−3, 3), (−4, 3)

   Describe fully the following transformations.

   a) $\Delta 1 \rightarrow \Delta 2$ **b)** $\Delta 1 \rightarrow \Delta 3$

   c) $\Delta 1 \rightarrow \Delta 4$ **d)** $\Delta 1 \rightarrow \Delta 5$

   e) $\Delta 1 \rightarrow \Delta 6$ **f)** $\Delta 3 \rightarrow \Delta 5$

   g) $\Delta 6 \rightarrow \Delta 2$

Stretch

The rectangle ABCD has been **stretched** in the direction of the y-axis so that A′B′ is twice AB.

A stretch is fully described if we know:

- the direction of the stretch and the invariant line
- the ratio of corresponding lengths, the stretch factor.

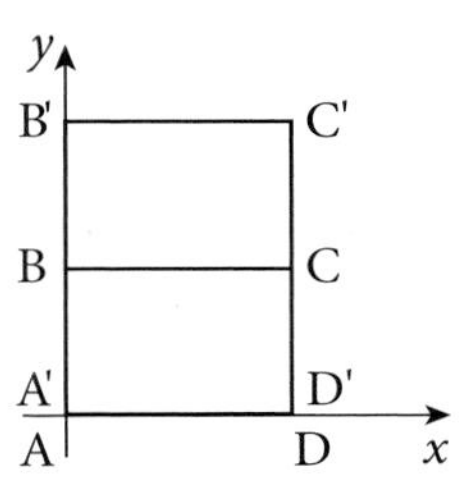

Stretch factor 2 parallel to the y-axis, invariant line $y = 0$.

Exercise 11.13

1. In each diagram the object is coloured green and AB is the invariant line. Calculate the scale factor for each of the stretches shown.

a)

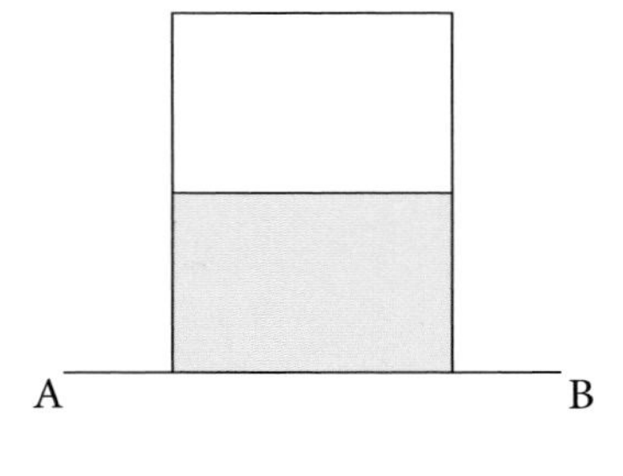

b)

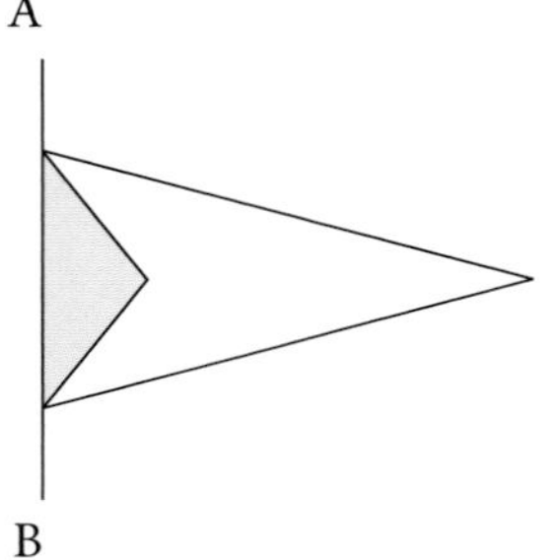

c)

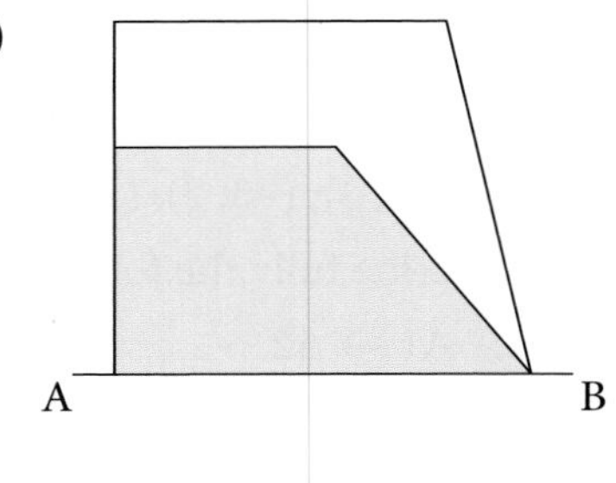

2. Triangle PQR is mapped onto P′Q′R′ by a stretch of scale factor 2 with invariant line AB.
Draw and label triangle P′Q′R′.

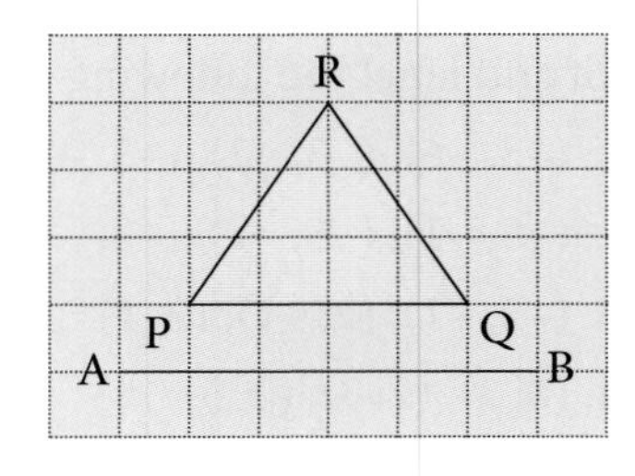

3.

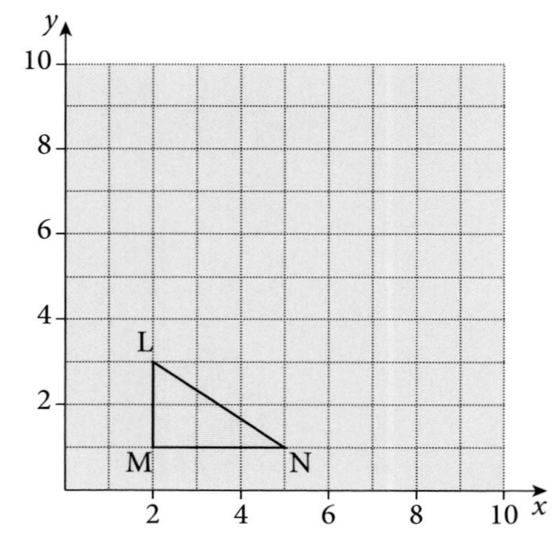

a) Draw and label triangle L′M′N′ which is the image of triangle LMN after a stretch of scale factor 3 with invariant line $y = 0$.

b) Draw and label triangle L′M′N′ which is the image of triangle LMN after a stretch of scale factor 2 with invariant line $x = 0$.

c) Write down the coordinates of the image of L in each case.

11.4 Combined transformations

It is convenient to denote transformations by a symbol.

Let **A** denote 'reflection in line $x = 3$' and

B denote 'translation $\begin{pmatrix} 2 \\ 1 \end{pmatrix}$'

Perform **A** on $\Delta 1$.

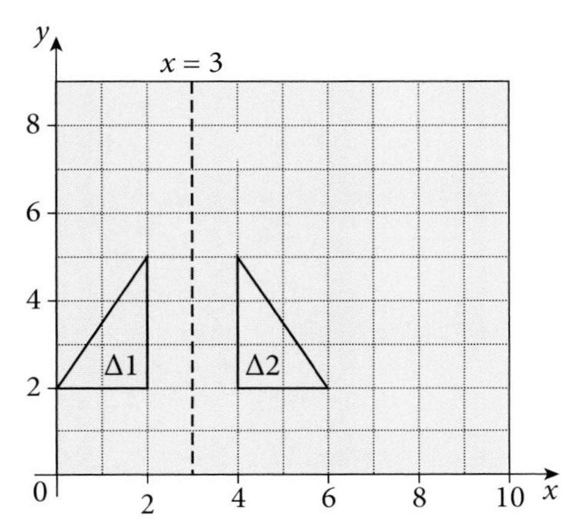

$\Delta 2$ is the image of $\Delta 1$ under the reflection in $x = 3$

i.e. $\mathbf{A}(\Delta 1) = \Delta 2$

$\mathbf{A}(\Delta 1)$ means 'perform the transformation **A** on triangle $\Delta 1$'

Perform **B** on $\Delta 2$.

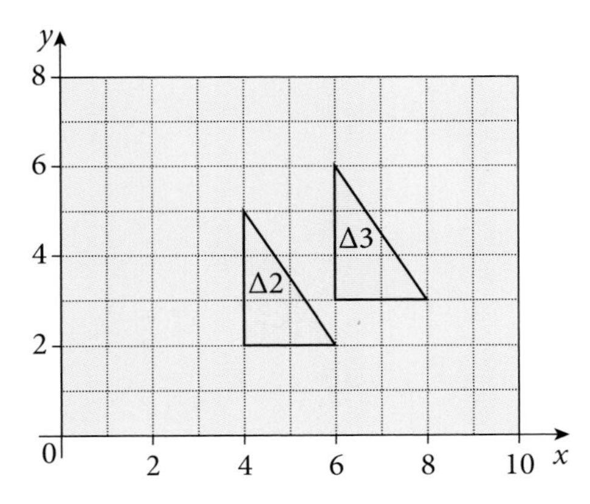

From Figure 2 we can see that

$\mathbf{B}(\Delta 2) = \Delta 3$

The effect of going from $\Delta 1$ to $\Delta 3$ may be written

$\mathbf{BA}(\Delta 1) = \Delta 3$

- It is very important to notice that $\mathbf{BA}(\Delta 1)$ means do **A** first and then **B**.

Inverse transformations

If translation **T** has vector $\begin{pmatrix} 3 \\ -2 \end{pmatrix}$, the translation which has the opposite effect has vector $\begin{pmatrix} -3 \\ 2 \end{pmatrix}$. This is written $\mathbf{T}^{-1}$.

If rotation **R** denotes a 90° clockwise rotation about (0, 0), then $\mathbf{R}^{-1}$ denotes a 90° *anti*clockwise rotation about (0, 0).

The **inverse** of a transformation is the transformation which takes the image back to the object.

Note:

For all reflections, the inverse is the same reflection.

e.g. if **X** is a reflection in $x = 0$, then $\mathbf{X}^{-1}$ is also a reflection in $x = 0$.

Exercise 11.14

In this exercise, transformations **A**, **B**, …, **H**, are as follows:

A denotes reflection in $x = 2$

B denotes 180° rotation, center (1, 1)

C denotes translation $\begin{pmatrix} -6 \\ 2 \end{pmatrix}$

D denotes reflection in $y = x$

E denotes reflection in $y = 0$

F denotes translation $\begin{pmatrix} 4 \\ 3 \end{pmatrix}$

G denotes 90° rotation clockwise, centre (0, 0)

H denotes enlargement, scale factor $+\frac{1}{2}$, centre (0, 0)

Draw x-and y -axes with values from −10 to +10.

1. Draw triangle LMN at L(2, 2), M(6, 2), N(6, 4).

Find the image of LMN under the following combinations of transformations. Write down the coordinates of the image of point L in each case:

a) **CA**(LMN) b) **ED**(LMN) c) **DB**(LMN)

d) **BE**(LMN) e) **EB**(LMN)

Remember

For the combined transformations **CA** you perform **A** first.

2. Draw triangle PQR at P(2, 2), Q(6, 2), R(6, 4). Find the image of PQR under the following combinations of transformations. Write down the coordinates of the image of point P in each case:

a) **AF**(PQR) b) **CG**(PQR)

c) **AG**(PQR) d) **HE**(PQR)

3. Draw triangle XYZ at X(−2, 4), Y(−2, 1), Z(−4, 1). Find the image of XYZ under the following combinations of transformations and state the equivalent single transformation in each case.

   a) $\mathbf{G}^2\mathbf{E}$(XYZ) **b)** **CB**(XYZ) **c)** **DA**(XYZ)

4. Draw triangle OPQ at O(0, 0), P(0, 2), Q(3, 2). Find the image of OPQ under the following combinations of transformations and state the equivalent single transformation in each case.

   a) **DE**(OPQ) **b)** **FC**(OPQ)

   c) **DEC**(OPQ) **d)** **DFE**(OPQ)

5. Draw triangle RST at R(−4, −1), $S\left(-2\frac{1}{2}, -2\right)$, T(−4, −4).

   Find the image of RST under the following combinations of transformations and state the equivalent single transformation in each case.

   a) **EAG**(RST) **b)** **FH**(RST) **c)** **GF**(RST)

6. Write down the inverses of the transformations **A**, **B**, . . . , **H**.

7. Draw triangle JKL at J(−2, 2), K(−2, 5), L(−4, 5). Find the image of JKL under the following transformations. Write down the coordinates of the image of point J in each case.

   a) $\mathbf{C}^{-1}$ **b)** $\mathbf{F}^{-1}$ **c)** $\mathbf{G}^{-1}$

   d) $\mathbf{D}^{-1}$ **e)** $\mathbf{A}^{-1}$

8. Draw triangle XYZ at X(1, 2), Y(1, 6), Z(3, 6).

   a) Find the image of XYZ under each of the transformations **BC** and **CB**.

   b) Describe fully the single transformation equivalent to **BC**.

   c) Describe fully the transformation **M** such that **MCB** = **BC**.

Revision exercise 11A

1.

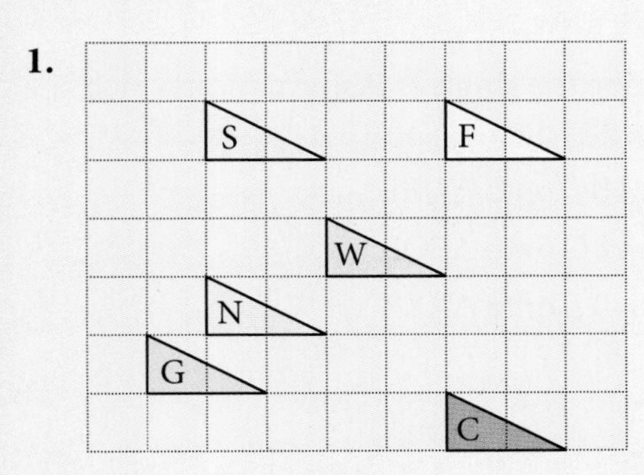

Write down the vector for each translation.

a) $\Delta N \rightarrow \Delta W$ **b)** $\Delta C \rightarrow \Delta S$

c) $\Delta F \rightarrow \Delta C$

d) Which triangle is the image of ΔW after translation by $\begin{pmatrix} 2 \\ -3 \end{pmatrix}$?

2. What vector is twice as long as and parallel to the vector $\begin{pmatrix} 3 \\ 5 \end{pmatrix}$?

3. What vector is twice as long as and opposite to the vector $\begin{pmatrix} 1 \\ -2 \end{pmatrix}$?

4. Write a vector which is perpendicular to the vector $\begin{pmatrix} 2 \\ 1 \end{pmatrix}$.

5. Under a certian translation, the image of the point (5, –2) is (3, –6). Find the image of the point (–2, 10) under the same translation.

6. Under a certian translation, the image of the point (–2, 0) is (6, –5). What point has (3, 3) as its image under the same translation?

7. a) Copy the diagram on squared paper.

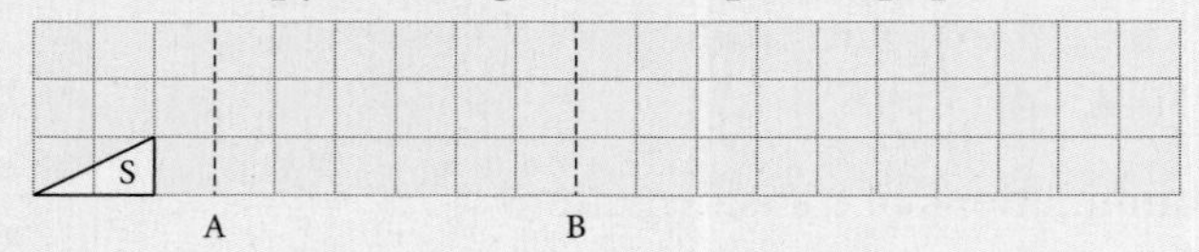

b) Draw the reflection of shape S in line A. Label the image S_1.

c) Draw the reflection of shape S_1 in line B. Label the image S_2.

d) Describe the single transformation from S to S_2.

8. Find the coordinates of the image of (1, 4) under:

a) a clockwise rotation of 90° about (0, 0)

b) a reflection in the line $y = x$

c) a translation which maps (5, 3) onto (1, 1).

9. Draw x- and y-axes with values from –8 to +8. Draw triangle ABC at A(1, –1), B(3, –1), C(1, –4). Find the image of ABC under the following enlargements.

a) scale factor 2, centre (5, –1)

b) scale factor 2, centre (0, 0)

c) scale factor $\frac{1}{2}$, centre (1, 3)

d) scale factor $-\frac{1}{2}$, centre (3, 1)

e) scale factor –2, centre (0, 0)

10.

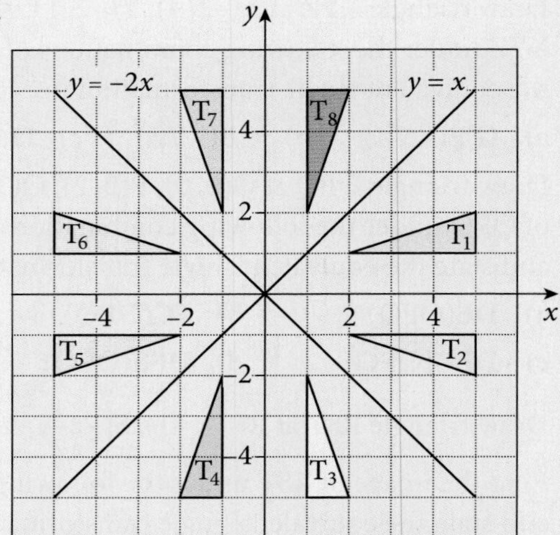

Using the diagram on the right, describe the transformations for the following.

a) $T_1 \rightarrow T_6$
b) $T_4 \rightarrow T_5$
c) $T_8 \rightarrow T_2$
d) $T_4 \rightarrow T_1$
e) $T_8 \rightarrow T_4$
f) $T_6 \rightarrow T_8$

11.

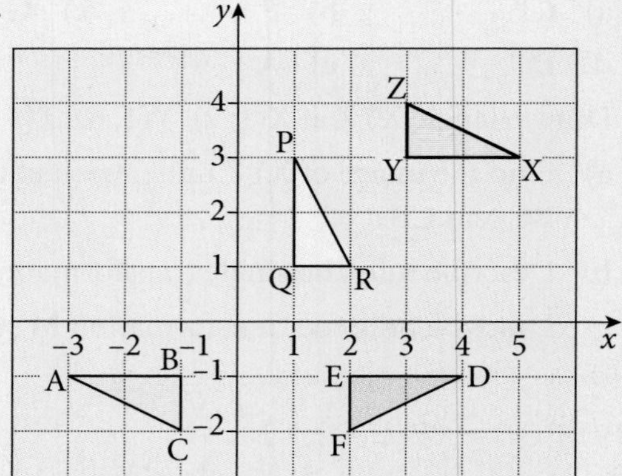

Describe the single transformation which maps:

a) ΔABC onto ΔDEF

b) ΔABC onto ΔPQR

c) ΔABC onto ΔXYZ

12. **M** is an reflection in the line $x + y = 0$.

> **Remember**
>
> For **MR**, perform **R** first.

R is a anticlockwise rotation of 90° about (0, 0).
T is a translation which maps (–1, –1) onto (2, 0).
Find the image of the point (3, 1) under.

a) **M** **b)** **R** **c)** **T**
d) **MR** **e)** **RT** **f)** **TMR**

13. How could ABCD be transformed onto A′B′C′D′ by

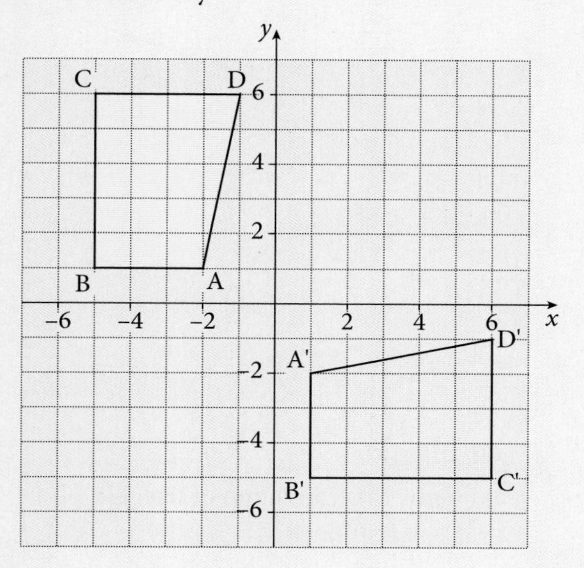

a) a single reflection
b) a reflection after a rotation of 90° anticlockwise about (0, 0)?
c) Would the image be the same if the order of the rotation and reflection were reversed?

14. Draw axes for both x and y between –8 and +8. Plot the points (1, 1), (3, 1), (3, 2), (2, 2), (2, 4) and (1, 4) and join up to make an 'L' shape.
This is mapped onto the points (–2, –2), (–2, –6), (–4, –6), (–4, –4), (–8, –4), (–8, –2) by *two* transformations; an enlargement with centre (0, 0) followed by a reflection. Describe these transformations as fully as possible.

15. Draw axes for both x and y between –5 and +5.
Plot and label the points A(1, 1), B(3, 1), C(5, 3) and D(1, 2). Join these up to make a quadrilateral.
The image of these points after a translation parallel to the y-axis followed by a reflection are A′(0, 3), B′(–2, 3), C′(–4, 5) and D′(0, 4). Plot A′B′C′D′ on the same set of axes.
Describe fully the translation and reflection required.

16. Three transformations are required to map ABC onto A′B′C′, an enlargement centre (0, 0) followed by a translation parallel to the x-axis, and lastly a rotation about (0, 0). Describe these transformations as fully as possible.

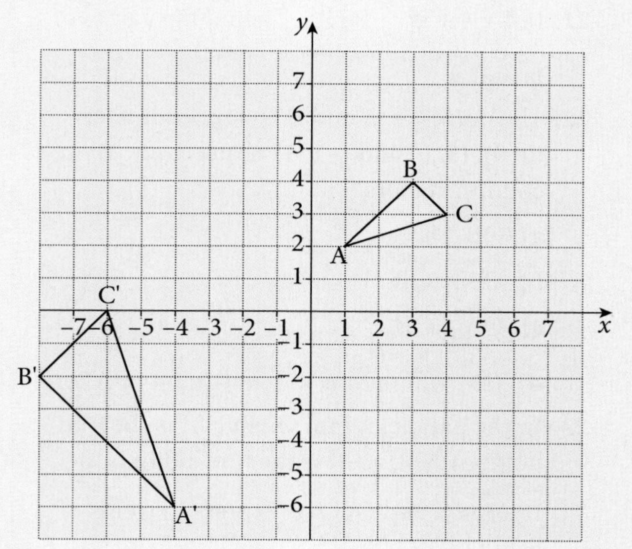

17. $\mathbf{a} = \begin{pmatrix} 5 \\ -1 \end{pmatrix}$ $\mathbf{b} = \begin{pmatrix} -3 \\ 4 \end{pmatrix}$

a) Find $2\mathbf{a} - \mathbf{b}$ **b)** Calculate $|\mathbf{b}|$

18. $\mathbf{p} = \begin{pmatrix} 3 \\ 1 \end{pmatrix}$ $\mathbf{q} = \begin{pmatrix} -2 \\ 3 \end{pmatrix}$

When $2\mathbf{p} + k\mathbf{q} = \begin{pmatrix} 0 \\ 11 \end{pmatrix}$ find the value of k.

19. In ΔOPR, the mid-point of PR is M.

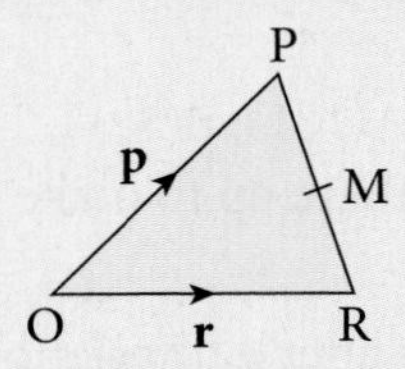

If $\overrightarrow{OP} = \mathbf{p}$ and $\overrightarrow{OR} = \mathbf{r}$, find in terms of **p** and **r**:

a) $\overrightarrow{PR}$

b) $\overrightarrow{PM}$

c) $\overrightarrow{OM}$

20. If $\mathbf{a} = \begin{pmatrix} 1 \\ 4 \end{pmatrix}$ and $\mathbf{b} = \begin{pmatrix} -3 \\ 4 \end{pmatrix}$, find:

a) $|\mathbf{b}|$ b) $|\mathbf{a} + \mathbf{b}|$ c) $|2\mathbf{a} - \mathbf{b}|$

21. If $4\begin{pmatrix} 1 \\ 3 \end{pmatrix} + 2\begin{pmatrix} 1 \\ m \end{pmatrix} = 3\begin{pmatrix} n \\ -6 \end{pmatrix}$, find the values of m and n.

22. The points O, A and B have coordinates (0, 0), (5, 0) and (−1, 4) respectively. Write as column vectors:

a) $\overrightarrow{OB}$

b) $\overrightarrow{OA} + \overrightarrow{OB}$

c) $\overrightarrow{OA} - \overrightarrow{OB}$

d) $\overrightarrow{OM}$ where M is the mid-point of AB.

23. In the parallelogram OABC, M is the mid-point of AB and N is the mid-point of BC.

If $\overrightarrow{OA} = \mathbf{a}$ and $\overrightarrow{OC} = \mathbf{c}$, express in trems of **a** and **c**:

a) $\overrightarrow{CA}$ b) $\overrightarrow{ON}$ c) $\overrightarrow{NM}$

Describe the relationship between CA and NM.

24. The vectors **a, b, c** are given by:

$\mathbf{a} = \begin{pmatrix} 1 \\ 5 \end{pmatrix}, \mathbf{b} = \begin{pmatrix} -2 \\ 1 \end{pmatrix}, \mathbf{c} = \begin{pmatrix} -1 \\ 17 \end{pmatrix}$

Find numbers m and n so that $m\mathbf{a} + n\mathbf{b} = \mathbf{c}$

Examination exercise 11B

1.

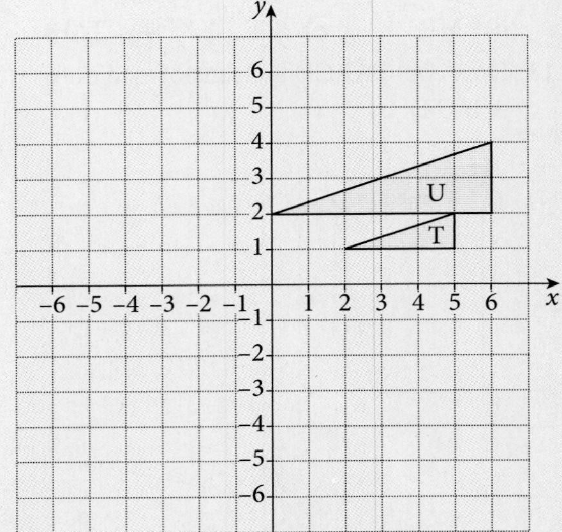

a) On the grid,

i) draw the translation of triangle T by $\begin{pmatrix} -6 \\ 3 \end{pmatrix}$, [2]

ii) draw the reflection of triangle T in the line $y = -x$. [2]

b) Describe fully the **single** transformation that maps triangle T onto triangle U. [3]

c) Write down the inverse of the transformation in **part a) i)**. [2]

Cambridge IGCSE International Mathematics 0607 Paper 4 Q8 October/November 2010

2.

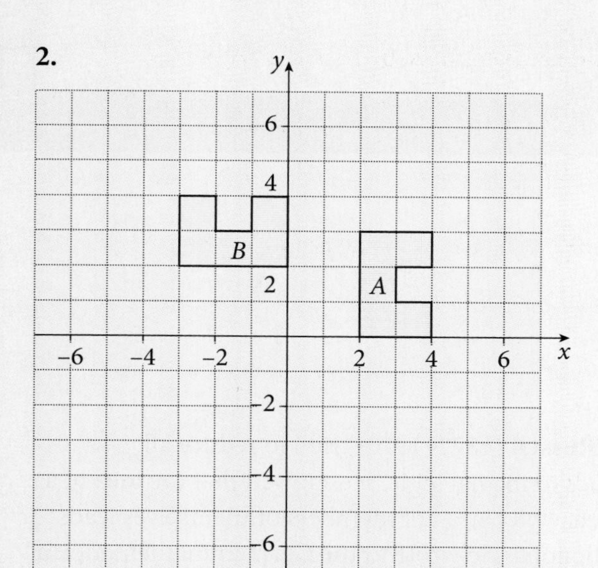

a) Describe fully the **single** transformation which maps shape A onto shape B. [3]

b) Draw the image of shape A after a stretch, with y–axis invariant and scale factor 2. [2]

Cambridge IGCSE International Mathematics 0607 Paper 2 Q9 May/June 2010

3. $\mathbf{p} = \begin{pmatrix} 2 \\ 3 \end{pmatrix}$ $\quad \mathbf{q} = \begin{pmatrix} -4 \\ 3 \end{pmatrix}$

a) Find $2\mathbf{p} - 3\mathbf{q}$. [2]

b) Calculate $|\mathbf{q}|$. [2]

Cambridge IGCSE International Mathematics 0607 Paper 21 Q4 May/June 2011

4. $\mathbf{p} = \begin{pmatrix} 5 \\ 1 \end{pmatrix}$ and $\mathbf{q} = \begin{pmatrix} -4 \\ 2 \end{pmatrix}$

a) Write $2\mathbf{p} - \frac{1}{2}\mathbf{q}$ as a column vector. [2]

b) Find $|\mathbf{q}|$ leaving your answer in surd form. [2]

Cambridge IGCSE International Mathematics 0607 Paper 2 Q7 May/June 2010

5. a) M and R are single transformations.

M is a reflection in the x-axis and R is an anti-clockwise rotation of 90° about the origin.

i) Find the image of the point (5, 7) under the transformation M. [1]

ii) Find the single transformation equivalent to M followed by R. [3]

b) T and U are translations represented by vectors **p** and **q**.

$\mathbf{p} = \begin{pmatrix} 1 \\ 2 \end{pmatrix}$ and $\mathbf{q} = \begin{pmatrix} -2 \\ 3 \end{pmatrix}$

$c\mathbf{p} + d\mathbf{q} = \begin{pmatrix} 0 \\ 21 \end{pmatrix}$.

Find the values of c and d. [4]

Cambridge IGCSE International Mathematics 0607 Specimen Paper 4 Q3 2010

12 Sets

Bertrand Russell (1872–1970) tried to reduce all mathematics to formal logic. He showed that the idea of a set of all sets which are not members of themselves leads to contradictions. He wrote to the German mathematician Gottlob Frege just as he was putting the finishing touches to a book that represented his life's work, pointing out that Frege's work was invalidated. Russell's elder brother, the second Earl Russell, showed great foresight in 1903 by queueing overnight outside the vehicle licensing office in London to have his car registered as A1.

9.1 Notation and meaning for:
is an element of ($\in$); is not an element of ($\notin$);
is a subset of ($\subseteq$); is a proper subset of ($\subset$);
universal set $\cup$, empty set $\varnothing$ or{ };
complement of A, (A'); number of elements in A, $n(A)$

9.2 Sets in descriptive form $\{ x \mid \quad \}$ or as a list

9.3 Venn diagrams with at most three sets

9.4 Intersection and union of sets

12.1 Set notation

1. $\cap$ 'intersection'

A $\cap$ B is shaded.

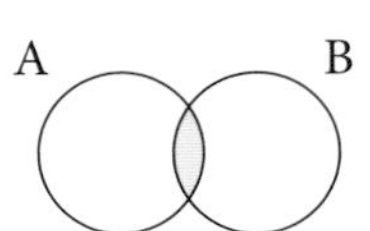

2. $\cup$ 'union'

A $\cup$ B is shaded.

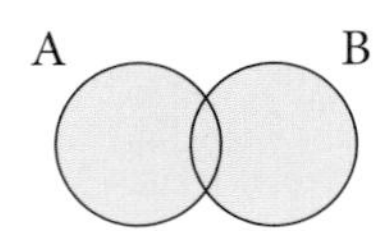

3. $\subset$ 'is a subset of'

   $A \subset B$

   [$B \not\subset A$ means 'B is *not* a subset of A']

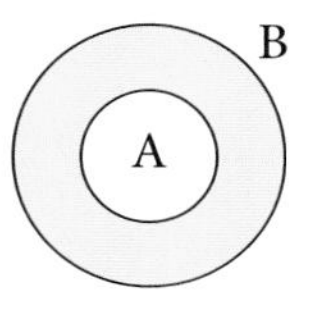

4. $\in$ 'is a member of'

   'belongs to'

   $b \in X$

   [$e \notin X$ means 'e is not a member of set X']

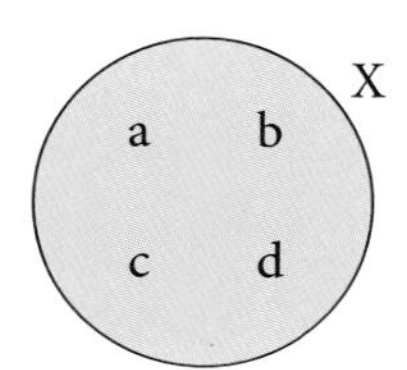

5. $\cup$ 'universal set'

   $\cup = \{a, b, c, d, e\}$

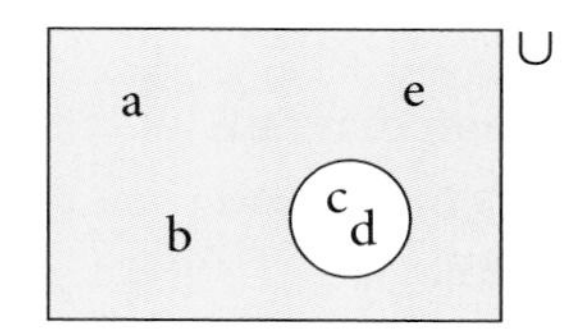

6. A' 'complement of'

   'not in A'

   A' is shaded

   $(A \cup A' = \cup)$

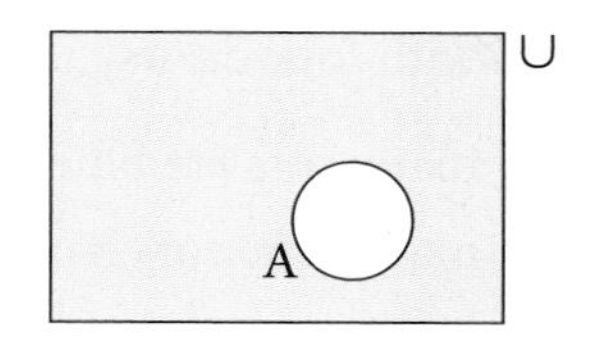

7. $n(A)$ 'the number of elements in set A'

   $n(A) = 3$

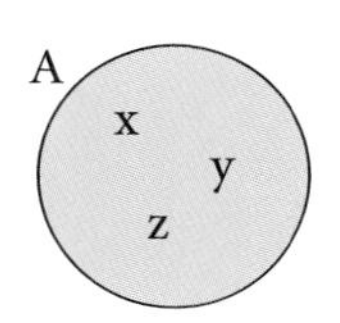

8. $A = \{x : x \text{ is an integer}, 2 \le x \le 9\}$

   A is the *set of* elements x *such that* x is an integer and $2 \le x \le 9$.

   The set A is $\{2, 3, 4, 5, 6, 7, 8, 9\}$.

9. $\varnothing$ or { } 'empty set'

$\varnothing \subset A$ for any set A

Exercise 12.1

1. In the Venn diagram,

   $\cup$ = {people in an hotel}

   T = {people who like toast}

   E = {people who like eggs}

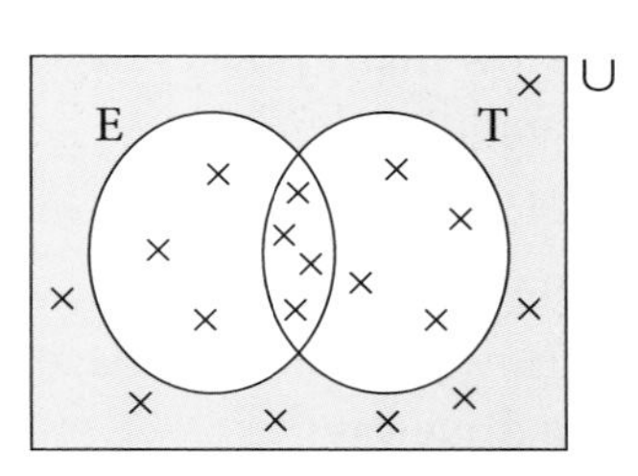

   a) How many people like toast?

   b) How many people like eggs but not toast?

   c) How many people like toast and eggs?

   d) How many people are in the hotel?

   e) How many people like neither toast nor eggs?

2. In the Venn diagram,
∪ = {boys in Year 10}
R = {members of the rugby team}
C = {members of the cricket team}

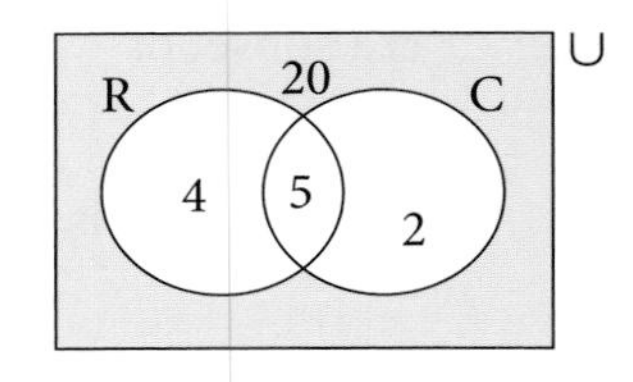

a) How many are in the rugby team?

b) How many are in both teams?

c) How many are in the rugby team but not in the cricket team?

d) How many are in neither team?

e) How many are there in Year 10?

3. In the Venn diagram,
∪ = {cars in a street}
B = {blue cars}
L = {cars with left-hand drive}
F = {cars with four doors}

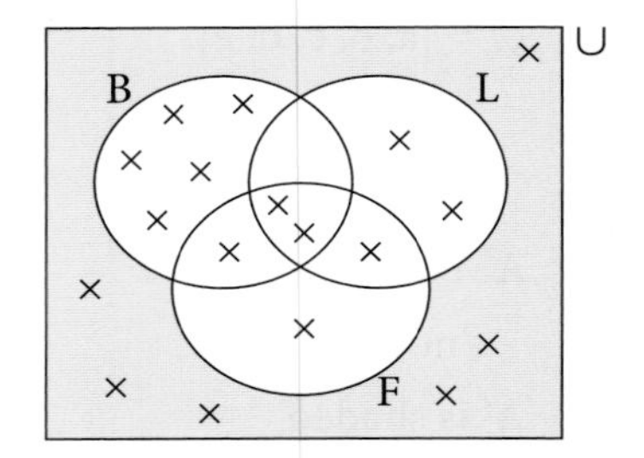

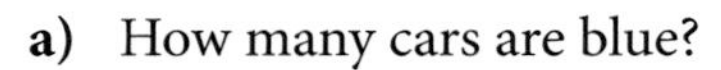

a) How many cars are blue?

b) How many blue cars have four doors?

c) How many cars with left-hand drive have four doors?

d) How many blue cars have left-hand drive?

e) How many cars are in the street?

f) How many blue cars with left-hand drive do not have four doors?

4. In the Venn diagram,
∪ = {houses in the street}
D = {houses with a driveway}
B = {houses with a basement}
G = {houses with a garden}

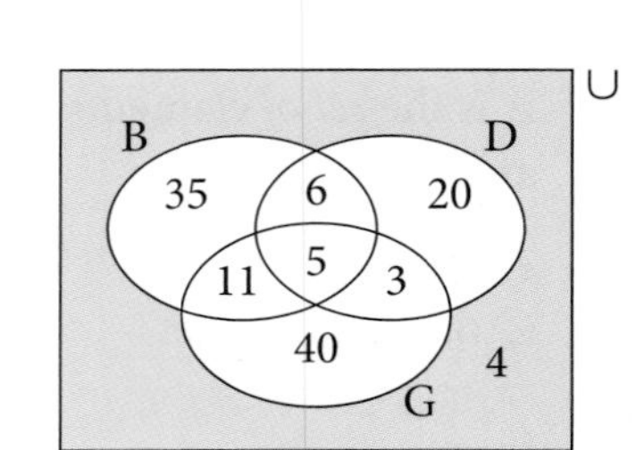

a) How many houses have gardens?

b) How many houses have a basement and a driveway?

c) How many houses have a basement and a driveway and a garden?

d) How many houses have a garden but not a basement or a driveway?

e) How many houses have a basement and a garden but not a driveway?

f) How many houses are there in the street?

5. In the Venn diagram,
   $\cup$ = {children in a mixed school}
   G = {girls in the school}
   S = {children who can swim}
   L = {children who are left-handed}

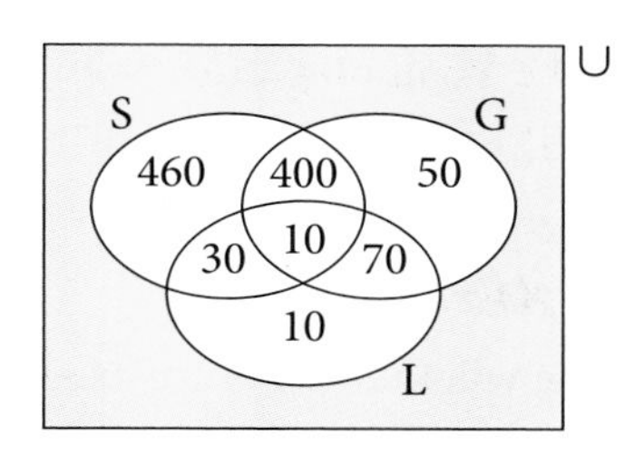

   a) How many left-handed children are there?
   b) How many girls cannot swim?
   c) How many boys can swim?
   d) How many girls are left-handed?
   e) How many boys are left-handed?
   f) How many left-handed girls can swim?
   g) How many boys are there in the school?

Example 1

$\cup = \{1, 2, 3 \ldots, 12\}$, $A = \{2, 3, 4, 5, 6\}$ and $B = \{2, 4, 6, 8, 10\}$.

a) $A \cup B = \{2, 3, 4, 5, 6, 8, 10\}$
b) $A \cap B = \{2, 4, 6\}$
c) $A' = \{1, 7, 8, 9, 10, 11, 12\}$
d) $n(A \cup B) = 7$
e) $B' \cap A = \{3, 5\}$

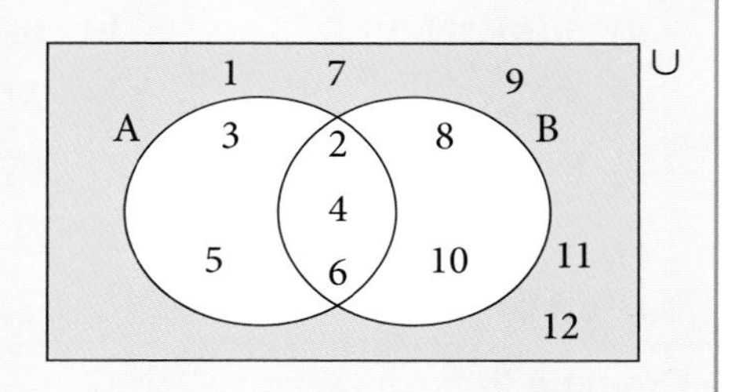

Exercise 12.2

In this exercise, be careful to use set notation only when the answer *is* a set.

1. If $M = \{1, 2, 3, 4, 5, 6, 7, 8\}$, $N = \{5, 7, 9, 11, 13\}$, find:
   a) $M \cap N$ **b)** $M \cup N$ **c)** $n(N)$ **d)** $n(M \cup N)$
   State whether these are 'true' or 'false':
   e) $5 \in M$ **f)** $7 \in (M \cup N)$ **g)** $N \subset M$ **h)** $\{5, 6, 7\} \subset M$
2. If $A = \{2, 3, 5, 7\}$, $B = \{1, 2, 3, \ldots, 9\}$, find:
   a) $A \cap B$ **b)** $A \cup B$ **c)** $n(A \cap B)$ **d)** $\{1, 4\} \cap A$
   State whether these are 'true' or 'false':
   e) $A \in B$ **f)** $A \subset B$ **g)** $9 \subset B$ **h)** $3 \in (A \cap B)$
3. If $X = \{1, 2, 3, \ldots, 10\}$, $Y = \{2, 4, 6, \ldots, 20\}$ and
   $Z = \{x : x \text{ is an integer}, 15 \le x \le 25\}$, find:
   a) $X \cap Y$ **b)** $Y \cap Z$ **c)** $X \cap Z$
   d) $n(X \cup Y)$ **e)** $n(Z)$ **f)** $n(X \cup Z)$
   State whether these are 'true' or 'false':
   g) $5 \in Y$ **h)** $20 \in X$ **i)** $n(X \cap Y) = 5$ **j)** $\{15, 20, 25\} \subset Z$.

4. If D = {1, 3, 5}, E = {3, 4, 5}, F = {1, 5, 10), find:

a) $D \cup E$ **b)** $D \cap F$

c) $n(E \cap F)$ **d)** $(D \cup E) \cap F$

e) $(D \cap E) \cup F$ **f)** $n(D \cup F)$

State whether these are 'true' or 'false':

g) $D \subset (E \cup F)$ **h)** $3 \in (E \cap F)$

i) $4 \notin (D \cap E)$

5. Find:

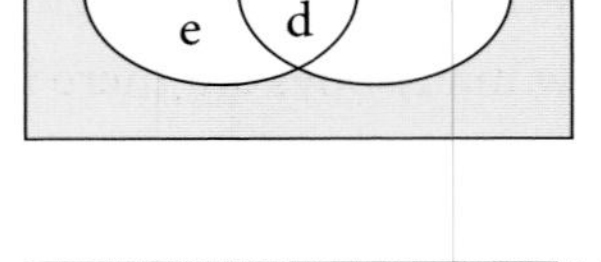

a) $n(E)$ **b)** $n(F)$

c) $E \cap F$ **d)** $E \cup F$

e) $n(E \cup F)$ **f)** $n(E \cap F)$

6. Find:

a) $n(M \cap N)$ **b)** $n(N)$

c) $M \cup N$ **d)** $M' \cap N$

e) $N' \cap M$ **f)** $(M \cap N)'$

g) $M \cup N'$ **h)** $N \cup M'$

i) $M' \cup N'$

Example 2

On a Venn diagram, shade the regions

a) $A \cap C$ **b)** $(B \cap C) \cap A'$

where A, B, C are intersecting sets.

a) $A \cap C$

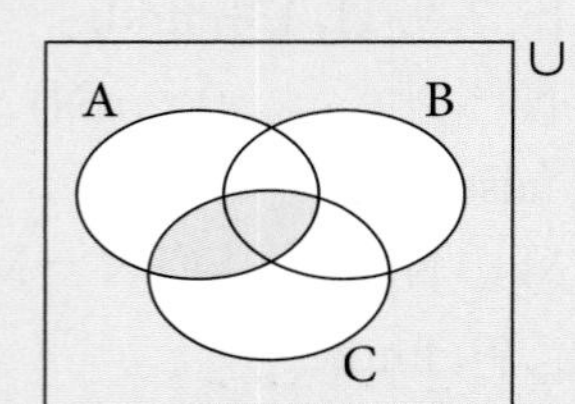

b) $(B \cap C) \cap A'$

[find $(B \cap C)$ first]

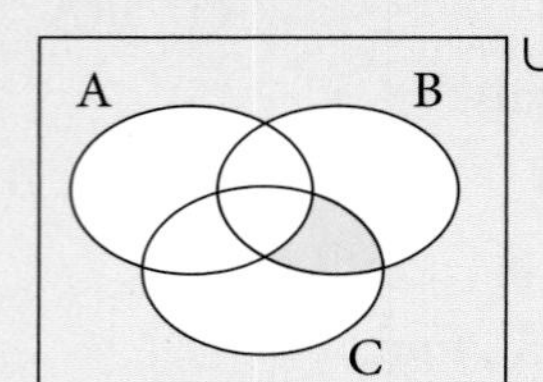

Exercise 12.3

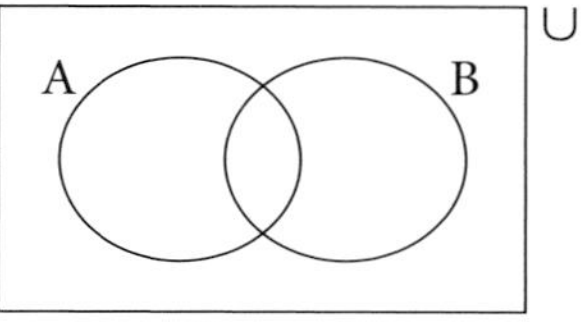

Figure 1

1. Draw six diagrams similar to Figure 1 and shade the following sets:
   a) $A \cap B$ b) $A \cup B$ c) A'
   d) $A' \cap B$ e) $B' \cap A$ f) $(B \cup A)'$

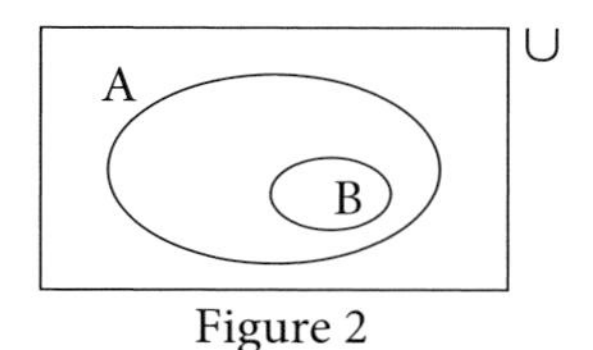

Figure 2

2. Draw four diagrams similar to Figure 2 and shade the following sets:
   a) $A \cap B$ b) $A \cup B$ c) $B' \cap A$ d) $(B \cup A)'$

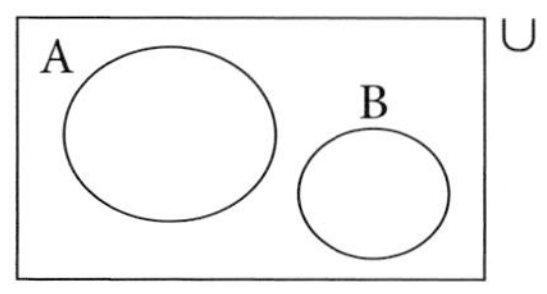

Figure 3

3. Draw four diagrams similar to Figure 3 and shade the following sets:
   a) $A \cup B$ b) $A \cap B$ c) $A \cap B'$ d) $(B \cup A)'$

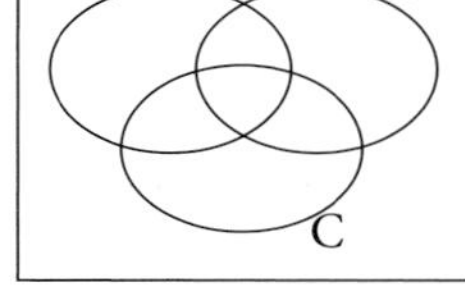

Figure 4

4. Draw eleven diagrams similar to Figure 4 and shade the following sets:
   a) $A \cap B$ b) $A \cup C$ c) $A \cap (B \cap C)$
   d) $(A \cup B) \cap C$ e) $B \cap (A \cup C)$ f) $A \cap B'$
   g) $A \cap (B \cup C)'$ h) $(B \cup C) \cap A$ i) $C' \cap (A \cap B)$
   j) $(A \cup C) \cap B'$ k) $(A \cup C) \cap (B \cap C)$

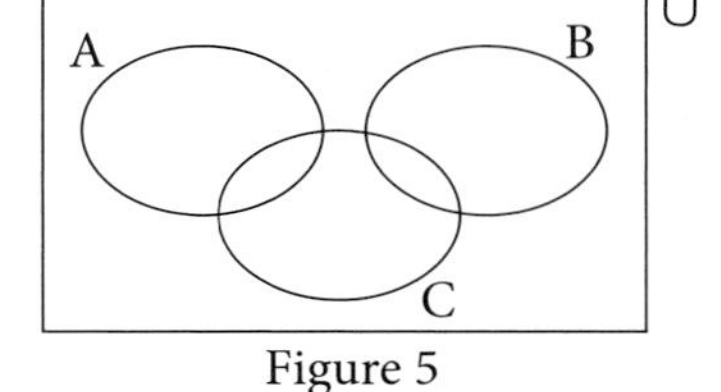

Figure 5

5. Draw nine diagrams similar to Figure 5 and shade the following sets:
   a) $(A \cup B) \cap C$ b) $(A \cap B) \cup C$
   c) $(A \cup B) \cup C$ d) $A \cap (B \cup C)$
   e) $A' \cap C$ f) $C' \cap (A \cup B)$
   g) $(A \cap B) \cap C$ h) $(A \cap C) \cup (B \cap C)$
   i) $(A \cup B \cup C)'$

6. Copy each diagram and shade the region indicated.

a) 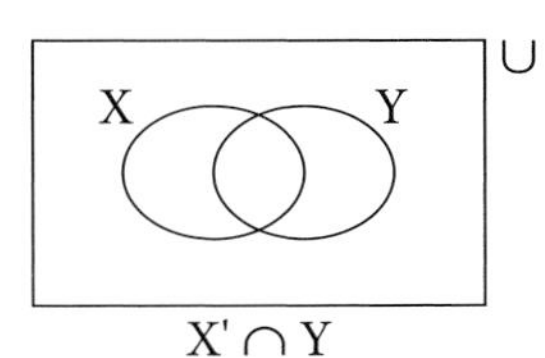

$X' \cap Y$

b) 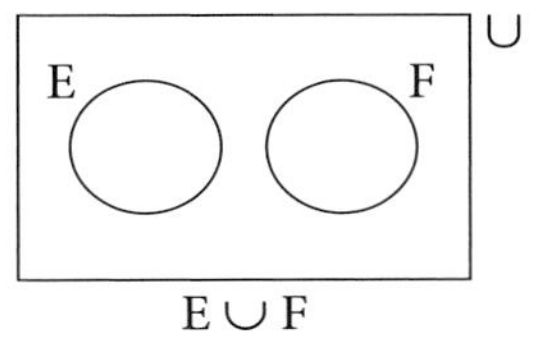

$E \cup F$

c) 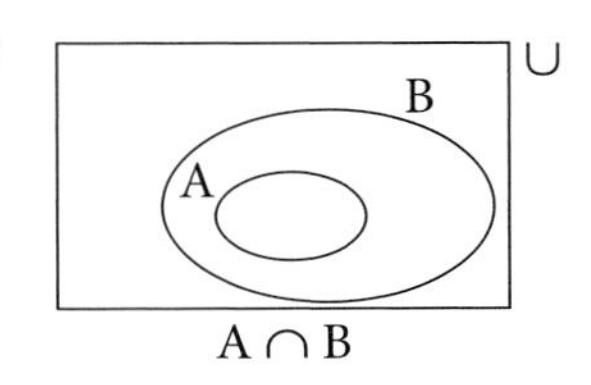

$A \cap B$

d) 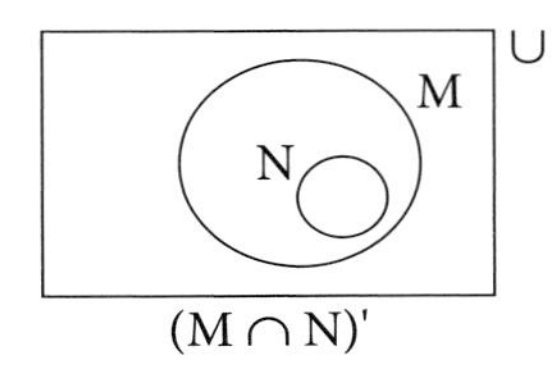

$(M \cap N)'$

7. Describe the shaded region.

a) 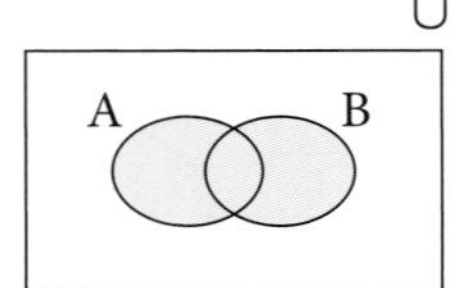

b) 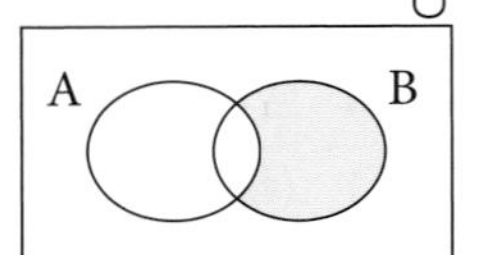

c) 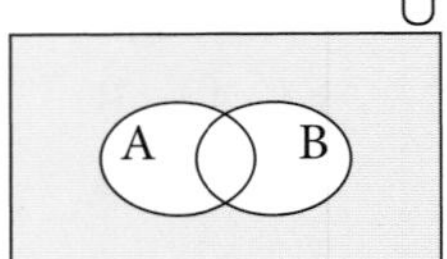

d) 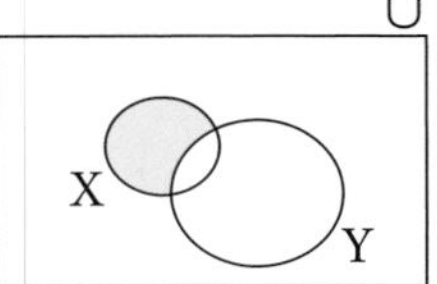

12.2 Logical problems

Example 3

In a form of 30 girls, 18 play netball and 14 play hockey, whilst 5 play neither. Find the number who play both netball and hockey.

Let U = {girls in the form}
N = {girls who play netball}
H = {girls who play hockey}
and x = the number of girls who play both netball and hockey

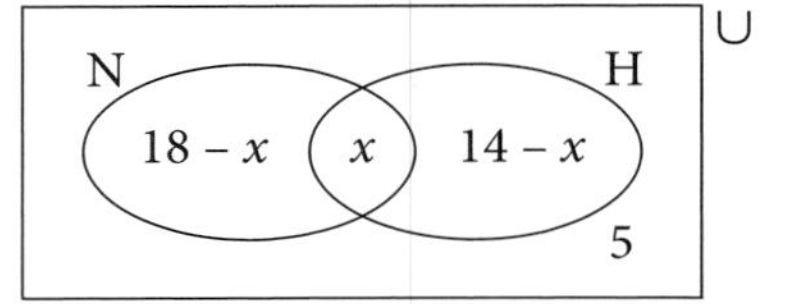

The number of girls in each portion of the universal set is shown in the Venn diagram.

Since $n(\mathrm{U}) = 30$

$$18 - x + x + 14 - x + 5 = 30$$
$$37 - x = 30$$
$$x = 7$$

∴ Seven girls play both netball and hockey.

Example 4

A = {sheep}
B = {horses}
C = {'intelligent' animals}
D = {animals which make good pets}

a) Express the following sentences in set language:

i) No sheep are 'intelligent' animals.

ii) All horses make good pets.

iii) Some sheep make good pets.

b) Interpret the following statements:

i) $\mathrm{B} \subset \mathrm{C}$

ii) $\mathrm{B} \cup \mathrm{C} = \mathrm{D}$

▶ Continued on next page

a) i) $A \cap C = \varnothing$

ii) $B \subset D$

iii) $A \cap D \neq \varnothing$

b) i) All horses are intelligent animals.

ii) Animals which make good pets are either horses or 'intelligent' animals (or both).

Exercise 12.4

1. In the Venn diagram $n(A) = 10$, $n(B) = 13$, $n(A \cap B) = x$ and $n(A \cup B) = 18$.

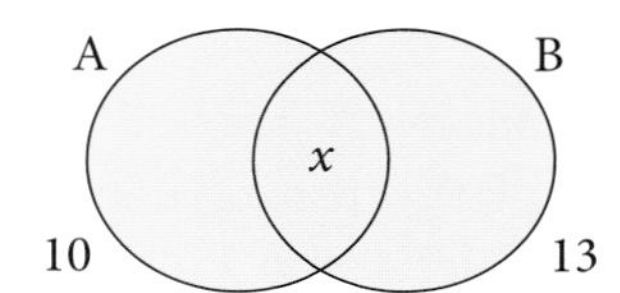

a) Write in terms of x the number of elements in A but not in B.

b) Write in terms of x the number of elements in B but not in A.

c) Add together the number of elements in the three parts of the diagram to obtain the equation $10 - x + 13 + x = 18$.

d) Hence find the number of elements in both A and B.

2. In the Venn diagram $n(A) = 21$, $n(B) = 17$, $n(A \cap B) = x$ and $n(A \cup B) = 29$.

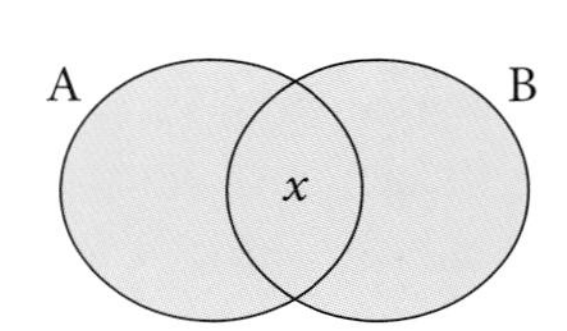

a) Write down in terms of x the number of elements in each part of the diagram.

b) Form an equation and hence find x.

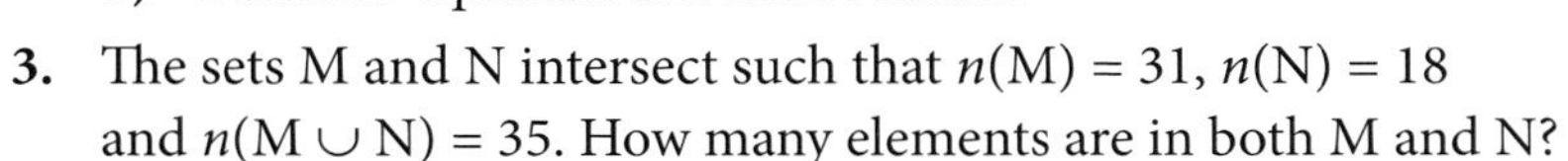

3. The sets M and N intersect such that $n(M) = 31$, $n(N) = 18$ and $n(M \cup N) = 35$. How many elements are in both M and N?

4. The sets P and Q intersect such that $n(P) = 11$, $n(Q) = 29$ and $n(P \cup Q) = 37$. How many elements are in both P and Q?

5. The sets A and B intersect such that $n(A \cap B) = 7$, $n(A) = 20$ and $n(B) = 23$. Find $n(A \cup B)$.

6. Twenty boys all play either football or basketball (or both). If thirteen play football and ten play basketball, how many play both sports?

7. Of the 53 staff at a school, 36 drink tea, 18 drink coffee and 10 drink neither tea nor coffee. How many drink both tea and coffee?

8. Of the 32 pupils in a class, 18 play golf, 16 play the piano and 7 play both. How many play neither?

9. Of the pupils in a class, 15 can spell 'parallel', 14 can spell 'Pythagoras', 5 can spell both words and 4 can spell neither. How many pupils are there in the class?

10. In a school, students must take at least one of these subjects: Maths, Physics or Chemistry. In a group of 50 students, 7 take all three subjects, 9 take Physics and Chemistry only, 8 take Maths and Physics only and 5 take Maths and Chemistry only. Of these 50 students, x take Maths only, x take Physics only and $x + 3$ take Chemistry only. Draw a Venn diagram, find x, and hence find the number taking Maths.

11. All of 60 different vitamin pills contain at least one of the vitamins A, B and C. Twelve have A only, 7 have B only, and 11 have C only. If 6 have all three vitamins and there are x having A and B only, x having B and C only and x having A and C only, how many pills contain vitamin A?

12. The exam results of the 30 members of a rugby squad were as follows: All 30 players passed at least two subjects, 18 players passed at least three subjects, and 3 players passed four subjects or more. Calculate:

    a) how many passed exactly two subjects,

    b) what fraction of the squad passed exactly three subjects.

13. In a group of 59 people, some are wearing hats, gloves or scarves (or a combination of these). 4 are wearing all three, 7 are wearing just a hat and gloves, 3 are wearing just gloves and a scarf and 9 are wearing just a hat and scarf. The number wearing only a hat is x, the number wearing only gloves is x, the number wearing only a scarf is $x - 2$, and the number wearing none of the three items is $(x - 2)$. Find x and hence the number of people wearing a hat.

14. In a street of 150 houses, three different newspapers are delivered: T, G and M. Of these houses, 40 receive T, 35 receive G, and 60 receive M. 7 receive T and G, 10 receive G and M and 4 receive T and M. 34 receive no paper at all. How many receive all three? (**Note**: If '7 receive T and G', this information does not mean 7 receive T and G *only*.)

15. If S = {Serbian men}, G = {good footballers}, express the following statements in words:

    a) $G \subset S$ **b)** $G \cap S = \varnothing$ **c)** $G \cap S \neq \varnothing$

    (Ignore the truth or otherwise of the statements.)

16. Given that $\mathscr{E}$ = {pupils in a school}, B = {boys}, H = {hockey players}, F = {football players}, express the following in words:

    a) $F \subset B$ **b)** $H \subset B'$

    c) $F \cap H \neq \varnothing$ **d)** $B \cap H = \varnothing$

    Express in set notation:

    e) No boys play football.

    f) All pupils play either football or hockey.

17. If $\cup$ = {living creatures}, S = {spiders}, F = {animals that fly}, T = {animals which taste nice}, express in set notation:

a) No spiders taste nice.

b) All animals that fly taste nice.

c) Some spiders can fly.

Express in words:

d) $S \cup F \cup T = \cup$ **e)** $T \subset S$

Revision exercise 12A

1. Given that $\cup = \{1, 2, 3, 4, 5, 6, 7, 8\}$, $A = \{1, 3, 5\}$, $B = \{5, 6, 7\}$, list the members of the sets:

a) $A \cap B$ **b)** $A \cup B$ **c)** A'

d) $A' \cap B'$ **e)** $A \cup B'$

2. The sets P and Q are such that $n(P \cup Q) = 50$, $n(P \cap Q) = 9$ and $n(P) = 27$. Find the value of $n(Q)$.

3.

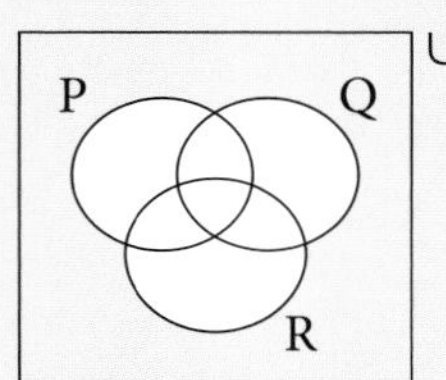

Figure 1

Draw three diagrams similar to Figure 1, and shade the following.

a) $Q \cap R'$ **b)** $(P \cup Q) \cap R$

c) $(P \cap Q) \cap R'$

4. Describe the shaded regions in Figures 2 and 3.

a)

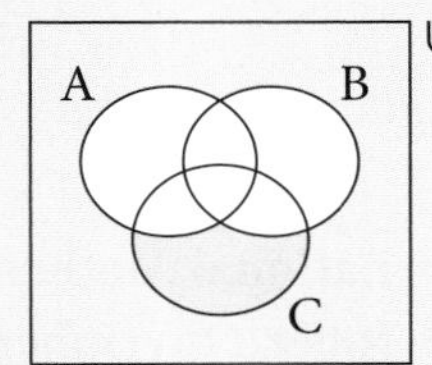

Figure 2

b)

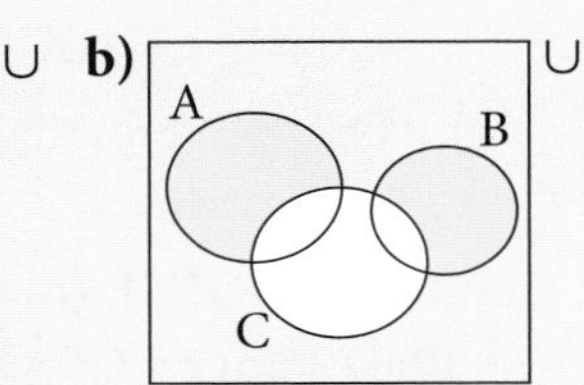

Figure 3

5. Given that $\cup$ = {people on a train}, M = {males}, T = {people over 25 years old} and S = {snooker players},

a) express in set notation:

i) all the snooker players are over 25

ii) some snooker players are women

b) express in words: $T \cap M' = \varnothing$

6.

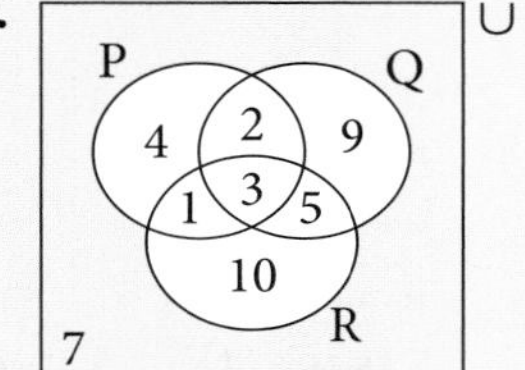

The figures in the diagram indicate the number of elements in each subset of $\cup$.

a) Find $n(P \cap R)$.

b) Find $n(Q \cup R)'$.

c) Find $n(P' \cap Q')$.

Examination exercise 12B

1. Using set notation describe the regions shaded on the Venn diagrams.

a)

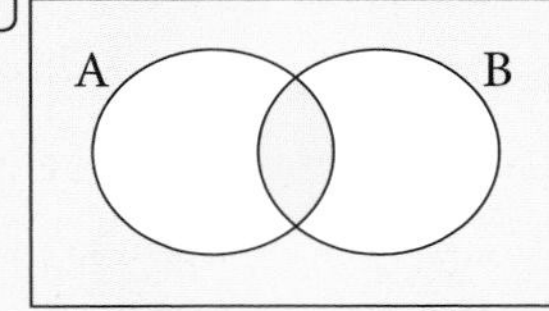

[1]

b) ∪ A B

[1]

Cambridge IGCSE International Mathematics 0607 Paper 21 Q6 May/June 2011

2. All the students in a class of 20 took tests in Mathematics and Chemistry.
The following table shows the results of these two tests.

	Pass	**Fail**
Mathematics	12	8
Chemistry	11	9

M is the set of students who passed the Mathematics test.
C is the set of students who passed the Chemistry test.
x is the number of students who passed both tests.

a) Write 3 expressions **in terms of x** to complete the Venn diagram.

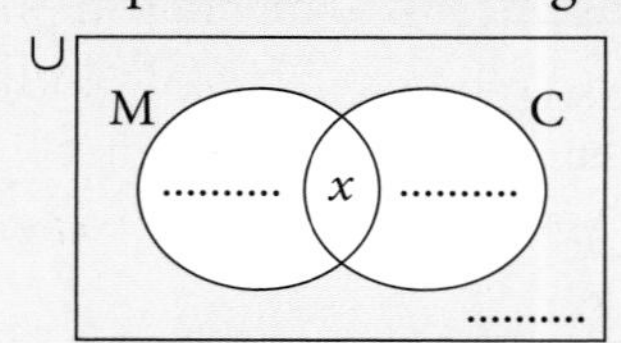

[3]

b) Two pupils failed both Mathematics and Chemistry.
Find the value of x, the number of students who passed both tests. [2]

Cambridge IGCSE International Mathematics 0607 Paper 21 Q10 May/June 2011

3. $U = \{x \mid 0 < x \leqslant 12, x \in \mathbb{Z}\}$
A = {multiples of 3} B = {factors of 30}
$C = \{x \mid 6 \leqslant x \leqslant 11, x \in \mathbb{Z}\}$

a) List the elements of the sets. [3]

A = { .... }

B = { .... }

C = { .... } [3]

b)

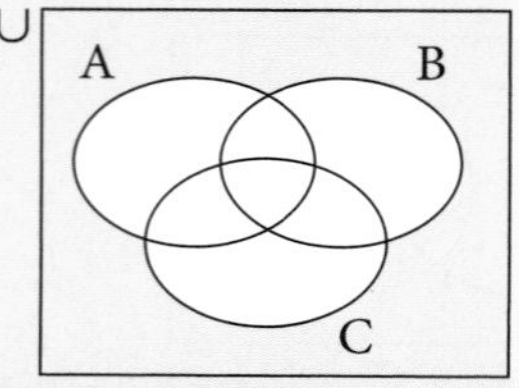

Put the 12 elements of U in the correct regions of the Venn diagram. [2]

c) Complete the following statements.

i) $A \cap B$ = { .... } [1]

ii) $A \cup C$ = { .... } [1]

iii) $(A \cup C) \cap B$ = { .... } [1]

iv) B' = { .... } [1]

v) $n(A \cap B \cap C)'$ = ... [1]

Cambridge IGCSE International Mathematics 0607 Paper 4 Q3 May/June 2010

4.

U = {Prime numbers less than 20}
A = {factors of 12}
B = {factors of 70}
C = {factors of 91}

a) List the 8 elements of set U.
(1 is **not** a prime number.) [1]

b) Write all the elements of U in the correct parts of the Venn diagram above. [3]

c) List the elements of $(B \cup C)'$. [1]

d) Write down the value of $n((B \cup C) \cap A')$. [1]

e) On the Venn diagram, shade the region $B \cap A' \cap C'$. [1]

Cambridge IGCSE International Mathematics 0607 Paper 4 Q9 October/November 2010

13 Probability

Blaise Pascal (1623–62) suffered the most appalling ill-health throughout his short life. He is best known for his work with Fermat on probability. Pascal's work on probability became of enormous importance and showed for the first time that absolute certainty is not a necessity in mathematics and science. He also studied physics, but his last years were spent in religious meditation and illness.

10.1 Probability $P(A)$ as a fraction, decimal or percentage
Significance of its value

10.2 Relative frequency as an estimate of probability

10.3 Expected frequency of occurrences

10.4 Combining events:
the addition rule $P(A \text{ or } B) = P(A) + P(B)$ (mutually exclusive)
the multiplication rule $P(A \text{ and } B) = P(A) \times P(B)$ (independent)

10.5 Tree diagrams including successive selection with or without replacement

10.6 Probabilities from Venn diagrams and tables

13.1 Probability

We can sometimes work out the probability of an event such as tossing a coin or rolling a dice. There are also many events which occur where it is very difficult (but not impossible) to estimate the probability of the event happening.

All major airlines regularly overbook aircraft because they can usually predict with accuracy the probability that a certain number of passengers will fail to arrive for the flight.

Relative frequency

Most people start to work out the probability of an 'event', like rolling a 3 on a dice by using equally likely outcomes. This is sometimes called the principle of **symmetry** (not to be confused with 'line symmetry' or 'rotational symmetry'). There are six equally likely outcomes so the probability of rolling a 3 is $\frac{1}{6}$.

> **Example 1**
>
> A 'trial' could be rolling a fair dice and a 'success' will be 'rolling a 4'.
>
> The probability of rolling a 4 is $\frac{1}{6}$

Suppose a 'trial' can have n equally likely results and suppose that a 'success' can occur in s ways (from the n). Then the probability of a 'success' $= \frac{s}{n}$.

- If an event *cannot* happen the probability of it occurring is 0.
- If an event is *certain* to happen the probability of it occurring is 1.
- All probabilities lie between 0 and 1.
  You write probabilities using fractions or decimals.

In many situations the principle of symmetry cannot be applied.

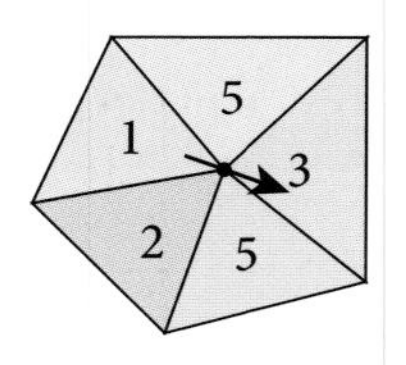

- It is not easy to work out the probability of spinning a 5 on the spinner shown, which has unequal sectors.
- When a drawing pin is dropped onto a hard surface, it will come to rest either point up or point down. You might think that the probability of the pin landing point up is $\frac{1}{2}$ because it must land either point up or point down. But a drawing pin is not a regular shape like a coin or a dice. Often the best way to find an estimate of the probability of an event, which cannot be predicted, is to perform an experiment.

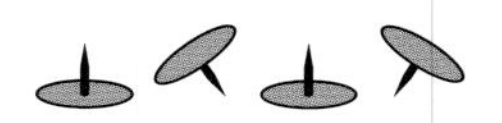

Experiment: 'Spinning a five'

A spinner, with unequal sectors, is spun 10 times and the number of 5s obtained is recorded. This procedure is repeated several times and the results are recorded in a table.

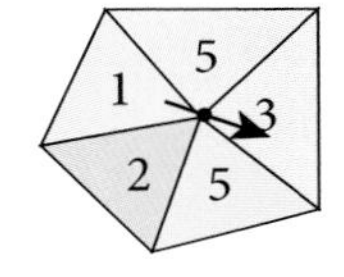

Relative frequency of spinning a 5 = $\left(\dfrac{\text{number of 5s obtained so far}}{\text{number of trials so far}}\right)$

For simplicity use ✓ for 'spinning a 5' and × for 'not spinning a 5'.

Results	Number of 5 s obtained so far	Number of trials so far	Relative frequency of spinning a 5
✓× × ✓× × × × × ×	2	10	= 0.2
× ✓ ✓ ✓ × ✓ ✓ ✓ ×✓	9	20	= 0.45
✓ × × ✓ ✓ × ✓ × × ✓	14	30	= 0.467(3 d.p.)
...	...	...	...

The experiment is continued until the number in the relative frequency column settles down to a fairly constant value.

A graph shows the progress of the experiment very clearly.

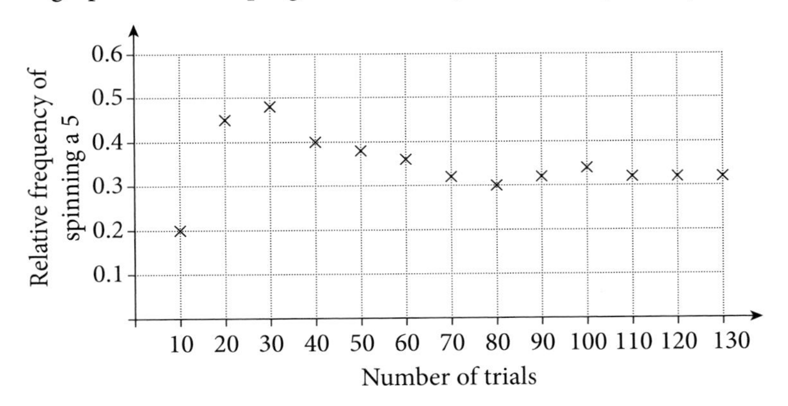

The graph may start by going up and down but after many trials the relative frequency settles down to a fairly constant number.

We can take the **relative frequency** of an event occurring as an estimate of the **probability** of that event occurring. The estimate improves as the number of trials is increased.

Exercise 13.1

1. Karim randomly selects a card from a pack and notes whether it is a Heart, Spade, Diamond or Club.
   Here are his results:

   S H D S S C H C C S D H H C S

   C H C D S D H C D S H S S C H

   a) What was the relative frequency of selecting a Heart?

   b) What was the relative frequency of selecting a Diamond?

2. In an experiment Tom drops 12 drawing pins onto a hard floor. He does the experiment 10 times and counts how many pins land 'point up'. His results were as follows.

Number of the 12 drawing pins that landed 'point up'									
3	5	6	2	4	7	3	3	4	5

   a) Use Tom's data to work out the probability that a *single* drawing pin will land point up.

   b) Tom continues the experiment until he has dropped the 12 drawing pins 100 times.
   About how many drawing pins in total would you expect to land point up?

3. Four friends are using a spinner for a game and they wonder if it is perfectly fair. They each spin the spinner several times and record the results.

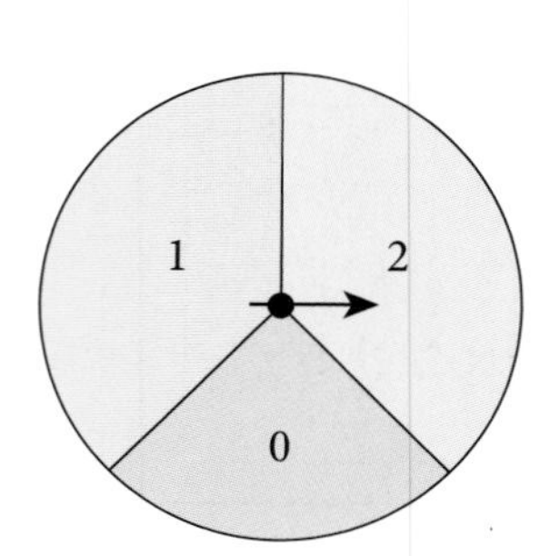

Name	**Number of spins**	**Results**		
		0	**1**	**2**
Andrea	30	12	12	6
Kamil	100	31	49	20
Bo	300	99	133	68
Anja	150	45	73	32

   a) Whose results are most likely to give the best estimate of the probability for getting each number?

   b) Make a table and collect together all the results. Use the table to decide whether you think the spinner is biased or unbiased.

   c) Use the results to work out the probability of the spinner getting a 2.

4. **Use your GDC**

Press MENU 5:Probability | 4:Random | 1:Number and press 1 ENTER (TI-Nspire)

Press MATH PRB 1:Rand ENTER (TI-84 Plus)

Press OPTN F6 OPTN F3 PROB F4 RAND F1 Ran# EXE (Casio)

You will generate a random number between 0.000... and 0.999... Take the first three digits to simulate tossing three coins. We could say any odd digit is a tail and any even digit is a head. So the number 0.346... represents THH.

Press the ENTER or EXE keys repeatedly to simulate tossing of three coins 100 or 200 times and work out the relative frequency of getting three heads.

Compare your result with the value that you would expect to get theoretically.

TI-Nspire	TI84-Plus	Casio
rand(1) {0.514702}	rand .8555803143	Ran# 0.2616534177
TTH	HTT	HHT

5. Find the relative frequency of a drawing pin landing pointing up.
Drop a drawing pin (or perhaps 10 drawing pins) onto a hard surface and count the number of times it lands point up. Record the results in groups of 10 trials in a table. Use ✓ for 'point up' and × for 'point down'.

Draw a graph of relative frequency against number of trials, and state the relative frequency of the event occurring after a large number (say 200) trials.
You could combine your results with other people in the class.

Working out the probability of an event occurring

Example 2

The numbers 1 to 20 are each written on a card.
The 20 cards are mixed together.
One card is chosen at random from the pack.
Find the probability that the number on the card is:

a) even **b)** a factor of 24 **c)** prime.

We will use '$p(x)$' to mean 'the probability of x'.

a) $p(\text{even}) = \frac{10}{20} = \frac{1}{2}$

b) $p(\text{factor of 24}) = p(1, 2, 3, 4, 6, 8, 12) = \frac{7}{20}$

c) $p(\text{prime}) = p(2, 3, 5, 7, 11, 13, 17, 19) = \frac{8}{20} = \frac{2}{5}$

In each case, we have counted the number of ways in which a 'success' can occur and divided by the number of possible results of a 'trial'.

Example 3

A black die and a white die are thrown at the same time. Display all the possible outcomes. Find the probability of obtaining:

a) a total of 5,

b) a total of 11,

c) a 2 on the black die and a 6 on the white die.

It is convenient to display all the possible outcomes on a grid.

There are 36 possible outcomes, shown where the lines cross.

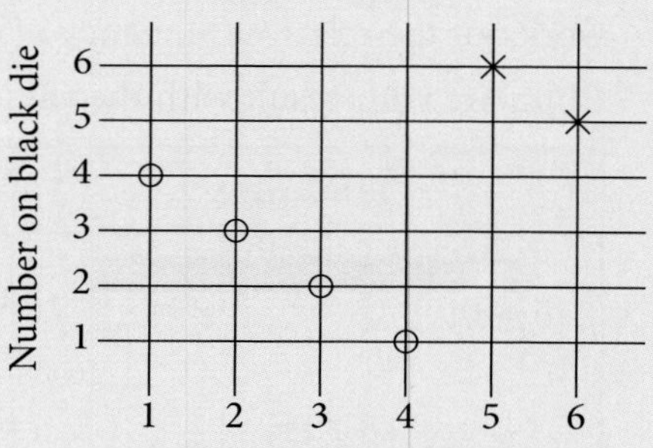

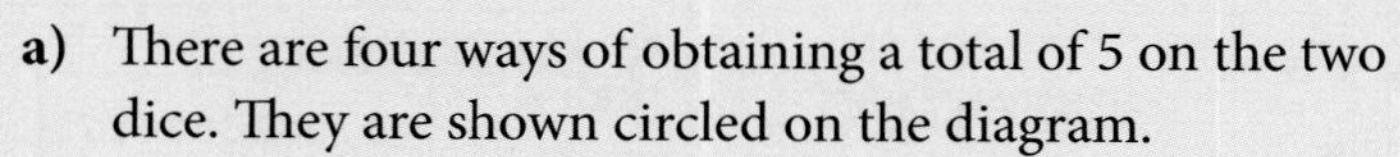

a) There are four ways of obtaining a total of 5 on the two dice. They are shown circled on the diagram.

$\therefore$ Probability of obtaining a total of 5 $= \frac{4}{36}$

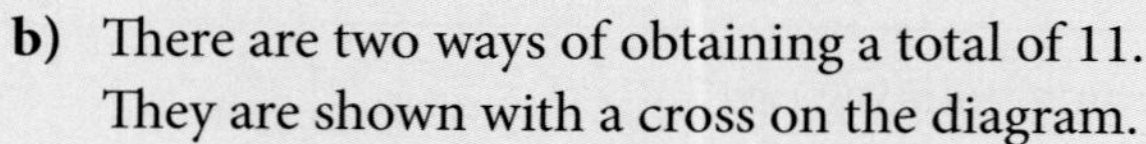

b) There are two ways of obtaining a total of 11. They are shown with a cross on the diagram.

$\therefore$ p (total of 11) $= \frac{2}{36} = \frac{1}{18}$

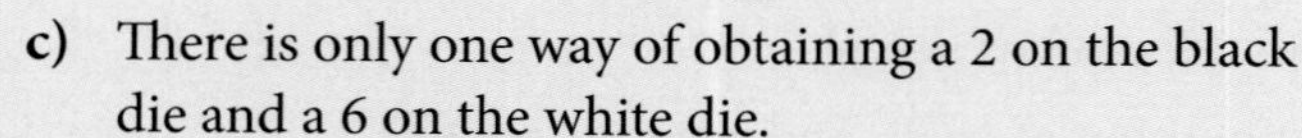

c) There is only one way of obtaining a 2 on the black die and a 6 on the white die.

$\therefore$ p (2 on black and 6 on white) $= \frac{1}{36}$

Exercise 13.2

In this exercise, all dice are normal cubic dice with faces numbered 1 to 6.

1. A fair die is thrown once. Find the probability of obtaining

a) a 6 **b)** an even number

c) a number greater than 3 **d)** a 3 or a 5.

2. The two sides of a coin are known as 'head' and 'tail'. A 10c and a 5c coin are tossed at the same time. List all the possible outcomes. Find the probability of obtaining

a) two heads **b)** a head and a tail.

3. A bag contains 6 red balls and 4 green balls.

a) Find the probability of selecting at random

i) a red ball **ii)** a green ball.

b) One red ball is removed from the bag. Find the new probability of selecting at random

i) a red ball **ii)** a green ball.

4. One letter is selected at random from the word UNNECESSARY. Find the probability of selecting

   a) an R
   b) an E
   c) an O
   d) a C

5. Three coins are tossed at the same time. List all the possible outcomes. Find the probability of obtaining

   a) three heads
   b) two heads and one tail
   c) no head
   d) at least one head.

6. A bag contains 10 red balls, 5 blue balls and 7 green balls. Find the probability of selecting at random

   a) a red ball
   b) a green ball
   c) a blue *or* a red ball
   d) a red *or* a green ball.

7. Cards with the numbers 2 to 101 are placed in a hat. Find the probability of selecting

   a) an even number
   b) a number less than 14
   c) a square number
   d) a prime number less than 20.

8. A red die and a blue die are thrown at the same time. List all the possible outcomes in a systematic way. Find the probability of obtaining

   a) a total of 10
   b) a total of 12
   c) a total less than 6
   d) the same number on both dice
   e) a total more than 9.

   What is the most likely total?

9. A die is thrown; when the result has been recorded, the die is thrown a second time. Display all the possible outcomes of the two throws. Find the probability of obtaining

   a) a total of 4 from the two throws
   b) a total of 8 from the two throws
   c) a total between 5 and 9 inclusive from the two throws
   d) a number on the second throw which is double the number on the first throw
   e) a number on the second throw which is four times the number on the first throw.

10. Find the probability of the following:

   a) throwing a number less than 8 on a single die
   b) obtaining the same number of heads and tails when five coins are tossed
   c) selecting a square number from the set

      A = {4, 9, 16, 25, 36, 49}

   d) selecting a prime number from the set A.

11. Four coins are tossed at the same time. List all the possible outcomes in a systematic way. Find the probability of obtaining:

a) two heads and two tails b) four tails

c) at least one tail d) three heads and one tail.

Remember

'Head' and 'tail' are the two sides of a coin.

12. Cards numbered 1 to 1000 were put in a box. Ali selects a card at random. What is the probability that Ali selects a card containing at least one 3?

13. One ball is selected at random from a bag containing 12 balls of which x are white.

a) What is the probability of selecting a white ball?

When a further 6 white balls are added the probability of selecting a white ball is doubled.

b) Find x.

14. Two dice and two coins are thrown at the same time. Find the probability of obtaining

a) two heads and a total of 12 on the dice

b) a head, a tail and a total of 9 on the dice

c) two tails and a total of 3 on the dice.

What is the most likely outcome?

15. A red, a blue and a green die are all thrown at the same time. Display all the possible outcomes in a suitable way. Find the probability of obtaining:

a) a total of 18 on the three dice b) a total of 4 on the three dice

c) a total of 10 on the three dice d) a total of 15 on the three dice

e) a total of 7 on the three dice f) the same number on each die.

13.2 Mutually exclusive and independent events

Two events are **mutually exclusive** if they cannot occur at the same time, e.g. selecting an even number or selecting a 1 from a set of numbers.

- The 'OR' rule:

  For mutually exclusive events A and B

  $p(\text{A or B}) = p(\text{A}) + p(\text{B})$

Two events are **independent** if the occurrence of one event is unaffected by the occurrence of the other, e.g. obtaining a 'head' on one coin, and a 'tail' on another coin when the coins are tossed at the same time.

- The 'AND' rule:

  $p(\text{A } \textit{and} \text{ B}) = p(\text{A}) \times p(\text{B})$

where $p(\text{A})$ = probability of A occurring, etc. This is the **multiplication law**.

Example 4

One ball is selected at random from a bag containing 5 red balls, 2 yellow balls and 4 white balls. Find the probability of selecting a red ball or a white ball.

The two events are mutually exclusive.

p (red ball *or* white ball) = p(red) + p(white)

$$= \frac{5}{11} + \frac{4}{11}$$

$$= \frac{9}{11}$$

Example 5

A fair coin is tossed and a fair die is rolled. Find the probability of obtaining a head and a 6.

The two events are independent.

p(head *and* 6) = p(head) × p(6)

$$= \frac{1}{2} \times \frac{1}{6}$$

$$= \frac{1}{12}$$

Exercise 13.3

1. A coin is tossed and a die is thrown. Write down the probability of obtaining

a) a head on the coin

b) an odd number on the die

c) a head on the coin and an odd number on the die.

2. A ball is selected at random from a bag containing 3 red balls, 4 black balls and 5 green balls. The first ball is replaced and a second is selected. Find the probability of obtaining

a) two red balls **b)** two green balls.

3. The letters of the word INDEPENDENT are written on individual cards are put into a box. A card is selected and then replaced and then a second card is selected. Find the probability of obtaining

a) the letter P twice **b)** the letter E twice.

4. Three coins are tossed and two dice are thrown at the same time. Find the probability of obtaining

a) three heads and a total of 12 on the dice

b) three tails and a total of 9 on the dice.

5. When a golfer plays any hole, he will take 3, 4, 5, 6, or 7 strokes with probabilities of $\frac{1}{10}, \frac{1}{5}, \frac{2}{5}, \frac{1}{5}$ and $\frac{1}{10}$ respectively.
He never takes more than 7 strokes. Find the probability of the following events:
   a) scoring 4 on each of the first three holes
   b) scoring 3, 4 and 5 (in that order) on the first three holes
   c) scoring a total of 28 for the first four holes
   d) scoring a total of 10 for the first three holes
   e) scoring a total of 20 for the first three holes.
6. A coin is biased so that it shows heads with a probability of $\frac{2}{3}$.
The same coin is tossed three times. Find the probability of obtaining
   a) two tails on the first two tosses **b)** a head, a tail and a head (in that order)
   c) two heads and one tail (in any order).

13.3 Tree diagrams

Example 6

A bag contains 5 red balls and 3 green balls. A ball is drawn at random and then replaced. Another ball is drawn.
What is the probability that both balls drawn are green?

The branch marked * involves the selection of a green ball twice.
The probability of this event is obtained by simply multiplying the fractions on the two branches.

$\therefore \quad p(\text{two green balls}) = \frac{3}{8} \times \frac{3}{8} = \frac{9}{64}$

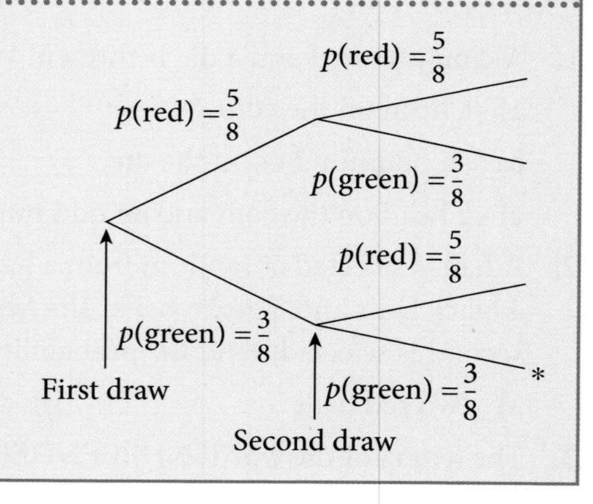

Example 7

A bag contains 5 red balls and 3 green balls. A ball is selected at random and not replaced. A second ball is then selected. Find the probability of selecting:

a) two green balls **b)** one red ball and one green ball.

a) $p(\text{two green balls}) = \frac{3}{8} \times \frac{2}{7}$

$= \frac{3}{28}$

b) $p(\text{one red, one green})$

$= \left(\frac{5}{8} \times \frac{3}{7}\right) + \left(\frac{3}{8} \times \frac{5}{7}\right)$

$= \frac{15}{28}$

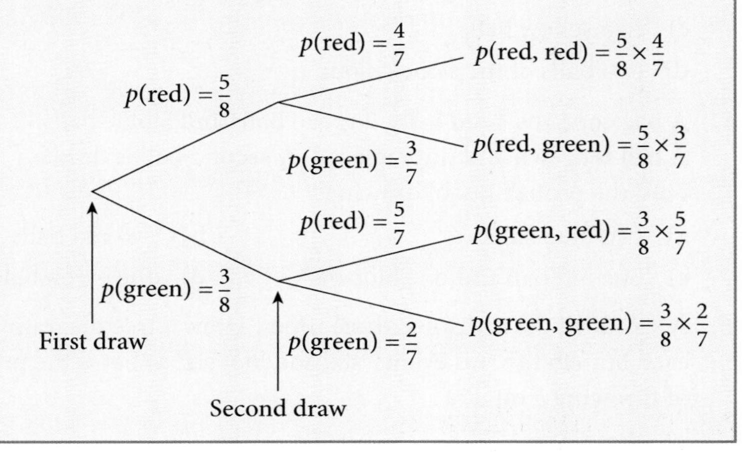

Exercise 13.4

1. A bag contains 10 discs; 7 are black and 3 white. A disc is selected, and then replaced. A second disc is selected. Copy and complete the tree diagram showing all the probabilities and outcomes.

   Find the probability of the following:

   a) both discs are black

   b) both discs are white.

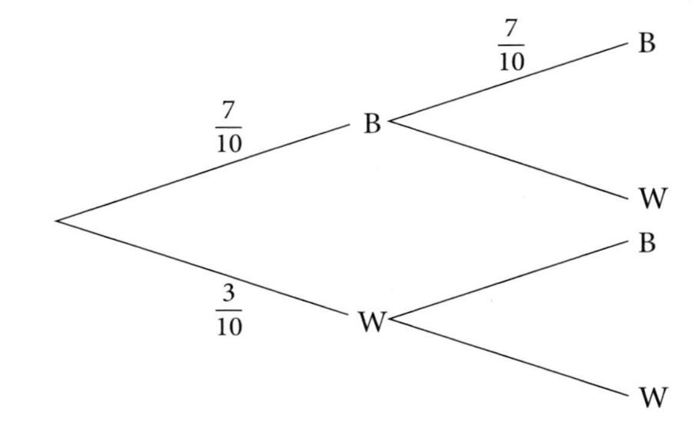

2. A bag contains 5 red balls and 3 green balls. A ball is drawn and then replaced before a ball is drawn again. Draw a tree diagram to show all the possible outcomes. Find the probability that:

   a) two green balls are drawn

   b) the first ball is red and the second is green.

3. A bag contains 7 green discs and 3 blue discs. A disc is drawn and *not* replaced. A second disc is drawn. Copy and complete the tree diagram.

   Find the probability that:

   a) both discs are green

   b) both discs are blue.

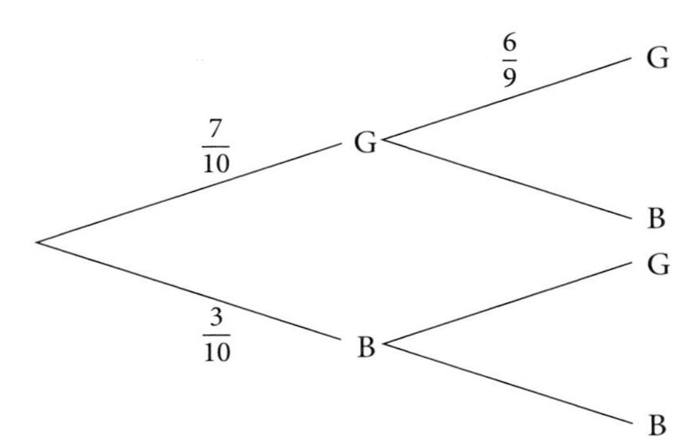

4. A bag contains 5 red balls, 3 blue balls and 2 yellow balls. A ball is drawn and not replaced. A second ball is drawn. Find the probability of drawing

   a) two red balls

   b) one blue ball and one yellow ball

   c) two yellow balls

   d) two balls of the same colour.

5. A bag contains 4 red balls, 2 green balls and 3 blue balls. A ball is drawn and not replaced. A second ball is drawn. Find the probability of drawing:

   a) two blue balls **b)** two red balls

   c) one red ball and one blue ball **d)** one green ball and one red ball.

6. A six-sided die is thrown three times. Draw a tree diagram, showing at each branch the two events 'six' and 'not six'. What is the probability of throwing a total of

   a) three sixes **b)** no sixes

   c) one six **d)** at least one six (use part **b**).

7. A bag contains 6 red marbles and 4 blue marbles. A marble is drawn at random and not replaced. Two further draws are made, again without replacement. Find the probability of drawing

   a) three red marbles **b)** three blue marbles

   c) no red marbles **d)** at least one red marble.

8. When a cutting is taken from a geranium the probability that it grows is $\frac{3}{4}$. Three cuttings are taken. What is the probability that

   a) all three grow **b)** none of them grow?

9. A die has its six faces marked 0, 1, 1, 1, 6, 6. Two of these dice are thrown together and the total score is recorded. Draw a tree diagram.

   a) How many different totals are possible?

   b) What is the probability of obtaining a total of 7?

10. A coin is biased so that the probability of a head is $\frac{3}{4}$. Find the probability that, when tossed three times, it shows

    a) three tails **b)** two heads and one tail

    c) one head and two tails **d)** no tails.

    Write down the sum of the probabilities in **a)**, **b)**, **c)** and **d)**.

11. A teacher decides to award exam grades A, B or C by a new method. Out of 20 children, three are to receive As, five Bs and the rest Cs. She writes the letters, A, B, and C on 20 pieces of paper and invites the pupils to draw their exam result, going through the class in alphabetical order. Find the probability that:

a) the first three pupils all get grade A

b) the first three pupils all get grade B

c) the first three pupils all get different grades

d) the first four pupils all get grade B

(Do not cancel down the fractions.)

12. The probability that an amateur golfer actually hits the ball is $\frac{1}{10}$. If four separate attempts are made, find the probability that the ball will be hit

a) four times **b)** at least twice **c)** not at all.

13. A box contains x milk chocolates and y plain chocolates. Two chocolates are selected at random. Find, in terms of x and y, the probability of choosing

a) a milk chocolate on the first choice

b) two milk chocolates

c) one of each sort

d) two plain chocolates.

14. If a hedgehog crosses a road before 7.00 am, the probability of being run over is $\frac{1}{10}$. After 7.00 am, the corresponding probability is $\frac{3}{4}$.

The probability of the hedgehog waking up early enough to cross before 7.00 am, is $\frac{4}{5}$.

What is the probability of the following events:

a) the hedgehog waking up too late to reach the road before 7.00 am

b) the hedgehog waking up early and crossing the road in safety

c) the hedgehog waking up late and crossing the road in safety

d) the hedgehog waking up early and being run over

e) the hedgehog crossing the road in safety.

15. Bag A contains 3 red balls and 3 blue balls. Bag B contains 1 red ball and 3 blue balls. A ball is taken at random from bag A and placed in bag B. A ball is then chosen from bag B. What is the probability that the ball taken from B is red?

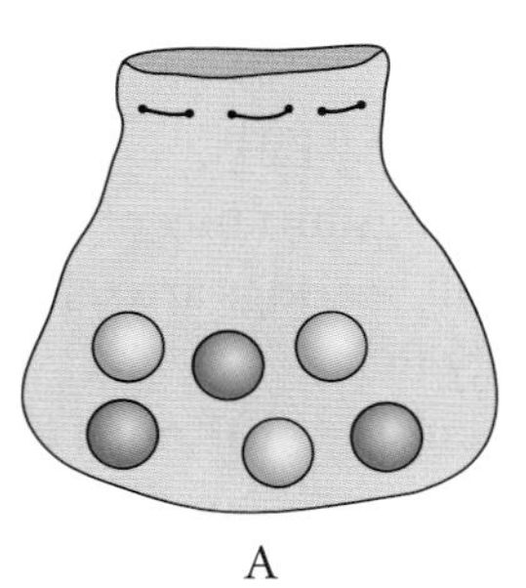

A

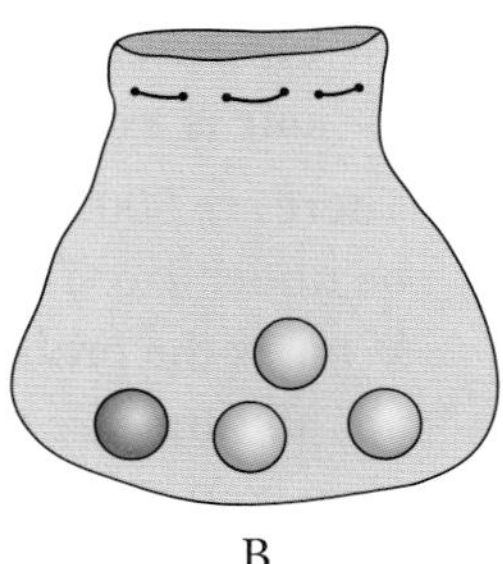

B

16. On a Monday or a Thursday, Ceren paints a 'masterpiece' with a probability of $\frac{1}{5}$. On any other day, the probability of producing a masterpiece is $\frac{1}{100}$. Find the probability that on one day chosen at random, she will paint a masterpiece.

17. Two dice, each with four faces marked 1, 2, 3 and 4, are thrown together.

a) What is the most likely total score on the faces pointing downwards?

b) What is the probability of obtaining this score on three successive throws of the two dice?

18. In the Venn diagram, $\mathcal{U}$ = {pupils in a class of 15}, G = {girls}, S = {swimmers}, F = {pupils who were born on a Friday}.
A pupil is chosen at random. Find the probability that the pupil:

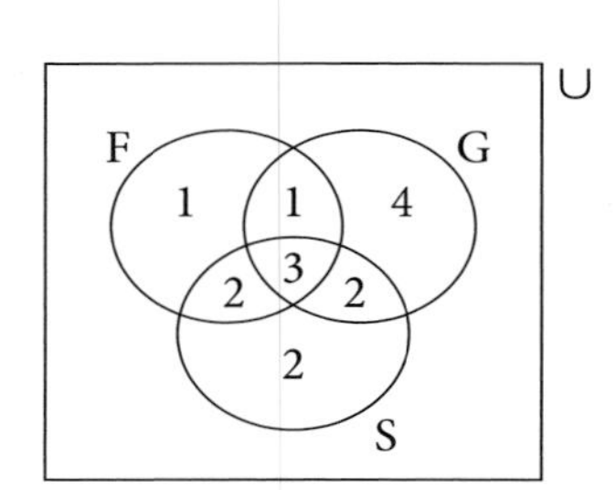

a) can swim **b)** is a girl swimmer

c) is a boy swimmer who was born on Friday.

Two pupils are chosen at random. Find the probability that

d) both are boys **e)** neither can swim

f) both are girls swimmers who were born on a Friday.

19. A bag contains 3 red, 4 white and 5 green balls. Three balls are selected without replacement. Find the probability that the three balls chosen are

a) all red **b)** all green

c) one of each colour.

If the selection of the three balls was carried out 1100 times, how often would you expect to choose

d) three red balls **e)** one of each colour?

20. There are 1000 components in a box of which 10 are known to be defective. Two components are selected at random.
What is the probability that

a) both are defective

b) neither are defective

c) just one is defective?

(Do *not* simplify your answers.)

21. There are 10 boys and 15 girls in a class. Two children are chosen at random. What is the probability that

a) both are boys **b)** both are girls

c) one is a boy and one is a girl?

22. There are 500 ball bearings in a box of which 100 are known to be undersize. Three ball bearings are selected at random.
What is the probability that

'Undersize' means 'smaller than usual'.

a) all three are undersize **b)** none are undersize?

Give your answers as decimals correct to three significant figures.

23. There are 9 boys and 15 girls in a class. Three children are chosen at random. What is the probability that

a) all three are boys **b)** all three are girls

c) one is a boy and two are girls?

Give your answers as fractions.

Revision exercise 13A

1. One card is picked at random from a pack of 52. Find the probability that it is

a) the Jack of diamonds

b) a two of any suit **c)** a diamond

2. A bag contains 9 balls: 3 red, 4 white and 2 yellow.

a) Find the probability of selecting a red ball.

b) The 2 yellow balls are replaced by 2 white balls.

Find the probability of selecting a white ball.

3. Ella and Kurt play 'rock, paper, scissors' 18 times. How many times would you expect Ella to win?

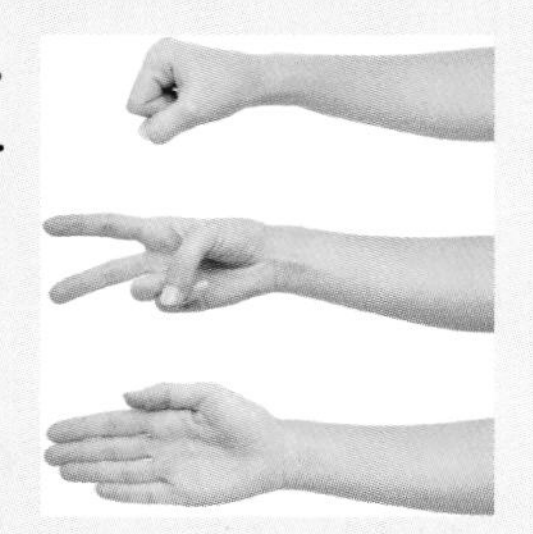

4. A fair dice is rolled 120 times. How many times would you expect to roll a two?

5. Neil played a card game with Phil. The cards were dealt so that both players received two cards. Neil's cards were a seven and a four. Phil's first card was a six.

Find the probability that Phil's second card was

a) a seven **b)** a King or a Queen of any suit.

6. The number of people visiting the Louvre Museum one day was 11 249. How many of these people would you expect to celebrate their birthdays on a Tuesday in the year 2021?

7. A coin is biased so that the probability of tossing a head is 56%.

a) What is the probability of tossing a tail with this coin?

b) How many tails would you expect when the coin is tossed 500 times?

8. One ball is selected from a bag containing n yellow balls, m red balls and 7 white balls. What is the probability of selecting a yellow ball?

9.

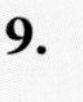

When two dice are thrown simultaneously, what is the probability of obtaining the same number on both dice?

10. A bag contains 20 discs of equal size of which 12 are red, x are blue and the rest are white.

a) If the probability of selecting a blue disc is $\frac{1}{4}$, find x.

b) A disc is drawn and then replaced. A second disc is drawn. Find the probability that neither disc is red.

11. Three dice are thrown. What is the probability that none of them shows a 1 or a 6?

12. A coin is tossed four times. What is the probability of obtaining at least three heads?

13. A bag contains 8 balls of which 2 are red and 6 are white. A ball is selected and not replaced. A second ball is selected. Find the probability of obtaining

    a) two red balls **b)** two white balls

    c) one ball of each colour.

14. A bag contains x green discs and 5 blue discs. A disc is selected. A second disc is drawn. Find, in terms of x, the probability of selecting

    a) a green disc on the first draw

    b) a green disc on the first and second draws, if the first disc is replaced

    c) a green disc on the first and second draws, if the first disc is *not* replaced.

15. In a group of 20 people, 5 cannot swim. If two people are selected at random, what is the probability that neither of them can swim?

16. **a)** What is the probability of winning the toss in five consecutive hockey matches?

    b) What is the probability of winning the toss in all the matches in the FA cup from the first round to the final (i.e. 8 matches)?

17. Mr and Mrs Singh have three children. What is the probability that

    a) all the children are boys

    b) there are more girls than boys?

    (Assume that a boy is as likely as a girl.)

18. The probability that it will rain today is $\frac{1}{6}$. If it is dry today, the probability that it will rain tomorrow is $\frac{1}{8}$. What is the probability that both today and tomorrow will be dry?

19. Two dice are thrown. What is the probability that the *product* of the numbers on top is:

    a) 12, **b)** 4, **c)** 11?

20. The probability of snow on 1st January is $\frac{1}{20}$. What is the probability that snow will fall on the next three 1st January?

Examination exercise 13B

1. In Hurghada the probability that the sun will shine on any day is 0.8. If the sun shines, the probability Ahmed will go to the beach is 0.9. If the sun does not shine, the probability he will go to the beach is 0.5.

    a) Complete the tree diagram.

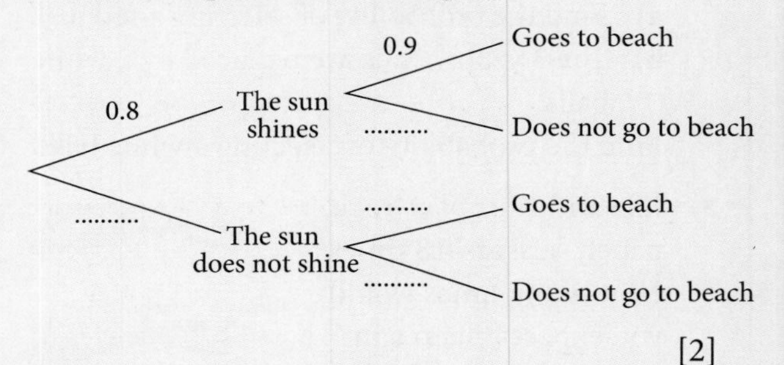

[2]

    b) Find the probability that Ahmed will go to the beach on a given day. [2]

Cambridge IGCSE International Mathematics 0607 Paper 2 Q10 October/November 2010

2. | 1 | 7 | 8 | 12 | 15 | 19 | 19 |
   |---|---|---|---|---|---|---|

    The diagram shows seven numbered cards.

    a) A card is chosen at random.

    i) Write down the probability that it is numbered 12 or 15. [1]

    ii) The probability that the number on the card is greater than x is $\frac{5}{7}$. Write down the value of x. [1]

b) Two cards are chosen at random, without replacement, from the seven numbered cards.

i) When the first card chosen is numbered 19, write down the probability that the second card is also numbered 19. [1]

ii)

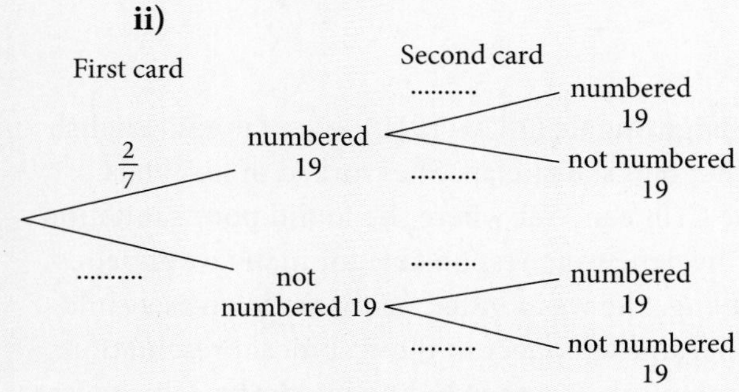

Complete the tree diagram, by writing the probabilities in the spaces. [2]

iii) Find the probability that both cards are numbered 19. [2]

iv) Find the probability that exactly one card is numbered 19. [3]

c) Cards are chosen at random, without replacement, from the seven numbered cards, until a card that is numbered 19 is chosen.
Find the probability that this happens with the third card. [2]

Cambridge IGCSE International Mathematics 0607 Paper 41 Q11 May/June 2011

3. A school bus picks up students at the town centre and takes them to the school.
On any day the probability that the bus is on time at the town centre is $\frac{5}{6}$.

a) Write down the probability that the bus is not on time at the town centre. [1]

b) If the bus is on time at the town centre, the probability that it is on time at the school is $\frac{7}{8}$.
If the bus is not on time at the town centre then the probability that it is on time at the school is $\frac{1}{4}$.

i) Draw a tree diagram and write the correct probability against each branch. [1]

ii) Calculate the probability that the bus is on time at the school. [3]

iii) Calculate the probability that the bus is never on time at the school in a week of 5 school days.
Give your answer as a decimal, correct to 2 significant figures. [2]

iv) There are 192 days in this school's year.
On how many days is the bus expected to be on time at the school? [1]

Cambridge IGCSE International Mathematics Paper 4 Q11 May/June 2010

14 Statistics

Florence Nightingale (1820–1910) was a famous English nurse, writer and statistician. She worked in hospitals during the Crimean War where she found poor sanitation and basic hygiene were responsible for many wounded soldiers dying. She was a gifted mathematician as a child and later became a pioneer in the graphical presentation of statistical information, including pie charts, which were not commonly used by others at that time. In 1859 she was elected the first female member of the Royal Statistical Society.

11.1 Reading and interpretation of graphs or tables of data

11.2 Discrete and continuous data

11.3 (Compound) bar chart, line graph, pie chart, stem and leaf diagram, scatter diagram

11.4 Mean, mode, median, quartiles and range from lists of discrete data
Mean, mode, median and range from grouped discrete data

11.5 Mean from continuous data

11.6 Histograms with frequency density on the vertical axis using continuous data (includes histograms with unequal class intervals)

11.7 Cumulative frequency table and curve
Median, quartiles, percentiles and inter-quartile range (read from curve)

11.8 Use of a graphics calculator to calculate mean, median, and quartiles for discrete data and mean for grouped data

11.9 Understanding and description of correlation (positive, negative or zero) with reference to a scatter diagram (the coefficient of correlation is not required)
Straight line of best fit (by eye) through the mean on a scatter diagram
Use a graphics calculator to find equation of linear regression

Syllabus objectives 11.8 and 11.9 are also covered in Chapter 1.

14.1 Reading and interpretation of data or graphs

The results of a statistical investigation are known as **data**.

Raw data are data in the form that they were collected, for example, the number of peas in 40 pods.

5	3	6	5	4	6	6	7	4	6
4	7	7	3	7	4	7	5	7	5
6	7	6	7	5	6	6	7	6	6
5	3	6	4	6	5	7	3	6	4

Data in this form are difficult to interpret.

Data can be presented in a variety of different ways that make it easier to interpret.

Frequency tables

This presentation is in the form of a tally chart. The tally provides a numerical value for the frequency of occurrence.

From the example above, there are 4 pea pods that contain 3 peas, 6 pods that contain 4 peas, etc.

Number of peas	Tally	Frequency (f)
3	||||	4
4	~~||||~~ |	6
5	~~||||~~ ||	7
6	~~||||~~ ~~||||~~ |||	13
7	~~||||~~ ~~||||~~	10

Display charts

You can use a pie chart, a bar chart or a frequency polygon to display data.

This pie chart illustrates the data about the number of peas in a pod.

Number of peas	Angle of pie chart
3	$\frac{4}{40} \times 360 = 36°$
4	$\frac{6}{40} \times 360 = 54°$
5	$\frac{7}{40} \times 360 = 63°$
6	$\frac{13}{40} \times 360 = 117°$
7	$\frac{10}{40} \times 360 = 90°$

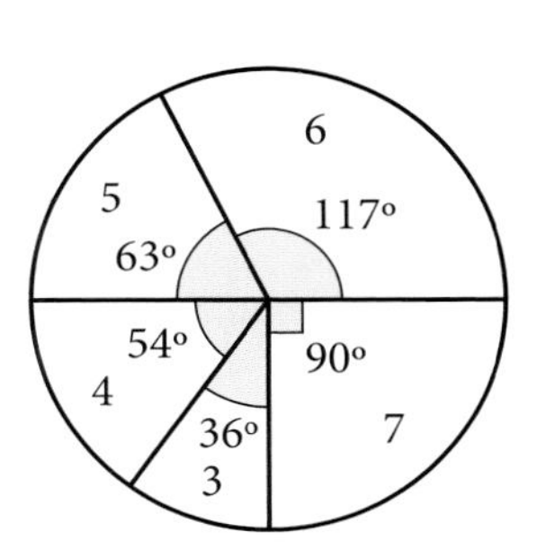

Here is the same data displayed in a bar chart and a frequency polygon.

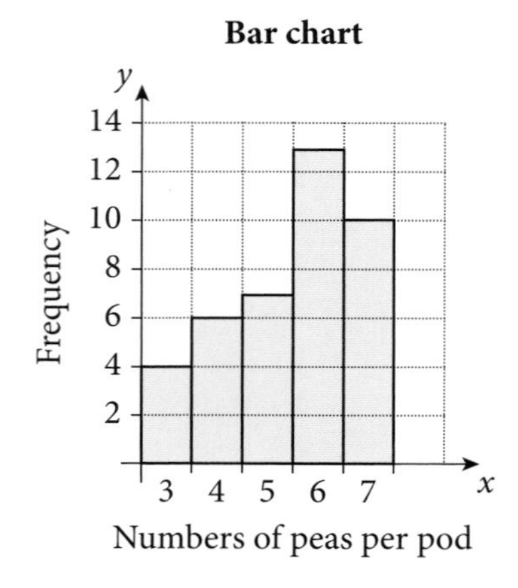

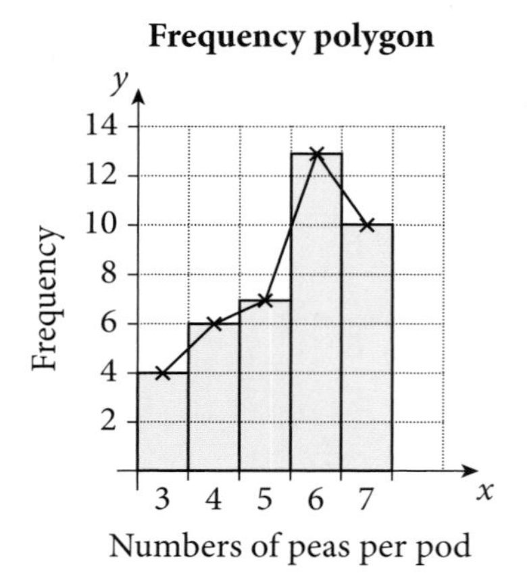

Note that in the bar chart and the frequency polygon, the vertical axis is always used to show frequency. It is the frequency that most clearly shows the mode, that is, that more pods contained 6 peas than any other number of peas.
In the frequency polygon, the boxes of the bar chart are replaced by a line joining the tops of their midpoints.

The example above relates to **discrete** data, that is, individually distinct number quantities. The next section is about **continuous** data, which are derived from measurements, for example, height, weight, age, time.

You collect discrete data by counting. You collect continuous data by measuring.

Continuous data

- Rounded data
  You round each measurement and record the frequency of this rounded quantity as for discrete data. For example, the lengths of 36 pea pods can be rounded to the nearest mm and then recorded as raw data – a pea measuring 59·2 mm is recorded as one that has a length of 59 mm.

  The table below shows rounded data: the lengths of 36 pea pods.

52	80	65	82	77	60	72	83	63
78	84	75	53	73	70	86	55	88
85	59	76	86	73	89	91	76	92
66	93	84	62	79	90	73	68	71

Grouped data

You group the measurements into classes with defined class boundaries. In this **grouped frequency table**, the data on lengths of pea pods are grouped into the classes shown in the left-hand column.

Length (l mm)	Tally	Frequency
$50 < l \leq 60$	||||	4
$60 < l \leq 70$	~~||||~~ |	6
$70 < l \leq 80$	~~||||~~ ~~||||~~ ||	12
$80 < l \leq 90$	~~||||~~ ~~||||~~	10
$90 < l \leq 100$	||||	4

So, the pea pod measuring 59·2 mm is one of 4 recorded in the class $50 < l \leq 60$.

Here are the grouped data displayed in a bar chart and a frequency polygon.

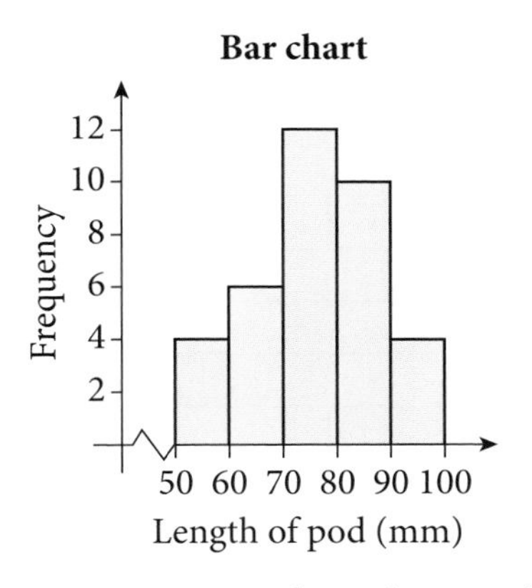

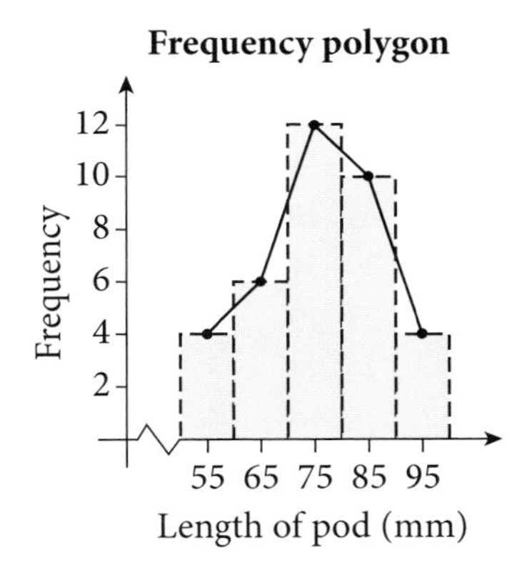

In the frequency polygon you mark the mid-points of the classes on the horizontal axis.

In some cases where there are lots of small classes, the midpoints of the frequency polygon are joined with a curve. The diagram is then called a **frequency curve**.

Exercise 14.1

1. Karine and Jackie intend to go skiing in February. They have information about the expected snowfall in February for two possible places.

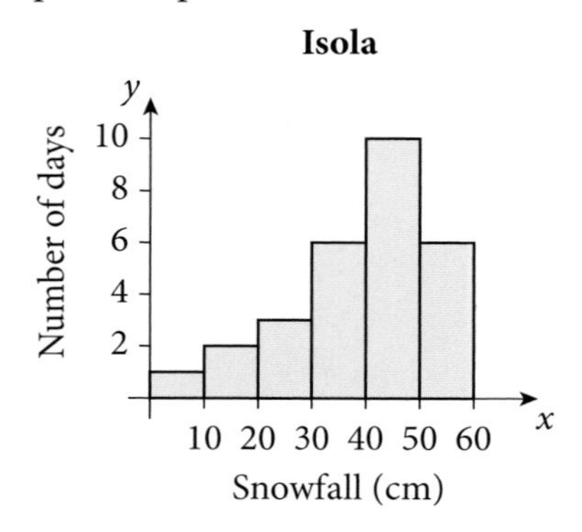

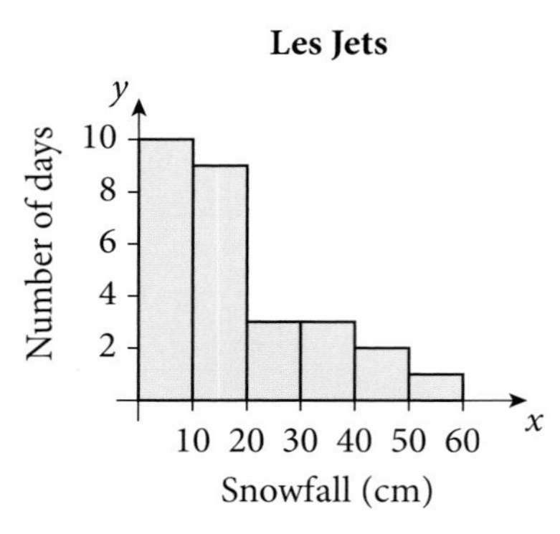

Decide where you think they should go. It doesn't matter where you decide, but you *must* say why, using the charts above to help you explain.

2. The chart shows information about people who use the Internet regularly.

   a) About what percentage of boys in the 5–9 age group used the Internet regularly?

   b) In what age groups did more women than men use the Internet?

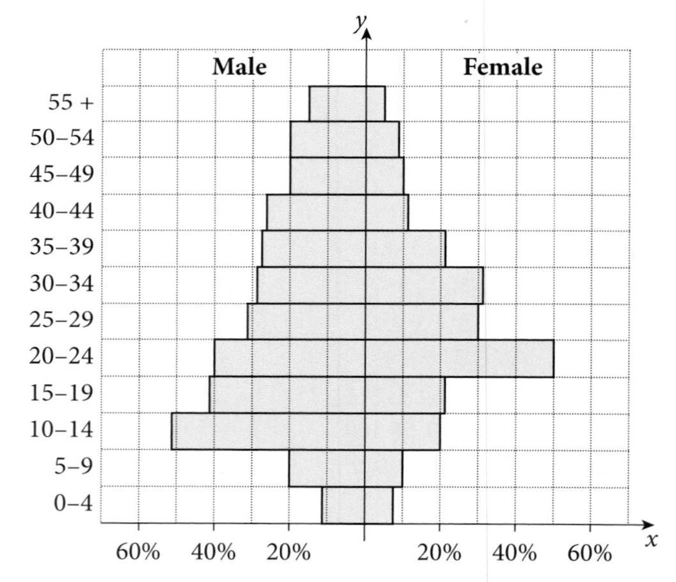

3. In a survey, Jake counted the number of people in 100 cars passing a set of traffic lights. Here are the results:

Number of people in car	0	1	2	3	4	5	6
Frequency	0	10	35	25	20	10	0

   a) Draw a bar chart to illustrate the data.

   b) On the same graph draw the frequency polygon.

Here the bar chart has been started. For frequency, use a scale of 1 cm for 5 units.

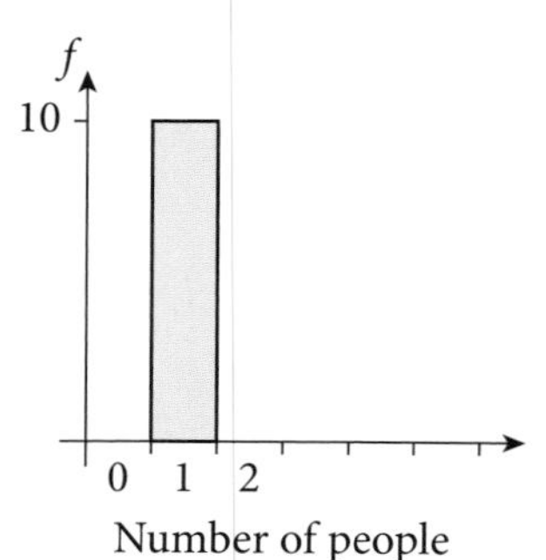

4. The diagram shows two frequency polygons giving the distribution of the weights of players in two different sports A and B.

   a) How many people played sport A?

   b) Comment on two differences between the two frequency polygons.

   c) Either for A or for B suggest a sport where you would expect the frequency polygon of weights to have this shape. Explain in one sentence why you have chosen that sport.

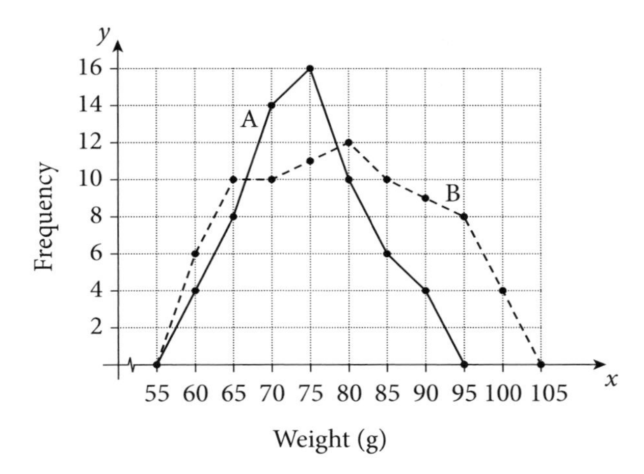

5. A scientist at an agricultural college is studying the effect of a new fertiliser for raspberries. She measures the heights of the plants and also the total weight of fruit collected. She does this for two sets of plants: one with the new fertiliser and one without it. Here are the frequency polygons:

------ with fertiliser

——— without fertiliser

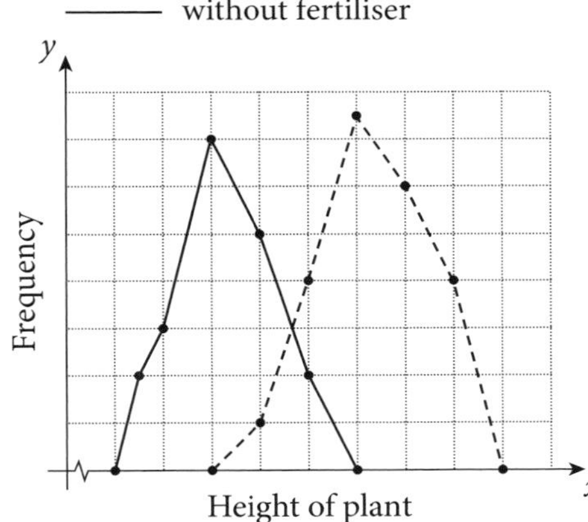

y

Frequency

Weight of raspberries

x

   a) What effect did the fertiliser have on the heights of the plants?

   b) What effect was there on the weights of fruit collected?

6. The diagram illustrates the production of apples in two countries. In what way could the pictorial display be regarded as misleading?

Brazil
470
thousand
tonnes

France
950
thousand
tonnes

7. The graph shows the performance of a company in the year in which a new manager was appointed. In what way is the graph misleading?

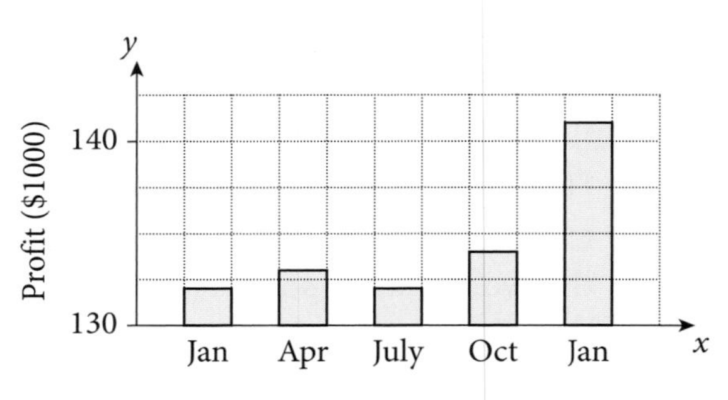

8. In one week Achim spends €120 on various items as shown in the pie chart.

   a) What fraction of his money did Achim spend on

i) food	**ii)** bus fare
iii) rent	**iv)** savings
v) entertainment	**vi)** clothes?

   b) How much of the €120 did Achim spend on

i) food	**ii)** fares
iii) rent	**iv)** savings?

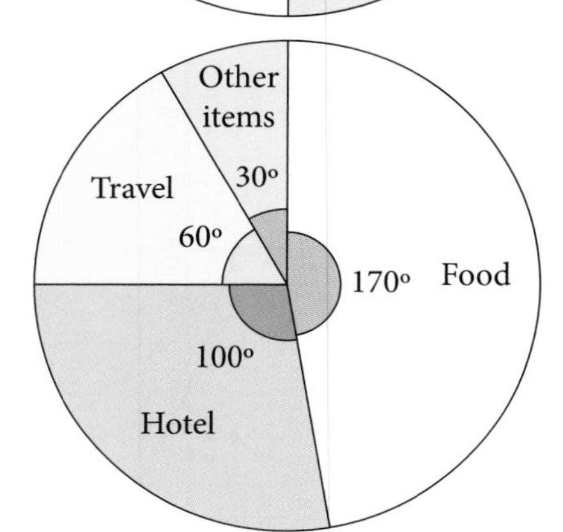

9. The total cost of a holiday was £900.
   The pie chart shows how this cost was made up.

   a) How much was spent on food?

   b) How much was spent on travel?

   c) How much was spent on the hotel?

   d) How much was spent on other items?

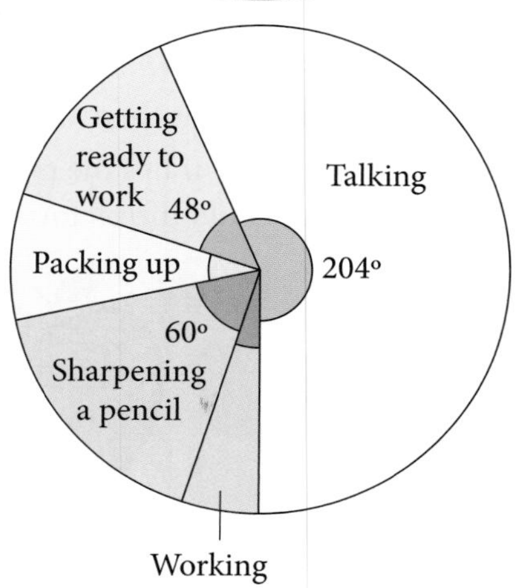

10. The pie chart shows how Maria spends her time in a maths lesson which lasts 60 minutes.

    a) How much time does Maria spend

    i) getting ready to work

    ii) talking

    iii) sharpening a pencil?

    b) Maria spends 3 minutes working. What is the angle on the pie chart for the time spent working?

11. A quantity of scrambled eggs is made using these ingredients.

Ingredient	eggs	milk	butter	cheese	salt/pepper
Mass	450 g	20 g	39 g	90 g	1 g

Calculate the angles on a pie chart corresponding to each ingredient.

12. The pie chart illustrates the sales of various brands of petrol.

a) What percentage of sales does Esso have?

b) If Texaco accounts for $12\frac{1}{2}\%$ of total sales, calculate the angles x and y.

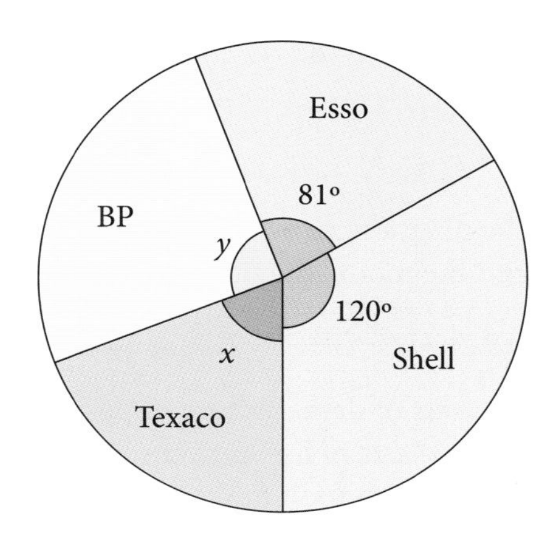

13. The cooking times for meals L, M and N are in the ratio $3 : 7 : x$. On a pie chart, the angle corresponding to L is 60°. Find x.

Stem-and-leaf diagrams

You can display data in groups in a **stem-and-leaf diagram**.

Here are the marks of 20 girls in a science test:

54	42	61	47	24	43	55	62	30	27
28	43	54	46	25	32	49	73	50	45

Put the marks into groups 20–29, 30–39, … 70–79.

Choose the tens digit as the 'stem' and the units as the 'leaf'.

The first four marks are shown [54, 42, 61, 47]

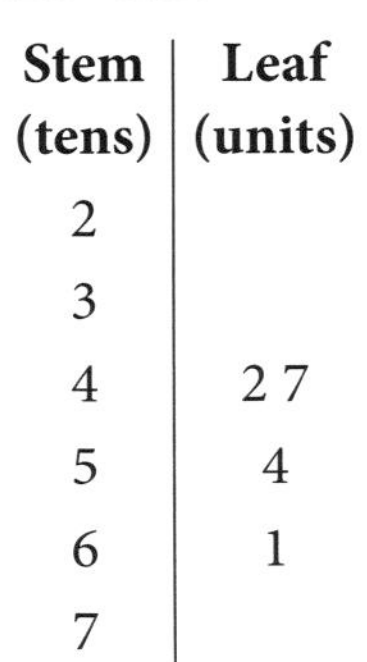

Stem (tens)	**Leaf (units)**
2	
3	
4	2 7
5	4
6	1
7	

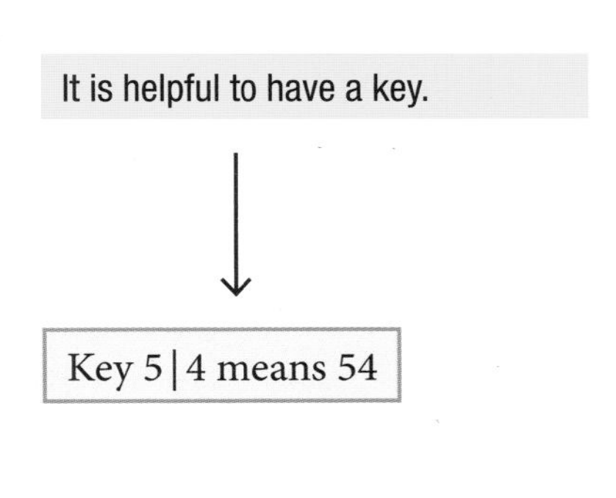

The complete diagram is below. The diagram on the right shows the leaves in numerical order:

Stem	Leaf
2	4 7 8 5
3	0 2
4	2 7 3 3 6 9 5
5	4 5 4 0
6	1 2
7	3

Stem	Leaf
2	4 5 7 8
3	0 2
4	2 3 3 5 6 7 9
5	0 4 4 5
6	1 2
7	3

The diagram shows the shape of the distribution. It is also easy to find the mode, the median and the range.

Back-to-back stem plots

Two sets of data can be compared using a **back-to-back stem plot**. The two sets of leaves share the same stem. Here are the marks of 20 boys who took the same science test as the girls:

33	55	63	74	20	35	40	67	21	38
51	64	57	48	46	67	44	59	75	56

Here are the boys' marks entered into a back-to-back stem plot:

Boys	Stem	Girls
1 0	2	4 5 7 8
8 5 3	3	0 2
8 6 4 0	4	2 3 3 5 6 7 9
9 7 6 5 1	5	0 4 4 5
7 7 4 3	6	1 2
5 4	7	3

Here is the key

Key (boys)
1|5 means 51

Key (girls)
5|4 means 54

You can see that the boys achieved higher marks than the girls in this test.

Exercise 14.2

1. The diagram shows the marks of 24 students in a test.

41	23	35	15	40	39	47	29
52	54	45	27	28	36	48	51
59	65	42	32	46	53	66	38

Stem	Leaf
1	
2	3
3	5
4	1
5	
6	

Key
2|3 means 23

a) Copy and complete the stem-and-leaf diagram. The first three entries are shown.

b) Write the range of the marks.

2. Draw a stem-and-leaf diagram for each set of data.

a)

24	52	31	55	40	37	58	61	25	46
44	67	68	75	73	28	20	59	65	39

b)

30	41	53	22	72	54	35	47
44	67	46	38	59	29	47	28

Stem	Leaf
2	
3	
4	
5	
6	
7	

3. Here is the stem-and-leaf diagram showing the masses, in kg, of some people in a lift.

a) Write the range of the masses

b) How many people were in the lift?

c) What is the median mass?

Stem (tens)	Leaf (units)
3	2 5
4	1 1 3 7 8
5	0 2 5 8
6	4 8
7	1
8	2

Key
3|2 means 32 kg

4. In this question the stem shows the units digit and the leaf shown the first digit after the decimal point.
Draw the stem and leaf diagram using these data.

2·4	3·1	5·2	4·7	1·4	6·2	4·5	3·3
4·0	6·3	3·7	6·7	4·6	4·9	5·1	5·5
1·8	3·8	4·5	2·4	5·8	3·3	4·6	2·8

Stem	Leaf
1	
2	
3	
4	
5	
6	

Key
3|7 means 3·7

a) What is the median?

b) Write the range.

5. Here is a back-to-back stem plot showing the pulse rates of several people.

a) How many men were tested?

b) What was the median pulse rate for the women?

c) Write a sentence to describe the main features of the data.

Men		Women
5 1	4	
7 4 2	5	3
8 2 0	6	2 1
5 2	7	4 4 5 8 9
2 6	8	2 5 7
4	9	2 8

Key (women)
5|3 means 53

Key (men)
1|4 means 41

Exercise 14.3

1. The bar chart shows the number of children playing various games on a given day.

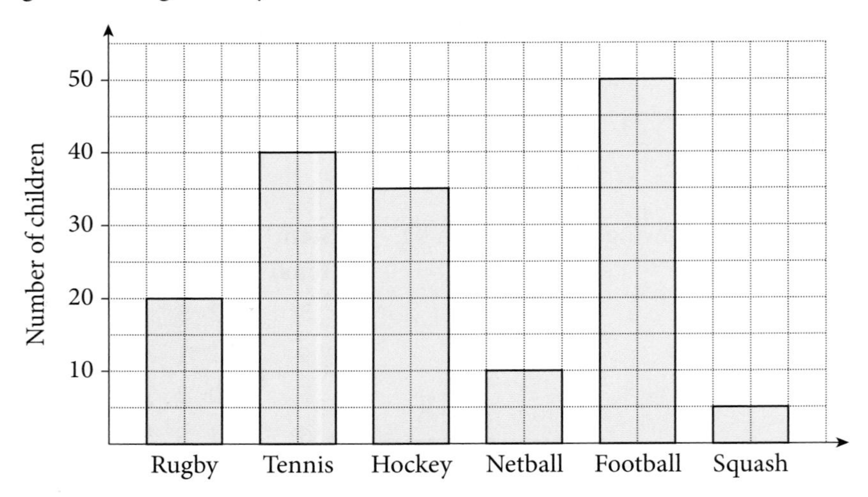

   a) Which game had the least number of players?

   b) What was the total number of children playing all the games?

   c) How many more footballers were there than tennis players?

2. The table shows the number of cars of different makes in a car park. Illustrate this data on a bar chart.

Make	Skoda	Renault	Saab	Kia	Subaru	Lexus
Number	14	23	37	5	42	18

3. The pie chart illustrates the values of various goods sold in one year by a certain shop. If the total value of the sales was $24 000, find the sales value of

   a) toys **b)** grass seed

   c) books **d)** food.

Food
Toys
45°
60°
Grass seed
90°
Books

4. The table shows the colours of a random selection of sweets. Calculate the angles on a pie chart corresponding to each colour.

Colour	red	green	blue	yellow	pink
Number	5	7	11	4	9

5. Calculate the angles on a pie chart corresponding to quantities A, B, C, D and E given in the tables.

a)

Quantity	A	B	C	D	E
Number	3	5	3	7	0

b)

Quantity	A	B	C	D	E
Mass	10 g	15 g	34 g	8 g	5 g

c)

Quantity	A	B	C	D	E
Length	7	11	9	14	11

6. A firm making office chairs sold its products in four countries:

5% were sold in Spain
15% were sold in France
15% were sold in Germany
65% were sold in the U.K.

What would be the angles on a pie chart drawn to represent this information?

7. The weights of A, B, C are in the ratio 2 : 3 : 4. Calculate the angles representing A, B, and C on a pie-chart.

8. The results of an opinion poll of 2000 people are represented on a pie chart. The angle corresponding to 'did not know' is 18°. How many people in the sample 'did not know'?

9.

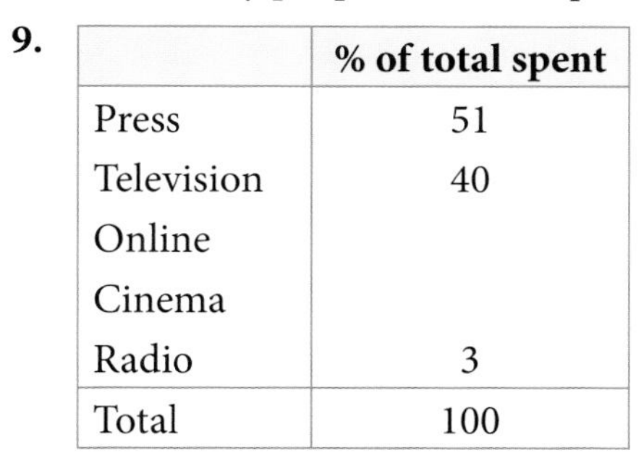

	% of total spent
Press	51
Television	40
Online	
Cinema	
Radio	3
Total	100

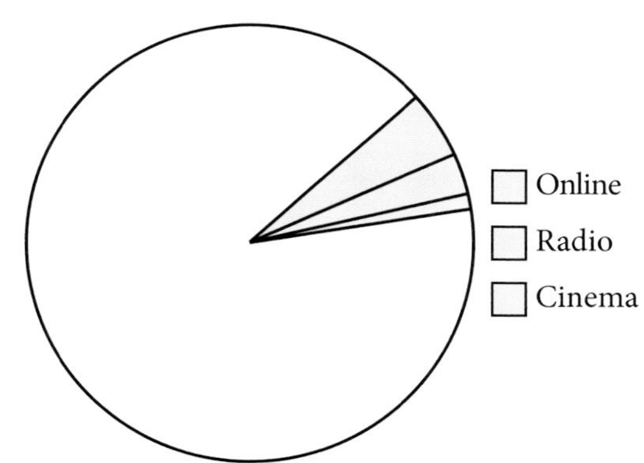

In 2012 in Spain, money was spent on advertisements in the press, TV, online, etc. The incomplete table and pie-chart show the way this was divided between the media.

a) Calculate the angle of the sector representing television, and complete the pie-chart.

b) The angle of the sector representing online advertising is 18°. Calculate the percentage spent on online advertising, and hence complete the table.

Frequency polygons

A frequency polygon can be drawn by joining the mid-points of the tops of the bars on a frequency chart.
Frequency polygons are used mainly to compare data.

- Here is a frequency chart showing the heights (or lengths) of the babies treated at a hospital one day.

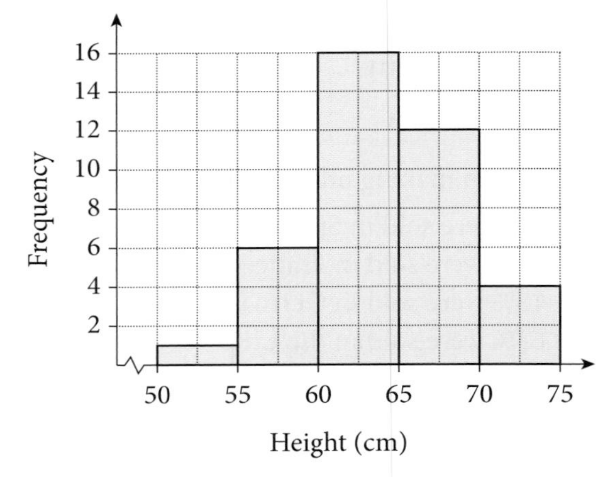

- Here is the corresponding frequency polygon, drawn by joining the mid-points of the tops of the bars.

  It is not necessary to draw the bars if you require only the frequency polygon.

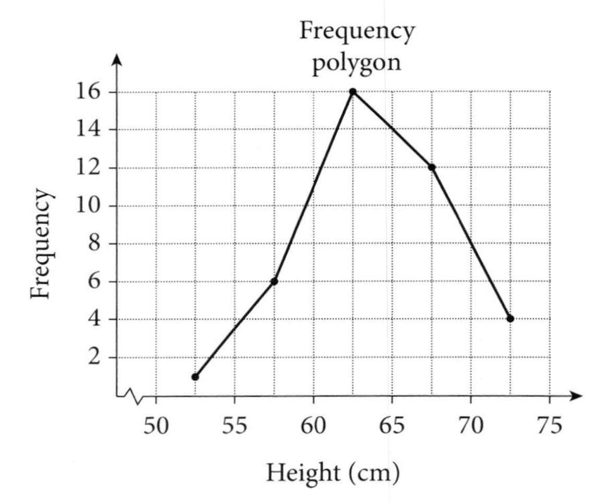

The diagram on the right shows the frequency polygons for the exam results of 34 pupils in two subjects, Maths and French. Two main differences are apparent:

1. The marks obtained in the Maths exam were significantly lower for most pupils.
2. The marks obtained in the French exam were more spread out than the Maths marks. The French marks were distributed fairly evenly over the range from 0 to 100% whereas the Maths marks were mostly between 0 and 40%.

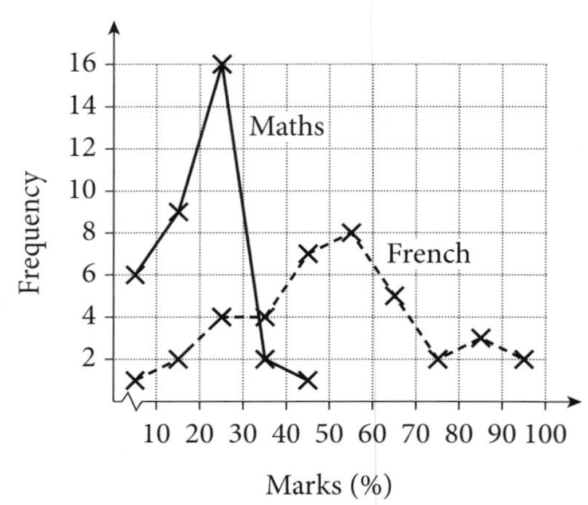

Exercise 14.4

1. Draw a frequency polygon for the distribution of masses of children drawn in the diagram.

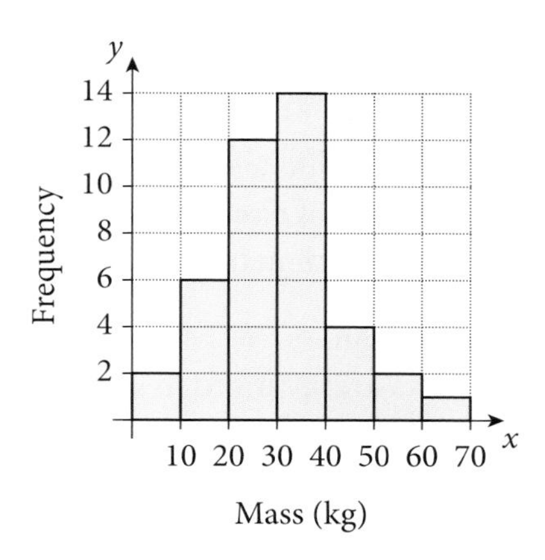

2. In a supermarket survey, shoppers were asked two questions as they left:
   a) How much have you just spent?
   b) How far away do you live?

   The results were separated into two groups: shoppers who lived less than 2 miles from the supermarket and shoppers who lived further away. The frequency polygons show how much shoppers in each group had spent. Decide which polygon, P or Q, is most likely to represent shoppers who lived less than 2 miles from the supermarket. Give your reasons.

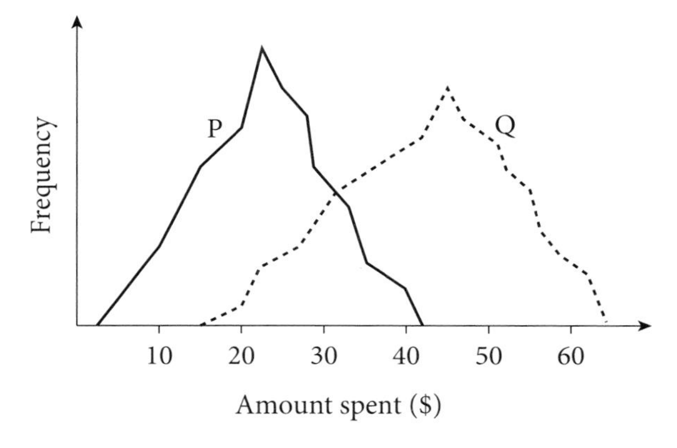

3. Scientists doing research in genetic engineering altered the genes of a certain kind of rabbit. Over a period of several years, measurements were made of the adult weight of the rabbits and their lifespans. The frequency polygons below show the results.

   What can you deduce from the two frequency polygons? Write one sentence about weight and one sentence about lifespan.

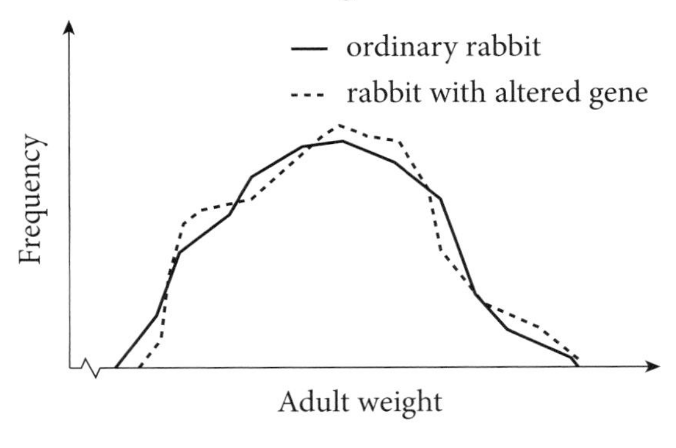

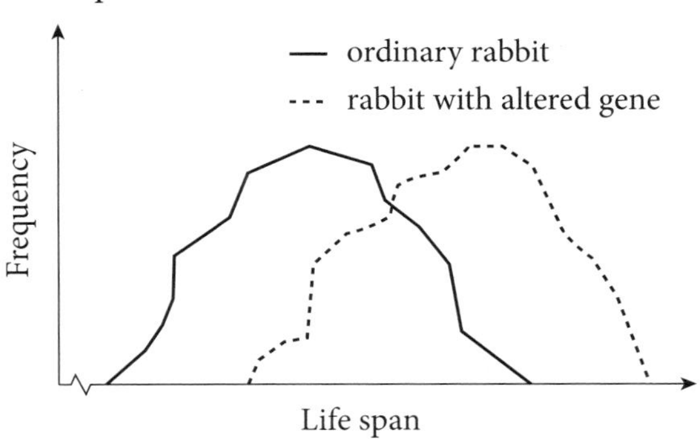

14.2 Averages, range and quartiles

Mean, median and mode

A single number called an average can be used to represent the whole range of a set of data, whether the data are discrete (for example, exam marks) or continuous (for example, heights).

The median
The data are arranged in order from the smallest to the largest; the middle number is then selected. This is really the central number of the range and is called the median.

If there are two 'middle' numbers, the median is in the middle of these two numbers.

The mean
All the data are added up and the total divided by the number of items. This is called the mean and is equivalent to sharing out all the data evenly.

The mode
The item which occurs most frequently in a frequency table is selected. This is the most popular value and is called the mode (from the French *à la mode* meaning 'fashionable').

Each average has its purpose and sometimes one is preferable to the other.

The median is fairly easy to find and has an advantage in being hardly affected by untypical values such as very large or very small values that occur at the ends of a distribution.

Look at these examination marks:

20	21	21	22	23	25	27	27	27	29	98	98
					↑						

Clearly the **median** value is 25.

The **mean** of the data is 35·5. It is easier to use in further work such as standard deviation, but clearly it does not give a true picture of the centre of distribution of the data.

The **mode** of these data is 27. It is easy to find and it eliminates some of the effects of extreme values. However it does have disadvantages, particularly in data which have two 'most popular' values, and it is not so widely used.

Range

In addition to knowing the centre of a distribution, it is useful to know the range or spread of the data.

- range = (largest value) − (smallest value)

For the examination marks, the range is 98 − 20 = 78.

Example 1

Find the median, mean, mode and range of this set of 10 numbers: 5, 4, 10, 3, 3, 4, 7, 4, 6, 5.

a) Arrange the numbers in order of size to find the median.

3, 3, 4, 4, 4, 5, 5, 6, 7, 10

↑

The median is the 'average' of 4 and 5.
So median = 4·5

b) $\text{mean} = \frac{(5+4+10+3+3+4+7+4+6+5)}{10}$

$= \frac{51}{10} = 5\cdot1$

c) mode = 4 because there are more 4s than any other number.

d) range = 10 − 3 = 7

GDC help

Using a GDC

TI-Nspire	TI84-Plus	Casio
=OneVar(7 n 10. 4 MinX 3. 6 Q_1X 4. 5 MedianX... 4.5 Q_3X 6. C10 =4.5	1-Var Stats ↑n=10 minX=3 Q_1=4 Med=4.5 Q_3=6 maxX=10	1-Variable Q1 =4 Med =4.5 Q3 =6 maxX =10 Mod =4 Mod:n=1
=OneVar(4 $\bar{x}$ 5.1 10 Σx 51. 3 Σx^2 301. 3 sx := sn-... 2.13177 4 σx := σnx... 2.02237 C2 =5.1	1-Var Stats $\bar{x}$=5.1 Σx=51 Σx^2=301 Sx=2.131770261 σx=2.022374842 ↓n=10	1-Variable $\bar{x}$ =5.1 Σx =51 Σx^2 =301 σx =2.02237484 sx =2.13177026 n =10

Exercise 14.5

1. Find the median, mean, mode and range of these sets of numbers:

   a) 3, 12, 4, 6, 8, 5, 4
   b) 7, 21, 2, 17, 3, 13, 7, 4, 9, 7, 9
   c) 12, 1, 10, 1, 9, 3, 4, 9, 7, 9
   d) 8, 0, 3, 3, 1, 7, 4, 1, 4, 4.

2. Copy and complete.

   a) The mean of 3, 5, 6 and □ is 6.
   b) The mean of 7, 8, □ and 4 is 8.

3. The frequency table shows the test results for a class of 30 students.

Mark	3	4	5	6	7	8
Frequency	2	5	4	7	6	6

   What was the modal mark?

4. a) Calculate the mean of 2, 5, 7, 7, 4. b) Hence find the mean of 32, 35, 37, 37, 34.

5. The total mass of five men is 380 kg. Calculate the mean mass of the men.

6. The temperature in a greenhouse was measured at midnight every day for a week. The results (in °C) were

   1 −2 4 0 5 2 −1

   What was the range of the temperatures?

7. The table shows the age, height and weight of seven children

	Mike	**Steve**	**Dora**	**Sam**	**Pat**	**Rayan**	**Gary**
Age (years)	16	15	17	15	16	15	16
Height (cm)	169	180	170	175	172	163	164
Weight (kg)	50	50	52	44	51	41	48

   a) What was the median age?
   b) What was the median height?
   c) What was the median weight?

8. The range for the eight numbers shown is 40.
   Find the **two** possible values of the missing number.

9. The mean weight of ten people in a lift is 70 kg. The weight limit for the lift is 1000 kg. Roughly how many more people can get into the lift?

10. There were ten cowboys in a saloon. The mean age of the men was 25 and the range of their ages was 6. Write each statement below and then write next to it whether it is true, possible or false.

   a) The youngest man was 18 years old.
   b) All the men were at least 20 years old.
   c) The oldest person was 4 years older than the youngest.
   d) Every man was between 20 and 26 years old.

11. These are the salaries of five employees in a small business.

Miss A : \$22 500   Mr B : \$17 900   Mr C : \$21 400

Mr D : \$22 500   Mrs E : \$85 300.

a) Find the mean, median and mode of their salaries.

b) Which does *not* give a fair average? Explain why in one sentence.

12. A farmer has 32 cattle to sell. Their weights in kg are as follows.

81	81	82	82	83	84	84	85
85	86	86	87	87	88	89	91
91	92	93	94	96	150	152	153
154	320	370	375	376	380	381	390

[Total weight = 5028 kg]

On the telephone, when talking to a potential buyer, the farmer describes the cattle and says the average weight is 'over 157 kg'.

a) Find the mean weight and the median weight.

b) Which average has the farmer used to describe his animals? Does this average describe the cattle fairly?

13. A gardening magazine sells seedlings of a plant through the post and claims that the average height of the plants after one year's growth will be 85 cm. A sample of 24 of the plants were measured after one year with these results (in cm)

6	7	7	9	34	56	85	89
89	90	90	91	91	92	93	93
93	94	95	95	96	97	97	99

(The sum of the heights is 1788 cm.)

a) Find the mean and the median height of the sample.

b) Is the magazine's claim about average height justified?

14. The mean weight of five people is 76 kg. The weights of four of the people are 72 kg, 74 kg, 75 kg and 81 kg. What is the weight of the fifth person?

15. The mean length of six rods is 44·2 cm. The mean length of five of them is 46 cm. How long is the sixth rod?

16. a) The mean of 3, 7, 8, 10 and x is 6. Find x.

b) The mean of 3, 3, 7, 8, 10, x and x is 7. Find x.

17. The mean height of 12 men is 1·70 m, and the mean height of 8 women is 1·60 m, Find

a) the total height of the 12 men

b) the total height of the 8 women

c) the mean height of the 20 men and women.

18. The total weight of 6 rugby players is 540 kg and the mean weight of 14 ballet dancers is 40 kg. Find the mean weight of the group of 20 rugby players and ballet dancers.

19. Write five numbers so that

    the mean is 6
    the median is 5
    the mode is 4.

    ? ? ? ? ?

20. Find five numbers so that the mean, median, mode and range are all 4.

21. The numbers 3, 5, 7, 8 and N are arranged in ascending order.
    If the mean of the numbers is equal to the median, find N.

22. The mean of five numbers is 11.

    The numbers are in the ratio 1 : 2 : 3 : 4 : 5.

    Find the smallest number.

23. The median of five consecutive integers is N.

    a) Find the mean of the five numbers.

    b) Find the mean and the median of the squares of the integers.

    c) Find the difference between the two values in part **b)**.

Calculating the mean from a frequency table

Example 2

The frequency table shows the weights of the eggs bought in a supermarket. Find the mean, median and model weight.

Weight	58 g	59 g	60 g	61 g	62 g	63 g
Frequency	3	7	11	9	8	2

a) Mean weight of eggs

$$= \frac{(58\times3)+(59\times7)+(60\times11)+(61\times9)+(62\times8)+(63\times2)}{(3+7+11+9+8+2)}$$

$$= \frac{2418}{40} = 60{\cdot}45\,\text{g}$$

b) There are 40 eggs so the median weight is the weight between the 20th and 21st numbers. By inspection, both the 20th and 21st weights are 60 g.
So median weight = 60 g

c) The modal weight = 60 g

60 g has the highest frequency.

▶ Continued on next page

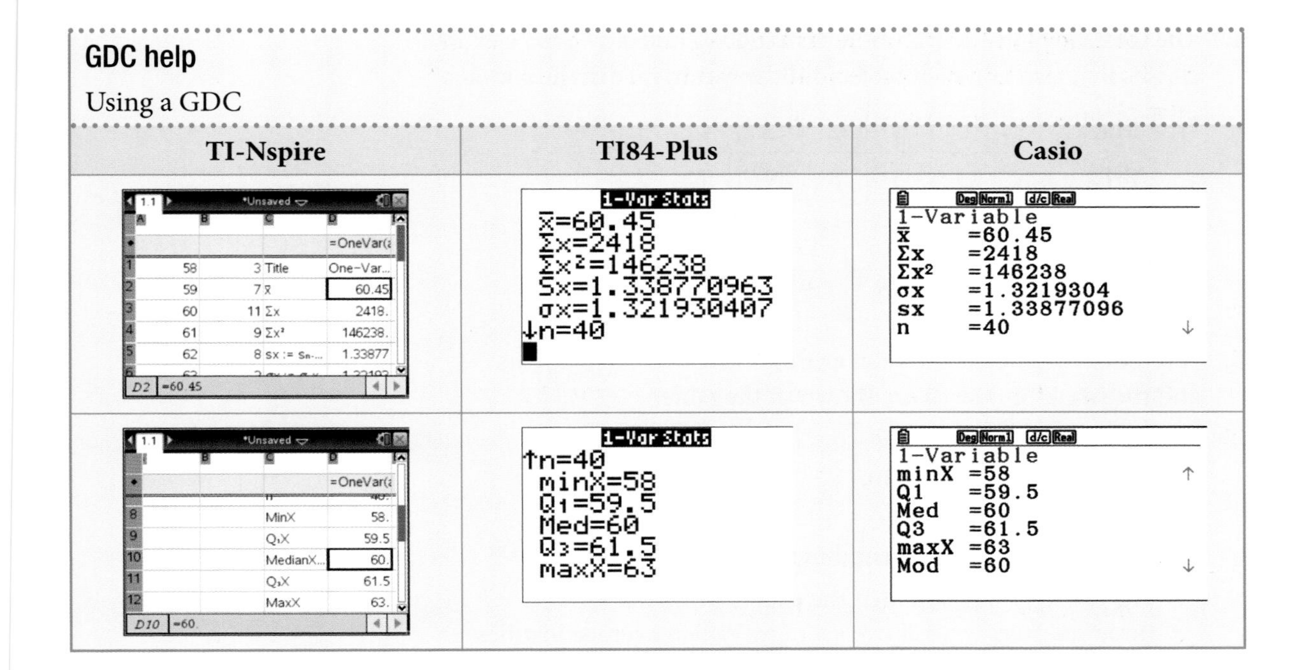

Exercise 14.6

1. The frequency table shows the weights of the 40 apples sold in a shop.

Weight	70 g	80 g	90 g	100 g	110 g	120 g
Frequency	2	7	9	11	8	3

Calculate the mean weight of the apples.

2. The frequency table shows the price of a packet of butter, in cents, in 30 different shops.

Price	49c	50c	51c	52c	53c	54c
Frequency	2	3	5	10	6	4

Calculate the mean price of a packet of butter.

3. A box contains 50 nails of different length as shown in the frequency table.

Length of nail	2 cm	3 cm	4 cm	5 cm	6 cm	7 cm
Frequency	4	7	9	12	10	8

Calculate the mean length of the nails.

4. The tables give the distribution of marks obtained by two classes in a test. For each table, find the mean, median and mode.

a)

Mark	0	1	2	3	4	5	6
Frequency	3	5	8	9	5	7	3

b)

Mark	15	16	17	18	19	20
Frequency	1	3	7	1	5	3

5. A teacher conducted a mental arithmetic test for 26 students and the marks out of 10 are shown in the table.

Mark	3	4	5	6	7	8	9	10
Frequency	6	3	1	2	0	5	5	4

a) Find the mean, median and mode.

b) The teacher congratulated the class saying that 'over three-quarters were above average'. Which average justifies this statement?

6. The table shows the number of goals scored in a series of football matches.

Number of goals	1	2	3
Number of matches	8	8	x

a) If the mean number of goals is 2·04, find x.

b) If the modal number of goals is 3, find the smallest possible value of x.

c) If the median number of goals is 2, find the largest possible value of x.

7. The table shows the results of a survey on the number of occupants per car.

Number of occupants	1	2	3	4
Number of cars	7	11	7	x

a) If the mean number of occupants is $2\frac{1}{3}$, find x.

b) If the mode is 2, find the largest possible value of x.

c) If the median is 2, find the largest possible value of x.

8. The marks obtained by the members of a class are summarised in the table.

Mark	x	y	z
Frequency	a	b	c

Calculate the mean mark in terms of a, b, c, x, y and z.

Example 3

The results of 51 students in a test are given in the frequency table.

Find **a)** the mean **b)** the median **c)** the mode.

Mark	30–39	40–49	50–59	60–69
Frequency	7	14	21	9

Don't forget the mean is only an **estimate** because you do not have the raw data and you have made an assumption with the mid-point of each interval.

In order to find the mean you approximate by saying each interval is represented by its mid-point. For the 30-39 interval you say there are 7 marks of 34·5 [that is (30 + 39) ÷ 2 = 34·5].

a) $\text{Mean} = \dfrac{(34{\cdot}5\times7)+(44{\cdot}5\times14)+(54{\cdot}5\times21)+(64{\cdot}5\times9)}{(7+14+21+9)}$

$= 50{\cdot}7745098$

$= 51$ (2 s.f.)

b) The median is the 26th mark, which is in the interval 50–59. You cannot find the exact median.

c) The **modal group** is 50–59. You cannot find an exact mode.

Later you will find out how to get an estimate of the median by drawing a cumulative frequency curve.

GDC help

Using a GDC

TI-Nspire	TI84-Plus	Casio
=OneVar(; 34.5, 7, Title, One-Var...; 44.5, 14, x̄, 50.7745; 54.5, 21, Σx, 2589.5; 64.5, 9, Σx², 135873.; sx := Sn-..., 9.37247; D2 =50.774509803922	1-Var Stats x̄=50.7745098 Σx=2589.5 Σx²=135872.75 Sx=9.372466978 σx=9.28012515 ↓n=51	1-Variable x̄ =50.7745098 Σx =2589.5 Σx² =135872.75 σx =9.28012514 sx =9.37246697 n =51

Exercise 14.7

1. The table gives the number of words in each sentence of a page in a book.

a) Copy and complete the table.

b) Work out an estimate for the mean number of words in a sentence.

Number of words	Frequency f	Midpoint x	fx
1–5	6	3	18
6–10	5	8	40
11–15	4		
16–20	2		
21–25	3		
Totals	20	–	

2. The results of 24 students in a test are given in the table.

Mark	40–54	55–69	70–84	85–99
Frequency	5	8	7	4

Find the mid-point of each group of marks and calculate an estimate of the mean mark.

3. The table shows the number of letters delivered to the 26 houses in a street.

Calculate an estimate of the mean number of letters delivered per house.

Number of letters delivered	**Number of houses (frequency)**
0–2	10
3–4	8
5–7	5
8–12	3

4. The histogram shows the heights of the 60 athletes in the 2012 Dutch athletics team.

   a) Calculate an estimate for the mean height of the 60 athletes.

   b) Explain why your answer is an **estimate** for the mean height.

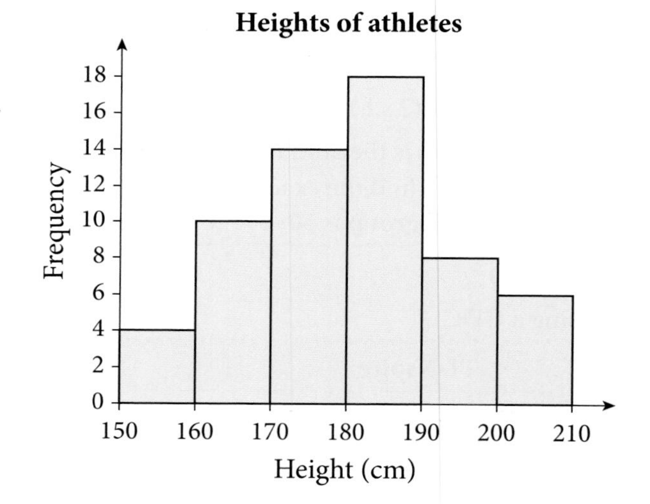

14.3 Histograms

Drawing a histogram

In a histogram, the **area** of each bar shows the frequency of the data. Histograms resemble bar charts but are not to be confused with them. In bar charts the height of each shows the frequency; histograms often have bars of varying widths. Because the area of the bar represents frequency, the height must be adjusted to correspond with the width of the bar. The vertical axis is not labelled 'frequency' but **frequency density**.

- Frequency density $= \dfrac{\text{frequency}}{\text{class width}}$

You can use histograms to represent both discrete data and continuous data, but their main purpose is for use with continuous data.

Example 4

Draw a histogram from the table for the distribution of ages of passengers travelling on a flight from Mexico City to New York.

Ages	Frequency
$0 \leq x < 20$	28
$20 \leq x < 40$	36
$40 \leq x < 50$	20
$50 \leq x < 70$	30
$70 \leq x < 100$	18

Note that the data have been collected into class intervals of different widths.

To draw the histogram, the heights of the bars must be adjusted by calculating frequency density.

Ages	Frequency	Frequency density (f.d.)
$0 \leq x < 20$	28	$28 \div 20 = 1{\cdot}4$
$20 \leq x < 40$	36	$36 \div 20 = 1{\cdot}8$
$40 \leq x < 50$	20	$20 \div 10 = 2{\cdot}0$
$50 \leq x < 70$	30	$30 \div 20 = 1{\cdot}5$
$70 \leq x < 100$	18	$18 \div 30 = 0{\cdot}6$

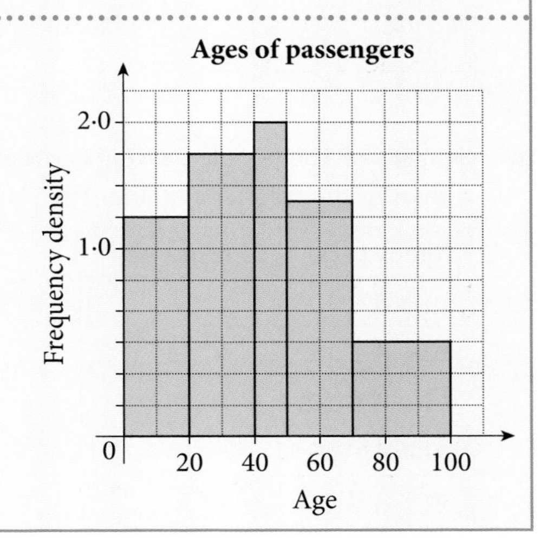

Exercise 14.8

1. Lakshmi measured the lengths of 20 copper nails. She recorded the results in a frequency table. Calculate the frequency densities and draw the histogram as started on the right.

Length l (in mm)	Frequency	Frequency density (f.d.)
$0 \leq l < 20$	5	$5 \div 20 = 0{\cdot}25$
$20 \leq l < 25$	5	
$25 \leq l < 30$	7	
$30 \leq l < 40$	3	

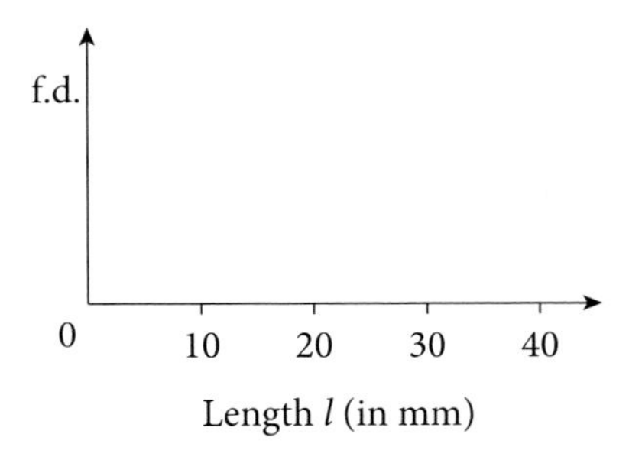

2. The frequency table has been started. It shows the volumes of 55 containers. Calculate the frequency densities and draw the histogram.

Volume (mm^3)	Frequency	Frequency density
$0 < V \leq 5$	5	$5 \div 5 = 1$
$5 < V \leq 10$	3	$3 \div$
$10 < V \leq 20$	12	
$20 < V \leq 30$	17	
$30 < V \leq 40$	13	
$40 < V \leq 60$	5	

3. Thirty students in a class are weighed on the first day of term. Draw a histogram to represent these data.
Note that the weights do not start at zero. This can be shown on the graph by a broken axis.

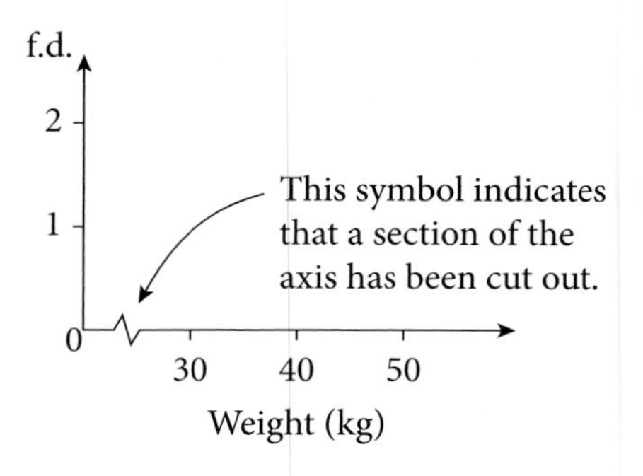

Weight w (kg)	Frequency
$30 < w \leq 40$	5
$40 < w \leq 45$	7
$45 < w \leq 50$	10
$50 < w \leq 55$	5
$55 < w \leq 70$	3

4. The ages of 120 people passing through a turnstile were recorded and are shown in the frequency table. The class boundaries are 0, 10, 15, 20, 30, 40.

Age A (yrs)	Frequency
$0 < A \leq 10$	18
$10 < A \leq 15$	46
$15 < A \leq 20$	35
$20 < A \leq 30$	13
$30 < A \leq 40$	8

Draw the histogram for the data.

5. The table shows the daily profit made at a beach resort sailing club.

Profit P (\$)	Frequency
$0 < P \leq 40$	16
$40 < P \leq 80$	48
$80 < P \leq 100$	30
$100 < P \leq 120$	40
$120 < P \leq 160$	36
$160 < P \leq 240$	16

Draw a histogram for the data.

6. Another common notation is used here for the masses of plums picked in an orchard.

The initial '20–' means '$20\,\text{g} \leq \text{mass} < 30\,\text{g}$'.
Draw a histogram with class boundaries at 20, 30, 40, 60, 80.

Mass (g)	20–	30–	40–	60–	80–
Frequency	11	18	7	5	0

Finding frequencies from histograms

You can draw up a frequency table from the data in a histogram. The you can find the actual size of the sample tested as well as the frequencies in each class.

Example 5

The histogram shows children's heights. Draw up the relevant frequency table and find the size of the sample which was tested.

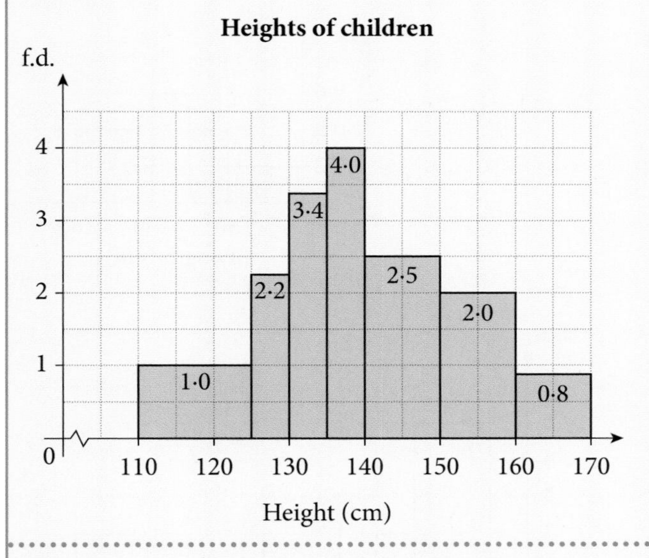

Use frequency density = $\frac{\text{frequency}}{\text{class width}}$

So frequency = (frequency density) × (class width)

Height x (cm)	f.d.	Frequency
$110 < x \leq 125$	1·0	1·0 × 15 = 15
$125 < x \leq 130$	2·2	2·2 × 5 = 11
$130 < x \leq 135$	3·4	3·4 × 5 = 17
$135 < x \leq 140$	4·0	4·0 × 5 = 20
$140 < x \leq 150$	2·5	2·5 × 10 = 25
$150 < x \leq 160$	2·0	2·0 × 10 = 20
$160 < x \leq 170$	0·8	0·8 × 10 = 8
		Total frequency = 116

There were 116 children in the sample measured.

Remember

The area of each bar represents frequency.

Exercise 14.9

1. From the histogram find
   a) how many of the lengths are in these intervals
      i) 40–60 cm
      ii) 30–40 cm
   b) the total frequency.

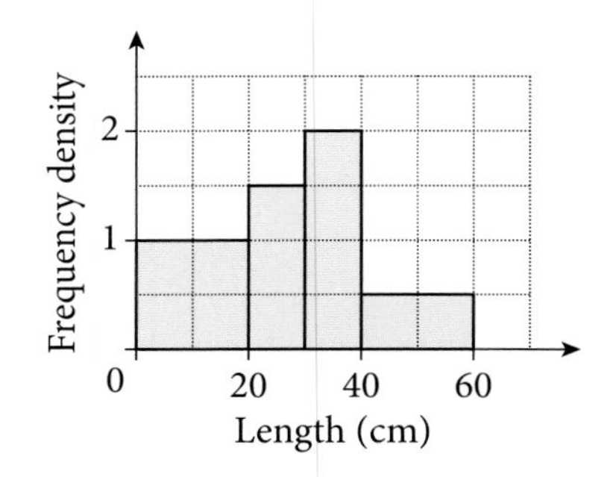

2. For this histogram find the total frequency.

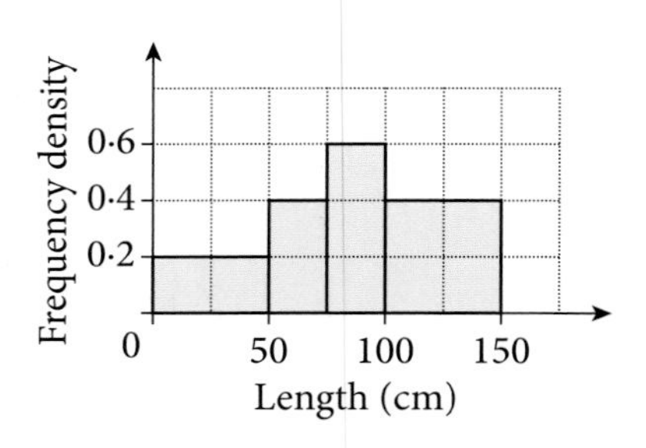

3. The histogram shows the ages of the trees in a small wood. How many trees were in the wood altogether?

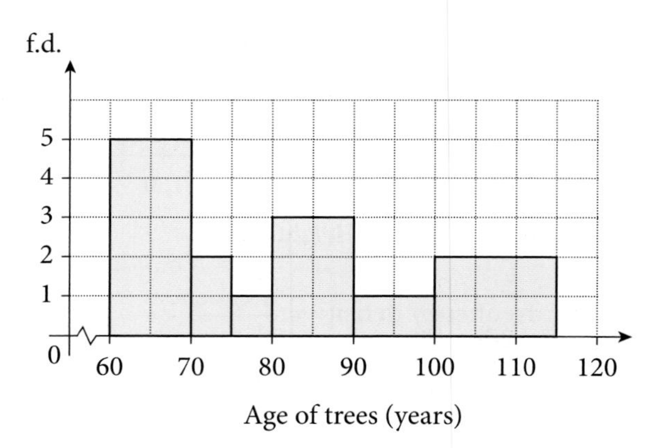

4. One day a farmer weighs all the hens' eggs which he collects. The results are shown in this histogram. There were 8 eggs in the class 50–60 g.
   Work out the numbers on the frequency density axis and hence find the total number of eggs collected.

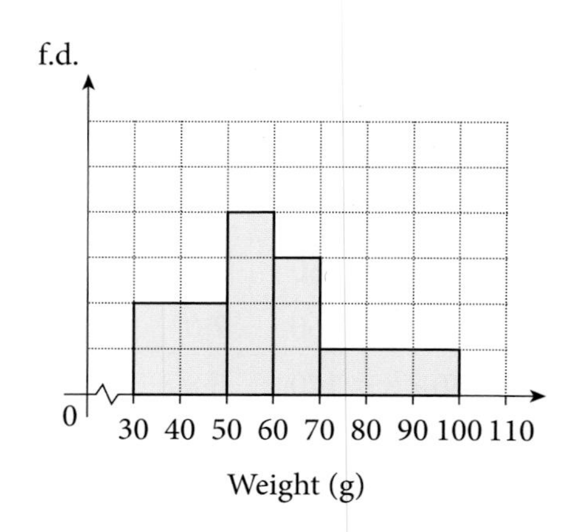

5. A different way of giving frequencies is to draw a key to show what a certain area represents. The histogram below has a guide to shows the size of the area which represents 4 people.

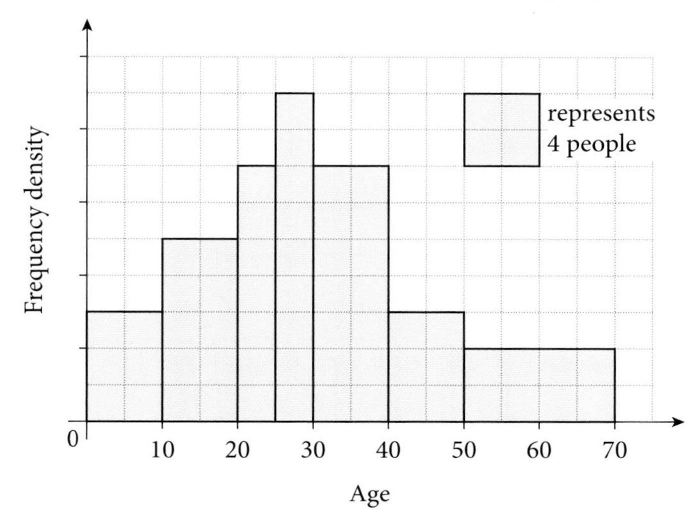

a) How many people are aged between 0 and 10?

b) How many people are aged between 10 and 20?

c) How many people are represented in the whole histogram?

6. A farmer takes a sample of 92 trees from a field and measures their heights.

Height (cm)	Frequency
100–120	16
120–140	24
140–150	18
150–160	10
160–200	24

He draws a bar 4 cm high to represent the trees of height 100–120 cm. How high will the bar be representing

a) 120–140 cm

b) 140–150 cm

c) 160–200 cm?

14.4 Cumulative frequency

Quartiles, interquartile range

The range is a simple measure of spread but one extreme (very high or very low) value can have a big effect.

The **interquartile range** is a better measure of spread.

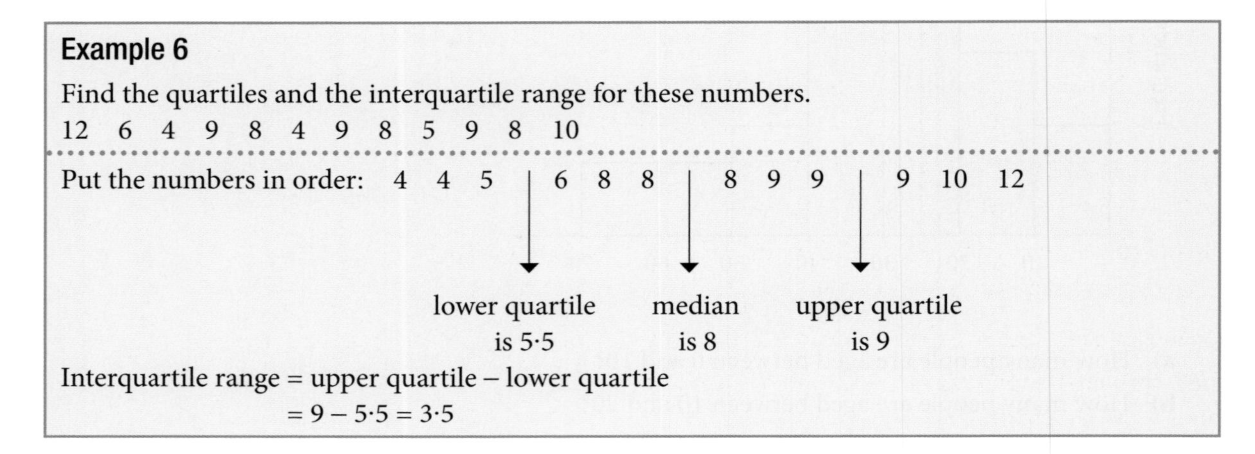

Example 6

Find the quartiles and the interquartile range for these numbers.

12 6 4 9 8 4 9 8 5 9 8 10

Put the numbers in order: 4 4 5 | 6 8 8 | 8 9 9 | 9 10 12

lower quartile is 5·5

median is 8

upper quartile is 9

Interquartile range = upper quartile – lower quartile

= 9 – 5·5 = 3·5

Exercise 14.10

For each set of data, work out

a) the lower and upper quartile

b) the interquartile range.

1. 1 1 4 4 5 8 8 8 9 10 11 11

2. 5 2 6 4 1 9 3 2 8 4 1 0

3. 7 9 11 15 18 19 23 27

4. 0 0 1 1 2 3 3 4 4 4 4 6 6 6 7 7

Cumulative frequency

- The total of the frequencies up to a particular value is called the **cumulative frequency**.

Data from a frequency table can be used to calculate cumulative frequencies. These new values, when plotted and joined, form a cumulative frequency curve, sometimes called an S-shaped curve.

It is a simple matter to find the median from the halfway point of a cumulative frequency curve.

Other points of location can also be found from this curve.
The cumulative frequency axis can be divided into 100 parts.

- The **upper quartile** is at the 75% point.
- The **lower quartile** is at the 25% point.

The quartiles are particularly useful in finding the central 50% of the range of the distribution; this is known as the **interquartile range**.

- Interquartile range = (upper quartile) – (lower quartile)

The interquartile range is an important measure of spread in that it shows how widely the data are spread.
Half the distribution is in the interquartile range. If the interquartile range is small, then the middle half of the distribution is bunched together.

Example 7

In a survey, 200 people were asked to state their weekly earnings. The results were plotted on a cumulative frequency curve.

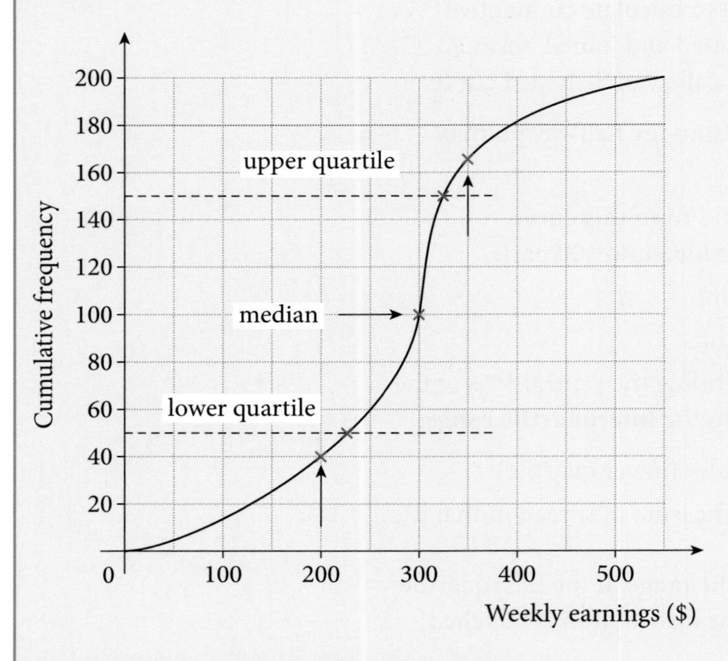

a) How many people earned up to \$350 a week?

b) How many people earned more than \$200 a week?

c) Find the interquartile range.

---

a) From the curve, about 165 people earned up to \$350 per week.

b) About 40 people earned up to \$200 per week.
There are 200 people in the survey, so 160 people earned more than \$200 per week.

c) Lower quartile (50 people) = \$235
Upper quartile (150 people) = \$325
∴ Interquartile range = 325 − 235
= \$90

Example 8

A pet shop owner likes to weigh all his mice every week as a check on their state of health. The table shows the weights of the 80 mice.

Weight w (g)	Frequency	Cumulative frequency	Weight represented by cumulative frequency
$0 < w \le 10$	3	3	≤ 10 g
$10 < w \le 20$	5	8	≤ 20 g
$20 < w \le 30$	5	13	≤ 30 g
$30 < w \le 40$	9	22	≤ 40 g
$40 < w \le 50$	11	33	≤ 50 g
$50 < w \le 60$	15	48	≤ 60 g
$60 < w \le 70$	14	62	≤ 70 g
$70 < w \le 80$	8	70	≤ 80 g
$80 < w \le 90$	6	76	≤ 90 g
$90 < w \le 100$	4	80	≤ 100 g

The table also shows the cumulative frequency.
Plot a cumulative frequency curve and hence estimate

a) the median **b)** the interquartile range.

From the cumulative frequency curve,

median = 55 g
lower quartile = 38 g
upper quartile = 68 g
interquartile range = 68 − 38
= 30 g

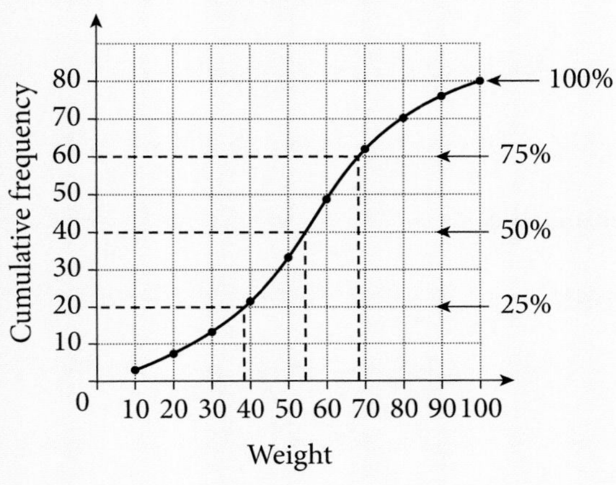

The points on the graph are plotted at the upper limit of each group of weights.

Exercise 14.11

1. The graph shows the cumulative frequency curve for the marks of 40 students in an examination.

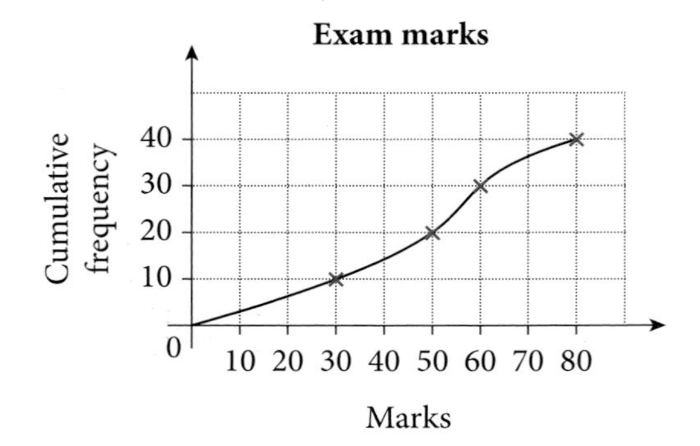

From the graph, estimate

a) the median mark

b) the mark at the lower quartile and at the upper quartile

c) the interquartile range

d) the pass mark if three-quarters of the students passed.

2. The graph shows the cumulative frequency curve for the marks of 60 students in an examination.

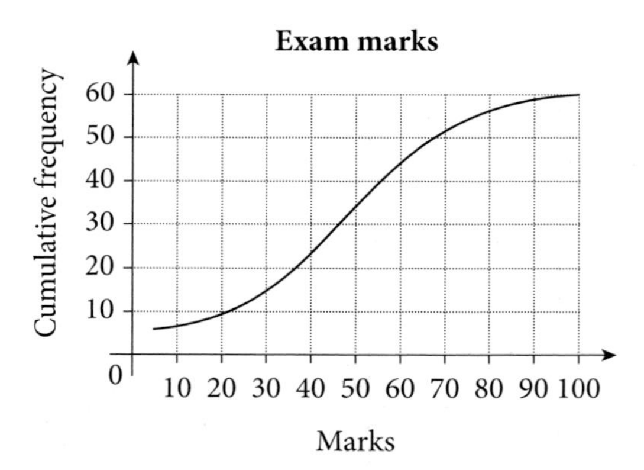

From the graph estimate

a) the median mark

b) the mark at the lower quartile and at the upper quartile

c) the interquartile range

d) the pass mark if two-thirds of the students passed.

3. The lifetimes of 500 electric light bulbs were measured in a laboratory. The results are shown in the cumulative frequency diagram.

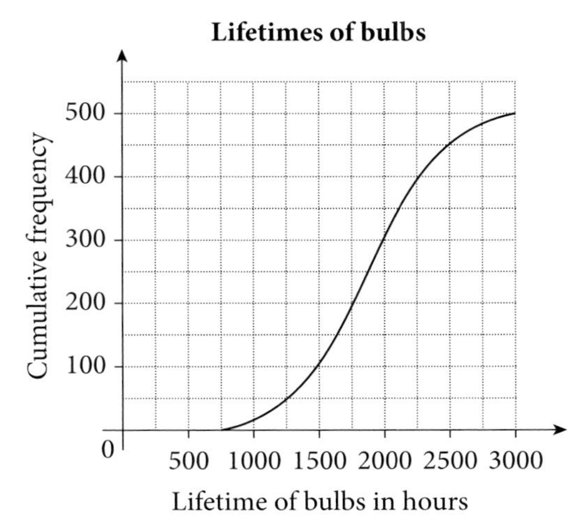

a) How many bulbs had a lifetime of 1500 hours or less?
b) How many bulbs had a lifetime of between 2000 and 3000 hours?
c) After how many hours were 70% of the bulbs dead?
d) What was the shortest lifetime of a bulb?

4. A photographer measures all the snakes required for a scene in a film involving a snake pit.

a) Draw a cumulative frequency curve for the results.

Length l (cm)	Frequency	Cumulative frequency	Upper limit
$0 < l \leq 10$	0	0	≤ 10
$10 < l \leq 20$	2	2	≤ 20
$20 < l \leq 30$	4	6	≤ 30
$30 < l \leq 40$	10	16	≤ 40
$40 < l \leq 50$	17		:
$50 < l \leq 60$	11		:
$60 < l \leq 70$	3		:
$70 < l \leq 80$	3		:

Use a scale of 2 cm for 10 units across the page for the lengths and 2 cm for 10 units up the page for the cumulative frequency. Remember to plot points at the **upper** end of the classes (10, 20, 30 etc).

b) Find

i) the median **ii)** the interquartile range.

5. As part of a medical inspection, a nurse measures the heights of 48 students in a school.

 a) Copy and complete the table, then draw a cumulative frequency curve for the results.

Height (cm)	Frequency	Cumulative frequency
$140 < h \leq 145$	2	2 [$\leq$ 145 cm]
$145 < h \leq 150$	4	6 [$\leq$ 150 cm]
$150 < h \leq 155$	8	14 [$\leq$ 155 cm]
$155 < h \leq 160$	9	
$160 < h \leq 165$	12	
$165 < h \leq 170$	7	
$170 < h \leq 175$	4	
$175 < h \leq 180$	2	

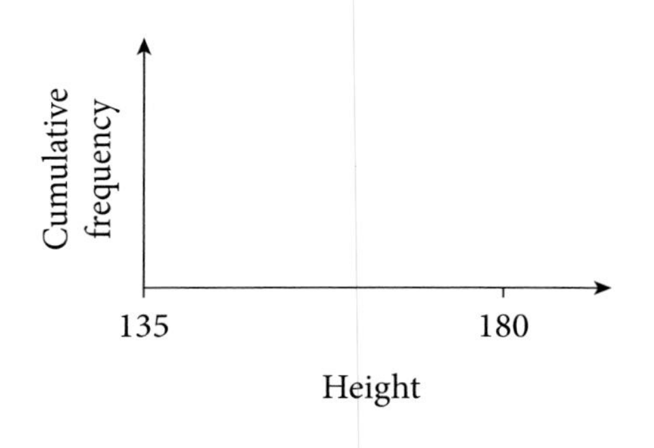

 Use a scale of 2 cm for 5 units across the page and 2 cm for 10 units up the page.

 b) Find

  i) the median

  ii) the interquartile range.

6. Henning and Boris are brilliant darts players. They recorded their scores over 60 throws. Here are Henning's scores:

Score (x)	$30 < x \leq 60$	$60 < x \leq 90$	$90 < x \leq 120$	$120 < x \leq 150$	$150 < x \leq 180$
Frequency	10	4	13	23	10

 a) Draw a cumulative frequency curve.

 Use a scale of 2 cm for 20 points across the page and 2 cm for 10 throws up the page.

 b) For Henning, find

  i) his median score

  ii) the interquartile range of his scores.

 For his 60 throws, Boris had a median score of 105 and an interquartile range of 20.

 c) Which of the players is more consistent? Give a reason.

7. The life of an unreliable photocopying machine is tested by timing how long it works before breaking down. The table shows the results for 50 machines.

Time, t (hours)	$2 < t \leq 4$	$4 < t \leq 6$	$6 < t \leq 8$	$8 < t \leq 10$	$10 < t \leq 12$
Frequency	9	6	15	15	5

a) Draw a cumulative frequency graph.
[Put t across the page, 1 cm = hour; cumulative frequency up the page]

b) Find

i) the median life of a machine

ii) the interquartile range.

c) The makers claim that their machines will work for at least 10 hours. What percentage of the machines do not match this description?

8. In an international competition 60 students from India and Switzerland did the same science test.

Marks	**India frequency**	**Switzerland frequency**	**India cum. freq.**	**Switzerland cum. freq.**
1–5	1	2	1 [≤5·5]	2 [≤5·5]
6–10	2	5	3 [≤10·5]	7 [≤10·5]
11–15	4	11	7 [≤15·5]	
16–20	8	16		
21–25	16	10		
26–30	19	8		
31–35	10	8		

The upper class boundaries for the marks are 5·5, 10·5, 15·5 etc.

The cumulative frequency graph should be plotted for values ≤5·5, ≤10·5, ≤15·5 and so on.

a) Using the same axes, draw the cumulative frequency curves for the Indian and Swiss results.

Use a scale of 2 cm for 5 marks across the page and 2 cm for 10 people up the page.

b) Find the median mark for each country.

c) Find the interquartile range for the India results.

d) Describe in one sentence the main difference between the two sets of results.

14.5 Scatter diagrams

Sometimes it is important to discover if there is a connection or relationship between two sets of data.

Examples:

- Are more ice creams sold when the weather is hot?
- Do tall people have higher pulse rates?
- Are people who are good at maths also good at science?
- Does watching TV improve examination rates?

If there is a relationship, it will be easy to spot if your data is plotted on a **scatter diagram** – that is a graph in which one set of data is plotted on the horizontal axis and the other on the vertical axis.

Here is a scatter graph showing the price of pears and the quantity sold.

We can see a *connection* – when the price was high the sales were low and when the price went down the sales increased.

This scatter graph shows the sales of the newspaper and the temperature. We can see there is *no connection* between the two variables.

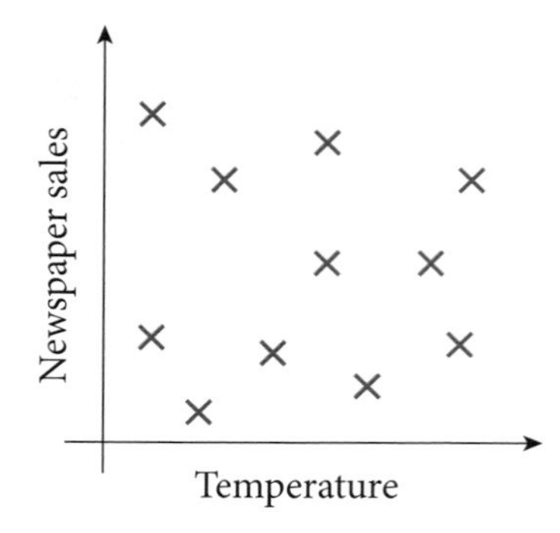

Correlation

The word **correlation** describes how things *co-relate*. There is correlation between two sets of data if there is a connection or relationship.

The correlation between two sets of data can be positive or negative and it can be strong or weak as indicated by the scatter graphs below.

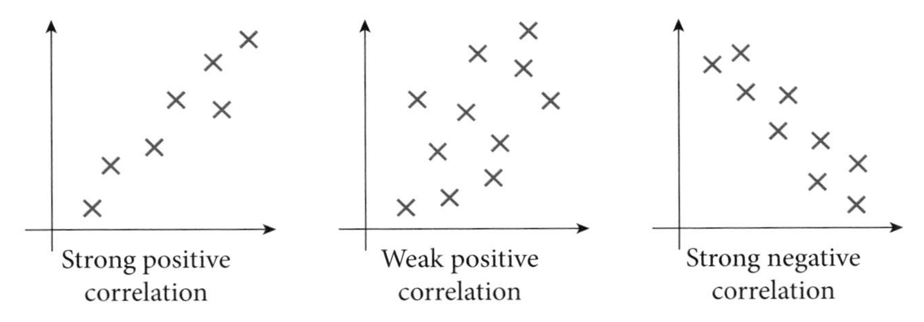

When the correlation is positive the points are around a line which slopes upward to the right. When the correlation is negative the 'line' slopes downwards to the right.

When the correlation is strong the points are bunched close to a line through their midst. When the correlation is weak the points are more scattered.

It is important to realise that often there is *no* correlation between two sets of data.

If, for example, we take a group of students and plot their Maths test results against their time to run 800 m, the graph might look like the one on the right. A common mistake in this topic is to 'see' a correlation on a scatter graph where none exists.

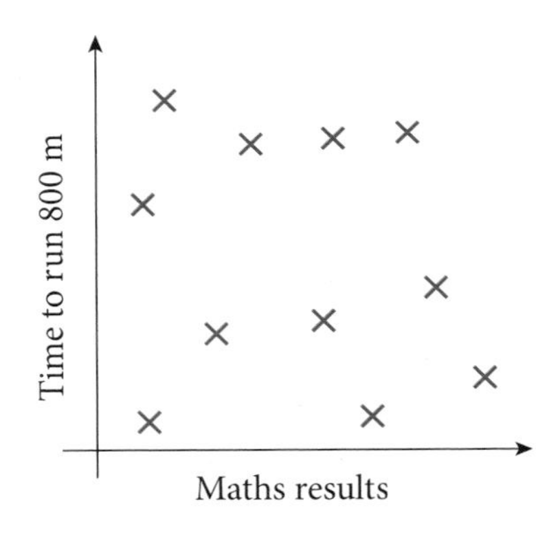

There is also *no* correlation in these two scatter graphs:

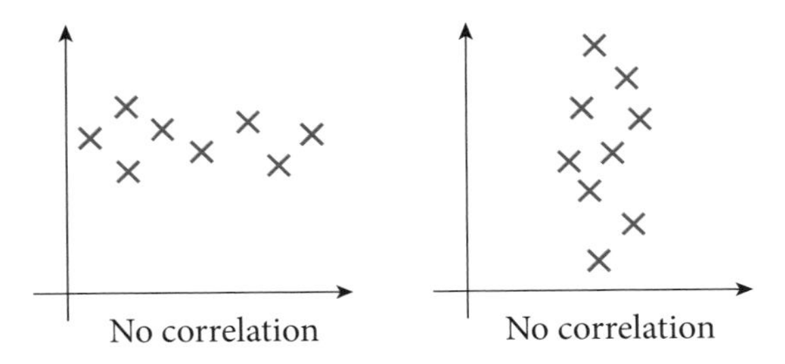

Line of best fit

When a scatter graph shows either positive or negative correlation, a **line of best fit** can be drawn. The sums of the distances to points on either side of the line are equal and there should be an equal number of points on each side of the line. The line should be drawn through the mean of the data and is easier to draw when a transparent ruler is used.

Here are the marks obtained in two tests by 9 students.

Student	A	B	C	D	E	F	G	H	I
Maths mark	28	22	9	40	37	35	30	23	?
Physics mark	48	45	34	57	50	55	53	45	52

A line of best fit can be drawn as there is strong positive correlation between the two sets of marks.

The line of best fit can be used to estimate the maths result of student J, who missed the Maths test but scored 52 in the Physics test.

We can estimate that student J would have scored about 33 in the maths test. It is not possible to be *very* accurate using scatter graphs. It is reasonable to state that student J ‘might have scored between 30 and 36’ in the maths test.

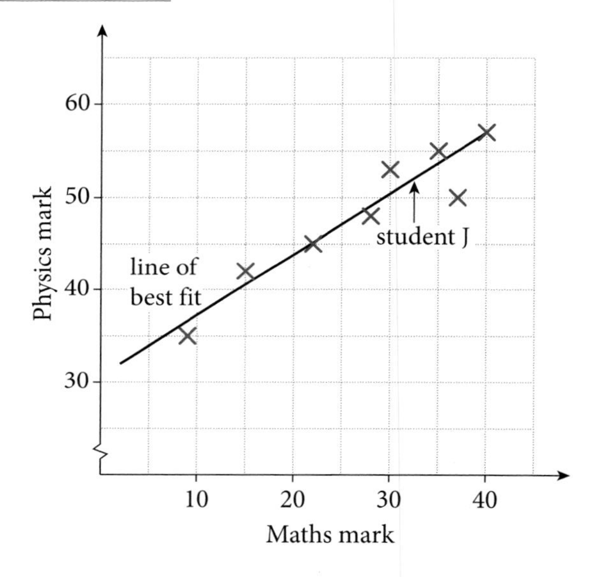

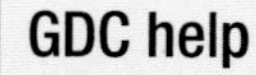

GDC help

The scatter graphs and line of best fit for the Maths and Physics marks can also be plotted on your GDC.

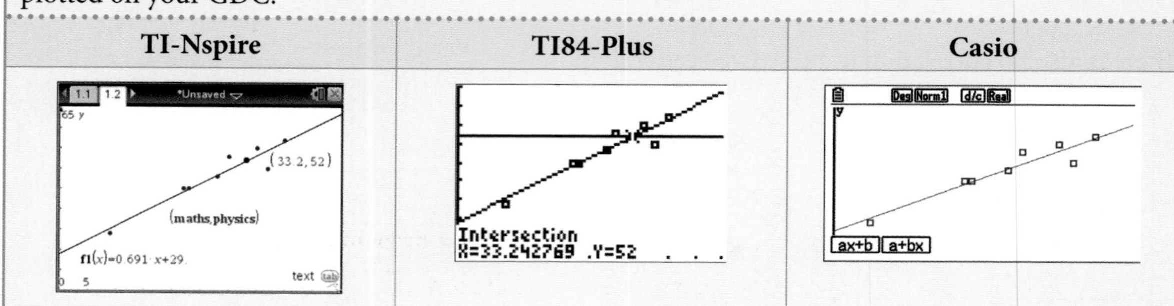

Here is a scatter graph in which the heights of boys of different ages is recorded. A line of best fit is drawn.

a) We can estimate that the height of an 8-year-old boy might be about 123 cm [say between 120 and 126 cm].

b) We can only predict a height within the range of values plotted. We could not extend the line of best and use it to predict the height of a 30 year old! Why not?

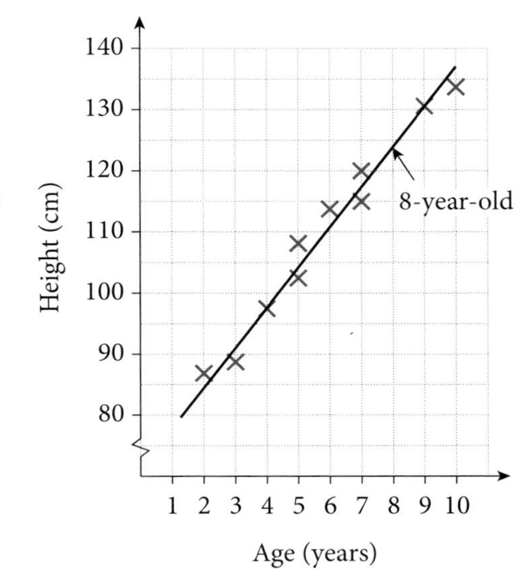

Exercise 14.12

1. Make the following measurements for everyone in your class:

- height (nearest cm)
- arm span (nearest cm)
- head circumference (nearest cm)
- hand span (nearest cm)
- pulse rate (beats/minute)

For greater consistency of measuring, one person (or perhaps two people) should do all the measurements of one kind (except on themselves!).

Enter all the measurements in a table, either on the board or on a sheet of paper.

Name	Height	Armspan	Head	
Raul	161	165	56	
Lara	150	148	49	
Anuradha				

a) Draw the scatter graphs shown below:

i)

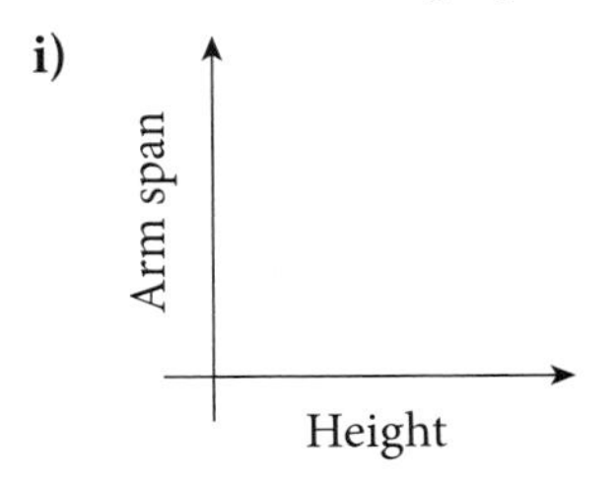

ii)

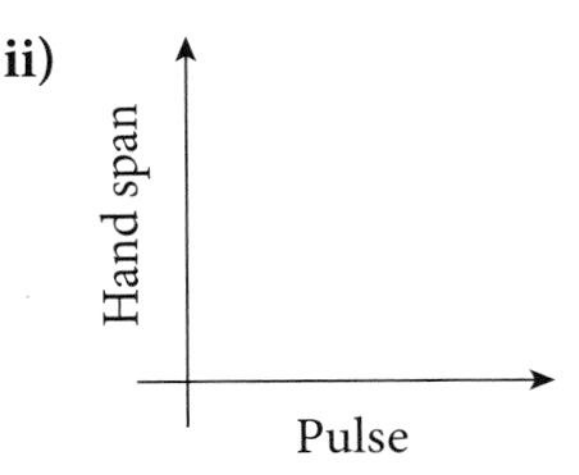

b) Describe the correlation, if any, in the scatter graphs you drew in part **a**).

c) i) Draw a scatter graph of two measurements where you think there might be positive correlation.

ii) Was there indeed a positive correlation?

2. Plot the points given on a scatter graph, with s across the page and p up the page. Draw axes with values from 0 to 20. Descirbe the correlation, if any, between the values of s and p (i.e. 'strong negative', 'weak positive' etc.)

a)

s	7	16	4	12	18	6	20	4	10	13
p	8	15	6	12	17	9	18	7	10	14

b)

s	3	8	12	15	16	5	6	17	9
p	4	2	10	17	5	10	17	11	15

c)

s	11	1	16	7	2	19	8	4	13	18
p	5	12	7	14	17	1	11	8	11	5

In questions **3**, **4** and **5** plot the points given on a scatter graph, with s across the page and p up the page.
Draw axes with the values from 0 to 20.
If possible draw a line of best fit on the graph.
Where possible estimate the value of p on the line of best fit where $s = 10$.

3.

s	2	14	14	4	12	18	12	6
p	5	15	16	6	12	18	13	7

4.

s	2	15	17	3	20	3	6
p	13	7	5	12	4	13	11

5.

s	4	10	15	18	19	4	19	5
p	19	16	11	19	15	3	1	9

6. The following data gives the marks of 11 students in a French test and in a German test.

French	15	36	36	22	23	27	43	22	43	40	26
German	6	28	35	18	28	28	37	9	41	45	17

a) Plot this data on a scatter graph, with French marks on the horizontal axis.

b) Draw the line of best fit.

c) Estimate the German mark of a student who got 30 in French.

d) Estimate the French mark of a student who got 45 in German.

7. The data below gives the petrol consumption figures of cars, with the same size engine, when driven at different speeds.

Speed (m.p.h.)	30	62	40	80	70	55	75
Petrol consumption (m.p.g.)	38	25	35	20	26	34	22

m.p.h. stands for 'miles per hour' and m.p.g. stands for 'miles per gallon'.

a) Plot a scatter graph and draw a line of best fit.

b) Estimate the petrol consumption of a car travelling at 45 m.p.h.

c) Estimate the speed of a car whose petrol consumption is 27 m.p.g.

Equation of linear regression

So far we have drawn a line of best fit on a scatter diagram by 'judgment'. Consequently, different lines will be drawn by different people for the same data. **Linear regression** is more complicated but has the advantage that it always produces the same line for a given set of data.

Briefly, we find the vertical distances of points d_1, d_2, d_3, ... from the line of best fit.
We then work out $d_1^2 + d_2^2 + d_3^2 + \ldots$
The **least squares regression line** in the line which makes this sum its lowest value.
The equation of the line of regression is found using a graphic display calculator (see page 33 in Chapter 1).

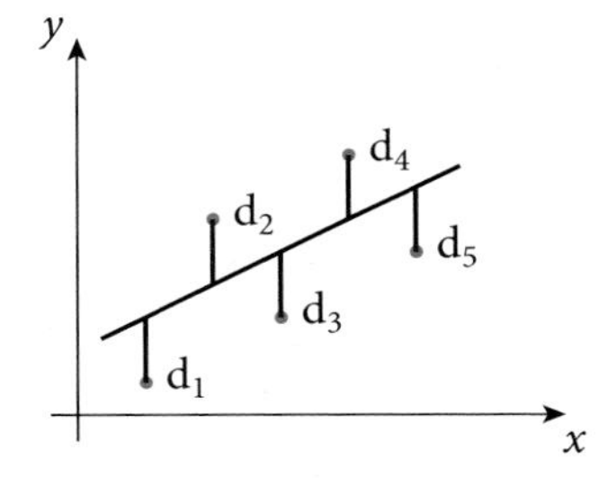

Exercise 14.13

1. Use a graphic display calculator to find the equation of the line of regression.

a)

x	2	14	14	4	12	18	12	6
y	5	15	16	6	12	18	13	7

b)

x	2	15	17	3	20	3	6
y	13	7	5	12	4	13	11

2. The following data gives the marks of 11 students in a French test and in a German test.

French, x	15	36	36	22	23	27	43	22	43	40	26
German, y	6	28	35	18	28	28	37	9	41	45	17

a) Plot this data on a scatter graph, with the French marks on the horizontal axis.

b) Find the equation of the line of regression.

c) Estimate the German mark of a student who got 30 in French.

d) Estimate the French mark of a student who got 45 in German.

3. The data below gives the petrol consumption figures of cars, with the same size engine, when driven at different speeds.

Speed (m.p.h.), x	30	62	40	80	70	55	75
Petrol consumption (m.p.g.), y	38	25	35	20	26	34	22

a) Find the equation of the line of regression.
b) Estimate the petrol consumption of a car travelling at 45 m.p.h.
c) Estimate the speed of a car whose petrol consumption is 27 m.p.g.

4. For a medical survey 12 children had the span of their right hand and the length of their right foot measured. These are the results:

Hand span (cm), x	16·5	15	14	12·5	12	11	24	24	21	21	20
Length of foot (cm), y	20	23	18	19	17	14	28	24·5	27·5	23	25

a) Draw a scatter graph, plotting handspan on the horizontal axis and length of foot on the vertical axis. Take values from 10 to 28 on both axes.
b) Describe the results in one sentence.
c) Find the equation of the line of regression.
d) One girl had her leg in plaster so her foot length could not be measured. Her hand span was 17·5 cm. Draw the line of regression on your scatter graph and use it to estimate the likely length of her foot.

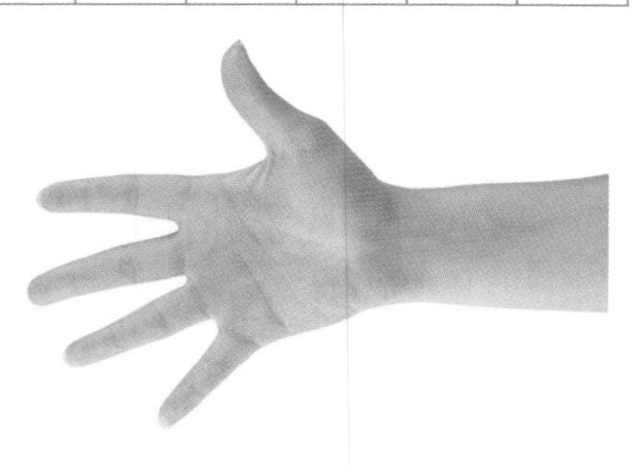

Revision exercise 14A

1. A pie chart is drawn with sectors to represent the following percentages:

   20%, 45%, 30%, 5%.

   What is the angle of the sector which represents 45%?

2. The pie chart shows the numbers of votes for candidates A, B and C in an election. What percentage of the votes were cast in favour of candidate C?

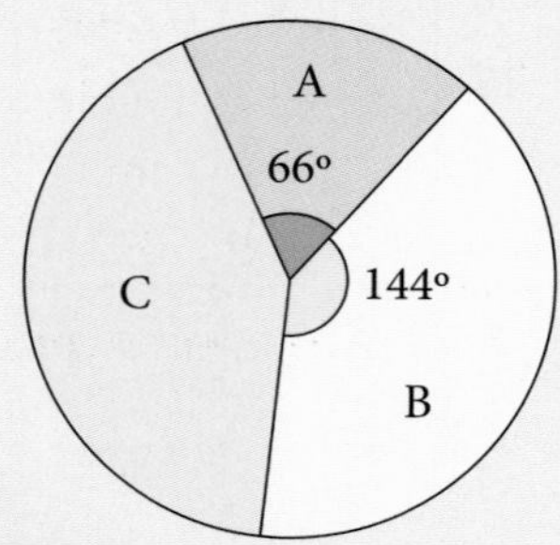

3. A pie chart is drawn from the following data to show the expenditure of a football club.

Wages	\$41 000
Travel	\$9000
Rates	\$6000
Miscellaneous	\$4000

   What is the angle of the sector showing the expenditure on travel?

4. The mean of four numbers is 21.
   a) Calculate the sum of the four numbers.

   Six other numbers have a mean of 18.

   b) Calculate the mean of the ten numbers.

5. Find:
   a) the mean
   b) the median
   c) the mode

   of the numbers 3, 1, 5, 4, 3, 8, 2, 3, 4, 1.

6.

Marks	3	4	5	6	7	8
Number of pupils	2	3	6	4	3	2

The table shows the number of pupils in a class who scored marks 3 to 8 in a test. Find

a) the mean mark **b)** the modal mark

c) the median mark.

7. The mean height of 10 boys is 1·60 m and the mean height of 15 girls is 1·52 m. Find the mean height of the 25 boys and girls.

8.

Mark	3	4	5
Number of pupils	3	x	4

The table shows the number of pupils who scored mark 3, 4 or 5 in a test. Given that the mean mark is 4·1, find x.

In questions **9** and **10**, draw a cumulative frequency curve, and find

a) the median **b)** the interquartile range.

9.

Mass (kg)	Frequency
1–5	4
6–10	7
11–15	11
16–20	18
21–25	22
26–30	10
31–35	5
36–40	3

10.

Length (cm)	Frequency
41–50	6
51–60	8
61–70	14
71–80	21
81–90	26
91–100	14
101–110	7
111–120	4

11. In a competition, 30 children had to pick up as many paper clips as possible in one minute using a pair of tweezers.
The results were as follows.

3	17	8	11	26	23	18	28	33	38
12	38	22	50	5	35	39	30	31	43
27	34	9	25	39	14	27	16	33	49

Construct a frequency table using intervals 1–10, 11–20, etc. and hence draw a cumulative frequency curve.

a) From the curve, estimate the median number of clips picked up.

b) From the frequency table, estimate the mean of the distribution using the mid-interval values 5·5, 15·5, etc.

c) Calculate the exact value of the mean using the original data.

d) Why is it possible only to estimate the mean in part **b)**?

12. The children in two schools took the same test in Mathematics and their results are shown.

IQR is shorthand for 'inter-quartile range'.

School A	School B
Median mark = 52%	Median mark = 51·8%
IQR = 7·2	IQR = 11·2

What can you say about these two sets of results?

13. As part of a health improvement programme, the heights of people from one town and one village in Gambia were measured. Here are the results:

People in town	People in village
Median height = 171 cm	Median height = 163 cm
IQR = 8·4	IQR = 3·7

What can you say about these two sets of results?

Examination exercise 14B

1. During one week a cafe records the number of hot drinks (x) and cold drinks (y) it sells each day. The table shows the results.

Day	Mon	Tues	Wed	Thurs	Fri	Sat	Sun
Number of hot drinks (x)	55	29	40	45	65	80	60
Number of cold drinks (y)	30	46	35	27	20	15	25

a) Complete the scatter diagram by plotting the points for Friday, Saturday and Sunday. The first four points have been plotted for you.

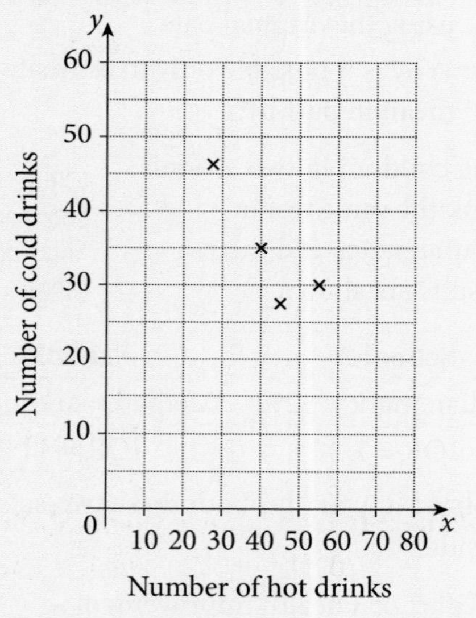

[2]

b) Describe any correlation between x and y. [1]

c) i) Find the equation of the line of regression, giving y in terms of x. [2]

ii) 50 hot drinks are sold on one day in the following week. How many cold drinks would you expect to be sold on this day? [2]

Cambridge IGCSE International Mathematics 0607 Paper 4 Q11 October/November 2010

2. For the set of data

1 2 4 5 6 8 9 9 10 12

find

a) the mean, [2]

b) the mode, [1]

c) the median, [1]

d) the lower quartile. [1]

Cambridge IGCSE International Mathematics 0607 Specimen Paper 2 Q8 2010

3. The cumulative frequency curve shows the heights of 200 plants measured correct to the nearest centimetre.

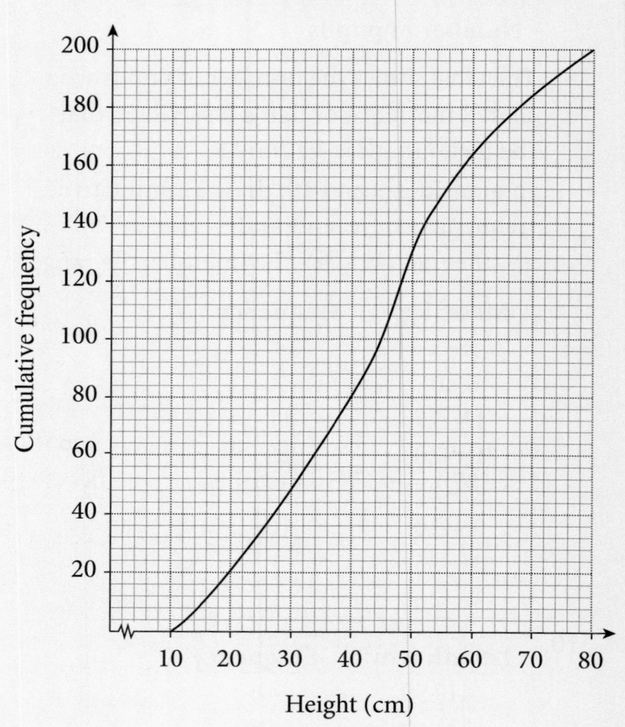

a) Use the graph to find

i) the median, [1]

ii) the interquartile range. [2]

b) Find the percentage of plants with heights greater than 50 cm. [2]

Cambridge IGCSE International Mathematics 0607 Paper 2 Q4 October/November 2010

4. The table shows the marks (x) gained by 100 students in an examination.

Mark (x)	Frequency
$0 \leqslant x < 10$	20
$10 \leqslant x < 20$	10
$20 \leqslant x < 40$	10
$40 \leqslant x < 45$	30
$45 \leqslant x < 60$	30

Use this information to draw a histogram on the grid below.

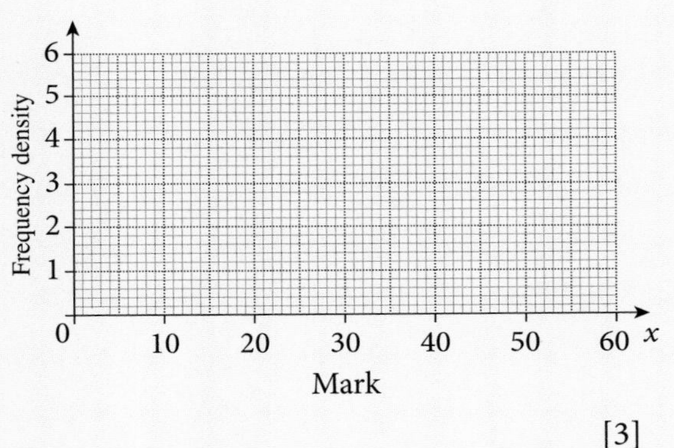

[3]

Cambridge IGCSE International Mathematics 0607 Paper 2 Q9 October/November 2010

5. The masses of 100 bags of flour are given in the table.

Mass (m grams)	Frequency
$980 \leqslant m < 990$	4
$990 \leqslant m < 1000$	10
$1000 \leqslant m < 1005$	50
$1005 \leqslant m < 1010$	20
$1010 \leqslant m < 1020$	8
$1020 \leqslant m < 1040$	8

a) Calculate an estimate of the mean mass of a bag of flour, correct to the nearest gram. [3]

b) i) Complete the frequency density column in this table.

Mass (m grams)	Frequency	Frequency density
$980 \leqslant m < 990$	4	
$990 \leqslant m < 1000$	10	
$1000 \leqslant m < 1005$	50	
$1005 \leqslant m < 1010$	20	
$1010 \leqslant m < 1020$	8	
$1020 \leqslant m < 1040$	8	

[3]

ii) On a grid with axes and scales as shown, draw an accurate histogram to show this information.

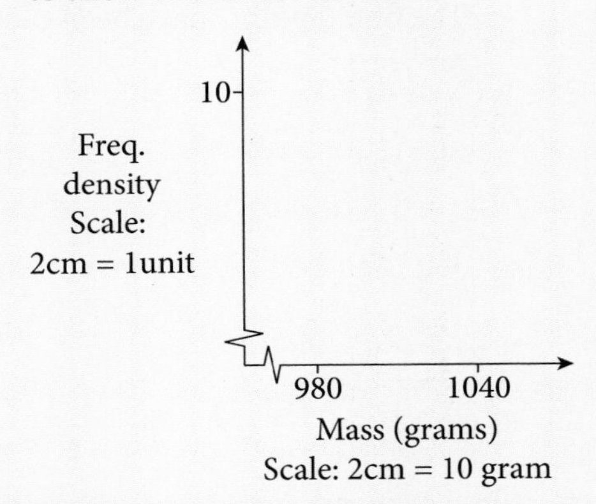

Cambridge IGCSE International Mathematics 0607 Paper 4 Q4 May/June 2010

6. Nyali sells cakes and ice creams.
She records the number of cakes (c) and the number of ice creams (i) she sells each day for 10 days. The results are shown in the table.

Number of cakes (c)	Number of ice creams (i)
48	50
60	18
52	38
40	50
60	30
36	54
70	14
20	70
44	46
50	50

a) Complete the scatter diagram.
The first 6 points have been plotted for you.

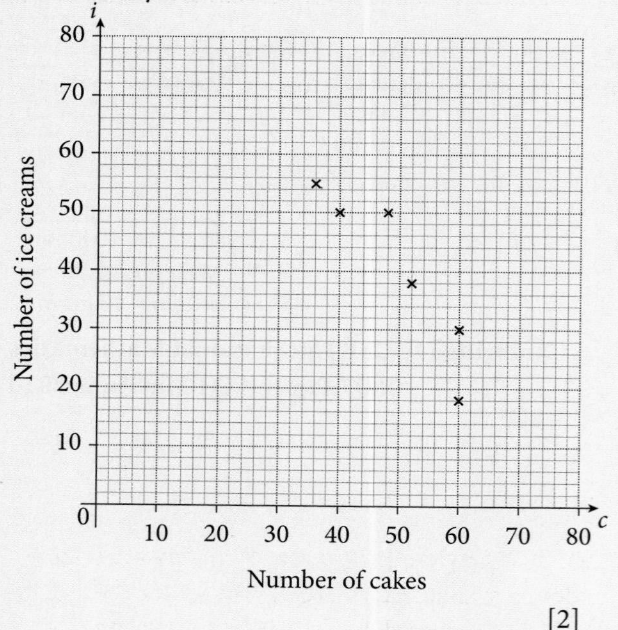

[2]

b) Write down one word to describe the correlation between c and i. [1]

c) Find the equation of the line of regression, writing i in terms of c. [2]

d) Use your equation to estimate the number of ice creams
Nyali sells on a day when she sells 67 cakes. [1]

Cambridge IGCSE International Mathematics 0607 Paper 41 Q7 May/June 2011

15 Investigations and mathematical modelling 2

Andrew Wiles (1953–) As a 10-year-old boy Andrew read about Fermat's Last Theorem which had not been proved since it was stated over 300 years ago. The theorem states that there are no integer solutions to the equation $x^n + y^n = z^n$ for $n > 2$.
Andrew spent nine years on the proof and eventually proved the theorem in 1995.

15.1 Investigations

There are a large number of possible starting points for these investigations so it may be possible for you to choose investigations which appeal to you. On other occasions, your teacher may set the same investigation for the whole class.

Here are a few guidelines for you:

- If the set problem is too complicated try an easier case.
- Draw your own diagrams.
- Make tables of your results and be systematic.
- Look for patterns.
- Is there a rule or formula to describe the results?
- Can you *predict* further results?
- Can you *prove* any rules which you may find?

Proof

- In the field of science a theory (like Newton's theory of gravitation) can never be proved. It can only be considered highly likely using all the evidence available at the time. The history of science contains many examples of theories which were accepted at the time but were later shown to be untrue when more accurate observation was possible.
- A mathematical proof is far more powerful. Once a theorem is proved mathematically it will *always* be true. Pythagoras proved his famous theorem over 2500 years ago and when he died he knew it would never be disproved.
  A proof starts with simple facts which are accepted. The proof then argues logically to the result which is required.

1. Diamonds and triangles

In this task you will draw rectangles on triangular dotty paper (make sure you have the paper the right way up!)

A

In each diagram you have:

- equilateral triangles (e),
- isosceles triangles (i),
- diamonds (d).

In all rectangles each corner must be on a dot.

This is correct:

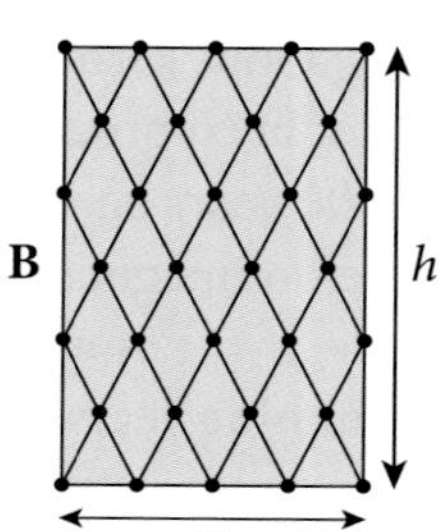

This is not allowed:

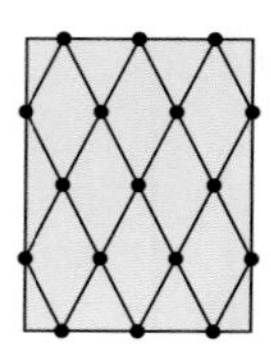

Each rectangle has width w and height h.

In rectangle A: $w = 3, h = 2$

In rectangle B: $w = 4, h = 3$

Both w and h must be whole numbers.

a) Count the number of equilateral triangles e in diagrams **A** and **B** and in rectangles of your own.

Can you find a rule connecting e and w or a rule connecting e and h?

b) Count the number of isosceles triangles i in your rectangles. Can you find a rule connecting i with either w or h?

c) The rule connecting d with w and h is more difficult to find. Be systematic by putting your results in tables. First keep w the same and change only h.

For example:

i)

w	h	d
2	2	4
2	3	?
2	4	?
2	5	?

ii) Change to $w = 3$

w	h	d
3	2	?
3	3	?
3	4	?
3	5	?

iii) Change to $w = 4$ etc.

As you look for a connection try the following:

- in part **i)** write a column in the table for $3h$
- in part **ii)** write a column in the table for $5h$
- in part **iii)** write a column in the table for $7h$

Can you see a rule connecting, w, h and d for each of the different widths?

Try to write '$d = \ldots$'

Now add a column for $w \times h \times 2$.

Can you now see a rule that works for any diagram you could draw?

2. Connect the transformations

The triangle ABC is mapped onto triangle A′B′C′ by a rotation followed by a translation.

For example: Rotation 90° anticlockwise about (1, 1)

followed by translation $\begin{pmatrix} -3 \\ 3 \end{pmatrix}$

or Rotation 90° anticlockwise about (3, 0)

followed by translation $\begin{pmatrix} -4 \\ 6 \end{pmatrix}$

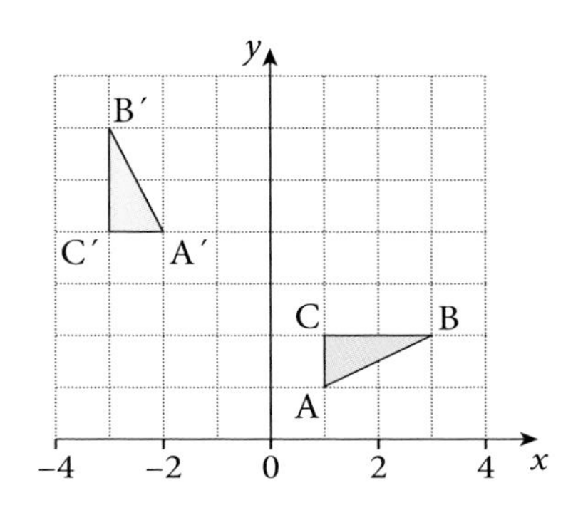

Is there a connection between the centre of the rotation and the vector of the translation?

This is not an easy question to answer. You need to work methodically and to record your results in a table.

Suggestions:

a) Take the following points as the centres of the rotation: (0, 0), (1, 1), (2, 2), (3, 3) etc.

Record the results in a table.

Write down any connection you notice between the centre of rotation and the vector of the translation.

Centre	Translation
(0, 0)	$\begin{pmatrix} -1 \\ 3 \end{pmatrix}$
(1, 1)	$\begin{pmatrix} -3 \\ 3 \end{pmatrix}$

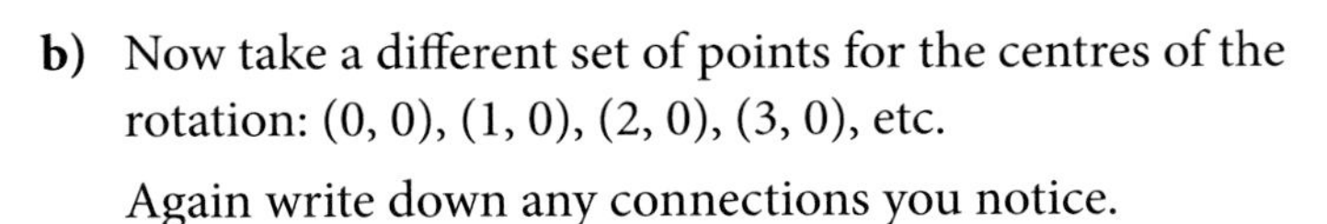

b) Now take a different set of points for the centres of the rotation: (0, 0), (1, 0), (2, 0), (3, 0), etc.

Again write down any connections you notice.

c) Can you use your results from **a)** and **b)** above to *predict* the vector of the translation if the centre of the rotation is (7, 7) or (10, 0)?

d) Investigate other sets of points for the centre of rotation. Can you find a general rule which predicts the vector of the translation for *any* centre of rotation?

3. In search of π

You have been using the fact that the circumference of a circle = $\pi \times$ diameter, where π is some value a bit more than 3. You may also have been told that π is an **irrational** number which means it does not have an exact decimal value, and that it has been calculated to many hundreds of decimal places. So how can we be sure this value is correct?

Firstly, find the perimeter of a regular pentagon where the distance from the centre to the vertices is 0·5 cm.

A pentagon is made up of five isosceles triangles.

$z = 360° \div 5 = 72°$

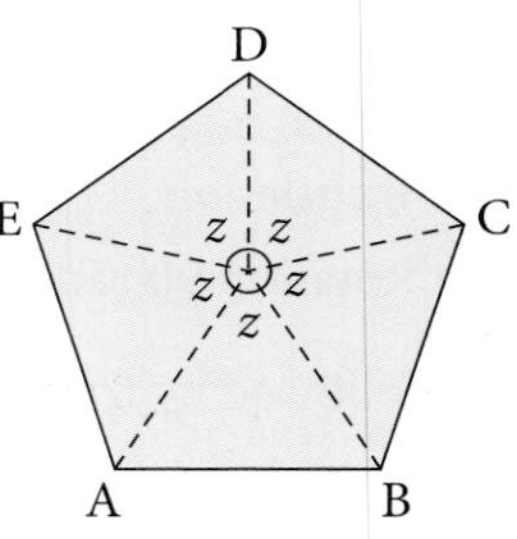

By splitting triangle AOB into two right-angled triangles we can find AX, where X is the mid-point of AB.

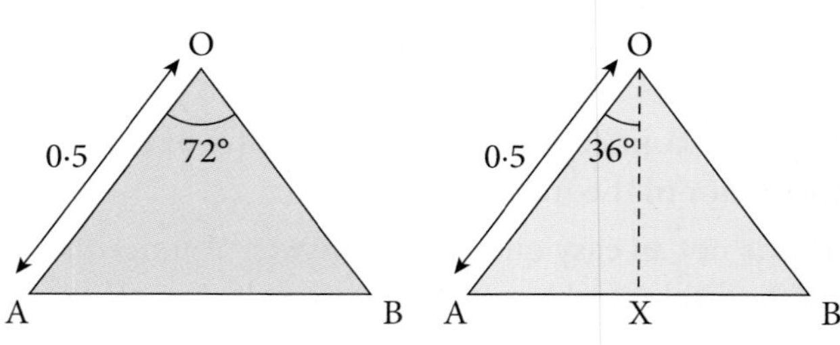

$AX = 0{\cdot}5 \sin 36°$

$AB = 2\,AX = 2(0{\cdot}5 \sin 36°)$

$AB = \sin 36° = 0{\cdot}5877852523$ (10 d.p.)

Perimeter of pentagon $= 5 \times 0{\cdot}5877852523$.

In a similar way, find the perimeters of the following regular polygons with the same distance from the centre to the vertices.

a) An octagon (8 sides) **b)** A decagon (10 sides)

c) A 20 sided polygon **d)** A 100 sided polygon

e) A 1000 sided polygon!

The more sides we take the closer our shape resembles a circle with a diameter of 1 cm and the nearer our perimeter gets to the circumference of a circle of diameter 1 cm.

4. Squares

For this investigation you need either dotty paper or squared paper.
The yellow square has an area of 1 unit.
Can you draw a square, with its corners on the dots, with an area of 2 units?
Can you draw a square with an area of 3 units?
Can you draw a square with an area of 4 units?
Investigate for squares up to 100 units.
For which numbers x can you draw a square of area x units?

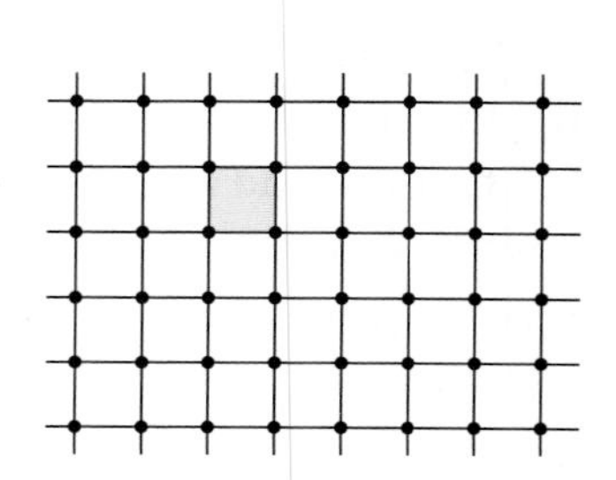

5. Painting cubes

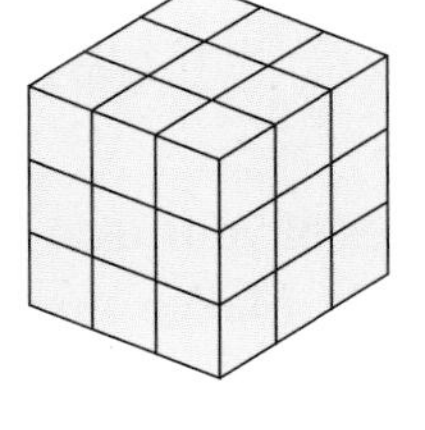

The large cube here consists of 27 unit cubes.
All six faces of the large cube are painted green.

How many unit cubes have 3 green faces?
How many unit cubes have 2 green faces?
How many unit cubes have 1 green face?
How many unit cubes have 0 green faces?

Suppose the large cube is $20 \times 20 \times 20$.
Answer the four questions above.
Answer the four questions for the cube which is $n \times n \times n$.

6. Find the connection

Work through the flow diagram several times, using a calculator.

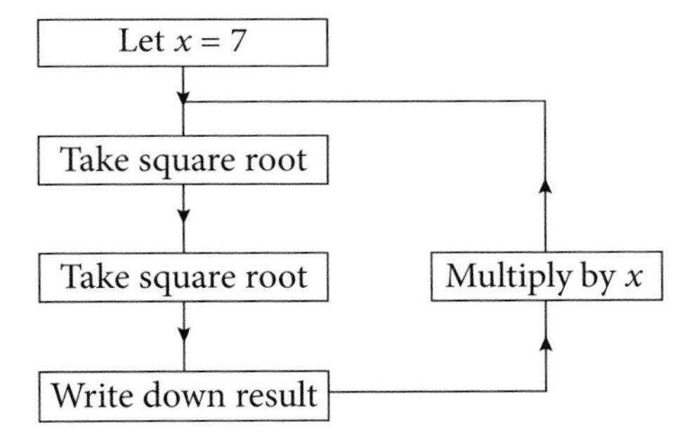

What do you notice?
Try different numbers for x (suggestions: 11, 5, 8, 27)
What do you notice?

What happens if you take the square root three times?

Suppose in the flow diagram you change
'Multiply by x' to 'Divide by x'. What happens now?

Suppose in the flow diagram you change
'Multiply by x' to 'Multiply by x^2'. What happens now?

7. Spotted shapes

For this investigation you need dotty paper. If you have not got any, you can make your own using a felt-tip pen and squared paper.

The rectangle in Diagram 1 has 10 dots on the perimeter ($p = 10$) and 2 dots inside the shape ($i = 2$). The area of the shape is 6 square units ($A = 6$).

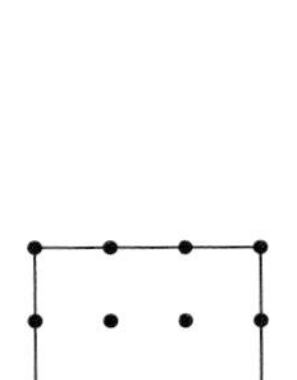
Diagram 1

The triangle in Diagram 2 has 9 dots on the perimeter ($p = 9$) and 4 dots inside the shape ($i = 4$). The area of the triangle is $7\frac{1}{2}$ square units $\left(A = 7\frac{1}{2}\right)$.

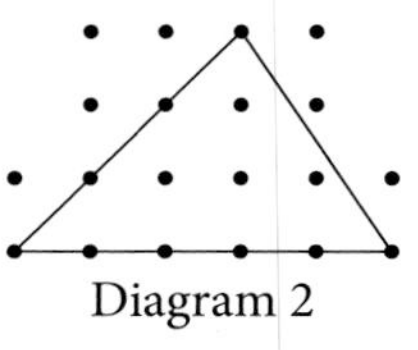

Diagram 2

Draw more shapes of your own design and record the values for p, i and A in a table. Make some of your shapes more difficult like the one in Diagram 3.

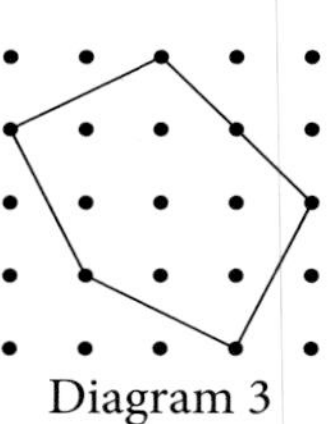

Diagram 3

Can you find a formula connecting p, i and A?

Try out your formula with some more shapes to see if it always works.

8. Diagonals

In a 4×7 rectangle the diagonal passes through 10 squares.

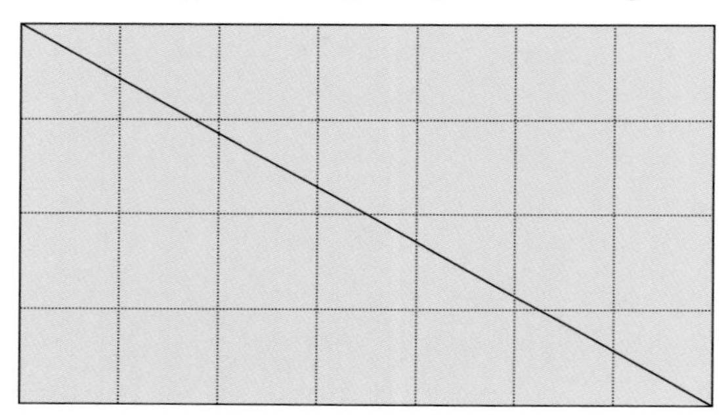

Draw rectangles of your own choice and count the number of squares through which the diagonal passes.

A rectangle is 640×250. How many squares will the diagonal pass through?

9. Biggest number

A calculator has the following buttons:

Also, the only digits buttons which work on this calculator are the '1', '2' and '3'.

a) You can press any button, but only once.
What is the biggest number you can get?

b) Now the '1', '2', '3' and '4' buttons are working.
What is the biggest number you can get?

c) Investigate what happens as you increase the number of digits which you can use.

10. Final score

The final score in a football match was 3–2. How many different scores were possible at half-time?

Investigate for other final scores where the difference between the teams is always one goal (1–0, 5–4, etc.). Is there a pattern or rule which would tell you the number of possible half-time scores in a game which finished 58–57?

Suppose the game ends in a draw. Find a rule which would tell you the number of possible half-time scores if the final score was 63–63.

Investigate for other final scores (3–0, 5–1, 4–2 etc).

11. Cutting paper

The rectangle ABCD is cut in half to give two smaller rectangles.

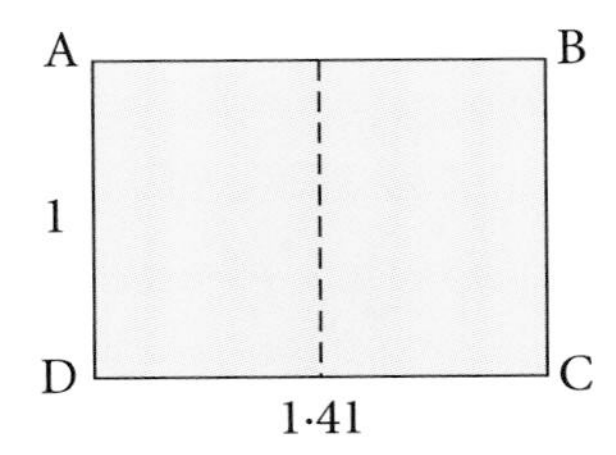

Each of the smaller rectangles is mathematically similar to the large rectangle. Find a rectangle which has this property.

What happens when the small rectangles are cut in half? Do they have the same property?

Why is this a useful shape for paper used in business?

12. Fibonacci sequence

Fibonacci was the nickname of the Italian mathematician Leonardo de Pisa (1170–1250 CE). The sequence which bears his name has fascinated mathematicians for hundreds of years. You can, if you like, join the Fibonacci Association which was formed in 1963.

Here is the start of the sequence:

1, 1, 2, 3, 5, 8, 13, 21, 34, 55, 89, 144,...

There are no prizes for working out the next term!

The sequence has many interesting properties to investigate. Here are a few suggestions:

a) Add three terms.
$1 + 1 + 2$, $1 + 2 + 3$, etc.
Add four terms.

b) Add squares of terms
$1^2 + 1^2$, $1^2 + 2^2$, $2^2 + 3^2$,...

c) Ratios
$\frac{1}{1} = 1$, $\frac{2}{1} = 2$, $\frac{3}{2} = 1{\cdot}5$,...

d) In fours | 2 3 5 8 |
$2 \times 8 = 16$, $3 \times 5 = 15$

e) In threes | 3 5 8 |
$3 \times 8 = 24$, $5^2 = 25$

f) Take a group of 10 consecutive terms. Compare the sum of the 10 terms with the seventh member of the group.

g) In sixes | 1 1 2 3 5 8 |
Square and add the first five numbers
$1^2 + 1^2 + 2^2 + 3^2 + 5^2 = 40$
$5 \times 8 = 40$
Now try seven numbers from the sequence, or eight...

13. Maximum cylinder

A rectangular piece of paper has a fixed perimeter of 40 cm. It could for example be 7 cm × 13 cm.
This paper can make a hollow cylinder of height 7 cm or of height 13 cm.
Work out the volume of each cylinder.

What dimensions should the paper have so that it can make a cylinder of the maximum possible volume?

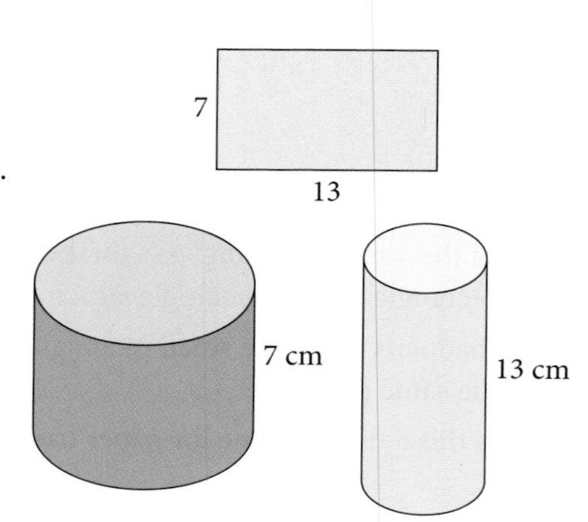

Past examination question paper

B MODELLING THE SOLAR SYSTEM **16 marks**

You are advised to spend no more than 35 minutes on this part.

Logarithms to base 10 are written as log.

1. The table below shows information about seven planets in the Solar System.

Planet	Distance from the Sun (S km)	Time to orbit the Sun (T days)	log S	log T
Mercury	5.79×10^7	88	7.8	1.9
Venus	1.08×10^8	225	8.0	2.4
Earth	1.50×10^8	365	8.2	2.6
Mars	2.28×10^8	687		
Jupiter	7.78×10^8	4330		
Saturn	1.43×10^9	10 800		
Pluto	5.91×10^9	90 800	9.8	5.0

Complete the table of values for log S and log T.
Give each value correct to 2 significant figures.

2. a) Draw a grid like the one shown and plot the seven points (log S, log T).
   b) Plot the mean point (8.6, 3.2) and use this to draw a line of best fit.
   (Do this by eye. Do not use your calculator.)

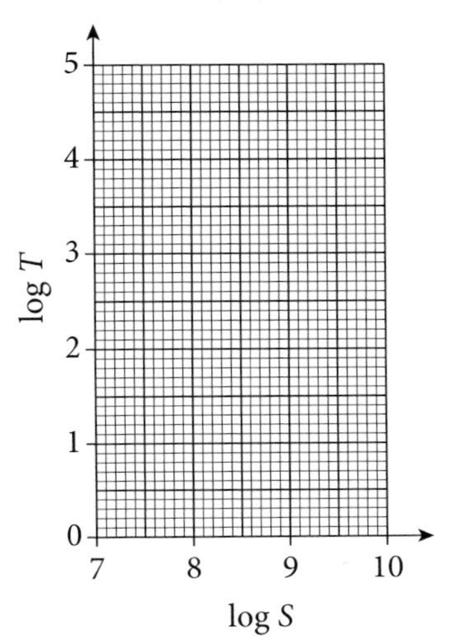

3. The time taken for the planet Uranus to orbit the Sun is 30 685 days.

   Use your graph to estimate the distance of Uranus from the Sun.
   Give your answer correct to 2 significant figures.

4. Let $x = \log S$ and $y = \log T$.
   The equation of the line of best fit is $y = mx + c$.

   Use your calculator to find the values of m and c.
   Give answer correct to 2 significant figures.

5. A model for this is $\log T = m \log S + c$.
   The distance of the planet Neptune from the Sun is 4.50×10^9 km.

   Use the model to find the time taken for Neptune to orbit the Sun.
   Give your answer in standard form correct to 2 significant figures.

6. Writing c as $\log k$, the model can be written as $\log T = m \log S + \log k$.
   a) Show that $T = kS^m$.
   b) Find the value of k.
   c) Write the model $T = kS^m$ using your values of k and m.
   Use the data for Earth to test this model.

Cambridge IGCSE International Mathematics 0607
Paper 6 QB October/November 2010

Answers

The answers given here are concise. However, when answering exam-style questions, you should show as many steps in your working as possible. Full worked solutions to the examination exercises can be found at www.oxfordsecondary.co.uk/ 0607. In addition, the CD contains step-by-step solutions to selected questions, given as a PowerPoint presentation.

Cambridge International Examinations bears no responsibility for the example answers to questions taken from its past paper questions which are contained in this publication.

2 Number

Exercise 2.1 page 40

1. 7·91 **2.** 22·22 **3.** 7·372 **4.** 0·066 **5.** 466·2 **6.** 1·22 **7.** 1·67
8. 1·61 **9.** 16·63 **10.** 24·1 **11.** 26·7 **12.** 3·86 **13.** 0·001 **14.** 1·56
15. 0·0288 **16.** 2·176 **17.** 0·02 **18.** 0·0001 **19.** 7·56 **20.** 0·7854 **21.** 360
22. 34 000 **23.** 18 **24.** 0·74 **25.** 2·34 **26.** 1620 **27.** 8·8 **28.** 1200
29. 0·00 175 **30.** 13·2 **31.** 200 **32.** 0·804 **33.** 0·8 **34.** 0·077 **35.** 0·0009
36. 0·01 **37.** 184 **38.** 20 **39.** 0·099 **40.** 3

Exercise 2.2 page 40

1. 20 **2.** 256; 65 536 **3.** (2 × 1·5) + (1 × 2) **4.** € 22 940 **5.** 250 m
6. 0·5 ÷ 10 **7.** $6^2+7^2+42^2=43^2;\ x^2+(x+1)^2+[x(x+1)]^2=(x^2+x+1)^2$
8. a) 54 × 9 = 486 **b)** 57 × 8 = 456 **c)** 52 × 2 = 104, 57 × 2 = 114 or 58 × 3 = 174
9. 5, 28; total 32 **11.** 12 units long, marks at 1, 4, 5 and 10 **12.** 21 **15.** 37

Exercise 2.3 page 42

1. $1\frac{11}{20}$ **2.** $\frac{11}{24}$ **3.** $1\frac{1}{2}$ **4.** $\frac{5}{12}$ **5.** $\frac{4}{15}$ **6.** $\frac{1}{10}$ **7.** $\frac{8}{15}$
8. $\frac{5}{42}$ **9.** $\frac{15}{26}$ **10.** $\frac{5}{12}$ **11.** $4\frac{1}{2}$ **12.** $1\frac{2}{3}$ **13.** $\frac{23}{40}$ **14.** $\frac{3}{40}$
15. $1\frac{7}{8}$ **16.** $1\frac{1}{12}$ **17.** $1\frac{1}{6}$ **18.** $2\frac{5}{8}$ **19.** $6\frac{1}{10}$ **20.** $9\frac{1}{10}$ **21.** $1\frac{9}{26}$
22. $\frac{1}{9}$ **23.** $\frac{2}{3}$ **24.** $5\frac{1}{4}$ **25.** $2\frac{2}{25}$ **26. a)** $\frac{1}{2}, \frac{7}{12}, \frac{2}{3}$ **b)** $\frac{2}{3}, \frac{3}{4}, \frac{5}{6}$ **c)** $\frac{1}{3}, \frac{5}{8}, \frac{17}{24}, \frac{3}{4}$
d) $\frac{5}{6}, \frac{8}{9}, \frac{11}{12}$ **27. a)** $\frac{1}{2}$ **b)** $\frac{3}{4}$ **c)** $\frac{17}{24}$ **d)** $\frac{7}{18}$ **e)** $\frac{3}{10}$ **f)** $\frac{5}{12}$
28. 5 **29.** Rome, Moscow **30.** $\frac{1}{2}+\frac{1}{3}+\frac{1}{6}$ **31.** $\frac{1}{2}+\frac{1}{4}+\frac{1}{5}+\frac{1}{20}$ **32.** $\frac{9}{40}$
33. $1\frac{5}{8}$ g **34.** $\frac{19}{48}$ **35. a)** $\frac{5}{24}$ **b)** 26 **c)** $\frac{3}{50}$
36. $\frac{7}{9}$ is missing fraction **37.** 11 **38.** 9 **39.** same

Exercise 2.4 page 44

1. 0·25 **2.** 0·4 **3.** 0·8 **4.** 0·75 **5.** 0·5 **6.** 0·375 **7.** 0·9 **8.** 0·625 **9.** $0{\cdot}41\dot{6}$ **10.** $0{\cdot}1\dot{6}$
11. $0{\cdot}\dot{6}$ **12.** $0{\cdot}8\dot{3}$ **13.** $0{\cdot}\dot{2}8571\dot{4}$ **14.** $0{\cdot}\dot{4}2857\dot{1}$ **15.** $0{\cdot}\dot{4}$ **16.** $0{\cdot}\dot{4}\dot{5}$ **17.** 1·2 **18.** 2·625 **19.** $2{\cdot}\dot{3}$
20. 1·7 **21.** 2·1875 **22.** $2{\cdot}\dot{2}8571\dot{4}$ **23.** $2{\cdot}\dot{8}571$ **24.** 3·19 **25.** $\frac{1}{5}$ **26.** $\frac{7}{10}$ **27.** $\frac{1}{4}$ **28.** $\frac{9}{20}$
29. $\frac{9}{25}$ **30.** $\frac{13}{25}$ **31.** $\frac{1}{8}$ **32.** $\frac{5}{8}$ **33.** $\frac{21}{25}$ **34.** $2\frac{7}{20}$ **35.** $3\frac{19}{20}$ **36.** $1\frac{1}{20}$ **37.** $3\frac{1}{5}$ **38.** $\frac{27}{100}$
39. $\frac{7}{1000}$ **40.** $\frac{11}{100000}$ **41.** 0·58 **42.** 1·42 **43.** 0·65 **44.** 1·61 **45.** 0·07 **46.** 0·16 **47.** 3·64 **48.** 0·60
49. $\frac{4}{15}$, 0·33, $\frac{1}{3}$ **50.** $\frac{2}{7}$, 0·3, $\frac{4}{9}$ **51.** $\frac{7}{11}$, 0·705, 0·71 **52.** $\frac{5}{18}$, 0·3, $\frac{4}{13}$

Exercise 2.5 page 45

1. $r=\frac{4}{9}$ **2.** $r=\frac{28}{99}$ **3.** $\frac{2}{9}$ **4.** $\frac{73}{99}$ **5.** $\frac{51}{99}$ **6.** $\frac{29}{99}$ **7.** $\frac{245}{999}$
8. $\frac{326}{999}$ **9.** $\frac{139}{333}$ **10.** $\frac{82}{99}$ **11. a)** $0{\cdot}1\dot{6}$ **b)** $0{\cdot}\dot{8}$ **c)** $0{\cdot}\dot{2}8571\dot{4}$ **d)** $0{\cdot}\dot{3}8461\dot{5}$

Exercise 2.6 *page 46*

1. 3, 11, 19, 23, 29, 31, 37, 47, 59, 61, 67, 73

2. **a)** 4, 8, 12, 16, 20 **b)** 6, 12, 18, 24, 30 **c)** 10, 20, 30, 40, 50 **d)** 11, 22, 33, 44, 55 **e)** 20, 40, 60, 80, 100

3. 12 and 24 **4.** 15 **5.** **a)** 1, 2, 3, 6 **b)** 1, 3, 9 **c)** 1, 2, 5, 10, **d)** 1, 3, 5, 15 **e)** 1, 2, 3, 4, 6, 8, 12, 24 **f)** 1, 2, 4, 8, 16, 32

6. **a)** Yes. Divide by 3, 5, 7, 11, 13, (i.e. odd prime numbers $<\sqrt{263}$) **b)** No **c)** Prime numbers $<\sqrt{1147}$

7. 2, 3, 5, 41, 67, 89 **8.** **a)** false **b)** true **c)** false **9.** **a)** $\frac{2}{3}$ **b)** $1\frac{5}{7}$ **10.** $\frac{11}{7}$ **11.** 50

12. **a)** 5 / 11 4 **b)** 7 / 8 9 **c)** 7 / 13 9 **13.** **a)** $-\sqrt{9}, -2, 1^3, \frac{7}{5}, \sqrt{25}, 3^2, |-11|$ **b)** $-(0{\cdot}2)^2, \left(\frac{1}{10}\right)^2, |x|, \sqrt{400}, (1 \div 0{\cdot}01)$

Exercise 2.7 *page 48*

1. Rational: $(\sqrt{17})^2$; 3·14; $\frac{\sqrt{12}}{\sqrt{3}}$; $3^{-1} + 3^{-2}$; $\frac{22}{7}$; $\sqrt{2{\cdot}25}$

3. **a)** A 4, B $\sqrt{41}$ **b)** A rational, B irrational **c)** Both rational **d)** B

4. **a)** 6π cm, irrational **b)** 6 cm, rational **c)** 36 cm^2, rational **d)** 9π cm^2, irrational **e)** $36 - 9\pi$ cm^2, irrational

5. **a)** true **b)** true **c)** true **8.** **a)** No **b)** Yes e.g. $\sqrt{8}\times\sqrt{2}=4$

10. **a)** 74 **b)** irrational **11.** rational **12.** (e.g.) 5·29

14. e.g. **a)** $\sqrt{7}$ **b)** $\sqrt{3}$ **c)** $\sqrt{2}$ **d)** π

Exercise 2.8 *page 50*

1. AF, EG, BD, CH

2. **a)** T **b)** T **c)** F **d)** T **e)** F **f)** T

3. **a)** $3+2\sqrt{2}$ **b)** $7-4\sqrt{3}$ **c)** 18 **d)** $14-6\sqrt{5}$ **e)** $6+4\sqrt{2}$ **f)** 8

4. **a)** T **b)** T **c)** F **d)** T **e)** F **f)** T

5. **a)** 4 **b)** $4\sqrt{2}$ **c)** $10\sqrt{3}$ **d)** $9\sqrt{2}$ **e)** $5\sqrt{5}$ **f)** $5\sqrt{3}$ **g)** $5\sqrt{5}$ **h)** $\sqrt{3}$ **i)** 2 **j)** $1\frac{1}{2}$ **k)** $2\frac{1}{2}$ **l)** $1\frac{1}{3}$

6. **a)** $2\sqrt{3}$ **b)** **i)** $4\sqrt{2}$ **ii)** $4\sqrt{3}$ **iii)** $3\sqrt{2}$ **iv)** $\frac{2}{3}\sqrt{3}$

7. **a)** $4\sqrt{2}-4$ **b)** **i)** $2\sqrt{3}-2$ **ii)** $\sqrt{5}-2$ **iii)** $\frac{10}{3}(\sqrt{7}+2)$

Exercise 2.9 *page 51*

1. **a)** $2n$ **b)** $10n$ **c)** $3n$ **d)** $11n$ **e)** $100n$ **f)** n^2 **g)** 10^n **h)** n^3

2. $n^2 + 4$ **3.** 9, 11, 13, 15, 17

4. **a)** 3, 4, 5, 6, 7 **b)** 5, 10, 15, 20, 25 **c)** 9, 19, 29, 39, 49 **d)** 97, 94, 91, 88, 85 **e)** $1, \frac{1}{2}, \frac{1}{3}, \frac{1}{4}, \frac{1}{5}$ **f)** 1, 4, 9, 16, 25

Exercise 2.10 *page 52*

1. $4n + 1$ **2.** $3n + 4$ **3.** $5n - 1$ **4.** $4n + 2$ **5.** $3n + 2$ **6.** $28 - 3n$ **7.** $5n$

8. 2^n **9.** $n(n + 2)$ **10.** $\frac{n}{n+1}$ **11.** $7n$ **12.** n^2 **13.** $\frac{5}{n^2}$ **14.** $\frac{n+2}{n}$

15. $4n - 1$ **16.** $2n + 3$ **17.** $9 - 2n$ **18.** $4n - 9$ **19.** **a)** $2n + 6$ **b)** $4n - 1$ **c)** $5n + 3$

20. **a)** $8n + 3$ **b)** $2n+\frac{1}{2}$ **c)** $3n - 10$ **21.** **a)** $3n + 1$ **b)** 3001

22. **a)** $4n$ **b)** $2n + 1$ **c)** $\frac{n}{n+1}$ **d)** $3n + 1$ **23.** **a)** 26 **b)** 10001

Exercise 2.11 *page 53*

1. **a)** 8 **b)** 8·17 **c)** 8·17 **2.** **a)** 20 **b)** 19·6 **c)** 19·62

3. **a)** 20 **b)** 20·0 **c)** 20·04 **4.** **a)** 1 **b)** 0·815 **c)** 0·81

5. **a)** 311 **b)** 311 **c)** 311·14 **6.** **a)** 0 **b)** 0·275 **c)** 0·28

7. **a)** 0 **b)** 0·00 747 **c)** 0·01 **8.** **a)** 16 **b)** 15·6 **c)** 15·62

9. **a)** 900 **b)** 900 **c)** 900·12 **10.** **a)** 4 **b)** 3·56 **c)** 3·56

11. **a)** 5 **b)** 5·45 **c)** 5·45 **12.** **a)** 21 **b)** 21·0 **c)** 20·96

13. **a)** 0 **b)** 0·0851 **c)** 0·09 **14.** **a)** 1 **b)** 0·515 **c)** 0·52

15. **a)** 3 **b)** 3·07 **c)** 3·07 **16.** 5·7 **17.** 0·8

18. 11·2 **19.** 0·1 **20.** 0·0 **21.** 11·1

Exercise 2.12 *page 54*

1. 70·56 **2.** 118·958 **3.** 451·62 **4.** 33 678·8 **5.** 0·6174 **6.** 1068 **7.** 19·53

8. 18 914·4 **9.** 38·72 **10.** 0·009 79 **11.** 2·4 **12.** 11 **13.** 41 **14.** 8·9

15. 4·7 **16.** 56 **17.** 0·0201 **18.** 30·1 **19.** 1·3 **20.** 0·31 **21.** 210·21

22. 294 **23.** 282·131 **24.** 35 **25.** 242

Exercise 2.13 page 55

1. \$2000 **2.** 6 kg **3.** \$600 **4.** 10 cm **5.** about 80 km
6. a) 900 **b)** 70 → 80 **c)** 0·2 kg **d)** 90 000 **e)** £2000
f) 0·5 **g)** 100 **h)** 300 **i)** £100 **7.** about 20 000 000
8. about 7 m

Exercise 2.14 page 56

1. 4×10^3 **2.** 5×10^2 **3.** 7×10^4 **4.** 6×10 **5.** 2.4×10^3 **6.** 3.8×10^2 **7.** 4.6×10^4
8. 4.6×10 **9.** 9×10^5 **10.** 2.56×10^3 **11.** 7×10^{-3} **12.** 4×10^{-4} **13.** 3.5×10^{-3} **14.** 4.21×10^{-1}
15. 5.5×10^{-5} **16.** 1×10^{-2} **17.** 5.64×10^5 **18.** 1.9×10^7 **19.** 1.1×10^9 **20.** 1.67×10^{-24} **21.** 5.1×10^8
22. 2.5×10^{-10} **23.** 6.023×10^{23} **24.** 3×10^{10} **25.** \$ 3.6×10^6

Exercise 2.15 page 57

1. 1.5×10^7 **2.** 3×10^8 **3.** 2.8×10^{-2} **4.** 7×10^{-9} **5.** 2×10^6
6. 4×10^{-6} **7.** 9×10^{-2} **8.** 6.6×10^{-8} **9.** 3.5×10^{-7} **10.** 1×10^{-16}
11. 8×10^9 **12.** 7.4×10^{-7} **13.** c, a, b **14.** 13 **15.** 16
16. i) 9×10^2, 4×10^2 **ii)** 1×10^8, 4×10^7 **17.** 50 min **18.** 6×10^2
19. a) 20·5s **b)** 6.3×10^{91} years

Exercise 2.16 page 59

1. 1 : 3 **2.** 1 : 6 **3.** 1 : 50 **4.** 1 : 1·6 **5.** 1 : 0·75
6. 1 : 0·375 **7.** 1 : 25 **8.** 1 : 8 **9.** 2·4 : 1 **10.** 2·5 : 1
11. 0·8 : 1 **12.** 0·02 : 1 **13.** \$15, \$25 **14.** \$36, \$84 **15.** 140 m 110 m
16. \$18, \$27, \$72 **17.** 15 kg, 75 kg, 90 kg **18.** 46 min, 69 min, 69 min
19. £39 **20.** 18 kg, 36 kg, 54 kg, 72 kg **21.** \$400, \$1000, \$1000, \$1600
22. 5 : 3 **23.** \$200 **24.** 3 : 7 **25.** 1 : 5 **26.** $\frac{1}{7}x$ **27.** 6
28. 12 **29.** \$120 **30.** 300 g **31.** 625

Exercise 2.17 page 60

1. \$1·68 **2.** \$252 **3.** 6 days **4.** $2\frac{1}{2}$ litres **5.** 60 km
6. 119 g **7.** \$68·40 **8.** $2\frac{1}{4}$ weeks **9.** 80 c **10. a)** 12; **b)** 2100
11. 4 **12.** 5·6 days **13.** \$175 **14.** 540° **15.** \$1·20
16. 190 m **17.** 1250 **18.** 11·2 h **19.** 57·1 min **20.** 243 kg
21. 12 days

Exercise 2.18 page 62

1. a) €14 **b)** £44·10 **c)** 758 pesos **d)** 69·75 rupees **e)** 209·76 yen
f) 0·26 dinars **2. a)** \$714·29 **b)** \$3968·25 **c)** \$0·16 **d)** \$3103·45
e) \$33·04 **f)** \$663·72 **3.** £3·39 **4.** Cheaper in UK by \$9·23
5. €494·67 **6.** Kuwait £325·86, France £495, Japan £635·53 **7.** €433·63

Exercise 2.19 page 64

1. a) 70 m **b)** 16 m **c)** 3·55 m **d)** 108·5 m
2. a) 5 cm **b)** 3·5 cm **c)** 0·72 cm **d)** 2·86 cm
3. a) 450 000 cm **b)** 4500 m **c)** 4·5 km **4.** 12·3 km
5. 4·71 km **6.** 1 : 20 000 000 **7.** 50 cm **8.** 64 cm **9.** 5·25 cm

Exercise 2.20 page 65

1. 40 m by 30 m; 12 cm²; 1200 m² **2.** 1 m², 6 m² **3.** 0·32 km² **4.** 50 cm²
5. 150 km² **6.** 18·75 hectares **7.** 240 cm² **8.** 1 : 50 000

Exercise 2.21 page 66

1. a) $\frac{3}{5}$ **b)** $\frac{6}{25}$ **c)** $\frac{7}{20}$ **d)** $\frac{1}{50}$
2. a) 25% **b)** 10% **c)** $87\frac{1}{2}$% **d)** $33\frac{1}{3}$% **e)** 72% **f)** 31%
3. a) 0·36 **b)** 0·28 **c)** 0·07 **d)** 0·134 **e)** 0·6 **f)** 0·875
4. a) 45%; $\frac{1}{2}$; 0·6 **b)** 4%; $\frac{6}{16}$; 0·38 **c)** 11%; 0·111; $\frac{1}{9}$ **d)** 0·3; 32%; $\frac{1}{3}$
5. a 85% **b)** 77·5% **c)** 23·75% **d)** 56% **e)** 10% **f)** 37·5%
6. a) 0·36 **b)** 0·67 **c)** 0·43 **d)** 0·48 **e)** 0·33
f) 0·11 **g)** 0·01 **7.** Yes

Exercise 2.22 page 67

1. a) \$15 b) 900 kg c) \$2·80 d) 125
2. \$32
3. 13·2c
4. 52·8 kg
5. a) \$1·02 b) \$21·58 c) \$2·22 d) \$0·53
6. \$243·28
7. \$26 182
8. 96·8%
9. 77·5%
10. \$71·48
11. 200
12. 29 000
13. 500 cm
14. \$6·30
15. 400 kg
16. 325
17. \$35·25
18. \$8425·60

Exercise 2.23 page 69

1. a) 25%, profit b) 25%, profit c) 10%, loss d) 20%, profit e) 30%, profit f) 7·5%, profit g) 12%, loss h) 54%, loss
2. 28%
3. $44\frac{4}{9}$%
4. 46·9%
5. 12%
6. $5\frac{1}{3}$%
7. a) \$50 b) \$450 c) \$800 d) \$12·40
8. \$50
9. \$12
10. \$5
11. 60c
12. \$2200
13. 14·3%
14. 20%
15. 8 : 11
16. 21%
17. 20%

Exercise 2.24 page 71

1. a) \$216 b) \$115·50 c) 2 years d) 5 years
2. \$2295, \$9045
3. 7·5%

Exercise 2.25 page 71

1. a) \$2180 b) \$2376·20 c) \$2590·06
2. a) \$5550 b) \$6838·16 c) \$8425·29
3. \$13 107·96
4. a) \$36 465·19 b) \$40202·87
5. a) \$9540 b) \$107 19·14 c) \$16117·63
6. a) \$14033·01 b) \$734·03 c) \$107946·25
7. \$9211·88
8. 8 years
9. 11 years
10. 13 years
11. \$30 000 at 8% (Although the total amount of money is greater for \$30 000 at 8% compound interest, the actual interest earned is greater for \$20 000 at 12% compound interest.)

Exercise 2.26 page 74

1. a) $2\frac{1}{2}$ h b) $3\frac{1}{8}$ h c) 75 s d) 4 h
2. a) 20 m/s b) 30 m/s c) $83\frac{1}{3}$ m/s d) 108 km/h e) 79·2 km/h f) 1·2 cm/s g) 90 m/s h) 25 mph i) 0·03 miles per second
3. a) 75 km/h b) 4·52 km/h c) 7·6 m/s d) 4×10^6 m/s e) $2·5 \times 10^8$ cm/s f) 200 km/h g) 3 km/h
4. a) 110 000 m b) 10 000 m c) 56 400 m d) 4500 m e) 50 400 m f) 80 m g 960 000 m
5. a) 3·125 h b) 76·8 km/h
6. a) 4·45 h b) 23·6 km/h c) 102·63 s d) 7·79 m/s
7. 46 km/h
8. a) 8 m/s b) 7·6 m/s
9. 1231 km/h
10. 3 h
11. 100s
12. $1\frac{1}{2}$ minutes
13. 600 m
14. $53\frac{1}{3}$ s
15. 5 cm/s
16. 60 s
17. 120 km/h

Exercise 2.27 page 76

1. $\frac{7}{25}$, 0·28, 28%; $\frac{16}{25}$, 0·64, 64%; $\frac{5}{8}$, 0·625, $62\frac{1}{2}$%
2. 12·4 m
3. 3·08 kg
4. $56\frac{1}{4}$ km
5. \$820
6. A
7. a) 19:17 b) 23:49
8. \$36
9. 1·32

Exercise 2.28 page 77

1. $1·08 \times 10^9$ km
2. 167 days
3. 3 h 21 min
4. 17
5. a) \$4·95 b) 25 c) \$11·75
6. 30 g zinc, 2850 g copper, 3000 g total

Exercise 2.29 page 77

1. 2 : 1
2. a) 9·85 b) 76·2 c) 223 512 d) 1678·1
3. \$252 000
4. a) 8 b) 24 c) 8 d) 8
5. 0·18 s
6. \$140
7. 5
8. THIS IS A VERY SILLY CODE
9. 29

Exercise 2.30 page 79

1. a) 1850, 1850, 92·5 b) 4592, 4592, 14 c) 50·4, 50·4, 63 d) 31·6, 31·6, 221·2 e) 42·3, 42·3, 384·93 f) 39·51, 39·51, 13.71 g) 21·2, 21·2, 95·4 h) 42·4, 42·4 i) 6·2449· · ·, 6·2449 · · · j) 29·63, 29·63
2. 1 kg
3. A–T, B–P, C–S, D–R, E–Q
4. a) 281 b) 36 c) 101 : 16
5. \$1000
6. 6 times
7. a) 35·99 b) 3·96 c) 316·8 d) 15·59 e) 198 f) 103·5
8. a) 20·56 b) 0·114 c) 1·23 d) 98·6 e) 198·9 f) 50:76
9. a) 5 b) 100 c) \$3000 d) 1 e) 0·2 kg f) 2 g) 100 h) £2000 i) 400

Revision exercise 2A page 80

1. a) 185 b) 150 c) 40 d) $\frac{11}{12}$ e) $2\frac{4}{5}$ f) $\frac{2}{5}$
2. 128 cm
3. $\frac{2}{5}$
4. $\frac{a}{b}$
5. a) 0·0547 b) 0·055 c) $5{\cdot}473 \times 10^{-2}$
6. 1·238
7. $750
8. a) $26 b) 6 : 5 c) 6
9. a) 3×10^7 b) $3{\cdot}7 \times 10^4$ c) $2{\cdot}7 \times 10^{13}$
10. a) i) 57·2% ii) $87\frac{1}{2}$% b) 40% c) 80 c

Revision exercise 2B page 81

1. 5%
2. a) $500 b) $37\frac{1}{2}$%
3. $357·88
4. 3·05
5. a) 2·4 km b) 1 km²
6. a) 300 m b) 60 cm c) 150 cm²
7. a) 1 : 50 000 b) 1 : 4 000 000
8. a) 22% b) 20·8% c) $240
9. $ 250
10. a) 0·005 m/s b) 1·6 s c) 172·8 km

Revision exercise 2C page 82

1. $33\frac{1}{3}$ km/h
2. a) 3 b) 10 c) 1, 9 d) 1, 8 e) $m = 3, n = 9$ f) $p = 1, q = 3, r = 9, s = 8, t = 10$
3. About 3
4. $2{\cdot}3 \times 10^9$
5. a) 600 b) 10 000 c) 3 d) 20
6. a) 0·5601 b) 3·215 c) 0·6161 d) 0·4743
7. a) 0·340 b) $4{\cdot}08 \times 10^{-6}$ c) 64·9 d) 0·119
8. 33·1%

Examination exercise 2D page 83

1. a) 250 km b) i) 20 m/s ii) 0·225 s
2. a) $11\frac{1}{9}$% b) 21 m c) i) 34·7 m ii) 6 years
3. a) $5\sqrt{3}$ b) $\frac{5+\sqrt{3}}{11}$
4. b) i) 54·02 km/n ii) 1918

3 Algebra 1

Exercise 3.1 page 85

1. 5° 2. −4° 3. −1° 4. 4° 5. −4° 6. 12°
7. −7° 8. −5° 9. −4° 10. 0° 11. a) C b) B 12. −17 m

Exercise 3.2 page 86

1. 13 2. 211 3. −12 4. −31 5. −66 6. 6·1 7. 9·1
8. −35 9. 18·7 10. −9 11. −3 12. 3 13. −2 14. −14
15. −7 16. 3 17. 181 18. −2·2 19. 8·2 20. 17 21. 2
22. −6 23. −15 24. −14 25. −2 26. −12 27. −80 28. −13·1
29. −4·2 30. 12·4 31. −7 32. 8 33. 4 34. −10 35. 11
36. 4 37. −20 38. 8 39. −5 40. −10 41. −26 42. −21
43. 8 44. 1 45. −20·2 46. −50 47. −508 48. −29 49. 0
50. −21 51. −0·1 52. −4 53. 6·7 54. 1 55. −850 56. 4
57. 6 58. −4 59. −12 60. −31

Exercise 3.3 page 87

1. −8 2. 28 3. 12 4. 24 5. 18 6. −35 7. 49
8. −12 9. −2 10. 9 11. −4 12. 4 13. −4 14. 8
15. 70 16. 6 17. $\frac{1}{4}$ 18. $-\frac{3}{5}$ 19. −0·01 20. 0·0002 21. 121
22. 6 23. −600 24. −1 25. −20 26. −2·6 27. −700 28. 18
29. −1000 30. 640 31. −6 32. −42 33. −0·4 34. −0·4 35. −200
36. −35 37. −2 38. $\frac{1}{2}$ 39. $-\frac{1}{4}$ 40. −90

Exercise 3.4 page 88

1. −10 2. 1 3. 12 4. −28 5. −2 6. 16 7. −3
8. 14 9. −28 10. 4 11. $-\frac{1}{6}$ 12. 9 13. −30 14. 24
15. −1 16. −2 17. −30 18. 7 19. 3 20. 16 21. 93
22. 2400 23. 10 24. 1 25. −4 26. 48 27. −1 28. 0
29. −8 30. 170 31. −3 32. 1 33. 1 34. 0 35. 15
36. 5 37. −2·4 38. −180 39. 5 40. −994 41. 2 42. −48
43. 2·5 44. −2·5 45. −32 46. 0 47. −0·1 48. −16 49. −4·3
50. $-\frac{1}{16}$
51.

−1	−2	3
4	0	−4
−3	2	1

3	2	−2
−4	1	6
4	0	−1

0	1	−4
−5	−1	3
2	−3	−2

Exercise 3.5 *page 89*

1. 21 2. 1·62 3. 396 4. 650 5. 63·8 6. 9×10^{12} 7. 800
8. $ac + ab - a^2$ 9. $r - p + q$ 10. 802; $4n + 2$ 11. $2n + 6$

Exercise 3.6 *page 91*

1. a) $4x + 1$ b) $4x - 2$ c) $x - 1$ d) $2x + 2$ e) $x + 3$ f) $3x - 4$ g) $x + 3$
h) $x + 7$ i) $x + 3$ j) $x + 4$ k) $x + 4$ l) $x + 4$

Exercise 3.7 *page 92*

1. 7 2. 13 3. 13 4. 22 5. 1 6. −1 7. 18
8. −4 9. −3 10. 37 11. 0 12. −4 13. −7 14. −2
15. −3 16. −8 17. −30 18. 16 19. −10 20. 0 21. 7
22. −6 23. −2 24. −7 25. −5 26. 3 27. 4 28. −8
29. −2 30. 2 31. 0 32. 4 33. −4 34. −3 35. −9 36. 4

Exercise 3.8 *page 92*

1. 9 2. 27 3. 4 4. 16 5. 36 6. 18 7. 1
8. 6 9. 2 10. 8 11. −7 12. 15 13. −23 14. 3
15. 32 16. 36 17. 144 18. −8 19. −7 20. 13 21. 5
22. −16 23. 84 24. 17 25. 6 26. 0 27. −25 28. −5
29. 17 30. $-1\frac{1}{2}$ 31. 19 32. 8 33. 19 34. 16 35. −16
36. 12 37. 36 38. 12 39. 2 40. 11 41. −23 42. −26
43. 5 44. 31 45. $4\frac{1}{2}$

Exercise 3.9 *page 94*

1. −20 2. 16 3. −42 4. −4 5. −90 6. −160 7. −2
8. −81 9. 4 10. 22 11. 14 12. 20 13. 1 14. $\sqrt{5}$
15. 4 16. $-6\frac{1}{2}$ 17. 54 18. 25 19. 4 20. 312 21. 45
22. 22 23. 14 24. −36 25. −7 26. 1 27. 901 28. −30
29. −5 30. $7\frac{1}{2}$ 31. −7 32. $-\frac{3}{13}$ 33. 7 34. −2 35. 0
36. $-4\frac{1}{2}$ 37. 6 38. 2 or −2 39. 26 40. $-\frac{4}{9}$ 41. $3\frac{1}{4}$ 42. $-\frac{5}{6}$
43. 4 44. $2\frac{2}{3}$ 45. $3\frac{1}{4}$ 46. $-2\frac{1}{6}$ 47. −3 48. 12 49. $1\frac{1}{3}$ 50. $-\frac{5}{36}$

Exercise 3.10 *page 95*

1. $3x + 11y$ 2. $2a + 8b$ 3. $3x + 2y$ 4. $5x + 5$ 5. $9 + x$ 6. $3 - 9y$ 7. $5x - 2y - x^2$
8. $2x^2 + 3x + 5$ 9. $-10y$ 10. $3a^2 + 2a$ 11. $7 + 7a - 7a^2$ 12. $5x$ 13. $\frac{10}{a} - b$ 14. $\frac{5}{x} - \frac{5}{y}$
15. $\frac{3m}{x}$ 16. $\frac{1}{2} - \frac{2}{x}$ 17. $\frac{5}{a} + 3b$ 18. $-\frac{n}{4}$ 19. $7x^2 - x^3$ 20. $2x^2$ 21. $x^2 + 5y^2$
22. $-12x^2 - 4y^2$ 23. $5x - 11x^2$ 24. $\frac{8}{x^2}$ 25. $5x + 2$ 26. $12x - 7$ 27. $3x + 4$ 28. $11 - 6x$
29. $-5x - 20$ 30. $7x - 2x^2$ 31. $3x^2 - 5x$ 32. $x - 4$ 33. $5x^2 + 14x$ 34. $-4x^2 - 3x$ 35. $5a + 8$
36. $a + 9$ 37. $ab + 4a$ 38. $y^2 + y$ 39. $2x - 2$ 40. $6x + 3$ 41. $x - 4$ 42. $7x + 5y$
43. $4x^2 - 11x$ 44. $2x^2 + 14x$ 45. $3y^2 - 4y + 1$ 46. $12x + 12$ 47. $4ab - 3a + 14b$ 48. $2x - 4$
49. a) $4ab$ b) $6a + 4b$ 50. a) $\times 6, + 1$ b) $\times 5, - 2, \times 3$ c) $\times 4, + 5, \div 3$ d) $\times 7, + 3, \times 2$
e) Square, + 7 f) −7, Square g) $\times 2, - 3$, Square, + 10 h) $\times 3, + 1$ Square, $+ 5, \div 7$
51. a) 2·5 b) 15 c) 7

Exercise 3.11 *page 96*

1. $x^2 + 4x + 3$ 2. $x^2 + 5x + 6$ 3. $y^2 + 9y + 20$ 4. $x^2 + x - 12$ 5. $x^2 + 3x - 10$ 6. $x^2 - 5x + 6$
7. $a^2 - 2a - 35$ 8. $z^2 + 7z - 18$ 9. $x^2 - 9$ 10. $k^2 - 121$ 11. $2x^2 - 5x - 3$ 12. $3x^2 - 2x - 8$
13. $2y^2 - y - 3$ 14. $49y^2 - 1$ 15. $9x^2 - 4$ 16. $6a^2 + 5ab + b^2$ 17. $3x^2 + 7xy + 2y^2$ 18. $6b^2 + bc - c^2$
19. $-5x^2 + 16xy - 3y^2$ 20. $15b^2 + ab - 2a^2$ 21. $2x^2 + 2x - 4$ 22. $6x^2 + 3x - 9$ 23. $24y^2 + 4y - 8$
24. $6x^2 - 10x - 4$ 25. $4a^2 - 16b^2$ 26. $x^3 - 3x^2 + 2x$ 27. $8x^3 - 2x$ 28. $3y^3 + 3y^2 - 18y$
29. $x^3 + x^2y + x^2z + xyz$ 30. $3za^2 + 3zam - 6zm^2$

Exercise 3.12 *page 97*

1. $x^2 + 8x + 16$ 2. $x^2 + 4x + 4$ 3. $x^2 - 4x + 4$ 4. $4x^2 + 4x + 1$ 5. $y^2 - 10y + 25$
6. $9y^2 + 6y + 1$ 7. $x^2 + 2xy + y^2$ 8. $4x^2 + 4xy + y^2$ 9. $a^2 - 2ab + b^2$ 10. $4a^2 - 12ab + 9b^2$
11. $3x^2 + 12x + 12$ 12. $9 - 6x + x^2$ 13. $9x^2 + 12x + 4$ 14. $a^2 - 4ab + 4b^2$ 15. $2x^2 + 6x + 5$

16. $2x^2 + 2x + 13$ 17. $5x^2 + 8x + 5$ 18. $2y^2 - 14y + 25$ 19. $10x - 5$ 20. $-8x + 8$
21. $-10y + 5$ 22. $3x^2 - 2x - 8$ 23. $2x^2 + 4x - 4$ 24. $-x^2 - 18x + 15$
25. a) $6x^2 + 12x$ b) $8x^2 + 12x$

Exercise 3.13 *page 98*

1. 8 2. 9 3. 7 4. 10 5. $\frac{1}{3}$ 6. 10 7. $1\frac{1}{2}$
8. -1 9. $-1\frac{1}{2}$ 10. $\frac{1}{3}$ 11. 35 12. 130 13. 14 14. $\frac{2}{3}$
15. $3\frac{1}{3}$ 16. $-2\frac{1}{2}$ 17. 3 18. $1\frac{1}{8}$ 19. $\frac{3}{10}$ 20. $-1\frac{1}{4}$ 21. 10
22. 27 23. 20 24. 18 25. 28 26. -15 27. $\frac{99}{100}$ 28. 0
29. 1000 30. $-\frac{1}{1000}$ 31. 1 32. -7 33. -5 34. $1\frac{1}{6}$ 35. 1
36. 2 37. -5 38. -3 39. $-1\frac{1}{2}$ 40. 2 41. 1 42. $3\frac{1}{2}$
43. 2 44. -1 45. $10\frac{2}{3}$ 46. 1·1 47. -1 48. 2 49. $2\frac{1}{2}$ 50. $1\frac{1}{3}$

Exercise 3.14 *page 99*

1. $-1\frac{1}{2}$ 2. 2 3. $-\frac{2}{5}$ 4. $-\frac{1}{3}$ 5. $1\frac{2}{3}$ 6. 6 7. $-\frac{2}{5}$
8. $-3\frac{1}{5}$ 9. $\frac{1}{2}$ 10. -4 11. 18 12. 5 13. 4 14. 3
15. $2\frac{3}{4}$ 16. $-\frac{7}{22}$ 17. $\frac{1}{4}$ 18. 1 19. 4 20. -11 21. $-7\frac{1}{3}$
22. $1\frac{1}{4}$ 23. -5 24. 6 25. 3 26. 6 27. 2 28. 3
29. 4 30. 3 31. $10\frac{1}{2}$ 32. 5 33. 2 34. -1 35. -17
36. $-2\frac{9}{10}$ 37. $2\frac{10}{21}$ 38. $\frac{1}{3}$ 39. 14 40. 15

Exercise 3.15 *page 100*

1. $\frac{1}{4}$ 2. -3 3. 4 4. $-7\frac{2}{3}$ 5. -43 6. 11 7. $-\frac{1}{2}$
8. 0 9. 1 10. $-1\frac{2}{3}$ 11. $\frac{1}{4}$ 12. 0 13. $-\frac{6}{7}$ 14. $1\frac{9}{17}$
15. $1\frac{22}{23}$ 16. $\frac{2}{11}$ 17. 4 cm 18. 5 m 19. 4

Exercise 3.16 *page 101*

1. $\frac{1}{3}$ 2. $\frac{1}{5}$ 3. $1\frac{2}{3}$ 4. -3 5. $\frac{5}{11}$ 6. -2 7. 6
8. $3\frac{3}{4}$ 9. -7 10. $-7\frac{2}{3}$ 11. 2 12. 3 13. 4 14. -2
15. -3 16. 3 17. $1\frac{5}{7}$ 18. $4\frac{4}{5}$ 19. 10 20. 24 21. 2

Exercise 3.17 *page 102*

1. 3 2. 5 3. -4 4. $6\frac{3}{4}$ 5. -3 6. 0 7. 3
8. $1\frac{1}{5}$ 9. 1 10. 2 11. 3 12. 4 13. $\frac{3}{5}$ 14. $1\frac{1}{8}$
15. -1 16. 1 17. 1 18. $\frac{1}{4}$ 19. $-\frac{1}{3}$ 20. $\frac{9}{10}$ 21. 1
22. 2 23. $-\frac{1}{7}$ 24. -4 25. 3

Exercise 3.18 *page 103*

1. 91, 92, 93 2. 21, 22, 23, 24 3. 57, 59, 61 4. 506, 508, 510 5. $12\frac{1}{2}$
6. $12\frac{1}{2}$ 7. $11\frac{2}{3}$ 8. $8\frac{1}{3}, 41\frac{2}{3}$ 9. $1\frac{1}{4}, 13\frac{3}{4}$ 10. $3\frac{1}{3}$ cm
11. 12 cm 12. 20 13. 5 cm 14. 7 cm 15. $18\frac{1}{2}, 27\frac{1}{2}$

Exercise 3.19 page 104

1. 20°, 60°, 100° 2. 45°, 60°, 75° 3. 5 4. 6, 8 5. 12, 24, 30 6. 5, 15, 8
7. 168·84 cm^2 8. $59\frac{2}{3}$ kg, $64\frac{2}{3}$ kg, $72\frac{2}{3}$ kg 9. 24, 22, 15 10. 48, 12 11. 40, 8
12. 6 13. 5 × 4 14. 14 15. \$45, \$31 16. \$21·50 17. $6\frac{2}{7}$

Exercise 3.20 page 106

1. \$3700 2. 3 3. $1\frac{3}{7}$ m 4. 80°, 100° 5. 30°, 60°, 90°, 120°, 150°, 270°
6. 26, 58 7. 2 km 8. 8 km 9. 400 m 10. 21 11. 23
12. \$3600 13. 15 14. $53\frac{1}{3}$, $56\frac{2}{3}$ 15. 2 km 16. 7, 8, 9 17. 2, 3, 4, 5

Exercise 3.21 page 108

1. $x = 2, y = 1$ 2. $x = 4, y = 2$ 3. $x = 3, y = 1$ 4. $x = -2, y = 1$ 5. $x = 3, y = 2$ 6. $x = 5, y = -2$
7. $x = 2, y = 1$ 8. $x = 5, y = 3$ 9. $x = 3, y = -1$ 10. $a = 2, b = -3$ 11. $a = 5, b = \frac{1}{4}$ 12. $a = 1, b = 3$
13. $m = \frac{1}{2}, n = 4$ 14. $w = 2, x = 3$ 15. $x = 6, y = 3$ 16. $x = \frac{1}{2}, z = -3$ 17. $m = 1\frac{15}{17}, n = \frac{11}{17}$ 18. $c = 1\frac{16}{23}, d = -2\frac{12}{23}$

Exercise 3.22 page 108

1. 1 2. −3 3. 2 4. 15 5. −2 6. −3
7. −2 8. −11 9. −21 10. 1 11. 0 12. 15
13. −10 14. 3 15. 6 16. −11 17. 2 18. 5
19. −19 20. −4 21. x 22. $-3x$ 23. $4x$ 24. $4y$
25. $9y$ 26. $3x$ 27. $-8x$ 28. $4x$ 29. $2x$ 30. $3y$

Exercise 3.23 page 110

1. $x = 2, y = 4$ 2. $x = 1, y = 4$ 3. $x = 2, y = 5$ 4. $x = 3, y = 7$ 5. $x = 5, y = 2$ 6. $a = 3, b = 1$
7. $x = 1, y = 3$ 8. $x = 1, y = 3$ 9. $x = -2, y = 3$ 10. $x = 4, y = 1$ 11. $x = 1, y = 5$ 12. $x = 0, y = 2$
13. $x = \frac{5}{7}, y = 4\frac{3}{7}$ 14. $x = 1, y = 2$ 15. $x = 2, y = -3$ 16. $x = 4, y = -1$ 17. $x = 3, y = 1$ 18. $x = 1, y = 2$
19. $x = 2, y = 1$ 20. $x = -2, y = 1$ 21. $x = 1, y = 2$ 22. $a = 4, b = 3$ 23. $x = -23, y = -78$
24. $x = 3, y = \frac{1}{2}$ 25. $x = 4, y = 3$ 26. $x = 5, y = -2$ 27. $x = \frac{1}{3}, y = -2$ 28. $x = 5\frac{5}{14}, y = \frac{2}{7}$ 29. $x = 3, y = -1$
30. $x = 5, y = 0{\cdot}2$

Exercise 3.24 page 111

1. $5\frac{1}{2}, 9\frac{1}{2}$ 2. 6, 3 or $2\frac{2}{5}, 5\frac{2}{5}$ 3. 4, 10 4. $a = 2, c = 7$ 5. $m = 4, c = -3$
6. $a = 1, b = -2$ 7. carrot seed 1c, lettuce seed 3c 8. TV \$200, DVD player \$450 9. 7, 3
10. 64°, 64°, 52° 11. white 2 g, brown $3\frac{1}{2}$ g 12. 120 cm, 240 cm 13. 150 m 350 m
14. 2c × 15, 5c × 25 15. 10c × 14, 50c × 7 16. 20

Exercise 3.25 page 113

1. teacher \$500, doctor \$700 2. current 4 m/s, fish 10 m/s 3. $\frac{5}{7}$ 4. $\frac{3}{5}$
5. boy 10, snake 3 6. 4, 7 7. $y = 3x - 2$ 8. walks 4 m/s, runs 5 m/s
9. \$1 × 15, \$5 × 5 10. 36, 9 11. A = 5, B = 7, C = 1 12. current $4\frac{1}{2}$ knots, submarine $20\frac{1}{2}$ knots
13. $a = 1, b = 2, c = 5, y = x^2 - 2x + 5$ 14. $y = 2x^2 - 3x + 5$ 15. $y = x^2 + 3x + 4$ 16. $y = x^2 + 2x - 3$

Exercise 3.26 page 114

1. $5(a + b)$ 2. $7(x + y)$ 3. $x(7 - x)$ 4. $y(y + 8)$ 5. $y(2y + 3)$ 6. $2y(3y - 2)$
7. $3x(x - 7)$ 8. $2a(8 - a)$ 9. $3c(2c - 7)$ 10. $3x(5 - 3x)$ 11. $7y(8 - 3y)$ 12. $x(a + b + 2c)$
13. $x(x + y + 3z)$ 14. $y(x^2 + y^2 + z^2)$ 15. $ab(3a + 2b)$ 16. $xy(x + y)$ 17. $2a(3a + 2b + c)$ 18. $m(a + 2b + m)$
19. $2k(x + 3y + 2z)$ 20. $a(x^2 + y + 2b)$ 21. $xk(x + k)$ 22. $ab^2(b + 2)$ 23. $bc(a - 3b)$ 24. $ae(2a - 5e)$
25. $ab(a^2 + b^2)$ 26. $x^2y(x + y)$ 27. $2xy(3y - 2x)$ 28. $3ab(b^2 - a^2)$ 29. $a^2b(2a + 5b)$ 30. $ax^2(y - 2z)$
31. $2ab(x + b + a)$ 32. $yx(a + x^2 - 2yx)$

Exercise 3.27 page 115

1. $(a + b)(x + y)$ 2. $(a + b)(y + z)$ 3. $(x + y)(b + c)$ 4. $(x + y)(h + k)$ 5. $(x + y)(m + n)$
6. $(a + b)(h - k)$ 7. $(a + b)(x - y)$ 8. $(m + n)(a - b)$ 9. $(h + k)(s + t)$ 10. $(x + y)(s - t)$
11. $(a - b)(x - y)$ 12. $(x - y)(s - t)$ 13. $(a - x)(s - y)$ 14. $(h - b)(x - y)$ 15. $(m - n)(a - b)$
16. $(x - z)(k - m)$ 17. $(2a + b)(x + 3y)$ 18. $(2a + b)(x + y)$ 19. $(2m + n)(h - k)$ 20. $(m - n)(2h + 3k)$
21. $(2x + y)(3a + b)$ 22. $(2a - b)(x - y)$ 23. $(x^2 + y)(a + b)$ 24. $(m - n)(s + 2t^2)$

Exercise 3.28 page 116

1. $(x + 2)(x + 5)$ 2. $(x + 3)(x + 4)$ 3. $(x + 3)(x + 5)$ 4. $(x + 3)(x + 7)$ 5. $(x + 2)(x + 6)$
6. $(y + 5)(y + 7)$ 7. $(y + 3)(y + 8)$ 8. $(y + 5)(y + 5)$ 9. $(y + 3)(y + 12)$ 10. $(a + 2)(a - 5)$
11. $(a + 3)(a - 4)$ 12. $(z + 3)(z - 2)$ 13. $(x + 5)(x - 7)$ 14. $(x + 3)(x - 8)$ 15. $(x - 2)(x - 4)$
16. $(y - 2)(y - 3)$ 17. $(x - 3)(x - 5)$ 18. $(a + 2)(a - 3)$ 19. $(a + 5)(a + 9)$ 20. $(b + 3)(b - 7)$
21. $(x - 4)(x - 4)$ 22. $(y + 1)(y + 1)$ 23. $(y - 7)(y + 4)$ 24. $(x - 5)(x + 4)$ 25. $(x - 20)(x + 12)$
26. $(x - 15)(x - 11)$ 27. $(y + 12)(y - 9)$ 28. $(x - 7)(x + 7)$ 29. $(x - 3)(x + 3)$ 30. $(x - 4)(x + 4)$

Exercise 3.29 page 117

1. $(2x + 3)(x + 1)$ 2. $(2x + 1)(x + 3)$ 3. $(3x + 1)(x + 2)$ 4. $(2x + 3)(x + 4)$ 5. $(3x + 2)(x + 2)$
6. $(2x + 5)(x + 1)$ 7. $(3x + 1)(x - 2)$ 8. $(2x + 5)(x - 3)$ 9. $(2x + 7)(x - 3)$ 10. $(3x + 4)(x - 7)$
11. $(2x + 1)(3x + 2)$ 12. $(3x + 2)(4x + 5)$ 13. $(3x - 2)(x - 3)$ 14. $(y - 2)(3y - 5)$ 15. $(4y - 3)(y - 5)$
16. $(2y + 3)(3y - 1)$ 17. $(2x - 5)(3x - 6)$ 18. $(5x + 2)(2x + 1)$ 19. $(6x - 1)(x - 3)$ 20. $(4x + 1)(2x - 3)$
21. $(6x + 5)(2x - 1)$ 22. $(16x + 3)(x + 1)$ 23. $(2a - 1)(2a - 1)$ 24. $(x + 2)(12x - 7)$ 25. $(x + 3)(15x - 1)$
26. $(8x + 1)(6x + 5)$ 27. $(16y - 3)(4y + 1)$ 28. $(15x - 1)(8x + 5)$ 29. $(3x - 1)(3x + 1)$ 30. $(2a - 3)(2a + 3)$

Exercise 3.30 page 118

1. $(y - a)(y + a)$ 2. $(m - n)(m + n)$ 3. $(x - t)(x + t)$ 4. $(y - 1)(y + 1)$ 5. $(x - 3)(x + 3)$
6. $(a - 5)(a + 5)$ 7. $\left(x-\frac{1}{2}\right)\left(x+\frac{1}{2}\right)$ 8. $\left(x-\frac{1}{3}\right)\left(x+\frac{1}{3}\right)$ 9. $(2x - y)(2x + y)$ 10. $(a - 2b)(a + 2b)$
11. $(5x - 2y)(5x + 2y)$ 12. $(3x - 4y)(3x + 4y)$ 13. $\left(x-\frac{y}{2}\right)\left(x+\frac{y}{2}\right)$ 14. $\left(3m-\frac{2}{3}n\right)\left(3m+\frac{2}{3}n\right)$
15. $\left(4t-\frac{2}{5}s\right)\left(4t+\frac{2}{5}s\right)$ 16. $\left(2x-\frac{z}{10}\right)\left(2x+\frac{z}{10}\right)$ 17. $x(x - 1)(x + 1)$ 18. $a(a - b)(a + b)$ 19. $x(2x - 1)(2x + 1)$
20. $2x(2x - y)(2x + y)$ 21. $3x(2x - y)(2x - y)$ 22. $2m(3m - 2n)(3m + 2n)$ 23. $5\left(x-\frac{1}{2}\right)\left(x+\frac{1}{2}\right)$
24. $2a(5a - 3b)(5a + 3b)$ 25. $3y(2x - z)(2x + z)$ 26. $4ab(3a - b)(3a + b)$ 27. $2a^3(5a - 2b)(5a + 2b)$
28. $9xy(2x - 5y)(2x + 5y)$ 29. 161 30. 404 31. 4400
32. 2421 33. 4329 34. 0·75 35. 4·8 36. −2469
37. 0·0761 38. −10 900 39. 53·6 40. 0·000 005

Exercise 3.31 page 119

1. −3, −4 2. −2, −5 3. 3, −5 4. 2, −3 5. 2, 6 6. −3, −7
7. 2, 3 8. 5, −1 9. −7, 2 10. $-\frac{1}{2}, 2$ 11. $\frac{2}{3}, -4$ 12. $1\frac{1}{2}, -5$
13. $\frac{2}{3}, 1\frac{1}{2}$ 14. $\frac{1}{4}, 7$ 15. $\frac{3}{5}, -\frac{1}{2}$ 16. 7, 8 17. $\frac{5}{6}, \frac{1}{2}$ 18. 7, −9
19. −1, −1 20. 3, 3 21. −5, −5 22. 7, 7 23. $-\frac{1}{3}, \frac{1}{2}$ 24. $-1\frac{1}{4}, 2$
25. 13, −5 26. $-3, \frac{1}{6}$ 27. $\frac{1}{10}, -2$ 28. 1, 1 29. $\frac{2}{9}, -\frac{1}{4}$ 30. $-\frac{1}{4}, \frac{3}{5}$

Exercise 3.32 page 120

1. 0, 3 2. 0, −7 3. 0, 1 4. $0, \frac{1}{3}$ 5. 4, −4 6. 7, −7
7. $\frac{1}{2}, -\frac{1}{2}$ 8. $\frac{2}{3}, -\frac{2}{3}$ 9. $0, -1\frac{1}{2}$ 10. $0, 1\frac{1}{2}$ 11. $0, 5\frac{1}{2}$ 12. $\frac{1}{4}, -\frac{1}{4}$
13. $\frac{1}{2}, -\frac{1}{2}$ 14. $0, \frac{5}{8}$ 15. $0, \frac{1}{12}$ 16. 0, 6 17. 0, 11 18. $0, 1\frac{1}{2}$
19. 0, 1 20. 0, 4 21. 0, 3 22. $\frac{1}{2}, -\frac{1}{2}$ 23. $1\frac{1}{3}, -1\frac{1}{3}$ 24. 3, −3
25. $0, 2\frac{2}{5}$ 26. $\frac{1}{3}, -\frac{1}{3}$ 27. $0, \frac{1}{4}$ 28. $0, \frac{1}{6}$ 29. $\frac{1}{4}, -\frac{1}{4}$ 30. $0, \frac{1}{5}$

Exercise 3.33 page 121

1. $-\frac{1}{2}, -5$ 2. $-\frac{2}{3}, -3$ 3. $-\frac{1}{2}, -\frac{2}{3}$ 4. $\frac{1}{3}, 3$ 5. $\frac{2}{5}, 1$ 6. $\frac{1}{3}, 1\frac{1}{2}$
7. −0·63, −2·37 8. −0·27, −3·73 9. 0·72, 0·28 10. 6·70, 0·30 11. 0·19, −2·69 12. 0·85, −1·18
13. 0·61, −3·28 14. $-1\frac{2}{3}, 4$ 15. $-1\frac{1}{2}, 5$ 16. 3·56, −0·56 17. 0·16, −3·16 18. $-\frac{1}{2}, 2\frac{1}{3}$
19. $-\frac{1}{3}, -8$ 20. $1\frac{2}{3}, -1$ 21. 2·28, 0·22 22. −0·35, −5·65 23. $-\frac{2}{3}, \frac{1}{2}$ 24. −0·58, 2·58
25. −2·69, 0·19 26. 0·22, −1·55 27. −0·37, 5·37 28. $-\frac{5}{6}, 1\frac{3}{4}$ 29. $-\frac{7}{9}, 1\frac{1}{4}$ 30. $1\frac{2}{5}, -2\frac{1}{4}$
31. $-4, 1\frac{1}{2}$ 32. $-3, 1\frac{2}{3}$ 33. $-2, 1\frac{2}{3}$ 34. $-3\frac{1}{2}, \frac{1}{5}$ 35. $-3, \frac{4}{5}$ 36. $-8\frac{1}{2}, 11$

Exercise 3.34 page 121

1. −3, 2
2. −3, −7
3. $-\frac{1}{2}, 2$
4. 1, 4
5. $-1\frac{2}{3}, \frac{1}{2}$
6. −0·39, −4·28
7. −0·16, 6·16
8. 3
9. $2, -1\frac{1}{3}$
10. −3 ,−1
11. 0·66, −22·66
12. −7, 2
13. $\frac{1}{4}, 7$
14. $-\frac{1}{2}, \frac{3}{5}$
15. $0, 3\frac{1}{2}$
16. $-\frac{1}{4}, \frac{1}{4}$
17. −2·77, 1·27
18. $-\frac{2}{3}, 1$
19. $-\frac{1}{2}, 2$
20. 0, 3
21. a) −1 b) 0·6258 c) 0·5961 d) 0·2210

Exercise 3.35 page 123

1. 8, 11
2. 11, 13
3. 12 cm
4. 6 cm
5. $x = 11$
6. 10 cm × 24 cm
7. 8 km north, 15 km east
8. 12 eggs
9. 4·96 × 11·96
10. 13 eggs
11. 4 or −1
12. 2, 5
13. $\frac{40}{x}$h, $\frac{40}{x-2}$h, 10 km/h
14. 4 km/h
15. 60 km/h
16. 5 km/h
17. 157 km
18. $x = 2$
19. $x = 3$ or 9·5
20. $\frac{3}{4}$
21. 9 cm or 13 cm

Revision exercise 3A page 125

1. a) $-2\frac{1}{2}$ b) $2\frac{2}{3}$ c) 0, −5 d) 2, −2 e) $-5, 2\frac{2}{3}$
2. a) 14 b) 18 c) 28
3. a) $(2x - y)(2x + y)$ b) $2(x + 3)(x + 1)$ c) $(2 - 3k)(3m + 2n)$ d) $(2x + 1)(x - 3)$
4. a) $x = 3, y = -2$ b) $m = 1\frac{1}{2}, n = -3$ c) $x = 7, y = \frac{1}{2}$ d) $x = -1, y = -2$
5. a) 8 b) 140 c) 29 d) 42 e) 6 f) −6
6. a) $2x - 21$ b) $(1 - 2x)(2a - 3b)$ c) 23
7. $5(a + 2b) + 3(n + 4m)$
8. a) 1 b) $10\frac{1}{2}$ c) $0, 3\frac{1}{2}$ d) −3, −2 e) 12
9. a) $z(z - 4)(z + 4)$ b) $(x^2 + 1)(y^2 + 1)$ c) $(2x + 3)(x + 4)$
10. $\frac{7}{8}$
11. a) $c = 5, d = -2$ b) $x = 2, y = -1$ c) $x = 9, y = -14$ d) $s = 5, t = -3$
12. a) $\frac{1}{2}, -\frac{1}{2}$ b) $\frac{7}{11}$ c) 3 d) 0, 5
13. a) 1·78, −0·28 b) 1·62, −0·62 c) 0·87, −1·54 d) 1·54, −4·54
14. a) $x = 9$ b) $x = 10$
15. a) 2 b) −3 c) 36 d) 0 e) 36 f) 4
16. speed = 5 km/h
17. 8 cm × 6·5 cm
18. a) −2, 4 b) 16 c) 6·19, 0·81
19. $-\frac{1}{5}, 3$
20. 8
21. $x = 13$
22. $a = 8, b = 12, c = 17$
23. 21
24. 18
25. $m = 60°, n = 50°$
26. 6 cm
27. −4

Examination exercise 3B page 126

1. $x = 3, y = -1$
2. −2, 2·5
3. $(2a - 5b)(c + 3)$
4. 3·56

4 Mensuration

Exercise 4.1 page 128

1. 10·2 m²
2. 22 cm²
3. 103 m²
4. 9 cm²
5. 31 m²
6. 6000 cm² or 0·6 m²
7. 26 m²
8. 18 cm²
9. 20 m²
10. 13 m
11. 15 cm
12. 56 m
13. 8 m, 10 m
14. 12 cm
15. 2500
16. $1 050 000
17. 6 square units
18. 14 square units
19. 1849
21. 1100 m

Exercise 4.2 page 130

1. 48·3 cm²
2. 28·4 cm²
3. 66·4 m²
4. 3·1 cm²
5. 18·2 cm²
6. 12·3 cm²
7. 2·78 cm²
8. 36·4 m²
9. 62·4 m²
10. 30·4 m²
11. 44·9 cm²
12. 0·28 m²
13. 63 m²
14. 70·7 m²
15. 14 m²
16. 65·8 cm²
17. 18·1 cm²
18. 8·0 m²
19. 52·0 cm²

Exercise 4.3 page 131

1. 124 m²
2. 10·7 cm
3. 50·9°
4. 4·85 m
5. 7·23 cm
6. 60°; 23·4 cm²
7. $\frac{11}{12}$ cm²
8. 110 cm²
9. 18·7 cm

Exercise 4.4 page 132

1. a) 570 b) 370 c) 2100 d) 9·9 e) 84 f) 0·2 g) 450 h) 7000, 70 000
2. a) 12 cm² b) 1200 mm²
3. a) 22 m² b) 220 000 cm²
4. a) 10 000 b) 8 c) 0·5 d) 0.026 d) 200

Exercise 4.5 page 134

1. a) 31·4 cm b) 78·5 cm²
2. a) 18·8 cm b) 28·3 cm²
3. a) 26·6 cm b) 49·1 cm²
4. a) 26·3 cm b) 33·3 cm²
5. a) 50·3 cm b) 174 cm²
6. a) 22·0 cm b) 10·5 cm²
7. a) 9·42 cm b) 6·44 cm²
8. a) 25·1 cm b) 25·1 cm²
9. a) 18·8 cm b) 12·6 cm²

Exercise 4.6 page 135

1. 2·19 cm 2. 30·2 m 3. 2·65 km 4. 9·33 cm 5. 14·2 mm 6. 497 000 km^2
7. a) 30:` b) 1508 cm^2 c) 508 cm^2 8. a) 40·8 m^2 b) 6 9. 5305
10. 60·9 cm 11. 29

Exercise 4.7 page 136

1. a) 80 b) 7 2. a) 98 cm^2 b) 14·0 cm^2 3. a) 33·0 cm b) 70·9 cm^2
4. 5.39 cm $(\sqrt{29})$ 5. 796 m^2 6. 28·0 cm 7. 57·5° 8. Yes

Exercise 4.8 page 138

1. a) 2·09 cm: 4·19 cm^2 b) 7·85 cm: 39·3 cm^2 c) 8·20 cm; 8·20 cm^2 2. 31·9 cm^2 3. 31·2 cm^2
4. a) 85·9° b) 57·3° c) 6·25 cm 5. a) 12 cm b) 30°
6. a) 3·98 cm b) 74·9° 7. a) 30° b) 10·5 cm 8. a) 18 cm
b) 38·2° 9. a) 10 cm b) 43·0° 10. 15·1 km^2 11. 26·5 cm

Exercise 4.9 page 141

1. a) 14·5 cm b) 72·6 cm^2 c) 24·5 cm^2 d) 48·1 cm^2
2. a) 5·08 cm^2 b) 82·8 cm^2 c) 5·14 cm^2 3. a) 60°, 9·06 cm^2 b) 106·3°, 11·2 cm^2
4. 3 cm 5. 3·97 cm 6. a) 13·5 cm^2 b) 405 cm^3
7. a) 129·9 cm^2 b) 184·3 cm^2 8. 19·6 cm^2 9. $0{\cdot}313r^2$
10. a) 8·37 cm b) 54·5 cm c) 10·4 cm

Exercise 4.10 page 143

1. a) 30 cm^3 b) 168 cm^3 c) 110 cm^3 d) 94·5 cm^3 e) 754 cm^3 f) 283 cm^3
2. a) 503 cm^3 b) 760 m^3 c) 12·5 cm^3 3. 3·98 cm 4. 6·37 cm 5. 1·89 cm
6. 91· 1 cm^3 7. 9·77 cm 8. 7·38 cm 9. 12·7 m 10. 4·24 litres 11. 106 cm/s
12. 1570 cm^3, 12·6 kg 13. 3 : 4 14. 5·37 cm 15. $118 750 16. No
17. 1·19 cm 18. 53 times 19. a) 229·2 cm^3 b) 382·0 cm^3 20. 191 cm

Exercise 4.11 page 147

1. 20·9 cm^3 2. 524 cm^3 3. 4189 cm^3 4. 101 cm^3 5. 268 cm^3 6. $4{\cdot}19x^3$ cm^3 7. 0·004 19 m^3
8. 3 cm^3 9. 93·3 cm^3 10. 48 m^3 11. 92·4 cm^3 12. 262 cm^3 13. 235 cm^3 14. 415 cm^3
15. 5 m 16. 2·43 cm 17. 23·9 cm 18. 6 cm 19. 3·72 cm 20. 1·93 kg

Exercise 4.12 page 148

1. 106 s 2. a) 125 b) 2744 c) $2{\cdot}7 \times 10^7$ 3. a) 0·36 cm b) 0·427 cm
4. a) 6·69 cm b) 39·1 cm 5. $10\frac{2}{3}$ cm^3 6. 1·05 cm^3 7. 488 cm^3 8. 53·6 cm^3
9. 74·5 cm^3 10. 4·24 cm 11. 764 cm 12. 123 cm^3 13. 54·5 litres
14. a) 16π b) 8 cm c) 6 cm 15. 471 cm^3 16. 2720 cm^3 17. 943 cm^3 18. 5050 cm^3

Exercise 4.13 page 150

1. a) 7000 b) 600 c) 1 d) 1 000 000 e) 3000 f) 8·5
2. a) 55 000 b) 13 000 c) 0·2 d) 600 e) 20 f) 1 000 000 000 3. 1085 g
4. a) 91·1 cm^3 5) 1·50 cm 6) 251 m^3

Exercise 4.14 page 151

1. a) 36π cm^2 b) 72π cm^2 c) 60π cm^2 d) $2{\cdot}38\pi$ m^2 e) 400π m^2 f) 65π cm^2 g) 192π mm^2
h) $10{\cdot}2\pi$ cm^2 i) $0{\cdot}000\,4\pi$ m^2 j) 98π cm^2, 147π cm^2
2. 1·64 cm 3. 2·12 cm 4. 3·46 cm 5. a) 3 cm b) 4 cm c) 3 cm d) 0·2 m
e) 6 cm f) 2·5 cm g) 7 m 6. 303 cm^2 7. $1179 8. $3870 9. 94·0 cm^3
10. 44·6 cm^2 11. 377 cm^2 12. 20 cm, 10 cm 13. 71·7 cm^2 14. 147 cm^2

Revision exercise 4A page 153

1. a) 14 cm^2 b) 54 cm^2 c) 50 cm^2 d) 18 cm^2
2. a) 56·5 m, 254 m^2 b) 10·8 cm c) 3·99 cm 3. a) 9π cm^2 b) 8 : 1
4. 3·43 cm^2, 4·57 cm^2 5. a) 12·2 cm b) 61·1 cm^2
6. a) 11·2 cm b) 10·3 cm c) 44·7 cm^2 d) 31·5 cm^2 e) 13·2 cm^2
7. 103·1° 8. 9·95 cm 9. a) 905 cm^3 b) 5·76 cm
10. 8·06 cm 11. 99·5 cm^3 12. 42 13. 4 cm
14. a) 15·6 cm^2 b) 93·5 cm^2 c) 3741 cm^3 15. 0·370 cm 16. 104 cm^2
17. 5·14 cm^2 18. 68c 19. 25 20. 20 cm^2 21. $ 7·2 billion

Examination exercise 4B page 155

1. a) i) 22 619 cm² ii) $ 5·43 b) i) 351858 cm² ii) 12 h 13 min
2. a) x^2y b) $2x^2 + 4xy$
3. a) prism b) i) 5·25 cm² ii) 810·6 g iii) 91·8 c) 24

5 Functions 1

Exercise 5.1 page 159

1. a) 5, 10, 1 b) 21, 101, −29
2. $x \to$ [×5] → [+4] → $5x + 4$
3. $x \to$ [−4] → [×3] → $3(x - 4)$
4. $x \to$ [×2] → [+7] → [square] → $(2x + 7)^2$
5. $x \to$ [×5] → [+9] → [÷4] → $\frac{5x+9}{4}$
6. $x \to$ [×3] → [subtract from 4] → [÷5] → $\frac{4-3x}{5}$
7. $x \to$ [square] → [×2] → [+1] → $2x^2 + 1$
8. $x \to$ [square] → [×3] → [÷2] → [+5] → $\frac{3x^2}{2} + 5$
9. $x \to$ [×4] → [−5] → [square root] → $\sqrt{(4x-5)}$
10. $x \to$ [square] → [+10] → [square root] → [×4] → $4\sqrt{x^2+10}$
11. $x \to$ [×3] → [subtract from 7] → [square] → $(7-3x)^2$
12. $x \to$ [×3] → [+1] → [square] → [×4] → [+5] → $4(3x + 1)^2 + 5$
13. $x \to$ [square] → [subtract from 5] → $5 - x^2$
14. $x \to$ [square] → [+1] → [square root] → [×10] → [+6] → [÷4] → $\frac{10\sqrt{(x^2+1)}+6}{4}$
15. $x \to$ [cube] → [÷4] → [+1] → [square] → [subtract 6] → $\left(\frac{x^3}{4}+1\right)^2 - 6$

16. a) −9, 11, $\frac{1}{2}$ b) 0·8, −2·7, $\frac{1}{80}$ c) 4, 1·2, 36
17. a) 0 b) 6 c) 12
18. a) 10 b) $\frac{1}{2}$ c) 2
19. a) 6, 24, 6 b) 0, $\sqrt{2}$, $\sqrt{6}$ c) −6, 6, $9\frac{3}{4}$
20. a) ±3 b) ±3 c) ±2 d) ±6
21. a) 10, 21 b) 111, 411, 990, 112
22. a) 7 b) 10 c) 5 d) 14 e) 7 f) 7
23. a) 3 b) 6 c) 8 d) 10
24. a) 11 b) 17 c) 7
25. a) 5 b) 17 c) $1\frac{1}{2}$ d) 3
26. $a = 3, b = 5$
27. $a = 2, b = -5$
28. $a = 7, b = 1$

Exercise 5.2 page 162

1. a) $x \mapsto 4(x+5)$ b) $x \mapsto 4x+5$ c) $x \mapsto (4x)^2$ d) $x \mapsto 4x^2$ e) $x \mapsto x^2+5$ f) $x \mapsto 4(x^2+5)$ g) $x \mapsto [4(x+5)]^2$
2. a) −2·5 b) $\pm\sqrt{\frac{5}{3}}$
3. a) $x \mapsto 2(x-3)$ b) $x \mapsto 2x-3$ c) $x \mapsto x^2-3$ d) $x \mapsto (2x)^2$ e) $x \mapsto (2x)^2-3$ f) $x \mapsto (2x-3)^2$
4. a) 2 b) 11 c) 6 d) 2 e) 1 f) 64
5. a) −3 b) 2 c) $1\frac{1}{2}$ d) 5
6. a) $x \to 2(3x-1)+1$ b) $x \to 3(2x+1)-1$ c) $x \to 2x^2+1$ d) $x \to (3x-1)^2$ e) $x \to 2(3x-1)^2+1$ f) $x \to 3(2x^2+1)-1$
7. a) 11 b) 9 c) 11 d) 14 e) 81 f) −1
8. a) 2 b) 0, 2 c) $\pm\sqrt{2}$
9. $x \mapsto \frac{x+2}{5}$
10. $x \mapsto \frac{x}{5}+2$
11. $x \mapsto \frac{x}{6}-2$
12. $x \mapsto \frac{3x-1}{2}$
13. $x \mapsto \frac{4x}{3}+1$
14. $x \mapsto \frac{x-2}{6}$
15. $x \mapsto \frac{2x-24}{5}$
16. $x \mapsto \frac{x-3}{-7}$
17. $x \mapsto \frac{3x-12}{-5}$
18. $x \mapsto 10-3x$
19. $x \mapsto \frac{4(5x+3)+1}{2}$
20. $x \mapsto \frac{7x-30}{-6}$
21. $x \mapsto 20x-164$
22. a) $\sqrt{x}$ b) log c) $x!$ d) x^2 e) $\frac{1}{x}$ f) tan g) $\frac{1}{x}$ h) $\sqrt{\ }$ i) cos j) log or ln k) tan l) $x!$ m) cos n) sin o) cos p) $\frac{1}{x}$ q) $x!$ r) log

Exercise 5.3 *page 165*

1. a) 6 **b)** 10 **c)** 2 **d)** 2 **e)** 10 **f)** 11

2. a)

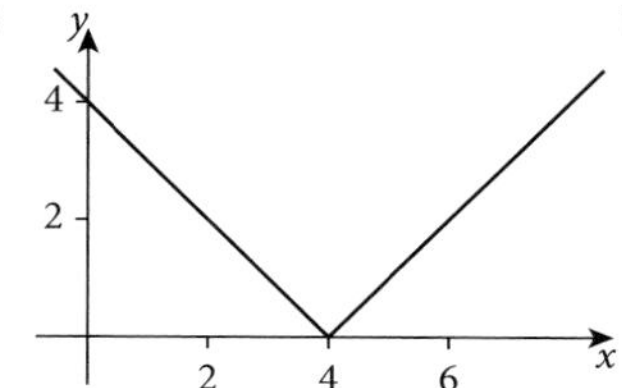

b)

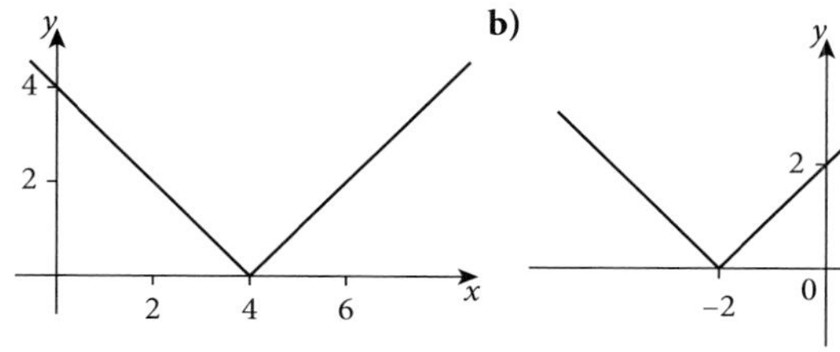

c)

d)

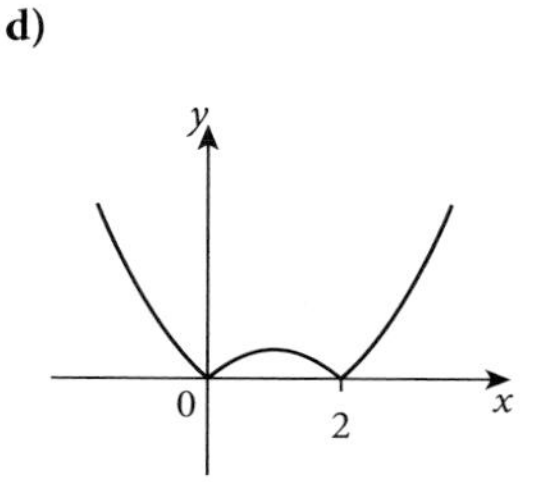

e)

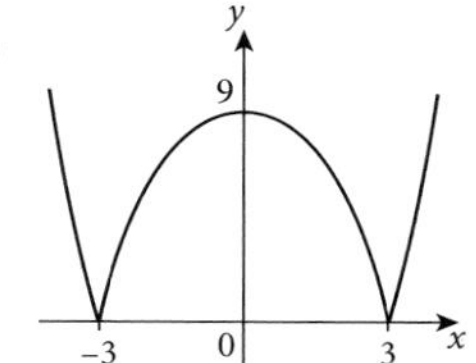

f)

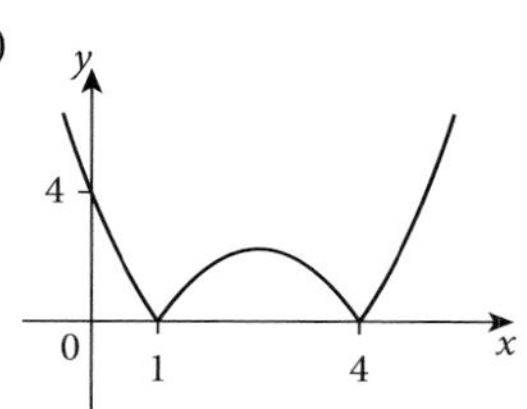

g)

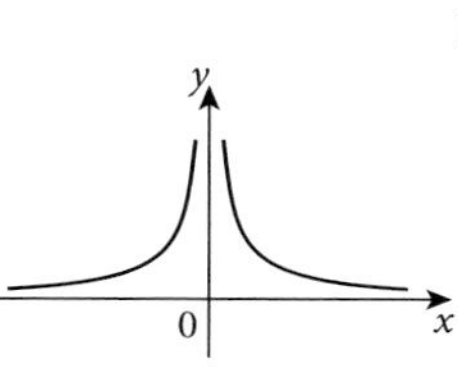

h)

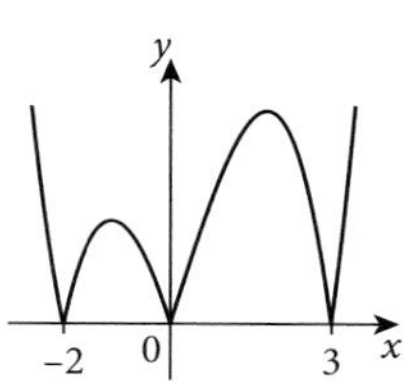

Exercise 5.4 *page 166*

1. a) $f(x)=ax^2+bx+c$ **b)** $f(x)=a^x$ **c)** $f(x)=\frac{a}{x}$ **d)** $f(x)=ax^3+bx^2+cx+d, a<0$

e) $f(x)=|ax+b|$ **f)** $f(x)=ax^3+bx^2+cx+d, a>0$

Exercise 5.5 *page 167*

In question **1** to **18** sketch graphs are on students' calculations.

The graphs of the following have asymptotes as follows:

Qu 3: $y=0, x=0$ Qu 4: $y=0$ Qu 8: $x=1, y=0$ Qu 10: $y=0$ Qu 11: $y=1$

Qu 18: $x=0, y=0$ Qu 19: **b)** smaller **c)** $x=0, y=x$

20. a)

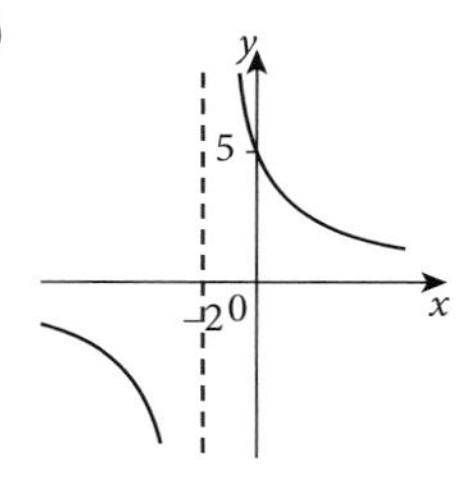

b)

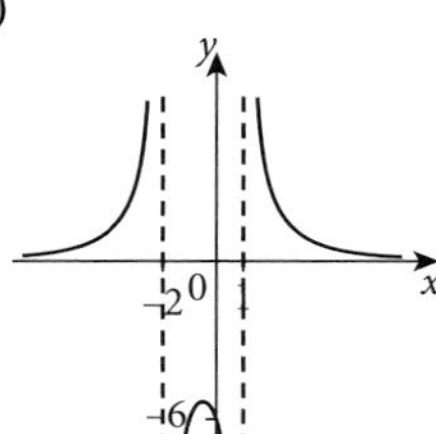

c)

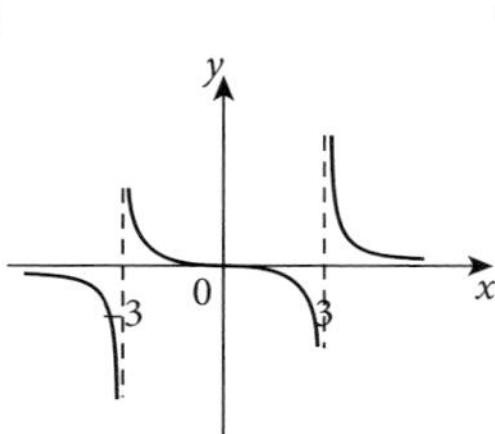

d)

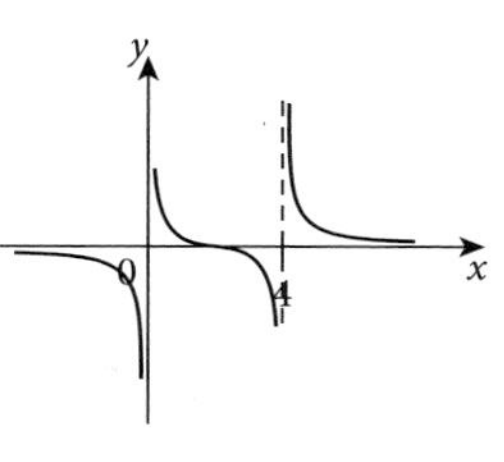

e)

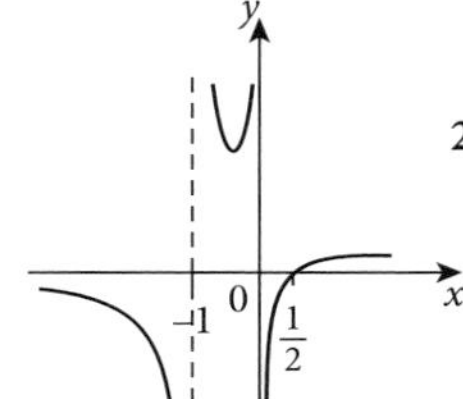

21. a) 1 **b)** 2 **c)** 2 **d)** 1 **e)** 2 **f)** 3

Exercise 5.6 *page 168*

1. a) i) 40 km **ii)** 24 km **iii)** 72 km **iv)** 8 km

b) i) 40 miles **ii)** 35 miles **iii)** 10 miles **iv)** 20 miles

2. a) i) €24 **ii)** €96 **iii)** €60 **b) i)** £47 **ii)** £70 **iii)** £117 **c)** £128

3. a) 180 **b)** $C = 0{\cdot}2x + 35$ **4. a)** 30 litres **b) i)** 8 km/litre **ii)** 6 km/litre; 30 litres **c)** $6\frac{2}{3}$ km/litre; 6 litres

5. a) 2000 **b)** 200 **c)** $1.6 \le x \le 2.4$ **6. a)** Yes **b)** No **c)** About \$250 – \$270

Exercise 5.7 *page 173*

1. a) $y=x^2-5x+4$ **b)** $y=x^2-x-6$ **c)** $y=5x-x^2$ **d)** $y=-(x^2+6x+5)$ **e)** $y=x^2-2x+1$

2. a) $y=2x^2-12x+16$ **b)** $y=x^2-3x-10$ **c)** $y=3-2x-x^2$

3. a) $y=3x(x-6)$ **b)** $y=3x^2-12$ **c)** $y=\frac{1}{5}(21-4x-x^2)$ **4. a)** (3, 5) **b)** (1, 11) **c)** (0, −6) **d)** (−2, 0)

Revision exercise 5A *page 174*

1. a) i) $\frac{x-4}{3}$ **ii)** $5x+2$ **b)** $5\frac{1}{3}$ **2. a)** -5 **b)** 0 **c)** -3 **d)** 8

3. a) 3 **b)** 0, 5 **4. a)** C **b)** A **c)** D **d)** E **5. a)** $x = 0, y = 0$ **b)** $y = 0$

6. a)

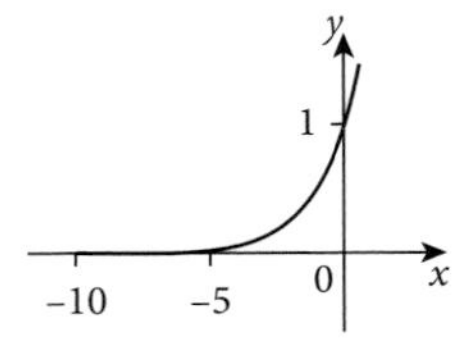

b)

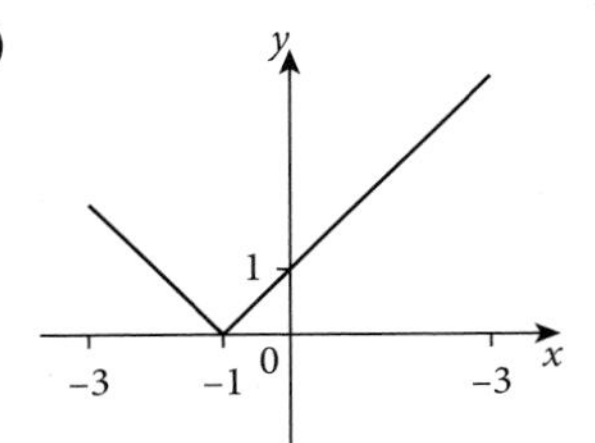

c)

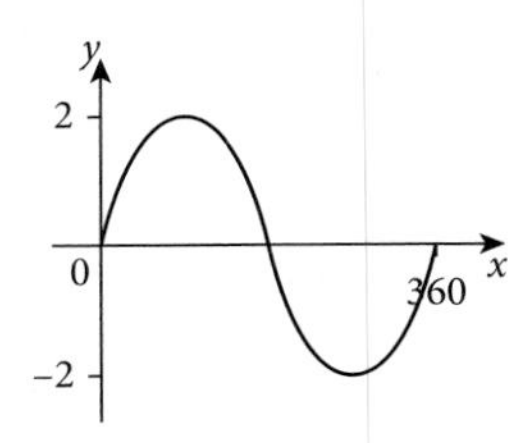

Examination exercise 5B *page 175*

1. a) A, **b)** C, **c)** F, **d)** E, **e)** H, **f)** G

2. a) 13 **b)** $3(2x-1)^2+1=12x^2-12x+4$ **c)** $\frac{x+1}{2}$ **3. a)**

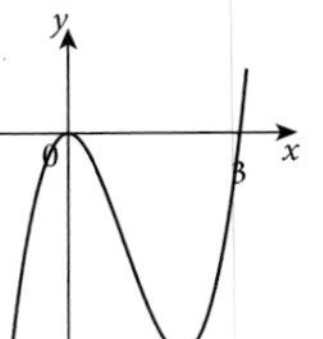

b) (0, 0), (3, 0)

c) (0, 0), (2, −4) **4. c) i)** 9 **ii)** 0 **iii)** $\sqrt[3]{x-1}$ **iv)** 28

5. a) $\frac{1}{2}$ **b)** −1·5  **c)** $\frac{x-5}{x}$

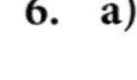

6. a)

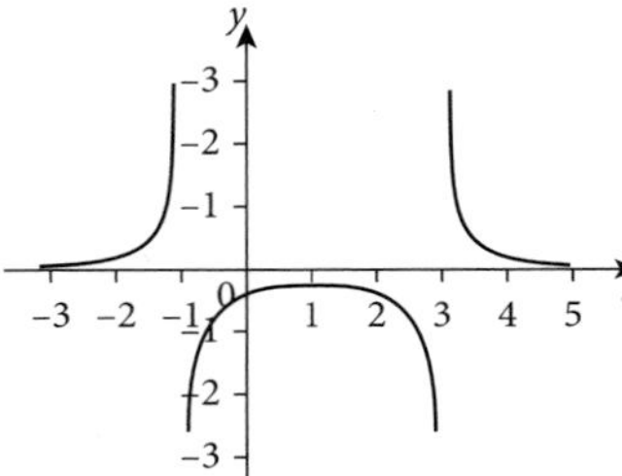

b) $x = -1, x = 3, y = 0$ **c)** (1, −0·25)

d) domain $(x \in \mathrm{R}), x \neq -1, x \neq 3$ range $y > 0, y \leq -0{\cdot}25$ **e) i)** 2 **ii)** 4

7. a)

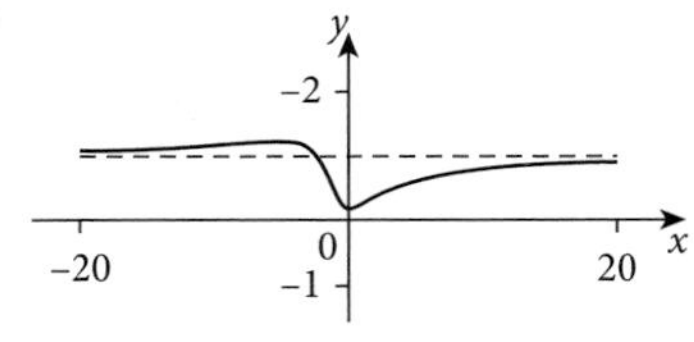

b) (−5·19, 1·24) **c)** 0·161 to 1·24 **d)** $y = 1$ **e)** −1·62

6 Investigations and mathematical modelling

6.1 Investigations *page 178*

Note: It must be emphasised that the *process* of obtaining reliable results is far more important than these few results· It is not suggested that 'obtaining a formula' is the only aim of these coursework tasks· The results are given here for some of the investigations merely as a check for teachers or students working on their own.

It is not possible to summarise the enormous number of variations which students might think of for themselves· Obviously some original thoughts will be productive while many others will soon 'dry up'.

1. With the numbers written in c columns, the difference for a $(n \times n)$ square is $(n-1)^2 \times c$.

5. a) For n blacks and n whites, numbers of moves $=\frac{n(n+1)}{2}$.

b) For n blacks and n whites, number of moves $=\frac{n(n+2)}{2}$.

c) For n of each colour, number of moves $=\frac{3}{2}n(n+1)$.

6. 4 × 4 : There are 30 squares (i·e· 16 + 9 + 4 + 1)

8 × 8 : There are 204 squares

(64 + 49 + 36 + 25 + 16 + 9 + 4 + 1)

n × n : Number of squares $= 1^2 + 2^2 + 3^2 + \cdots + n^2$

$=\frac{n}{6}(n+1)(2n+1)$

This could be found using method of differences or from the standard result for $\sum_1^n r^2$.

6.2 Mathematical modelling page 183

1. For part (d) state clearly the assumption you have made about the increased braking distance in bad weather.
2. For a square card, corner cut out $= \frac{1}{6}$ (size of card).

   For a rectangle $a \times 2a$, corner cut out $\cong \frac{a}{4{\cdot}732}$.

7 Geometry

Exercise 7.1 page 187

1. 95° 2. 49° 3. 100° 4. 77° 5. 129° 6. 95°
7. $a = 30°$ 8. $e = 30°, f = 60°$ 9. 110° 10. $x = 54°$ 11. $a = 40°$
12. $a = 36°, b = 72°, c = 144°, d = 108°$ 13. 105° 14. $a = 30°, b = 120°, c = 150°$ 15. $x = 20°, y = 140°$
16. $a = 120°, b = 34°, c = 26°$ 17. $a = 68°, b = 58{\cdot}5°$ 18. 25°
19. 44° 20. $a = 30°, b = 60°, c = 150°, d = 120°$ 21. $a = 10°, b = 76°$ 22. $e = 71°, f = 21°$
23. 144° 24. 70° 25. 41°, 66° 26. 46°, 122° 27. 36°

Exercise 7.2 page 189

1. $a = 72°, b = 108°$ 2. $x = 60°, y = 120°$ 3. $(n - 2)180°$ 4. 110° 5. 60°
6. $128\frac{4}{7}°$ 7. 15 8. 12 9. 9 10. 18
11. 12 12. 36° 13. 42° 14. 113° (293° is an alternative answer for a concave decagon)
15. a) 135° b) 45° c) 22·5° 16. 105°
17. $a = 18°, b = 81°, c = 18°, d = 162°$ 18. 36 19. $x = 36°, y = 144°$

Exercise 7.3 page 192

1. $a = 116°, b = 64°, c = 64°$ 2. $a = 64°, b = 40°$ 3. $x = 68°$
4. $a = 40°, b = 134°, c = 134°$ 5. $m = 69°, y = 65°$ 6. $t = 48°, u = 48°, v = 42°$
7. $a = 118°, b = 100°, c = 62°$ 8. $a = 34°, b = 76°, c = 70°, d = 70°$ 9. 72°, 108°
10. 26° 11. 60°

Exercise 7.4 page 194

1. 10 cm 2. 4·12 cm 3. 4·24 cm 4. 12·7 cm 5. 8·72 cm 6. 5·66 cm
7. 6·63 cm 8. 5 cm 9. 17 cm 10. 4 cm 11. 9·85 cm 12. 7·07 cm 13. 3·46 m

Exercise 7.5 page 195

1. 5·29 cm 2. 6·16 cm 3. 5·14 cm² 4. 72·7 cm² 5. 103 cm²
6. 18 cm² 7. 14·3 cm² 8. 5·39 units 9. a) 4·47 b) 6·08 c) 7·21 (all units)

Exercise 7.6 page 197

1. 40·3 km 2. 13·6 cm 3. 6·34 m 4. 4·58 cm 5. 84·9 km
6. 24 cm 7. 9·80 cm 8. 5, 4, 3; 13, 12, 5; 25, 24, 7; 41, 40, 9; 61, 60, 11
9. $x = 4$ m, 20·6 m 10. 9·49 cm 11. 18·5 km 12. 11·40 cm 13. 11·36 cm
14. 8·43 cm 15. a) 7·55 b) 12·5 c) 14·9 16. 10 feet

Exercise 7.7 page 200

1. a) 1, 1 b) 1, 1 c) 2, 2 d) 2, 2 e) 4, 4 f) 0, 2 g) 5, 5
   h) 0, 1 i) 1, 1 j) 0, 2 k) 0, 2 l) 0, 2 m) ∞, ∞ n) 0, 4
4. square 4, 4; rectangle 2, 2; parallelogram 0, 2; rhombus 2· 2; trapezium 0, 1; kite 1, 1; equilateral triangle 3, 3; regular hexagon 6, 6
5. 34°, 56° 6. 35°, 35° 7. 72°, 108°, 28° 8. 40°, 30°, 110° 9. 116°, 32°, 58°
10. 55°, 55° 11. 26°, 26°, 77° 12. 52°, 64°, 116° 13. 70°, 40°, 110° 14. 54°, 72°, 36° 15. 60°, 15°, 75°, 135°

Exercise 7.8 page 202

1. $a = 2\frac{1}{2}$ cm, $e = 3$ cm 2. $x = 6$ cm, $y = 10$ cm 3. $x = 12$ cm, $y = 8$ cm
4. m = 10 cm, $a = 16\frac{2}{3}$ cm 5. $y = 6$ cm 6. $x = 4$ cm, $w = 1\frac{1}{2}$ cm
7. $e = 9$ cm, $f = 4\frac{1}{2}$ cm 8. $x = 13\frac{1}{3}$ cm, $y = 9$ cm 9. $m = 6$ cm, $n = 6$ cm
10. $m = 5\frac{1}{3}$ cm, $z = 4\frac{4}{5}$ cm 11. $v = 5\frac{1}{3}$ cm, $w = 6\frac{2}{3}$ cm 12. No 13. AO = 2 cm, DO = 6 cm 14. 16 m
15. a) Yes b) No c) No d) Yes e) Yes f) No g) No h) Yes

Exercise 7.9 page 204

1. Q = 25 cm, R = 45 cm, S = 70 cm **2.** BDA, CBD, CBA (any two), $x = 3{\cdot}75$ cm **3.** 180 cm² **4.** No **5.** 3·5 **6.** $4\frac{2}{3}$ cm or $1\frac{13}{14}$ cm **7.** $\frac{x}{x-1} = \frac{x-1}{\frac{x}{2}}$, $x = 3{\cdot}4$ **8.** $y^2 = qz$ **10.** $x = 0{\cdot}618$, $1{\cdot}618 : 1$

Exercise 7.10 page 207

1. A and G; B and E.

Exercise 7. 11 page 209

1. 16 cm² **2.** 27 cm² **3.** $11\frac{1}{4}$ cm² **4.** $14\frac{1}{2}$ cm² **5.** 128 cm² **6.** 12 cm² **7.** 8 cm **8.** 18 cm **9.** $4\frac{1}{2}$ cm **10.** $7\frac{1}{2}$ cm **11.** $2\frac{1}{2}$ cm **12.** 6 cm

Exercise 7.12 page 210

1. A = 32 cm² **2.** B = 27·9 cm² **3.** C = 40 cm² **4.** d = 225 cm² **5. a)** $16\frac{2}{3}$ cm² **b)** $10\frac{2}{3}$ cm² **6. a)** 25 cm² **b)** 21 cm² **7.** 8 cm² **8.** 6 cm **9.** 24 cm² **10. a)** $1\frac{4}{5}$ cm **b)** 3 cm **c)** 3 : 5 **d)** 9 : 25 **11.** 150 **12.** 360 **13.** Small potatoes (for the same weight)

Exercise 7.13 page 213

1. 480 cm³ **2.** 540 cm³ **3.** 160 cm³ **4.** 4500 cm³ **5.** 81 cm³ **6.** 11 cm³ **7.** 16 cm³ **8.** $85\frac{1}{3}$ cm³ **9.** 4 cm **10.** 21 cm **11.** 4·6 cm **12.** 9 cm **13.** 6·6 cm **14.** $4\frac{1}{2}$ cm **15.** $168\frac{3}{4}$ cm³ **16.** 106·3 cm³ **17.** 12 cm **18. a)** 2 : 3 **b)** 8 : 27 **19.** 8 : 125 **20.** $x_1^3 : x_2^3$ **21.** 54 kg **22.** 240 m² **23.** $9\frac{3}{8}$ litres **24.** $2812\frac{1}{2}$ cm²

Exercise 7.14 page 216

1. $a = 27°$, $b = 30°$ **2.** $c = 20°$, $d = 45°$ **3.** $c = 58°$, $d = 41°$, $e = 30°$ **4.** $f = 40°$, $g = 55°$, $h = 55°$ **5.** $a = 32°$, $b = 80°$, $c = 43°$ **6.** $x = 34°$, $y = 34°$, $z = 56°$ **8.** $a = 92°$ **9.** $b = 42°$ **10.** $c = 46°$, $d = 44°$ **11.** $e = 49°$, $f = 41°$ **12.** $g = 76°$, $h = 52°$ **13.** $x = 48°$ **14.** $y = 32°$ **15.** $x = 22°$ **16.** $a = 36°$, $x = 36°$

Exercise 7.15 page 219

1. $a = 94°$, $b = 75°$ **2.** $c = 101°$, $d = 84°$ **3.** $x = 92°$, $y = 116°$ **4.** $c = 60°$, $d = 45°$ **5.** 37° **6.** 118° **7.** $e = 36°$, $f = 72°$ **8.** 35° **9.** 18° **10.** 90° **11.** 30° **12.** $22\frac{1}{2}°$ **13.** $n = 58°$, $t = 64°$, $w = 45°$ **14.** $a = 32°$, $b = 40°$, $c = 40°$ **15.** $a = 18°$, $c = 72°$ **16.** 55° **17.** $e = 41°$, $f = 41°$, $g = 41°$ **18.** 8° **19.** $x = 30°$, $y = 115°$ **20.** $x = 80°$, $z = 10°$

Exercise 7.16 page 221

1. $a = 18°$ **2.** $x = 40°$, $y = 65°$, $z = 25°$ **3.** $c = 30°$, $e = 15°$ **4.** $f = 50°$, $g = 40°$ **5.** $h = 40°$, $i = 40°$ **6.** $n = 36°$ **7.** $k = 50°$, $m = 50°$, $n = 80°$, $p = 80°$ **8.** $n = 16°$, $p = 46°$ **9.** $x = 70°$, $y = 20°$, $z = 55°$

Revision exercise 7A page 223

2. 80° **3. a)** 30° **b)** $22\frac{1}{2}°$ **c)** 12 **4. a)** 40° **b)** 100° **5.** 4·12 cm **6. i)** 3 cm **ii)** 5·66 cm **7. c)** $2\frac{4}{5}$ cm **8. b)** 6 cm **9.** $3\frac{2}{3}$ cm, $1\frac{1}{11}$ cm **10.** 6 cm **11.** 250 cm³ **12. a)** $3\frac{1}{3}$ cm **b)** 1620 cm³ **13. a)** 1 m² **b)** 1000 cm³ **14. a)** 50° **b)** 128° **c)** $c = 50°$, $d = 40°$ **d)** $x = 10°$, $y = 40°$ **15. a)** 55° **b)** 45° **16. a)** $\sqrt{89} \approx 9{\cdot}43$ cm **b)** $\sqrt{98} \approx 9{\cdot}90$ cm **17.** $a = 45°$, $b = 75°$, $c = 15°$ **18.** 10 cm, 15·3 cm² **19.** 125 **20.** 32·6 cm **21.** 2·5 cm **22.** 15·3 cm² **23.** 92°

Examination exercise 7B page 225

1. a) 35° **b)** 125° **c)** 15°
2. a) i) 116° **ii)** 32° **iii)** 61° **iv)** 7° **b)** sum of opposite angles = 180°
3. a) i) 60° **ii)** 135° **iii)** 110° **b) i)** 75° **ii)** 105°
4. a) i) $A\hat{B}X = D\hat{C}X$, $A\hat{X}B = D\hat{X}C$, the triangles have the same three angles **ii)** 18 cm² **b) i)** 50° **ii)** 98° **iii)** 5·14 cm **iv)** 4 cm

8 Algebra 2

Exercise 8.1 page 228

1. < **2.** > **3.** > **4.** = **5.** < **6.** < **7.** = **8.** > **9.** < **10.** > **11.** < **12.** > **13.** > **14.** > **15.** = **16.** F **17.** F **18.** T **19.** F **20.** F **21.** T **22.** T **23.** F **24.** F **25.** $x > 13$ **26.** $x < -1$ **27.** $x < 12$ **28.** $x \le 2\frac{1}{2}$ **29.** $x > 3$ **30.** $x \ge 8$ **31.** $x < \frac{1}{4}$ **32.** $x \ge -3$ **33.** $x < -8$ **34.** $x < 4$ **35.** $x > -9$ **36.** $x < 8$ **37.** $x > 3$ **38.** $x \ge 1$ **39.** $x < 1$ **40.** $x > 2\frac{1}{3}$

Exercise 8.2 page 229

1. $x > 5$ **2.** $x \leq 3$ **3.** $x > 6$ **4.** $x \geq 4$ **5.** $x < 1$ **6.** $x < -3$ **7.** $x > 0$
8. $x > 4$ **9.** $x > 2$ **10.** $x < -3$ **11.** $1 < x < 4$ **12.** $-2 \leq x \leq 5$ **13.** $1 \leq x < 6$ **14.** $0 \leq x < 5$
15. $-1 \leq x \leq 7$ **16.** $0{\cdot}2 < x < 2$ **17.** $x > 80$ **18.** $x > 10$ **19.** $x < -2{\cdot}5$ **20.** $0 < x < 4$ **21.** $5 \leq x \leq 9$
22. $-1 < x < 4$ **23.** $5{\cdot}5 < x < 6$ **24.** $\frac{1}{2} < x < 8$ **25.** $-8 < x < 2$ **26.** $\{1, 2, 3, 4, 5, 6\}$ **27.** $\{7, 11, 13, 17, 19\}$
28. $\{2, 4, 6, 8, 10\}$ **29.** $\{4, 9, 16, 25, 36, 49\}$ **30.** $\{5, 10\}$ **31.** $\{-4, -3, -2, -1\}$
32. $\{2, 3, 4, \cdots 12\}$ **33.** $\{1, 4, 9\}$ **34.** $\{2, 3, 5, 7, 11\}$ **35.** $\{2, 4, 6, \cdots 18\}$

Exercise 8.3 page 230

1. $-\frac{1}{2} < x < 4$ **2.** $0 < x < 4$ **3.** $-2 < x < -1$ **4.** $9 < x < 11$ **5. a)** $n > 3$ **b)** $n < 9$ **6. a)** 8 **b)** 1
7. a) 6 **b)** 35 **c)** 15 **8.** 1, 2, 3, 4 **9.** 7 **10.** 10
11. $\frac{1}{2}$ or others **12.** 1, 2, 3, ····· 14 **13.** 19 **14.** $\frac{1}{2}$ or others **15.** 19
16. a) $-3 \leq x < 6$ **b)** $-2 < x < 2$ **c)** $-3 \leq x \leq 2$ **d)** $-3 \leq x < 7$ **17.** 17

Exercise 8.4 page 232

1. $x > 3$ **2.** $x < 7, y < 5$ **3.** $y \geq x$ **4.** $y \leq x, x \leq 8, y \geq -2$
5. a) $x + y \leq 7, x \geq 0, y \geq x$ **b)** $x + y \leq 6, y \geq 0, y \leq x + 2$

6.
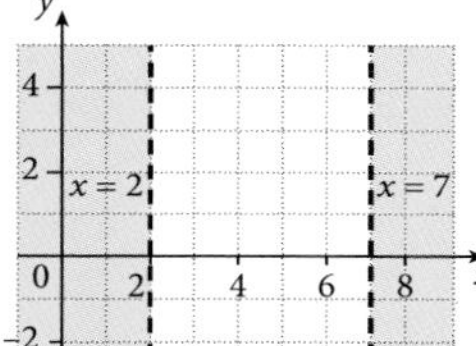

7.
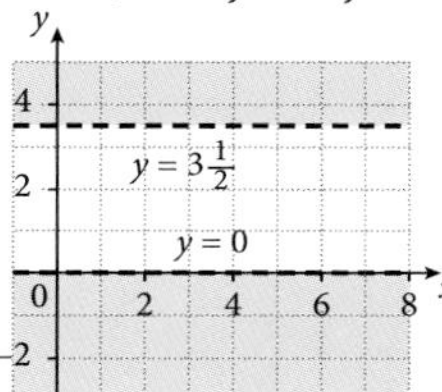

8.
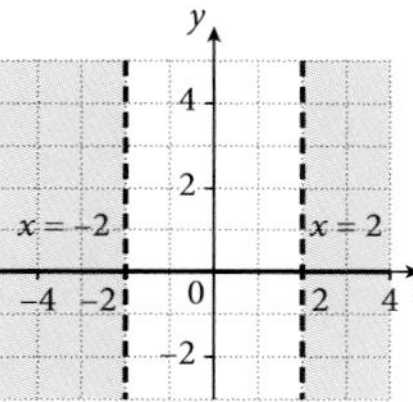

9.
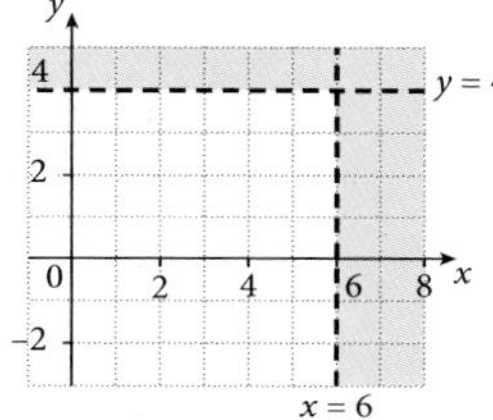

10.
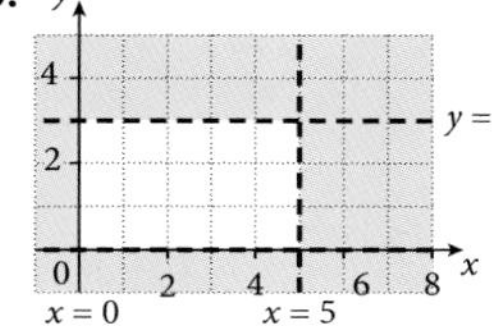

11.
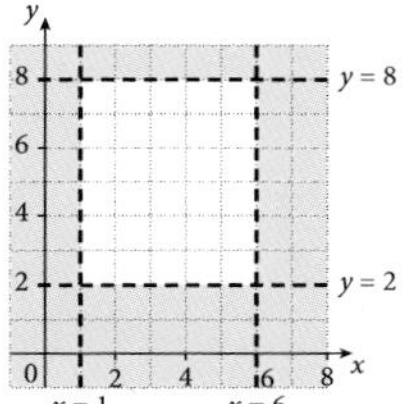

12.
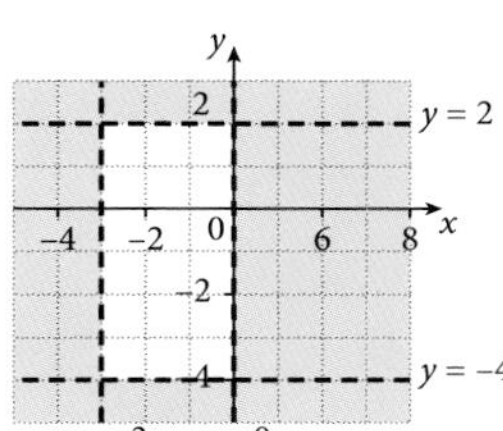

13.
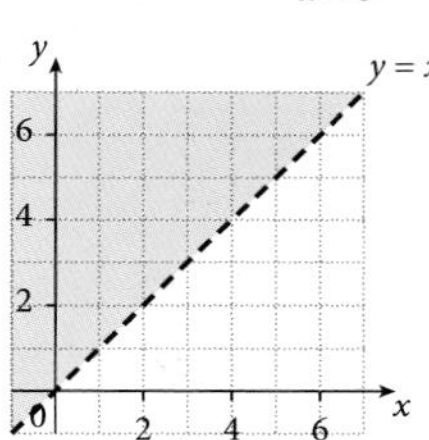

14.
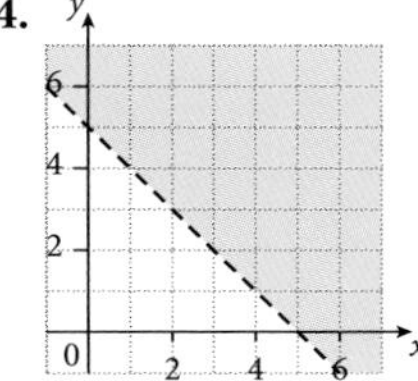

15.
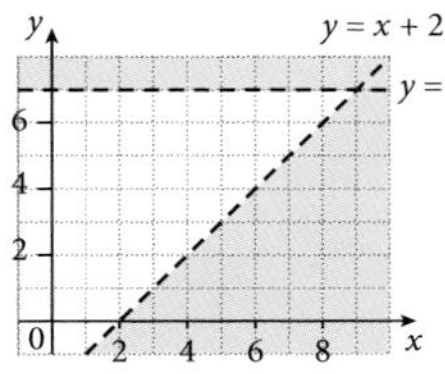

Exercise 8.5 page 234

1. a) $-1 < x < 1$ **b)** $x > 4, x < -4$ **c)** $-3 \leq x \leq 3$
d) $-4 < x < 4$ **e)** $x > 5, x < -5$ **f)** $x \geq 4, x \leq -4$
2. a) $x \geq 6$ or $x \leq -2$ **b)** $-4 < x < -2$ **c)** $x \geq 6$ or $x \leq -1$
d) $-3 < x < -2$ **e)** $x \geq 4$ or $x \leq 3$ **f)** $1 \leq x \leq 2$
g) $\frac{3}{2} < x < 4$ **h)** $-\frac{2}{3} < x < 5$ **i)** $-4\frac{1}{2} < x < \frac{1}{2}$
j) $x \geq 3$ or $x \leq \frac{4}{3}$ **k)** $x \leq -\frac{1}{2}, x \geq \frac{3}{2}$ **l)** $-2 < x < \frac{13}{3}$ **m)** $2 < x < \frac{5}{2}$ **n)** $x > \frac{5}{2}$ or $x < \frac{1}{3}$
3. $x > 4$ **4.** $x > 5$ **5.** $x < -3$ **6.** $p \leq 4$ and $p \geq 20$

Exercise 8.6 page 235

1. 2^4 **2.** 7^6 **3.** $3^2 \times 2^5$ **4.** a^5 **5.** p^3 **6.** $5^4 \times 8^2$ **7.** 3^{10} **8.** 4^5 **9.** 8^8 **10.** 7^{44}
11. 9^{24} **12.** n^5 **13.** a^{10} **14.** n^{10} **15.** y^8 **16.** n^5 **17.** a^8 **18.** x^{10} **19.** 3^3 **20.** 2^5
21. 4^6 **22.** $5^3 \times 5^2 = 5^5$ **23.** $7^5 \div 7^3 = 7^2$ **24.** $3^2 \times 3^{10} = 3^{12}$ **25.** $7^{13} \div 7^{10} = 7^3$ **26.** $n^3 \times n^3 = n^6$
27. $3^{100} \div 3^{80} = 3^{20}$ **28.** $3^6 \div 3 = 3^5$ **29.** $10^{10} \div 10^9 = 10$ **30.** 2^4 **31.** 25 **32.** 4 **33.** 100
34. 27 **35.** 1 **36.** 16 **37.** 16 **38.** 1000 **39.** 1

Exercise 8.7 page 236

1. 3^4 **2.** $4^2 \times 5^3$ **3.** 3×7^3 **4.** $2^3 \times 7$ **5.** 10^{-3} **6.** $2^{-2} \times 3^{-3}$ **7.** $15^{\frac{1}{2}}$ **8.** $3^{\frac{1}{3}}$ **9.** $10^{\frac{1}{5}}$ **10.** $5^{\frac{3}{2}}$
11. x^7 **12.** y^{13} **13.** z^{10} **14.** z^{100} **15.** m **16.** e^{-5} **17.** y^2 **18.** w^6 **19.** y **20.** x^{10}
21. 1 **22.** w^{-5} **23.** w^{-5} **24.** x^7 **25.** a^8 **26.** k^3 **27.** 1 **28.** x^{29} **29.** y^2 **30.** x^6
31. z^4 **32.** t^{-4} **33.** $4x^6$ **34.** $16y^{10}$ **35.** 1 **36.** $21w^{-3}$ **37.** $2n^4$ **38.** $2x$

Exercise 8.8 page 237

1. 27 **2.** 1 **3.** $\frac{1}{9}$ **4.** 25 **5.** 2 **6.** 4 **7.** 9 **8.** 2 **9.** 27 **10.** 3

11. $\frac{1}{3}$ **12.** $\frac{1}{2}$ **13.** 1 **14.** $\frac{1}{5}$ **15.** 10 **16.** 8 **17.** 32 **18.** 4 **19.** $\frac{1}{9}$ **20.** $\frac{1}{8}$

21. 18 **22.** 10 **23.** 1000 **24.** $\frac{1}{1000}$ **25.** $\frac{1}{9}$ **26.** 1 **27.** $1\frac{1}{2}$ **28.** $\frac{1}{25}$ **29.** $\frac{1}{10}$ **30.** $\frac{1}{4}$

31. $\frac{1}{4}$ **32.** 100 000 **33.** 1 **34.** $\frac{1}{32}$ **35.** 0·1 **36.** 0·2 **37.** 1·5 **38.** 1 **39.** 9 **40.** $1\frac{1}{2}$

41. $\frac{3}{10}$ **42.** 64 **43.** $\frac{1}{100}$ **44.** $1\frac{2}{3}$ **45.** $\frac{1}{100}$ **46.** 1 **47.** 100 **48.** 6 **49.** 750 **50.** −7

51. a) 0 **b)** $\frac{2}{5}$

Exercise 8.9 page 238

1. $25x^4$ **2.** $49y^6$ **3.** $100a^2b^2$ **4.** $4x^2y^4$ **5.** $2x$ **6.** $\frac{1}{9y}$ **7.** x^2 **8.** $\frac{x^2}{2}$ **9.** 1 **10.** $\frac{2}{x}$

11. $36x^4$ **12.** $25y$ **13.** $16x^2$ **14.** $27y$ **15.** 25 **16.** 1 **17.** 49 **18.** 1 **19.** $8x^6y^3$ **20.** $100x^2y^6$

21. $\frac{3x}{2}$ **22.** $\frac{2}{x}$ **23.** x^3y^5 **24.** $12x^3y^2$ **25.** $10y^4$ **26.** $3x^3$ **27.** $x^3y^2z^4$ **28.** x **29.** $3y$ **30.** $27x^{\frac{3}{2}}$

31. $10x^3y^5$ **32.** $32x^2$ **33.** $\frac{5}{2}x^2$ **34.** $\frac{9}{x^2}$ **35.** $2a^2$ **36.** $a^3b^3c^6$ **37. a)** 2^5 **b)** 2^7 **c)** 2^6 **d)** 2^0

38. a) 3^{-3} **b)** 3^{-4} **c)** 3^{-1} **d)** 3^{-2} **39.** 16 **40.** $\frac{1}{4}$ **41.** $\frac{1}{6}$ **42.** 1 **43.** $16\frac{1}{8}$ **44.** $\frac{3}{8}$

45. $\frac{1}{4}$ **46.** $\frac{5}{256}$ **47.** $1\frac{1}{16}$ **48.** 0 **49.** $\frac{1}{4}$ **50.** $\frac{1}{4}$ **51.** 3 **52.** 4 **53.** −1 **54.** −2

55. 3 **56.** 3 **57.** 1 **58.** $\frac{1}{5}$ **59.** 0 **60.** −4 **61.** 2 **62.** −5 **63.** 1 **64.** $\frac{1}{18}$

65. a) 3·60 **b)** 5·44 **66. a)** 1024 **b)** 30 **c)** 31

Exercise 8.10 page 239

1. 2^{19} **2.** 6^6 **3.** 10^3 **4.** 10^{21} m³ **5.** 10^5 **6.** 10^7 seconds, less than a year

7. a) 3 **b)** 4 **c)** 10 **d)** 12 **e)** 2 **f)** 3 **g)** 10 **h)** 2

8. Assuming the number are labelled a, b, c, d, e, f the order of size is d, b, a, c, f, e

Exercise 8.11 page 240

1. about 2·5 **3. a)** 2·9 **b)** 1·5, 4·9 **4. a)** 1 **b)** $\sqrt[n]{n} \to 1$ **5. a)** 3 **b)** 4·56

Exercise 8.12 page 242

1. $2\frac{1}{2}$ **2.** 3 **3.** $\frac{B}{A}$ **4.** $\frac{T}{N}$ **5.** $\frac{K}{M}$ **6.** $\frac{4}{y}$ **7.** $\frac{C}{B}$ **8.** $\frac{D}{4}$ **9.** $\frac{T+N}{9}$ **10.** $\frac{B-R}{A}$

11. $\frac{R+T}{C}$ **12.** $\frac{N-R^2}{L}$ **13.** $\frac{R-S^2}{N}$ **14.** 2 **15.** −7 **16.** L + B **17.** N + T **18.** 2 **19.** $4\frac{1}{2}$ **20.** $\frac{N-C}{A}$

21. $\frac{L-D}{B}$ **22.** $\frac{F-E}{D}$ **23.** $\frac{H+F}{N}$ **24.** $\frac{T+Z}{Y}$ **25.** $\frac{B+L}{R}$ **26.** $\frac{Q-m}{V}$ **27.** $\frac{n+a+m}{t}$ **28.** $\frac{s-t-n}{q}$ **29.** $\frac{t+s^2}{n}$ **30.** $\frac{c-b}{V^2}$

Exercise 8.13 page 243

1. 12 **2.** 10 **3.** BD **4.** TB **5.** RN **6.** bm **7.** 26

8. $BT + A$ **9.** $AN + D$ **10.** $B^2N - Q$ **11.** $ge + r$ **12.** $4\frac{1}{2}$ **13.** $\frac{DC-B}{A}$ **14.** $\frac{pq-m}{n}$

15. $\frac{vS+t}{r}$ **16.** $\frac{qt+m}{z}$ **17.** $\frac{bc-m}{A}$ **18.** $\frac{AE-D}{B}$ **19.** $\frac{nh+f}{e}$ **20.** $\frac{qr-b}{g}$ **21.** 4 **22.** −2

23. 2 **24.** $A - B$ **25.** $\frac{v^2-r}{r}$ **26.** $\frac{w-t^2}{n}$ **27.** $\frac{n-2}{q}$ **28.** $\frac{1}{4}$ **29.** $-\frac{1}{7}$

30. $\frac{B-DE}{A}$ **31.** $\frac{D-NB}{E}$ **32.** $\frac{h-bx}{f}$ **33.** $\frac{v^2-Cd}{h}$

Exercise 8.14 page 243

1. $\frac{1}{2}$ **2.** $1\frac{2}{3}$ **3.** $\frac{B}{C}$ **4.** $\frac{T}{X}$ **5.** $\frac{M}{B}$ **6.** $\frac{n}{m}$ **7.** $\frac{v}{t}$

8. $\frac{n}{\sin 20°}$ **9.** $\frac{7}{\cos 30°}$ **10.** $\frac{B}{x}$ **11.** $6\frac{2}{3}$ **12.** $\frac{ND}{B}$ **13.** $\frac{HM}{N}$ **14.** $\frac{et}{b}$

15. $\frac{vs}{m}$ **16.** $\frac{mb}{t}$ **17.** 1 **18.** $\frac{5}{6}$ **19.** $\frac{A}{C-B}$ **20.** $\frac{V}{H-G}$ **21.** $\frac{r}{n+t}$

22. $\frac{b}{q-d}$ **23.** $\frac{m}{t+n}$ **24.** $\frac{b}{d-h}$ **25.** $\frac{d}{C-e}$ **26.** $\frac{m}{r-e^2}$ **27.** $\frac{n}{b-t^2}$ **28.** $\frac{d}{mn-b}$

Exercise 8.15 page 244

1. 4 2. 24 3. 11 4. $B^2 - A$ 5. $D^2 - C$ 6. $H^2 + E$ 7. $\frac{c^2 - b}{a}$
8. $a^2 + m$ 9. $\frac{b^2 + t}{g}$ 10. $b - r^2$ 11. $d - t^2$ 12. $b^2 + d$ 13. $n - c^2$ 14. $b - f^2$
15. $c - g^2$ 16. $\frac{M - P^2}{N}$ 17. $\frac{D - B}{A}$ 18. $A^4 + D$ 19. $\pm\sqrt{g}$ 20. ± 4 21. $\pm\sqrt{B}$
22. $\pm\sqrt{(B - A)}$ 23. $\pm\sqrt{(M + A)}$ 24. $\pm\sqrt{(b - a)}$ 25. $\pm\sqrt{(C - m)}$ 26. $\pm\sqrt{(d - n)}$ 27. $\pm\sqrt{\frac{n}{m}}$ 28. $\pm\sqrt{\frac{b}{a}}$ 29. $\frac{at}{z}$
30. $\pm\sqrt{\left(\frac{m + t}{a}\right)}$ 31. $\pm\sqrt{(a - n)}$ 32. $\pm\sqrt{40}$ 33. $\pm\sqrt{(B^2 + A)}$ 34. $\pm\sqrt{(x^2 - y)}$ 35. $\pm\sqrt{(t^2 - m)}$ 36. 8
37. $\frac{M^2 - A^2B}{A^2}$ 38. $\frac{M}{N^2}$ 39. $\frac{N}{B^2}$ 40. $a - b^2$ 41. $\pm\sqrt{(a^2 - t^2)}$ 42. $\pm\sqrt{(m - x^2)}$ 43. $\frac{4}{\pi^2} - t$
44. $\frac{B^2}{A^2} - 1$ 45. $\pm\sqrt{\left(\frac{C^2 + b}{a}\right)}$ 46. $\pm\sqrt{\left(\frac{b^2 + a^2x}{a^2}\right)}$ 47. $\pm\sqrt{(x^2 - b)}$ 48. $\pm\sqrt{(c - b)a}$ 49. $\frac{c^2 - b^2}{a}$ 50. $\pm\sqrt{\left(\frac{m}{a + b}\right)}$

Exercise 8.16 page 245

1. $3\frac{2}{3}$ 2. 3 3. $\frac{D - B}{2N}$ 4. $\frac{E + D}{3M}$ 5. $\frac{2b}{a - b}$ 6. $\frac{e + c}{m + n}$ 7. $\frac{3}{x + k}$
8. $\frac{C - D}{R - T}$ 9. $\frac{z + x}{a - b}$ 10. $\frac{nb - ma}{m - n}$ 11. $\frac{d + xb}{x - 1}$ 12. $\frac{a - ab}{b + 1}$ 13. $\frac{d - c}{d + c}$ 14. $\frac{M(b - a)}{b + a}$
15. $\frac{n^2 - mn}{m + n}$ 16. $\frac{v}{3}$ 17. $\frac{a(b + c)}{b - 2a}$ 18. $\frac{5x}{3}$ 19. $-\frac{4z}{5}$ 20. $\frac{mn}{p^2 - m}$ 21. $\frac{mn + n}{4 + m}$

Exercise 8.17 page 246

1. $-\left(\frac{by + c}{a}\right)$ 2. $\pm\sqrt{\left(\frac{e^2 + ab}{a}\right)}$ 3. $\frac{n^2}{m^2} + m$ 4. $\frac{a - b}{1 + b}$ 5. $3y$ 6. $\frac{a}{e^2 + c}$ 7. $-\left(\frac{a + lm}{m}\right)$
8. $\frac{t^2g}{4\pi^2}$ 9. $\frac{4\pi^2d}{t^2}$ 10. $\pm\sqrt{\frac{a}{3}}$ 11. $\pm\sqrt{\left(\frac{t^2e - ba}{b}\right)}$ 12. $\frac{1}{a^2 - 1}$ 13. $\frac{a + b}{x}$ 14. $\pm\sqrt{(x^4 - b^2)}$
15. $\frac{c - a}{b}$ 16. $\frac{a^2 - b}{a + 1}$ 17. $\pm\sqrt{\left(\frac{G^2}{16\pi^2} - T^2\right)}$ 18. $-\left(\frac{ax + c}{b}\right)$

Exercise 8.18 page 247

1. 47 2. 12 3. 440 4. $\frac{2}{3}$ or $\frac{1}{2}$ 5. **a)** 440 **b)** 8 days 6. 11·6

Exercise 8.19 page 248

1. $\frac{5}{7}$ 2. $\frac{7}{8}$ 3. $5y$ 4. $\frac{1}{2}$ 5. 4 6. $\frac{x}{2y}$ 7. 2 8. $\frac{a}{2}$
9. $\frac{2b}{3}$ 10. $\frac{a}{5b}$ 11. a 12. $\frac{7}{8}$ 13. $\frac{5 + 2x}{3}$ 14. $\frac{3x + 1}{x}$ 15. $\frac{32}{25}$ 16. $\frac{4 + 5a}{5}$
17. $\frac{3}{4 - x}$ 18. $\frac{b}{3 + 2a}$ 19. $\frac{5x + 4}{8x}$ 20. $\frac{2x + 1}{y}$ 21. $\frac{x + 2y}{3xy}$ 22. $\frac{6 - b}{2a}$ 23. $\frac{2b + 4a}{b}$ 24. $x - 2$

Exercise 8.20 page 249

1. $\frac{x + 2}{x - 3}$ 2. $\frac{x}{x + 1}$ 3. $\frac{x + 4}{2(x - 5)}$ 4. $\frac{x + 5}{x - 2}$ 5. $\frac{x + 3}{x + 2}$ 6. $\frac{x + 5}{x - 2}$
7. $\frac{x + 2}{x}$ 8. $\frac{3x}{x + 5}$ 9. $\frac{1}{2}$ 10. $\frac{3x}{x - 5}$ 11. $\frac{3x - 5}{x}$ 12. $\frac{x - 2}{x - 1}$

Exercise 8.21 page 250

1. $\frac{3}{5}$ 2. $\frac{3x}{5}$ 3. $\frac{3}{x}$ 4. $\frac{4}{7}$ 5. $\frac{4x}{7}$ 6. $\frac{4}{7x}$ 7. $\frac{7}{8}$
8. $\frac{7x}{8}$ 9. $\frac{7}{8x}$ 10. $\frac{5}{6}$ 11. $\frac{5x}{6}$ 12. $\frac{5}{6x}$ 13. $\frac{23}{20}$ 14. $\frac{23x}{20}$
15. $\frac{23}{20x}$ 16. $\frac{1}{12}$ 17. $\frac{x}{12}$ 18. $\frac{1}{12x}$ 19. $\frac{5x + 2}{15}$ 20. $\frac{7x + 2}{12}$ 21. $\frac{9x + 13}{10}$
22. $\frac{1 - 2x}{12}$ 23. $\frac{2x - 9}{15}$ 24. $\frac{-3x - 12}{14}$ 25. $\frac{3x + 1}{x(x + 1)}$ 26. $\frac{7x - 8}{x(x - 2)}$ 27. $\frac{8x + 9}{(x - 2)(x + 3)}$ 28. $\frac{4x + 11}{(x + 1)(x + 2)}$
29. $\frac{-3x - 17}{(x + 3)(x - 1)}$ 30. $\frac{7x - 5}{(x + 1)(x - 2)}$

Exercise 8.22 page 250

1. $\frac{3n}{5}$ 2. $\frac{2x}{5}$ 3. $\frac{3n}{4}$ 4. $\frac{x}{10}$ 5. $\frac{5x}{8}$ 6. $\frac{x}{12}$ 7. $\frac{9x}{20}$
8. $\frac{3x}{20}$ 9. $\frac{x^2}{12}$ 10. $\frac{t}{5}$ 11. $\frac{ab}{8}$ 12. $\frac{9x^2}{16}$ 13. $\frac{y^2}{9}$ 14. $\frac{8t}{5}$
15. $3\frac{1}{3}$ 16. $\frac{15x^2}{4}$ 17. $\frac{8}{x}$ 18. $\frac{6}{t}$ 19. $\frac{21}{p}$ 20. $\frac{12}{x^2}$

Exercise 8.23 page 251

1. a) $1\frac{1}{10}$ b) $1\frac{1}{10}x$ c) $\frac{3}{10}$ d) $1\frac{1}{5}$ 2. a) $\frac{1}{12}$ b) $\frac{m}{12}$ c) $\frac{m^2}{12}$ d) $1\frac{1}{3}$
3. a) $\frac{17}{24}$ b) $\frac{17s}{24}$ c) $\frac{s}{8}$ d) 3 4. a) $\frac{3}{14}$ b) $\frac{3t}{14}$ c) $\frac{5t^2}{14}$ d) $1\frac{3}{7}$
5. a) $\frac{n}{2}, \frac{1}{2}n$ b) $\left(\frac{n}{2}\right)^2, \frac{n}{2}\times\frac{n}{2}$ c) $\frac{n}{2}-\frac{n}{4}$ 6. $\frac{3y+2x}{xy}$ 7. $\frac{3}{t}$ 8. $\frac{4q-5p}{pq}$ 9. r^2 10. $\frac{p+q}{x}$
11. $\frac{11}{2x}$ 12. $\frac{n}{m}$ 13. $\frac{3q}{p}$ 14. $\frac{y}{a^3}$ 15. $\frac{16x}{15}$ 16. $\frac{5a-2b}{10}$ 17. $\frac{9t+4z}{12}$
18. a) $\frac{101}{2}$ b) $\frac{n+1}{2}$

Exercise 8.24 page 252

1. 12 2. 10 3. 20 4. −14 5. −25 6. $1\frac{1}{2}$ 7. $2\frac{2}{3}$ 8. $\frac{4}{5}$ 9. $\frac{6}{7}$ 10. $\frac{4}{9}$
11. 2 12. 12 13. $2\frac{2}{3}$ 14. 15 15. $-\frac{8}{11}$ 16. −50

Exercise 8.25 page 252

1. 12 2. 18 3. −5 4. 12 5. 22 6. 51 7. 3 8. $1\frac{1}{4}$ 9. $\frac{1}{4}$ 10. 3
11. $\frac{5}{11}$ 12. −2 13. $\frac{1}{2}$ 14. $-1\frac{3}{7}$ 15. −2 16. $\frac{1}{2}$ 17. $3\frac{8}{21}$

Exercise 8.26 page 253

1. $3n + 2$ 2. a) $4n + 1$ b) $6n - 4$ 3. $5n + 1$ 4. $4n + 2$
5. a) $2n + 3$ b) $4n - 1$ c) $6n - 4$ 6. a) $8n - 6$ b) $3n + 4$ c) $9n + 12$
7. $2n + 1$ 8. $4n + 1$ 9. a) $h = 3n + 4$ b) $p = 5n - 3$ c) $s = \frac{n}{2} + 3$

Exercise 8.27 page 255

1. $w = 4n + 6$ 2. $w = 6n + 4$ 3. $d = 3h + 2$ 4. a) $z = 6n - 5$ b) $z = 15 - 3n$ 5. $p = 12 - 2c$
6. a) $s = 2h - 1$ b) 171 7. a) 21 b) $4n + 1$ 8. a) $\frac{3}{5}, \frac{3}{10}, \frac{3}{17}$
b)

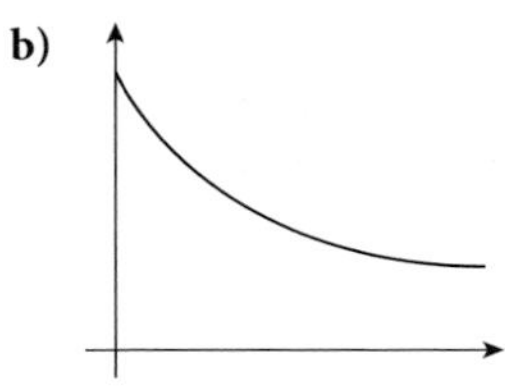

9. a) 22 b) 3001 c) $3n + 1$

Exercise 8.28 page 258

1. a) $3n + 1$ b) $5n + 4$ c) $11n - 3$ 2. $n^2 + 2n - 1$ 3. $n^2 + 3n + 1$ 4. $n^2 - 2n + 3$
5. $2n^2 + n + 1$ 6. $3n^2 + 4$ 7. $n^2 + 5n - 4$ 8. $\frac{n^2+n}{2}$ 9. $n^2 - 2n + 5$ 10. c) $n^2 + 3n$

Exercise 8.29 page 259

1. $n^3 - n^2 + 2n + 1$ 2. $n^2 + 5n - 3$ 3. $n^3 - 5n - 1$ 4. $n^2 - 2n + 9$ 5. $7n - 17$
6. $3n^3 + 3n - 2$ 7. $3n^2 - n + 1$ 8. $\frac{n^3+3n}{2}$ 9. $n^3 - 4n^2 + 3n - 5$ 10. $a = -4$
11. $a = -1$ 12. $a = 19$, $f(x) = (x + 4)(x + 3)(x + 1)$

Exercise 8.30 page 259

1. $n^3 + 3n + 1$ 2. $n^2 + 5n - 3$ 3. $n^3 - 5n + 2$
4. $2n^3 + n^2 - 2$ 5. $n^3 + n + 2$ 6. $3n^3 - n^2 + 5$
7. a) 4, 12, 24, 40 b) $2n^2 + 2n$ c) 144
8. a) $a = -4$ b) 2, 0, 2, 14 9. b) 2870 10. a) $2n$ b) $2n + 2$

Exercise 8.31 page 264

1 **a)** $3 \times 2^{n-1}$ **b)** $2 \times 5^{n-1}$ **c)** $20 \times \left(\frac{1}{2}\right)^{n-1}$ **d)** $2000 \times \left(\frac{1}{5}\right)^{n-1}$
e) $\frac{1}{8} \times 2^{n-1}$ ($= 2^{n-4}$) **f)** $4 \times 3^{n-1}$ **g)** $5 \times (-3)^{n-1}$
2. **a)** 1023 **b)** 2186 **c)** 195 312 **d)** 615
3. **a)** 79·9 **b)** 80·0 **c)** 607·5 **d)** 1706·7 **e)** 66·7
4. $a = 7, r = 3$ **5.** $a = \frac{35}{32}, r = 2$ **6.** **a)** $7r$ **b)** 5 **7.** **b)** $-4, 3$
8. **a)** 1·05 **b)** £1276·28 **9.** **a)** $r = -5, 4$ **b)** 9, −45, 225; 9, 36,144
10. **c)** $r = 3, a = 7$ **11.** **b)** 15 **12.** **a)** −177147 **c)** 11 **13.** 5
14. **a)** $\frac{5}{2}(3^n - 1)$ **b)** 1328600 **c)** 16 **15.** 7 **16.** 4, −12

Exercise 8.32 page 268

1. **a)** $S = ke$ **b)** $v = kt$ **c)** $x = kz^2$ **d)** $y = k\sqrt{x}$
e) $T = k\sqrt{L}$ **f)** $C = kr$ **g)** $A = kr^2$ **h)** $V = kr^3$
2. **a)** 9 **b)** $2\frac{2}{3}$ **3.** **a)** 35 **b)** 11 **4.** **a)** 75 **b)** 4

5.

x	1	3	4	$5\frac{1}{2}$
z	4	12	16	22

6.

r	1	2	4	$1\frac{1}{2}$
V	4	32	256	$13\frac{1}{2}$

7.

h	4	9	25	$2\frac{1}{4}$
w	6	9	15	$4\frac{1}{2}$

8. **a)** 18 **b)** 2 **9.** **a)** 42 **b)** 4 **10.** $333\frac{1}{3}$ N/cm^2 **11.** 180 m; 2 s
12. 675 J; $\sqrt{\frac{4}{3}}$ cm **13.** 4 cm; 49 h **14.** $15\frac{5}{8}$ h **15.** 9000 N; 25 m/s **16.** $15^4 : 1$ (50 625 : 1)

Exercise 8.33 page 270

1. **a)** $x = \frac{k}{y}$ **b)** $s = \frac{k}{t^2}$ **c)** $t = \frac{k}{\sqrt{q}}$ **d)** $m = \frac{k}{w}$ **e)** $z = \frac{k}{t^2}$
2. **a)** 1 **b)** 4 **3.** **a)** $2\frac{1}{2}$ **b)** $\frac{1}{2}$
4. **a)** 36 **b)** ±4 **5.** **a)** 1·2 **b)** ±2
6. **a)** 16 **b)** ±10 **7.** **a)** 6 **b)** 16 **8.** **a)** $\frac{1}{2}$ **b)** $\frac{1}{20}$

9.

y	2	4	1	$\frac{1}{4}$
z	8	4	16	64

10.

t	2	5	20	10
y	25	4	$\frac{1}{4}$	1

11.

x	1	4	256	36
r	12	6	$\frac{3}{4}$	2

12. **a)** 6 **b)** 50 **13.** **a)** 0·36 **b)** 6
14. $k = 100, n = 3$

x	1	2	4	10
z	100	$12\frac{1}{2}$	1·5625	$\frac{1}{10}$

15. $k = 12, n = 2$

v	1	4	36	10 000
y	12	6	2	$\frac{3}{25}$

16. 2·5 m^3; 200 N/m^2 **17.** 3 h; 48 men **18.** 2 days; 200 days **19.** 6 cm

Revision exercise 8A page 273

1. **a)** $\frac{9x}{20}$ **b)** $\frac{7}{6x}$ **c)** $\frac{5x-2}{6}$ **d)** $\frac{5x+23}{(x-1)(x+3)}$ **2.** **a)** $(x - 2)(x + 2)$ **b)** $\frac{3}{x+2}$
3. **a)** $s = t(r + 3)$ **b)** $r = \frac{s-3t}{t}$ **c)** $t = \frac{s}{r+3}$ **4.** **a)** $z = x - 5y$ **b)** $m = \frac{11}{k+3}$ **c)** $z = \frac{T^2}{C^2}$
5. **a)** 50 **b)** 50 **6.** **a)** 16 **b)** ±4
7. **a)** **i)** 3 **ii)** 4 **iii)** $\frac{1}{4}$ **b)** **i)** 4 **ii)** 0
8. **a)** 9, 10 **b)** 2, 3, 4, 5 **9.** $\frac{t^2}{k^2} - 5$ **10.** $\frac{z+2}{z-3}$ **11.** **a)** $\frac{3}{5}$ **b)** $\frac{k(1-y)}{y}$

12. 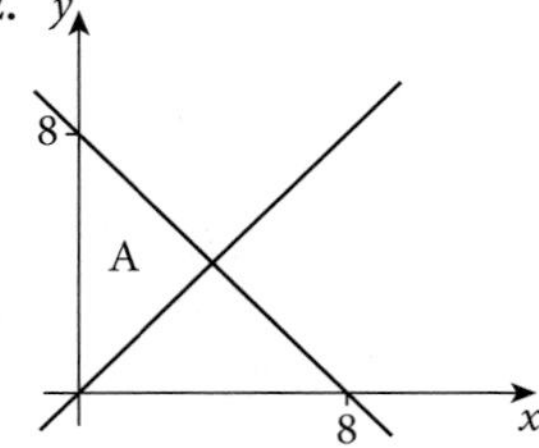

13. a) $1\frac{5}{6}$ b) 0·09 14. 21

15. a) $\frac{5+a^2}{2-a}$ b) $-\left(\frac{cz+b}{a}\right)$ c) $\frac{a^2+1}{a^2-1}$

16. a) $\frac{7}{2x}$ b) $\frac{3a+7}{a^2-4}$ c) $\frac{x-8}{x(x+1)(x-2)}$

17. $p=\frac{10t^2}{s}$ 18. 6 or 7 19. $y \geq 2,\ x+y \leq +6,\ y \leq 3x$

20. $x \geq 0,\ y \geq x-2,\ x+y \leq 7,\ y \geq 0$

21. a) $3n+7$ b) n^2+n+4 c) n^3+3n-1

22. a) 512 b) 6 h c) 2^{21} 23. a) $x > 5$ or $x < -5$ b) $-4 < x < -2$ 24. n^3+n+5

25. a) $5 \times 2^{n-1}$ b) $2 \times 3^{n-1}$ c) $100 \times \left(\frac{1}{2}\right)^{n-1}$ 26. 765 27. $a=2\frac{1}{2},\ r=2$

28. 4 29. 4, −5 30. a) $x+1$ b) $x(x+1)^{99}$

Examination exercise 8B *page 274*

1. a) $\frac{1}{49}$ b) 4 2. $x=\pm\sqrt{c^2+d^2}$ 3. a) i) 1 ii) 6 b) 7

4. a) $F=\frac{36}{d^2}$ b) 4 5. a) 62, 79 b) n^2+2n-1

6. a) −0·32, 1·19 b) $-0{\cdot}32 < x < 1{\cdot}19$ 7. a) 24, 35 b) n^2-1

9 Functions 2

Exercise 9.1 *page 278*

11. (0, 0), (1, 4), (1·6, 1·6) 12. (0, 1), (2·25, 1), (4·5, 10) 13. (−1·5, 1·5), (0·67, 8), (3·5, 8), (3·5, −3·5)
14. (4, −2), (0·33, 5·33), (−2·29, −5·14) 15. (−2, 3), (0·6, 8·2), (2·5, 2·5), (1·33, 1·33)

Exercise 9.2 *page 279*

1. $1\frac{1}{2}$ 2. 2 3. 3 4. $1\frac{1}{2}$ 5. $\frac{1}{2}$ 6. $-\frac{1}{6}$ 7. −7

8. −1 9. 4 10. −4 11. 5 12. $-1\frac{3}{7}$ 13. 6 14. 0

15. 0 16. infinite 17. infinite 18. −8 19. $5\frac{1}{3}$ 20. 0 21. $\frac{b-d}{a-c}$ or $\frac{d-b}{c-a}$

22. $\frac{n+b}{m-a}$ 23. $\frac{2f}{a}$ 24. −4 25. 0 26. $-\frac{6d}{c}$ 27. a) $-1\frac{1}{5}$ b) $\frac{1}{10}$ c) $\frac{4}{5}$

28. a) infinite b) $-\frac{3}{10}$ c) $\frac{3}{10}$ 29. $3\frac{1}{2}$ 30. a) $\frac{n+4}{2m-3}$ b) $n=-4$ c) $m=1\frac{1}{2}$

31. b) 7·2 c) (4, 4) 32. b) yes, PQ = PR c) (3, 2·5)

Exercise 9.3 *page 281*

1. 1, 3 2. 1, −2 3. 2, 1 4. 2, −5 5. 3, 4 6. $\frac{1}{2}$, 6 7. 3, −2

8. 2, 0 9. $\frac{1}{4}$, −4 10. −1, 3 11. −2, 6 12. −1, 2 13. −2, 3 14. −3, −4

15. $\frac{1}{2}$, 3 16. $-\frac{1}{3}$, 3 17. 4, −5 18. $1\frac{1}{2}$, −4 19. 10, 0 20. 0, 4

Exercise 9.4 *page 282*

1. $y=3x+7$ 2. $y=2x-9$ 3. $y=-x+5$ 4. $y=2x-1$ 5. $y=3x+5$

6. $y=-x+7$ 7. $y=\frac{1}{2}x-3$ 8. $y=2x-3$ 9. $y=3x-11$ 10. $y=-x+5$

11. $y=\frac{1}{3}x-4$

Exercise 9.5 *page 282*

1. A: $y=3x-4$ B: $y=x+2$ 2. C: $y=\frac{2}{3}x-2$ D: $y=-2x+4$ 3. a) $y=2x+5$ b): $y=-x+3$

4. a) $y=3x+1$ b) $y=x-2$ 5. a) (0, 4) b) $2y=x+8$ 6. $n=2$

Exercise 9.6 page 284

1. 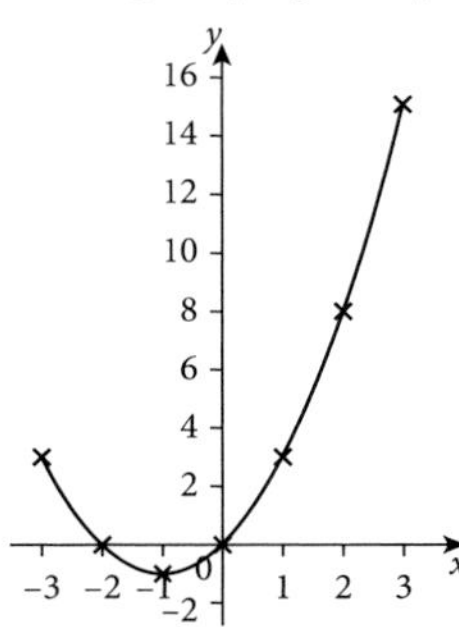

2. 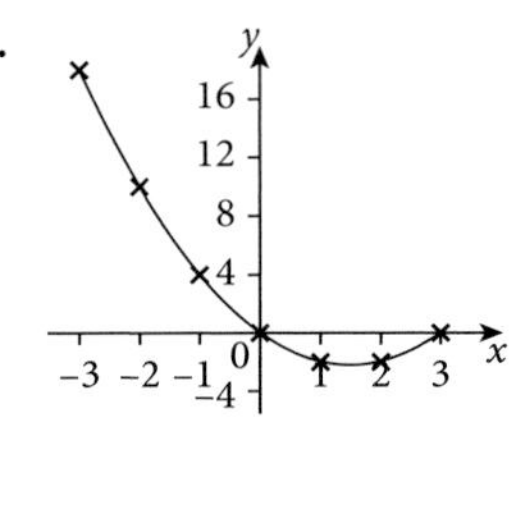

3. 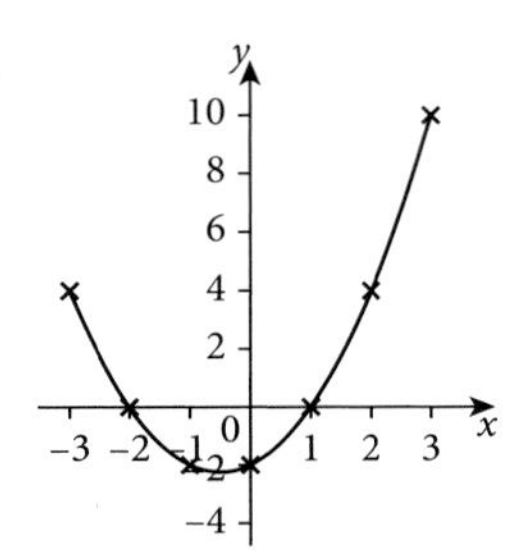

4. 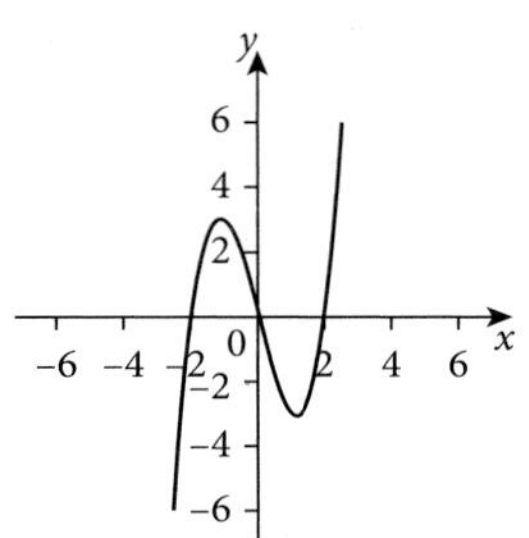

5. 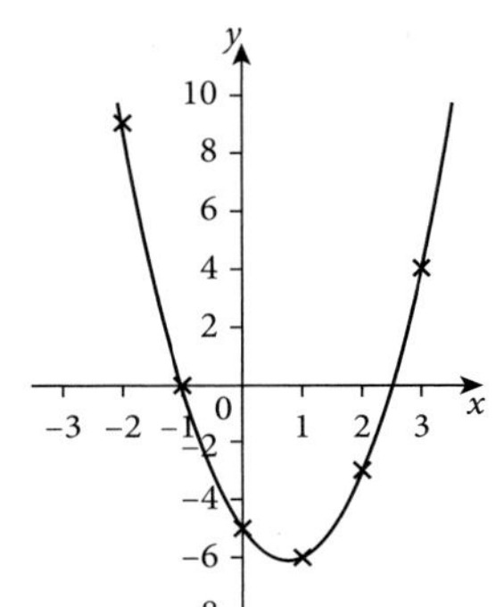

6. 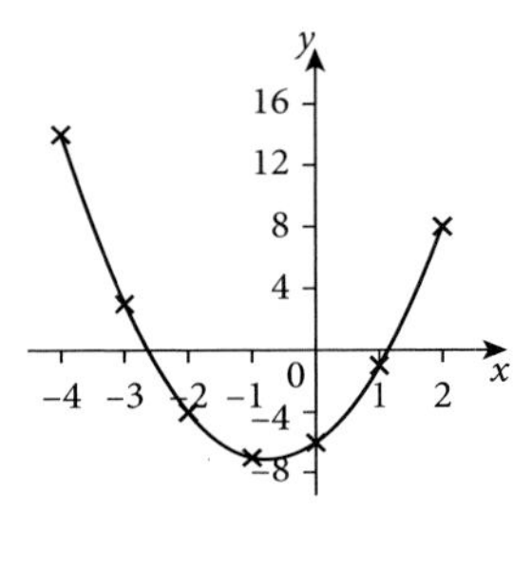

7. 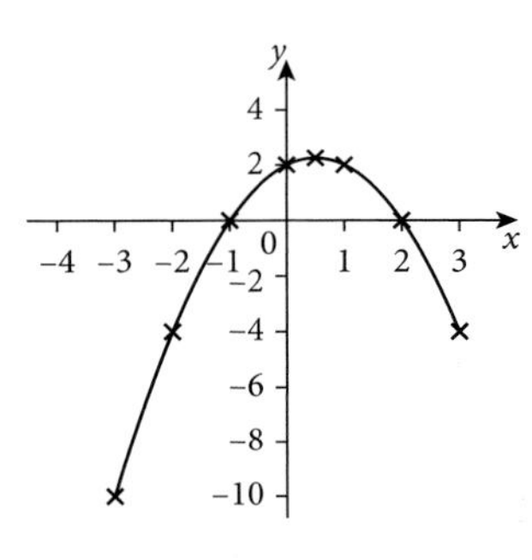

8. 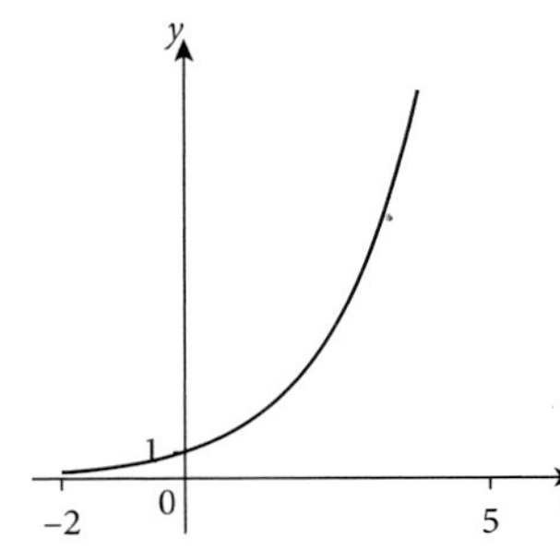

9. 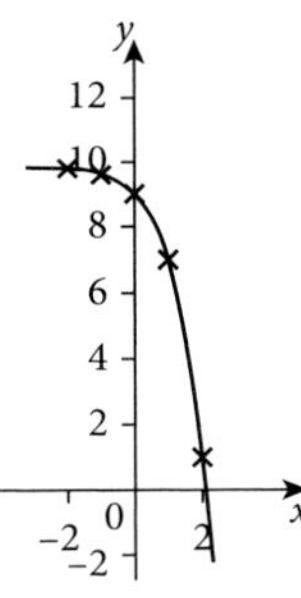

10. 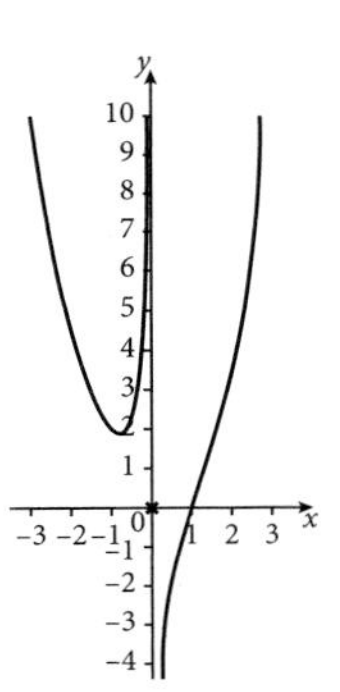

Exercise 9.7 page 285

1. a) 4 **b)** 8 **c)** 10·6 **2. a)** 3 **b)** −5 **c)** 1·5
3. a) 7·25 **b)** −2 **c)** −0·8, 3·8 **6.** 0·25 **7.** $y = 1, x = -1$
8. $y = x, x = 0$ **9. a)** 0·75 **b)** 1·23 **10. a)** −2·47 **b)** 1·4

Exercise 9.8 page 289

1. a) −0·4, 2·4 **b)** −0·8, 3·8 **c)** −1, 3 **d)** −0·4, 2·4 **2.** −0·3, 3·3 **3.** 0·6, 3·4 **4.** 0·3, 3·7
5. a) $y = 3$ **b)** $y = -2$ **c)** $y = x + 4$ **d)** $y = x$ **e)** $y = 6$
6. a) $y = 6$ **b)** $y = 0$ **c)** $y = 4$ **d)** $y = 2x$ **e)** $y = 2x + 4$
7. a) $y = -4$ **b)** $y = 2x$ **c)** $y = x - 2$ **d)** $y = -3$ **e)** $y = 2$
8. a) $y = 5$ **b)** $y = 2x$ **c)** $y = 0{\cdot}2$ **d)** $y = 3 - x$ **e)** $y = 3$
9. a) $y = 0$ **b)** $y = -2\frac{1}{2}$ **c)** $y = -8x$ **d)** $y = -3$ **e)** $y = -5\frac{1}{2}x$
10. a) −1·65, 3·65 **b)** −1·3, 2·3 **c)** −1·45, 3·45 **11. a)** 1·7, 5·3 **b)** 0·2, 4·8
12. a) −3·3, 0·3 **b)** −4·6, − 0·4 **13. a)** −2·35, 0·85 **b)** −2·8, 1·8
14. a) i) −0·4, 2·4 **ii)** −0·5, 2 **b)** $-1{\cdot}3 \le x \le 2{\cdot}8$ **15. a)** 3·4 **b)** 2·4, 7·6 **c)** 4·2
16. a) ±3·7 **c)** ±2·8 **17. a)** 1·72 **b)** 0, ± 1·4
18. a) $1{\cdot}6 \le x \le 7{\cdot}4$ **b)** 6·9 **19. a)** 2·6 **b)** 0.46, 3.31 **c)** 0.64
20. a) −1·6, 0·6 **b)** $-\frac{1}{2}$, 1

Exercise 9.9 page 292

1. **a)** quadratic, negative x^2 **b)** cubic, positive x^3 **c)** reciprocal **d)** cubic, negative x^3

e) quadratic, positive x^2 **f)** exponential

2. **a)**

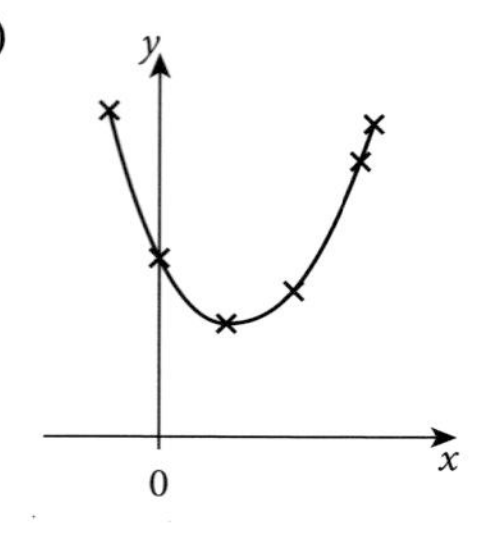

b)

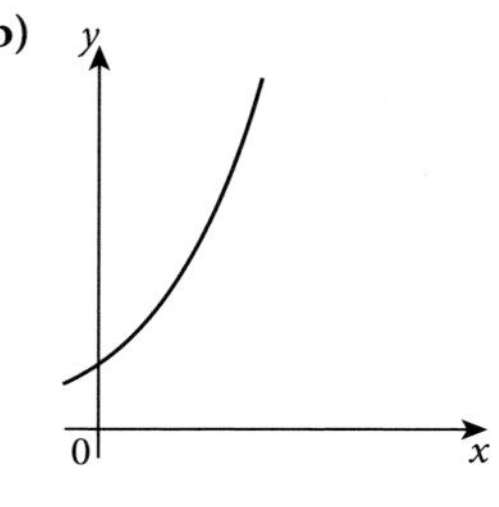

c)

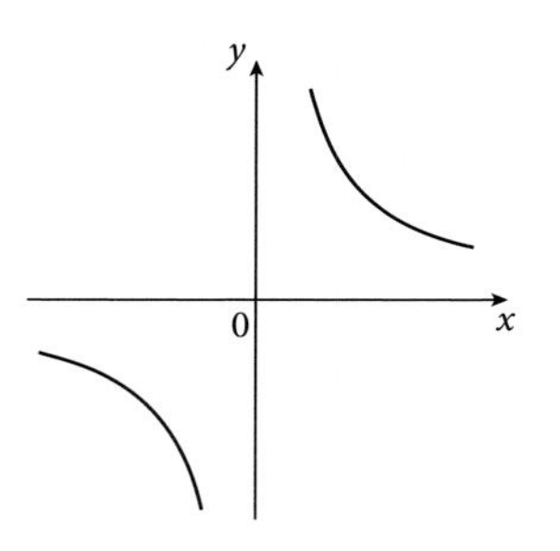

d)

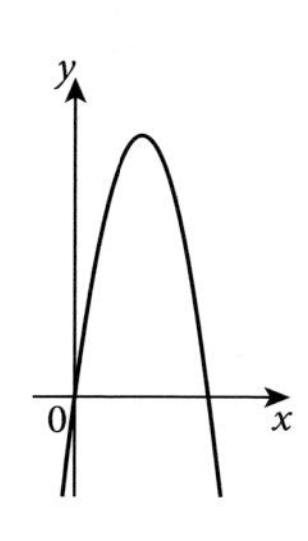

e)

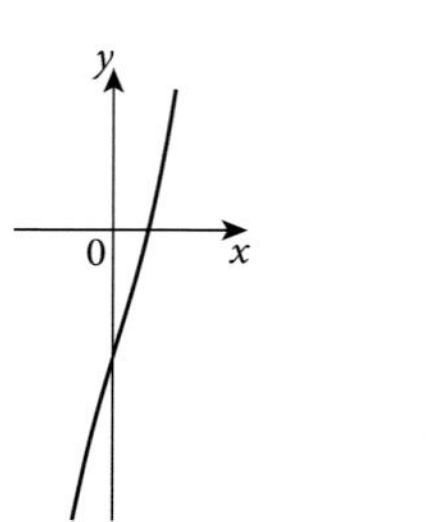

f)

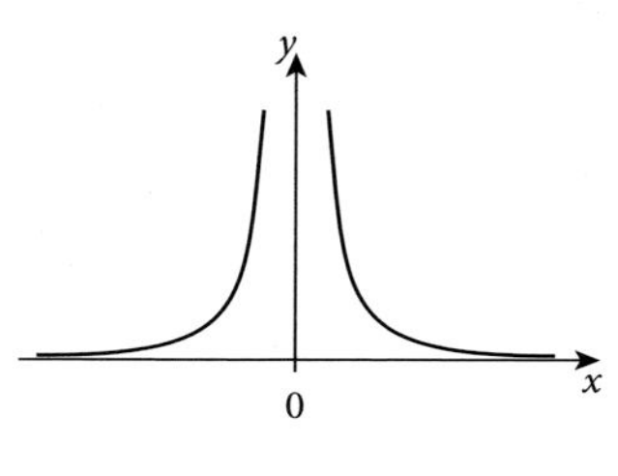

3. **i)** c **ii)** b **iii)** a

iv) e **v)** f **vi)** d

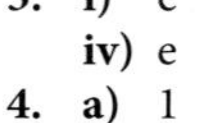

4. **a)** 1 **b)** 2

c) 1 **d)** 3

5.

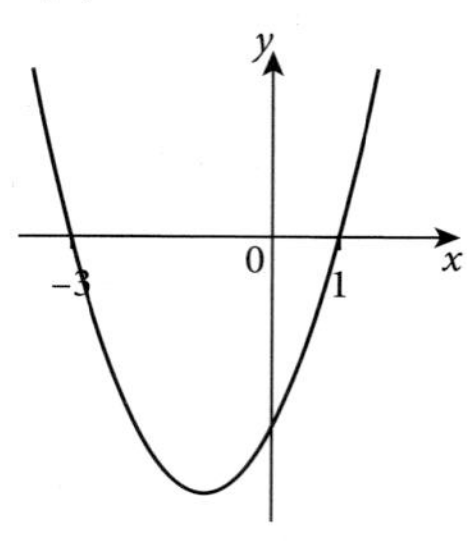

6. **a)**

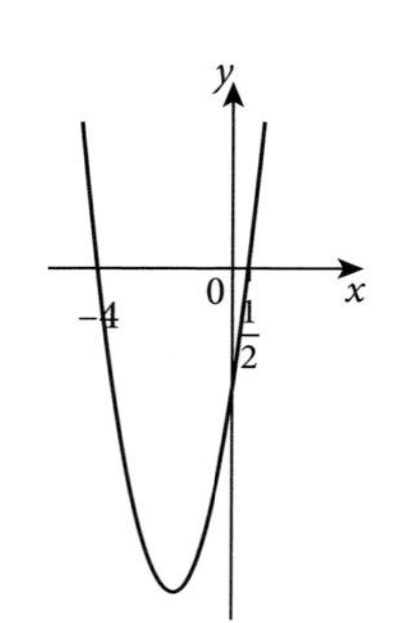

b)

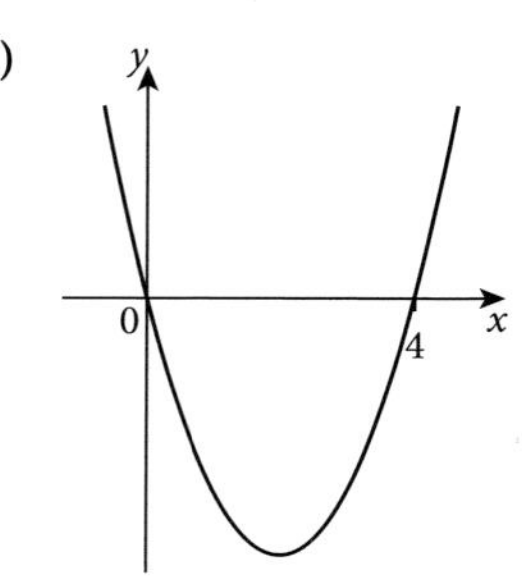

c)

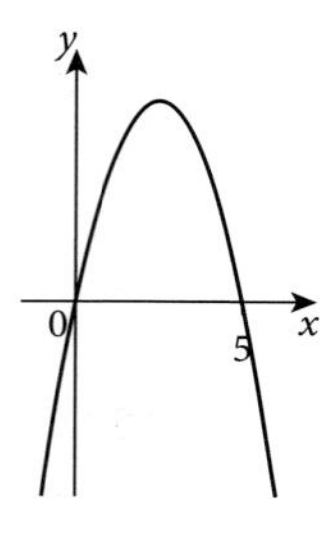

7. **a)**

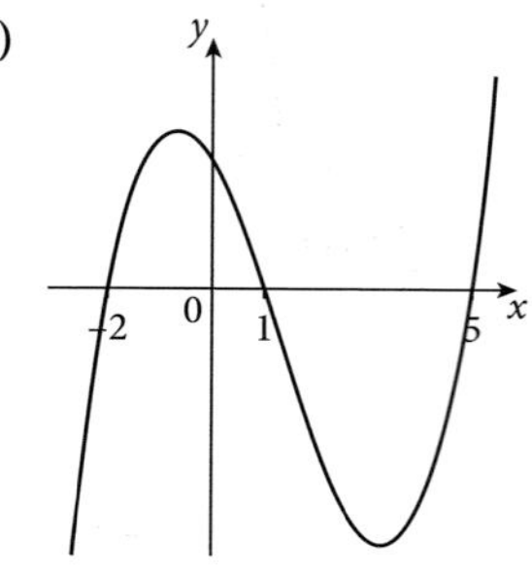

8. **a)**

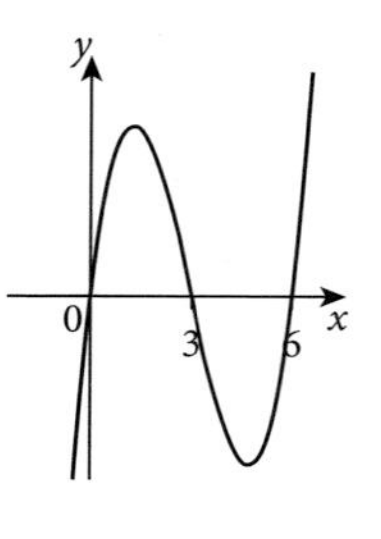

b)

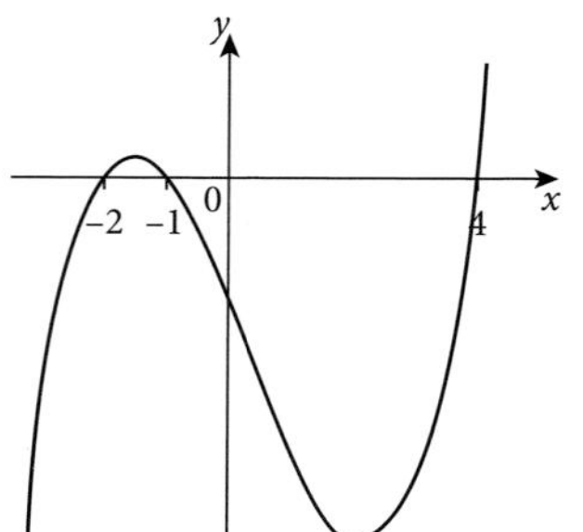

c)

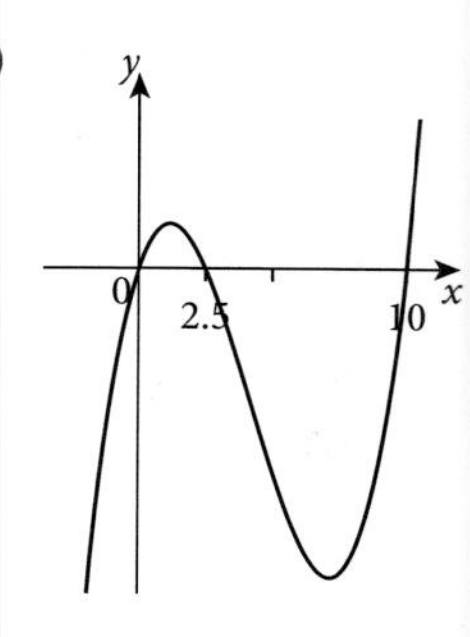

Exercise 9.10 page 296

1.

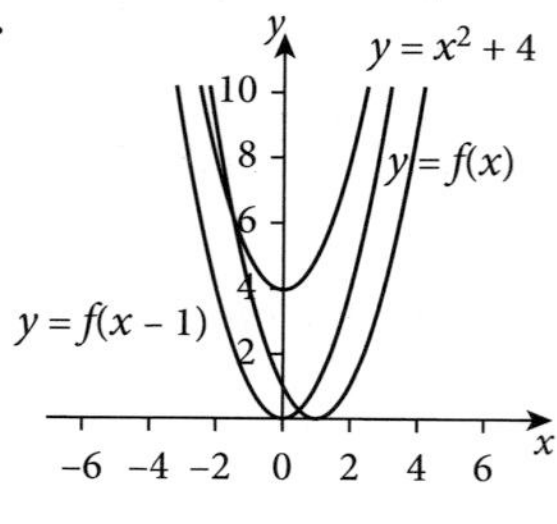

2.

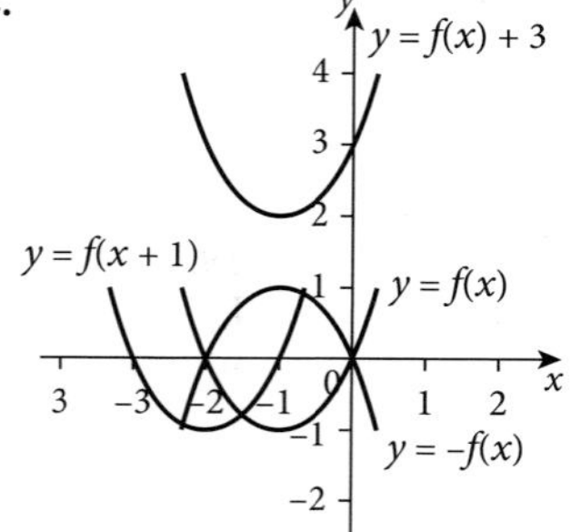

3. **a)** Translation 2 units ↓, A'(0, – 2)

b) Translation 7 units →, A'(7, 0)

4. **a)** A'(−2, −1), B'(0, −3), C'(2, 0), reflection in $y = 0$

b) A'(0, 1), B'(2, 3), C'(4, 0), translation 2 units →

5.

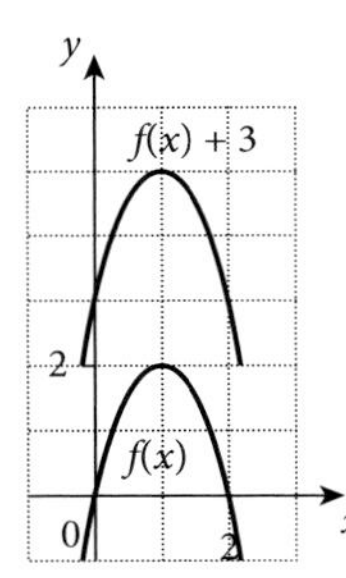

6.

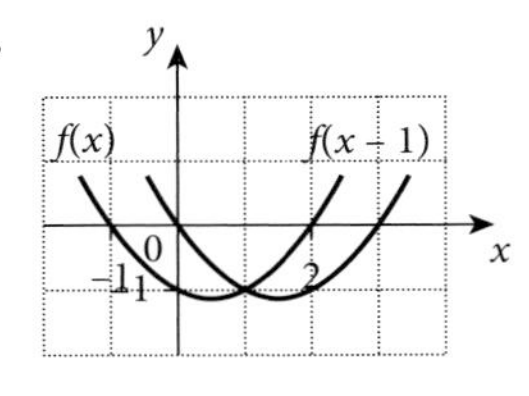

7.

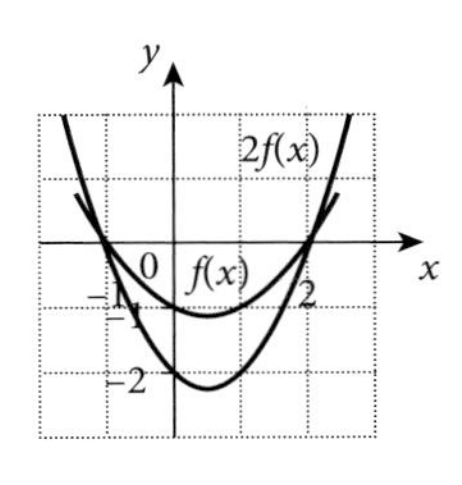

8.

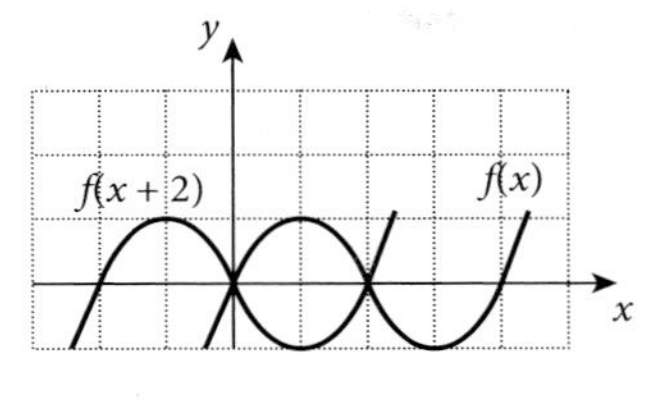

9.

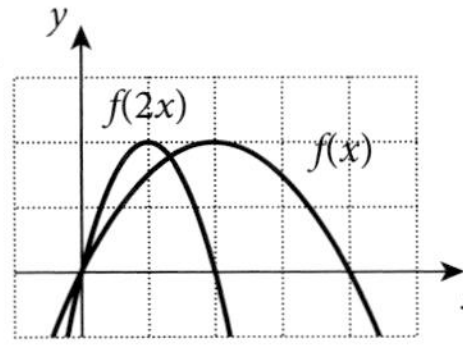

10.

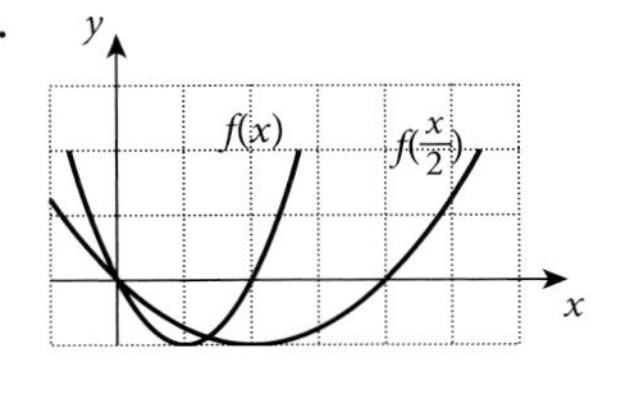

11.

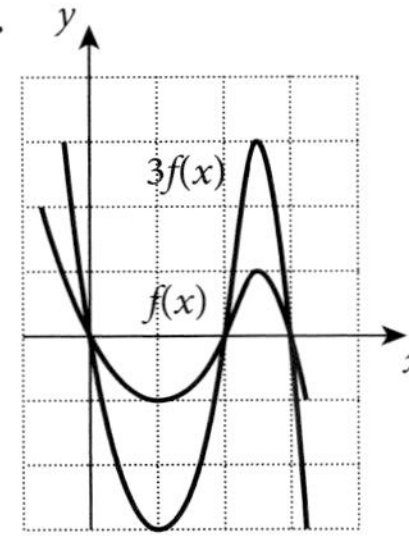

12.

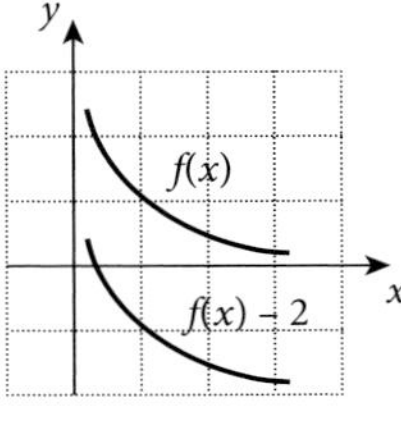

13.

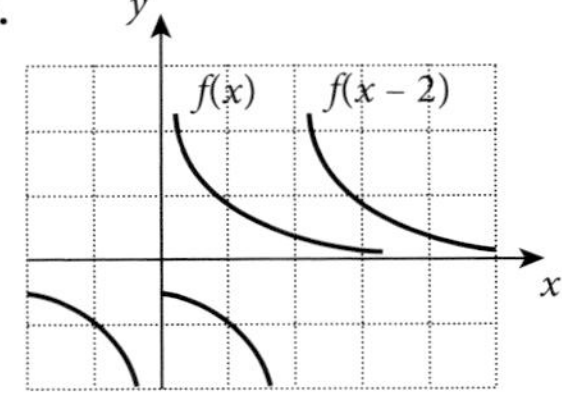

14.

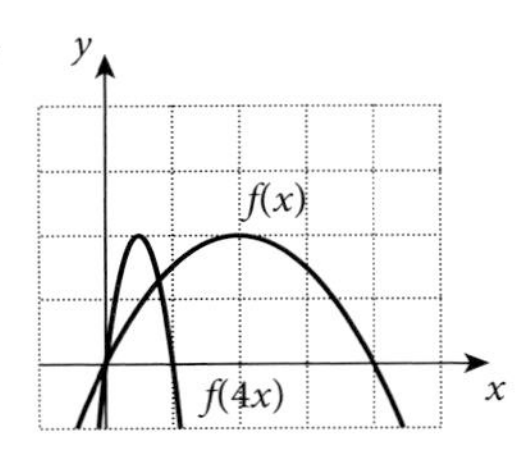

15.

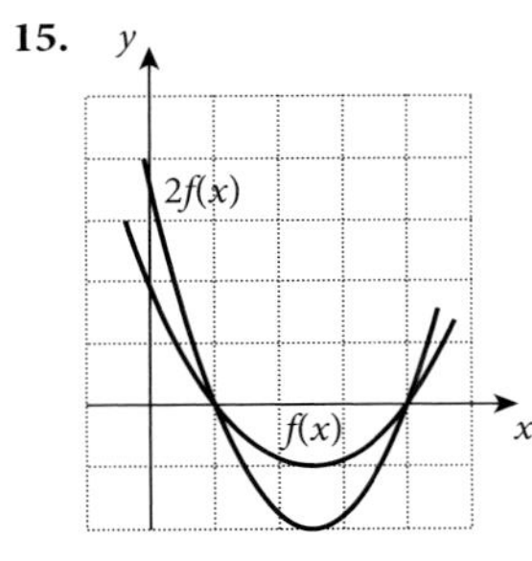

16.

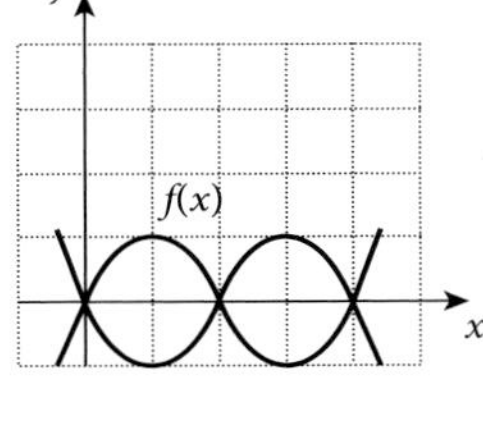

Exercise 9.11 page 297

1.

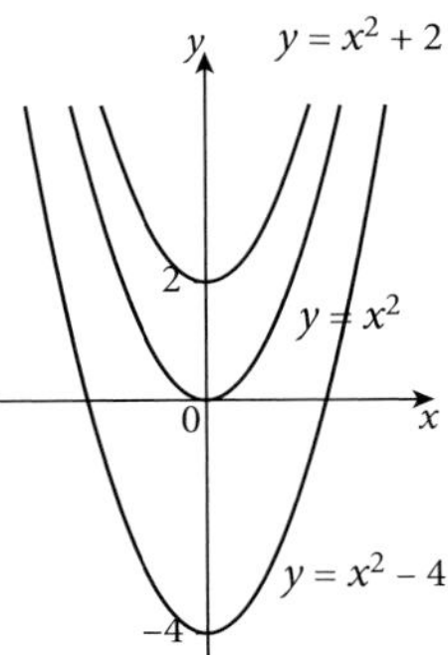

2.

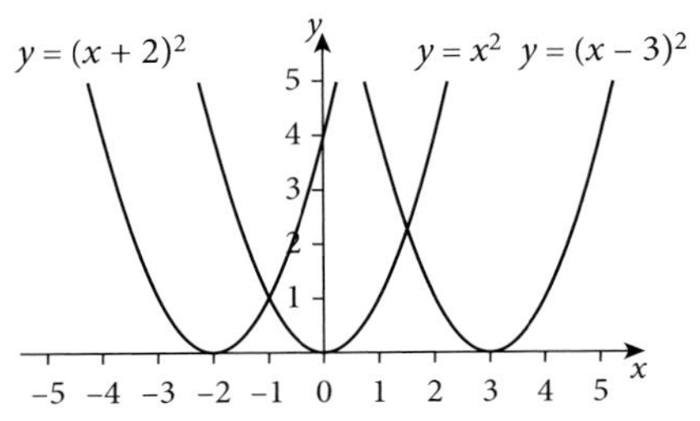

3.

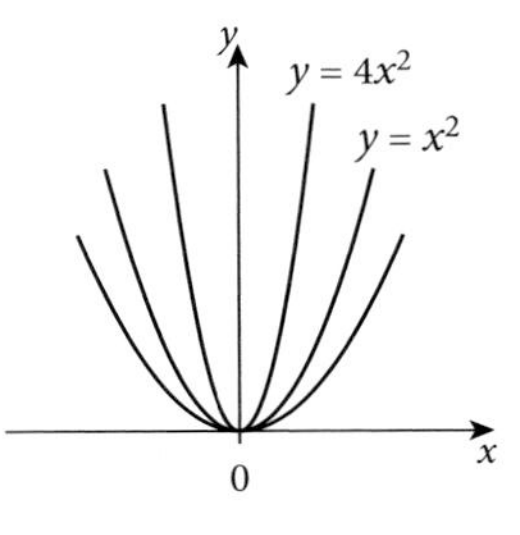

4.

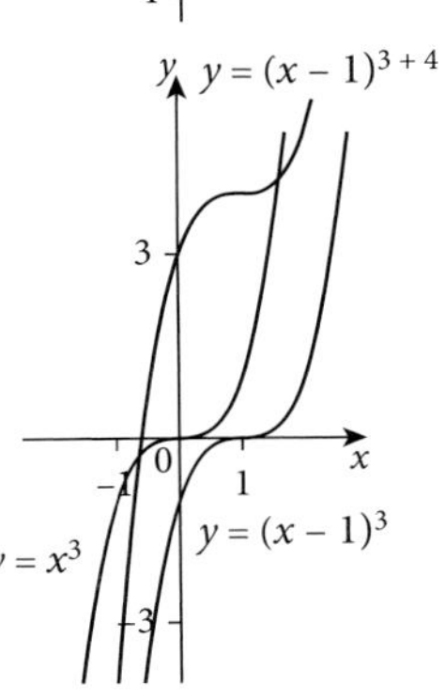

5.

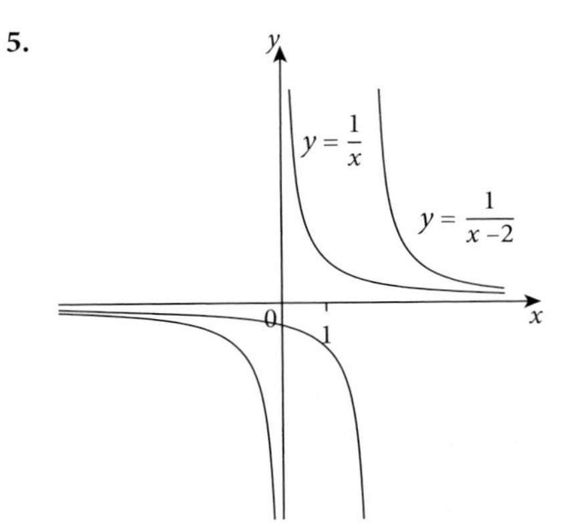

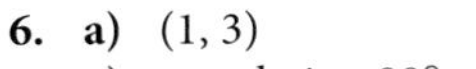

6. **a)** (1, 3) **b)** (−1, 1)

7. **a)** translation 90° ←

b) stretch parallel to y axis × 2

8. **a)** translation 1 unit ↑

b) translation 90 units →

c) stretch parallel to y axis × 2

Exercise 9.12 *page 301*

1. **a)** $\log_5 125 = 3$ **b)** $\log_2 32 = 5$ **c)** $\log_{10} 10000 = 4$ **d)** $\log_2 8 = 3$ **e)** $\log_7 49 = 2$ **f)** $\log_x 10 = n$

2. **a)** 2 **b)** 3 **c)** 2 **d)** 5 **e)** 0 **f)** −1 **g)** −2 **h)** $\frac{1}{2}$ **i)** $-\frac{1}{2}$ **j)** 3 **k)** $\frac{3}{2}$ **l)** 2

3. **a)** $x = \log_{10} 7$ **b)** $x = \log_5 3$ **c)** $x = \log_8 2$

4. **a)** log 21 **b)** log 20 **c)** log 64 **d)** log 72 **e)** log 121 **f)** log 2 **g)** log 8 **h)** log 162

5. **a)** $3 \log a + \log b$ **b)** $\log b + 2 \log c$ **c)** $\frac{1}{2}\log a$ **d)** $5 \log c$ **e)** $\log a + 2 \log b - 3 \log c$ **f)** $4\log a + 3\log b - \frac{1}{2}\log c$

6. **a)** 5 **b)** 4 **c)** −2 **d)** 0 **e)** 2 **f)** 3

7. **a)** 1·23 **b)** 2·40 **c)** 1·58 **8.** **a)** 4·09 **b)** 5·22 **c)** 2·05 **d)** 2·84 **e)** 0·850 **f)** 0·880 **g)** 1·25 **h)** 4·51 **i)** 1·11 **j)** −1·10 **9.**

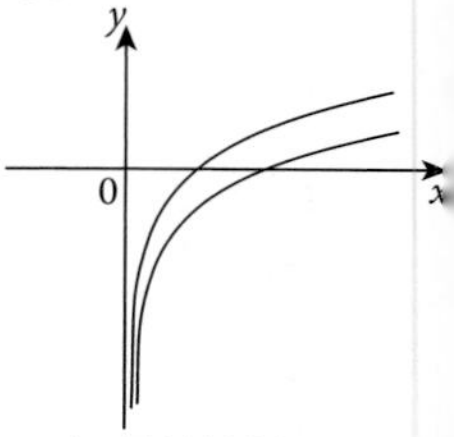

Exercise 9.13 *page 303*

1. **a)** 5 **b)** 2 **c)** $\frac{1}{2}$ **d)** 0 **e)** $\frac{1}{2}$ **f)** $\frac{1}{4}$ **g)** −3 **h)** $-\frac{1}{2}$ **i)** $1\frac{1}{2}$

2. **a)** $\log\left(\frac{x+4}{x}\right)$ **b)** $\log\left(\frac{x+1}{x+2}\right)$ **c)** $\log_3 \frac{(x+3)(x+5)}{(x+1)}$ **d)** $\log_7 (3x+2)(x-3)$ **e)** $\log_2 (x-1)(x+2)$

3. **a)** $\log_3\left(\frac{x+4}{x}\right)$ **b)** $\frac{1}{2}$ **4.** **a)** $\frac{1}{12}$ **b)** $\frac{1}{6}$ **c)** $\frac{10}{11}$

5. **a)** 36 **b)** 16 **c)** 6 **6.** **a)** 1 **b)** 2·01 **c)** 200 000 **d)** 33 **e)** 5764804 **f)** $1.495\left(=5^{\frac{1}{4}}\right)$ **g)** $2.47\left(15^{\frac{1}{3}}\right)$ **h)** 100 **i)** $2.24\left(5^{\frac{1}{2}}\right)$ **7.** 1·89 **8.** $k = \frac{\log 3}{\log 7}$

9. $k = \frac{\log 5}{\log 8}$ **10.** **a)** $3 \log x + \log y = \log 3$ **b)** $y = \frac{3}{x^3}$ **11.** **b)** $y = 5x^3$ **12.** 1·53

Revision exercise 9A *page 304*

1. **a)** $y = x - 7$ **b)** $y = 2x + 5$ **c)** $y = -2x + 10$ **d)** $y = \frac{x+1}{2}$

2. **a)** 2 **b)** 1 **c)** $-3\frac{1}{2}$ **d)** 0 **e)** 10

3. **a)** 2, −7 **b)** −4, 5 **c)** $\frac{1}{2}$, 4 **d)** $-\frac{1}{2}$, 5 **e)** −2, 12 **f)** $-\frac{2}{3}$, 8

4. A: $y = 6$; B: $y = \frac{1}{2} - 3$; C: $y = 10 - x$; D: $y = 3x$ **5.** A: $4y = 3x - 16$; B: $2y = x - 8$; C: $2y + x = 8$; D: $4y + 3x = 16$

6. **a)** $y = 2x - 3$ **b)** $y = 3x + 4$ **c)** $y = 10 - x$ **d)** $y = 7$

7. **a)** A (0, −8), (4, 0) **b)** 2 **c)** $y = 2x - 8$ **8.** 25 sq· units **9.** −3 **10.** 219

11. **a)** −0.73, 2.73 **b)** 0·75, 2·75 **c)** −2, 2 **d)** 0·7

12. **a)** −1·65, 3·65 **b)** −1·3, 2·3 **c)** −1·45, 3·45

13. **a)** 1 **b)** 2 **c)** 2 **d)** 1 **e)** 2 **f)** 3

14. **a)** 3·35 **b)** 2·4, 7·6 **c)** 4·24 **15.** $0 < k < 4$ **16.** (1, 2)

17. **a)** 4·16, 0·84 **b)** $0.65 < x < 3.85$ **c)** 3·3

18. **a)** 9·2 **b)** about 1·4 **c)** 1·7 **19.** **a)** log 128 **b)** log 9

20. **a)** 1·5 **b)** 1·431 **21.** **a)** $\frac{x+3}{x}$ **b)** $\frac{1}{5}$

Examination exercise 9B *page 306*

1. $a = 7$ **2.** **a)** $y = 2x + 2$ **b)** $2x + 4y = 23$ **3.** $a = 4, b = 2$ **4.** **a)** $y = x^2 + 3$ **b)** $y = (x - 3)^2$

5. **a)** real numbers **b)** 3, 90 **c)** **i)** stretch, factor 2, x- axis invariant **ii)** translation $\begin{pmatrix} -60 \\ 0 \end{pmatrix}$

6. **a)** 72 **b)** 2

7. **a)** **i)**

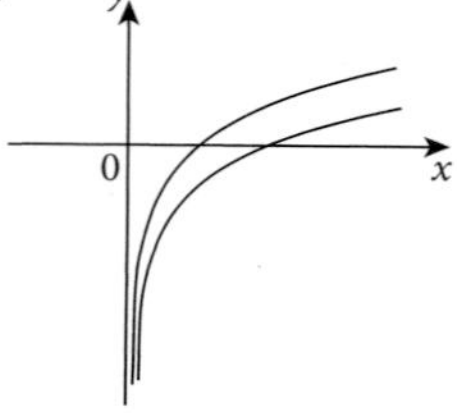

ii) (1, 0) **iii)** $x = 0$ **iv)** $f(x) \le 1$ **b)** 2·40 **c)** $0 \le g(x) \le 1$

d) translation of $y = f(x)$ by $\begin{pmatrix} -1 \\ 0 \end{pmatrix}$

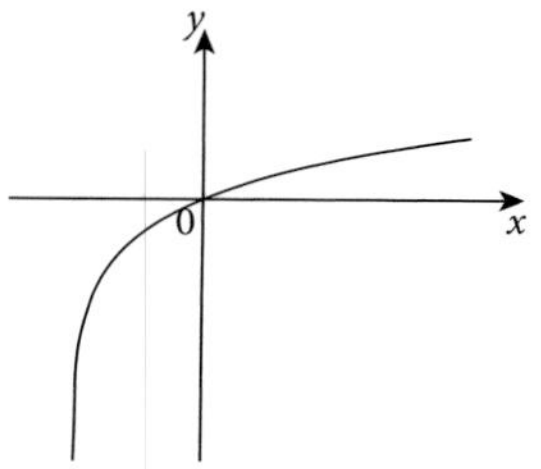

10 Trigonometry

Exercise 10.2 *page 312*

1. 4·54 **2.** 3·50 **3.** 3·71 **4.** 6·62 **5.** 8·01 **6.** 31·9 **7.** 45·4
8. 4·34 **9.** 17·1 **10.** 13·2 **11.** 38·1 **12.** 3·15 **13.** 516 **14.** 79·1
15. 5·84 **16.** 2·56 **17.** 18·3 **18.** 8·65 **19.** 11·9 **20.** 10·6 **21.** 119
22. 10·1 **23.** 3·36 cm **24.** 4·05 cm **25.** 4·10 m **26.** 11·7 m **27.** 9·48 cm **28.** 5·74 m
29. 9·53 cm **30.** 100 m **31.** 56·7 m **32.** 16·3 cm **33.** 0·952 cm **34.** 8·27 m

Exercise 10.3 *page 314*

1. 5, 5·55 **2.** 13·1, 27·8 **3.** 34·6, 41·3 **4.** 20·4, 11·7 **5.** 94·1, 94·1 **6.** 15·2, 10, 6·43 **7.** 4·26
8. 3·50 **9.** 26·2 **10.** 8·82 **11. a)** 17·4 cm **b)** 11·5 cm **c)** 26·5 cm
12. a) 6·82 cm **b)** 6·01 cm **c)** 7·31 cm

Exercise 10.4 *page 316*

1. 36·9° **2.** 44·4° **3.** 48·2° **4.** 60° **5.** 36·9° **6.** 50·2° **7.** 29·0°
8. 56·4° **9.** 38·9° **10.** 43·9° **11.** 41·8° **12.** 39·3° **13.** 60·3° **14.** 50·5°
15. 13·6° **16.** 34·8° **17.** 60·0° **18.** 42·0° **19.** 36·9° **20.** 51·3° **21.** 19·6°
22. 17·9° **23.** 32·5° **24.** 59·6° **25.** 54·8° **26.** 46·3°

Exercise 10.5 *page 319*

1. 19·5° **2.** 4·1 m **3. a)** 26·0 km **b)** 23·4 km **4. a)** 88·6 km **b)** 179·3 km **5.** 4·1 m
6. 8·6 m **7. a)** 484 km **b)** 858 km **c)** 985 km, 060·6° **8.** 954 km, 133° **9.** 56·3°
10. 35·5° **11.** 71·6° **12.** 91·8° **13.** 180 m **14.** 36·4° **15.** 10·3 cm

Exercise 10.6 *page 320*

1. 9·51 cm **2.** 71·1° **3.** 53:7 m **4.** 67·1 m **5.** 138 m **6.** 83·2 km **7.** 60°
8. 241 cm **9.** 13·9 cm **10.** 29·0 cm **11.** Yes **12.** 11·09 m, 11·09; 222 m **13.** 13·2 cm
14. 4·4 m **15.** (all cm) $x = 16{\cdot}4$, $y = 113.6$, width = 255 **16.** 3·13 m **17.** 25·6 cm **18.** 27·1 cm

Exercise 10.7 *page 323*

1. a) $\frac{1}{2}$ **b)** $\sqrt{3}$ **c)** $\frac{1}{2}$ **d)** $\frac{\sqrt{3}}{2}$ **e)** 1 **f)** 1 **g)** $\frac{1}{\sqrt{3}}$ **h)** $\frac{\sqrt{3}}{2}$
5. a) 6 cm **b)** $2\sqrt{3}$ cm **c)** 3 cm **d)** 2·41 cm **e)** $7\sqrt{3}$ cm **f)** $8\sqrt{3}$ cm

Exercise 10.8 *page 325*

1. a) 13 cm **b)** 13·6 cm **c)** 17·1° **2. a)** 4·04 m **b)** 38·9° **c)** 11·2 m **d)** 19·9°
3. a) 8·49 cm **b)** 8·49 cm **c)** 10·4 cm **d)** 35.3° **e)** 35·3°
4. a) 14·1 cm **b)** 18·7 cm **c)** 69·3° **d)** 29·0° **e)** 41·4°
5. a) 4·47 m **b)** 7·48 m **c)** 63·4° **d)** 74·5° **e)** 53·3° **6.** 10·8 cm; 21·8°
7. a) $h \tan 65°$ or $\frac{h}{\tan 25°}$ **b)** $h \tan 57°$ or $\frac{h}{\tan 33°}$ **c)** 22·7 m **8.** 22·6 m **9.** 55·0 m
10. 7·26 m **11.** 43·3°

Exercise 10.9 *page 328*

3. 162° **4.** 153° **5. a)** 140° **b)** 110° **c)** 50° **6.** 110° **7.** 135°
8. 160° **9.** 82° **10.** 58°, 122° **11.** 20·5°, 159·5° **12.** 60°, 120°
13. a) 41°, 139° **b)** 30°, 150° **14. a) i)** 16°, 111° **ii)** 153° **b)** 2·2 **c)** 63°

Exercise 10.10 *page 329*

1. a) amplitude = 1, period = 180° **b)** amplitude = 1, period = 120°
2. a) 180° **b)** infinite **3.** a is amplitude **4. a)** 2 **b)** $\frac{1}{2}$ **c)** 3

5.

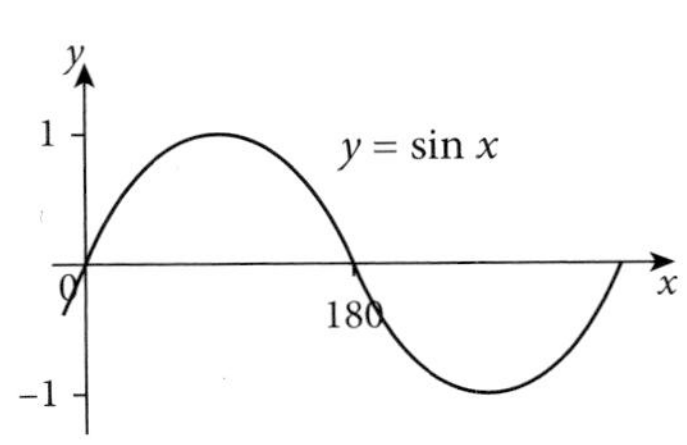

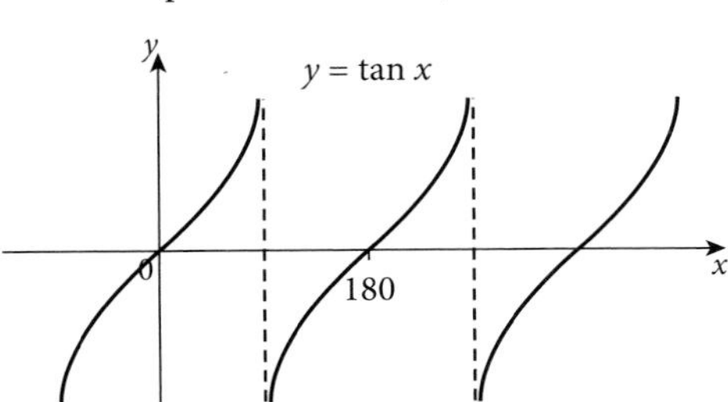

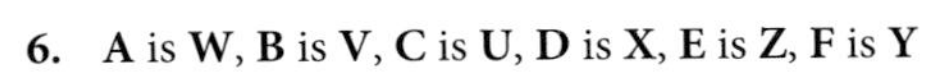
6. **A** is **W**, **B** is **V**, **C** is **U**, **D** is **X**, **E** is **Z**, **F** is **Y**

Exercise 10.11 *page 331*

1. 6·38 m **2.** 12·5 m **3.** 5·17 cm **4.** 40·4 cm **5.** 7·81 m, 7·10 m **6.** 3·55 m, 6·68 m
7. 8·61 cm **8.** 9·97 cm **9.** 8·52 cm **10.** 15·2 cm **11.** 35·8° **12.** 42·9°
13. 32·3° **14.** 37·8° **15.** 35·5°, 48·5° **16.** 68.8°, 80.0° or 111.2°, 37.6° **17.** 64.6° **18.** 34.2°
19. 50.6° **20.** 39.1° **21.** 39·5° **22.** 21·6°

Exercise 10.12 *page 335*

1. 6·24 **2.** 6·05 **3.** 5·47 **4.** 9·27 **5.** 10·1 **6.** 8·99 **7.** 5·87
8. 4·24 **9.** 11·9 **10.** 154 **11.** 25·2° **12.** 78·5° **13.** 115·0° **14.** 111·1°
15. 24·0° **16.** 92·5° **17.** 99·9° **18.** 38·2° **19.** 137·8° **20.** 34·0° **21.** 60·2°
22. 8.72 **23.** 1.40° **24.** 7.38°

Exercise 10.13 *page 337*

1. 6·7 cm **2.** 10·8 m **3.** 35·6 km **4.** 25·2 m **5.** 38·6°, 48·5°, 92·9° **6.** 40·4 m
7. 9·8 km; 085·7° **8.** **a)** 29·6 km **b)** 050·5° **9.** **a)** 10·8 m **b)** 72·6° **c)** 32·6°
10. 378 km, 048·4° **11.** **a)** 62·2° **b)** 2·33 km **12.** 9·64 m **13.** 8·6°

Revision exercise 10A *page 338*

1. **a)** 45·6° **b)** 58·0° **c)** 3·89 cm **d)** 33·8 m **2.** **a)** 1·75 **b)** 60·3°
3. **a)** 12·7 cm **b)** 5·92 cm **c)** 36·1° **4.** 5·39 cm **5.** **a)** 220° **b)** 295° **6.** 0·335 m
7. **a)** 6·61 cm **b)** 12·8 cm **c)** 5·67 cm **8.** **a)** 86·9 cm **b)** 53·6 cm **c)** 133 cm **9.** 10·1 cm^2
10. **a)** 14·1 cm **b)** 35·3° **c)** 35·3° **11.** **a)** 6·63 cm **b)** 41·8° **12.** **a)** 11·3 cm **b)** 8·25 cm
c) 55·6° **13.** 45·2 km, 33·6 km **14.** 73·4° **15.** 8·76 m, 9·99 m **16.** 0·539
17. 4·12 cm, 9·93 cm **18.** 26·4° **19.** 30·5 cm **20.** 919 km h^{-1} **21.** 8·23 cm
22. **a)** $\frac{1}{2}$ **b)** $\sqrt{3}$ **c)** $\frac{1}{\sqrt{2}}$ **d)** $\frac{3}{2}$ **e)** $\frac{3}{4}$ **f)** 2 **g)** 1

Examination exercise 10B *page 340*

1. **a)** North B 210° A **b)** 25 km **2.** 30°, 150° **3.** 120°, 240°
4. **a)** **i)** 29·6 cm^2 **ii)** 13·2 cm **iii)** 120° **b)** 45·1°
5. **a)** 1190 km **b)** 1199 – 1200 km **c)** 568500 – 569400 km^2
d) 1490 – 1500 km **e)** 28·2 cm
6. **a)** 0·8333 **b)** **ii)** 56·4°, 123·6°
iii) 67·2°

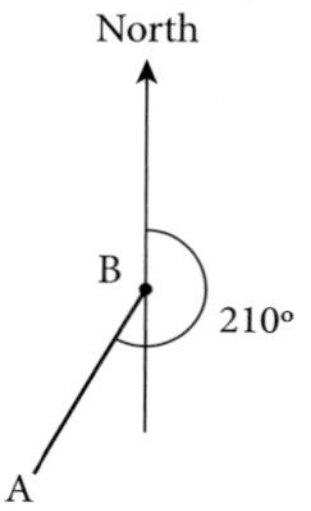

11 Vectors and transformations

Exercise 11.1 *page 344*

1. **d** **2.** **2c** **3.** **3c** **4.** **3d** **5.** **5d** **6.** **3c** **7.** **−2d**
8. **−2c** **9.** **−3c** **10.** **−c** **11.** **c + d** **12.** **c + 2d** **13.** **2c + d** **14.** **3c + d**
15. **2c + 2d** **16.** **2c + 3d** **17.** **2c − d** **18.** **3c − d** **19.** **−c + 2d** **20.** **−c + 3d** **21.** **−c + d**
22. **−c − 2d** **23.** **−2c − 2d** **24.** **−3c − 6d** **25.** **−2c + 3d** **26.** **c + 6d** **27.** $\overrightarrow{QI}$ **28.** $\overrightarrow{QU}$
29. $\overrightarrow{QH}$ **30.** $\overrightarrow{QB}$ **31.** $\overrightarrow{QF}$ **32.** $\overrightarrow{QJ}$ **33.** $\overrightarrow{QZ}$ **34.** $\overrightarrow{QL}$ **35.** $\overrightarrow{QE}$
36. $\overrightarrow{QX}$ **37.** $\overrightarrow{QW}$ **38.** $\overrightarrow{QK}$ **39.** **a)** **−a** **b)** **b − a** **c)** **−b** **d)** **a + b**
40. **a)** **a + b** **b)** **a − 2b** **c)** **−a + b** **d)** **−a − b** **41.** **a)** **−a − b** **b)** **3a − b** **c)** **2a − b**
d) **−2a + b** **42.** **a)** **a − 2b** **b)** **a − b** **c)** **2a** **d)** **−2a + 3b** **43** **a) or d)**
44. **a)** **2a − c** **b)** **2a − c** **c)** **3a** **d)** **a + b + c** **e)** **−3a − b** **45.** **a)** **b − c** **b)** **2b + 2c**
c) **a + 2b + 2c** **d)** **−a − b** **e)** **c − a − b** **46.** **a)** **a + b** **b)** **−a + c** **c)** **a + b + c** **d)** **b − c**
e) **−a + 2c**

Exercise 11.2 *page 347*

1. **a)** **a + c** **b)** **− a + b** **c)** **2b** **d)** **−2a** **e)** **−2a + 2b** **f)** **−a + b** **g)** **a + b**
h) **b** **i)** **−b + 2a** **j)** **−2b + a** **2.** **a)** **a** **b)** **−a + b** **c)** **3b** **d)** **−2a**
e) **−2a + 3b** **f)** $-\mathbf{a} + \frac{3}{2}\mathbf{b}$ **g)** $\mathbf{a} + \frac{3}{2}\mathbf{b}$ **h)** $\frac{3}{2}\mathbf{b}$ **i)** **−b + 2a** **j)** **−3b + a**
3. **a)** **2a** **b)** **−a + b** **c)** **2b** **d)** **−3a** **e)** **−3a + 2b** **f)** $-\frac{3}{2}\mathbf{a} + \mathbf{b}$ **g)** $\frac{3}{2}\mathbf{a} + \mathbf{b}$

h) $\frac{1}{2}\mathbf{a}+\mathbf{b}$ i) $-\mathbf{b}+3\mathbf{a}$ j) $-2\mathbf{b}+\mathbf{a}$ **4.** a) $\frac{1}{2}\mathbf{a}$ b) $-\mathbf{a}+\mathbf{b}$ c) $4\mathbf{b}$ d) $-\frac{3}{2}\mathbf{a}$

e) $-\frac{3}{2}\mathbf{a}+4\mathbf{b}$ f) $-\mathbf{a}+\frac{8}{3}\mathbf{b}$ g) $\frac{1}{2}\mathbf{a}+\frac{8}{3}\mathbf{b}$ h) $-\frac{1}{2}\mathbf{a}+\frac{8}{3}\mathbf{b}$ i) $\frac{3}{2}\mathbf{a}-\mathbf{b}$ j) $\mathbf{a}-4\mathbf{b}$

5. a) $5\mathbf{a}$ b) $\mathbf{b}-\mathbf{a}$ c) $\frac{3}{2}\mathbf{b}$ d) $-6\mathbf{a}$ e) $\frac{3}{2}\mathbf{b}-6\mathbf{a}$ f) $\mathbf{b}-4\mathbf{a}$ g) $2\mathbf{a}+\mathbf{b}$

h) $\mathbf{a}+\mathbf{b}$ i) $6\mathbf{a}-\mathbf{b}$ j) $\mathbf{a}-\frac{3}{2}\mathbf{b}$ **6.** a) $4\mathbf{a}$ b) $\mathbf{b}-\mathbf{a}$ c) $3\mathbf{b}$ d) $-5\mathbf{a}$

e) $3\mathbf{b}-5\mathbf{a}$ f) $\frac{3}{4}\mathbf{b}-\frac{5}{4}\mathbf{a}$ g) $\frac{15}{4}\mathbf{a}+\frac{3}{4}\mathbf{b}$ h) $\frac{11}{4}\mathbf{a}+\frac{3}{4}\mathbf{b}$ i) $5\mathbf{a}-\mathbf{b}$ j) $\mathbf{a}-3\mathbf{b}$ **7.** $\frac{1}{2}\mathbf{s}-\frac{1}{2}\mathbf{t}$

8. $\frac{1}{3}\mathbf{a}+\frac{2}{3}\mathbf{b}$ **9.** $\mathbf{a}+\mathbf{c}-\mathbf{b}$ **10.** $2\mathbf{m}+2\mathbf{n}$ **11.** a) $\mathbf{b}-\mathbf{a}$ b) $\mathbf{b}-\mathbf{a}$ c) $2\mathbf{b}-2\mathbf{a}$ d) $\mathbf{b}-2\mathbf{a}$

e) $\mathbf{b}-2\mathbf{a}$ f) $2\mathbf{b}-3\mathbf{a}$

12. a) $\mathbf{y}-\mathbf{z}$ b) $\frac{1}{2}\mathbf{y}-\frac{1}{2}\mathbf{z}$ c) $\frac{1}{2}\mathbf{y}+\frac{1}{2}\mathbf{z}$ d) $-\mathbf{x}+\frac{1}{2}\mathbf{y}+\frac{1}{2}\mathbf{z}$ e) $-\frac{2}{3}\mathbf{x}+\frac{1}{3}\mathbf{y}+\frac{1}{3}\mathbf{z}$ f) $\frac{1}{3}\mathbf{x}+\frac{1}{3}\mathbf{y}+\frac{1}{3}\mathbf{z}$

Exercise 11.3 *page 350*

1.

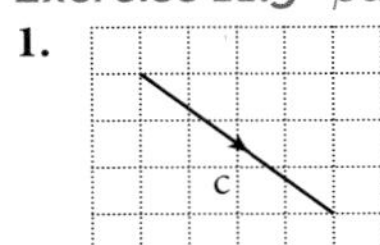

2.

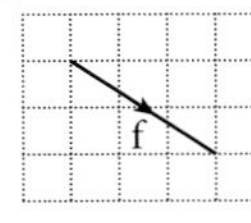

3.

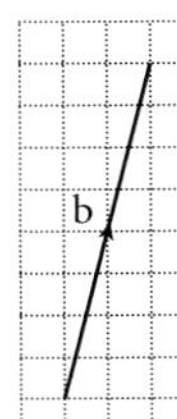

4.

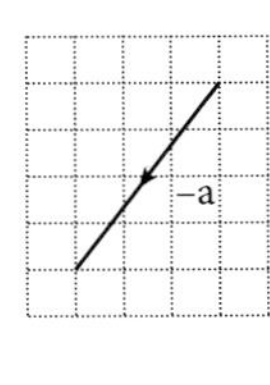

5.

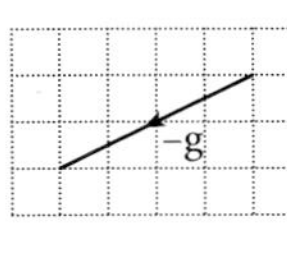

6.

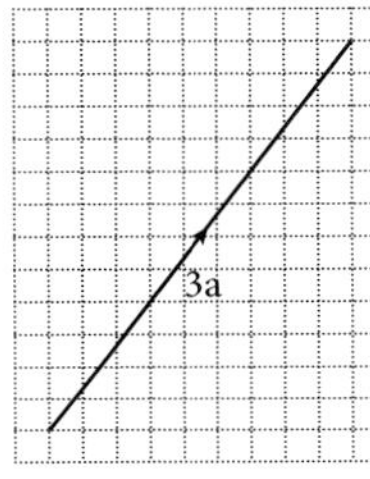

7.

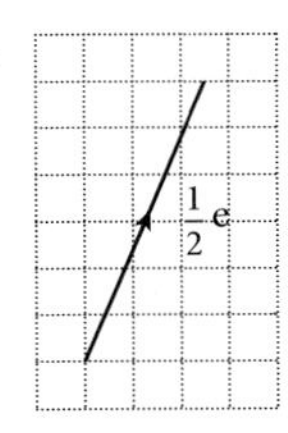

8.

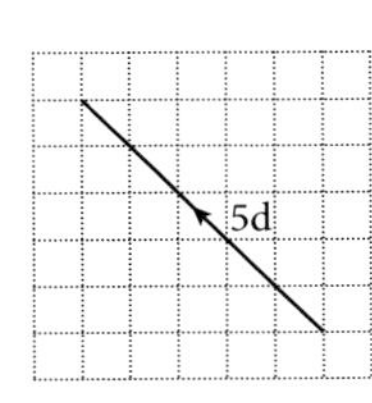

9.

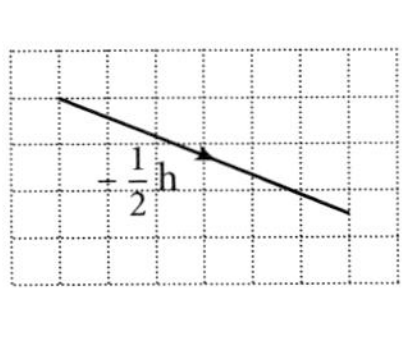

10.

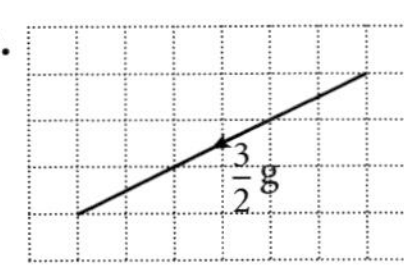

11.

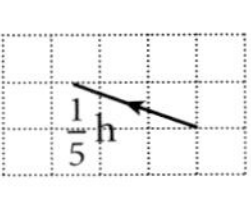

12.

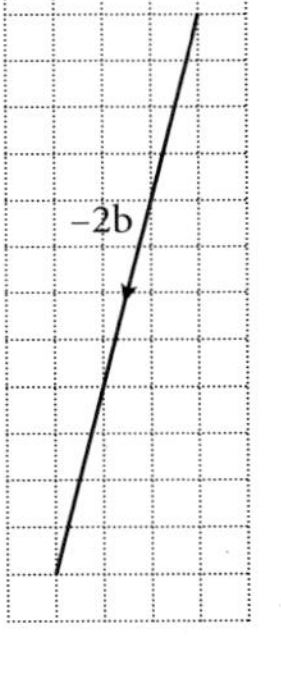

13. $\begin{pmatrix}-11\\9\end{pmatrix}$ **14.** $\begin{pmatrix}-1\\-4\end{pmatrix}$ **15.** $\begin{pmatrix}4\\8\end{pmatrix}$ **16.** $\begin{pmatrix}4\\3\end{pmatrix}$ **17.** $\begin{pmatrix}8\\-7\end{pmatrix}$ **18.** $\begin{pmatrix}18\\-1\end{pmatrix}$ **19.** $\begin{pmatrix}7\\-4\end{pmatrix}$

20. $\begin{pmatrix}-18\\-16\end{pmatrix}$ **21.** $\begin{pmatrix}13\\19\end{pmatrix}$ **22.** $\begin{pmatrix}8\\5\end{pmatrix}$ **23.** $\begin{pmatrix}10\\-13\end{pmatrix}$ **24.** $\begin{pmatrix}17\\35\end{pmatrix}$ **25.** $\begin{pmatrix}4\\8\end{pmatrix}$ **26.** $\begin{pmatrix}4\\3\end{pmatrix}$

27. $\begin{pmatrix}-1\\1\end{pmatrix}$ **28.** $\begin{pmatrix}-16\\3\end{pmatrix}$ **29.** $\begin{pmatrix}-2\frac{1}{2}\\-3\end{pmatrix}$ **30.** $\begin{pmatrix}-3\frac{1}{2}\\\frac{1}{2}\end{pmatrix}$ **31.** $\begin{pmatrix}-3\\-7\end{pmatrix}$ **32.** $\begin{pmatrix}-2\\12\end{pmatrix}$ **33.** $\begin{pmatrix}4\\8\end{pmatrix}$

34. $\begin{pmatrix}-5\frac{1}{2}\\4\frac{1}{2}\end{pmatrix}$ **35.** $\begin{pmatrix}0\\0\end{pmatrix}$ **36.** $\begin{pmatrix}-2\frac{1}{2}\\-6\end{pmatrix}$ **37.** b) **l** and **s**; **n** and **r**; **p** and **t**; **m** and **q** **38.** a) true b) true

c) true d) true e) false f) false

39. a)

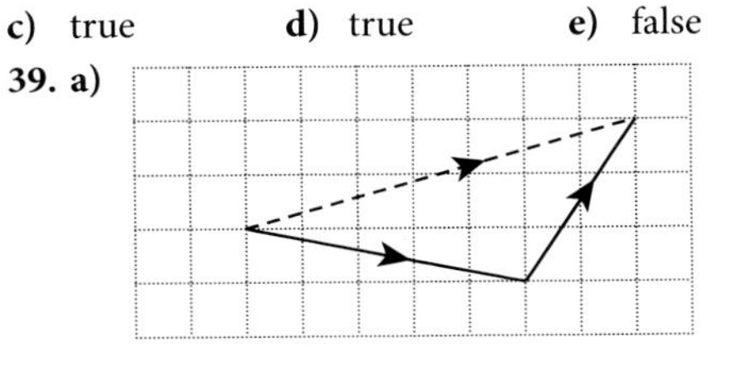

b)

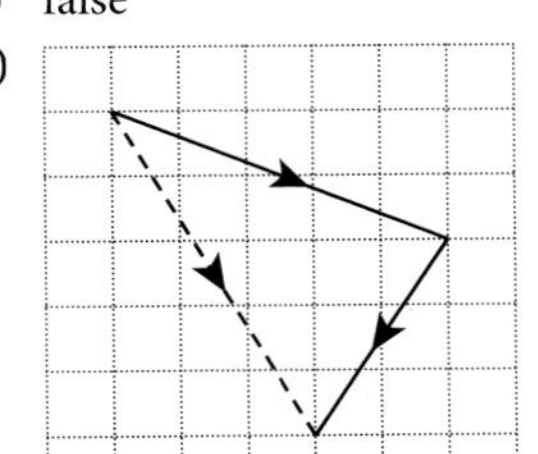

40. a)

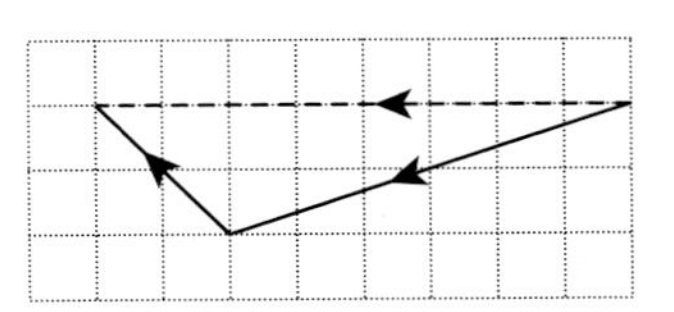

b)

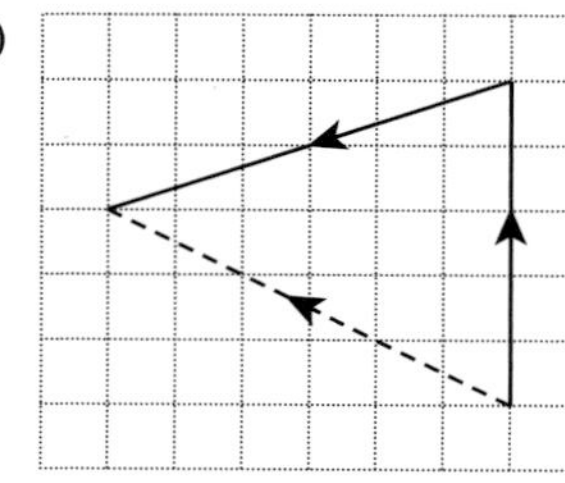

c)

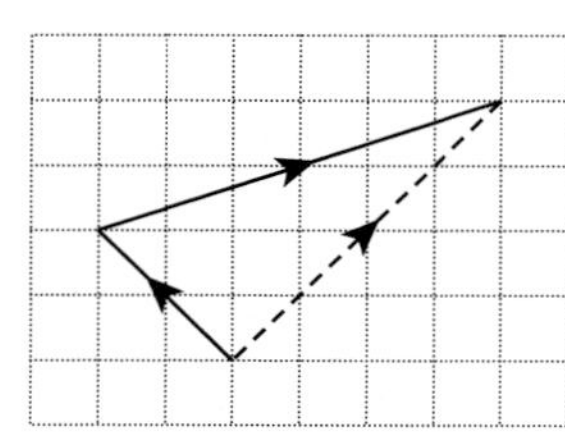

d)

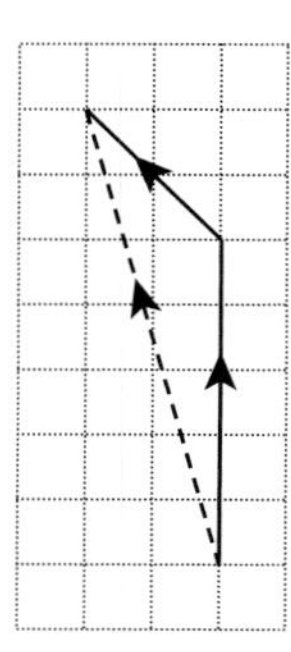

Exercise 11.4 *page 352*

1. $\begin{pmatrix} 2 \\ -2 \end{pmatrix}$ **2.** $\begin{pmatrix} 6 \\ -2 \end{pmatrix}$ **3. b)** $\begin{pmatrix} 0 \\ 3 \end{pmatrix}$; $\begin{pmatrix} -5 \\ -5 \end{pmatrix}$ **4. b)** $\begin{pmatrix} 4 \\ 2 \end{pmatrix}$; $\begin{pmatrix} -7 \\ 0 \end{pmatrix}$

5. a) $\begin{pmatrix} 3 \\ -3 \end{pmatrix}$ **b)** $\begin{pmatrix} 1\frac{1}{2} \\ -1\frac{1}{2} \end{pmatrix}$ **c)** $\begin{pmatrix} 3\frac{1}{2} \\ 3\frac{1}{2} \end{pmatrix}$; $M\left(3\frac{1}{2},\ 3\frac{1}{2}\right)$ **6. a)** $\begin{pmatrix} -1 \\ 6 \end{pmatrix}$ **b)** $\begin{pmatrix} -\frac{1}{2} \\ 3 \end{pmatrix}$ **c)** $\begin{pmatrix} 5\frac{1}{2} \\ -4 \end{pmatrix}$; $M\left(5\frac{1}{2},\ -4\right)$

7. a) i) $\begin{pmatrix} -6 \\ 3 \end{pmatrix}$ **ii)** $\begin{pmatrix} -2 \\ 1 \end{pmatrix}$ **iii)** $\begin{pmatrix} 2 \\ 3 \end{pmatrix}$ **b) i)** $\begin{pmatrix} 0 \\ -9 \end{pmatrix}$ **ii)** $\begin{pmatrix} 0 \\ -3 \end{pmatrix}$ **iii)** $\begin{pmatrix} -2 \\ 2 \end{pmatrix}$

8. $\begin{pmatrix} -1 \\ 2 \end{pmatrix}$ or $\begin{pmatrix} -1 \\ 2 \end{pmatrix}$ **9.** $\begin{pmatrix} 0 \\ 2 \end{pmatrix}$ or $\begin{pmatrix} 0 \\ -2 \end{pmatrix}$ **10. a)** $\mathbf{q} - \mathbf{p}$ **b)** $\mathbf{q} + 2\mathbf{p}$ **c)** $\mathbf{p} + \mathbf{q}$

11. a) $\begin{pmatrix} 1 \\ -3 \end{pmatrix}$ **b)** $\begin{pmatrix} -1 \\ 3 \end{pmatrix}$ **c)** $\begin{pmatrix} 3 \\ 1 \end{pmatrix}$ **d)** $\begin{pmatrix} -3 \\ -1 \end{pmatrix}$

Exercise 11.5 *page 353*

1. 5 **2.** $\sqrt{17}$ **3.** 13 **4.** 3 **5.** 5 **6.** $\sqrt{45}$ **7.** $\sqrt{74}$
8. $\sqrt{208}$ **9.** 10 **10.** $\sqrt{89}$ **11. a)** $\sqrt{320}$ **b)** no **12. a)** $\sqrt{148}$ **b)** no
13. $\sqrt{29}$ **14.** $\sqrt{26}$ **15.** $\sqrt{10}$ **16. a)** 5 **b)** $n = \pm 4$ **17. a)** 13 **b)** $m = \pm 13$
18. a) 5 **b)** $p = 0$ **19. a)** 9 **b)** 6 **c)** 5 **20. a)** 30 **b)** 5 **c)** $\sqrt{50}$ **d)** 4

Exercise 11.6 *page 355*

1

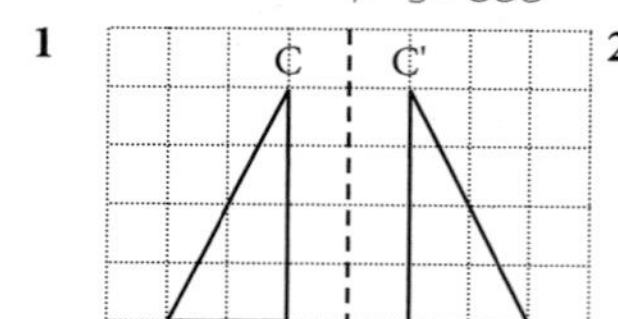

2

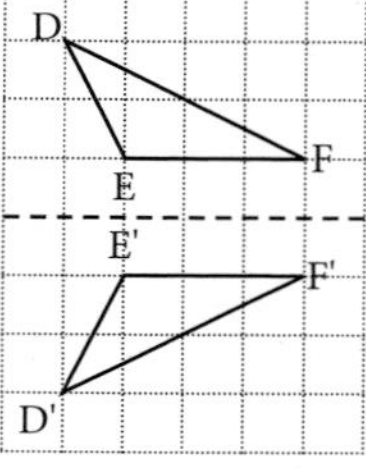

3

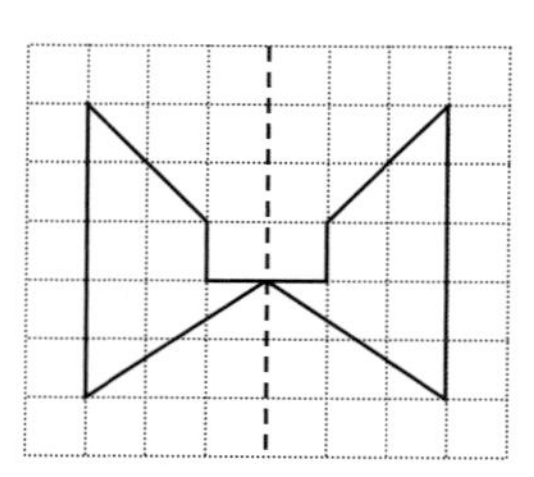

4

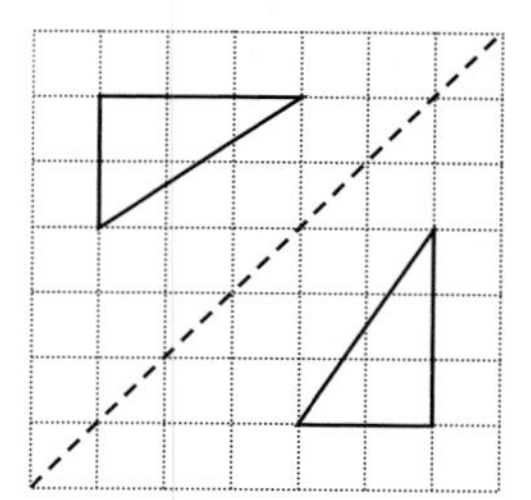

5

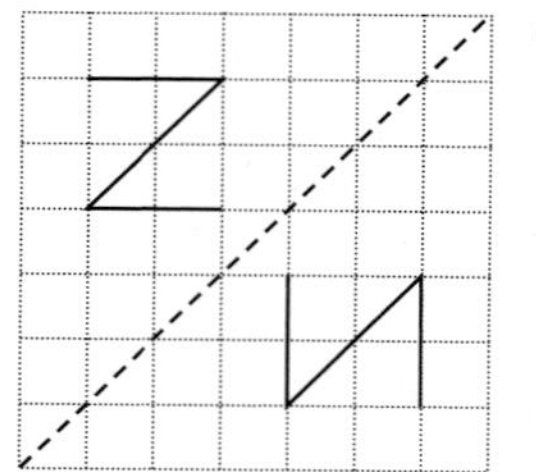

6

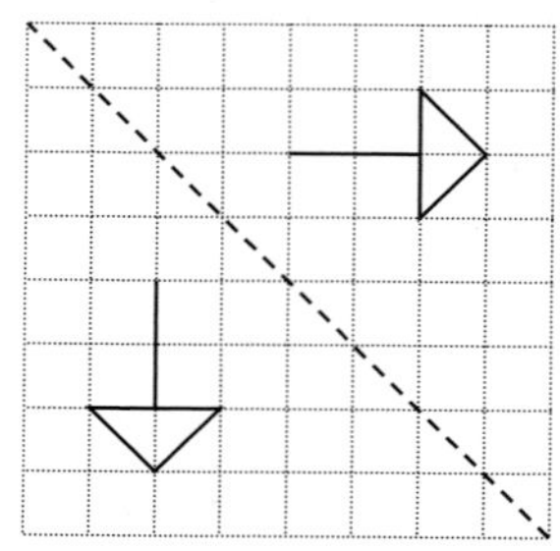

7

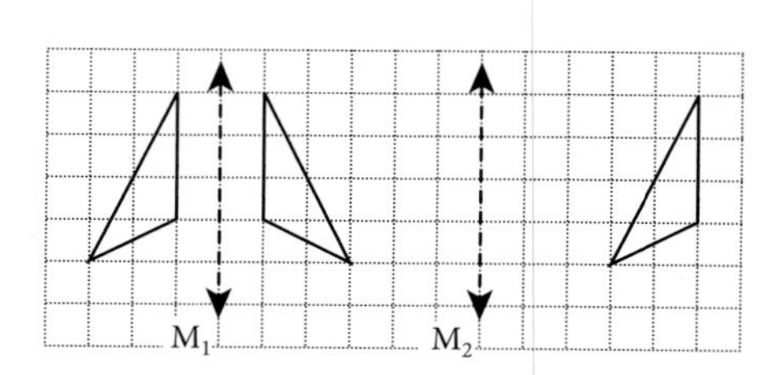

8

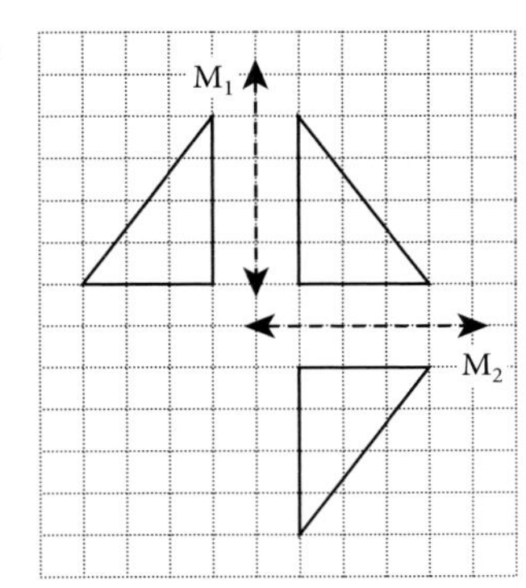

9

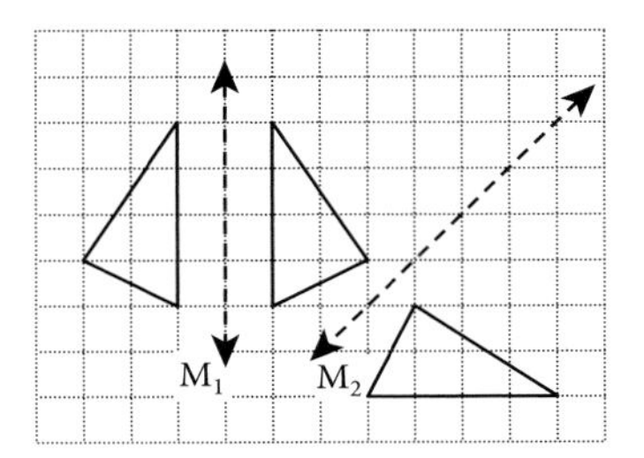

Exercise 11.7 page 356

1. c) i) (−6, 8) **ii)** (6, −4) **iii)** (8, 6)
2. c) i) (8, 8) **ii)** (8, −6) **iii)** (−8, 6)
3. c) i) (3, −1) **ii)** (4, 2) **iii)** (−1, 1)
4. b) i) $y = 1$ **ii)** $y = x$ **iii)** $y = -x$
iv) $y = 2$
5. f) (1, −1), (−3, −1), (−3, −3)
6. f) (8, −2), (8, −6), (6, −6)

Exercise 11.8 page 358

1.

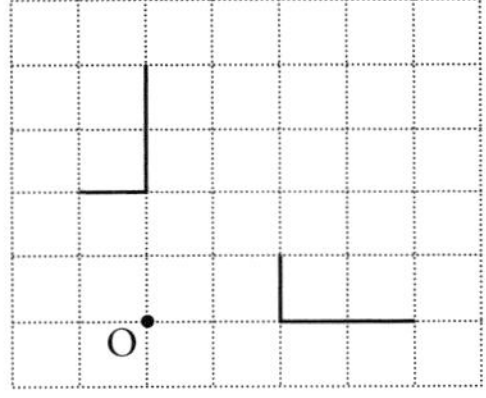

2

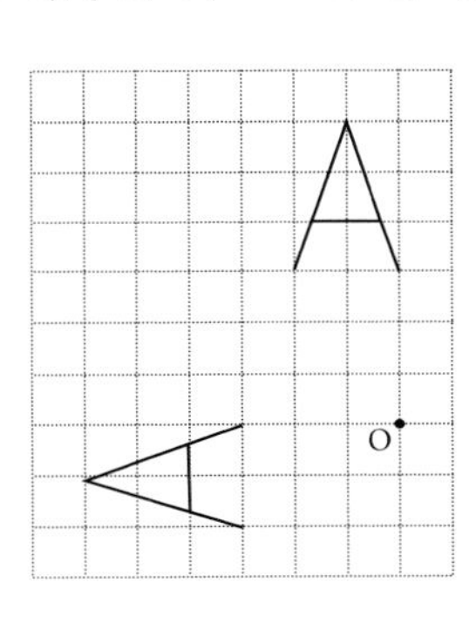

3

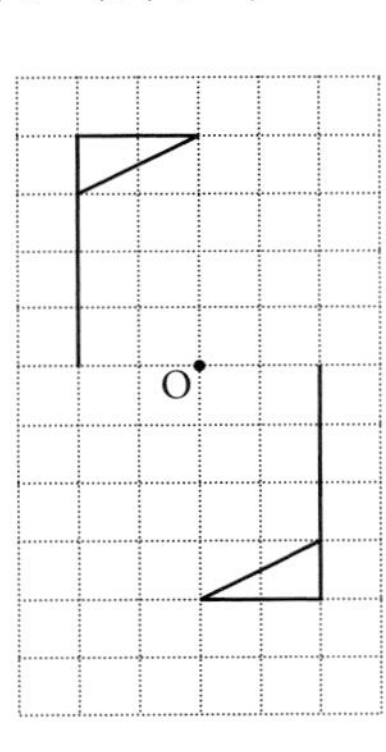

4

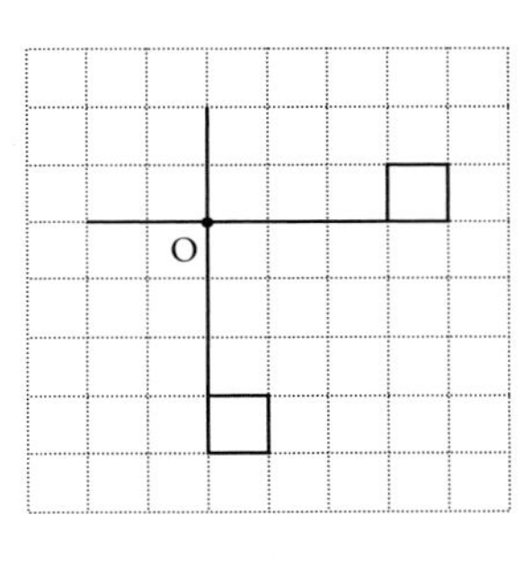

5

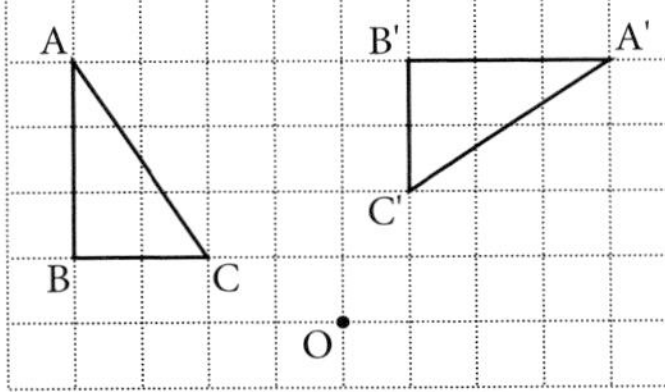

90° clockwise

6

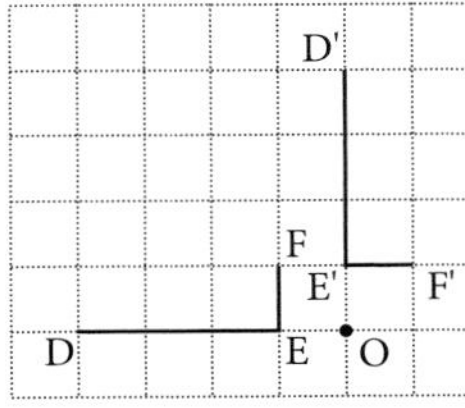

90° clockwise

7

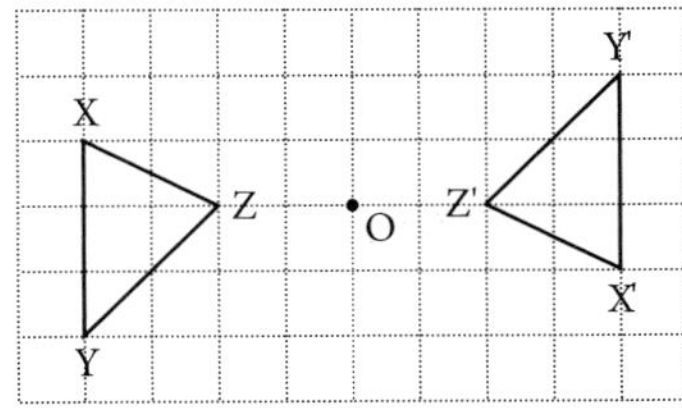

180°

8

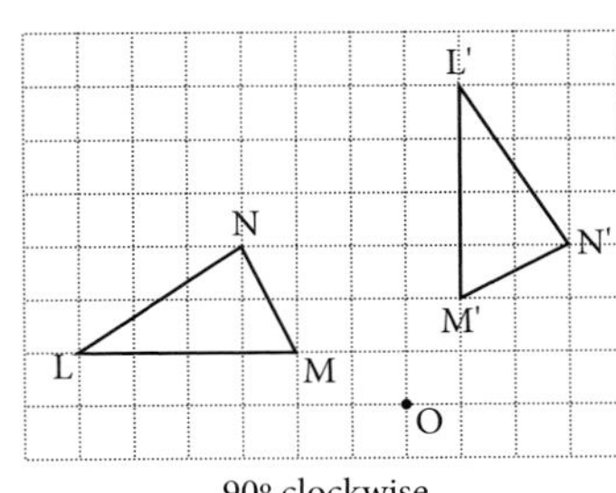

90° clockwise

Exercise 11.9 page 359

1. a) A′(3, −1) B′(6, −1) C′(6, −3)
b) D′(3, −3) E′(3, −6) F′(1, −6)
c) P′(−7, −4) Q′(−5, −4) R′(−5, −1)
2. b) i) (4, −1), (7, −1), (7, −3)
ii) (−1, −4), (−1, −7), (−3, −7)
iii) (−4, 1), (−7, 1), (−7, 3)
3. c) (−2, 1), (−2, −1), (1, −2)
4. e) (−5, 2), (−5, 6), (−3, 5)
5. b) i) 90° anticlockwise, centre (0, 0)
ii) 180°, centre (2, 1)
iii) 90° clockwise, centre (2, 0)
iv) 180°, centre $\left(3\frac{1}{2},\ 2\frac{1}{2}\right)$
v) 90° anticlockwise, centre (6, 1)
vi) 90° clockwise, centre (1, 3)
6. e) i) 180°, centre $\left(\frac{1}{2}, \frac{1}{2}\right)$
ii) 90°, anticlockwise, centre (−2, 4)

Exercise 11.10 page 361

1. a) $\begin{pmatrix} 7 \\ 3 \end{pmatrix}$ **b)** $\begin{pmatrix} 0 \\ -9 \end{pmatrix}$ **c)** $\begin{pmatrix} 9 \\ 10 \end{pmatrix}$ **d)** $\begin{pmatrix} -10 \\ 3 \end{pmatrix}$ **e)** $\begin{pmatrix} -1 \\ 13 \end{pmatrix}$ **f)** $\begin{pmatrix} 10 \\ 0 \end{pmatrix}$ **g)** $\begin{pmatrix} -9 \\ -4 \end{pmatrix}$ **h)** $\begin{pmatrix} -10 \\ 0 \end{pmatrix}$

2. (5, 2) **3.** (5, 6) **4.** (8, −5) **5.** (0, 6) **6.** (4, −7) **7.** (−3, 4)
8. (−3, −5) **9.** (−1, −8) **10.** (5, 2) **11.** (−2, 1)

Exercise 11.11 page 363

1.

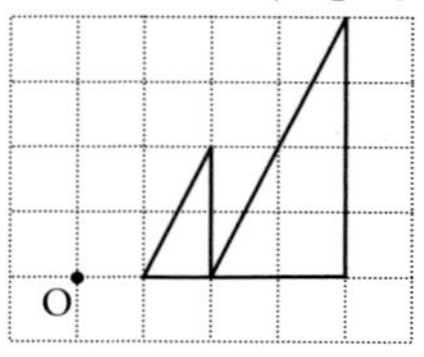

2.

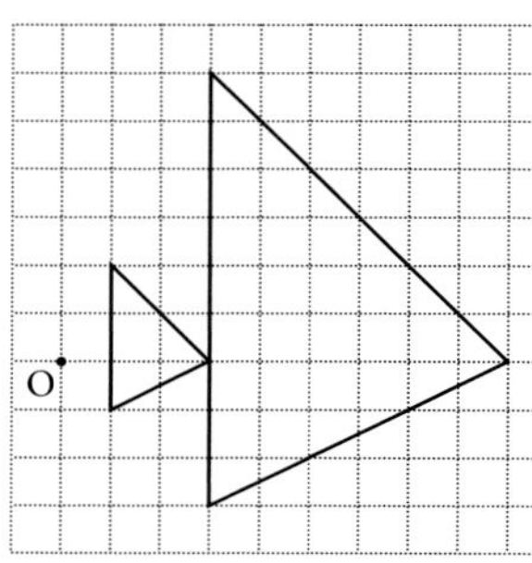

3.

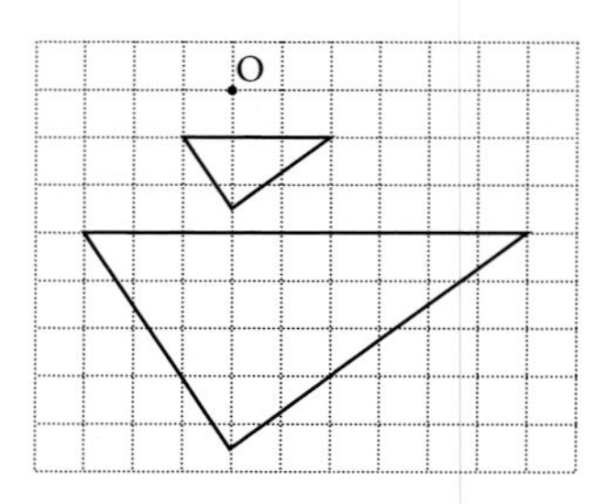

4.

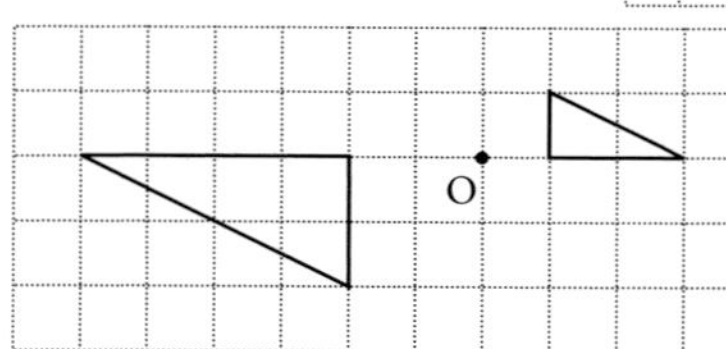

5.

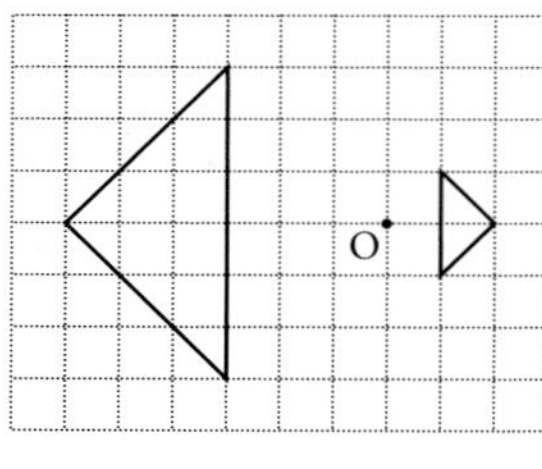

6.

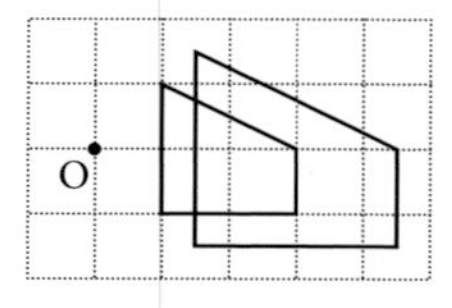

7. (4, 8), (8, 4), (10, 10) **8.** (3, 6), (7, 2), (9, 8) **9.** (1, 1), (10, 4), (4, 7) **10.** (1, 4), (7, 8), (11, 2)

11. (2, 1) +3 **12.** (11, 9), $\frac{1}{2}$ **13.** (5, 4), −2 **14.** (6, 6), −1 **15.** $\left(\frac{1}{2}, 1\right), \left(6\frac{1}{2}, 1\right), \left(\frac{1}{2}, 5\right)$ **16.** (3, 4), (3, 3), (5, 3)

17. (3, 7), (1, 7), (3, 2) **18.** (10, 7), (6, 7), (6, 5) **19.** (6, 5), $\left(3\frac{1}{2}, 5\right), \left(3\frac{1}{2}, 3\right)$

Exercise 11.12 page 364

1. b) i) Rotation 90° clockwise, centre (0, −2) **ii)** Reflection in $y = x$ **iii)** Translation $\begin{pmatrix} 3 \\ 7 \end{pmatrix}$

iv) Enlargement, scale factor 2, centre (−5, 5) **v)** Translation $\begin{pmatrix} -7 \\ -3 \end{pmatrix}$ **vi)** Reflection in $y = x$

2. a) Rotation 90° clockwise, centre (4, −2) **b)** Translation $\begin{pmatrix} 8 \\ 2 \end{pmatrix}$ **c)** Reflection in $y = x$

d) Enlargement, scale factor $\frac{1}{2}$, centre (7, −7) **e)** Rotation 90° anticlockwise, centre (−8, 0)

f) Enlargement, scale factor 2, centre (−1, −9) **g)** Rotation 90° anticlockwise, centre (7, 3)

3. a) Enlargement, scale factor $1\frac{1}{2}$, centre (1, −4) **b)** Rotation 90° clockwise, centre (0, −4)

c) Reflection $y = -x$ **d)** Translation $\begin{pmatrix} 11 \\ 10 \end{pmatrix}$ **e)** Enlargement, scale factor $\frac{1}{2}$, centre (−3, 8)

f) Rotation 90° anticlockwise, centre $\left(\frac{1}{2}, 6\frac{1}{2}\right)$ **g)** Enlargement, scale factor 3, centre (−2, 5)

Exercise 11.13 page 366

1. a) 2 **b)** 3 **c)** 1·5

2.

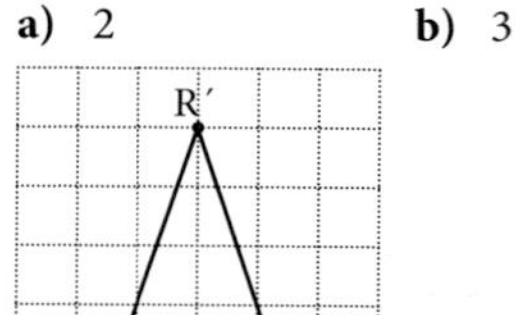

3.

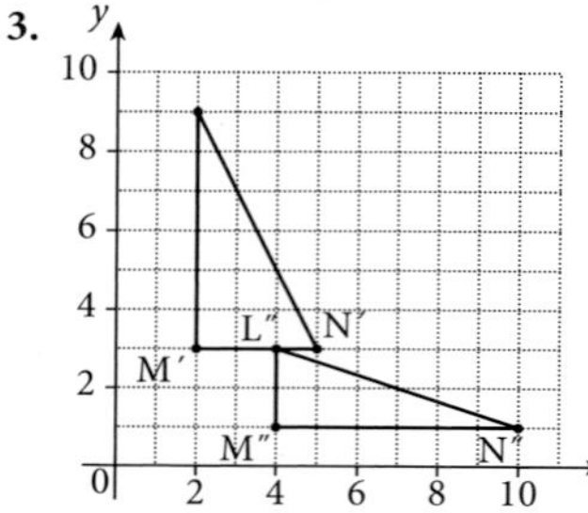

Exercise 11.14 page 368

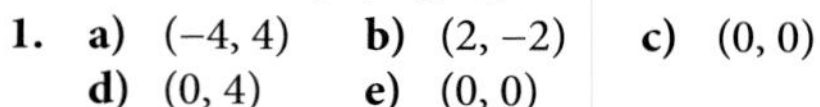

1. a) (−4, 4) **b)** (2, −2) **c)** (0, 0)
d) (0, 4) **e)** (0, 0)

2. a) (−2, 5) **b)** (−4, 0) **c)** (2, −2)
d) (1, −1)

3. a) reflection in y-axis
b) rotation 180°, centre (−2, 2)
c) rotation 90° clockwise, centre (2, 2)

4. a) rotation 90° anticlockwise, centre (0, 0) **b)** translation $\begin{pmatrix} -2 \\ 5 \end{pmatrix}$
c) rotation 90° anticlockwise, centre (2, −4)
d) rotation 90° anticlockwise, centre $\left(-\frac{1}{2}, 3\frac{1}{2}\right)$ **5. a)** rotation 90° anticlockwise, centre (2, 2)
b) enlargement, scale factor $\frac{1}{2}$, centre (8, 6) **c)** rotation 90° clockwise, centre $\left(-\frac{1}{2}, -3\frac{1}{2}\right)$

6. A^{-1} : reflection in $x = 2$ B^{-1} : B C^{-1} : translation $\begin{pmatrix} 6 \\ -2 \end{pmatrix}$ D^{-1} : D E^{-1} : E

F^{-1} : translation $\begin{pmatrix} -4 \\ -3 \end{pmatrix}$ G^{-1} : 90° rotation anticlockwise, centre (0, 0) H^{-1} : enlargement, scale factor 2, centre (0, 0)

7. a) (4, 0) **b)** (−6, −1) **c)** (−2, −2) **d)** (2, −2) **e)** (6, 2)

8. b) rotation =, 180°, centre (4, 0) **c)** translation $\begin{pmatrix} 12 \\ -4 \end{pmatrix}$

Revision exercise 11A page 369

1. a) $\begin{pmatrix} 2 \\ 1 \end{pmatrix}$ **b)** $\begin{pmatrix} -4 \\ 5 \end{pmatrix}$ **c)** $\begin{pmatrix} 0 \\ -5 \end{pmatrix}$ **d)** C **2.** $\begin{pmatrix} 6 \\ 10 \end{pmatrix}$ **3.** $\begin{pmatrix} -2 \\ 4 \end{pmatrix}$

4. various e·g· $\begin{pmatrix} 1 \\ -2 \end{pmatrix}, \begin{pmatrix} -1 \\ 2 \end{pmatrix} \cdots$ **5.** $\begin{pmatrix} -4 \\ 6 \end{pmatrix}$ **6.** (−5, 8) **7. d)** translation $\begin{pmatrix} 12 \\ 0 \end{pmatrix}$

8. a) (4, −1) **b)** (4, 1) **c)** (−3, 2)

9. a) A′(−3, −1) B′(1, −1) C′(−3, −7) **b)** A′(2,−2) B′(6,−2) C′(2,−8) **c)** A′(1, 1)B′(2, 1) C′$\left(1, -\frac{1}{2}\right)$

d) A′(4, 2) B′(3, 2) C′$\left(4, 3\frac{1}{2}\right)$ **e)** A′(−2, 2) B′(−6, 2) C′(−2, 8)

10. a) reflection in y-axis **b)** reflection in $y = x$ **c)** rotation, −90°, centre (0, 0)
d) reflcrion in $y = -x$ **e)** rotation, 180°, centre (0, 0) **f)** rotation, −90°, centre (0, 0)

11. a) reflection in $x = \frac{1}{2}$ **b)** reflection in $y = -x$ **c)** rotation, 180°, centre (1, 1) **12. a)** (−1, −3)
b) (−1, 3) **c)** (6, 2) **d)** (−3, 1) **e)** (−2, 6) **f)** (0, 2)

13. a) reflection in $y = x$ **b)** reflection in $x = 0$ **c)** No

14. a) enlargement s.f.2, reflection in $y = -x$

15. translation $\begin{pmatrix} 0 \\ 2 \end{pmatrix}$, reflection in $x = \frac{1}{2}$

16. enlargement s.f.2, translation $\begin{pmatrix} -8 \\ 0 \end{pmatrix}$, rotation 90° antioclockwise, centre (0, 0) **17. a)** $\begin{pmatrix} 13 \\ -6 \end{pmatrix}$ **b)** 5

18. $k = 3$ **19. a)** $\mathbf{r} - \mathbf{p}$ **b)** $\frac{1}{2}\mathbf{r} - \frac{1}{2}\mathbf{p}$ **c)** $\frac{1}{2}\mathbf{r} + \frac{1}{2}\mathbf{p}$

20. a) 5 **b)** $\sqrt{68}$ **c)** $\sqrt{41}$ **21.** $n = 2, m = -15$

22. a) $\begin{pmatrix} -1 \\ 4 \end{pmatrix}$ **b)** $\begin{pmatrix} 4 \\ 4 \end{pmatrix}$ **c)** $\begin{pmatrix} 6 \\ -4 \end{pmatrix}$ **d)** $\begin{pmatrix} 2 \\ 2 \end{pmatrix}$

23. a) $\mathbf{a} - \mathbf{c}$ **b)** $\frac{1}{2}\mathbf{a} + \mathbf{c}$ **c)** $\frac{1}{2}\mathbf{a} - \frac{1}{2}\mathbf{c}$ CA is parallel to NM and CA = 2 NM· **24.** $m = 3, n = 2$

Examination exercise 11 B page 372

1. a) i) triangle at (−4, 4), (−1, 4), (−1, 5) **ii)** triangle at (−1, −2), (−1, −5), (−2, −5)
b) enlargement, factor 2, centre (4, 0) **c)** translation $\begin{pmatrix} 6 \\ -3 \end{pmatrix}$

2. a) rotation, centre (0, 0) 90° anticlockwise **b)**

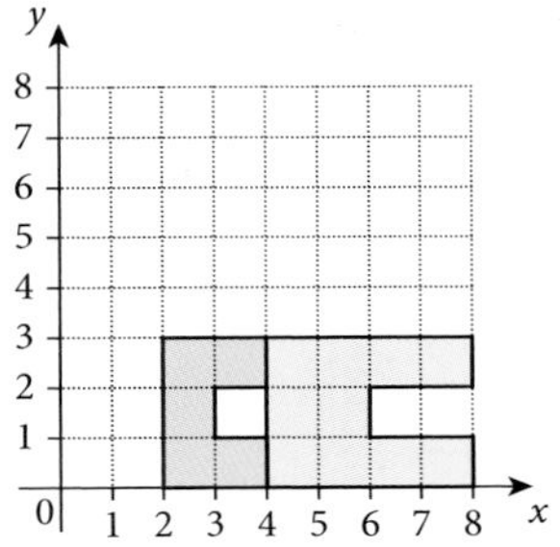

3. a) $\begin{pmatrix} 16 \\ -3 \end{pmatrix}$ **b)** 5

4. a) $\begin{pmatrix} 12 \\ 1 \end{pmatrix}$ **b)** $\sqrt{20}$ or $2\sqrt{5}$

5. a) i) (5, −7) **ii)** reflection in $y = x$ **b)** $c = 6, d = 3$

12 Sets

Exercise 12.1 page 375

1. a) 8 **b)** 3 **c)** 4 **d)** 18 **e)** 7 **2. a)** 9 **b)** 5 **c)** 4
d) 20 **e)** 31 **3. a)** 8 **b)** 3 **c)** 3 **d)** 2 **e)** 18 **f)** 0
4. a) 59 **b)** 11 **c)** 5 **d)** 40 **e)** 11 **f)** 124 **5. a)** 120
b) 120 **c)** 490 **d)** 80 **e)** 40 **f)** 10 **g)** 500

Exercise 12.2 *page 377*

1. a) {5, 7} **b)** {1, 2, 3, 4, 5, 6, 7, 8, 9, 11, 13} **c)** 5 **d)** 11 **e)** true
f) true **g)** false **h)** true **2. a)** {2, 3, 5, 7} **b)** {1, 2, 3, ⋯ 9}
c) 4 **d)** Ø **e)** false **f)** true **g)** false **h)** true
3. a) {2, 4, 6, 8, 10} **b)** {16, 18, 20} **c)** Ø **d)** 15 **e)** 11 **f)** 21 **g)** false
h) false **i)** true **j)** true **4. a)** {1, 3, 4, 5} **b)** {1, 5}
c) 1 **d)** {1, 5} **e)** {1, 3, 5, 10} **f)** 4 **g)** true **h)** false
i) true **5. a)** 4 **b)** 3 **c)** {b, d} **d)** {a, b, c, d, e} **e)** 5
f) 2 **6. a)** 2 **b)** 4 **c)** {1, 2, 4, 6, 7, 8, 9} d) {7, 9}
e) {1, 2, 4} **f)** {1, 2, 4, 7, 9} **g)** {1, 2, 4, 6, 8} **h)** {6, 7, 8, 9} **i)** {1, 2, 4, 7, 9}

Exercise 12.3 *page 379*

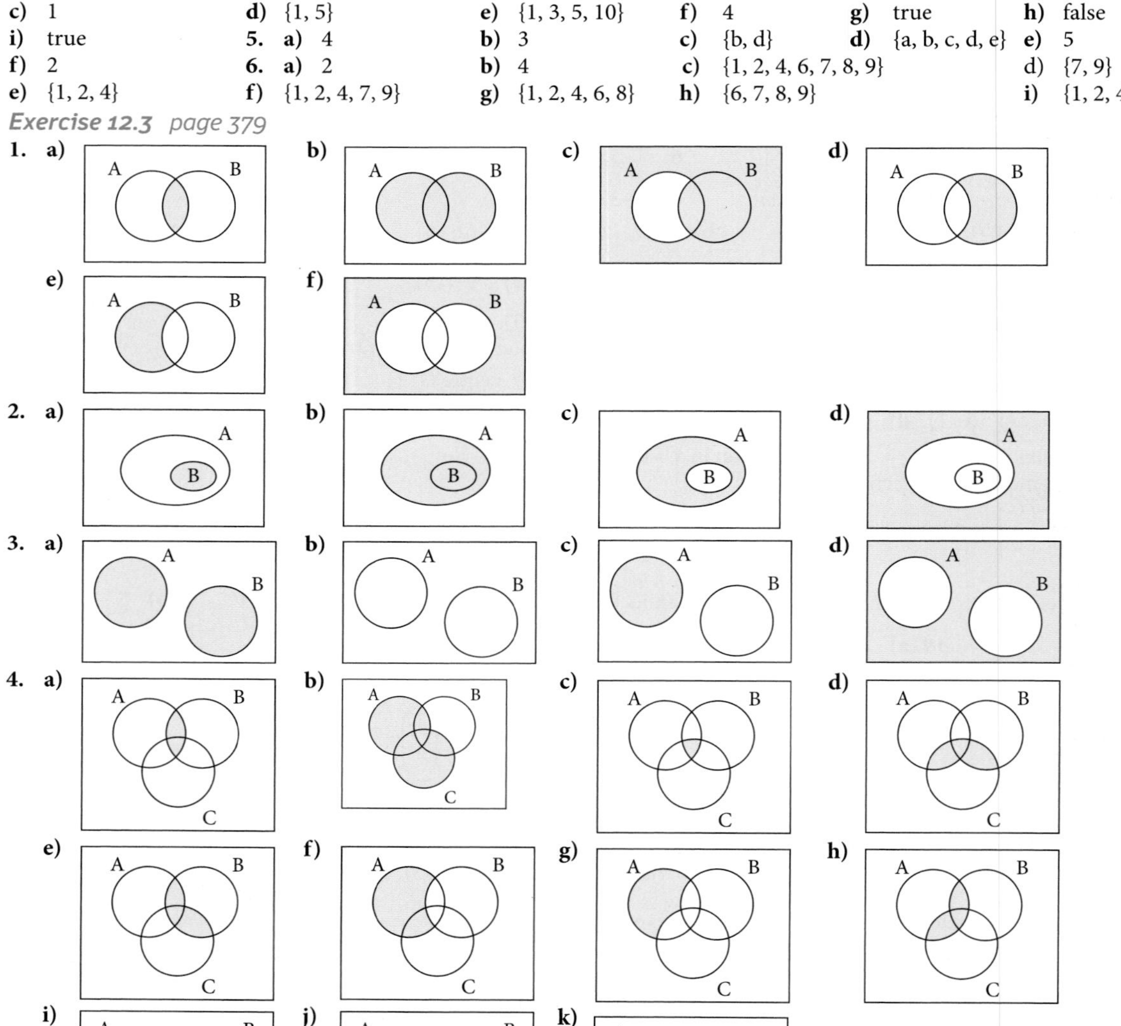

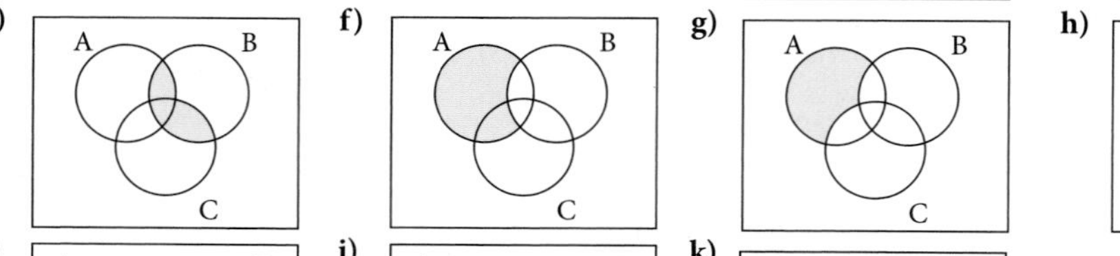

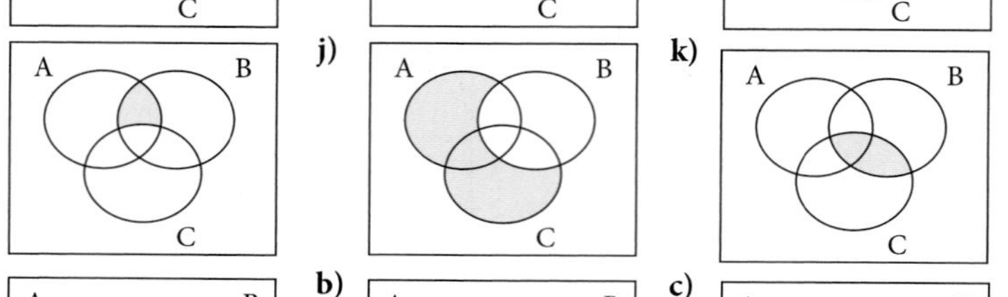

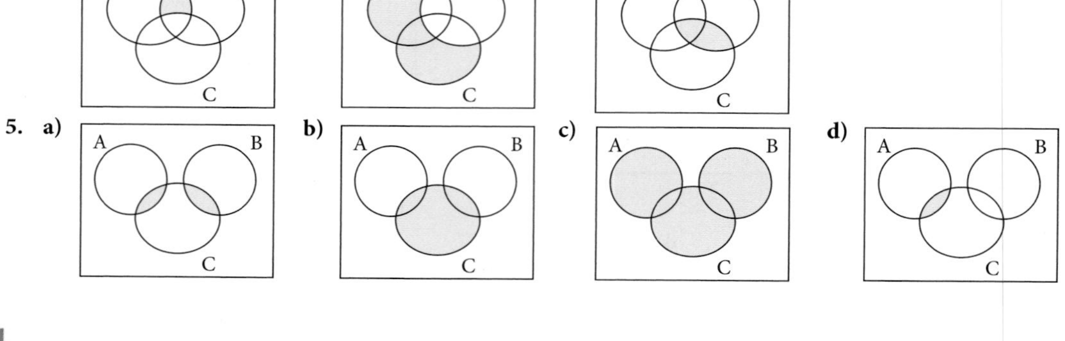

e) 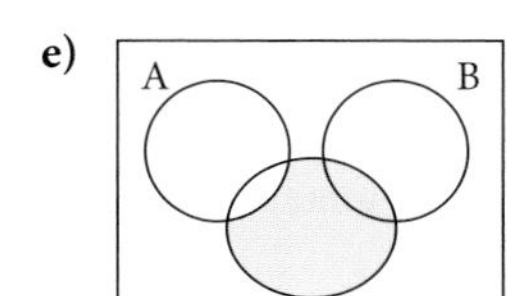

f)

g) 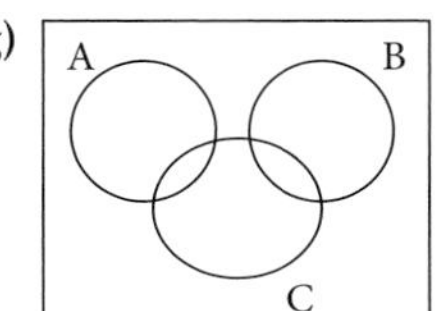

h) 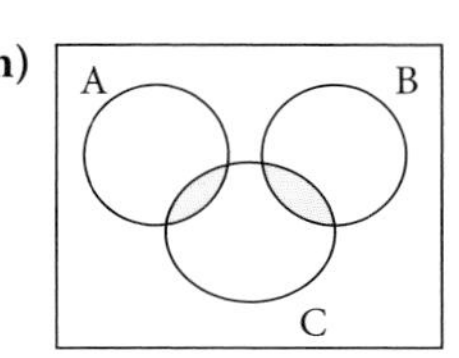

i) 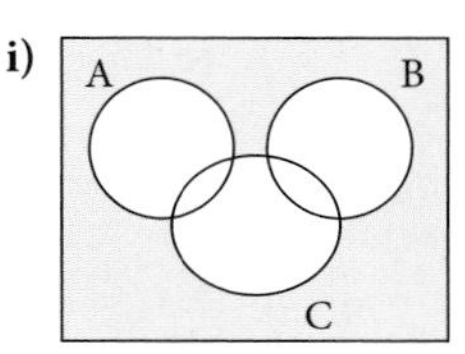

6. a) 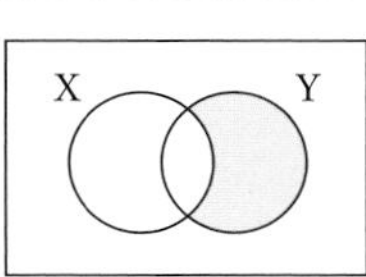

b)

c) 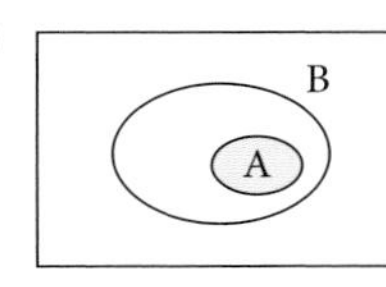

d)

7. a) $A \cup B$ **b)** $A' \cap B$ **c)** $(A \cup B)'$ **d)** $X' \cap Y'$

Exercise 12.4 *page 381*

1. a) $10 - x$ **b)** $13 - x$ **d)** 5 **2.** 9 **3.** 14 **4.** 3
5. 36 **6.** 3 **7.** 11 **8.** 5 **9.** 28 **10.** $x = 6$; 26
11. 34 **12. a)** 12 **b)** $\frac{1}{2}$ **13.** $x = 10$; 30 **14.** 2

15. a) All good footballers are Serbian men. **b)** No good footballers are Serbian men.
c) There are some good footballers who are Serbian men.

16. a) The football players are all boys. **b)** The hockey players are all girls.
c) There are some people who play both football and hockey. **d)** There are no boys who play hockey.
e) $B \cap F = \emptyset$ **f)** $H \cup F = \cup$ **17. a)** $S \cap T = \emptyset$ **b)** $F \subset T$ **c)** $S \cap F \neq \emptyset$
d) All living creatures are either spiders, animals that fly or animals which taste nice.
e) Animals which taste nice are all spiders.

Revision Exercise 12 A *page 383*

1. a) {5} **b)** {1, 3, 5, 6, 7} **c)** {2, 4, 6, 7, 8} **d)** {2, 4, 8} **e)** {1, 2, 3, 4, 5, 8} **2.** 32

3. a) 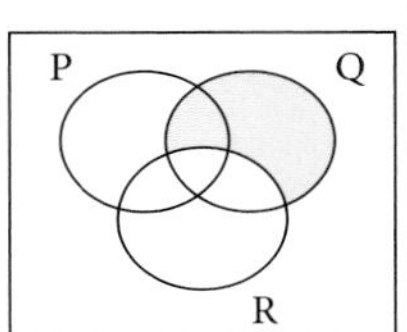

b) 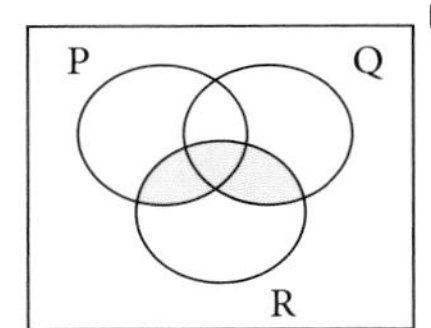

c) 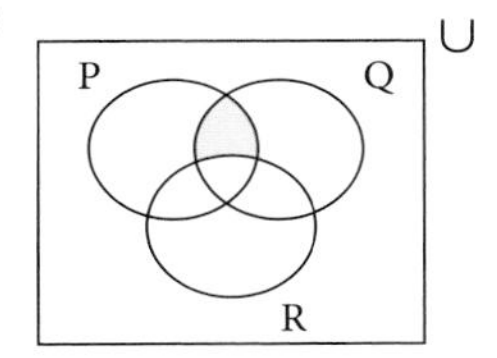

4. a) $(A \cup B)' \cap C$ **b)** $(A \cup B) \cap C'$
5. a) i) $S \subset T$ **ii)** $S \cap M' \neq \emptyset$ **b)** There are no women on the train over 25 years old.
6. a) 4 **b)** 11 **c)** 17

Examination exercise 12 B *page 383*

1. a) $A \cap B$ **b)** $A' \cap B$ **2. a)** $12 - x$, $11 - x$, $x - 3$ **b)** 5 **3. a)** {3, 6, 9, 12}, {1, 2, 3, 5, 6, 10}, {6, 7, 8, 9, 10, 11}

b) 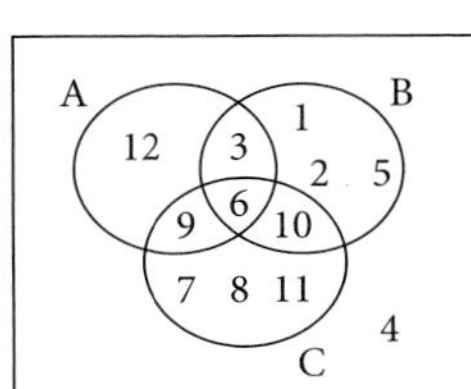

c) i) 3, 6 **ii)** 3, 6, 7, 8, 9, 10, 11, 12 **iii)** 3, 6, 10
iv) 4, 7, 8, 9, 11, 12 **v)** 11

4. a) 2, 3, 5, 7, 11, 13, 17, 19 **b)**
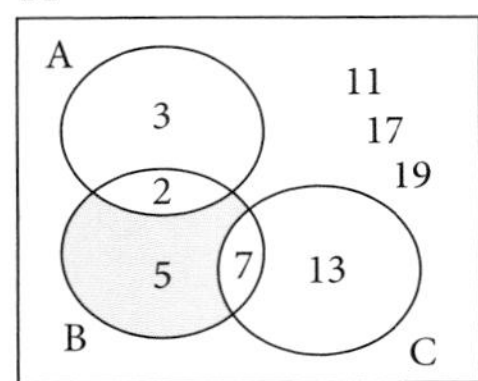

c) 3, 11,17,19 **d)** 3
e) B only shaded (i.e. parts in A and C not shaded)

13 Probability

Exercise 13.1 *page 388*

1. a) $\frac{4}{15}$ **b)** $\frac{1}{6}$ **2. a)** $\frac{7}{20}$ **b)** 420 **3. a)** Bo (most trials) **b)** biased **c)** $\frac{63}{290}$

Exercise 13.2 *page 390*

1. a) $\frac{1}{6}$ **b)** $\frac{1}{2}$ **c)** $\frac{1}{2}$ **d)** $\frac{1}{3}$ **2. a)** $\frac{1}{4}$ **b)** $\frac{1}{2}$ **3. a) i)** $\frac{3}{5}$ **ii)** $\frac{2}{5}$
b) i) $\frac{5}{9}$ **ii)** $\frac{4}{9}$ **4. a)** $\frac{1}{11}$ **b)** $\frac{2}{11}$ **c)** 0 **d)** $\frac{1}{11}$ **5. a)** $\frac{1}{8}$ **b)** $\frac{3}{8}$

c) $\frac{1}{8}$ d) $\frac{7}{8}$ **6.** a) $\frac{5}{11}$ b) $\frac{7}{22}$ c) $\frac{15}{22}$ d) $\frac{17}{22}$ **7.** a) $\frac{1}{2}$ b) $\frac{3}{25}$

c) $\frac{9}{100}$ d) $\frac{2}{25}$ **8.** a) $\frac{1}{12}$ b) $\frac{1}{36}$ c) $\frac{5}{18}$ d) $\frac{1}{6}$ e) $\frac{1}{6}$; most likely total = 7.

9. a) $\frac{1}{12}$ b) $\frac{5}{36}$ c) $\frac{2}{3}$ d) $\frac{1}{12}$ e) $\frac{1}{36}$ **10.** a) 1 b) 0 c) 1 d) 0

11. a) $\frac{3}{8}$ b) $\frac{1}{16}$ c) $\frac{15}{16}$ d) $\frac{1}{4}$ **12.** $\frac{271}{1000}$ **13.** $\frac{x}{12}$, 3 **14.** a) $\frac{1}{144}$ b) $\frac{1}{18}$

c) $\frac{1}{72}$, head, tail, and total of 7·

15. a) $\frac{1}{216}$ b) $\frac{1}{72}$ c) $\frac{1}{8}$ d) $\frac{5}{108}$ e) $\frac{5}{72}$ f) $\frac{1}{36}$

Exercise 13.3 *page 393*

1. a) $\frac{1}{2}$ b) $\frac{1}{2}$ c) $\frac{1}{4}$ **2.** a) $\frac{1}{16}$ b) $\frac{25}{144}$

3. a) $\frac{1}{121}$ b) $\frac{9}{121}$ **4.** a) $\frac{1}{288}$ b) $\frac{1}{72}$

5. a) $\frac{1}{125}$ b) $\frac{1}{125}$ c) $\frac{1}{10000}$ d) $\frac{3}{500}$ e) $\frac{3}{500}$

6. a) $\frac{1}{9}$ b) $\frac{4}{27}$ c) $\frac{4}{9}$

Exercise 13.4 *page 395*

1. a) $\frac{49}{100}$ b) $\frac{9}{100}$ **2.** a) $\frac{9}{64}$ b) $\frac{15}{64}$ **3.** a) $\frac{7}{15}$

b) $\frac{1}{15}$ **4.** a) $\frac{2}{9}$ b) $\frac{2}{15}$ c) $\frac{1}{45}$ d) $\frac{14}{45}$

5. a) $\frac{1}{12}$ b) $\frac{1}{6}$ c) $\frac{1}{3}$ d) $\frac{2}{9}$ **6.** a) $\frac{1}{216}$

b) $\frac{125}{216}$ c) $\frac{25}{72}$ d) $\frac{91}{216}$ **7.** a) $\frac{1}{6}$ b) $\frac{1}{30}$

c) $\frac{1}{30}$ d) $\frac{29}{30}$ **8.** a) $\frac{27}{64}$ b) $\frac{1}{64}$ **9.** a) 6

b) $\frac{1}{3}$ **10.** a) $\frac{1}{64}$ b) $\frac{27}{64}$ c) $\frac{9}{64}$ d) $\frac{27}{64}$; Sum = 1

11. a) $\frac{3}{20}\times\frac{2}{19}\times\frac{1}{18}\left(=\frac{1}{1140}\right)$ b) $\frac{1}{4}\times\frac{4}{19}\times\frac{1}{6}\left(=\frac{1}{114}\right)$ c) $\left(\frac{3}{20}\times\frac{5}{19}\times\frac{12}{18}\right)\times 6$

d) $\frac{5}{20}\times\frac{4}{19}\times\frac{3}{18}\times\frac{2}{17}$ **12.** a) $\frac{1}{10000}$ b) $\frac{523}{10000}$ c) $\frac{9^4}{10^4}$

13. a) $\frac{x}{x+y}$ b) $\frac{x(x-1)}{(x+y)(x+y-1)}$ c) $\frac{2xy}{(x+y)(x+y-1)}$ d) $\frac{y(y-1)}{(x+y)(x+y-1)}$

14. a) $\frac{1}{5}$ b) $\frac{18}{25}$ c) $\frac{1}{20}$ d) $\frac{2}{25}$ e) $\frac{77}{100}$

15. $\frac{3}{10}$ **16.** $\frac{9}{140}$ **17.** a) 5 b) $\frac{1}{64}$ **18.** a) $\frac{3}{5}$

b) $\frac{1}{3}$ c) $\frac{2}{15}$ d) $\frac{2}{21}$ e) $\frac{1}{7}$ f) $\frac{1}{35}$

19. a) $\frac{1}{220}$ b) $\frac{1}{22}$ c) $\frac{3}{11}$ d) 5 e) 300

20. a) $\frac{10\times 9}{1000\times 999}$ b) $\frac{990\times 989}{1000\times 999}$ c) $\frac{2\times 10\times 990}{1000\times 999}$ **21.** a) $\frac{3}{20}$ b) $\frac{7}{20}$

c) $\frac{1}{2}$ **22.** a) 0·00781 b) 0·511 **23·** a) $\frac{21}{506}$ b) $\frac{455}{2024}$ c) $\frac{945}{2024}$

Revision exercise 13A *page 399*

1. a) $\frac{1}{52}$ b) $\frac{1}{13}$ c) $\frac{1}{4}$ **2.** a) $\frac{1}{3}$ b) $\frac{2}{3}$

3. 9 **4.** 20 **5.** a) $\frac{3}{49}$ b) $\frac{8}{49}$ **6.** About 1607

7. a) 44% b) 220 **8.** $\frac{n}{n+m+7}$ **9.** $\frac{1}{6}$ **10.** a) 5

b) $\frac{4}{25}$ 11. $\frac{8}{27}$ 12. $\frac{5}{16}$ 13. a) $\frac{1}{28}$ b) $\frac{15}{28}$

c) $\frac{3}{7}$ 14. a) $\frac{x}{x+5}$ b) $\left(\frac{x}{x+5}\right)^2$; $\frac{x}{x+5}$, $\frac{x(x-1)}{(x+5)(x+4)}$ 15. $\frac{1}{19}$

16. a) $\frac{1}{32}$ b) $\frac{1}{256}$ 17. a) $\frac{1}{8}$ b) $\frac{1}{2}$ 18. $\frac{35}{48}$

19. a) $\frac{1}{9}$ b) $\frac{1}{12}$ c) 0 20. $\frac{1}{20^3}$

Examination exercise 13B page 400

1. a)

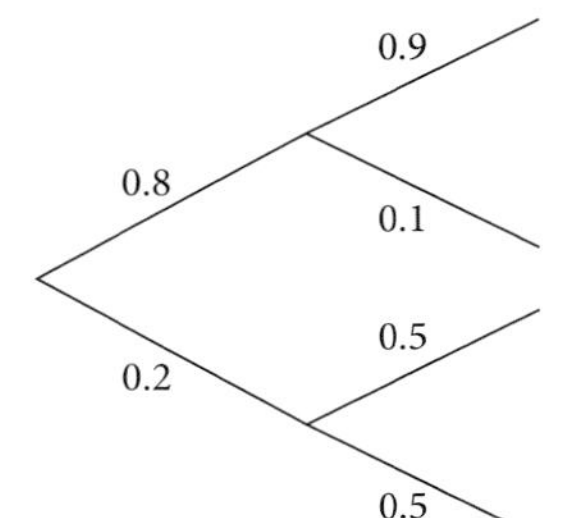

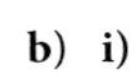

b) 0·82

2. a) i) $\frac{2}{7}$ ii) 7 b) i) $\frac{1}{6}$

ii) $\frac{5}{7}$, $\frac{1}{6}$, $\frac{5}{6}$, $\frac{2}{6}$, $\frac{4}{6}$ iii) $\frac{2}{42}$ iv) $\frac{20}{42}$ c) $\frac{40}{210}$

3. a) $\frac{1}{6}$ b) i)

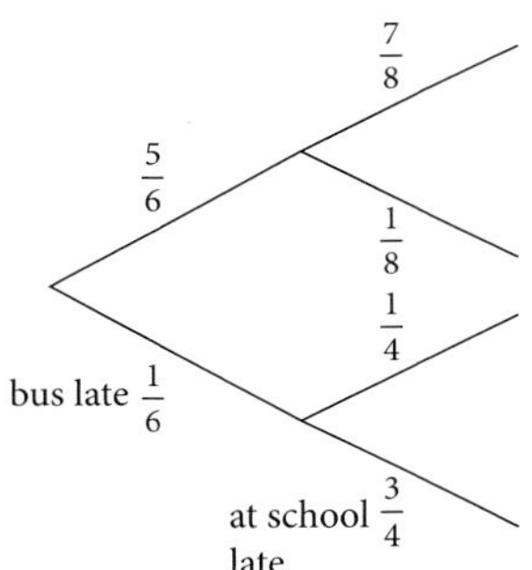

ii) $\frac{37}{48}$ iii) 0·00063 iv) 148

14. Statistics

Exercise 14.1 page 406

1. For example, Isola, because the snowfall is greater on average.
2. a) 20% b) 20–24 and 30–34
3. a–b

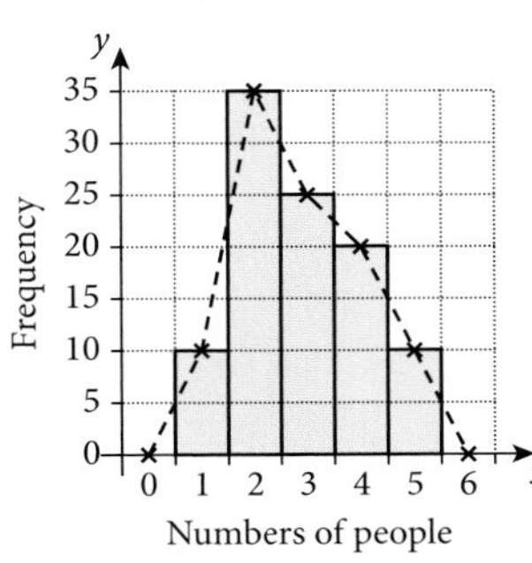

4. a) 62
   b) The mean weight is lower for A. The range is higher for B.
   c) For example, B could be rugby, as weight of player varies with position.
5. a) The fertilizer made the plants taller, on average.
   b) The fertilise had no effect on the weights.
6. Height has been used to represent production, but the width of the picture also varies, making the difference look much greater.
7. The profit axis does not start at zero, so the increase in profit looks much greater than it is.

8. a) i) $\frac{1}{8}$ ii) $\frac{1}{6}$ iii) $\frac{5}{12}$ iv) $\frac{1}{12}$ v) $\frac{1}{8}$ vi) $\frac{1}{12}$

b) i) € 15 ii) € 20 iii) € 50 iv) € 10 9. a) £425 b) £150

c) £250 d) £75 10. a) i) 8 min ii) 34 min iii) 10 min b) 18°

11. 270°, 12°, 23°, 54°, 1° 12. a) 22·5% b) 45°, 114° 13 8

Exercise 14.2 page 410

1. a) Key: 2|3 means 23 **b)** 51

Stem	Leaf
1	5
2	3 7 8 9
3	2 5 6 8 9
4	0 1 2 5 6 7 8
5	1 2 3 4 9
6	5 6

2. a) Key: 3|1 means 31

Stem	Leaf
2	0 4 5 8
3	1 7 9
4	0 4 6
5	2 5 8 9
6	1 5 7 8
7	3 5

b) Key: 2|2 means 22

Stem	Leaf
2	2 8 9
3	0 5 8
4	1 4 6 7 7
5	3 4 9
6	7
7	2

3. a) 50 kg **b)** 15 **c)** 50 kg

4. a) 4·5 **b)** 5·3 Key: 3|7 means 3·7

Stem	Leaf
1	4 8
2	4 4 8
3	1 3 3 7 8
4	0 5 5 6 6 7 9
5	1 2 5 8
6	2 3 7

5. a) 13 **b)** 78

c) The women's pulse rates were higher on average (median 78 compared with 62).
The spreas was greater for men (range 53 compared with 45).

Exercise 14.3 page 412

1. a) Squash **b)** 160 **c)** 10 **3. a)** \$3000 **b)** \$4000 **c)** \$6000 **d)** \$11 000

4. red 50°; green 70°; blue 110°; yellow 40°; pink 90°

5. a) A 60°; B 100°; C 60°; D 140°; E 0° **b)** A 50°; B 75°; C 170°; D 40°; E 25° **c)** A 48·5°; B 76·2°; C 62·3°; D 96·9°; E 76·2°

6. 18°, 54°, 234° **7.** 80°, 120°, 160° **8.** 100 **9. a)** 144° **b)** posters 5%, cinema 1%

Exercise 14.4 page 415

2. Polygon P is most likely to represent shoppers who lived nearest as they would possibly shop more often and spend less each time.

3. Adult weight unaffected by gene change· Life span greater with altered gene.

Exercise 14.5 page 418

1. a) Median = 5, Mean = 6, Mode = 4, Range = 9 **b)** Median = 7, Mean = 9, Mode = 7, Range = 19
c) Median = 8, Mean = 6·5, Mode = 9, Range = 11 **d)** Median = 3·5, Mean = 3·5, Mode = 4, Range = 8

2. a) 10 **b)** 13 **3** 6 **4. a)** 5 **b)** 35 **5.** 76 kg **6.** 7° C

7. a) 16 years **b)** 170 cm **c)** 50 kg **8.** 2, 45 **9.** 4 people

10. a) False **b)** True **c)** False **d)** Possible **11. a)** Mean = \$33 920, Median = \$22 500, Mode = \$22 500 **b)** Mean

12. a) Mean = 157·125 kg, Median = 91 kg **b)** The mean. No, it's too high.

13. a) Mean = 74·5 cm· Median = 91 cm
b) Yes. It is more appropriate to use the median, as the mean has been distorted by a few short plants.

14. 78 kg **15.** 35·2 cm **16. a)** 2 **b)** 9 **17. a)** 20·4 m
b) 12·8 m **c)** 1·66 m **18.** 55 kg **19.** For example: 4, 4, 5, 8, 9

20. For example: 2, 4, 4, 4, 6 **21.** 12 **22.** $3\frac{2}{3}$ **23. a)** N **b)** Mean = $N^2 + 2$, Median = N^2 **c)** 2

Exercise 14.6 page 421

1. 96·25 g **2.** 51·9 c **3.** 4·82 cm

4. a) Mean = 3·025, Median = 3, Mode = 3 **b)** Mean = 17·75, Median = 17, Mode = 17

5. a) Mean = 6·62, Median = 8, Mode = 3 **b)** Mode

6. a) 9 **b)** 9 **c)** 15 **7. a)** 5 **b)** 10 **c)** 10 **8.** $\frac{ax+by+cz}{a+b+c}$

Exercise 14.7 page 423

1. a

Number of words	Frequency f	Midpoint x	fx
1–5	6	3	18
6–10	5	8	40
11–15	4	13	52
16–20	2	18	36
21–25	3	23	69
Totals	20	—	215

b) 10·75 **2.** 68·25 **3.** 3·77

4. a) 181 cm **b)** The raw data is unavailable and an assumption has been made with the midpoint of each interval.

Exercise 14.8 *page 425*

1.

Length / (in mm)	Frequency	Frequency density (f.d.)
$0 \le l < 20$	5	$5 \div 20 = 0{\cdot}25$
$20 \le l < 25$	5	$5 \div 5 = 1$
$25 \le l < 30$	7	$7 \div 5 = 1{\cdot}4$
$30 \le l < 40$	3	$3 \div 10 = 0{\cdot}3$

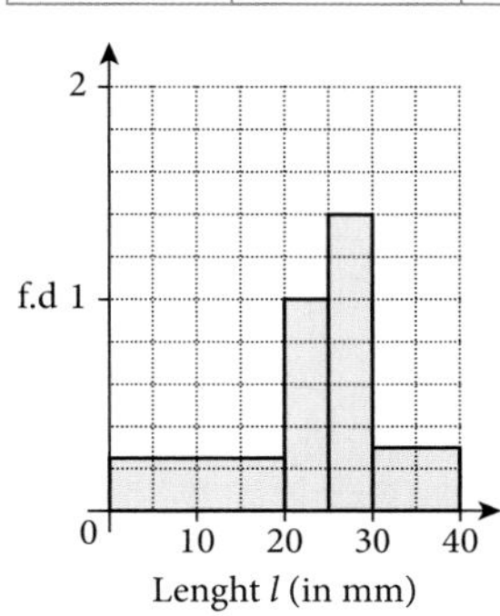

2.

Volume (mm³)	Frequency	Frequency density
$0 < V \le 5$	5	$5 \div 5 = 1$
$5 < V \le 10$	3	$3 \div 5 = 0{\cdot}6$
$10 < V \le 20$	12	$12 \div 10 = 1{\cdot}2$
$20 < V \le 30$	17	$17 \div 10 = 1{\cdot}7$
$30 < V \le 40$	13	$13 \div 10 = 1{\cdot}3$
$40 < V \le 60$	5	$5 \div 20 = 0{\cdot}25$

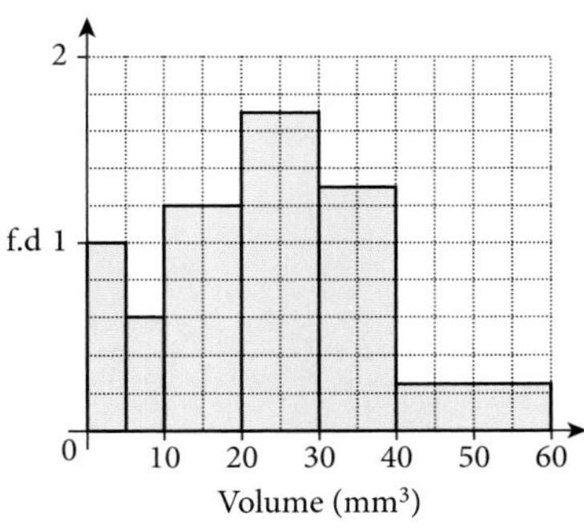

3

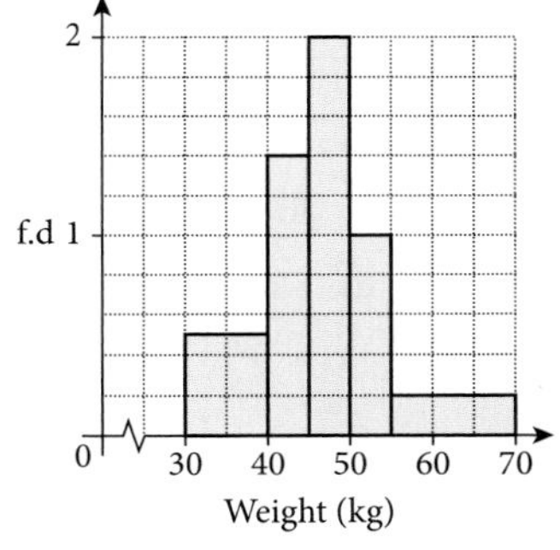

4

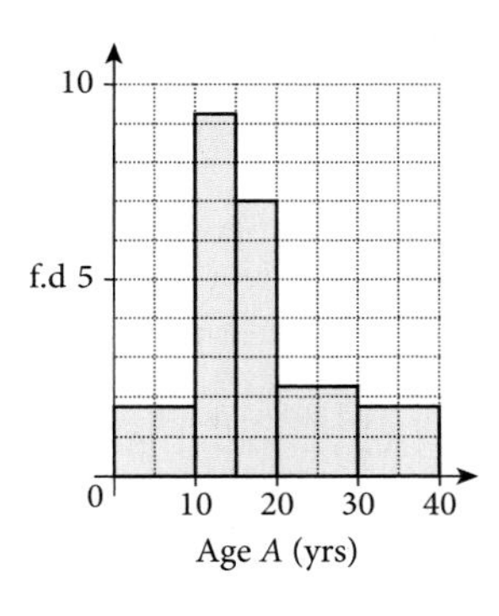

5

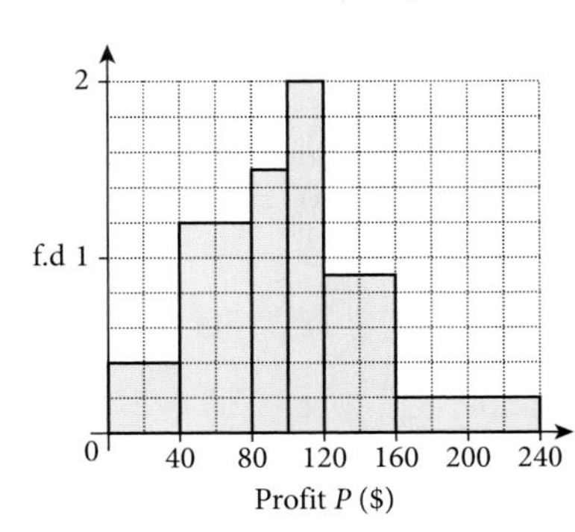

6

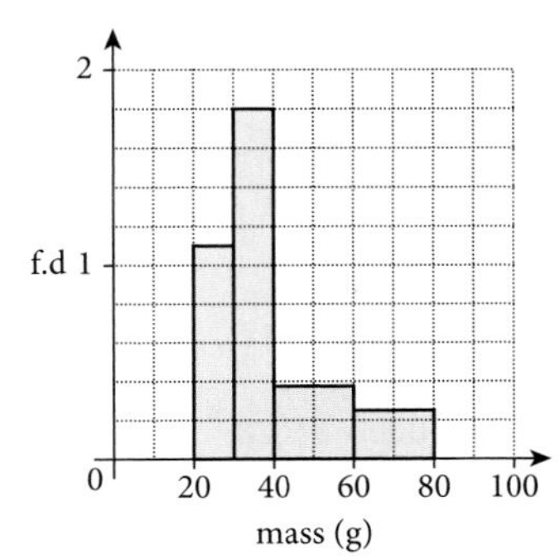

Exercise 14.9 *page 428*

1. a) i) 10 **ii)** 20 **b)** 65 **2.** 55 **3.** 135

4. 28 eggs **5. a)** 6 **b)** 10 **c)** 60 **6. a)** 6 cm

b) 9 cm **c)** 3 cm

Exercise 14.10 *page 430*

1. a) Q1 = 4, Q3 = 9·5 **b)** 5·5 **2. a)** Q1 = 1·5, Q3 = 5·5

b) 4 **3. a)** Q1 = 10, Q3 = 21 **b)** 11

4. a) Q1 = 1·5, Q3 = 6 **b)** 4·5

Exercise 14.11 page 434

1. a) 50 b) 30, 60 c) 30 d) 30
   b) 30, 61 c) 31 d) 36
   c) 2100 hours d) 750 hours
2. a) 46 b) 200
3. a) 100

4. a)

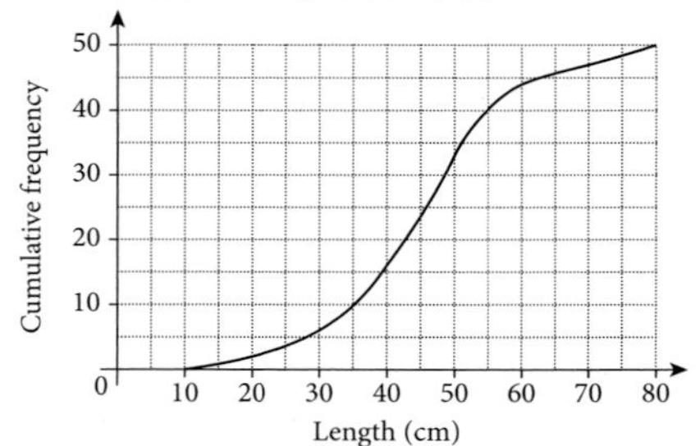

b) i) 46 cm ii) 17 cm

5. a)

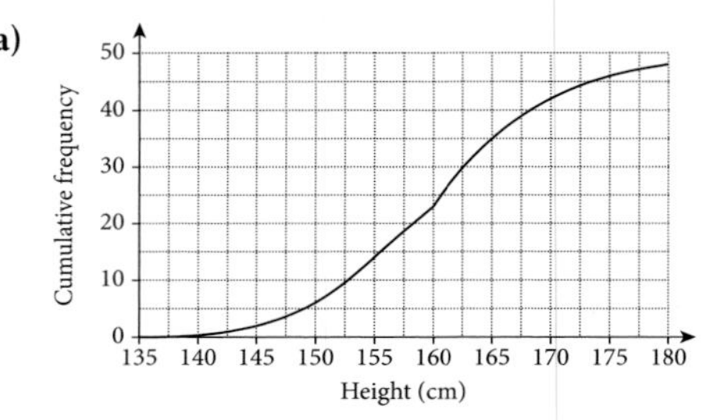

b) i) 160·5 cm ii) 11·5 cm

6. a)

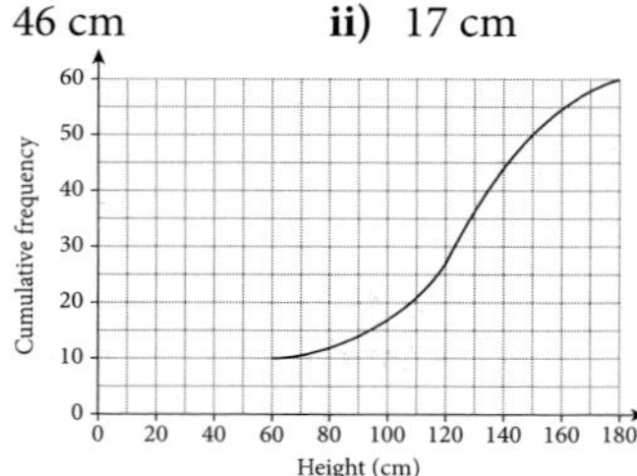

b) i) 125 ii) 52
c) Boris, his interquartile range is smaller.

7. a)

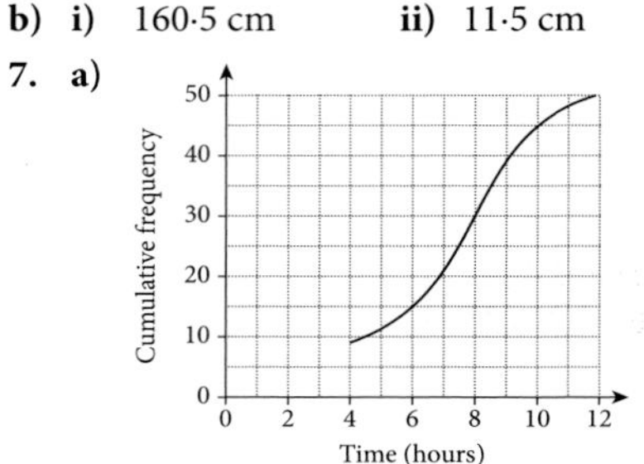

b) i) 7·4 hours ii) 3·5 hours c) 90%

8. a)

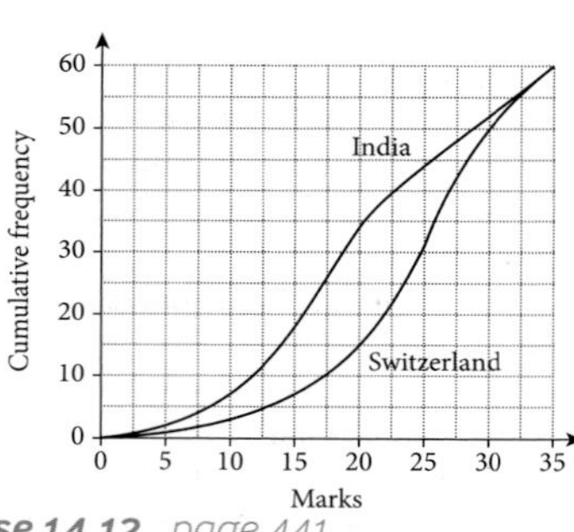

b) India 25, Switzerland $18\frac{1}{2}$
c) India $8\frac{1}{2}$
d) The Indian results are higher on average, and less spread out.

Exercise 14.12 page 441

2. a) strong positive correlation b) no correlation c) weak negative correlation
3. 11
4. 9
5. no correlation
6. c) about 26 d) about 46
7. b) 33 m.p.g. c) 63 m.p.h.

Exercise 14.13 page 443

1. a) $y = 0{\cdot}87\,x + 2{\cdot}58$ b) $y = -0{\cdot}50\,x + 14$
2. a) $y = 1{\cdot}18\,x - 9{\cdot}3$ b) about 26 c) about 46
3. a) $y = -\,0{\cdot}36\,x + 49{\cdot}9$ b) 33·7 m.p.g. c) 63·6 m.p.h.
4. b) positive correlation c) $y = 0{\cdot}83x + 7{\cdot}28$ d) 21·8 → 22 cm

Revision exercise 14A page 444

1. 162°
2. 41·7%
3. 54°
4. a) 84 b) 19·2
5. a) 3·4 b) 3 c) 3
6. a) 5·45 b) 5 c) 5
7. 1·552 m
8. 3
9. a) 20 kg b) 10·5 kg
10. a) 80·5 cm b) 20 cm
11. a) 26 b) 25·8 c) 26·1
12. The median mark is about the same for both schools but the results are more spread out for school B.
13. The people in the town are taller and their heights are more spread out than the heights of the people in the village.

Examintation exercise 14B page 446

1. b) negative correlation c) i) $y = -\,0{\cdot}565x + 58{\cdot}5$ ii) about 30
2. a) 6·6 b) 9 c) 7 d) 4
3. a) i) 45 ii) 25 b) 34 to 36
4. frequency densties 2, 1, 0·5, 6, 2
5. a) 1005 b) 0·4, 1, 10, 4, 0·8, 0·4
6. b) negative c) $i = -1{\cdot}14\,c + 96{\cdot}8$ d) 20 or 21

15 Investigations and mathematical modelling 2

1. *Diamonds and triangles* page 450

a) $e = 2w$

b) $i = 2h$

c) When, $w = 2$, $d = 3h - 2$ [or $d = 3h - w$];
When $w = 3$, $d = 5h - 3$ [or $d = 5h - w$].
When $w = 4$, $d = 7h - 4$ [or $d = 7h - w$]·
For any diagram, $d = (2w - 1)h - w$ or $d = 2wh - h - w$.

3. *In search of π* page 452

a) 8 sides 3·061467459

b) 10 sides 3·090169944

c) 20 sides 3·128689301

d) 100 sides 3·141075908

e) 1000 sides 3·141587486

5. *Painting cubes* page 453

For $n \times n \times n$: 3 green faces = 8
2 green faces = $12\ (n - 2)$
1 green face = $6(n - 2)^2$
0 green face = $(n - 2)^3$

7. *Spotted shapes* page 453

$$A = i + \frac{1}{2}p - 1$$

12. *Fibonacci sequence* page 456

a) Another Fibonnacci sequence.

b) Terms are alternate terms of original sequence.

c) Ratio tends towards 1·618 (to 4 s.f.), the 'Golden Ratio'

d) (first × fourth) = (second × third) + 1

e) (first × third) ± 1 = (second)2, alternating + and –

f) sum of 10 terms = 11 × seventh term.

g) For six terms $a\ b\ c\ d\ e\ f$ let
$x = e \times f - (a^2 + b^2 + c^2 + d^2 + e^2)$

The numbers in the first difference column are the squares of the terms in the original Fibonacci sequence.

	x	first difference
first six	0	1
second six	1	1
third six	2	4
fourth six	6	9
fifth six	15	25
sixth six	40	64
seventh six	104	169
eighth six	273	

2. *Connect the transformations* page 451

a) Suppose centre of rotation has equal x and y coordinates, say (a, a)
Then, for the vector of the corresponding translation:
Top number = $-(2a + 1)$
Bottom number = 3

b) Suppose centre of rotation is $(a, 0)$,
Then the vector for the corresponding translation has:
Top number = $-(a + 1)$
Bottom number = $(a + 3)$

c) For centre (7, 7), vector is $\begin{pmatrix} -15 \\ 3 \end{pmatrix}$ i·e· $\begin{pmatrix} -(2\times7+1) \\ 3 \end{pmatrix}$

For centre (10, 0), vector is $\begin{pmatrix} -11 \\ 13 \end{pmatrix}$ i·e· $\begin{pmatrix} -(10+1) \\ 10+3 \end{pmatrix}$

6. *Find the connection* page 453

Square root twice, 'multiply by x' gives cube root of x.
Square root three times, 'multiply by x' gives seventh root of x.
Square root twice, 'devide by x' gives $1/(\sqrt[3]{x})$
Square root twice, 'multiply by x^2' gives $(\sqrt[3]{x})^2$

8. *Diagonals* page 454

Consider three cases of rectangles $m \times n$

a) m and n have no common factor. Number of squares $= m + n - 1$.
e.g. 3×7, number $= 3 + 7 - 1 = 9$

b) n is a multiple of m. Number of squares $= n$ e.g. 3×12, number = 12

c) m and n share a common factor a
so $m \times n = a(a \times m') \times (a \times n')$
number of squares $= a(m' + n' - 1)$
e.g. $640 \times 250 = (10 \times 64) \times (10 \times 25)$, number of squares $= 10(64 + 25 - 1) = 880$.

Past examination paper
The Solar System page 457

1.

8·4	2·8
8·9	3·6
9·2	4·0

2. b) mean (8·6, 3·2). Line of best fit drawn through mean.

3. $2·8 \times 10^2$ km / $3·2 \times 10^9$ km

4. $m = 1·5$, $c = -9·6$

5. $7·6 \times 10^4$ days / $5·4 \times 10^4$ days

6. b) $k = 2·0 \times 10^{-10}$ / $2·5 \times 10^{-10}$

c) The model produces a value of T of about 365, which shows that the data for Earth fit the model well.

Index